W9-CMD-975

AR

ANNUAL REVIEW OF MICROBIOLOGY

ANNUAL REVIEW OF MICROBIOLOGY

VOLUME 46, 1992

L. NICHOLAS ORNSTON, *Editor*
Yale University

ALBERT BALOWS, *Associate Editor*
Centers for Disease Control, Atlanta

E. PETER GREENBERG, *Associate Editor*
University of Iowa, Iowa City

ANNUAL REVIEWS INC. 4139 EL CAMINO WAY P.O. BOX 10139 PALO ALTO, CALIFORNIA 94303-0897

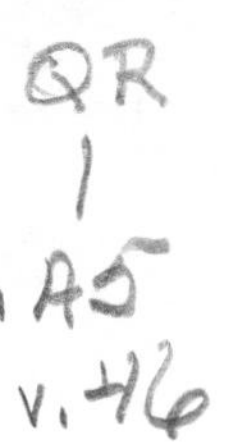

AR ANNUAL REVIEWS INC.
Palo Alto, California, USA

International Standard Serial Number: 0066–4227
International Standard Book Number: 0–8243–1146-9
Library of Congress Catalog Card Number: 49-432

♾ The paper used in this publication meets the minimum requirements of American National Standard for Information Sciences—Permanence of Paper for Printed Library Materials, ANSI Z39.48-1984.

Typesetting by Kachina Typesetting Inc., Tempe, Arizona; John Olson, President; Jeannie Kaarle, Typesetting Coordinator; and by the Annual Reviews Inc. Editorial Staff

PRINTED AND BOUND IN THE UNITED STATES OF AMERICA

PREFACE

A knight errant, returning from a quest for knowledge, stated that it had taken him a quarter of a century to learn that truth is an illusion. It follows that a well-crafted illusion can become truth. We shouldn't place too much emphasis on one knight's stand, but his comments deserve consideration as we reflect upon how we learn and how we teach.

By learning, we become acquainted with the unknown, and through teaching we communicate what we hope is the known. In both activities, our efforts are guided by a sense of anticipation that has been shaped by our experience. We swiftly scan for reinforcement of established associations and move more slowly when our assumptions come under fire. However we move, excessive information becomes a burden. We need to know what is important, and we rely upon the reviewer of a subject to select the information required for effective thought.

Rarely is the drive toward scholarship leisurely. The lecture must be given or the grant request must be prepared before a rapidly approaching deadline. In these situations, back collections of the *Annual Review of Microbiology* receive bursts of attention. Members of the editorial committee are among those who engage in these flurries of activity, and expectation of such needs influences suggestions for the topics and authors that will be represented in the next volume. We seek authors who are unafraid of emphasis to review topics of enduring interest. The challenge to the reviewer is to distill the essence of a subject into an article of limited length. This constraint must be honored so that each volume represents the diversity of present research in microbiology. The result is a book that, fun to open for the first time, becomes a reliable companion when we are under pressure to produce graceful scholarship.

For this volume, Abdul Matin and Graham Walker served as guests and aided the deliberations of the editorial board. The authors, the creators in the Annual Reviews process, met the expectations that accompanied their selection, and the volume materialized under the capable guidance of its production editor, Amanda Suver. Mistakes that might have been introduced into the process by the editor were prevented by his tactful administrative associate, Rosemarie Hansen.

L. NICHOLAS ORNSTON
EDITOR

Annual Review of Microbiology
Volume 46, 1992

CONTENTS

 (*continued*)

INDEXES

OTHER REVIEWS OF INTEREST TO MICROBIOLOGISTS

From the *Annual Review of Biochemistry,* Volume 61 (1992)

DNA Looping, Robert Schleif
Chromosome and Plasmid Partition in Escherichia coli, Sota Hiraga
Prokaryotic DNA Replication, Kenneth J. Marians
Proton Transfer in Reaction Centers from Photosynthetic Bacteria, M. Y. Okamura and G. Feher
Transpositional Recombination: Mechanistic Insights from Studies of Mu and Other Elements, Kiyoshi Mizuuchi
Pheromone Response in Yeast, Janet Kurjan

From the *Annual Review of Cell Biology,* Volume 8 (1992)

ABC Transporters: from Microorganisms to Man, Christopher F. Higgins
The Interaction of Bacteria with Mammalian Cells, Stanley Falkow, Ralph R. Isberg, and Daniel A. Portnoy
Retroviral Reverse Transcription and Integration: Progress and Problems, J. M. Whitcomb and S. H. Hughes

From the *Annual Review of Entomology,* Volume 37 (1992)

Polydnaviruses: Mutualists and Pathogens, Jo-Ann G. W. Fleming
The Mode of Action of Bacillus thuringiensis *Endotoxins,* Sarjeet S. Gill, Elizabeth A. Cowles, and Patricia V. Pietrantonio

From the *Annual Review of Fluid Mechanics,* Volume 24 (1992)

Hydrodynamic Phenomena in Suspensions of Swimming Microorganisms, T. J. Pedley and J. O. Kessler

From the *Annual Review of Genetics,* Volume 26 (1992)

Fungal Senescence, Anthony J. F. Griffiths
Genetics of the Fission Yeast Schizosaccharomyces pombe, Jacqueline Hayles and Paul Nurse
Translational Accuracy and the Fitness of Bacteria, C. G. Kurland
Genetics of Legionella pneumophila *Virulence,* A. Marra and H. A. Shuman
Communication Modules in Bacterial Signaling Proteins, John S. Parkinson and Eric C. Kofoid

 (*continued*)

Developmentally Regulated Gene Rearrangements in Prokaryotes, Robert Haselkorn
Genetics and Biogenesis of Bacterial Flagella, Robert M. Macnab
Genetics and Intermediary Metabolism, Daniel G. Fraenkel
DNA Replication of Phages and Plasmids, Sue Wickner
Genetic Analysis Using the Polymerase Chain Reaction, Henry A. Erlich and Norman Arnheim
Genetics of Retroviral Integration, Steven P. Goff

From the *Annual Review of Immunology,* Volume 10 (1992)

Roles of $\alpha\beta$ and $\gamma\delta$ T Cell Subsets in Viral Immunity, Peter C. Doherty, William Allan, Maryna Eichelberger, and Simon R. Carding
Regulation of Immunity to Parasites by T Cells and T Cell–Derived Cytokines, A. Sher and R. L. Coffman

From the *Annual Review of Medicine,* Volume 43 (1992)

Complications of Lyme Borreliosis, W. Donald Cooke and Raymond J. Dattwyler
Helicobacter pylori *and Gastroduodenal Disease,* Timothy L. Cover and Martin J. Blaser
Detection of Microbial Nucleic Acids for Diagnostic Purposes, N. Cary Engelberg and Barry I. Eisenstein
Chlamydia pneumoniae, *Strain TWAR Pneumonia,* J. Thomas Grayston
Acute Progressive Epstein-Barr Virus Infections, Stephen E. Straus
The Resurgence of Measles in the United States, 1989–1990, William L. Atkinson, Walter A. Orenstein, and Saul Krugman
Quinolone Antimicrobial Agents, Harold C. Neu

From the *Annual Review of Plant Physiology and Plant Molecular Biology,* Volume 43 (1992)

Chronicles from the Agrobacterium–*Plant Cell DNA Transfer Story,* Patricia C. Zambryski
Cell Biology of Pathogenesis, Adrienne R. Hardham
Host-Range Determinants of Plant Viruses, William O. Dawson and Mark E. Hill
Regulation of the Vesicular-Arbuscular Mycorrhizal Symbiosis, Roger T. Koide and R. Paul Schreiner

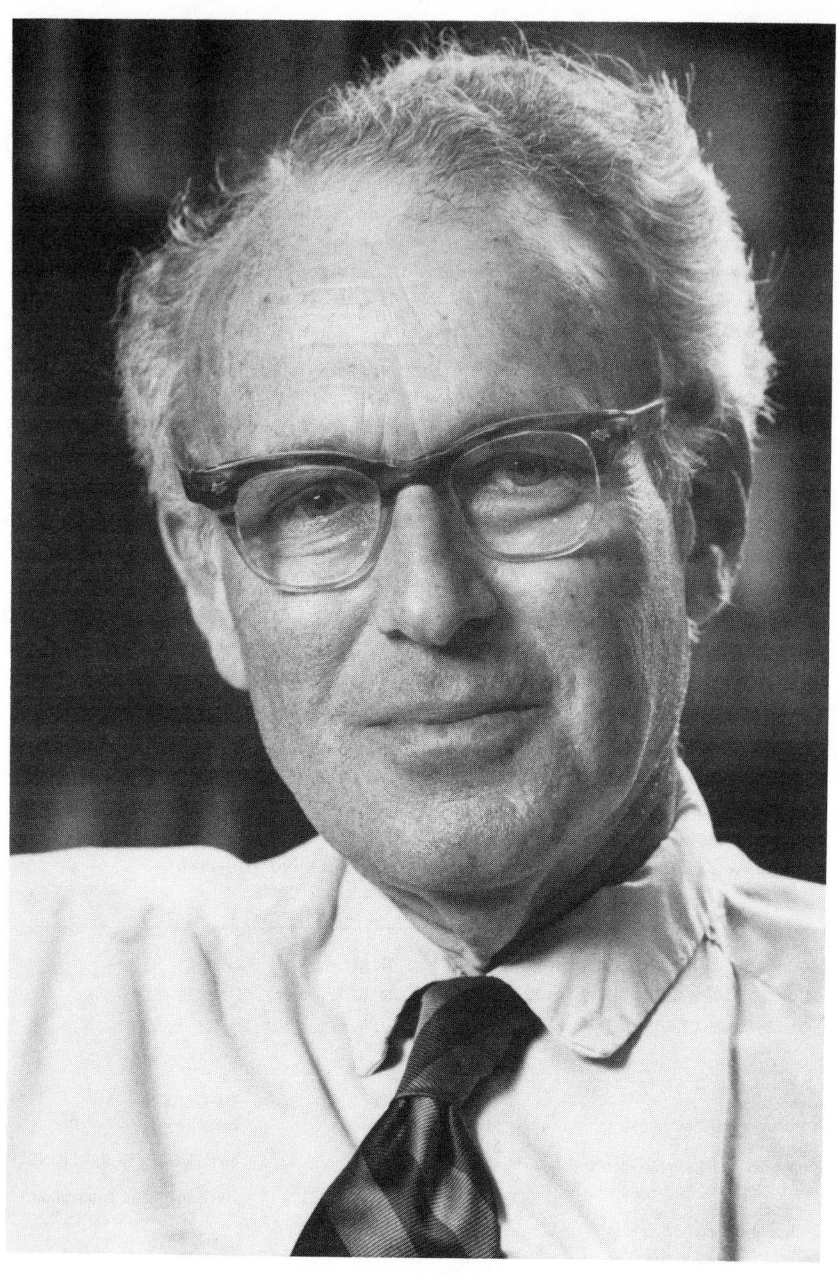

Bernard D. Davis

Annu. Rev. Microbiol. 1992. 46:1–33

SCIENCE AND POLITICS: Tensions Between the Head and the Heart

Bernard D. Davis

Bacterial Physiology Unit, Harvard Medical School, Boston, Massachusetts 02115

KEY WORDS: autobiography, science and society, genetic revolution

---

CONTENTS

Evolution, Genetics, and Society

I consider myself very fortunate in having chosen a career in science at such a favorable time. When the study of genetics finally penetrated into microbiology in the early 1940s, it revolutionized the life sciences, and by natural inclination, I was swept up in this venture.

In this essay, I summarize a career that encountered few obstacles and describe the influence of a few individuals on its shape. But in the final section, I take up another, far more sweeping and controversial domain, the relations between science and society. These increasingly drew my attention in the 1970s, a pivotal period in my life, during which I no longer found full-time preoccupation with my research entirely satisfying.

One reason for this restlessness was that the greatest excitement in the life sciences had moved from microbial to molecular genetics, but I had not

0066-4227/92/1001-0001$02.00

moved with it. But perhaps more important, the world around me was struggling with agonizing questions about justice and race and the meaning of equality and affirmative action. At the time, I was much impressed by a little book by the distinguished humanist and evolutionary biologist T. Dobzhansky (30). He wrote that social equality is a moral and political goal: it is weakened, rather than strengthened, when we tie it to vulnerable assumptions of biological equality, rather than recognizing the reality of biological diversity.

In those turbulent times, Dobzhansky's book fell like a stone. But I shared his concern, as I saw unsound assumptions about human biology being used to distort affirmative action, shifting its aim from equal opportunity to equal numerical results. I therefore decided to spend 1974–1975 at the Center for Advanced Study in the Behavioral Sciences, in Palo Alto, Calif., studying behavioral genetics with the aim of writing a book on human diversity. This work would emphasize the evolutionary aspects of the subject more than the controversial data on IQ. More important, it would articulate the principle that the scientific facts do not prescribe policy, but they should improve it by testing the underlying assumptions.

I did not write the book, partly because the skeptical responses of most of the behavioral scientists at the Center to my nagging about the importance of genetics made it clear that it would be extremely difficult to convince a wider public. During that year, I did publish a transcript of a conference on human diversity, which I organized under the auspices of the American Academy of Arts and Sciences (26). Currently, the academic left is less effective in discouraging study of human behavioral genetics; the public certainly no longer needs convincing that genes are important; and political discussion of problems of affirmative action no longer goes on only under the table.

During the year at the Center, I enjoyed frequent contact with a scholar who had virtually founded the field of the sociology of science, Robert Merton. Toward the end of the year, he remarked that he had never met anyone who had more internalized the canons of science. At the time, this seemed a great compliment. But a decade later, at my retirement party, I offered a different interpretation. It stemmed from my growing fascination with the centrality of evolution, not only for biology but also for our perspective on the nature of the universe and of human beings. This fascination had led me to teach a course for nonscientist undergraduates at Harvard College on evolution, genetics, and society.

Initially, this new and expansive interest led me to share the widely held view that the crowning glory of evolution was the emergence of the human mind, capable of understanding its own evolution. But I gradually became impressed by an additional key principle of evolution: it selects not for maximizing but for optimizing a trait. Hence, if I had internalized the canons of science maximally—focusing as sharply as possible on logical, analytical approaches to problems, and regarding their emotional aspects as diversions—then I was perhaps suffering from too much emphasis on objectivity, at the expense of what we often vaguely call wisdom.

In fact, in trying to give something of a portrait here, I must recognize that my genes seem to have steered me toward a rather skewed dedication to objective knowledge and truth, and to the importance of building on reality rather than on hope. But I recognize that moral values are more important in our daily lives, and they have competed strongly for my interest.

On the other hand, I have been less interested in esthetic and artistic values. This weakness has made me rather unresponsive to some major areas of cultural achievement. For example, in principle, I can recognize beauty in the multiple levels of meaning and the ambiguities of poetry, yet for me, these conflict with the search for clarity and are less interesting. So too with the visual arts—though here my limited aesthetic responses to details may involve a hereditary problem in dealing with spatial relations; as a student I did badly with morphological subjects.

My interest in moral introspection, in contrast to the classical Greek emphasis on beauty as the major goal of culture, fits the traditions of my Jewish background. And as for many other scientists, for me the extraordinarily abstract beauties and mysteries of the branch of mathematics called music present no conflict with the objective external world, and music has been an important part of my life.

In trying to provide some coherence to the picture, I oversimplify it by focusing on personal tensions between the heart and the head. Our whole society exhibits a similar antinomy as it struggles to deal with the explosive growth of our scientific knowledge and powers.

Personal Background

I was born in 1916 in Franklin, Massachusetts, a town of about 7000 located 30 miles from Boston. My parents had met in this country after coming from Lithuania in their late teens. The first immigrant from a region usually was followed by neighbors, and the first generation from a given shtetl usually arranged to end up in the same burial ground. Boston became a center for people from the Vilna province. In the U.S., our not-very-Jewish family name, Davis, was created from the middle name David by an immigration officer because he found the surname on the passport, Borukhovitch, too difficult to Anglicize.

I heard little from my parents about their earlier life in Europe. They focused instead on the remarkable opportunities afforded by this country, on the long and peculiar history of the Jewish people, and on the difficulty of maintaining the religious bonds that had held this people together for so long.

My parents could not arrange for me to learn the Hebrew that had contributed so much to this continuity, so they adopted a widespread American compromise: for instance, I was tutored to enunciate, without understanding the language, the long prayer that is required for Bar Mitzvah. This irrational solution so disturbed me that many years later, on a sabbatical in Israel, I found myself reacting equally irrationally: trying to learn Hebrew proved too unpleasant, though I had learned several other languages with pleasure.

Another experience turned me against religion very early. My maternal grandmother came as a dowry with her youngest daughter, my mother. She never learned English, though our town had only half a dozen Jewish families. Her main concern was strict adherence to religious rituals (including peasant superstitions that she thought were religious). She would not turn on the electric light on the Sabbath, because it was equivalent to making a fire and therefore was forbidden work. But she allowed us children to turn lights on for her convenience. I recall vividly my hand on the switch at the stairway, at the age of perhaps 12, thinking that if this system would help her to get into Heaven I wanted none of it. Thus began, however primitively, my shift to atheism, which I later decided not to soften with the euphemism "agnosticism."

Much has been written about the Jews of the lower East side in New York and their intense political, religious, and literary conflicts, but a second form of the Diaspora, scattered in small towns, has been neglected. As was typical, my father learned English at night school, was lent enough capital for a peddler's pack, and was helped to find a town that did not yet have a peddler. He eventually came to own a retail store—a mode of assimilation that became widespread in the small towns of America.

These scattered Jewish families lacked the social solidarity of those in the urban community, and so they were inevitably more alienated from their earlier culture. We children certainly suffered a good deal of teasing for our differences, but it was a pretty mild anti-Semitism. On the other hand, the relative isolation in small towns also provided direct benefits, by diluting competition. Arthur Kornberg's autobiography describes his bitter experience at City College in New York, where only one out of 200 Jewish premedical students got into any medical school. My experience at Franklin was quite different. The principal of the high school followed up my application to Harvard by taking me to visit the director of admissions and describing me as a promising student. This seemed to assure my admission. My subsequent record in the college made admission to Harvard Medical School seem almost automatic.

My father was intensely dedicated to providing the best educational opportunities for his children. He succeeded, but at great personal cost. He had built a flourishing store, but it was destroyed by a fire, with little insurance, in

1928; he borrowed to rebuild, and in 1929 came the depression. My brother, the valedictorian of his high-school class, had just started at Harvard. We three younger children also all became valedictorians, and my father took on the expense of sending us all through Harvard or Radcliffe—and then me through medical school. After his death a decade later, the family was moved by the vivid evidence of what sustained him: the first item in each of his account books was a report on some child's academic achievement.

I later felt guilty for not appreciating his sacrifices—and at having criticized his preoccupation with the perpetual problem of money. (I found asking for money very uncomfortable; later every grant application was a burden.) As high-school students, we children clerked in the store. I recall the agony of adult customers, in the depression years, deciding whether to spend ten cents or fifteen cents on a Christmas present. My father honed our skills in arithmetic by practicing sums during our family's Sunday drives. After I entered Harvard, I proudly demonstrated the slide rule to him that I had had to buy for a course. "You set 2 here, and 3 there, and here you see the answer, 6." I will never forget his expression as he asked, "For this you paid $2.00?"

While my mother encouraged pleasant relations with non-Jewish neighbors, she also emphasized that for real trust "your own is your own." But many years later, when she was more assimilated, and also desperate that I was not yet married at 39, she wholeheartedly accepted a non-Jewish daughter-in-law.

My father tried to persuade my mother to work in the store, but she insisted on caring for the children full time. I also chose a wife whose primary aim was raising a family, and I followed my father's pattern of being excessively preoccupied by work. Although I felt that time spent in scholarly activities was more justified than time spent making money, the result was similar: less time spent with the growing children, which I now regret.

Premature death of my grandparents in the old country, followed by remarriage, led to my having a number of half-aunts and half-uncles. As a child, I noticed that the set who shared a particular stepmother were kindly, sweet people, while the direct children of that mother all had difficult personalities. I cannot help wondering whether these striking differences might not have initiated a sensitivity to the role of genes in behavior.

For some years, my brother aspired to a career as a violin virtuoso. I played piano, less well, and chamber music became a major avocation, shared with my brother and later with my wife and son (a professional cellist). I was a poor athlete, and withdrawal from competition with my skillful older brother probably helped drive me to the excellent town library. The science teacher in our high school was hardly inspiring, and so medicine seemed the natural outlet for my curiosity about how things work.

I began school in a one-room schoolhouse that served both the first grade (which stayed until lunch) and the second grade (which left earlier and returned after lunch). One day, in the first grade, the teacher asked me to return after lunch—and so I had skipped a grade and was a young member of my classes thereafter. (That school was recently reported to be the oldest functioning one-room schoolhouse in this country.)

An interest in language showed up early in my life, along with stubborn independence and excessive confidence in the power of logic. I learned to read before I began school, and when my mother helped me take out my first book from the library, I had great difficulty with the beginning—realizing only years later that I had struggled with the introduction for teachers. Also, I can still picture my nose pressed against a window pane, staring at "Union" on the sign at the street corner, knowing how the letter U is pronounced, and trying to figure out by logic how one could spell the word for onion if not with a "u."

In graduating from high school, I learned a lesson in future grantsmanship. The topic I chose for my valedictory address was Creative Chemistry, but the teacher insisted that all the essays have a central theme: the bicentenary, that year, of the birth of George Washington. Linking these two themes seemed impossible, until inspiration struck: "Little did George Washington dream that chemistry. . . ." The teacher was satisfied with this opening sentence, and the audience received a buoyant 1930s message on progress through chemistry.

College: Medicine, Chemistry, Biology

On entering college in 1932, I planned to seek a broad education in history and literature before moving on into a career in medicine. But I soon switched to concentration in biochemistry, for I had no difficulty in obtaining a grade of A in courses in science, mathematics, or language, but I could not get a better grade than C in the introductory history course, with its relatively large volume of reading matter, no matter how I tried.

In a curious episode connected with that switch, a brilliant upperclassman concentrating in history, who had just been elected to Phi Beta Kappa as one of the top eight in the class, had decided to go to medical school, and so he was taking the introductory physics course. He was having difficulty with the material and he asked me to study with him to help get him oriented. It soon became clear that one of us could assimilate the content of science easily, while for the other it had a different Gestalt. He returned to history (and particularly the history of discovery); years later he became Librarian of Congress. It seems to me that each of us develops early a few key ideas that spring up repeatedly in our subsequent intellectual life. It therefore may not be too farfetched to suggest that this episode, like that involving my aunts, further encouraged my later growing interest in the importance of genetic differences for our behavior.

Another theme, prominent in much of my scientific life, turned up in my undergraduate honors thesis: an interest in interactions of proteins more complex than the Michaelis-Menten relation between enzyme and substrate. This field became much more important later, when the atypical functional interactions could be correlated with shifts in shape (i.e. allostery). My research in this area, on the hemoglobin of a species of fish, was guided by a very kindly and encouraging teacher, D. Bruce Dill.

The oxygen dissociation curve of most hemoglobins, measured by equilibration with the gas at various tensions, has a sigmoid curve rather than the hyperbolic one predicted for a first-order reversible association. This property greatly increases the amount of oxygen that the hemoglobin transports between the lungs and the tissues in each round of circulation, and physiologists had long regarded it as a triumph of evolution in making circulatory systems efficient. Many years later, at the Cold Spring Harbor Symposium in 1961, an allosteric enzyme within the bacterial cell was shown to have a similar sigmoid relation between concentration of substrate and the amount bound (which in turn regulates the catalytic activity of the enzyme). Jacques Monod emphasized the physiological value of this mechanism for amplifying the sensitivity of the regulatory response. Having benefited from the breadth provided by a medical education, I was delighted to be able to point out the close parallel between the already well-explored hemoglobin system and the newly discovered allosteric regulation of an enzyme.

I was particularly interested in this parallel because it is striking how often the specialized products required for the physiology of higher organisms are not new in the sense of a fundamental new property of a molecule: the three billion years of prokaryotic evolution provided a wide array of nuts and bolts, which the additional 600 million years of evolution of multicellular organisms could then combine and permutate. Their novelty lies in the variety of ecological niches that they allow organisms to fill, and the mechanisms evolved for these purposes, more than in unusual features of the molecules. Moreover, I am amused by my own sense of gratification when one of the presumably late marvels of evolution is found to have originated in bacteria—"we" got there first. This identification with one's material is probably widespread among scientists, defining their turf.

At the same time, it is humbling to realize that the biomedical sciences have generated a rather constricted notion of the range of properties of living matter. If the familiar forms of life at the earth's surface should disappear, evolution would not necessarily start from scratch again, at the prebiotic stage. Some remnants of the DNA pool could still be maintained in what we regard as "peculiar" organisms: for example, those that thrive at the pressures encountered at the deep-sea bottom, and at temperatures above 100°C.

Clearly, our notion of the normal range of properties of living matter tends to be anthropocentric. But it is broadening now, as an interest in the whole panorama of evolution, starting from its prebiotic phase (the "origin of life"), increasingly replaces medicine's domination of the life sciences. But an evolutionary perspective came to me much later. When I was an undergraduate, the curriculum paid little attention to genetics and evolutionary biology compared with physiology.

I did my thesis research in the Fatigue Laboratory, a curious organization where Dill had done most of the experimental work underlying L. J. Henderson's theoretical studies on blood as a physiological system. Henderson's search for broad principles fascinated me, especially as synthesized in his imaginative book on *The Fitness of the Environment*. But by that time he had become infatuated with the scientism of Pareto, whom he saw as creating a scientifically rigorous sociology. Henderson therefore had less influence on my intellectual development than one might have expected. Nevertheless, I certainly absorbed from him the principle that one should seek reciprocal interactions of multiple components—a system—rather than simply linear causal relationships. Also, his views might have contributed, subliminally, to my own later interest in interactions of science and society.

During college, I spent a summer at Woods Hole with my biochemistry tutor, Ancel Keys, doing analyses of sea water. This experience began a life-long association with that unusual community. I taught in the Marine Biological Laboratory, wrote for many summers in its library, and enjoyed activities, as a regular summer resident, that have greatly enriched the lives of my family members.

By the end of college I had difficulty deciding whether to go on to medical school or to seek a PhD in chemistry instead. I was attracted by the intellectual elegance of physical chemistry, which extracted general principles from empirical details. I also had been inspired by George Kistiakowsky's graduate course in physical chemistry. The deciding factor for me was probably the realization that chemistry then offered very limited career opportunities for a Jew, while a medical degree provided a broader set of options—including that of going into business for oneself if necessary. In those days, virtually no one sought the double MD/PhD degree, and I finally pursued the MD.

In making this choice, I was excessively confident that a good grounding in the powerful tools of physical chemistry, which I had formally mastered in Kistiakowsky's course, virtually guaranteed success in solving the presumably easier problems of medicine. I did not conceal this overconfidence, and it resulted in a painful comeuppance. I had the impression that I was expected to end up with highest honors. But having read every paper on hemoglobin, I incorrectly assumed that my thesis should correct the many errors that I found in that literature. The chairman of the oral examination, Professor Edwin J.

Cohn from the medical school, was neither impressed nor amused, and he had strong views on how people should behave. Hence, I ended up with a disappointing magna cum laude rather than a summa.

My interest in music came alive in college. Previously I had dutifully taken piano lessons, but teachers in that era discouraged sight reading. Now, playing piano again for pleasure, and without a teacher, I found that reading at sight came easily to me. For a long period I studied until midnight and then climbed seven flights of stairs to a room in the tower of Eliot House, where there was a Steinway that one could play at any hour.

To round out the description of my college experience: I was shy, and awkward with girls, and I did not take much advantage of the opportunities for social development. Of course I had a circle of friends, but with little money to spend we focused more on serious discussions than on seeking fun.

I was indeed a very earnest student. Perhaps I was responding to my father's hopes and sacrifices more than I realized. In 1936, I moved on to Harvard Medical School, dreaming that I could become a scientist, but also with a sense of obligation to help patients directly.

Medical School and Internship

Fairly soon after entering medical school, I arranged for part-time research, pushed by a fatherly upperclassman. He pointed out that E. J. Cohn had one of the few laboratories that was studying proteins in depth and that if he got to know me better he would surely overcome his unfavorable reaction to my smart-aleck thesis. My friend suggested that I should therefore try to work directly under him. That is what I did.

I spent a great deal of time in Cohn's laboratory throughout medical school, almost like a graduate student. At that time Cohn and some other investigators were struggling to move the study of proteins from the morass of colloid chemistry into the world of definite chemical structures. Cohn believed that study of the electrical properties of the whole molecules, correlated with similar studies on peptides, was the most rewarding feature to pursue at that time.

The studies of Cohn and John T. Edsall in this area did indeed add a great deal of useful knowledge. However, it seems clear now (though it may not have been so clear to me then) that Cohn's forceful direction of all the efforts of his highly coordinated laboratory along one line caused him to miss a more fruitful approach that emerged later: the development by Stein & Moore, and by Sanger and others, of the analytical tools for studying a purified protein as an organic molecule with a definable sequence.

Cohn set me to work studying the electrophoretic mobility of hemoglobin with the traditional cylindrical U-tube, using the color of the protein to locate the boundary. Electrophoresis had not then been used much as a tool for studying proteins. Tiselius in Sweden scooped us by introducing sophisticated new technical developments, including an optical system that depended on refractive index and hence could be used to analyze mixtures of any proteins.

What Cohn and I reported, in my first publication (how important to the budding scientist!), now seems very obvious: electrophoretic mobility varies not only with pH (as was already known) but also with ionic strength (23). Developing the theory that would explain the quantitative features of this effect was beyond me, and so I was not unhappy to drop the problem at that stage—the beginning of my shift of interest from the chemical toward the biological aspects of biochemistry. I did not foresee how powerful electrophoresis would subsequently become as an analytical and preparative tool.

Cohn had an extraordinary personality, and the members of his laboratory competed with anecdotes about his harshness—yet with appreciation for his intellectual standards. As an example of his need to control, he renamed me "Ben," and others in the laboratory followed his lead. But they also depended on him for a job, while as a medical student I was more independent. When we finally sent off a joint manuscript he said "Ben, you've fought me every step of the way, and I respect you for it." Moreover, I am sure he had a hand in the decision to award me highest honors on graduation from medical school.

Cohn shrewdly foresaw that World War II would bring in our country, and that it would create valuable opportunities for research. He built up perhaps the first sizable research program in the basic biomedical sciences to be funded by the government: fractionation of bovine plasma. During my last year of medical school (1939–1940) I participated in this project with enthusiasm, under the guidance of Thomas McMeekin. The project did not succeed, for it was based on the false assumption that purified albumin might serve as a blood substitute without causing the immunological reactions encountered with whole foreign plasma. But Cohn showed his usual capacity for extracting valuable dividends from a bold and even perhaps rash program, and the fractionation of blood that he began has continued to be a fruitful enterprise.

I learned a good deal from E. J. Cohn about how to direct a laboratory—including more than I realized about why one should not direct people too closely. My success in standing up to him no doubt reinforced my tendency to rebel against authority.

Of course many other activities in medical school aroused my interest, and they had a strong enough attraction to keep me in clinical work a bit longer. Many scientists have proceeded directly from medical school into research without an internship, but I did not lose easily the sense of obligation to try to be a warm physician as well as a scientist.

The choice of hospital was easy. The chairman of the Department of

Medicine at Johns Hopkins Hospital, learning that the Tiselius apparatus was a powerful new tool for analyzing proteins, offered a fellowship, combined with a part-time internship, so that I could build such an apparatus and use it to seek novel abnormalities in plasma proteins. Unfortunately, despite my substantial laboratory experience, I did not have the maturity to make the most of the opportunity. I had a technician who analyzed a random plasma from the clinical laboratory each day, and we encountered many new patterns; but I did not follow through with most of them. For example, buried among them are the first agammaglobulinemia and the extraordinarily high concentrations of a novel protein in multiple myeloma. But evidently such descriptive findings, however novel, did not seize my interest, probably because they lacked an accompanying mechanistic explanation.

Meanwhile, I made my first discovery that seemed to me quite original. The sulfonamide drugs were receiving a great deal of attention at Johns Hopkins, and at a department meeting someone reported that sulfathiazole, unlike sulfanilamide, did not reach nearly as high levels in the cerebrospinal fluid (CSF) as in the plasma. The accepted explanation was a limited ability of sulfathiazole to penetrate into the CSF. But it seemed to me that the problem was one in the physical chemistry of a distribution at equilibrium: both drugs might penetrate freely, yielding the same concentration of free compound in the two fluids, but additional sulfathiazole might bind to something in plasma (very likely a protein). To test this hypothesis, as an intern without a laboratory, I equilibrated plasma in a cellophane bag against buffer containing either drug and then sent the samples to the clinical laboratory as though they came from a patient. Sure enough, sulfathiazole was extensively bound to plasma proteins but sulfanilamide was not (3).

This finding led me to add to a small literature that demonstrated the ability of serum albumin to bind a remarkable variety of compounds. I published a review, of which I was quite proud (4), emphasizing the physiological importance of this binding—not only in influencing the distribution of drugs in the body, but even more in protecting cells from many toxic compounds, both endogenous and foreign.

As an intern, I was conscientious but bored by many details, except for those patients whose problems suggested an interesting scientific challenge. The bulk of medical practice does not fit this bill. So I was not a very good intern, and I was not invited to continue into a residency.

In addition, the social attitude toward African Americans in Baltimore, and at Johns Hopkins Hospital, disturbed me. The white and the "colored" patients, though receiving identical care, had separate wards, blood banks, and toilets. I was shocked to learn that a nurse was not allowed to take orders on a colored patient if the intern referred to him as Mister rather than by his first name. I have been pleased to see how well Hopkins now deals with the problems of integration and of recruiting black students—perhaps with more balance than my own school.

At the end of my internship, I entered the U.S. Public Health Service. I might not have known about this relatively obscure alternative to the regular armed forces, except that the wife of the Surgeon-General had been my patient. Also my part-time internship included routine duties on the metabolic ward at Hopkins, a research unit that was concentrating on study of the newly available adrenal steroids, and on the basis of rumors that German aviators were using these hormones to increase altitude tolerance, the Hopkins unit had initiated a cooperative research with a small aviation medicine facility at the NIH in nearby Bethesda. On entering the USPHS I joined that unit, where I completed a study begun by another research fellow. The results did indeed show that rats given deoxycorticosterone survived a higher simulated altitude.

I might note an additional interest that arose during medical school: the study of foreign languages. Leaving dormitory life, I moved to a rooming house run by an impecunious Boston dowager, in which most of the other tenants were German refugees. I enjoyed the chance to learn conversational German, and in later years, I created a similar opportunity to learn Russian.

U.S. Public Health Service; Tuberculosis

After I spent a brief period in aviation medicine, the USPHS assigned me to set up a study of biological false-positive serological tests for syphilis, which were exempting significant numbers from the military draft. The Service hoped that I could discover some properties of the serum proteins that would distinguish the false reactions. I felt that I needed guidance in immunology, which I was fortunately able to obtain in the laboratory of Elvin Kabat, at Columbia University College of Physicians and Surgeons. Thus began a long friendship, as well as the first of a series of appointments in various research institutions of New York City.

This research failed to solve the problem, but it taught me a good deal. Elvin handled his part of the arrangement very conscientiously and modestly, but I must confess that I, being accustomed to being on top of my problem, found it uncomfortable at times to be working with a master of the field who was always one jump ahead of me. Among our findings we obtained and characterized a pure solution of the Wassermann antibody (at a time when any pure antibody was hard to obtain) by simple ether extraction of the antigen (cardiolipin) from the precipitate that it formed with positive sera (29). The antibody turned out to be a larger molecule than most antibodies, and it was later identified as IgM. We also found that the pure antibody (or any isolated gamma globulin fraction) destroyed the activity of complement in the complement fixation test, and the effect was prevented by restoring the albumin that would normally be present in the serum in the test (27). I am not sure that this interaction has yet been explained.

I was also invited to review the literature on biological false-positive tests for syphilis, and in the medical-student tradition of being thorough and dutiful, I took pride in tracking down every reference. I later lost interest in erudition as a goal, compared with the discovery of significant novelty. But respect for historical continuity is another matter: I still enjoy introducing material within a historical framework, and I have regretted the need to cut down the history in the successive editions of a textbook.

Though the Heidelberger-Kabat school of quantitative immunochemistry was then at the leading edge of immunology, I did not enjoy doing repetitious nitrogen analyses. After this project, the USPHS offered another opportunity in immunology, which might have fit my interests better, in the laboratory of Jules Freund (at the Public Health Research Institute of New York City). His war project aimed at using his powerful new immuno-adjuvant technique to try to develop a vaccine against malaria. My assignment was to develop a complement fixation diagnostic test based on fractions of the blood of our infected ducks and monkeys (6). But my stay was short because I found it difficult to accept Freund's insistence on secrecy. My position was very uncomfortable, but I fortunately received strong support from Robert Loeb, a forthright person whom I had come to admire while at Columbia.

Meanwhile, World War II drew to a close. I had expected to return to academia, but I had already seen that one could achieve great flexibility and independence, and excellent facilities for research, within the theoretically bureaucratic confines of government. The USPHS created a Tuberculosis Control Division at the end of the war, and the director of its research program, Carroll Palmer, invited me to set up a basic science research laboratory, at whatever site I thought most suitable. This opportunity seemed too good to pass up. Cohn, however, was furious, professing that his training had not been designed to be wasted in a second-rate institution. In fact, even though Carroll Palmer was disappointed that he could not arouse in me a deep interest in the intellectual challenges of his field, epidemiology, he provided a budget for my research on virtually no basis except personal confidence, and he gave me extraordinary freedom to pursue any leads.

To direct a tuberculosis research laboratory effectively, I needed a background in bacteriology. In medical school, this subject had been largely descriptive. Moreover, research on tuberculosis was carried out mostly by convalescents in sanatoria and in general was not very exciting. But Rene Dubos was then enthusiastically describing a new approach to tuberculosis, in an outstanding institution, the Rockefeller Institute. I therefore sought training in his laboratory.

My interactions with Dubos were a very important part of my development. It was an intellectually and culturally inspiring period for me—even though the science was not as rigorous, in retrospect, as what I had encountered with Cohn and with Kabat. As I have already described this experience in some detail, in a book celebrating Dubos' contributions to the development of antibiotics (19), I cover it here only very briefly.

Reminiscing about the contrast between Dubos and his teacher Selman Waksman encouraged me, belatedly, to rethink a rather snobbish attitude of admiring only those scientists with brilliant, often romanticized, ideas, while undervaluing those whose systematic, persistent, and even pedestrian approach laid solid foundations or provided practical benefits. Today, the growth of biotechnology has broken down the earlier barrier between biologists in academia and in industry. Though this development has created conflicts of interest, the benefits, including faster discovery and marketing of useful products, seem to be far outweighing the costs.

It further struck me that even though Waksman received his Nobel Prize for the discovery of the first effective drug against tuberculosis, his most valuable discovery was the principle that a persistent search for useful antibiotics will pay off. Dubos, in contrast, was too impatient to continue the search when his early antibiotic proved to be too toxic. His charismatic style subsequently led him increasingly to activities with wide public appeal, especially in the environmentalist movement, so we regrettably lost contact.

Incidentally, Dubos immensely admired the very different, profound style of Oswald Avery, and I am sorry that he did not introduce us while I was at Rockefeller. Among the friendships that I did develop there, I particularly treasure that with the highly original Rollin Hotchkiss, who was continuing the Avery work on genetic transformation.

My main discovery during the period with Dubos stemmed logically from my earlier work. By adding both a nonionic detergent and Cohn's Fraction V (albumin), Dubos developed culture media that provided somewhat faster and more dispersed cultivation of *Mycobacterium tuberculosis*. I found that the detergent was slowly hydrolyzed, and the released free fatty acids were very toxic to mycobacteria. The albumin neutralized that toxicity by its tight binding of the fatty acid (24). This finding solidly established the scavenging property of albumin that I had discussed earlier, and also the principle of a nonnutrient, protective growth factor for bacteria.

In this work, I further noted that normal serum contains a low concentration of free fatty acid—but as a few sentences inserted in a paper on methods, this finding was buried. Later, other workers rediscovered this fraction and showed that despite its small size its rapid turnover gives it an important role in lipid metabolism.

The period with Dubos had an additional, unexpected impact on my life: infection with the tubercle bacillus, with a minimal surface lesion that caused a persistent pneumothorax. Perhaps persons who were tuberculin-negative (as I was), and hence lacked the partial immunity of those with a positive test, should not have undertaken work with virulent tubercle bacilli. And we certainly should not have employed careless techniques, as we did in a macho manner, proud of the medical tradition of taking risks for the benefit of others.

Physicians were then unwilling to use the toxic drug streptomycin on minimal cases, and the pinhole of my pneumothorax finally required surgery. Meanwhile, the traditional treatment, prolonged rest, gave me a sort of early sabbatical for more than a year. I think it was very valuable, at a time when I was ruminating about the program that I expected to launch on recovery. Much romantic literature has been written about the predilection of tuberculosis for people of talent and even genius; but I suspect that if there is any correlation it is because the prolonged rest gave the victims the chance to meditate about what they really valued.

My reading during this period turned more to philosophy than to current science, and I read Bertrand Russell's *History of Western Philosophy* twice, marveling at the clarity of his style. I intended to continue to read philosophy after recovery, but that interest did not persist. Philosophy has increasingly struck me as mostly an intellectual game, enjoyable for those who choose it but of declining importance in an age of science.

The scientific reading that most fascinated me was a review by George Beadle on the use of biochemical mutants of the mold *Neurospora* (i.e. those blocked in a biosynthetic step) as tools for genetic and biochemical studies. It seemed to me that such work, on universally distributed biosynthetic pathways, should be deeply satisfying because it was near the trunk of the evolutionary tree, while attempts to grow bigger and better tubercle bacilli were only twigs.

In developing the new tuberculosis laboratory, I was probably reluctant to expose persons without any immunity to virulent tubercle bacilli. So we launched a Tuberculosis Research Laboratory without any tubercle bacilli! It was located in a New York City health facility near the Rockefeller. The city lent the space to the Department of Preventive Medicine at nearby Cornell, and in turn, the chairman, an epidemiologist with no use for labs, lent them to us.

Bacterial Genetics and Amino Acid Biosynthesis; New York University Medical School

At first I floundered in searching for a focus for our new laboratory. But Beadle's review had already planted a seed in my mind, and it suddenly germinated during a seminar, when the speaker noted that biochemical mutants of *Escherichia coli* had technical advantages over those of *Neurospora* but were more difficult to isolate. Recalling that penicillin kills bacterial cells only when they are growing, I realized that in a culture growing in minimal medium it should kill off the wild-type *E. coli* while allowing any rare mutants with an additional growth requirement to survive.

The same idea occurred to Lederberg & Zinder, who submitted a letter to the *Journal of Biological Chemistry*. Lederberg generously offered to ask the journal to hold up their manuscript if I wished to send one immediately. The journal rejected both papers, on grounds of insufficient biochemical interest. We demonstrated our annoyance by publishing the two short papers in a chemical journal (5), but they really were papers on a bacteriological method. I am more pleased to recall the spirit that led us to have the two reprints bound under a single cover. Because independent discoveries in science do not represent a zero sum game, it does not seem to me that simultaneous announcement diminishes credit to either party.

I soon had a more extensive collection of mutants than the *Neurospora* group had accumulated in years (7). I named this class "auxotrophic" (for the additional nutritional requirement). I also introduced the terms "phenome" and "phenomic lag" to explain why my initial experiments, in which cells were exposed to penicillin immediately after mutagenic irradiation, had failed: any alteration in the genome would not yet have been expressed in the phenome. The cells required some growth after the irradiation, to allow phenotypic expression, before killing by penicillin could be selective.

Because I had never studied genetics, I arranged to give a summer course in Beadle's department at Caltech, as a way to learn some genetics and to become acquainted with that outstanding group, as well as to tout the virtues of bacteria. I also discussed with Lederberg a variety of challenging problems that I had thought of in the wide-open new field of bacterial genetics. But almost invariably he had already ruled out the value of each or already done it. I decided that competition with him in genetics would be much less profitable than mining the intermediates in biosynthesis, which auxotrophic mutants accumulated in large quantities.

Succumbing to the easy prosperity afforded by this field was probably a mistake for me, in the light of my later interest in more complex biological mechanisms. This was all the more true because I depended on associates for the chemical identification of our novel intermediates. And though I took the phage course at the Cold Spring Harbor Laboratory—which initiated a treasured friendship with Max Delbruck—I still did not get deeply into bacterial genetics itself. Van Niel's famous summer course at Pacific Grove on general microbiology, which I audited, had a stronger influence on my interests and my approach to problems.

Nevertheless, it seems to me that the role of the explosive advances in bacterial genetics, as a major foundation for much of the early work in molecular genetics, was taken for granted all too quickly. I made one contribution to the early phase of bacterial genetics: use of a U-tube with a sintered glass barrier to separate two of Lederberg's conjugating strains but allow the surrounding medium to be pumped back and forth. The results showed that conjugation requires cell contact and hence is not a process of transformation by a labile substance in the medium (9).

In studying biosynthesis, I did undertake one prolonged program: working out many of the steps in a common pathway of aromatic biosynthesis, leading to tyrosine, phenylalanine, tryptophan, and *p*-aminobenzoate. This path also led to a previously unknown growth factor, *p*-hydroxybenzoate (8), which others later found to be a precursor of a quinone cofactor. The first intermediate that we identified in the common pathway was an already known but obscure natural plant product, shikimic acid. This intermediate was accumulated by mutants blocked immediately after its production, and it supported the growth of those blocked earlier (10). It is gratifying that the shikimic pathway has given rise to several books and to a review of the past decade with over 500 references.

I will not dwell on my early contributions to this pathway, but I would like to acknowledge my debt to Roger Stanier. He was studying the breakdown of aromatic compounds by soil bacteria, and he suggested, and supplied, the shikimic acid that turned out to be an intermediate in my pathway (but not in his). The sample was prepared for him by H. O. L. Fischer from the dried fruit of the shikimi tree, obtained from a Chinese pharmacist. The fruit, which contains an alkaloid as well as shikimic acid, is used as a laxative, and the pharmacist originally pretended not to know of it. Stanier fortunately learned why and persisted: the fruit is apparently also used traditionally, in larger doses, to poison one's mother-in-law (11).

My associates identified many intermediates and enzymes in the aromatic pathway, as well as in pathways to several other amino acids. These investigators included Ulrich Weiss, Ivan Salamon, Susumu Mitsuhashi, Charles Gilvarg, Elijah Adams, Edwin Kalan, and Henry Vogel. Our approach could not tell us how the aromatic pathway branched off from the central metabolic pathways (which I named amphibolic, for both catabolic and anabolic). To solve this problem, David Sprinson at Columbia College of Physicians & Surgeons initiated a long, enjoyable collaboration. Using precursors radioactively labeled in specific atoms (some at 5 counts per minute above background!), and then enzymes, he showed that three of the atoms of shikimic acid come from phosphoenolpyruvate and the other four from erythrose-4-phosphate to yield 3-deoxyarabinoheptulosonic acid-7-P (37). I admired his patient and thorough approach, as an organic biochemist, because I tended to seek problems with intellectual challenges but easy technical solutions.

Identifying the sequence of a pathway opens up the possibility of studying its regulatory mechanisms. Mitsuhashi worked on this problem, but we used bioassays that were not very satisfactory and we did not publish the results. I was deeply interested in problems of regulation (14), but I failed to follow through and clearly establish the role of feedback in repressing as well as in inducing enzyme synthesis (which Vogel and I did note in an abstract). This was probably the largest mistake in my choice of directions. One factor may have been my propensity to intermittent depression at that stage in my life.

My group also made interesting contributions in some other areas. The role of the tricarboxylic acid (TCA) cycle in *E. coli* was then very much in dispute because the organism could not metabolize citrate, supplied exogenously, and yet it could oxidize acetate. Gilvarg settled the matter by showing that a glutamate auxotroph lacked citrate synthase, and it could not oxidize acetate (31); hence the oxidation in the wild-type proceeds via endogenous citrate. This is an excellent example of the sharp tools provided by mutants.

Carl Hirsch, Hans Kornberg, and Chandra Amarasingham further pursued work on the TCA cycle in the lab to which I subsequently moved. The results included the finding that succinate dehydrogenase is distinct from the anaerobically induced fumarate reductase. Chandra encountered something quite surprising. Under anaerobic conditions, the organism does not oxidize acetate, and as we expected, it does not make the superfluous α-ketoglutarate dehydrogenase. But the enzyme was also absent from aerobic cells growing happily on glucose. Instead the acetate accumulates until it reaches a critical, quite high concentration of free acetic acid, which induces formation of α-ketoglutarate dehydrogenase and thus completes the cycle (1). Thus, under optimal conditions, the culture initially extracts only one-third of the available energy from glucose and stores the remainder in the medium. The teleonomic significance of this regulatory response is not clear. Unfortunately Chandra died prematurely, and this finding has not been pursued by students of the cycle, though it seems to me quite fundamental.

At NYU, Werner Maas was primarily responsible for the first strong evidence that a gene can affect the structure, rather than only the quantity, of its product. He showed that a temperature-sensitive pantothenate auxotroph formed a temperature-sensitive enzyme (34). We often discussed ways of generalizing this finding, but we failed to formulate clearly the later concept of conditionally lethal mutants. Any scientist can look back and see boats he should not have missed, but this was a large one!

My chief in the USPHS finally decided that he could not justify my work in his tuberculosis program. At the same time I was offered chairmanship of the Department of Pharmacology at New York University Medical School, a school that was more willing than most to make unorthodox appointments based on future promise. I knew little pharmacology, and I have been amused to meet former students who recall with pleasure that I emphasized principles. They did not know that I had little else to offer!

Just when I was moving from the USPHS to academia, in 1954, FBI agents visited me and unsuccessfully pressed me to identify Communists whom I might have met as a moderately radical student. The experience was frightening, and it sensitized me to the heavy hand of government. My flirtation with a communist cell in New York reflects the astounding success of the Party in securing converts among people distressed by the economic injustices so prominent in depression years, but the secrecy and deception required by this approach to politics made me very uncomfortable. My experience with committed Marxists also sensitized me to their frequently tricky techniques in academic disputes.

Before starting at New York University, I spent two months in the laboratory of Jacques Monod in Paris. It was an exciting period there; the work on gene regulation was gathering momentum. But I was an observer more than a participant, and I failed to build on this exposure. I think my research has been limited by a conservative reluctance to use new techniques.

I enjoyed very much becoming a teacher at NYU, though more in the theatrical role of the lecturer than that of a teacher relating to individual students or small groups. But I am impressed by how much a teacher can influence students, even as a lecturer, if he is willing to go beyond conveying facts and offer guidance on principles and values.

My research during the three years at NYU wound up our contributions on biosynthesis. I enjoyed extracting from the wealth of detail a broad generalization about intermediary metabolism: "On the Importance of Being Ionized" (13).

I also discovered the first specific transport system in a bacterial membrane and its inducibility (12). *E. coli* ordinarily cannot use citrate as a carbon source, but we had shown that it does metabolize endogenous citrate via the TCA cycle within the cells. The block therefore had to result from impermeability. Hence, when we encountered conditions that permitted the cell to use exogenous citrate, without any general increase in permeability, I had to conclude that we were inducing a specific transport system for uptake of the compound.

At that time, the idea of adaptively changing the composition of a morphological unit such as a membrane, and even more the idea that the tiny bacterium could have a wide variety of specific transport systems, seemed wild. Indeed, in publishing this work (13), I could only credit with a footnote some important data produced by a subsequently very distinguished postdoctoral fellow, Howard Green, who feared that such a claim might destroy his reputation. I also had difficulty persuading Monod, who had just discovered accumulation of nonmetabolized β-galactosides in bacteria, that he must be dealing with specific active transport rather than with binding to intracellular constituents. Monod soon went on to develop the "permease" as a major contribution, while my further studies on citrate were not fruitful. But I thought he was excessively ungenerous in utterly ignoring the logic of my pioneer finding.

In moving from Paris to NYU, I invited Luigi Gorini, who had encountered an undefined aromatic growth factor, to join us. A collaboration between him and Werner Maas led to an important contribution to our understanding of regulation: feedback in the arginine pathway not only provides economy in response to an exogenous supply of arginine, but it ensures a proper level of endogenous synthesis, over a wide range of conditions. In steady-state growth, this level is usually far below the cell's capacity, but the reserve capacity allows a rapid adaptation to shifts in the supply of nutrients.

I saw a good deal of Leo Szilard, whose main interest had moved from physics to microbial genetics. He lived in a hotel in New York and circulated through nearby laboratories, critically evaluating their latest experiments and generously offering ideas that led to many valuable publications. He was the cleverest and most cerebral person I have ever met, and like his other young friends, I was overawed by him. But his character could present problems. He applied to the National Science Foundation for a lifetime salary, with no restrictions. As a member of the Committee that ruled on the application, I supported it as an appropriate way for society to recognize and assist this genius. But I failed to persuade the Committee to approve the grant—I believe because I had failed to persuade Szilard to make even a token commitment to the several sponsoring institutions. Our contacts subsequently evaporated.

Szilard had an extraordinary capacity for realistic and farsighted analysis of political as well as scientific issues. But his behavior sometimes reflected an insensitivity to the rules of behavior expected by others—which may help to explain why he was excluded from development of the atomic bomb after he had done much to initiate the program.

While at NYU, I married Elizabeth Menzel, who has brought a great deal of balance to my life. We had our first child in New York and then moved to Boston, where we both had roots and where it would be easier to raise a family. We had two more children in Boston, and the three have added a great deal to life's pleasures. Their striking differences in temperament, apparent from the moment of birth, reinforced my ideas about the importance of genetic diversity. It has been very gratifying to watch their progress, as a computer software engineer, an independent filmmaker, and a graduate student working on protein structure (and also conducting an orchestra). I am strongly dedicated to the nuclear family as the most natural pattern for our species.

Though I admired the spirit at NYU, and what it has been able to accomplish with limited resources, I could not resist an invitation to return to my alma mater, to a more comfortable city for raising a family, and to chairing a department in my field, microbiology.

Aminoglycosides and the Ribosome; Harvard

The move back to Boston in 1957 renewed ties with the university where I had studied for eight years, and it shifted my teaching from pharmacology to microbiology. The Dean at Harvard, George Berry, was intensely dedicated to the school. I tended to rebel at his forceful, authoritarian style, but I recognized that he, like E. J. Cohn, also could respect independence and put talent to good use.

Our microbiology department was then responsible for teaching immunology. A decade had passed since I had worked in that field, but it still looked to me like only a minor branch on the tree of science. The introduction of molecular genetics, which made immunology such a powerful model system to cell biology, was still a few years off.

In my new role, I took the teaching and administrative duties too seriously and delegated too little. Hence I became increasingly distanced from the details of my research, a common problem for senior scientists. Nevertheless, several excellent associates were highly productive, focusing mostly on the interactions of the ribosome with antibiotics and on the mechanisms of drug resistance. I viewed these as challenging problems in their own right and also as useful tools for studying ribosomal function.

During that period a leading microbiologist, W. Barry Wood, Jr., invited me to join him, along with Renato Dulbecco, Herman Eisen, and Harold Ginsberg, in writing a new kind of microbiology text for medical students (25). We proposed that in order to prepare students for the future advances in medical science microbiologists should now emphasize much more the use bacteria as model cells in the new genetic and molecular biology. This view was quite different from the widespread belief that microbiology was no longer an exciting area of medical research because the empirical triumphs of antibiotics had eliminated many of the challenges of infectious disease.

In fact, it was a great time for a microbiology department with the new orientation. Faculty from other departments would come to the lectures to learn about the latest developments at the Pasteur Institute. Later, I noted that at the 25th reunion of a medical class of that era the customary symposium on research by its members was devoted entirely to molecular studies of microbial pathogenesis. It struck me as being no coincidence that our earlier teaching about exciting scientific advances, even though not close to medicine, had stimulated some medical students to apply the same approach later to pathogenic microbes.

My role in the book gradually expanded because I turned out to be more interested than the other coauthors in its style. As a young student I disliked writing, partly because teachers then emphasized belles lettres rather than expository prose. But the final product of a scientist is usually a paper, and so I was a professional writer willy-nilly. I therefore decided to undertake a course in self-improvement, primarily by reading Fowler's *Modern English Usage* from A to Z. Precision in communication then became almost as interesting a challenge for me as precision in the data.

In fact, I may have developed a time-wasting habit of reading scientific papers with too much attention to the language. But it seems to me unfortunate that current pressures encourage fast publication, for as the literature grows it seems ever more important that it be made more accessible by being well written. Dubos told me that Avery stored each of his manuscripts (a few per year) in a drawer for a month or more before final revision—a difficult model to follow today.

The five coauthors of the text discussed drafts of their chapters in great detail during two summers at Woods Hole. Because my comments on style were so extensive the others began to comment only on the content. In addition, Barry Wood developed a health problem and so I became director of the project. I then found myself trying to achieve a homogeneous style in a book by five independent senior scientists. Their reactions to my heavy editing were often not happy, but we remained good friends and ended up agreeing that the final product benefited. I regard my work on this book, through four editions and 25 years, as one of my most important contributions to microbiology.

At one stage of work on the first edition, I took off two months for full-time writing, in a colonial town in Mexico, San Miguel Allende. It was the most idyllic period in our family life. We had no contact with phone, radio, or periodicals, but we did not feel bored, even though our news was mostly about such matters as the foaling of our neighbor's burro. When we returned to Boston, I found that what I had missed in world affairs could be assimilated in an hour or so, yet I resumed the lifelong habit of compulsive reading of the trivia that form so much of our news.

From the start of my connection with tuberculosis research, I had tried, intermittently, to understand the action of streptomycin. I did not recognize that I was making one of the commonest and most serious mistakes in scientific investigation: choosing a problem that was not ripe for solution. Streptomycin and the other aminoglycoside antibiotics irreversibly block protein synthesis; the nonviable cell remains grossly intact, and almost anything that one can measure then changes. It was, therefore, difficult to decide which effects were important.

Our first breakthrough came from the work of Nitya Anand, a exceptionally idealistic person and excellent pharmaceutical chemist from India who joined us for a year to broaden his background for drug development. He subsequently became Director of the Central Drug Research Institute, responsible for virtually all drug development in India. In our lab, he discovered an effect of streptomycin that was quite unexpected, because it was not obviously connected to the inhibition of protein synthesis: bacteria growing in its presence become permeable, in both directions, to a variety of small molecules, including increased uptake of streptomycin itself. Moreover, the effect is prevented when protein synthesis is reversibly inhibited by chloramphenicol (2). We inferred that streptomycin acts directly on the growing cytoplasmic membrane, causing nonspecific damage, and we suggested that this effect, rather than the block in protein synthesis, might be the lethal step. Donald Dubin added detailed data on its kinetics that seemed to support this hypothesis.

However, the mechanism of protein synthesis was just then beginning to open up, and Roger Stanier suggested, on the basis of indirect evidence, that streptomycin blocks protein synthesis by binding to the ribosome. This conclusion was soon directly confirmed by others. But Stanier's intellectually powerful paper also had an unfortunate influence on the field, for he emphasized the importance of separating the key step in streptomycin action from the epiphenomena. Only 25 years later did we realize that there is no key step. Moreover, his paper caused all workers in the field to neglect, for decades, the membrane damage. I in particular was embarrased at having overinterpreted this effect as the cause of cell death.

Building on Anand's discovery, Paul Plotz, then a medical student and now a distinguished immunologist at the NIH, explained the known synergism in the bactericidal action of a β-lactam plus an aminoglycoside—an effect of considerable clinical value. By exposing the cells sequentially to sublethal concentrations of the two antibiotics, he showed that pretreatment of growing cells with penicillin, which distorted cell-wall synthesis, evidently induced membrane damage, because it increased sensitivity to subsequent killing by streptomycin. In the reverse order, the two agents were not synergistic. Others later confirmed the sensitization by penicillin directly by measuring uptake of radioactively labeled aminoglycosides. The brutally direct approach to most problems with labeled reagents generally enables deeper analysis, but I have always had sympathy for the biologist who makes the initial discovery by a primitive but ingenious experiment and then is quickly forgotten.

Meanwhile, Luigi Gorini, whom I had brought from NYU to Harvard, discovered that in the cell streptomycin at sublethal levels causes misreading, rather than blockage, in protein synthesis. This discovery was independent of my program, except that Julian Davies in my group, working with Walter Gilbert, confirmed the misreading in vitro, and he later defined some of the specific misreadings of one base as another.

Unfortunately, as Gorini acquired fame for this important discovery he grew increasingly resentful of our relationship, and then he began to compete directly with a post-doc of mine, Fred Sparling, who was well along in a study of the ribosomes in cells heterozygous for streptomycin sensitivity and resistance (36). Gorini's insistence on continuing the competition destroyed our friendship, and it brought to a head tensions in the department. I can recognize now that the egalitarian revolt against authority in 1968 was only part of the reason for the tensions, since I had failed to change the style of strong administration of the department that the school had initially encouraged. I withdrew from administration and set up a small, independent unit. While the result was painful, the change greatly improved my research.

Studies by Juan Modelell (from Spain), though very carful, failed to identify the links between the ribosome and other chages in streptomycin-killed cells. In a later major advance P.-C Tai and Brian Wallace (from Australia) solved one mystery: how the ribosome can have two different, mutually exclusive responses, blockade or misreading, though it binds tightly only one molecule of streptomycin. These investigators separated initiating free ribosomes from chain-elongation polysomes, which lack initiation factors and hence can only complete the already growing chains. They found that several classes of antiribosomal antibiotics (including aminoglycosides) block further synthesis only when they bind to an initiating ribosome. However, on chain-elongating ribosomes aminoglycosides have a unique effect: they cause misreading and slowing but not blockade, while the other classes (such as spectinomycin) do not cause misreading and some indeed have no apparent effect (29b). Since inhibition of growth is reversible with the latter classes and irreversible (i.e. lethal) with the aminoglycosides, it appeared that the characteristic misreading effect of the aminoglycosides is probably involved in their lethal action.

The nature of this involvement, as well as the mechanism of the membrane damage and its relation to the ribosomes, remained unexplained. Accordingly, when I retired I felt that I had lost the battle with streptomycin. However, an invitation to review antibiotic actions on the ribosome, at a meeting of the Society of General Microbiology in England, led to the answer—which could have been recognized many years earlier. I brought some reprints with me to study before the meeting. Assimilating a bundle of facts at such a time is quite different from more or less remembering them over the years, and doing so in a new atmosphere may encourage new associations. Sitting in a London hotel room, I suddenly had a Eureka: if misreading affects all proteins being synthesized, it would include those that will be incorporated in the membrane. An abnormal protein there might create nonspecific leakiness, thus providing the missing link between ribosome and membrane.

With this link we only needed to recognize that killing depends not on one key step but on a cycle of multiple steps, each equally important (17). First, a few molecules of antibiotic stray into the essentially impermeable cell, perhaps through transient imperfections in the growing membrane. Because they first encounter mostly chain-elongating ribosomes, which predominate in growing cells, they cause misreading. This causes membrane damage, which results in increased uptake of antibiotic—far beyond the amount needed to saturate the ribosome population. All the ribosomes are thus fixed in initiation complexes, blocking protein synthesis. Recent work has confirmed this mechanism, and it has also explained why the block is irreversible: the abnormal membrane protein is rapidly destroyed, thus caging the antibiotic in the cells (2a).

Additional, particularly strong evidence for the proposed bactericidal mechanism is its ability to explain a remarkable paradoxical effect observed earlier by others: low concentrations of puromycin accelerate killing by streptomycin, while high concentrations block it. Because puromycin releases incomplete chains, which is a form of misreading, membrane damage provides an obvious explanation. At low puromycin concentrations, the released chains would be long enough to cause such damage, while at high concentrations, they would be too short (17).

The extensive literature on the action of aminoglycosides in recent years has focused mostly on quantitating their uptake. But the final uptake of streptomycin may reach 100 times the molar concentration of ribosomes in the cell. Because the cell has already been killed by uptake too low to be detected against the background of surface adsorption, the measured uptake may be irrelevant, except as an extrapolation revealing membrane damage.

Among the other problems taken up by my graduate students or post-docs at Harvard, Loretta Leive studied the competition between exogenous and endogenous sources of diaminopimelate, which we had shown to be a precursor of lysine. It is also a component of the peptidoglycan of *E. coli*. She demonstrated that, in these incorporations, the two corresponding diaminopimelate sources, differentially labeled, exhibit a gradient or compartmentalization (32). This problem is still not well understood and does not fit the picture of the cytoplasm as a homogeneous solution.

David Smith discovered that novobiocin interferes with DNA metabolism—an effect that became useful when others later identified the enzyme on which it acts, DNA gyrase. Eliora Ron, from Israel, found that growth at elevated temperatures is limited not by melting of the membrane, which is what we expected, but by reversible inactivation of the first enzyme in the biosynthesis of methionine. She has since found this curious property in all the bacterial species tested.

Porter Anderson found that inosine can pair with adenosine in the translation of polyinosinic acid as an artificial messenger, and I suggested that this fit of two purines in a double helix, and in Crick's wobble hypothesis, could best be explained if one of the purines was in the *syn* rather than the usual *anti* configuration (22). This was one more bright idea, subsequently established by others, that gave me pleasure but had no visible impact as an isolated contribution based on indirect evidence.

Elizabeth Mingioli was a most effective technician and my virtual right hand for many years. Unfortunately, having such an effective assistant accelerated my withdrawal from doing experiments myself. Among many contributions, she showed that vitamin B_{12} could replace methionine in certain mutants blocked in its methylation step.

I also became involved with the ribosome cycle, at a time when there was much controversy over the distribution of the ribosomal particles (polysomes, single ribosomes, and native 30S and 50S subunits) in the lysates of *E. coli* prepared in different ways. We distinguished initiating monosomes from free ribosomes, which accumulate after polysome runoff, by their difference in dissociability at low Mg^{2+}. Others had shown that the three initiation factors (IF1,2,3) are not free in the lysates but are bound to 30S ribosomal subunits. It seemed to me logical to conclude that one or more of the attached IFs must serve as a dissociation factor, preventing the 30S and 50S subunits from pairing and thus stabilizing them as a reservoir awaiting initiation.

This idea came shortly after I began a wonderful half-year sabbatical actually working in the lab with Pnina Elson at the Weitzmann Institute in Israel. I instructed Eliora Ron and Robert Kohler, back at home, to prepare a mixture of IFs and see whether it would cause purified free ribosomes to dissociate. It failed, so I cabled to ask them to try 10 times as much. They thought I had been touched by the hot sun in Israel—but the experiment worked. A. R. Subramanian identified the dissociation factor as IF3 (38).

Subramanian further found, unexpectedly, that at the end of translation the ribosome is released as a 70S particle rather than as subunits. Tai later showed (as did Kaji elsewhere) that this release is accelerated by a protein ribosome-release factor, whose significance is still not clear. The ribosome cycle was now quite complete. Michael Gottlieb, Robert Beller, and S. Ramagopal further added to our understanding of the ribosome, and Nicolette Lubsen, a graduate student from the Netherlands, studied the complex initiation factors of rabbit reticulocytes.

In one of our most important contributions, Robert Thompson provided the first experimental confirmation of Hopfield's suggestion that the recognition step in protein synthesis involves proofreading, which greatly increases its accuracy (39). Our approach was very direct: forming ternary aa-tRNA-EFTu complexes and showing that the ratio of GTP hydrolyzed to the amount of incorporated amino acid varied as predicted for a cognate, near-cognate, or distant tRNA.

We also became interested in a different and neglected aspect of the ribosome: its possible role in cell death. The reported extensive loss of total RNA in starving cells suggested to me that in starved *E. coli* cultures the breakdown of ribosomes might proceed to the stage where death could result from complete loss of ribosomes. An undergraduate, Selina Luger, showed that this was indeed true (28), though others have obtained quite different results under other conditions.

Our findings support what seems to me an interesting theoretical generalization: in a starving cell, complete elimination of any species of protein is not lethal if the cell retains or can restore the capacity to transcribe and translate its messenger when supplied with the necessary building blocks and energy; but if any protein required for protein synthesis (including ribosomes) is exhausted, it cannot be regenerated. I am glad to see breakdown of ribosomes during starvation now being studied in depth in other laboratories, because in the cycles in nature bacteria face famine much more often than feasting.

My last major area of research was protein transport across cell membranes. In a dense, short paper, Cesar Milstein, at Cambridge, England, had shown that a special, cleavable sequence, which he called a signal sequence, initiates transfer of an immunoglobulin chain across the membrane of a lymphoid cell. Walter Smith in our laboratory demonstrated such cotranslational transfer in bacteria more directly: chains protruding from the surface could be chemically labeled or enzymatically cleaved, with the inner terminus still attached to membrane-bound ribosomes in the cell (35). Milstein did not continue work on this problem, having meanwhile discovered how to grow monoclonal antibodies in hybridomas, and he seemed to be losing credit for the work on secretion. I wrote with Tai a review that was aimed in part at straightening out the history of the signal sequence (29a) for it seems to me important to assign credit properly for original discoveries, both to satisfy a sense of fairness and to provide motivation for investigators.

Further pursuing protein transport, David Rhoads showed that incorporation of protein into membrane vesicles in vitro requires a membrane potential. Seikoh Horiuchi & Michael Caulfield, using the gram-positive *Bacillus subtilis* to avoid outer membrane fragments, isolated a complex of four proteins, attached to cytoplasmic membrane or to ribosomes, that appear to be involved in protein secretion (2b). Others have identified this complex (also in another gram-positive organism, *Staphylococcus aureus*) as pyruvate dehydrogenase, and so its function in secretion is not certain.

We also examined the problem of how gram-negative bacteria excrete proteins to the exterior even though they lack any evident source of energy for moving them from the periplasm across the outer membrane. Stephen Lory showed that when this excretion is inhibited by ethanol, which distorts membrane organization, the protein flows through the junctions between the inner and outer membranes and is then held up on the external surface of the outer membrane; none is found in the periplasm. If the ethanol is removed, the bound protein can be released to the exterior by cleavage of its signal sequence (33). More recent work, however, on other systems, as well as by Lory using a mutation to block the pathway, has made it uncertain that the normal pathway bypasses the periplasm.

One of my last scientific papers, far from microbiology, was on what seems to me a major mystery in biology: the function of sleep, during which the brain is about as active as during waking hours. Francis Crick published in *Nature* a theory that all this work serves primarily to correct errors encoded in memory during that day. I offered an opposite view, based on the theory that memories are encoded by changes in proteins that occur when synapses are fired. Since these proteins inevitably fade, through the normal process of turnover, I suggested that during sleep we consolidate waning memories by firing sets of neurons in a process that systematically scans the brain (15). This mechanism does not seem to me to bear on the separate problem of why we dream, and it would not be surprising if a process of scanning memories were influenced by the cognitive and emotional content of the most recently recorded items. My paper was rejected by *Nature* and appeared in a less prominent journal. But it is gratifying to see that in the currently expanding research on sleep, the idea of consolidation of memories seems to be acquiring increasing acceptance.

Another speculative publication was stimulated by a paper by John Cairns, who found that a substrate, lactose, increased the rate of mutation from lac^- to lac^+. He offered a provocative Lamarckian interpretation that challenged a fundamental principle in genetics and evolutionary biology: that mutations occur in nature at random rather than being directed. I suggested an alternative mechanism that did not contradict this principle. In my mechanism, the mutagenic effect of the substrate depends on induction of transcription of its operon, which creates a short region of single-stranded DNA. This region, moving along the operon, is more mutable than double-stranded DNA, and so it would introduce a bias in the mutation rate—but a bias is not a directed mutation (18).

On retirement in 1984, I turned over my laboratory to P.-C. Tai, who had worked with me for over 15 years, and I have been delighted to see his success in further dissecting the problems of protein export. I subsequently enjoyed periods of teaching at the University of Tel Aviv and at the National Taiwan University Medical School. In addition, Carlos Chagas of Brazil had converted the ancient Pontifical Academy of Sciences into a useful advisory group to the Pope, and at a workshop on genetics I predicted that the Church would find it increasingly difficult to refuse to use our growing knowledge of genetics to prevent human misery by prenatal diagnosis and elective abortion. I was encouraged by the free and open discussion with the bishop in charge of family policy, but the response at other levels remains to be seen.

I often felt that I might not be guiding my students and fellows in enough detail. However, I have been pleased to see that nearly all of them have found new and interesting directions in their later careers, rather than remaining specialists building on their earlier training. This flexibility may have been encouraged by the freedom to pursue their problems with a good deal of independence in my laboratory.

To try to encapsulate the portrait of me as a scientist: I clearly have "internalized the canons of science," emphasizing rationality and reality, more than most. I think my strongest suit in science has been critical, logical analysis, leading to a simple but decisive experiment. And although a systematic program, pursuing the shikimate pathway, has probably contributed most to my scientific reputation, I have tended not to pursue programs at length but to skim the cream from a variety of problems. My greatest satisfaction in science has come from unexpected associations, such as those leading to the penicillin method, the ribosome dissociation factor, or the multistep action of streptomycin. I have perhaps been more willing than most serious scientists to publish ideas that seemed bright and even playful but lacked proof. But my predictions on genetic engineering have been too conservative.

In dealing with social issues I have focused on defects in current policies and in the underlying assumptions, rather than on supporting laudable aims and achievements that are already widely accepted. I have functioned more as a Socratic gadfly than as a member of committees or an active participant in political organizations.

Science and Society: the Genetic Revolution

Virtually a second career, in science and society, began in 1969, quite unexpectedly.

The Biology Department at Boston University was sponsoring a symposium, at a national meeting, where Peter Medawar was to discuss the ethical impacts of molecular genetics, but meanwhile he suffered a stroke. With great diffidence I filled the breach. I then spent a sabbatical year, as already noted, at the Center for Advanced Study of the Behavioral Sciences. This rather deep involvement in an unfamiliar subject was stimulated in part by my reaction to what seemed to me irresponsible attacks on behavioral genetics by a group called Science for the People, and particularly by a brilliant population geneticist in my university, Richard Lewontin. After he had made some especially outrageous statements on a television program, I enjoyed challenging him to a formal debate, like Thomas Huxley debating Bishop Wilberforce in the 19th Century.

Soon I was publishing on various aspects of science and society, and in 1986 I collected the articles in a book, *Storm Over Biology* (16). Its topics included evolution and sociobiology, genetics and racism, and genetic engineering. I summarize here two additional items from the book.

The first, a guest editorial in the *New England Journal of Medicine,* applauded efforts to increase the number of minority physicians but emphasized that especially in medical education we must not sacrifice standards, because lives are at stake. A very similar statement that I had drafted had been favorably received by the administration in my school, and because other medical schools also had problems with affirmative action, it seemed useful to publish the piece in a professional journal. I did not anticipate how the news media would portray me as a racist, with the result that students picketed me and Dean Robert Ebert denounced me in a letter to all other medical deans. My colleagues offered no public support, though I suspect that most of them agreed with me. Many years later, I received an unexpected rehabilitation when an annual report of the next Dean, Daniel Tosteson, commended *Storm Over Biology*. But the charge of racism has undoubtedly influenced the reception of my subsequent publications.

I determined to avoid further involvement with the topic. However, when the suit by Bakke over racial preference in medical school admissions subsequently reached the Supreme Court, the extensive public discussion all supported one or the other of the two extremes: de facto quotas or color-blind admission. I finally felt obligated to call attention to a third possibility: that we stretch standards but reject quotas, in order to increase numbers but still have some control over the cutoff. The court adopted this position, though on different, legal grounds.

The second piece in *Storm Over Biology,* on objectivity in science, emphasizes that the word science is used with three different meanings in different contexts: a methodology, the activities of people using that methodology, and the resulting body of knowledge. Because the results are tested against nature, they can be objective even though the activities are highly subjective. A great deal of confused criticism of science seems to me to have arisen from failure to distinguish these three meanings of the word.

The areas of my interest in the interactions of science and society continued to include public ambivalence about genetic engineering. I edited a book for a general audience, *The Genetic Revolution* (20), by a group of experts on scientific and social issues.

I have also become concerned by the increasing politicization of science. An early example was the attack on human behavioral genetics. More recently, the Human Genome Project has thrust on biomedical research (21) the precedent of a centralized, large-scale organization. A third example, with much greater menace, has arisen from outside our community: increasing public ambivalence about science, and increasing intrusion of government in the style of research, as a consequence of exaggeration of the problem of misconduct.

It is particularly challenging to try to foresee the future impacts of the genetic revolution. Current discussion focuses mostly on hazards and problems arising from our increased power to manipulate DNA. However, I suspect that our greatest problems will arise from our increased insights into our genetic diversity. These insights not only may give us more prognostic information than we can handle comfortably, they will surely have major implications before long for social policy, including education, the distribution of jobs and of rewards, and the definition of racial justice. And perhaps the more distant future will bring an irresistible pressure to guide human evolution.

Finally, my interest in social implications of evolutionary biology has led me to defend sociobiology, which focuses on cooperative instincts as well as on the competitive drives that dominated early evolutionary thinking. Moreover, while premature or grossly distorted applications of genetics and evolutionary biology were often used in the past to rationalize conservative and even racist political views, modern genetics has had the opposite effect, for it has replaced the false typological notion of races with a populational view, which recognizes races as groups in which prolonged reproductive separation has inevitably led to different, but overlapping, gene pools. But a review of my thoughts and writings on the aspects of science and society that I have listed would take us too far from the central theme of this autobiography.

We scientists are probably more idiosyncratic than we realize in viewing the search for truth as a paramount human goal. Most people prefer a belief because it is expected to make them, or others, feel or act better, rather than because it is based on evidence. But with the mounting global crises, an effective set of responses would have to be anchored in the reality that science elucidates. So we come back to the antinomy of the heart and the head—and the hope that we will be guided by each only in its proper realm.

Literature Cited

1. Amarasingham, C. R., Davis, B. D. 1965. Regulation of α-ketoglutarate dehydrogenase formation in *Escherichia coli*. *J. Biol. Chem.* 240:3664–70
2. Anand, N., Davis, B. D. 1960. Damage by streptomycin to the cell membrane of *Escherichia coli*. *Nature* 185:22–23; Anand, N., Davis, B. D., Armitage, A. K. 1960. Uptake of streptomycin by *Escherichia coli*. *Nature* 185:23–24

2a. Busse, H.-J., Wostmann, C., Bakker, E. P. 1992. The bactericidal action of streptomycin: membrane permeabilization caused by the insertion of mistranslated proteins into the cytoplasmic membrane of *Escherichia coli* and subsequent caging of the antibiotic inside due to degradation of these proteins. *J. Gen. Microbial.* In press

2b. Caulfield, M. P., Horiuchi, S., Tai, P.-C., Davis, B. D. 1984. The 64-kilodalton membrane protein of *Bacillus subtilis* is also present as a multiprotein complex on membrane-free ribosomes. *Proc. Natl. Acad. Sci. USA* 81:7772–76

3. Davis, B. D. 1942. Binding of sulfonamides by plasma proteins. *Science* 95: 78–81
4. Davis, B. D. 1946. Physiological significance of the binding of molecules by plasma proteins. *Am. Sci.* 34:611–18
5. Davis, B. D. 1948. Isolation of biochemically deficient mutants of bacteria by penicillin. *J. Am. Chem. Soc.* 70:4267; Lederberg, J., Zinder, N. 1948. Concentration of biochemical mutants of bacteria with penicillin. *J. Am. Chem Soc.* 70:4267; 1987. Citation classic. *Curr. Contents* 30(33):17
6. Davis, B. D. 1948. Complement fixa-

tion with soluble antigens of *Plasmodium knowlesi* and *Plasmodium lophurae*. *J. Immunol.* 58:269–75

7. Davis, B. D. 1950. Studies on nutritionally deficient bacterial mutants isolated by means of of penicillin. *Experientia* 6:41–48
8. Davis, B. D. 1950. *p*-Hydroxybenzoic acid: a new bacterial vitamin. *Nature* 166:1120–21
9. Davis, B. D. 1950. Nonfiltrability of the agents of genetic recombination in *Escherichia coli*. *J. Bacteriol.* 60:507–8
10. Davis, B. D. 1951. Aromatic biosynthesis. I. The role of shikimic acid. *J. Biol. Chem.* 191:315–25
11. Davis, B. D. 1954–55. Biochemical explorations with bacterial mutants. *Harvey Lect.* 50:230–44
12. Davis, B. D. 1956. Relations between enzymes and permeability (membrane transport) in bacteria. In *Enzymes: Units of Biological Structure and Function*, ed. O. H. Gaebler, pp. 509–22. New York: Academic
13. Davis, B. D. 1958. On the importance of being ionized. *Arch. Biochem. Biophys.* 78:497–509; 1986. Reprinted in *The Biologist (England)* 33:291–95
14. Davis, B. D. 1961. The teleonomic significance of biosynthetic control mechanisms. *Cold Spring Harbor Symp.* 26: 1–10
15. Davis, B. D. 1985. Sleep and the maintenance of memory. *Perspect. Biol. Med.* 28:457–64
16. Davis, B. D. 1986. *Storm Over Biology: Essays on Science, Sentiment, and Public Policy*. Buffalo: Prometheus Books
17. Davis, B. D. 1987. Mechanism of bactericidal action of aminoglycosides. *Microbiol. Rev.* 51:341–50
18. Davis, B. D. 1989. Transcriptional bias: a non-Lamarckian mechanism for substrate-induced mutations. *Proc. Natl. Acad. Sci. USA* 36:5005–9

18a. Davis, B. D. 1989. Evolutionary principles and the regulation of engineered bacteria. *Genome* 31:864–69

19. Davis, B. D. 1990. Two perspectives: on Rene Dubos, and on antibiotic action. In *Launching the Antibiotic Era*, ed. C. L. Moberg, Z. A. Cohn, pp. 69–80. New York: Rockefeller Univ. Press; 1991. Reprinted in *Perspect. Biol. Med.* 35:37–48
20. Davis, B. D., ed. 1991. *The Genetic Revolution: Scientific Prospects and Public Perceptions*, Baltimore: Johns Hopkins Univ. Press
21. Davis, B. D. 1992. Sequencing the human genome: a faded goal. *Bull. N. Y. Acad. Med.* 68:115–25
22. Davis, B. D., Anderson P., Sparling, P. F. 1973. Pairing of inosine with adenosine in codon position two in the translation of polyinosinic acid. *J. Mol. Biol.* 76:223–32
23. Davis, B. D., Cohn, E. J. 1939. The influence of ionic strength and pH on electrophoretic mobility. *J. Am. Chem. Soc.* 61:2092–58
24. Davis, B. D., Dubos, R. J. 1947. The binding of fatty acids by serum albumin, a protective growth factor in bacteriological media. *J. Exp. Med.* 86:215–28
25. Davis, B. D., Dulbecco, R., Eisen, H. N., Ginsberg, H., Wood, W. B. Jr. 1967. *Microbiology*. New York: Harper and Row. 1st ed.
26. Davis, B. D., Flaherty, P., eds. 1976. *Human Diversity*. Cambridge, MA: Ballinger
27. Davis, B. D., Kabat, E. A., Harris, A., Moore, D. H. 1944. The anticomplementary activity of serum gamma globulin. *J. Immunol.* 49:223–28
28. Davis, B. D., Luger, S. M., Tai, P.-C. 1986. Role of ribosome degradation in the death of starved *Escherichia coli* cells. *J. Bacteriol.* 166:439–45
29. Davis, B. D., Moore, D. H., Kabat, E. A., Harris, A. 1945. Electrophoretic, ultracentrifugal, and immmunochemical studies on Wassermann antibody. *J. Immunol.* 50:1–8

29a. Davis, B. D., Tai, P.-C. 1980. The mechanism of protein secretion across membranes. *Nature* 283:433–38

29b. Davis, B. D., Tai, P.-C., Wallace, B. J. 1974. Complex interactions of antibiotics with the ribosome. *Ribosomes*, pp. 771–89. Cold Spring Harbor: Cold Spring Harbor Laboratories

30. Dobzhansky, T. 1973. *Genetic Diversity and Human Equality*. New York: Basic Books
31. Gilvarg, C., Davis, B. D. 1956. The role of the tricarboxylic acid cycle in acetate oxidation in *Escherichia coli*. *J. Biol. Chem.* 222:307–19
32. Leive, L., Davis, B. D. 1965. Evidence for a gradient of exogenous and endogenous diaminopimelate in *Escherichia coli*. *J. Biol. Chem.* 240: 4370–78
33. Lory, S., Tai, P.-C,, Davis, B. D. 1983. Mechanism of excretion of protein by gram-negative bacteria: *Pseudomonas aeruginosa* exotoxin A. *J. Bacteriol.* 156:695–702
34. Maas, W. K., Davis, B. D. 1952. Production of an altered pantothenate-synthesizing enzyme by a temperature-sensitive mutant of *Escherichia coli*. *Proc. Natl. Acad. Sci. USA* 38:785–91

35. Smith, W. P., Tai. P.-C., Thompson, R. C., Davis, B. D. 1977. Extracellular labeling of nascent polypeptides traversing the membrane of *Escherichia coli*. *Proc. Natl. Acad. Sci USA* 74:2830–34

36. Sparling, P. F., Modolell, J., Takeda, Y., Davis, B. D. 1963. Ribosomes from *Escherichia coli* merodiploids heterozygous for resistance to streptomycin and to spectinomycin. *J. Mol. Biol.* 37:407–17

37. Srinivasan, P. R., Shigeura, H. T., Sprecher, M., Sprinson, B., Davis, B. D. 1956. The biosynthesis of shikimic acid from D-glucose. *J. Biol. Chem.* 220:477–87

38. Subramanian, A. R., Davis B. D., Beller, R. J. 1969. The ribosome dissociation factor and the ribosome-polysome cycle. *Cold Spring Harbor Symp.* 34:223–30

39. Thompson, R. C., Stone, P. J. 1977. Proofreading of the codon-anticodon interaction on ribosomes. *Proc. Natl. Acad. Sci. USA* 74:198–202

Annu. Rev. Microbiol. 1992. 46:35–64

GENETICS OF *CAMPYLOBACTER* AND *HELICOBACTER*

Diane E. Taylor

Departments of Microbiology and Medical Microbiology and Infectious Diseases, University of Alberta, Edmonton, Alberta, Canada T6G 2H7

KEY WORDS: genetic transformation, bacterial plasmids, cloning vectors, genome mapping, antibiotic resistance

CONTENTS

0066-4227/92/1001-0035$02.00

Abstract

This article reviews the current state of genetic analysis of *Campylobacter* and *Helicobacter*. Chromosomal genes cloned from *Campylobacter* and *Helicobacter* species are listed along with the method used to identify the cloned gene. *Campylobacter* plasmid genes that have been cloned and expressed in *Escherichia coli* and that specify resistance to tetracycline, kanamycin, or chloramphenicol are presented.

This review also examines our current knowledge of genetic exchange in *Campylobacter,* including conjugative plasmid transfer, natural transformation, electrotransformation, and bacteriophage transduction. In *Helicobacter,* natural transformation has been described and both plasmids and bacteriophages have been observed. Plasmid cloning vectors have been constructed for *Campylobacter*. Available vectors are discussed and restriction maps of some useful vectors that we have constructed are included.

The genome sizes of *C. jejuni* and *C. coli* are approximately 1.7 megabases (Mb), whereas the genome size of *H. pylori* ranges from 1.60 to 1.73 Mb. The positions of various genes on the *C. jejuni* and *C. coli* genome maps have been determined using both homologous and heterologous DNA probes. Genomic maps of these organisms are presented.

INTRODUCTION

This review deals with genetic developments in the genera *Campylobacter* and *Helicobacter*. Bacteria in these genera have a spiral or S-shaped morphology and are 0.5 to 8.0 μm long and 0.2 to 0.5 μm wide. They are gram negative and microaerophilic, with a G+C content that varies from 28 to 44 mol%. Species of *Campylobacter* have a single polar unsheathed flagellum at one or sometimes both ends of the cell, whereas species of *Helicobacter* have four to six sheathed flagella at one or sometimes both poles. The decision to transfer *Campylobacter pylori* (formerly *C. pyloridis*) to the new genus *Helicobacter* (41) was based on ribosomal RNA sequencing (118), fatty acid profiles, biochemical reactions, and morphological characteristics. The taxonomic position of several members of the *Campylobacter* genus is currently in a state of flux (149).

Campylobacter species are pathogens or commensals in a wide range of animal species. *Campylobacter jejuni* is a common cause of human diarrheal illness (125), and several other *Campylobacter* species, *C. coli, C. fetus* subsp. *fetus, C. hyointestinalis, C. lari,* and *C. upsaliensis* (46), can also cause similar disease manifestations. In contrast, *Helicobacter pylori* is frequently isolated or observed in preparations from gastric biopsies of patients with gastritis and/or ulcers (82, 83). Other recently named *Helicobacter* species include *H. mustelae, H. felis,* and *H. nemestrinae* isolated, respectively, from stomach tissue of ferrets (37), cats (107), and the pigtailed macaque (13).

The considerable recent interest in both *Campylobacter* and *Helicobacter* has resulted in several excellent review articles about these organisms. For information on the association of gastric disease and *H. pylori,* the reader is referred to other reviews (8, 16, 31, 32, 42, 44, 110). Recent reviews of *Campylobacter* have stressed taxonomy (109), epidemiology (12), association with human disease (46), antibiotic resistance (133) and pathophysiology, and early genetic studies (152).

GENES CLONED FROM *CAMPYLOBACTER* AND *HELICOBACTER*

Campylobacter *House-Keeping Genes*

Table 1 lists chromosomal genes that have been cloned from *Campylobacter*. Of those that have been cloned from *C. jejuni* or *C. coli* and successfully expressed in *Escherichia coli,* all can be classified as general house-keeping genes. Housekeeping genes are those that appear to be highly conserved across species boundaries, and encode similar functions required for the maintenance and growth of many different bacterial species, e.g. those required for amino acid biosynthesis.

In 1985, Lee and coworkers (72) reported that they had cloned in *E. coli* two genes from *C. jejuni* required for proline biosynthesis by selection for complementation in a *proAproB* mutant of *E. coli*. Although the DNA sequence of the DNA fragment from the *C. jejuni* gene library has not been determined (P. Guerry, personal communication), a similar strategy has been used to clone the *proA* gene and determine its sequence (V. L. Chan, personal communcation). Investigators have used this complementation procedure to clone several genes from amino acid pathways (see Table 1). These strategies have probably been successful because they rely on strong selective pressure: the requirement for growth of *E. coli* in absence of a particular amino acid to maintain the *Campylobacter* gene(s). Completion of the DNA sequences of these genes is important for comparison with homologous genes found in other species and to determine if they are stable in *E. coli* in the absence of selective pressure. Answers to these questions may help explain why difficulties have been encountered in cloning certain *Campylobacter* genes (see below).

Other house-keeping genes cloned from *C. jejuni* include 16S and 23S ribosomal RNA genes (59, 115), which were identified by hybridization of ribosomal RNA from *C. jejuni*. Two transfer RNA genes (those for alanine and leucine) were also cloned and were identified by their proximity to a 16S rRNA gene (116). The DNA sequences of these tRNA genes have been determined.

Table 1 Genes cloned from *Campylobacter* species and *Helicobacter* species

Microorganism	Gene cloned[a]	Cloning strategy[b]	Location	DNA sequence published	Reference
C. jejuni	γ Glutamyl kinase *(proA)*	Reverse genetics[c]	Chromosome	No	72; V. L. Chan[d]
	γ Glutamyl phosphate reductase *(proB)*		Chromosome	No	72
C. jejuni	Serine hydroxylmethyltransferase *(glyA)*	Reverse genetics[c]	Chromosome	Yes	21, 22
C. jejuni	β-Isopropylmalate (IPM), dehydrogenase *(leuB)*, IPM isomerases *(leuC, leuD)*	Reverse genetics[c]	Chromosome	No	A. Labigne[d]
C. jejuni	Acetylornithinase *(argE)*	Reverse genetics[c]	Chromosome	No	D. E. Taylor & M. Bussiere[e]
C. jejuni	Arginosuccinase *(argH)*	Reverse genetics[c]	Chromosome	No	V. L. Chan[d]
C. jejuni	Flagellin *(flaAflaB)*	λgt11 + antibody[f]	Chromosome	Yes	98, 99
C. jejuni	Flagellin *(flaA)*	PCR[g]	Chromosome	Yes	35
C. coli	Flagellin *(flaAflaB)*	Oligonucleotide probe[h]	Chromosome	Yes	48, 77
C. jejuni	Major outer membrane protein (MOMP)	λgt11 + antibody[f]	Chromosome	No[i,j]	138
C. jejuni	Ribosomal RNA: 16S, 23S	Hybridization of rRNA	Chromosome	No[j]	59, 115
C. jejuni	tRNA (Ala), tRNA (Leu)	Located next to 16S rRNA gene	Chromosome	Yes	116
C. fetus	Surface-array protein *(sapA)*	λgt11 + antibody[f]	Chromosome	Yes	9

C. coli	Chloramphenicol acetyltransferase *(cat)*	Direct selection[k]	Plasmid	Yes	120, 155
CLO[l]	Aminoglycoside phosphotransferase *(aphA-1)*	Direct selection[k]	Chromosome	Yes	104
C. coli	Aminoglycoside phosphotransferase *(aphA-3)*	Direct selection[k]	Plasmid	Yes	148
C. coli	Aminoglycoside phosphotransferase *(aphA-7)*	Direct selection[k]	Plasmid	Yes	144
C. coli	Tetracycline resistance *(tetO)*	Direct selection[k]	Plasmid	Yes	128
C. jejuni	Tetracycline resistance *(tetO)*	Direct selection[k]	Plasmid	Yes	81, 131, 140
H. pylori	Urease *(ureA, ureB)*	λgt11 + antibody[f]	Chromosome	Yes	28, 29
H. pylori	Urease *(ureA, ureB, ureC, ureD)*	Urease production[m]	Chromosome	Yes	68
	(ureE, ureF, ureG, ureH)			No	29a
H. pylori	26,000-Dalton protein[n]	Oligonucleotide probe[h]	Chromosome	Yes	103

[a] The complete name of the genetic property that has been cloned is given. The genotypic designation, where available, is included in parentheses.

[b] Refers to the strategy used to obtain the gene in question.

[c] Reverse genetics refers to complementation of an *E. coli* mutant by cloned gene(s) from *Campylobacter* species, e.g. complementation of *proA* and *proB* mutations.

[d] Personal communication.

[e] In preparation.

[f] Refers to the construction of gene libraries using phage λgt11 and subsequent screening with antibody to the protein of interest (see 55).

[g] PCR, sequence determined using polymerase chain reaction and oligonucleotide primers.

[h] Selected by hybridization of an oligonucleotide probe based on the amino acid sequence of the protein of interest.

[i] Complete open reading frame of MOMP has not yet been cloned.

[j] Partial DNA sequence of MOMP available (K. Hiratsuka & D. E. Taylor, unpublished data); sequences of 16S and 23S rDNA genes available from Dr. S. Cohen, Gene Track Inc., Boston, MA.

[k] Direct selection for expression of an antibiotic-resistance phenotype after cloning in *E. coli*.

[l] CLO, *Campylobacter*-like organism.

[m] Genes *(ureA, ureB, ureC, ureD)* were identified by transient urease production in *C. jejuni* (68); genes (*ureE, ureF, ureG* and *ureH*) are required for expression in *E. coli* (29a).

[n] Identity of the 26-kDa protein is as yet unknown.

Campylobacter *Virulence Genes*

C. jejuni and *C. coli* cause much gastrointestinal morbidity. Therefore, genetic studies have focused on structural proteins and other products believed to play a role in pathogenesis. These studies have mainly examined flagella of *C. jejuni* and *C. coli;* however, an enterotoxin and major outer membrane protein of *C. jejuni,* as well as a surface protein of *C. fetus fetus,* have also been investigated.

FLAGELLAR GENES Flagella play an important role in virulence because the bacterium needs them to colonize the intestines. Aflagellate mutants and nonmotile strains cannot colonize in animal models (1, 89, 96). In addition, the flagella are highly immunogenic and patients produce antibody to flagella soon after infection (7, 92, 158).

The genetics of flagellar production have been investigated extensively in both *C. jejuni* 81116 (98, 99, 156) and *C. coli* VC167 (47–49, 77, 146) and to a lesser extent in *C. jejuni* IN1 (34). *C. jejuni* 81116 and *C. coli* VC167 contain two copies of the flagellin genes, designated *flaA* and *flaB,* located adjacent to one another in a head-to-tail configuration (48, 99). In *C. jejuni* 81116, both genes comprise 1731 base pairs that are 95% identical. Primer extension studies demonstrated that *flaA* mRNA is transcribed from a σ^{28} promotor in *C. jejuni* 81116 (97). In *E. coli,* this promoter transcribes genes involved in chemotaxis, motility, and flagella function (2). The *flaB* gene lacks any recognizable promoter sequence, and initial studies suggested that it is not transcribed (97). However, more recent work suggests that a very low level of transcription from *flaB* may occur (156). In *C. coli* VC167, both *flaA* and *flaB* genes possess promoters. Although the former is a typical σ^{28} promotor, the latter appears to resemble a (*nif*) σ^{54} promoter (47). This promotor is usually activated in response to nitrogen starvation (67). The *flaB* gene is expressed at low levels in *C. coli* VC167 (47).

The calculated molecular weights of flagellin A and B estimated from the *flaA* and *flaB* genes of *C. jejuni* 81116 are 59,538 and 59,909, respectively. However, the observed molecular weight of the flagellin in polyacrylamide gels is 62,000, although sometimes a second flagellar protein of 60,000 is also observed (95). These two observed flagellins of different sizes do not represent products of the *flaA* and *flaB* genes, because a monoclonal antibody that recognizes both proteins reacted with expressed *flaA* fragments but not with homologous *flaB* fragments (100). These size differences may result from posttranslational modification of amino acid residues within the flagellar protein (99).

The availability of the deduced amino acid sequences derived from DNA sequencing of three *flaA* genes enables the identification of common and variable regions within the flagellin proteins. Two common regions consisting of the 170 amino acid N-terminal region (C1) and the 100 amino acid C-terminal region (C2) exhibit 94 and 96% identity, respectively. A variable region, V1, occupies the middle of the flagellin. Several areas within the V1 region are predicted to be surface-exposed and probably correspond to areas with surface epitopes (35).

Deletion mutant analysis utilizing a gene replacement technique (69) showed that *flaA*$^+$*flaB*$^-$ derivatives of both *C. jejuni* 81116 and *C. coli* VC167 have normal flagella, whereas *flaA*$^-$*flaB*$^-$ mutants are aflagellate (47, 156). Mutants of *flaA*$^-$*flaB*$^+$ type in both strains had short truncated filaments and much reduced motility compared with wild-type. Mutants of VC167 that are *flaA*$^+$*flaB*$^-$ were slightly less motile than wild-type cells, although they produced a flagellar filament indistinguishable in length from wild-type (47). Antiserum specific for the *flaB* gene product reacted sparsely along the entire length of the filament in *flaA*$^+$*flaB*$^+$ cells. These results suggest that the flagellar filament of *C. coli* VC167 is composed of both *flaA* and *flaB* gene products and that the *flaB* gene product constitutes less than 20% of the wild-type filament (47). Such an organization with the *flaB* subunit intertwined among *flaA* subunits is reminiscent of that seen in the complex flagellar filaments of *Rhizobium meliloti* (111), *Caulobacter crescentus* (88), and also apparently in *H. pylori* (65). In contrast, the *C. jejuni* 81116 flagellum contains only the *flaA* gene product (99, 100, 156).

Both *flaA* and *flaB* genes appear to be present in most *Campylobacter* strains examined using DNA hybridization analysis (146). These genes apparently have been maintained throughout evolution and presumably confer some selective advantage on the host bacterium. The data of Guerry et al (47) suggest that the *flaB* gene product confers increased motility on the flagella, but this is apparently not the case with all strains. The two copies of *fla* genes may be maintained within the *Campylobacter* genome to provide a duplicate in case the expression copy undergoes a mutation or deletion event rendering it nonfunctional. Mistakes in the expression copy could be corrected by a recombinational event within the alternate copy. In addition, genetic exchange among flagellar genes could involve natural transformation (see below) in which DNA from lysed cells in the population may be taken up. Such a process could play an important role in generating antigenic diversity in *Campylobacter* species.

PHASE VARIATION Flagellar expression in *Campylobacter* spp. is subject to both phase and antigenic variation (47, 49, 51, 76). Phase variation refers to the ability of some strains of *Campylobacter* to switch on and off flagellar production. Caldwell et al (19) showed that *C. jejuni* 81116 cells undergo a bidirectional transition between flagellated and aflagellated phenotypes. The $Fla^+ \rightarrow Fla^-$ transition occured at rates approximately 3×10^{-3} per cell per generation whereas the $Fla^- \rightarrow Fla^+$ transition occurred at the rate of approximately 4×10^{-7} per cell per generation. Passage through the rabbit intestine selected for emergence of the Fla^+ phenotype. Although many strains of *C. jejuni* and *C. coli* undergo phase variation (158, 159) and aflagellate variants are simple to select in vitro, the genetic events involved are far from clear. Presumably some rearrangement could occur in the upstream region of *flaA* that turns off flagellar synthesis. However, this event must be reversible because Fla^+ variants can be selected either by passage through an animal model or in the laboratory. Other alternatives could include a repressor that turns off flagellar synthesis or an activator required to turn it on. The situation may be similar to that seen in other pathogens, such as *Bordetella pertussis* (64) and *Vibrio cholerae* (87), that have coordinate regulation of virulence gene expression. Research may yet show that other virulence traits in *Campylobacter* spp. are controlled at the genetic level by such a system.

ANTIGENIC VARIATION Antigenic variation refers to the ability of *Campylobacter* species to reversibly express flagella of different antigenic specificities (51). In *C. coli* VC167, antigenic variation corresponds to the production of two flagellins of different molecular weights, 61,500 (T1) and 59,500 (T2) (47, 51). Isoelectric focusing of purified flagellar filaments from several *C. jejuni* serotypes also showed multiple charged flagellins (91). These size and charge differences may result from posttranslational modification of amino acid residues within the flagellar protein(s) (76, 77). Guerry et al (48, 49) showed that *C. coli* VC167 undergoes a DNA rearrangement associated with the flagellar antigenic variation. The rearrangements were detected using a 700–base pair probe from VC167 that contains homology to members of the *Enterobacteriaceae*. Further studies suggest that the flagellar variation was associated with an uncharacterized rearrangement adjacent to a 23S rDNA locus (P. Guerry, personal communication). The molecular events underlying antigenic variation are complex and unclear.

EXPRESSION OF *CAMPYLOBACTER* FLAGELLIN GENES IN *E. COLI* Molecular analysis of the flagellin genes has been hampered by the inability of *E. coli* to express detectable levels of the proteins. The *flaA* genes from both *C. jejuni* 81116 and *C. coli* VC167 possess a strong σ^{28} type promotor, and fusion of the *flaA* promotor to a promotorless chloramphenicol acetyltransferase gene has demonstrated that the promotor can function in *E. coli* (48). Other reasons for lack of expression in *E. coli* include codon utilization differences (124) and lack of posttranslational modification of amino acids in *E. coli*. Although *flaA* genes specify mRNA containing suboptimal codons for peptide synthesis in *E. coli* (48), this is probably not the major factor in lack of expression. Fusion of the *flaA* and *flaB* proteins with the *cro-lacZ* protein was obtained using the pEX series of vectors (100). In this system, the higher proportion of rare codons in *C. jejuni* DNA compared with *E. coli* did not appear to be a limiting factor for expression. The advantage of this fusion protein expression system is that the fusion proteins precipitate inside the cell and are protected from proteolysis.

ENTEROTOXIN GENE Some strains of *C. jejuni* produce an enterotoxin (56, 61–63, 85, 119). Klipstein et al (62) found a correlation between enterotoxin production by strains of *Campylobacter* and watery diarrhea; however, many strains of *C. jejuni* and *C. coli* isolated from stool specimens do not appear to produce enterotoxin (38). Studies with DNA probes for cholera toxin (CT) and *E. coli* heat-labile enterotoxin (LT) genes could not demonstrate homology at the molecular level between *C. jejuni* and *C. coli* DNA and CT or LT genes (5, 102, 152). *Campylobacter* enterotoxin genes are chromosomal and not plasmid mediated (141). Using an oligonucleotide probe similar to the coding region for a postulated ganglioside GM1-binding site on the *toxB* gene from *Vibrio cholerae* and the *eltB* gene from *E. coli,* Calva et al located *C. jejuni* homologous chromosomal sequences (20). The oligonucleotide hybridized to a *Sau*3A digest of total DNA of all tested *C. jejuni* isolates. Because not all *C. jejuni* strains produced the enterotoxin, the authors concluded that some of the enterotoxin genes are probably inactive.

Work on enterotoxins has been complicated by disagreements arising partly from the use of different cell lines, different culture conditions, and different methods of observation, so it is important to clone the putative enterotoxin gene both for DNA sequencing and expression studies. However, Calva et al report difficulties in cloning and maintaining this and other *C. jejuni* sequences in *E. coli* hosts (20). More recently, increased stability of *C. jejuni* DNA recombinants was obtained by cloning in a *mutL* derivative (deficient in methyl-directed DNA repair) (E. Calva, personal communication).

MAJOR OUTER MEMBRANE PROTEIN Fusion proteins containing a portion of the major outer membrane protein (MOMP), were obtained using a polyclonal antibody directed against the MOMP from *C. jejuni* UA580 to screen λgt11 libraries. Two different DNA fragments of 147 base pairs (bp) and 1845 bp were identified (138). One fragment hybridized only with *C. jejuni* DNA and can be used as a *C. jejuni*–specific probe, whereas the other hybridized with all *C. jejuni* and *C. coli* strains tested, as well as with some *C. lari* strains (139). DNA sequencing (K. Hiratsuka & D. E. Taylor, unpublished data) demonstrated that both of these fragments lacked the complete open reading frame for the MOMP believed to be a porin (54). Although porins are not strictly virulence determinants, the MOMPs in *Campylobacter* spp. act as major immunogens and may be involved in the uptake and exclusion of various antibiotics (53, 106). Work is in progress to complete the sequence of these putative MOMP genes and to compare the deduced protein sequences to porins of other species.

SURFACE-ARRAY PROTEIN OF *C. FETUS* Surface arrays (S-layers), which consist of regularly arranged protein subunits that are self-assembling and form two-dimensional paracrystalline arrays of protein monomers, have been observed in *C. fetus* subsp. *fetus* (10, 36). Wild-type *C. fetus* strains that possess S-layers, but not spontaneous mutants that lack them, resist killing by normal and immune serum and resist phagocytosis because of defective binding of host C3b to bacterial cell surfaces (11).

Polyclonal antibody to surface-array proteins (SAP) of *C. fetus* with molecular weights of 97,000–149,000 were used to select clones from a library of λgt11 into which *C. fetus* chromosomal fragments of 1.0–6.5 kilobases (kb) had been ligated (9). A clone with a 4.0-kb insert was subcloned in pUC9, and expression of a protein of 98,000 Daltons was obtained in *E. coli*. The protein was not fused with β-galactosidase nor was expression inducible by isopropyl-thiogalactoside (IPTG). These studies failed to identify a promoter sequence, and expression was assumed to depend on a flanking promotor present within the vector. The complete sequence of the *C. fetus sapA* gene was determined; it encodes a 933–amino acid polypeptide with a calculated molecular weight of 96,758. Because the first 20 amino acids matched exactly those determined from N-terminal sequencing of the SAP protein, this polypeptide is apparently secreted without a leader sequence. The *C. fetus* SAP contains a small but distinctive region within a hydrophobic region that is homologous with other S-layer proteins from *E. coli, Klebsiella pneumoniae,* and *Yersinia* and *Leishmania* species. There is little overall homology among either primary or secondary structures of the S-layer proteins for which structural genes have been cloned, as would be expected because S-layer proteins appear to be biochemically diverse. Further study of the *C. fetus* SAP protein is eagerly awaited, because this protein appears to contain some interesting functional domains including sites for C3b interaction, procoagulant activity, and calcium-binding.

Campylobacter *Antibiotic Resistance Genes*

Genes encoding resistance to three different antibiotics in *Campylobacter* species have been cloned and sequenced (Table 1). Resistance to the antibiotics kanamycin, tetracycline, and chloramphenicol is usually plasmid mediated, and some plasmids carry more than one antibiotic-resistance determinant (133). Several of these antibiotic-resistance determinants are believed to have been acquired outside the *Campylobacter* genus and to have spread to it by heterologous genetic exchange. All the resistance determinants from

Campylobacter also confer antibiotic resistance in *E. coli*. Several have been useful in the construction of plasmid vectors (see below).

KANAMYCIN RESISTANCE Three different genetic determinants specifying kanamycin resistance (Km^r) have been identified in *Campylobacter* spp. However, all three act via a similar mechanism, namely the production of a 3'-O-aminoglycoside phosphotransferase. The *Campylobacter*-like organism (CLO) strain BM2196 contains a chromosomally located gene, *aphA-1*, which is almost identical to the Km^r determinant in Tn*903* originally derived from *E. coli* (104). The insertion sequence IS*15*-Δ, which is widespread in gram-negative bacteria, was adjacent to the Km^r gene in BM2196. This result suggests that this Km^r determinant was acquired by *Campylobacter* spp. from a member of the *Enterobacteriaceae* (104).

In contrast, a Km^r determinant from a *C. coli* plasmid pIP1433 of 48 kb (143) specifies a 3'-aminoglycoside phosphotransferase of type III encoded by *aphA-3*, a gene found previously only in gram-positive cocci (148). Therefore, Km^r in *C. coli* could also result from the acquisition of a gene from a gram-positive coccus (71, 148).

A third Km^r phosphotransferase gene, *aphA-7*, was cloned from a 14-kb *C. jejuni* plasmid, pS1178 (144). The DNA sequence of the *aphA-7* gene was most closely related to that in *Streptococcus faecalis;* however, the G+C ratio of the open reading frame of *aphA-7* was 32.8%, which suggested that the *aphA-7* gene may be indigenous to *Campylobacter* (144).

TETRACYCLINE RESISTANCE Tetracycline resistance (Tc^r) determinants from the conjugative *C. coli* plasmid pIP1433 and the conjugative *C. jejuni* plasmids pUA466 and pFKT1025 have been cloned (128, 131, 140, 144), and DNA sequences were determined. Tc^r determinants from both *C. jejuni* and *C. coli* plasmids are highly homologous at the nucleotide level (81, 128, 133) and have been given the designation *tet(O)* to conform to the current nomenclature for Tc^r determinants (75). The *tet(O)* genes demonstrate 75–76% homology with the *tet(M)* gene of *Streptococcus pneumoniae* (133), and Tet(O) has also been identified in both *Streptococcus* and *Enterococcus* spp. (164), leading to the conclusion that the *tet(O)* determinant has been acquired by *Campylobacter* spp., probably from a gram-positive coccus (128, 133).

The mechanism of Tet(O) resistance remains something of a mystery. The 69-kilodalton (kDa) Tet(O) protein acts at the level of protein synthesis to counteract the inhibitory effects of tetracycline (80). The amino acid sequence of the Tet(O) protein shows considerable homology at the amino terminal end with elongation factor (EF)-Tu and even greater homology with EF-G (80). Homologies extending throughout the entire length of EF-G and Tet(M) (17) and EF-G and Tet(O) (E. K. Manavathu & D. E. Taylor, unpublished data) have been noted. Tet(O) doubtless functions as a GTPase. However, why Tet(O) resembles EF-G is not yet clear. Neither EF-Tu nor EF-G have been shown to be inhibited directly by tetracycline.

CHLORAMPHENICOL RESISTANCE Sagara et al (120) cloned a resistance determinant specifying chloramphenicol resistance (Cm^r) from a *C. coli* plasmid (pNR9589) isolated in Japan. We sequenced the Cm^r determinant and identified a *cat* gene that specified a chloramphenicol acetyltransferase (155). This *cat* gene is most closely related to *cat* genes from *Clostridium perfringens* and *Clostridium difficile* with which it shows 67% identity (4). Because Cm^r is very rare in *Campylobacter* species, *C. coli* may have acquired the *cat* gene from *Clostridium* spp.

Helicobacter *Genes*

UREASE GENES An unusual feature of all *H. pylori* isolates is the production of a large quantity of urease responsible for hydrolysis of urea to ammonia and carbon dioxide (15). The urease is believed to be an important factor in the colonization by *H. pylori* cells of the gastric mucosa and in their ability to cause damage to mucosal tissue (52, 127). Therefore, genetic studies of *H. pylori* have focused on cloning the genes responsible for urease production.

Clayton et al first reported the cloning of *H. pylori* urease genes using λgt11 and detection of 66-kDa and 31-kDa antigens with antiserum raised against the purified *H. pylori* urease (29). The DNA sequence of the cloned DNA fragment corresponded to two polypeptides, UreA (26.7 kDa) and UreB (60.5 kDa) (28). However, no urease activity was detected in *E. coli* (29).

Using a different strategy, Labigne et al (68) identified a 44-kb portion of the *H. pylori* genome by cosmid cloning that permitted temporary biosynthesis of urease when transferred by conjugation to *C. jejuni*. They took this approach because of the failure of *H. pylori* urease expression in *E. coli*. Subcloning led to localization of the urease gene cluster to a 4.2-kb region of DNA. Four open reading frames in the order *ureC, ureD, ureA, ureB* were identified with predicted molecular weights of 49.2, 15.0, 26.5, and 61.6 kDa, respectively. Only a single copy of the gene cluster is present in *H. pylori*. Polypeptides corresponding to products of *ureA, ureB,* and *ureC* but not *ureD* were identified in *E. coli* minicells. The UreA and UreB polypeptides, which correspond to the two structural subunits of urease, appear phylogenetically more closely related to jack bean urease than to other bacterial ureases, which are composed of three subunits. Upstream of both the *ureA* and *ureD* genes (310 bp in each case) is a sequence resembling a σ^{54} promoter (67), which suggests that the expression of *ureA, ureB,* and *ureD* genes is under the same transcriptional control and may be subject to nitrogen regulation. In contrast, upstream of the *ureC* gene is an *E. coli* consensus promoter (σ^{70}). The roles of the UreC and UreD polypeptides are not clear. The latter possesses features typical of a transmembrane protein and may function to transport or anchor the enzyme (68). Recently, expression in *E. coli* of urease activity from *H. pylori* was obtained using nitrogen-limitation growth conditions. However, at least four additional genes, designated *ureE, ureF, ureG,* and *ureH,* cloned from *H. pylori* were required for urease expression in *E. coli* (29a).

M_r 26,000 SURFACE PROTEIN In an unsuccessful attempt (T. J. Trust, personal communication) to purify the fibrillar hemagglutinin (34, 53), O'Toole and coworkers (103) purified a surface protein of unknown function with an apparent M_r of 26,000 from *H. pylori* by extraction with 0.2 M glycine hydrochloride or mild detergent extraction with 0.6% octylglucoside. The N-terminal amino acid sequence for the first 46 residues was determined, and synthetic oligonucleotides capable of encoding amino acid residues 22–26 were synthesized. Those oligonucleotides were used to clone the gene encoding the M_r 26,000 protein within a 900-bp fragment in the vector pK18. Expression of the M_r 26,000 protein in *E. coli* could not be detected in immunoblots but was detected after the fragment was recloned in the expression vector pKK233-2. The size of the protein deduced from the DNA sequence was M_r 22,000. However, the protein produced in *E. coli* and identified using immunoblotting comigrated with that purified from *H. pylori,* i.e. M_r 26,000. The protein produced in *E. coli* appears to be a fusion protein with an additional 45 amino acid residues expressed from within the *E. coli* vector (103). Although several questions remain about the role of this protein in *H. pylori* and its expression in *E. coli,* the gene appears to reside in all *H. pylori* strains tested.

Difficulties Encountered in Cloning and Gene Expression

Examples of the difficulties experienced during the cloning or attempted cloning of *Campylobacter* and *Helicobacter* chromosomal genes are numerous, both in the published literature and in verbal communication. Problems encountered include: failure to express the gene of interest, as for example the *Campylobacter* flagellin genes (48, 99); instability of cloned genes, e.g. the putative *Campylobacter* enterotoxin (20); identification of a peptide of the correct molecular weight but failure to detect enzymic activity, e.g. urease from *H. pylori* (29, 68); and cloning only a portion of the gene, e.g. MOMP from *C. jejuni* (138; K. Hiratsuka & D. E. Taylor, unpublished data) and the M_r 26,000 protein from *H. pylori* (103).

Possible explanations for these difficulties include: (*a*) the presence in *Campylobacter* of unusual promotor sequences that are not recognized or are recognized much less efficiently in *E. coli;* (*b*) the high A-T content of *C. jejuni* and *C. coli* (32 mol% G+C) and *H. pylori* (36–38 mol% G+C) that may cause DNA sequences from these organisms to be recognized in *E. coli* as strong promotors. Also, regions of 70–80% A-T content that are also rich in static bends can serve as upstream activators of promotors (90). Such strong promotors are stable only in vectors in which efficient terminator signals protect plasmid-control elements from excessive transcription (14). Other possibilities include: (*c*) failure of *E. coli* to process some gene products because of a lack of accessory genes; (*d*) differences in methylation of DNA between *E. coli* and *Campylobacter* that may result in instability; and (*e*) limitation of expression of *Campylobacter* and *Helicobacter* genes due to the presence of suboptimal codons for peptide synthesis in *E. coli*.

First, difficulties related to promotor efficiency can be addressed by using an expression vector, thereby supplying an efficient promotor, rather than relying on the indigenous promotor or on one located within the cloning vector. As DNA sequences of additional *Campylobacter* and *Helicobacter* genes become available, it will be helpful to identify the promoters used and to compare them with those of other bacterial genera. Second, the use of special cloning vectors containing termination signals may be helpful. Such a vector (pJDC9) has been developed for cloning genes from the pneumococcus, which also has a low G+C content, and (as shown by early studies) cannot generate stable DNA fragments greater than 2 kb when cloned in common *E. coli* vectors (27). Third, when lack of accessory proteins for processing presents a problem, one should be able to identify the protein with an antibody, or by its size, without looking for expression directly. Fourth, the use of an *E. coli* strain deficient in methyl-directed DNA repair (*mutL*) may be helpful in cloning some genes. Finally, although some genes from *Campylobacter* specify mRNA with a high percentage of rare codons, these do not necessarily limit expression, at least in the case of *flaA*-specified peptides (100) and the *cat* gene product (155).

PLASMID VECTORS

Shuttle Vectors

The first shuttle vector to be constructed for *E. coli*-to-*Campylobacter* transfer was pILL550, which confered resistance to kanamycin in both *Campylobacter* and *E. coli* (70). This plasmid contains an origin of replication derived from the *C. coli* plasmid pIP445 (70) that functions in *Campylobacter* species as well as one that functions in *E. coli*. The presence of an *oriT* sequence from the IncPα plasmid RK2 (50) enables the vector to be mobilized by a transfer-competent P-group plasmid into *Campylobacter* species. More recently, Wang & Taylor (154) constructed several more shuttle vectors based on a strategy similar to that devised by Labigne-Roussel et al (70). Figure 1 shows the restriction maps of three shuttle vectors, and Table 2 lists characteristics of these and several additional vectors. All these vectors contain the LacZ' determinant that can complement a defective β-galactosidase in *E. coli* and is useful for selection of clones by their blue and white color on plates containing Xgal (150). Various antibiotic resistance determinants, Cm^r, Km^r, and Tc^r, which consist of the *cat, aphA-3,* and *tet(O)* genes, respectively (see section on *Campylobacter* antibiotic-resistance genes), are used as markers for plasmid selection in *Campylobacter,* although all are also expressed in *E. coli*.

Suicide Vectors

Suicide vectors that can be introduced into *Campylobacter* but cannot replicate in these species have also been constructed (69, 153) (see Table 2). This approach has been used to mutagenize 16S rRNA genes (69), to construct a *leuB* mutant of *C. jejuni* by shuttle transposon mutagenesis (A. Labigne, personal communication), and to inactivate flagellar genes in both *C. jejuni* and *C. coli* (47, 156). To be successful, this approach requires that the cloned *Campylobacter* gene be available and that the original copy of the gene be disrupted by insertional mutation, usually with a resistance determinant. Once the suicide vector is mobilized into *C. jejuni,* homologous recombination occurs between the cloned *Campylobacter* gene and the chromosome. In some cells, the original copy of the gene is replaced by the mutated allele to generate the required mutant.

Campylobacter *Vectors*

Plasmids that replicate only in *Campylobacter* have also been constructed (155) (see Table 2). They consist of the origin of replication from a *C. coli* plasmid, a resistance determinant that functions in *Campylobacter* spp., and a

Table 2 *Campylobacter* cloning vectors

Plasmid (size)[a]	Marker[b]	Replication origin *E. coli*	Replication origin *Campylobacter*	*oriT*[c]	Reference
pUOA13 (8.7)	*aphA-3, bla, lacZ'*	+	+	+	154
pUOA14 (8.2)	*aphA-3, bla, cat*	+	+	+	155
pUOA15 (11.1)	*bla, lacZ', tet(O)*	+	+	+	154
pUOA17 (8.2)	*aphA-3, lacZ'*	+	+	+	154
pUOA18 (7.4)	*cat*	+	+	+	150
pUOA19 (5.0)	*aphA-3*	−	+	−	150
pUOA20 (4.8)	*cat*	−	+	−	150
pUOA22 (4.1)	*aphA-3, bla*	+	−	−	153
pUOA23 (3.8)	*cat, bla*	+	−	−	153

[a] Plasmid size in kilobases.
[b] The markers *bla* and *lacZ'* are not expressed in *Campylobacter* species.
[c] *oriT* is the origin of transfer from a broad-host-range IncP plasmid (50).

multiple cloning site. These vectors should be useful for subcloning *Campylobacter* DNA fragments into the multiple cloning site with subsequent transfer into *Campylobacter* species by electroporation or natural transformation (see next section). They are being used in our laboratory to search for DNA sequences that enhance uptake of the natural transformation process (154).

GENETIC EXCHANGE MECHANISMS

Plasmid Transfer

CAMPYLOBACTER Conjugative plasmids encoding Cm^r, Km^r, and/or Tc^r are found in some *C. jejuni* but more often in *C. coli* strains (130–132, 134, 135, 145). These plasmids usually range in size from 45 to 50 kb with a G+C content of 31–33 mol%, or approximately equivalent to those of the host species (136). Restriction maps of Tc^r and Km^r plasmids have been constructed (131, 132, 143, 145), but only the resistance determinants have been located on the plasmids. Nothing is known about the arrangements of genes involved in plasmid conjugative transfer or replication. Plasmid transfer frequencies ranged from about 1×10^{-5} to 1×10^{-3} transconjugants per recipient cell in a 24-h mating period. All plasmids tested transferred more efficiently on a solid surface than in liquid medium (135, 136), and their host range was restricted to closely related *Campylobacter* species (130, 135).

HELICOBACTER Plasmid transfer has not been reported in *Helicobacter* species, although plasmids have been visualized in *H. pylori* (108, 147). Tjia

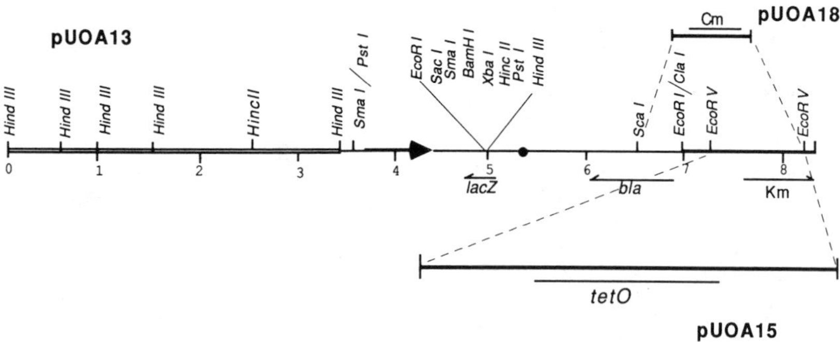

Figure 1 Restriction maps of the shuttle vectors pUOA13, pUOA15, and pUOA18. Position of resistance determinants: Km, kanamycin resistance; Cm, chloramphenicol resistance; *tetO*, tetracycline resistance (all of which are expressed in both *E. coli* and *Campylobacter* species); and *bla*, ampicillin resistance (β lactamase production), which is only expressed in *E. coli*. (*Double bar*) DNA sequence from the Campylobacter plasmid pIP1445; (*arrow*) *oriT* DNA; (*single bar*) pUC13 DNA. Numbers represent kilobase pairs.

et al (147) found that 58% of strains contained one or more plasmids ranging in size from 1.8 kb to 40 kb; whereas Penfold et al (108) found that 48% of strains examined yielded plasmids ranging in size from 3.7 kb to >148 kb. Nucleotide sequence analysis of a 1.5-kb plasmid from *H. pylori* (60a) demonstrated significant homology between it and plasmids that replicated via the "rolling-circle" replication mechanism. This suggests that the plasmid has been acquired by *H. pylori* from a gram-positive coccus since these plasmids have not previously been found in gram-negative bacteria. It has not yet been possible to ascribe a phenotype to any of the plasmids.

No strains of *H. pylori* are as yet reported to be resistant to any of the common antibiotics that are plasmid-mediated in *C. jejuni* and *C. coli,* i.e. Cm^r, Km^r, Tc^r (58, 137). Because these determinants are frequently carried on conjugative plasmids in *Campylobacter* as well as other species, we should not be surprised that conjugative plasmids have not yet been identified in *H. pylori*. Whether or not the increasing use of antibiotics to treat patients with *H. pylori* results in the emergence of plasmid-mediated resistance remains to be seen. However, these organisms, hidden presumably under the mucus layer, have been present in the stomach of patients treated orally with antibiotics for other infections, and have not yet developed resistance. Other factors may be important in the development of resistance, such as the proximity of other bacterial species carrying resistance determinants and the ability of *H. pylori* to take up available DNA.

Natural Transformation

CAMPYLOBACTER Wang & Taylor (154) observed that strains of both *C. coli* and *C. jejuni* could take up DNA without any special treatment, such as $CaCl_2$ or heat shock. Natural-transformation frequencies from streptomycin resistance (Str^r) and nalidixic acid resistance (Nal^r) were approximately 1×10^{-3} transformants per recipient cell for *C. coli* and 1×10^{-4} transformants per recipient cell for *C. jejuni*. Cotransformation frequencies for Str^r and Nal^r were 2×10^{-7} for *C. coli* UA585, which suggests that these two markers are not closely linked. Incubation with DNaseI prevents transformation. Although some strains of *Campylobacter* produce extracellular DNase, there is no association between the DNase-producing ability of the recipient strain and its capacity to take up endogenous DNA. All five *C. coli* strains tested were naturally competent, whereas only three out of six *C. jejuni* strains were competent.

The competence process is not understood in *Campylobacter* species. It is almost completely independent of growth phase; early-log-phase bacteria are slightly more competent than late-log-phase cells. The maximum transformation frequency obtained with *C. coli* UA585, with 0.1 μg DNA/ml, was 4×10^5 transformants per microgram of DNA. DNA from *C. jejuni* UA466R

(Nal^r Str^r) could transform Nal^r to *C. coli* UA585 at about 20% efficiency compared with homologous DNA, but these interspecies Nal^r transformants grew more slowly than either parent. *C. jejuni* UA466 could be transformed to Str^r by *C. coli* UA417R DNA at 1% efficiency; however, these transformants exhibited normal growth rates. Therefore, some cross-species transformation apparently occurs between *C. jejuni* and *C. coli*. Unlabeled *C. jejuni* DNA can also effectively compete with ^{32}P-labeled *C. coli* DNA for uptake into *C. coli* UA585 (154).

Transformation of *Campylobacter* with plasmid DNA is much less efficient than with chromosomal DNA. Small plasmids such as those shown in Table 2 transform *C. coli* UA585 at a frequency 1000-fold lower than that of chromosomal DNA markers. However, when a recipient strain such as *C. coli* UA585 contains a homologous plasmid, transformation frequencies are increased 100-fold. The plasmid present in the recipient seems to act as a rescue plasmid by recombining with the incoming plasmid (154).

If campylobacters are similar to other microorganisms such as *Haemophilus influenzae* and *Neisseria gonorrhoeae,* then their DNA probably contains a specific sequence necessary for binding to and uptake into campylobacter cells. Uptake sequences of 11 base pairs (30) and 10 base pairs (40) have been identified in *H. influenzae* and *N. gonorrhoeae,* respectively. These sequences, which are presumably present throughout the DNA of a particular *Campylobacter* species, may not be present at all on plasmids or perhaps are not present in enough copies to trigger uptake, especially if part of the plasmid has been acquired from an unrelated bacterium that does not possess the relevant uptake sequence.

HELICOBACTER Natural transformation in *H. pylori* has been reported (94). Of 25 clinical isolates, 22 were naturally competent for transformation of Str^r, including the *H. pylori* type strain NCTC 11637. The transformation frequency of *H. pylori* NCTC 11637 was 5×10^{-4}. DNA from a Str^r mutant of *C. jejuni* could not transform competent *H. pylori*. We have confirmed that *H. pylori* strains are naturally competent for transformation. Both Str^r and rifampicin-resistance markers were used to demonstrate DNA uptake (Y. Wang & D. E. Taylor, in preparation).

Both *Campylobacter* and *Helicobacter* spp., therefore, can take up DNA from other individuals in the population. This behavior may be important in the spread of antibiotic resistance such as high level erythromycin and quinolone resistance in *Campylobacter* spp. (33, 43, 142, 163) and perhaps metronidazole resistance in *H. pylori* (6, 39).

Electrotransformation

Electrotransformation, or electroporation, refers to the use of high voltages to induce the uptake of plasmid DNA into cells. Miller et al (86) reported electroporation of the shuttle vector pILL521 into *C. jejuni* C31 at frequencies as high as 1.2×10^6 transformants per microgram of DNA. Using the commercially available Gene Pulsar apparatus (BioRad), a field strength of 21.5 kV/cm, and a time constant of 2 ms, Yan (160) obtained maximum transformation frequencies of 5×10^3 transformants per microgram of DNA, which was equivalent to approximately 1×10^{-6} transformants per viable cell using *C. jejuni* C31. *Campylobacter* spp. appear to tolerate exposure to high-voltage electric fields without difficulty (86, 160). However, although *C. jejuni* C31 has been used successfully for electroporation, some other strains of *C. jejuni* and *C. coli* do not act as efficient strains in electroporation studies (160).

Bacteriophage Transduction

CAMPYLOBACTER Ritchie et al (117), Grajewski et al (45), and Salama et al (121, 122) have described bacteriophages specific for *C. jejuni* and *C. coli*. These workers have been concerned with developing bacteriophage typing schemes for epidemiological studies of *Campylobacter* infections. None of the Preston phages obtained by Salama and coworkers were lysogenic, and treatment with mitomycin-C was not useful for phage recovery (121). Nevertheless, we have successfully used bacteriophage $\phi 3$ of the Preston typing phages to transduce erythromycin resistance (Ery^r) from *C. coli* UA733 to *C. coli* UA585 (S. Salama & D. E. Taylor, unpublished data). Further studies are required to determine if transduction of the Ery^r marker represents specialized or generalized transduction and to determine if other markers can be transduced by $\phi 3$ or by other *C. jejuni* and *C. coli* bacteriophages.

In an early report, Chang & Ogg described phage-mediated transduction of a Str^r marker between strains of *C. fetus* subsp. *fetus* and from *C. fetus fetus* to *C. fetus* subsp. *venerealis* (23). They were able to effect transduction of glycine tolerance from *C. fetus fetus* to *C. fetus venerealis* using the same phage, VFP-11 (24). These authors pointed out that glycine tolerance may not be a reliable means of separating these two subspecies, as it can be transduced in a single step. Using phage VFP-13, Ogg & Chang (101) demonstrated bacteriophage conversion of a serotype V strain of *C. fetus fetus* to a serotype I strain. Also, some isolates reacted with antiserum to both serotype I and V, showing that both serospecific antigens were produced. Unfortunately, no further work has been published on transduction in these subspecies of *C. fetus*.

HELICOBACTER A single study reports visualization of a bacteriophage with a head size of 70×60 nm and a tail of at least 120 nm (123). Lysogeny was maintained during subculture in the laboratory for more than three months. Thus, *H. pylori* may also be capable of bacteriophage transduction.

GENETIC MAPPING

Genome Size and Mapping of Campylobacter *Species*

GENOME SIZES OF *C. JEJUNI* AND *C. COLI* Pulsed-field gel electrophoresis (PFGE) has facilitated the determination of genome sizes and the construction of physical maps of the chromosomes of several bacterial species (66). Bacterial genome sizes vary from *Myxococcus xanthus* (26), the largest, with a genome size of 9.45 Mb, to *Mycoplasma genitalium,* the smallest, with a genome size of only 585 kb (129).

Both *C. jejuni* UA580 (NCTC 1168) and *C. coli* UA417 have genome sizes of approximately 1.7 Mb, as determined using PFGE after *Sal*I and *Sma*I digestion (25). Nuijten et al (97) also determined that *C. jejuni* has a genome size of 1.7 Mb, although others (60) have obtained a slightly higher estimate. Therefore, *Campylobacter* species have genomes that are slightly smaller than *Haemophilus influenzae* Rd at 1.9 Mb (73, 74) and that are only about one-third of the size of the *E. coli* chromosome (126). The small genome size of *Campylobacter* spp. is consistent with their small and delicate nature, requirement for supplemented medium for growth, failure to ferment carbohydrates or degrade complex substances, and their biochemical inertness (57).

GENOME MAPS OF *C. JEJUNI* AND *C. COLI* The genomes of *C. jejuni* UA580 and *C. coli* UA417 consist of a single circular DNA molecule (Figures 2 and 3). One unusual feature of both the *C. jejuni* UA580 and *C. coli* UA417 genome maps is the arrangement of ribosomal RNA genes. Others have demonstrated that *C. jejuni* strains contain three copies of rRNA genes (59, 69, 97). Our data confirm this result, but we found that for both *C. jejuni* UA580 and *C. coli* UA417, at least two of the 16S rRNA genes are located apart from the 23S rRNA genes. Therefore, in the *C. jejuni* and *C. coli* strains we have examined, the rRNA genes do not appear to reside in operons as they do in *E. coli* or *H. influenzae,* but assume an arrangement more characteristic of organisms, such as *Mycoplasma galisepticum, Leptospira interrogans,* and *Thermus thermophilus* (66). The *flaAflaB* genes in *C. jejuni* 81116 reside together approximately 300 kb from one 16S rRNA gene and 700 kb from another (97). Similarly, flagellar genes in *C. jejuni* UA580 are located about 700 kb and 170 kb away from the two closest 16S rRNA genes (see Figure 2). These results suggest that the relative locations of flagellar genes and 16S rRNA genes are at least partially conserved in two different *C. jejuni* strains.

Ribosomal protein genes (*rplJ* and *rplL*) that encode, respectively, the large ribosomal subunit proteins L7/L12 and L10 reside together at 90 min on the *E. coli* map (3). A DNA probe from *E. coli* carrying these genes hybridized to

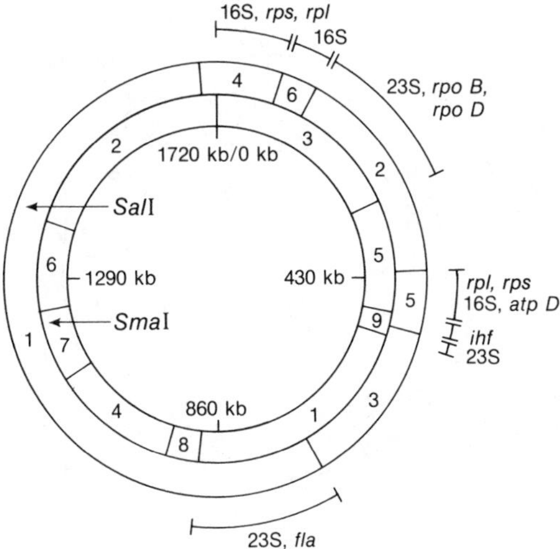

Figure 2 Genome map of *C. jejuni* UA580. Map was constructed from partial digestion patterns obtained with *Sma*I and *Sal*I, and by hybridization of ^{32}P-labeled DNA fragments after extraction from low melting point agarose. Both homologous and heterologous DNA probes were used for mapping. These were *C. jejuni flaAflaB* (pIVB3-300) (99); *C. jejuni* 16S rDNA (pAR140) (115); *E. coli* 23S rDNA (pCW1) (157); *E. coli rplJrplL* (pNO2016) (113); *E. coli rpsGrpsL* (pNO2005) (112); *E. coli hiphimA* (*lhf*) (pPLhip-himA-5) (93); *Bacillus subtilis* RNA polymerase σ^{43} subunit (*rpoD*) (pCP522) (114); *H. influenzae* RNA polymerase β subunit (*rpoB*) (pRIF2) (18); *E. coli atpD* (pBJC505) (151).

two regions in both the *C. jejuni* and *C. coli* maps separated by about 250 kb. The same two regions also hybridized to an *E. coli* probe for the *rpsG* and *rpsL* genes, which encode two small-subunit ribosomal proteins, S7 and S12, and are located at 73 min on the *E. coli* genetic map (3). Therefore, both *C. jejuni* and *C. coli* appear to have two separate clusters comprising a single 16S rRNA gene and both large and small ribosomal protein genes. The Str^r and Ery^r mutations probably also map to ribosomal protein genes (163). However, these genes are not located close to either of the ribosomal gene protein clusters so far identified on the *C. coli* UA417 genome (Figure 3).

Genome Size and Mapping of Helicobacter *Species*

The genome sizes of 30 *H. pylori* strains range from 1.60 to 1.73 Mb (D. E. Taylor, M. Eaton, N. Chang & S. Salama, submitted). The genome of *H. pylori* UA802 is a single circle with a size of 1.71 Mb. *H. mustelae* has

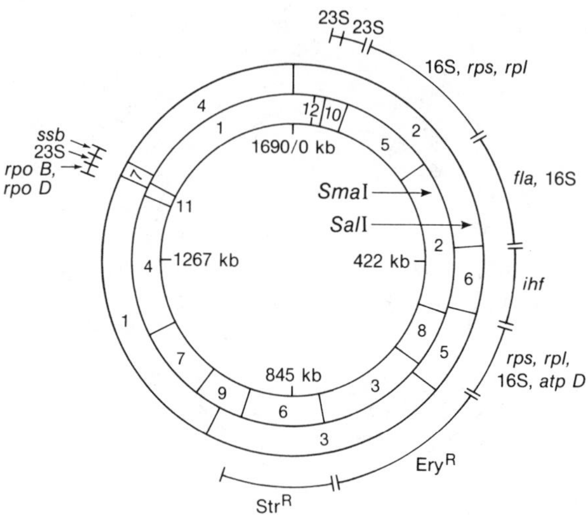

Figure 3 Genome map of *C. coli* UA417. See legend to Figure 2 for map construction and probes used. Additional probe was *E. coli* (*ssb*) (pTL119A 5) (78). Positions of Str^r and Ery^r markers were mapped by natural transformation of extracted DNA fragments (162).

approximately the same genome size as *H. pylori* (D. E. Taylor, unpublished data).

PFGE analysis of *H. pylori* strains from gastric biopsies would indicate that their genomes are highly variable and are far more variable than those of *C. jejuni* or *C. coli*. Yet digest patterns of a single strain maintained in the laboratory remained constant over time, and identical patterns were obtained from three *H. pylori* isolates obtained from antrum, fundus, and body of the stomach of a single patient (D. E. Taylor, M. Eaton, N. Chang & S. Salama, submitted). These results agree with findings from conventional agarose gel electrophoresis of *H. pylori* genome DNA using enzymes with more frequent cut sites (79, 105).

Restriction endonucleases useful for PFGE analysis of *C. jejuni* and *C. coli,* i.e. *Sal*I and *Sma*I, are not useful for analysis of *H. pylori*. Instead, *Not*I and *Nru*I cut most but not all *H. pylori* strains into a reasonable number of fragments for PFGE analysis. For *H. mustelae, Sfi*I and *Sal*I appear to be the enzymes of choice (D. E. Taylor, unpublished data).

As more genes from *Campylobacter* and *Helicobacter* are cloned and sequenced, it will be useful to extend genome mapping studies of individual strains to try to determine if gene arrangements are conserved. The amount of repetitive DNA appears to be small within the *C. jejuni, C. coli,* and *H. pylori* strains mapped so far because hybridization studies of restriction fragments give clear-cut rather than equivocal results. There is, however, a small degree of variability among the genomes of unrelated isolates of both *C. jejuni* and *C. coli* (161), and this phenomenon is even more pronounced in *H. pylori* (D. E. Taylor, M. Eaton, N. Chang & S. Salama, submitted). Bacterial genomes that constantly undergo genetic exchange via transformation, such as members of *Campylobacter* and *Helicobacter* species, may be prone to rapid genomic rearrangements. Although gene arrangement is usually conserved at the origins and terminus of replication in many bacteria (66), these regions have yet to be identified in *Campylobacter* or *Helicobacter* species.

CONCLUDING REMARKS

The number of investigators interested in the genetics of *Campylobacter* and *Helicobacter* species is increasing. Over the next few years, more genes will be cloned and sequenced from both of these genera. The relative ease with which genes involved in amino acid biosynthesis can be cloned should increase our understanding of these biosynthetic pathways in *Campylobacter* spp. The problems associated with cloning and expression of more specialized *Campylobacter* genes in *E. coli* will probably be overcome by using a variety of different approaches. Unusual features of these organisms that require further investigation include their genome variability, which has been documented in both *C. jejuni* and *C. coli,* but is even more pronounced in *H. pylori,* and the unusual arrangement of their rRNA genes. Studies of regulation of virulence traits in these species should lead to a clearer understanding of the way in which both *Campylobacter* and *Helicobacter* infect their human hosts.

ACKNOWLEDGMENTS

I acknowledge financial support from the Natural Sciences and Engineering Research Council of Canada, the Canadian Bacterial Diseases Network, and Glaxo Canada. I thank all those who supplied unpublished information including E. Calva, V. L. Chan, N. Chang, S. Cohen, P. Guerry, A. Labigne, I. Nachamkin, T. Trust, S. Salama, M. Eaton, W. Yan, and particularly Y. Wang for Figure 1. The help of M. B. Skirrow and C. S. Goodwin in critically reading the manuscript and of Heather Mitchell in typing it is also gratefully acknowledged.

Literature Cited

1. Aguero-Rosenfeld, M. E., Yang, X.-H., Nachamkin, I. 1990. Infection of adult syrian hampsters with flagellar variants of *Campylobacter jejuni*. *Infect. Immunol*. 58:2214–19
2. Arnosti, D. N., Chamberlin, M. J. 1989. Secondary sigma factor controls transcription of flagellar and chemotaxis genes in *Escherichia coli*. *Proc. Natl. Acad. Sci. USA* 86:830–34
3. Bachmann, B. J. 1987. Linkage map of *Escherichia coli* K-12, edition 7. In Escherichia coli *and* Salmonella typhimurium. *Cellular and Molecular Biology,* ed. F. C. Neidhardt, pp. 807–76. Washington, DC: Am. Soc. Microbiol.
4. Bannam, T. I., Rood, J. I. 1991. Relationship between the *Clostridium perfringens catQ* gene product and chloramphenicol acetyltransferases of other bacteria. *Antimicrob. Agents Chemother*. 35:471–76
5. Barg, B. H., Wachsmuth, I. K., Morris, G. K., Hill, W. E. 1986. Probing of *Campylobacter jejuni* with DNA coding for *Escherichia coli* heat-labile enterotoxin. *J. Infect. Dis*. 154:542
6. Becx, M. C. J. M., Janssen, A. J. H. M., Clasener, H. A. L., DeKoning, R. W. 1990. Metronidazole-resistant *Helicobacter pylori*. *Lancet* 335:539–40
7. Black, R. E., Levine, M. M., Clements, M. L., Hughes, T. P., Blaser, M. J. 1988. Experimental *Campylobacter jejuni* infection in humans. *J. Infect. Dis*. 157:472–79
8. Blaser, M. J. 1987. Gastric *Campylobacter*-like organisms, gastritis, and peptic ulcer disease. *Gastroenterology* 93:371–83
9. Blaser, M. J., Gotschlich, E. C. 1990. Surface array protein of *Campylobacter fetus:* cloning and gene structure. *J. Biol. Chem*. 265:14529–35
10. Blaser, M. J., Smith, P. F., Hopkins, J. A., Bryner, J., Heinzer, I., Wang, W.-L.-L. 1987. Pathogenesis of *Campylobacter fetus* infections: serum resistance associated with high-molecular-weight surface proteins. *J. Infect. Dis*. 155:696–706
11. Blaser, M. J., Smith, P. F., Repine, J. C., Joiner, K. A. 1988. Pathogenesis of *Campylobacter fetus* infections: failure of encapsulated *Campylobacter fetus* to bind C3b explains serum and phagocytosis resistance. *J. Clin. Invest*. 81: 1434–44
12. Blaser, M. J., Taylor, D. N., Feldman, R. A. 1983. Epidemiology of *Campylobacter jejuni* infections. *Epidemiol. Rev*. 5:157–76
13. Bronsdon, M. A., Goodwin, C. S., Sly, L. I., Chilvers, T., Schoenknecht, F. D. 1991. *Helicobacter nemestrinae* sp. nov., a spiral bacterium found in the stomach of a pigtailed macaque (*Macaca nemestrina*). *Int. J. Syst. Bacteriol*. 41:148–53
14. Brosius, J. 1984. Toxicity of an overproduced foreign gene product in *Escherichia coli* and its use in plasmid vectors for the selection of transcription terminators. *Gene* 27:161–72
15. Buck, G. E., Gourley, W. K., Lee, W. K., Subramanyam, K., Latimer, J. M., DiNuzzo, A. R. 1986. Relation of *Campylobacter pylori* to gastritis and peptic ulcer. *J. Infect. Dis*. 153:664–69
16. Buck, G. E. 1990. *Campylobacter pylori* and gastroduodenal disease. *Clin. Microbiol. Rev*. 3:1–12
17. Burdett, V. 1991. Purification and characterization of Tet(M), a protein that renders ribosomes resistant to tetracycline. *J. Biol. Chem*. 266:2872–77
18. Butler, P. D., Moxon, R. 1990. A physical map of the genome of *Haemophilus influenzae* type b. *J. Gen. Microbiol*. 136:2333–42
19. Caldwell, M. B., Guerry, P., Lee, E. C., Burans, J. P., Walker, R. I. 1985. Reversible expression of flagella in *Campylobacter jejuni*. *Infect. Immunol*. 50:941–43
20. Calva, E., Torres, J., Vazquez, M., Angeles, V., de la Vega, H., Ruiz-Palacios, G. M. 1989. *Campylobacter jejuni* chromosomal sequences that hybridize to *Vibrio cholera* and *Escherichia coli* LT enterotoxin genes. *Gene* 75:243–51
21. Chan, V. L., Bingham, H. 1991. Complete sequence of the *Campylobacter jejuni glyA* gene encoding serine hydroxymethyltransferase. *Gene* 101:51–58
22. Chan, V. L., Bingham, H., Kibue, A., Nayudu, P. R. V., Penner, J. L. 1988. Cloning and expression of the *Campylobacter jejuni glyA* gene in *Escherichia coli*. *Gene* 73:185–91
23. Chang, W.-J., Ogg, J. E. 1970. Transduction in *Vibrio fetus*. *Am. J. Vet. Res*. 31:919–24
24. Chang, W.-J., Ogg, J. E. 1971. Transduction and mutation to glycine tolerance in *Vibrio fetus*. *Am. J. Vet. Res*. 32:649–53
25. Chang, N., Taylor, D. E. 1990. Use of pulsed-field agarose gel electrophoresis

to size *Campylobacter* spp. genomes and to construct a *Sal*I map of *Campylobacter jejuni* UA580. *J. Bacteriol.* 172: 5211–17

26. Chen, H., Keseler, I. M., Shimkets, L. J. 1990. The genome size of *Myxococcus xanthus* determined by pulsed-field gel electrophoresis. *J. Bacteriol.* 172: 4206–13

27. Chen, J.-D., Morrison, D. A. 1987. Cloning of *Streptococcus pneumoniae* DNA fragments in *Escherichia coli* requires vectors protected by strong transcriptional termination. *Gene* 64:179–87

28. Clayton, C. L., Pallen, M. J., Kleanthous, H., Wren, B. W., Tobaqchali, S. 1989. Nucleotide sequence of two genes from *Helicobacter pylori* encoding for urease subunits. *Nucleic Acids Res.* 18:362

29. Clayton, C. L., Wren, B. W., Mullany, P., Topping, A., Tobaqchali, S. 1989. Molecular cloning and expression of *Campylobacter pylori* species-specific antigens in *Escherichia coli* K-12. *Infect. Immunol.* 57:623–29

29a. Cussac, V., Ferrero, R., Labigne, A. 1991. Expression of *Helicobacter pylori* urease activity in *Escherchia coli* host strains. *Microb. Ecol. Health Dis.* 4:5139

30. Danner, D. B., Deich, R. A., Sisco, K.-L., Smith, H. O. 1980. An eleven-base-pair sequence determines the specificity of DNA uptake in *Haemophilus* transformation. *Gene* 11:311–18

31. Dick, J. D. 1990. *Helicobacter (Campylobacter) pylori:* a new twist to an old disease. *Annu. Rev. Microbiol.* 44:249–69

32. Dooley, C. P., Cohen, H. 1988. The clinical significance of *Campylobacter pylori*. *Ann. Intern. Med.* 108:70–79

33. Endtz, H. Ph., Ruijs, G. J., van Klingeren, B., Jansen, W. H., van der Reyden, T., Mouton, R. P. 1991. Quinolone resistance in campylobacter isolated from man and poultry following the introduction of fluoroquinolones in veterinary medicine. *J. Antimicrob. Chemother.* 27:199–208

34. Evans, D. G., Evans, D. J., Jr., Moulds, J. J., Graham, D. Y. 1988. N-acetylneuraminyllactose-binding fibrillar hemagglutinin of *Campylobacter pylori:* a putative colonization factor antigen. *Infect. Immunol.* 56:2896–2906

35. Fischer, S. H., Nachamkin, I. 1991. Common and variable domains of the flagellin gene, *flaA* in *Campylobacter jejuni*. *Mol. Microbiol.* 5:1151–58

36. Fogg, G. C., Yang, L., Wang, E., Blaser, M. J. 1990. Surface array proteins of *Campylobacter fetus* block lectin-mediated binding to type A lipopolysaccharide. *Infect. Immunol.* 58:2738–94

37. Fox, J. G., Taylor, N. S., Edmons, P., Brenner, D. J. 1988. *Campylobacter pylori* subsp. *mustelae* subsp. nov. isolated from the gastric mucosa of ferrets (*Mustela putorius furo*), and an emended description of *Campylobacter pylori*. *Int. J. Syst. Bacteriol.* 38:367–70

38. Fricker, C. R., Park, R. W. A. 1989. A two-year study of the distribution of 'thermophilic' campylobacters in human, environmental and food samples from the Reading area with particular reference to toxin production and heat-stable serotype. *J. Appl. Bacteriol.* 66: 477–90

39. Glupczynski, Y., Burette, A., Dekoster, E., Nyst, J.-F., Deltenre, M., et al. 1990. Metronidazole resistance in *Helicobacter pylori*. *Lancet* 335:976–77

40. Goodman, S. D., Scocca, J. J. 1988. Identification and arrangement of the DNA sequence recognized in specific transformation of *Neisseria gonorrhoeae*. *Proc. Natl. Acad. Sci. USA* 85:6982–86

41. Goodwin, C. S., Armstrong, J. A., Chilvers, R., Peters, M., Collins, M. D., et al. 1989. Transfer of *Campylobacter pylori* and *Campylobacter mustelae* to *Helicobacter* gen. nov. as *Helicobacter pylori* comb. nov. and *Helicobacter mustelae* comb. nov., respectively. *Int. J. Syst. Bacteriol.* 39:397–405

42. Goodwin, C. S., Armstrong, J. A., Marshall, B. J. 1986. *Campylobacter pyloridis,* gastritis, and peptic ulceration. *J. Clin. Pathol.* 39:353–65

43. Gootz, T. D., Martin, B. A. 1991. Characterization of high-level quinolone resistance in *Campylobacter jejuni*. *Antimicrob. Agents Chemother.* 35:840–45

44. Graham, D. Y. 1989. *Campylobacter pylori* and peptic ulcer disease. *Gastroenterology* 96:615–25 (Suppl.)

45. Grajewski, B. A., Kusek, J. W., Gelfand, H. W. 1985. Development of a bacteriophage typing system for *Campylobacter jejuni* and *Campylobacter coli*. *J. Clin. Microbiol.* 22:13–18

46. Griffiths, P. L., Park, R. W. A. 1990. Campylobacters associated with human diarrhoeal disease. *J. Appl. Bacteriol.* 69:281–301

47. Guerry, P., Alm, R. A., Power, M. E., Logan, S. M., Trust, T. J. 1991. Role of two flagellin genes in *Campylobacter* motility. *J. Bacteriol.* 173:4757–64

48. Guerry, P., Logan, S. M., Thornton, S.,

Trust, T. J. 1990. Genomic organization and expression of *Campylobacter* flagella genes. *J. Bacteriol.* 172:1853–60
49. Guerry, P., Logan, S. M., Trust, T. J. 1988. Genomic rearrangements associated with antigenic variation in *Campylobacter coli. J. Bacteriol.* 170: 316–19
50. Guiney, D. G., Yakobson, E. 1983. Location and nucleotide sequence of the transfer origin of the broad host range plasmid RK2. *Proc. Natl. Acad. Sci. USA* 80:3595–98
51. Harris, L. A., Logan, S. M., Guerry, P., Trust, T. J. 1987. Antigenic variation of *Campylobacter* flagella. *J. Bacteriol.* 169:5066–71
52. Hazell, S., Lee, A. 1986. *Campylobacter pyloridis* urease, hydrogen ion back diffusion and gastric ulcers. *Lancet* 2: 15–17
53. Huang, J., Smyth, C. J., Kennedy, N. P., Arbuthnott, J. P., Keeling, P. W. N. 1988. Haemagglutinating activity of *Campylobacter pylori. FEMS Microbiol. Lett.* 56:109–12
54. Huyer, M., Parr, T. R. Jr., Hancock, R. E. W., Page, W. J. 1986. Outer membrane porin protein of *Campylobacter jejuni. FEMS Microbiol. Lett.* 37: 247–50
55. Huynh, T. V., Young, R. A., Davis. R. W. 1985. Construction and screening cDNA libraries in lambda gt10 and lambda gt11. In *DNA Cloning Techniques: a Practical Approach,* ed. D. Glover, 2:49–78. Eynsham, UK: IRL Press
56. Johnson, W. M., Lior, H. 1984. Toxins produced by *Campylobacter jejuni* and *Campylobacter coli. Lancet* 1:229–30
57. Karmali, M. A., Skirrow, M. B. 1984. Taxonomy of the genus *Campylobacter*. In Campylobacter *Infection in Man and Animals,* ed. J. Butzler, pp. 1–20. Boca Raton, FL: CRC Press
58. Kasper, G., Dickgiesser, N. 1984. Antibiotic sensitivity of *"Campylobacter pylori." Eur. J. Clin. Microbiol.* 3:444
59. Kim, N. W., Chan, V. L. 1989. Isolation and characterization of the ribosomal RNA genes of *Campylobacter jejuni. Curr. Microbiol.* 19:247–52
60. Kim, N. W., Chan, V. L. 1991. Genomic characterization of *Campylobacter jejuni* by field inversion gel electrophoresis. *Curr. Microbiol.* 22:123–27
60a. Kleanthous, K., Clayton, C. L., Tobaqchali, S. 1991. Characterization of a plasmid from *Helicobacter pylori* encoding a replicating protein common to plasmids in gram-negative bacteria. *Mol. Microbiol.* 5:2377–89
61. Klipstein, F. A., Engert, R. F. 1984. Properties of crude *Campylobacter jejuni* heat-labile enterotoxin. *Infect. Immunol.* 45:314–19
62. Klipstein, F. A., Engert, R. F. 1985. Immunological relationship of the B subunits of *Campylobacter jejuni* and *Escherichia coli* heat-labile enterotoxins. *Infect. Immunol.* 48:629–33
63. Klipstein, F. A., Engert, R. F., Short, H. B. 1986. Enzyme-linked immunosorbent assays for virulence properties of *Campylobacter jejuni* clinical isolates. *J. Clin. Microbiol.* 23:1039–43
64. Knapp, S., Mekalanos, J. J. 1988. Two trans-acting regulatory genes (*vir* and *mod*) control antigenic modulation in *Bordetella pertussis. J. Bacteriol.* 170: 5059–66
65. Kostrzynska, M., Betts, J. D., Austin, J. W., Trust, T. J. 1991. Identification, characterization, and spacial localization of two flagellin species of *Helicobacter pylori* flagella. *J. Bacteriol.* 173:937–46
66. Krawiec, S., Riley, M. 1990. Organization of the bacterial chromosome. *Microbiol. Rev.* 54:502–39
67. Kustu, S., Santero, E., Keener, J., Popham, D., Weiss, D. 1989. Expression of σ^{54} (*ntrA*)-dependent genes is probably united by a common mechanism. *Microbiol. Rev.* 53:367–76
68. Labigne, A., Cussac, V., Courcoux, P. 1991. Shuttle cloning and nucleotide sequences of *Helicobacter pylori* genes responsible for urease activity. *J. Bacteriol.* 173:1920–31
69. Labigne-Roussel, A., Courcoux, P, Tompkins, L. 1988. Gene disruption and replacement as a feasible approach for mutagenesis of *Campylobacter jejuni. J. Bacteriol.* 170:1704–8
70. Labigne-Roussel, A., Harel, J., Tompkins, L. 1987. Gene transfer from *Escherichia coli* to *Campylobacter* species: development of shuttle vectors for genetic analysis of *Campylobacter jejuni. J. Bacteriol.* 169:5320–23
71. Lambert, T., Gerbaud, G., Trieu-Cuot, P., Courvalin, P. 1985. Structural relationship between genes encoding 3'-aminoglycoside phosphotransferases in *Campylobacter* and gram-positive cocci. *Ann. Inst. Pasteur (Paris)* 136B:135–50
72. Lee, E. C., Walker, R. I., Guerry, P. 1985. Expression of *Campylobacter* genes for proline biosynthesis in *Escherichia coli. Can. J. Microbiol.* 31: 1064–67
73. Lee, J. J., Smith. H. O. 1988. Sizing of the *Haemophilus influenzae* Rd genome

by pulsed-field agarose gel electrophoresis. *J. Bacteriol.* 170:4402–5

74. Lee, J. J., Smith, H. O., Redfield, R. J. 1989. Organization of the *Haemophilus influenzae* Rd genome. *J. Bacteriol.* 171:3016–24
75. Levy, S. B., McMurry, L. M., Burdett, V., Courvalin, P., Hillen, W., et al. 1989. Nomenclature for tetracycline resistance determinants. *Antimicrob. Agents Chemother.* 33:1373–74
76. Logan, S. M., Guerry, P., Rollins, D. M., Burr, D. H., Trust, T. J. 1989. In vivo antigenic variation of *Campylobacter* flagellin. *Infect. Immunol.* 57:2583–85
77. Logan, S. M., Trust, T. J., Guerry, P. 1989. Evidence for posttranslational modification and gene duplication of *Campylobacter* flagellin. *J. Bacteriol.* 171:3031–38
78. Lohman, T. M., Green, J. M., Beyer, R. S. 1986. Large scale overproduction and rapid purification of the *Escherichia coli ssb* gene product: expression of the *ssb* gene under λPL control. *Biochemistry* 25:21–25
79. Majewski, S. I. H., Goodwin, C. S. 1988. Restriction endonuclease analysis of the genome of *Campylobacter pylori* with a rapid extraction method: evidence for considerable genomic variation. *J. Infect. Dis.* 157:465–71
80. Manavathu, E. K., Fernandez, C. L., Cooperman, B. S., Taylor, D. E. 1990. Molecular studies on the mechanism of tetracycline resistance mediated by Tet(O). *Antimicrob. Agents Chemother.* 34:71–77
81. Manavathu, E. K., Hiratsuka, K., Taylor, D. E. 1988. Nucleotide sequence analysis and expression of a tetracycline resistance gene from *Campylobacter jejuni*. *Gene* 62:17–26
82. Marshall, B. J. 1983. Unidentified curved bacilli on gastric epithelium in chronic active gastritis. *Lancet* 1:1273–75
83. Marshall, B. J. 1989. History of the discovery of *C. pylori*. In Campylobacter pylori *in Gastritis and Peptic Ulcer Disease,* ed. M. J. Blaser, pp. 7–23. New York: Igaku-Shoin
84. Martin, P., Trieu-Cuot, P., Courvalin, P. 1986. Nucleotide sequence of the *tetM* tetracycline resistance determinant of the streptococcal conjugative shuttle transposon Tn*1545*. *Nucleic Acids Res.* 14:7047–58
85. McCardell, B. A., Madden, J. M., Lee, E. C. 1984. Production of cholera-like toxin by *Campylobacter jejuni/coli*. *Lancet* 1:448–49
86. Miller, J. F., Dower, W. J., Tompkins, L. S. 1988. High voltage electroporation of bacteria: genetic transformation of *Campylobacter jejuni* with plasmid DNA. *Proc. Natl. Acad. Sci. USA* 85: 856–60
87. Miller, V. L., Taylor, R. K., Mekalanos, J. J. 1987. Cholera toxin transcriptional activator Tox^R is a transmembrane DNA binding protein. *Cell* 48:271–79
88. Minnich, S. A., Ohta, N., Taylor, N., Newton, N. 1988. Role of the 25-, 27- and 29-kilodalton flagellins in *Caulobacter crescentus* cell motility: method for construction of detection and Tn*5* insertion mutants by gene replacement. *J. Bacteriol.* 170:3953–60
89. Morooka, T., Umeda, A., Amako, K. 1985. Motility as an intestinal colonization factor for *Campylobacter jejuni*. *J. Gen. Microbiol.* 131:1973–80
90. Morrison, D. A., Jaurin, B. 1990. *Streptococcus pneumoniae* possesses canonical *Escherichia coli* (sigma 70) promotors. *Mol. Microbiol.* 4:1143–52
91. Nachamkin, I., Yang, X.-H. 1988. Isoelectric focusing of *Campylobacter jejuni* flagellin: microheterogeneity and restricted antigenicity of changed species with a monoclonal antibody. *FEMS Microbiol. Lett.* 49:235–38
92. Nachamkin, I, Yang, X.-H. 1989. Human antibody response to *Campylobacter jejuni* flagellin protein and synthetic N-terminal flagellin peptide. *J. Clin. Microbiol.* 27:2195–98
93. Nash, H. A., Robertson, C. A., Flamm, E., Weisberg, R. A., Miller, H. I. 1987. Overproduction of *Escherichia coli* integration host factor, a protein with nonidentical subunits. *J. Bacteriol.* 169:4124–27
94. Nedenskov-Sorensen, P., Bukholm, G., Bovre, K. 1990. Natural competence for genetic transformation in *Campylobacter pylori*. *J. Infect. Dis.* 161:365–66
95. Newell, D. G. 1986. Monoclonal antibodies directed against the flagella of *Campylobacter jejuni:* cross-reacting and serotypic specificity and potential role in diagnosis. *J. Hyg.* 96:377–84
96. Newell, D. G., McBride, H., Dolby, J. M. 1985. Investigations on the role of the flagella in the colonization of infant mice with *Campylobacter jejuni* and attachment of *Campylobacter jejuni* to human epithelial cell lines. *J. Hyg.* 95:217–27
97. Nuijten, P. J. M., Bartels, C., Bleumink-Pluym, N. M. C., Gaastra, W., van der Zeijst, B. A. M. 1990. Size and physical map of the *Campylobacter*

jejuni chromosome. *Nucleic Acids Res.* 18:6211–14

98. Nuijten, P. J. M., Bleumink-Pluym, N. M. C., Gaastra, W., van der Zeijst, B. A. M. 1989. Flagellin expression in *Campylobacter jejuni* is regulated at the transcriptional level. *Infect. Immunol.* 57:1084–88
99. Nuijten, P. J. M., van Asten, F. J. A. M., Gaastra, W., van der Zeijst, B. A. M. 1990. Structural and functional analysis of two *Campylobacter jejuni* flagellin genes. *J. Biol. Chem.* 265: 17798–17804
100. Nuijten, P. J. M., van der Zeijst, B. A. M., Newell, D. G. 1991. Localization of immunogenic regions on the flagellin proteins of *Campylobacter jejuni* 81116. *Infect. Immunol.* 59:1100–5
101. Ogg, J. E., Chang, W.-J. 1972. Phage conversion of serotypes in *Vibrio fetus*. *Am. J. Vet.* 33:1023–29
102. Olsvik, O., Wachsmuth, K., Morris, G., Feeley, J. C. 1984. Genetic probing of *Campylobacter jejuni* for cholera toxin and *Escherichia coli* heat-labile enterotoxin. *Lancet* 1:449
103. O'Toole, P. W., Logan, S. M., Kostrzynska, M., Wadström, T., Trust, T. J. 1991. Isolation and biochemical and molecular analyses of a species-specific protein antigen from the gastric pathogen *Helicobacter pylori*. *J. Bacteriol.* 173:505–13
104. Ouellette, M., Gerbaud, G., Lambert, T., Courvalin, P. 1987. Acquisition by a *Campylobacter*-like strain of *aphA-1*, a kanamycin resistance determinant from members of the family *Enterobacteriaceae*. *Antimicrob. Agents Chemother.* 31:1021–26
105. Owen, R. J., Bickley, J., Moreno, M., Costas, M., Morgan, D. R. 1991. Biotype and macromolecular profiles of cytotoxin-producing strains of *Helicobacter pylori* from antral gastric mucosa. *FEMS Microbiol. Lett.* 79:199–204
106. Page, W. J., Huyer, G., Huyer, M., Worobec, E. A. 1989. Characterization of the porins of *Campylobacter jejuni* and *Campylobacter coli* and implications for antibiotic susceptibility. *Antimicrob. Agents Chemother.* 33:297–303
107. Paster, B. J., Lee, A., Fox, J. G., Dewhirst, F. E., Tordoff, L. A., et al. 1991. Phylogeny of *Helicobacter felis*. *Int. J. Syst. Bacteriol.* 41:31–38
108. Penfold, S. S., Lastovica, A. J., Elisha, B. G. 1988. Demonstration of plasmids in *Campylobacter pylori*. *J. Infect. Dis.* 157:850–51
109. Penner, J. L. 1988. The genus *Campylobacter:* a decade of progress. *Clin. Microbiol. Rev.* 1:157–72
110. Peterson, W. L. 1991. *Helicobacter pylori* and peptic ulcer disease. *New Eng. J. Med.* 324:1043–48
111. Pleier, E., Schmitt, R. 1991. Expression of two *Rhizobium meliloti* flagellin genes and their contribution to the complex filament structure. *J. Bacteriol.* 173:2077–85
112. Post, L. E., Nomura, M. 1980. DNA sequence of the *str* operon of *Escherichia coli*. *J. Biol. Chem.* 255: 4660–66
113. Post, L. E., Strycharz, G. D., Nomura, M., Lewis, H., Dennis, P. P. 1979. Nucleotide sequence of the ribosomal protein gene cluster adjacent to the gene for RNA polymerase β in *Escherichia coli*. *Proc. Natl. Acad. Sci. USA* 76:1697–1701
114. Price, C. W., Doi, R. H. 1985. Genetic mapping of *rpoD* implicates the major sigma factor of *Bacillus subtilis* RNA polymerase in sporulation initiation. *Mol. Gen. Genet.* 201:88–95
115. Rashtchian, A., Abbott, M. A., Shaffer, M. 1987. Cloning and characterization of genes coding for ribosomal RNA in *Campylobacter jejuni*. *Curr. Microbiol.* 14:311–17
116. Rashtchian, A., Shaffer, M. 1986. The nucleotide sequences of two tRNA genes from *Campylobacter jejuni*. *Nucleic Acids Res.* 14:5560
117. Ritchie, A. E., Bryner, J. H., Foley, J. W. 1983. Role of DNA and bacteriophage in *Campylobacter* auto-agglutination. *J. Med. Microbiol.* 16:333–40
118. Romaniuk, P. J., Zoltowska, B., Trust, T. J., Lane, D. J., Olsen, G. J., et al. 1987. *Campylobacter pylori,* the spiral bacterium associated with human gastritis, is not a true *Campylobacter* sp. *J. Bacteriol.* 169:2137–41
119. Ruiz-Palacios, G. M., Torres, J., Torres, N. I., Escamilla, E., Ruiz-Palacios, B. R., Tamayo, J. 1983. Cholera-like enterotoxin produced by *Campylobacter jejuni:* characterization and clinical significance. *Lancet* 2:250–52
120. Sagara, K., Mochizuki, A., Okamura, N., Nakaya, R. 1987. Antimicrobial resistance of *Campylobacter jejuni* and *Campylobacter coli* with special reference to plasmid profiles of Japanese clinical isolates. *Antimicrob. Agents Chemother.* 31:713–19
121. Salama, S., Bolton, F. J., Hutchinson, D. N. 1989. Improved method for the isolation of *Campylobacter jejuni* and

Campylobacter coli bacteriophages. *Lett. Appl. Microbiol.* 8:5–7

122. Salama, S. M., Bolton, F. J., Hutchinson, D. N. 1990. Application of a new phage typing scheme to campylobacters isolated during outbreaks. *Epidemiol. Infect.* 104:405–11
123. Schmid, E. N., Von Recklinghausen, G., Ansorg, R. 1990. Bacteriophages in *Helicobacter (Campylobacter) pylori*. *J. Med. Microbiol.* 32:101–4
124. Sharp. P. M., Li, W.-H. 1987. The codon adaptation index—a measure of directional synonymous codon usage bias, and its potential applications. *Nucleic Acids Res.* 15:1281–95
125. Skirrow, M. B. 1977. *Campylobacter enteritis*—a "new" disease. *Br. Med. J.* 2:9–11
126. Smith, C. L., Econome, J. G., Schutt, A., Klco, S., Cantor, C. R. 1987. A physical map of the *Escherichia coli* K-12 genome. *Science* 236:1448–53
127. Smoot, D. T., Mobley, H. L. T., Chippendale, G. R., Lewison, J. F., Resau, J. F. 1990. *Helicobacter pylori* urease activity is toxic to human gastric epithelial cells. *Infect. Immunol.* 58:1992–94
128. Sougakoff, W., Papadopoulou, B., Nordmann, P., Courvalin, P. 1987. Nucleotide sequence and distribution of gene *tetO* encoding tetracycline resistance in *Campylobacter coli*. *FEMS Microbiol. Lett.* 44:153–59
129. Su, C. J., Baseman, J. B. 1990. Genome size of *Mycoplasma genitalium*. *J. Bacteriol.* 172:4705–7
130. Taylor, D. E. 1984. Plasmids from *Campylobacter*. In Campylobacter *Infection in Man and Animals,* ed. J. P. Butzler, pp. 87–96. Boca Raton, FL: CRC Press
131. Taylor, D. E. 1986. Plasmid-mediated tetracycline resistance in *Campylobacter jejuni:* expression in *Escherichia coli* and identification of homology with streptococcal class M determinant. *J. Bacteriol.* 165:1037–39
132. Taylor, D. E., Chang, N., Garner, R. S., Sherburne, R., Mueller, L. 1986. Incidence and antibiotic resistance and characterization of plasmids in *Campylobacter jejuni* isolated from clinical sources in Alberta, Canada. *Can. J. Microbiol.* 32:28–32
133. Taylor, D. E., Courvalin, P. 1988. Mechanisms of antibiotic resistance in *Campylobacter* species. *Antimicrob. Agents Chemother.* 32:1107–12
134. Taylor, D. E., DeGrandis, S. A., Karmali, M. A., Fleming, P. C. 1980. Transmissible tetracycline resistance in *Campylobacter jejuni*. *Lancet* 2:797
135. Taylor, D. E., DeGrandis, S. A., Karmali, M. A., Fleming, P. C. 1981. Transmissible plasmids from *Campylobacter jejuni*. *Antimicrob. Agents Chemother.* 19:831–35
136. Taylor, D. E., Garner, R. S., Allan, B. J. 1983. Characterization of tetracycline resistance plasmids from *Campylobacter jejuni* and *Campylobacter coli*. *Antimicrob. Agents Chemother.* 24:930–35
137. Taylor, D. E., Hargreaves, J. A., Ng, L.-K., Sherbaniuk, R. W., Jewell, L. D. 1987. Isolation and characterization of *Campylobacter pyloridis* from gastric biopsies. *Am. J. Clin. Path.* 87:49–54
138. Taylor, D. E., Hiratsuka, K. 1990. Use of non-radioactive DNA probes for detection and *Campylobacter jejuni* and *Campylobacter coli* in stool specimens. *Mol. Cell. Probes* 4:261–71
139. Taylor, D. E., Hiratsuka, K., Mueller, L. 1989. Isolation and characterization of catalase-negative and catalase-weak strains of *Campylobacter* species, including "*Campylobacter upsaliensis*" from humans with gastroenteritis. *J. Clin. Microbiol.* 27:2042–45
140. Taylor, D. E., Hiratsuka, K., Ray, H., Manavathu, E. K. 1987. Characterization and expression of a cloned tetracycline resistance determinant from *Campylobacter jejuni* plasmid pUA466. *J. Bacteriol.* 169:2984–89
141. Taylor, D. E., Johnson, W. M., Lior, H. 1987. Cytotoxic and enterotoxic activities of *Campylobacter jejuni* are not specified by the tetracycline resistance plasmids pMAK175 and pUA466. *J. Clin. Microbiol.* 25:150–51
142. Taylor, D. E., Ng, L.-K., Lior, H. 1985. Susceptibility of *Campylobacter* species to nalidixic acid, enoxacin, and other DNA gyrase inhibitors. *Antimicrob. Agents Chemother.* 28:708–10
143. Taylor, D. E., Yan, W., Ng, L.-K., Manavathu, E. K., Courvalin, P. 1988. Genetic characterization of kanamycin resistance in *Campylobacter coli*. *Ann. Inst. Pasteur/Microbiol.* 139:665–76
144. Tenover, F. C., Gilbert, T., O'Hara, P. 1989. Nucleotide sequence of a novel kanamycin resistance gene, *aphA-7,* from *Campylobacter jejuni* and comparison with other kanamycin phosphotransferase genes. *Plasmid* 22:52–58
145. Tenover, F. C., LeBlanc, D. J., Elvrum, P. 1987. Cloning and expression of a tetracycline resistance determinant from *Campylobacter jejuni* in *Escherichia coli*. *Antimicrob. Agents Chemother.* 31:1301–6
146. Thornton, S. A., Logan, S. M., Trust, T. J., Guerry, P. 1990. Polynucleotide

sequence relationships among flagellin genes of *Campylobacter jejuni* and *Campylobacter coli*. *Infect. Immunol.* 58:2686–89

147. Tjia, T. N., Harper, W. E. S., Goodwin, C. S., Grubb, W. B. 1987. Plasmids in *Campylobacter pyloridis*. *Microbios Lett.* 36:7–11
148. Trieu-Cuot, P., Gerbaud, G., Lambert, T., Courvalin, P. 1985. *In vivo* transfer of genetic information between gram-positive and gram-negative bacteria. *EMBO J.* 4:3583–87
149. Vandamme, P., Falsen, E., Rossau, R., Hoste, B., Seegers, P., et al. 1991. Revision of *Campylobacter, Helicobacter* and *Wolinella* taxonomy: emendation of generic descriptions and proposal of *Arcobacter* gen. nov. *Int. J. Syst. Bacteriol.* 41:88–103
150. Vieira, J., Messing, J. 1982. The pUC plasmids, an M13 mp7-derived system for insertion mutagenesis and sequencing with synthetic universal primers. *Gene* 19:259–68
151. von Mayenburg, K., Jorgensen, B. B., Nielsen, J., Hansen, F. G. 1982. Promotors of the *atp* operon coding for the membrane-bound ATP synthetase of *Escherichia coli* mapped by Tn*10* insertion mutations. *Mol. Gen. Genet.* 188:240–48
152. Walker, R. I., Caldwell, M. B., Lee, E. C., Guerry, P., Trust, T. J., Ruiz-Palacios G. M. 1986. Pathophysiology of *Campylobacter enteritis*. *Microbiol. Rev.* 50:81–94
153. Wang, Y. 1991. *Transformation and antibiotic resistance in* Campylobacter *species*. PhD thesis. Univ. Alberta, Edmonton, Canada
154. Wang, Y., Taylor, D. E. 1990. Natural transformation of *Campylobacter* species. *J. Bacteriol.* 172:949–55
155. Wang, Y., Taylor, D. E. 1990. Chloramphenicol resistance in *Campylobacter coli:* nucleotide sequence, expression, and cloning vector construction. *Gene* 94:23–28
156. Wassenaar, T. M., Bleumink-Pluym, N. M. C., van der Zeijst, B. A. M. 1991. Inactivation of *Campylobacter jejuni* flagellin genes by homologous recombination demonstrates that *flaA* but not *flaB* is required for invasion. *EMBO J.* 8:2055–61
157. Weitzmann, C. J., Cunningham, P. R., Ofengand, J. 1990. Cloning in vitro transcription, and biological activity of *Escherichia coli* 23S ribosomal RNA. *Nucleic Acids Res.* 18:3515–20
158. Wenman, W. M., Chai, J., Louie, T. J., Goudreau, C., Lior, H., et al. 1985. Antigenic analysis of flagellar protein and other proteins. *J. Clin. Microbiol.* 21:108–12
159. Wenman, W. M., Taylor, D. E., Lior, H. 1985. The flagellar protein determines *Campylobacter jejuni* serotype. In *Recent Advances in Chemotherapy. Proc. 14th Int. Congr. Chemotherapy,* ed. J. Ishigama, pp. 361–62. Kyoto, Japan: Univ. Tokyo Press
160. Yan, W. 1990. *Characterization of erythromycin resistance in* Campylobacter *spp*. PhD thesis. Univ. Alberta, Edmonton, Alberta, Canada
161. Yan, W., Chang, N., Taylor, D. E. 1991. Pulsed-field gel electrophoresis of *Campylobacter jejuni* and *Campylobacter coli* genomic DNA and its epidemiologic application. *J. Infect. Dis.* 163:1068–72
162. Yan, W., Taylor, D. E. 1991. Sizing and mapping of the genome of *Campylobacter coli* strain UA417R using pulsed field gel electrophoresis. *Gene* 101:121–25
163. Yan, W., Taylor, D. E. 1991. Characterization of erythromycin resistance in *Campylobacter jejuni* and *Campylobacter coli*. *Antimicrob. Agents Chemother.* 35:1989–96
164. Zilhao, R., Papadopoulou, B., Courvalin, P. 1988. Occurrence of the *Campylobacter* resistance gene *tetO* in *Enterococcus* and *Streptococcus* spp. *Antimicrob. Agents Chemother.* 32: 1793–96

Annu. Rev. Microbiol. 1992. 46:65–94

THE LIPOPHOSPHOGLYCAN OF *LEISHMANIA* PARASITES

Salvatore J. Turco and Albert Descoteaux

Department of Biochemistry, University of Kentucky Medical Center, Lexington, Kentucky 40536

KEY WORDS: glycolipid, glycoconjugates, glycosylphosphatidyl inositol, leishmaniasis, macrophages

---

CONTENTS

0066-4227/92/1001-0065$02.00

Abstract

Protozoan parasites of the genus *Leishmania* have the remarkable ability to avoid destruction in the hostile environments they encounter throughout their life cycle. The molecular details of how these pathogens persevere with impunity under harsh conditions are beginning to be understood. The fact that *Leishmania* parasites have adapted to not only survive, but to proliferate probably is due to the protection conferred by specialized molecules on the parasite's cell surface. One such macromolecule is a novel glycoconjugate called lipophosphoglycan. This heterogeneous, lipid-containing polysaccharide is the major surface molecule of the parasite and has been implicated in a surprisingly large number of functions that may contribute the the parasite's pathogenesis. This review emphasizes the structural aspects of lipophosphoglycan and its possible functions and biosynthesis.

INTRODUCTION

Leishmaniasis

Protozoan parasites of the genus *Leishmania* are the causative agents of leishmaniasis, a disease that annually afflicts millions of people world-wide. Depending on the pathogenic species involved, leishmaniasis is clinically divided into three types: cutaneous, mucocutaneous, and visceral. The etiologic agents of cutaneous disease are *L. major, L. tropica, L. mexicana,* and *L. aethiopica,* which restrict their infection to dermal tissues. Mucocutaneous disease (espundia) is attributed to *L. braziliensis,* which replicates primarily in mucous tissues and causes gross disfigurements. *L. donovani* is responsible for visceral leishmaniasis or kala-azar, a chronic and often fatal disease. Detailed descriptions of the clinical aspects and geographical distribution of the various leishmaniases are reviewed elsewhere (22, 52, 107).

Current treatment of leishmaniasis is rather crude and relies on chemotherapy. The various types of leishmaniases are treated with a repertoire of antimonial drugs (e.g. Pentostam) or diamidines (e.g. Pentamidine). Disadvantages of these antiquated drugs include their numerous side effects, the possibility of being mutagenic, the requirement of prolonged treatment, and the lack of knowledge about their precise mode of action. Because researchers are only now gaining fundamental knowledge of the molecular aspects that distinguish the parasites from their mammalian hosts, innovative drug design is in its infancy. Once the distinguishing features are elucidated, new and effective therapeutic protocols will undoubtedly evolve.

The Life Cycle of Leishmania *Parasites*

The life cycle of of *Leishmania* parasites consists of two stages. In one stage, the parasite lives as an extracellular, flagellate promastigote form in the alimentary tract of its insect vector, the sandflies *Phlebotomus* and *Lutzomyia* spp. While attached to the epithelial cells lining the midgut of the sandfly, the promastigotes multiply and are avirulent. In a process called metacyclogenesis (127), promastigotes eventually cease dividing, detach from the epithelial cells, and migrate to the mouthparts of the insect. These metacyclic promastigotes are virulent. This sequential development of promastigotes from a dividing, noninfectious stage to a resting, infective stage has been observed for promastigotes growing in the sandfly alimentary tract and in axenic culture (127).

Metacyclic promastigotes are inoculated into the microscopic wound produced in a human or other suitable vertebrate when a sandfly feeds. The parasites then attach to the macrophages attracted to the wound and enter these host cells by a receptor-mediated process. Upon entry into the phagolysosome of the macrophage, promastigotes differentiate to nonflagellate amastigotes. Surprisingly, these amastigotes not only survive, they proliferate in this usually hostile environment. Ultimately, the infected macrophages lyse, releasing the amastigotes into the surrounding environment where they can infect other macrophages. The life cycle is completed when a feeding sandfly bites an infected host and the parasite is present in the bloodmeal.

Importance of Cell-Surface Glycoconjugates

In order for a *Leishmania* parasite to propagate, it must avoid destruction (*a*) in the gut of the sandfly where the organism could be susceptible to digestive enzymes that degrade the bloodmeal, (*b*) in the bloodstream of the vertebrate host where the parasite transiently exists and would be exposed to the lytic complement pathway; and (*c*) in the phagolysosome of host macrophages where the parasite would be vulnerable to hydrolytic enzymes and the microbicidal oxidative burst. The molecular details of how this pathogen perseveres with impunity in obviously hostile conditions is beginning to be understood. Cell-surface glycoconjugates undoubtedly play a key role in the survival of the *Leishmania* parasite throughout its existence. One of the major glycoconjugates synthesized by the parasite is lipophosphoglycan (LPG), a heterogeneous, lipid-containing polysaccharide. This review emphasizes the structural aspects of LPG and its possible functions and biosynthesis.

THE STRUCTURE OF LPG

Lipophosphoglycan of Promastigotes

THE GENERALIZED STRUCTURE OF LPG The promastigote form of all *Leishmania* parasites synthesizes LPG. LPG is the major glycoconjugate of the promastigote and is localized over all its surface including the flagellum

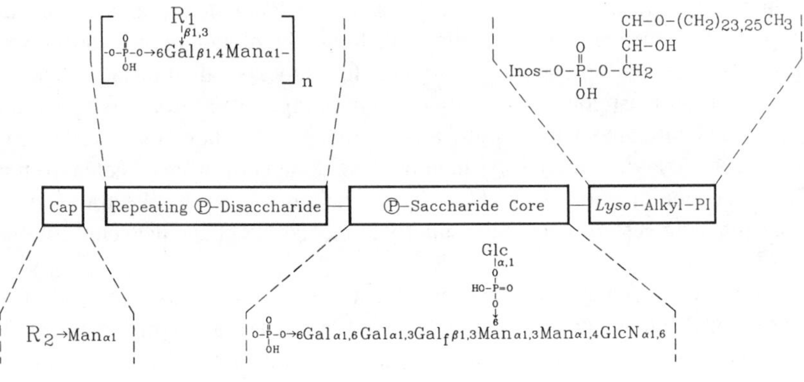

	R_1	R_2
L. donovani ($n_{avg.}$=16)	= H	= $\text{Man}_{\alpha1,2}$–
		= $\text{Gal}_{\beta1,4}$–
		= $\text{Man}_{\alpha1,2}[\text{Gal}_{\beta1,4}]$–
		= $\text{Man}_{\alpha1,2}\text{Man}_{\alpha1,2}$–
		= $\text{Man}_{\alpha1,2}\text{Man}_{\alpha1,2}[\text{Gal}_{\beta1,4}]$–
L. major ($n_{avg.}$=27)	= H	= $\text{Man}_{\alpha1,2}$–
	= Gal	
	= $\text{Ara}_{\alpha1,2}\text{Gal}$–	
	= $\text{Gal}_{\beta1,3}\text{Gal}$–	
	= $\text{Glc}_{\beta1,3}\text{Gal}$–	
	= $\text{Ara}_{\alpha1,2}\text{Gal}_{\beta1,3}\text{Gal}$–	
	= $\text{Gal}_{\beta1,3}\text{Gal}_{\beta1,3}\text{Gal}$–	
	= $\text{Ara}_{\alpha1,2}\text{Gal}_{\beta1,3}\text{Gal}_{\beta1,3}\text{Gal}$–	
L. mexicana ($n_{avg.}$=16)	= H	= $\text{Man}_{\alpha1,2}$–
	= Glc	= $\text{Man}_{\alpha1,2}\text{Man}_{\alpha1,2}$–
		= $\text{Man}_{\alpha1,2}[\text{Gal}_{\beta1,4}]$–

Figure 1 Structures of LPG from three promastigote species of *Leishmania,* grown in log to late-log phase of growth.

(72, 109). Each parasite cell contains several million molecules of LPG (82, 84, 104). As shown in Figure 1, LPG consists of four domains: (*a*) a phosphatidylinositol lipid anchor, (*b*) a phosphosaccharide core, (*c*) a repeating phosphorylated saccharide region, and (*d*) a small oligosaccharide cap structure. Structural analyses of LPG from several species of *Leishmania* indicate complete conservation of the lipid anchor, extensive conservation of the phosphosaccharide core, and variability of sugar composition and sequence in the repeating phosphorylated saccharide units and the cap structure (62a, 84, 104, 141a).

THE LIPID ANCHOR OF LPG The polysaccharide portion of LPG is anchored by the unusual phospholipid derivative 1-*O*-alkyl-2-*lyso*-phosphatidyl(*myo*)inositol. In the LPGs from *L. donovani* (104), *L. major* (81), and *L. mexicana* (62a), and probably in those from all species of *Leishmania,* the aliphatic chain consists of either a C_{24} or C_{26} saturated, unbranched hydrocarbon. As can many glycosylphosphatidylinositol (GPI)-anchored proteins (reviewed in 23), LPG can be hydrolyzed by bacterial phosphatidylinositol-specific phospholipase C, producing 1-*O*-alkylglycerol and the entire polysaccharide chain as products.

THE PHOSPHOSACCHARIDE CORE OF LPG Attached to the inositol of the lipid anchor of LPG is the phosphosaccharide core region. In *L. donovani* (147), *L. major* (84), and *L. mexicana* (62a), the glycan core consists of an unacetylated glucosamine, two mannoses, a galactose-6-phosphate, a galactopyranose, and a galactofuranose. The presence of the latter is extremely unusual in eukaryotic glycoconjugates, especially because the furanose is internal in a carbohydrate chain. As with all other reported glycosylphosphatidylinositol-anchored proteins reported thus far (23), LPG possesses the Man(α1,4)GlcN(α1,6)*myo*-inositol-1-PO_4 motif. The LPG cores of *L. donovani* (141a) and *L. mexicana* (62a) possess a glucosyl-α1-phosphate attached in phosphodiester linkage to the C6 hydroxyl of the proximal mannose residue. A substantial percentage of the *L. major* LPG also contains the identical glucosyl-α1-phosphate substitution, while the remainder does not (84). Another interesting sequence in the core region is the Gal(α1,3)Gal unit, which is reportedly the epitope for circulating antibodies in patients with leishmaniasis (3, 4, 144).

THE REPEATING UNITS OF LPG The salient feature of LPG is the repeating phosphorylated saccharide region. All LPG molecules reported thus far (62a, 82, 84, 146) contain multiple units of a backbone structure of PO_4-6Gal(β1,4)Man(α1). One of the noteworthy features of the backbone is the 4-*O*-substituted mannose residue, which is not present in any other known eukaryotic glycoconjugate. The *L. donovani* LPG (82, 146) contains no other substitutions of the backbone sequence, whereas in the *L. mexicana* LPG, approximately 25% of galactose residues are substituted at the C3 hydroxyl with βGlc residues (62a). The repeating units of the *L. major* LPG are the most complex in that approximately 87% of the galactose residues are further substituted with small saccharide side chains containing one to four residues of galactose, glucose, or the pentose arabinose (84). The presence of the common Gal-Man disaccharide backbone and the species-specific substitutions on the galactose residue could account for common and species-specific epitopes reported in serological observations (135). The number of repeating units per LPG molecule directly depends on the growth stage of the promastigote and is discussed below.

THE CAPPING OLIGOSACCHARIDES OF LPG LPG is terminated at the nonreducing end with one of several small neutral oligosaccharides containing galactose or mannose. Although the LPG of *L. major* possesses the most complicated series of repeating units, it is capped with the simplest structure (84), consisting exclusively of the disaccharide Man(α1,2)Man(α1). The most abundant terminal oligosaccharide of the *L. donovani* LPG is the branched trisaccharide Gal(β1,4)[Man(α1,2)]Man(α1) (141a), which is also a capping structure of the LPG of *L. mexicana* (62a).

THE TERTIARY STRUCTURE OF LPG The three-dimensional solution structure of the repeating PO_4-6Gal(β1,4)Man(α1) disaccharide units of the LPG derived from *L. donovani* has been recently determined by use of a combination of homo- and heteronuclear NMR spin coupling constant measurements together with restrained molecular mechanical minimization and molecular dynamics simulations (61). The repeating units, as expected, have limited mobility in solution about the Gal(β1,4)Man linkages. In contrast, a variety of stable rotamers exist about the Man(α1)-PO_4-6Gal linkages. An important feature of each of these low energy conformers is that the C3 hydroxyl of each galactose residue is exposed and freely accessible. This is the particular position for which glucose is substituted in the LPG from *L. mexicana* (62a), or for which galactose, glucose, and arabinose are substituted in the LPG from *L. major* (84). Thus, these additional units could be accommodated without major conformational changes to the repeat backbone.

Another intriguing finding from the molecular modeling studies (61) is that each of the stable conformers of the Man(α1)-PO_4-6Gal linkages may exist in a different configuration within the same LPG molecule. These torsional oscillations allow the LPG molecule to contract or expand in a manner reminiscent of a slinky spring, resulting in a molecule whose length can range from 90 Å when fully contracted to 160 Å when fully extended, assuming an average of 16 repeat units.

Developmental Modification of LPG During Promastigote Metacyclogenesis

A peculiar and significant observation of LPG is that modifications in its structure accompany the process of metacyclogenesis (127, 129, 130). These changes in LPG were initially detected in studies using the peanut agglutinin lectin and using stage-specific monoclonal antibodies (129, 130). Recent investigations have focused on the structural comparison of the LPG isolated from *L. major* promastigotes grown logarithmically and from stationary phase of growth. Analysis of the LPG derived from these growth stages of *L. major* revealed conservation of the *lyso*-1-*O*-alkylphosphatidylinositol lipid anchor and the phosphosaccharide core. The most striking difference, however, was an approximate doubling in size displayed by the metacyclic form of LPG, resulting from an approximate doubling in the number of repeating phosphorylated saccharide units (128). The relative increase in size of metacyclic LPG is consistent with freeze-fracture electron microscopic studies of the surface of metacyclic promastigotes (108). Their cell surfaces contain densely packed filamentous structures not present on logarithmically grown, noninfectious promastigotes. Furthermore, a greater than twofold thickening of the surface glycocalyx could be specifically labeled with the monoclonal antibody directed against the metacyclic LPG.

A second and more subtle difference in the two versions of LPG from *L. major* have been noted in the repeating phosphorylated saccharide composition (128). The repeat units of LPG obtained from logarithmically grown *L. major* contain terminal β-galactose residues that branch off the disaccharide backbone as a side chain. These galactose residues account for the agglutinability of such parasites to peanut agglutinin (130). Upon differentiation to metacyclic promastigotes, the repeat units terminate predominately with α-arabinose and, to a lesser extent, β-glucose residues, which are not ligands for the lectin. These compositional differences undoubtedly explain the expression of a novel epitope on metacyclic LPG (129). Such changes in LPG structure may have profound implications on function and suggest that there are important points of regulation of the glycosyltransferases involved in LPG biosynthesis.

Similarly, the LPG of *L. donovani* also undergoes an approximate twofold elongation during differentiation of the parasite to the metacyclic state (D. L. Sacks & S. J. Turco, unpublished observations). A determination of an average number of 16 repeat units from LPG isolated from *L. donovani* grown for extended time in culture supports this observation (146). LPG prepared from promastigotes freshly differentiated from hamster-derived amastigotes has an average repeat unit number of approximately 30 (82). Furthermore, a compositional change occurs, but one unlike that observed with the *L. major* LPG. Although the repeating units of LPG remain as PO_4-6Gal (β1,4)Man(α1) in metacyclic *L. donovani,* alterations of the cap oligosaccharides emerge. The sole terminal galactose residue present in the cap oligosaccharide of LPG from noninfectious parasites is absent in metacyclic LPG. This absence of terminal galactose would explain the loss of the ability of peanut agglutinin to agglutinate infectious *L. donovani* promastigotes (D. L. Sacks & S. J. Turco, unpublished observation).

LPG of Amastigotes

All of the structural information known about LPG has been obtained from LPG isolated from the *Leishmania* promastigotes. Until recently, there had been no information regarding the occurrence of LPG in amastigotes. The presence of LPG has been examined in the amastigotes of both *L. donovani* and *L. major,* and distinct results were obtained. *L. donovani* amastigotes apparently cannot synthesize LPG; at least a 10^4-fold down-regulation in LPG was found, corresponding to less than 100 molecules/cell (82). On the other hand, evidence has shown that *L. major* amastigotes express an LPG that is both biochemically (148) and antigenically (95) distinct from promastigotes. Although LPGs from both forms of *L. major* have much in common, the amastigote stage-specific LPG contains subtle carbohydrate differences that are not yet known (148). The temporal regulation of LPG expression during parasite differentiation was studied immunologically in vitro (45). During amastigote-to-promastigote transformation, the amastigote-specific form of LPG disappeared after subculture for 48 h, whereas during promastigote-to-amastigote transformation, the amastigote-specific form of LPG was detected in 12 h. The quantities of the *L. major* amastigote version of LPG appear to be much less than the promastigote counterpart. A stage-specific LPG has not been reported in other species of *Leishmania*.

Glycosylphosphoinositides

The various parts of LPG have been shown to exist in *Leishmania* promastigotes as components of proteins or as distinct entities. Regarding the latter, a family of molecules termed glycosylphosphatidylinositol antigens (GPIs) (119, 120, 137) or glycosylinositolphospholipids (GIPLs) (36, 79, 80, 82, 83) is abundant in *L. major* and *L. donovani*. Structural analyses of these glycolipids have indicated that they closely resemble the phosphosaccharide core–phosphatidylinositol region of LPG. In particular, the *L. major* GIPLs consist of a small mannose- and galactose-containing glycan that is glycosidically linked via an unacetylated glucosamine residue to either 1-*O*-alkyl-2-acyl-PI or *lyso*-1-*O*-alkyl-PI. The glycan parts of these molecules are completely identical to the analogous portions of LPG, including the salient Man(α1,4)GlcN(α1,6)*myo*-inositol and, in the larger GIPLs, the rare galactofuranose. Unlike those of *L. major,* the GIPLs of promastigotes of *L. donovani* are not galactosylated and synthesize abundant GIPLs containing one to four mannose residues (82, 137). Although *L. donovani* amastigotes do not synthesize LPG, they continue to synthesize GIPLs in quantities comparable to that reported in promastigotes (82). The *L. donovani* amastigote GIPLs, containing one to three mannose residues, were structurally different from promastigote GIPLs and appear to be precursors to glycolipid anchors of proteins. In addition, ceramidephosphoinositides have been found in *L. donovani* (69). Whether these particular phosphosphingolipids can be further substituted with carbohydrate residues as observed in *Trypanosoma cruzi* (77, 110), yeast, fungi, and plants (75) has not been established.

Extracellular LPG-Like Glycoconjugates

LPG-like substances, collectively termed excreted factor (EF), are reportedly present in conditioned medium from *Leishmania* parasites (38, 72, 139). The components of the EF can be organized into three categories. In one, LPG can form very tight complexes with albumin in the medium (72). Analysis of this form of LPG in the medium indicates that it is identical in all respects to the cell-associated LPG (72). One probable interpretation is that the lipid portion of LPG interacts with the hydrophobic binding pocket of albumin, facilitating its release from the surface of the promastigote.

In a second category, the repeating phosphorylated saccharide units of LPG comprise a carbohydrate chain of an acid phosphatase secreted by *L. donovani* (6), *L. tropica* (64, 65), and *L. mexicana* (63). The number of repeating units per carbohydrate chain and the nature of the linkage to the polypeptide are unknown. The latter is not believed to involve a typical N-glycosidic linkage to an Asn residue because the repeating units were not removed from acid phosphatase following N-glycanase digestion (6).

The third category of LPG-like substances is an extracellular phosphoglycan (exPG), which was purified from conditioned medium of *L. donovani* and then characterized (51a). Structural analysis indicated that the exPG consisted of the following structure: (CAP)$\rightarrow$[PO_4-6Gal(β1,4)Man(α1)]$_{10-11}$. The cap was found to be one of several small neutral oligosaccharides; the most abundant was the branched trisaccharide Gal(β1,4)[Man(α1,2)]Man(α1). Thus, the exPG is identical to cell-derived LPG except that it lacks a lipid anchor, the phosphosaccharide core, and several repeating units. Results from surface labeling with galactose oxidase/Na[^{3}H]$_4$ led to the conclusion that the exPG originates from surface LPG. The mechanism of release of the exPG as well as its possible function, if any, are not known.

FUNCTIONS OF LPG

The uniqueness of the overall structure of LPG and its highly unusual domains indicates that LPG might have one, or possibly several, important functions for the *Leishmania* parasite in its life cycle. Indeed, evidence has been provided for a surprisingly large number of potential functions that enable the promastigote to survive in the hydrolytic environments it encounters.

Functions in Sandfly-Leishmania *Interactions*

Differentiation and multiplication of the promastigote form of *Leishmania* parasites take place in the midgut of the sandflies *Phlebotomus* and *Lutzomyia* spp. soon after the ingestion of amastigotes during a bloodmeal on an infected host (19). Metacyclogenesis, or development from dividing noninfective forms into a resting, infective form, is believed to take place temporally in the midgut of the invertebrate host as the parasites move toward the anterior end of the gut (19, 131, 132). As described above (Developmental Modification of LPG During Metacyclogenesis), one of the major physical changes that occurs during metacyclogenesis is a structural modification of LPG (130). Although the importance of these modifications with respect to the successful survival of the parasite within the mammalian host has been largely documented (127, 145), information is beginning to appear regarding their role during metacyclogenesis itself. The possibility that these developmentally regulated structural changes may control attachment and detachment of maturing promastigotes from midgut epithelial cells, and hence their migration toward the mouthparts (145), was recently substantiated (26). In this study, expression of both LPG and the major cell-surface glycoprotein gp63 during the development of *L. major* in the gut of the sandfly *Phlebotomus papatasi* was examined in situ. Large amounts of nonmetacyclic, parasite-free LPG were detected on the surface of epithelial cells from the sandfly gut wall. In contrast, despite its abundant expression on promastigotes, gp63 was not detected on gut cells. The absence of metacyclic LPG suggests that the modified version of the glycoconjugate does not bind to these cells and, consequently, would allow infective parasites to move forward as they mature.

LPG may also protect the promastigotes against hydrolytic activity within the sandfly gut. Indeed, Schlein et al (133) increased survival of a foreign *L. major* strain in the stomach of *P. papatasi* by supplementing infective bloodmeal with LPG derived from an indigenous *L. major* strain (133). Clearly, additional studies are required to understand the functions of LPG during the development of the parasite within the sandfly.

Functions in Bloodstream

COMPLEMENT ACTIVATION AND RESISTANCE TO COMPLEMENT-MEDIATED DAMAGES Between the time of inoculation and infection of macrophages, promastigotes are exposed to the potential lytic effects of normal serum. Several studies were aimed at understanding the mechanisms by which promastigotes avoid destruction by the host's complement system (reviewed in 68). It appears that the developmentally regulated modifications of LPG represent the major resistance mechanism.

Promastigotes of all *Leishmania* species from log-phase cultures (noninfective) are extremely sensitive to fresh serum (43). On the other hand, stationary-phase cultures, which contain metacyclic promastigotes, display an increased resistance to lysis (43, 113). Activation of complement kills promastigotes because heat-inactivated serum and ethylenediaminetetraacetic acid (EDTA)-chelated serum fail to lyse promastigotes. Using the peanut agglutinin lectin, which does not agglutinate infective metacyclic promastigotes (130), Puentes et al (113) generated populations of pure log-phase and metacyclic *L. major* promastigotes and used them to further study complement resistance (113). Surprisingly, failure to bind the complement component C3 does not account for resistance to killing, inasmuch as both log-phase and metacyclic promastigotes activate complement rapidly and bind radiolabeled C3 after incubation in serum (113). These observations confirm the conclusions of an earlier study on resistance to complement-mediated lysis of amastigotes (98). The deposition of C3b on *L. major* metacyclics occurs through the classical pathway, and C3 is not covalently linked. On the other hand, C3 binding on log-phase promastigotes is mediated by efficient activation of the alternative pathway. Interestingly, LPG is the C3 acceptor molecule on both log-phase growth and metacyclic promastigotes, as determined by immunoprecipitation of LPG from promastigotes previously incubated with ^{125}I-C3. In contrast, regardless of the growth phase, *L. donovani* promastigotes bind C3 mainly as hemolytically inactive iC3b through activation of the alternative pathway (112). Half of the bound C3 is rapidly released from the parasite as a consequence of an unusual proteolytic cleavage, raising the possibility that C3 cleavage may be modulated by the major surface protease gp63. However, LPG does not mediate C3 binding on *L. mexicana* promastigotes, as it occurs solely on the gp63 (121).

The possibility that larger LPG molecules on metacyclic promastigotes are responsible for their resistance to complement-mediated lysis is supported by the observation that in *L. major,* most of C5b-9 complexes are spontaneously released from the metacyclic promastigote surface. This action precludes their insertion into the membrane and death of the parasite (111). Consistent with these observations, LPG may provide a barrier against the elevated titers of antileishmanial antibodies associated with kala-azar. Pooled kala-azar serum shows a strong reactivity with a LPG-deficient mutant of *L. donovani.* In contrast, little reactivity is observed with wild-type promastigotes (70). This observation is in agreement with the notion that the humoral response associated with kala-azar does not contribute to immunity. In addition, this masking effect of LPG may protect developing promastigotes in the insect's gut from antibodies present in the bloodmeal.

Collectively, these studies support the critical role of LPG, as well as the developmentally associated modifications, in the initial contact of *Leishmania* parasites with their mammalian hosts.

ATTACHMENT TO HOST MACROPHAGES Since *Leishmania* parasites infect primarily mononuclear phagocytic cells, attachment to potential host cells undoubtedly requires specific recognition molecules on the surface of both parasites and macrophages. Several *Leishmania* and macrophage cell-surface molecules have been implicated in the attachment of the parasite to its host cell (reviewed in 123). The glycoprotein gp63 present on all *Leishmania* species studied (9, 10, 40) and LPG can be considered parasite ligands (55, 118, 124, 125, 153), whereas CR1, CR3, and the mannose-fucose receptor represent the corresponding macrophage receptors (8, 25, 96, 141, 154, 155). Although binding occurs in the absence of serum, the presence of C3 dramatically increases the survival of metacyclic promastigotes (25, 97, 156). This phenomenon may be explained by the conversion of C3 into C3b through activation of the classical pathway and the subsequent CR1-mediated binding and internalization. The use of both CR1 and CR3 may favor the survival of *Leishmania* promastigotes because they reportedly promote phagocytosis without triggering the oxidative burst (157). Another report, however, stated that CR3, when suitably ligated, participates in macrophage activation (32).

Attachment of *L. major* but not *L. mexicana* promastigotes to macrophages is inhibited by the Fab fragment of an anti-*L. major* LPG antibody, suggesting that LPG is a parasite receptor for macrophages (55). Binding of purified *L. major* LPG to macrophage and nonmacrophage cells appears to be temperature dependent. Two mechanisms of binding may be involved: a specific mechanism, by which the carbohydrate part of the molecule binds to a macrophage receptor, and a nonspecific mechanism in which the lipid of LPG interacts with the membrane of the cells, probably through insertion into the lipid bilayer. In contrast, *L. donovani* LPG and its delipidated derivative bind to a variety of different cell types in a temperature-independent manner (143).

A study aimed at identifying the receptor(s) responsible for recognition of LPG on macrophages revealed that in the presence of serum, the binding site for *L. mexicana* LPG is contained, or conferred, by the α-chains of both CR3 and p150,95 (141), two members of the CD18 family of integrins. Competition experiments revealed that *L. mexicana* LPG and *Escherichia coli* lipopolysaccharide share the same binding site on the CD18 family of integrins. Interestingly, the binding site of LPG on CR3 is distinct from the binding site of C3bi.

These binding studies strongly suggest a role for LPG in the attachment of promastigotes to macrophages. However, in the absence of complement, phagocytosis of LPG-deficient mutants of *L. donovani* and *L. major* is similar or even superior to that of wild-type promastigotes (35, 57, 73, 88). LPG-deficient variants have exposed glycoproteins on their surface, such as gp63. Because these glycoproteins contain polymannose chains (89), the LPG-deficient variants may enter macrophages via a receptor not normally utilized.

Such a receptor may be the mannose-fucose receptor; the exact contribution of LPG in the recognition and attachment processes still remains to be determined.

Intracellular Functions

INTRACELLULAR SURVIVAL IN HOST PHAGOLYSOSOMES Subsequent to attachment of promastigotes to their receptors, internalization is achieved through the formation of a phagosome. Secondary lysosomes then fuse with the parasitophorous vacuole to form a phagolysosome in which the parasite transforms and multiplies as amastigotes (7, 18). This implies that *Leishmania* parasites have adapted to survive in the highly destructive environment of the phagolysosome, where they encounter degradative enzymes and toxic oxygen products.

Handman & Greenblatt (54) provided the first evidence that an excreted factor (presumably containing LPG) may be important for the intracellular survival of *Leishmania* parasites. They noticed that addition of concentrated excreted factor from *L. enrietti* promastigote cultures promoted the growth of this parasite in mouse peritoneal macrophages, which under normal circumstances are nonpermissive. Medium conditioned by *L. tropica* promastigotes was without effect. The possibility that the preparation of excreted factor contained other parasite molecules, however, cannot be excluded.

Use of *Leishmania* strains deficient in the biosynthesis of LPG clearly demonstrated that LPG is required for intracellular survival of promastigotes. An avirulent clone of *L. major* isolated from a rodent and lacking LPG was phagocytized by macrophages but was killed within 18 hours (57). Passive transfer of purified LPG from a virulent strain of *L. major* into the avirulent promastigotes confered the ability to survive in macrophages.

Several LPG-deficient variants of *L. donovani* selected for resistance to the ricin agglutinin lectin (73) are phagocytized but cannot survive in human monocytes. As in the avirulent strain of *L. major,* passive transfer of purified LPG significantly prolonged *L. donovani* mutant survival in monocytes (88). Because one particular ricin-resistant variant synthesizes lower amounts (20%) of a truncated form of LPG (86) and still could not sustain an infection (88), these results suggest that a minimum number of intact LPG molecules might be necessary for successful intracellular survival. Isolation and characterization of additional LPG-variants may be helpful to address this issue. Selection of ricin agglutinin-resistant mutants of *L. major* and determination of their infectivity both in vitro and in vivo (35) confirmed that LPG is a parasite factor determining infectivity and virulence, essential for intracellular survival. Interestingly, these *L. major* LPG-variants express normal or elevated amounts of glycosylinositolphospholipids, which are related to the phosphosaccharide core–phosphatidylinositol region of LPG (see Glycosylphosphoinositides).

The mechanisms by which LPG protects the parasite against the microbicidal activities displayed by the macrophage is an important area of research. Much hinges on the fate of LPG upon entry of the parasite into macrophages. LPG epitopes can be visualized by immunofluorescence with anti-LPG monoclonal antibodies on the surface of macrophages as early as five to ten min postinfection and are localized to the immediate area of internalization of the promastigote (143). The epitopes are evenly distributed over the entire macrophage surface by 25 min postinfection. The epitopes are maximally discernable one to two days postinfection, and by five to six days, the LPG epitopes can no longer be detected. Thus, intracellular functions for the promastigote form of LPG would have to be attributed within the first several days postinfection. The various properties of LPG with respect to intracellular survival are reviewed in the next sections.

INHIBITION OF HYDROLYTIC ENZYMES Adaptation to life in a phagolysosome requires the ability to resist, inhibit, or inactivate host hydrolytic enzymes. It has been suggested that survival of *Leishmania* parasites may depend on their ability to inhibit lysosomal enzymes (2). Resistance to lysosomal enzymic digestion was also suggested, based on surface properties of the parasite (18).

In an investigation (37) of the effect of partially purified excreted factor (LPG) upon the activity of four hydrolytic enzymes from peritoneal macrophages of mice, LPG did not affect acid phosphatase, β-glucuronidase, and N-acetyl-β-glucosaminidase. But β-galactosidase activity was highly inhibited after 3 h of incubation. The strong negative charge of the LPG molecule may account for the observed inhibitory effect. Competitive inhibition by the abundant phosphorylated disaccharide Gal(β1,4)Man in LPG provides an alternative explanation. Whether these in vitro results could be applied to the in vivo situation remains to be determined.

The role of LPG in protecting the parasite from digestion by lysosomal enzymes was further examined by measuring the rate of cytolysis of erythrocytes coated and uncoated with LPG (34). LPG coating significantly diminished the rate of cytolysis by macrophages, suggesting that this glycoconjugate may indeed enable *Leishmania* parasites to survive in the presence of hydrolytic enzymes.

CHELATOR OF CALCIUM Calcium plays an important role in the regulation of cellular functions, chiefly as an intracellular second messenger and as an enzyme cofactor. Interestingly, *L. major*-infected macrophages contain approximately 40% more exchangeable calcium than uninfected controls

(34). Similarly, macrophages engulfing LPG-coated erythrocytes have increased levels of calcium compared to macrophages engulfing control erythrocytes, possibly as a consequence of calcium binding by LPG (34). Recent studies using NMR examined the effect of calcium on the tertiary structure of the glycan moiety of LPG. The investigators concluded that calcium does not perturb the tridimensional structure of the glycan and that it binds to LPG in the vicinity of the phosphate groups (61). Therefore, the ability of LPG to chelate calcium may have important implications with respect to the ability of *Leishmania* parasites to survive within macrophages.

LPG might be a chelator of other important divalent cations, such as ferrous iron. Chelation of the latter metal presumably would prevent production, via the Fenton reaction, of the destructive hydroxyl radicals during activation of macrophages.

INHIBITOR OF HOST PROTEIN KINASE C Protein kinase C (PKC) is a multifunctional protein kinase that specifically phosphorylates serine and threonine residues (reviewed in 101). This enzyme is characterized by a catalytic domain containing an ATP-binding site and a regulatory domain that contains the sites involved in calcium, diacylglycerol (the physiological activator), and phospholipid binding (102). By virtue of its pivotal role in transmembrane signaling (42, 62), PKC modulates a wide variety of cellular functions. In phagocytes, one of the events mediated by PKC is the initiation of the oxidative burst. Phosphorylation and membrane association of the NADPH oxidase complex components represent the first steps of this process (5, 30). The active NADPH oxidase complex catalyzes the one-electron transfer from NADPH to oxygen, generating superoxide anion released at the outer surface of the plasma membrane into the extracellular space or into phagocytic vacuoles. Further reductions of the superoxide anions results in formation of hydrogen peroxide, hydroxyl radicals, and singlet oxygen. These events are involved in one of the main physiological functions of macrophages, the elimination of microbes. Products of the oxidative burst are deleterious for *Leishmania* promastigotes, in particular hydrogen peroxide (100, 105, 116). Susceptibility of promastigotes to this oxygen product may be explained by a deficiency in catalase and glutathione peroxidase, which are both scavengers of hydrogen peroxide (100). In contrast, amastigotes display a superior ability to scavenge hydrogen peroxide (20), probably as a result of increased catalase production (21).

Attenuation or inhibition of the host cell's respiratory burst may represent a critical factor for the survival of *Leishmania* parasites. *Leishmania* parasites can impair the stimulation of the macrophage oxidative burst stimulated by zymosan, bacterial lipopolysaccharide, or lymphokines (14, 106). This impairment is parasite specific because inert latex beads do not block the respiratory burst, and appears to be a function of the number of parasites per macrophage.

Inhibitor of the oxidative burst A role for LPG in the impairment of the respiratory burst was suggested by the demonstration that it is a potent inhibitor of purified rat-brain PKC activity in vitro (87). This study revealed that LPG is a competitive inhibitor ($K_I < 1\ \mu M$) with respect to diolein (a diacylglycerol), and a noncompetitive inhibitor with respect to phosphatidylserine. Inhibition of PKC is selective, since LPG does not affect the catalytic fragment of PKC and the cAMP-dependent protein kinase. Additional studies revealed that the 1-*O*-alkylglycerol fragment exhibits the most potent inhibitory activity, although the phosphoglycan portion also significantly inhibits purified PKC activity (85). These results strongly suggest that LPG interacts with the regulatory domain of PKC, which contains the binding sites for diacylglycerol, calcium, and phospholipids. Decrease in the calcium-dependent activity of PKC was explained by chelation of calcium.

The inhibitory property of LPG on purified PKC may be of major significance, considering the role of PKC in the induction of the respiratory burst. In this regard, phagocytosis of LPG-coated beads inhibits oxygen consumption in monocytes stimulated with phorbol myristate acetate, a synthetic activator of PKC (88). Using indirect immunofluorescence, LPG or a LPG fragment was detected in the outer membrane of monocytes containing LPG-coated beads. Thus, LPG may effectively interact with the monocyte's PKC or another factor involved in production of the oxidative burst. Whether it is the intact LPG molecule or a processed form that mediates the inhibition of the respiratory burst is still unknown. Nevertheless, these findings are consistent with the hypothesis that inhibition of PKC activity represents an important, if not essential, event for successful establishment of *Leishmania* parasites within their host cell.

LPG may not be the only leishmanial molecule involved in suppressing the oxidative burst. A leishmanial-surface acid phosphatase inhibits the neutrophil-derived burst upon stimulation with a chemotactic peptide, but not with a phorbol ester (46).

Inhibitor of c-fos *gene expression* Activation of PKC also results in the expression of several genes, including several protooncogenes such as the c-*fos* gene (50, 74, 93). The Fos protein is a *trans*-acting transcription factor that forms a stable transcriptional complex with the product of another oncogene, c-*jun* (reviewed in 24). According to the prevailing hypothesis, Fos functions as a nuclear third messenger molecule that regulates gene expression in response to environmental signals. In macrophages, c-*fos* gene expression is inducible through a PKC- or a cAMP-dependent pathway.

Activation of PKC results in a rapid and transient increase in c-*fos* mRNA levels (114), whereas elevation of cAMP stimulates a stable and long-lasting expression of the c-*fos* gene (11).

Macrophages infected with *L. donovani* display an impaired ability to express the c-*fos* gene in response to lipopolysaccharide or diacylglycerol, suggesting that the parasite can interfere with a PKC-dependent gene expression pathway (28). Incubation of macrophages with LPG or its delipidated version resulted in the inhibition of PKC-dependent c-*fos* gene expression. In contrast, LPG did not impair the ability of macrophages to express the c-*fos* gene in response to cAMP (29). This observation is in agreement with the selective inhibitory effect of LPG on purified enzyme activity (87). However, the exact molecular mechanism by which LPG interacts with the components of this PKC-dependent signal transduction pathway is not known.

Because macrophage-activating cytokines such as interferon (IFN)-γ and TNF-α act through PKC-dependent signal transduction pathways (41, 53, 136), impairment of PKC-dependent gene expression would attenuate the impact of external activating signals and therefore be beneficial for intracellular *Leishmania* parasites.

Inhibitor of chemotaxis Rapid accumulation of blood monocytes at sites of inflammation represents an important step of the inflammatory response. Binding of a chemoattractant to its specific cell-surface receptor initiates a cascade of intracellular events (140, 151), which involves activation of PKC (76). *L. donovani* LPG and its delipidated counterpart are both potent inhibitors of monocyte and neutrophil chemotactic locomotion (44). Inhibition of inflammatory reactions in the lesion sites may contribute to the chronicity of the disease.

SCAVENGERS OF TOXIC OXYGEN METABOLITES In addition to inhibiting PKC-mediated enzymatic induction of the oxidative burst (88), LPG may protect promastigotes by scavenging toxic oxygen metabolites generated during the burst (17). Indeed, LPG is highly effective in scavenging hydroxyl radicals and superoxide anions, as determined in vitro using electron spin resonance spectroscopy and spin-trapping assays. The scavenging activity of LPG is largely conferred by the repeating phosphorylated disaccharide units. This property can be explained by oxidation of the hydroxyl groups of sugars to ketones (reviewed in 48).

Addition of LPG to monocytes following induction of the oxidative burst results in an immediate, dose-dependent reduction of chemiluminescence, suggesting an effective scavenging of the superoxide anions already secreted into the culture medium (88). Inhibition of chemiluminescence is also observed with monocytes treated with LPG but not with the glycosylinositolphospholipid antigens (44), which confirms the scavenging property of the repeating phosphorylated disaccharides. Incubation of *L. major* promastigotes and their excreted factors (which contain LPG) with polymorphonuclear leukocytes also inhibits chemiluminescence (39). Thus, LPG may protect the parasite from the damaging effects of the oxidative burst through at least two distinct mechanisms: (*a*) attenuation of the PKC-mediated induction of the burst, and (*b*) scavenging of the cytocidal products of the burst.

INHIBITOR OF IL-1 PRODUCTION Macrophages and other antigen-presenting cells play a crucial modulatory role during the establishment of an effective immune response through the activation of T lymphocytes (149). Two macrophage accessory functions regulate T helper lymphocyte activation: (*a*) the expression of major histocompatibility complex class II molecules and (*b*) the production of interleukin (IL)-1 (149). In *L. donovani*-infected macrophages, these two functions are defective (103, 115, 117). Interestingly, incubation of monocytes with purified *L. donovani* LPG inhibits lipopolysaccharide (LPS)-induced IL-1 secretion (44). Whether or not LPG is responsible for the inhibition of IL-1 in infected macrophages remains to be determined.

MODULATOR OF TNF RECEPTORS Tumor necrosis factor (TNF) is a cytokine that pleiotropically affects the immune response (138). Macrophages produce TNF in response to a broad range of stimuli, including several microorganisms and derived molecules. In turn, TNF activates macrophage cytotoxic functions and the subsequent destruction of the invader. *E. coli* LPS, a potent inducer of TNF, stimulates macrophages to rapidly internalize their TNF receptors (31). This phenomenon may reflect a microbial evasion strategy of host defense or a host strategy to prevent autotoxicity. As does bacterial LPS, LPG causes an important reduction in the ability of macrophages to bind TNF (29). Although the *Leishmania* parasite stimulates macrophages to secrete TNF (49), a down-regulation of TNF receptors may play a role in the survival of *Leishmania,* particularly during the initiation of infection. The fate of TNF receptors following phagocytosis of *Leishmania* parasites, however, is not known.

ANALOGOUS FUNCTIONS BY GLYCOLIPIDS FROM MYCOBACTERIA *Mycobacterium leprae* and *Mycobacterium tuberculosis,* the etiologic agents of leprosy and tuberculosis, respectively, survive and multiply in mononuclear phagocytes. Production of large amounts of cell wall–associated glycolipids is a feature common to both *M. leprae* and *M. tuberculosis* (12, 47). Recent studies indicate that these glycolipids may represent virulence factors that contribute to the intracellular survival of mycobacteria. Indeed, in a manner similar to LPG, mycobacterial glycolipids effectively scavenge oxygen radicals (16, 17), suppress the oxidative burst in monocytes (13, 150), and inhibit PKC activity both in vitro and in vivo (13, 16) as well as PKC-dependent gene expression in macrophages (16). From an evolutionary point of view, it is striking that a protozoan and a bacterial intracellular parasite have both evolved structurally and functionally similar virulence factors to counter the microbicidal activities of macrophages.

BIOSYNTHESIS OF LPG

Because LPG is essential for the survival of *Leishmania* as it progresses through its life-cycle, chemotherapeutic drugs conceivably could be designed to inhibit LPG synthesis; the parasite should then be vulnerable to normal host defense mechanisms. Detailed knowledge of LPG biosynthesis might provide a specific target in the search for more efficaceous drugs against leishmaniasis.

Synthesis of the PI Anchor Region

Although LPG possesses a *lyso*-1-*O*-alkyl PI anchor, the synthesis probably involves formation of a 1-*O*-alkyl-2-acyl-PI precursor. *Leishmania* parasites possess significant amounts of 1-*O*-alkyl-2-acyl-phospholipids, but not *lyso*-alkylphospholipids (152). The latter are cytotoxic for *Leishmania* (1) as well as other eukcaryotic cells (60). Based on information from other eukaryotic systems, the possible pathway that leads to the assembly of the phosphatidylinositol anchor of LPG can be postulated. Ether phospholipid synthesis might be initiated by acylation of C1 of glycolytic intermediate dihydroxyacetonephosphate. A key enzyme believed to be involved in ether lipid synthesis (dihydroxyacetonephosphate acyltransferase) has been reported (58) in *Leishmania* glycosomes (organelles resembling peroxisomes). The acyl group may then be replaced with a fatty alcohol. *Leishmania* parasites incorporate fatty alcohols in ether phospholipids (59). The resultant alkyl dihydroxyacetonephosphate might then be reduced with NADPH and acylated at the *sn*-2 position forming 1-*O*-alkyl-2-acyl phosphatidic acid. The latter may be activated to the CDP-derivative by CTP and then condensed with *myo*-inositol to form 1-*O*-alkyl-2-acyl-PI. However, this pathway has not been established in *Leishmania* species, and another route of lipid anchor assembly for LPG may exist.

Assembly of the Core-PI Region

Predictions on the pathway of core-PI synthesis can be derived from the recently elucidated structures of the GIPLs isolated from *Leishmania* spp. (79, 80, 82, 83) and from details on GPI anchor assembly reported in African trypanosomes (reviewed in 33). The biosynthetic pathway of the trypanosomal GPI anchor contains a transfer of N-acetylglucosamine from UDP-GlcNAc to PI forming GlcNAc-PI. Deacetylation of the product results in GlcN-PI. Mannose residues are then added by using mannosylphosphoryldolichol as the mannosyl donor (91). Analogous reactions in *Leishmania* spp. have not yet been demonstrated.

In *Leishmania* spp., addition of the first mannose residue to GlcN-PI presumably would yield Man(α1,4)GlcN-PI, a precursor to LPG and GPI-anchored proteins, such as the glycoprotein gp63 (134). The addition of the second mannose residue is at a branch point in the biosynthetic pathway of leishmanial GPI anchors. In the synthesis of the GPI anchor of gp63 (134), the second mannose would form Man(α1,6)Man(α1,4)GlcN-PI, whereas in LPG, it would yield Man(α1,3)Man(α1,4)GlcN-PI. In LPG biosynthesis, three galactose residues then would be added, one of which is the galactofuranose; the donor of this sugar is unknown. Three of the GIPLs that have been isolated and characterized from *L. major* (GIPLs 1–3) contain one to three galactose residues, respectively (83). Because their complete glycan structures are consistent with the LPG core structure including the galactofuranose residue, GIPLs 1–3 most likely are intermediates in LPG biosynthesis. Of significance, the lipid portions of GIPLs 1–3 are 1-*O*-alkyl-2-acyl-PI. Two other GIPLs have been isolated and are the *lyso*-derivatives of GIPLs 2 and 3. Thus, the pathway may involve assembly of the glycan core on 1-*O*-alkyl-2-acyl-PI, which is followed by deacylation of the fatty acid near completion of the core. The glucose(α1)-phosphate probably is added sometime after the deacylation step because, as reported so far, no GIPL contains this substituent.

As discussed above (LPG of Amastigotes), amastigotes of *L. donovani* do not synthesize LPG (82). These intracellular forms apparently do not synthesize Man(α1,3)Man(α1,4)GlcN-PI, which would be an intermediate in LPG biosynthesis. This observation led to the proposal that down-regulation of a putative α1,3-mannosyltransferase in amastigotes contributes to a lack of LPG expression (82).

Polymerization of Repeating Units

In contrast to the virtual lack of enzymological information regarding the synthesis of the other portions of LPG, data have been reported on repeating unit polymerization. An in vitro membrane system from *L. donovani* capable of synthesizing LPG repeating units has been developed (15). The galactose and mannose residues of the LPG repeating units of *L. donovani* are believed to be added from their respective nucleotide-sugar donors sequentially and directly to LPG (Figure 2). While mannosylphosphoryldolichol might participate as a mannosyl donor in core-PI synthesis, no evidence was found for the possible involvement of the mannolipid in repeating unit assembly. Additional information indicates that guanosine diphosphate (GDP)-Man donates mannose-1-phosphate (15), thereby conserving the α-anomeric configuration of the mannosylphosphate bond. Thus, the repeating units of LPG appear to be polymerized by the individual alternating transfer of galactose and mannose-1-phosphate residues from their nucleotide derivatives. Experiments using monensin, an inhibitor of Golgi function, have produced evidence that repeating-unit assembly occurs in the Golgi (6). The addition of the hexoses that comprise side chains of the LPG repeating units of *L. major* and *L. mexicana* has not yet been examined. One of the key glycosyltransferases to be investigated is a putative arabinosyltransferase from *L. major*. As mentioned earlier (Developmental Modification of LPG During Promastigote Metacyclogenesis), the LPGs from metacyclic forms of *L. major* contain arabinose residues not contained in terminal side chains in the repeating units and in noninfectious forms. Thus, induction of the arabinosyltransferase in *L. major* may be an important regulatory enzyme in metacyclogenesis in that species.

Synthesis of Capping Oligosaccharides

All but one of the cap structures elucidated from various leishmanial LPGs contain a Man(α1,2)Man(α1) at the reducing end. It is, therefore, tempting to

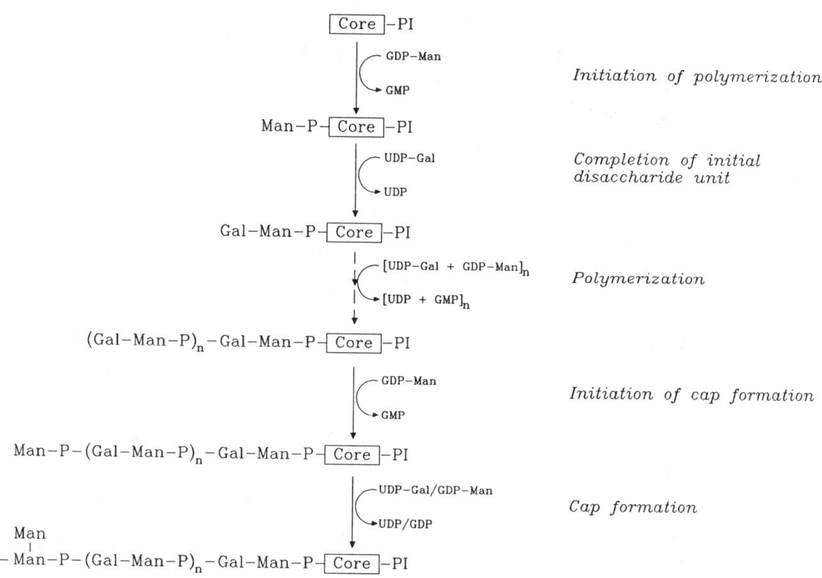

Figure 2 Proposed pathway of assembly of the repeating units and capping oligosaccharides of the *L. donovani* LPG. The core structure is Gal(α1,6)Gal(α1,3)Gal$_f$(β1,3)- Man-(α1,3)Man(α1,4)GlcN(α1,6), and PI is *lyso*-1-*O*-alkylphosphatidylinositol. The sequence of monosaccharide addition in cap formation is not yet known.

speculate on the existence of a Man(α1,2)mannosyltransferase. The activity of such a putative enzyme would result in the signal for cessation of LPG elongation with the formation of a chain-terminating Man(α1,2)Man-containing cap oligosaccharide. Because metacyclogenesis is accompanied by an approximate doubling in the size of LPG, a chain-terminating mannosyltransferase may prove to be one of the key regulatory enzymes in LPG biosynthesis. In the in vitro glycosylating system from *L. donovani* that can generate repeating units (15), several small neutral oligosaccharides were also observed upon fragmentation and analysis of the LPG product. These oligosaccharides may be capping structures, but are not characterized as yet.

IMMUNOLOGICAL ASPECTS OF LPG

LPG as a Vaccine Candidate

Because LPG is the most abundant surface molecule of *Leishmania* promastigotes, several studies have examined this molecule's potential as a chemically defined vaccine. When administered to genetically resistant mice, in which cutaneous lesions resolve spontaneously, purified *L. major* LPG induced full protection against a challenge with promastigotes, whereas partial protection was achieved in the susceptible mice (56). This observation suggested that LPG might be considered as a candidate vaccine antigen against cutaneous leishmaniasis. Immunization with the water-soluble carbohydrate portion (phosphoglycan), however, had no effect on the development of lesions (56). Further studies revealed that the phosphoglycan is in fact a disease-promoting antigen (92).

Incorporation of LPG into liposomes proved to be an efficient way to immunize mice against *L. mexicana* without causing any exacerbation of the disease (122). Adoptive transfer of T-cells isolated from immunized mice into syngeneic mice provided protection against a challenge with *L. mexicana* promastigotes, indicating that protection is a function of antigen-specific T-cells.

Activation of specific T helper subsets of lymphocytes represents one of the key cellular events in the establishment of an effective immune response. Several lines of evidence support the notion that T lymphocytes play a dominant role in the acquired resistance to *Leishmania* parasites (78, 99). It was therefore of interest to determine whether LPG can be recognized by T-cells. Mice vaccinated with *L. major* LPG contained an increased frequency of *L. major*-reactive T-cells (94). Furthermore, LPG induced a specific delayed-type hypersensitivity in *L. major*-infected mice, which also produced T-cell-dependent IgG to LPG. The authors suggested the possibility that T-cells can recognize and respond to LPG although the T-cells did not respond to LPG in vitro. Indeed, LPG stimulates a weak proliferation of T lymphocytes from patients with active cutaneous leishmaniasis caused by *L.*

major (66). Surprisingly, these T-cells did not respond to gp63, which is nonetheless considered as a potential vaccine candidate. Similarly, T lymphocytes from cured kala-azar patients proliferate and produce IFN-γ in response to *L. major* LPG (71).

In another observation, Mendonça and coworkers (90) reported that T lymphocytes from cutaneous leishmaniasis patients responded to highly purified *L. braziliensis* LPG, whereas proteinase K–treated LPG did not stimulate any response. This important observation indicates that stimulation of T-cell responses by LPG may, in fact, be induced by tightly associated protein contaminants.

Purification and partial characterization of the LPG-associated protein contaminants revealed the presence of several proteins. Subsequent T-cell proliferation studies showed that these peptides were potent stimulators of T-cells from leishmaniasis patients (126), as well as from mice immunized with protein-contaminated LPG (67), whereas protein-free LPG failed to stimulate any T-cell responses. While LPG does not appear to be able to elicit T-cell responses, the glycoconjugate may act as a natural adjuvant for the proteins with which it is tightly complexed.

Use for Serotyping Leishmania *Strains*

Several polyclonal and monoclonal antibodies generated against *Leishmania* strains react with epitopes present on LPG (27, 51, 135). These antibodies proved useful for serotyping *Leishmania* strains, with respect to their ability to selectively precipitate the excreted form of LPG. However, monoclonal antibodies that were initially believed to recognize the core region of *L. donovani* LPG (142) were subsequently found to react with protein contaminants (67). Therefore, we can infer that some of the monoclonal and polyclonal antibodies used for serotyping may, in fact, recognize LPG-associated proteins.

CONCLUDING REMARKS

The list of functions proposed for LPG is extensive and surprisingly large. However, the uniqueness of the overall structure of LPG with its unusual domains could account for the multifunctional aspects of the LPG molecule. Whether the structural polymorphisms of LPG contribute to the various pathologies associated with the different leishmaniases remains to be established.

Acknowledgments

This laboratory is funded by the United Nations Development Program/World Bank/World Health Organization Special Program for Research and Training in Tropical Diseases and by National Institutes of Health Grant AI20941. S. J. T. is a Burroughs Wellcome Scholar in Molecular Parasitology. A. D. is supported by a Postdoctoral fellowship from the Medical Research Council of Canada.

Literature Cited

1. Acterberg, V., Gercken, G. 1987. Cytotoxicity of ester and ether lysophospholipids on *Leishmania donovani* promastigotes. *Mol. Biochem. Parasitol.* 23:117–22
2. Alexander, J., Vickerman, K. 1975. Fusion of host cell secondary lysosomes with the parasitophorous vacuoles of *Leishmania mexicana* infected macrophages. *J Protozool.* 22:502–8
3. Avila, J. L., Rojas, M., Garcia, L. 1988. Persistence of elevated levels of galactosyl-α(1,3)galactose antibodies in sera from patients cured of visceral leishmaniasis. *J. Clin. Microbiol.* 26: 1842–47
4. Avila, J. L., Rojas, M., Towbin, H. 1988. Serological activity against galactosyl-α(1,3)galactose in sera from patients with several kinetoplastida infections. *J. Clin. Microbiol.* 26:126–32
5. Baggiolini, M., Wyman, M. P. 1990. Turning on the respiratory burst. *TIBS* 15:69–72
6. Bates, P. A., Hermes, I., Dwyer, D. M. 1990. Golgi-mediated posttranslational processing of secretary acid phosphatase by *Leishmania donovani* promastigotes. *Mol. Biochem. Parasitol.* 38:247–56
7. Berman, J. D., Dwyer, D. M., Wyler, D. J. 1979. Multiplication of *Leishmania* in human macrophages in vitro. *Infect. Immunol.* 26:375–79
8. Blackwell, J. M., Ezekowitz, R. A. B., Roberts, M. B., Channon, J. Y., Sim, R. B., et al. 1985. Macrophage complement and lectin-like receptors bind *Leishmania* in the absence of serum. *J. Exp. Med.* 162:324–31
9. Bouvier, J., Etges, R. J., Bordier, C. 1985. Identification and purification of membrane and soluble forms of the major surface protein of *Leishmania* promastigotes. *J. Biol. Chem.* 260:15504–9
10. Bouvier, J., Etges, R. J., Bordier, C. 1987. Identification of the promastigote surface protease in seven species of *Leishmania. Mol. Biochem. Parasitol.* 24:73–79
11. Bravo, R., Neuberg, M., Burckhardt, J., Almendral, J., Wallich, R., et al. 1987. Involvement of common and cell type–specific pathways in c-fos gene control: stable induction by cAMP in macrophages. *Cell* 48:251–60
12. Brennan, P. J. 1989. Structure of mycobacteria: recent developments in defining cell wall carbohydrates and proteins. *Rev. Infect. Dis.* 11:S420–30
13. Brozna, J. P., Horan, M., Rademacher, J. M., Pabst, K. M., Pabst, M. J. 1991. Monocyte responses to sulfatide from *Mycobacterium tuberculosis:* inhibition of priming for enhanced release of superoxide, associated with increased secretion of interleukin-1 and tumor necrosis factor alpha, and altered protein phosphorylation. *Infect. Immunol.* 59: 2542–48
14. Buchmuller-Rouiller, V., Mauël, J. 1987. Impairment of the oxidative metabolism of mouse peritoneal macrophages by intracellular *Leishmania* spp. *Infect. Immunol.* 55:587–93
15. Carver, M. A., Turco, S. J. 1991. Cell-free biosynthesis of lipophosphoglycan from *Leishmania donovani:* characterization of microsomal galactosyltransferase and mannosyltransferase activities. *J. Biol. Chem.* 266:10974–81
16. Chan, J., Fan, X., Hunter, S. W., Brennan, P. J., Bloom, B. R. 1991. Lipoarabinomannan, a possible virulence factor involved in persistence of *Mycobacterium tuberculosis* within macrophages. *Infect. Immunol.* 59:1755–61
17. Chan, J., Fujira, T., Brennan, P., McNeil, M., Turco, S. J., et al. 1989. Microbial glycolipids: possible virulence factors that scavenge oxygen radicals. *Proc. Natl. Acad. Sci. USA* 86:2453–57
18. Chang, K. P., Dwyer, D. M. 1976. Multiplication of a human parasite *Leishmania donovani* in phagolysosomes of hamster macrophages in vitro. *Science* 193:678–80
19. Chang, K. P., Fong, D., Bray, R. S. 1985. Biology of *Leishmania* and leishmaniasis. In *Leishmaniasis,* ed. K. P. Chang, R. S. Bray, pp. 1–30. New York: Elsevier
20. Channon, J. Y., Blackwell, J. M. 1985. A study of the sensitivity of *Leishmania donovani* promastigotes and amastigotes to hydrogen peroxide. I. Differences in

sensitivity correlate with parasite-mediated removal of hydrogen peroxide. *Parasitology* 91:197–206

21. Channon, J. Y., Blackwell, J. M. 1985. A study of the sensitivity of *Leishmania donovani* promastigotes and amastigotes to hydrogen peroxide. II. Possible mechanisms involved in protective H_2O_2 scavenging. *Parasitology* 91:207–17
22. Channon, J. Y., Blackwell, J. M. 1986. Molecular biology of *Leishmania*. *Parasitol. Today* 2:45–53
23. Cross, G. A. M. 1990. Glycolipid anchoring of plasma membrane proteins. *Annu. Rev. Cell. Biol.* 6:1–39
24. Curran, T., Franza, B. R. Jr. 1988. Fos and Jun: the AP-1 connection. *Cell* 55:395–97
25. Da Silva, R. P., Hall, B. F., Joiner, K. A., Sacks, D. L. 1989. CR1, the C3b receptor, mediates binding of infective *Leishmania major* metacyclic promastigotes to human macrophages. *J. Immunol.* 143:617–22
26. Davies, C. R., Cooper, A. M., Peacock, C., Lane, R. P., Blackwell, J. M. 1990. Expression of LPG and GP63 by different developmental stages of *Leishmania major* in the sandfly *Phlebotomus papatasi*. *Parasitology* 101:337–43
27. de Ibarra, A. A. L., Howard, J. G., Snary, D. 1982. Monoclonal antibodies to *Leishmania tropica major:* specificities and antigen location. *Parasitology* 85:523–31
28. Descoteaux, A., Matlashewski, G. 1989. C-*fos* and TNF gene expression in *Leishmania donovani*-infected macrophages. *Mol. Cell. Biol.* 9:5223–27
29. Descoteaux, A., Turco, S. J., Sacks, D. L., Matlashewski, G. 1991. *Leishmania donovani* lipophosphoglycan selectively inhibits signal transduction in macrophages. *J. Immunol.* 146:2747–53
30. Dewald, B., Thelen, M., Baggiolini, M. 1988. Two transduction sequences are necessary for neutrophil activation by receptor agonists. *J. Biol. Chem.* 263: 16179–84
31. Ding, A. H., Sanchez, E., Srimal, S., Nathan, C. F. 1989. Macrophages rapidly internalize their tumor necrosis factor receptors in response to bacterial lipopolysaccharide. *J Biol. Chem.* 264: 3924–29
32. Ding, A. H., Wright, S. D., Nathan, C. F. 1987. Activation of mouse peritoneal macrophages by monoclonal antibodies to MAC-1 (complement receptor type 2). *J. Exp. Med.* 165:733–49
33. Doering, T. L., Masterson, W. J., Hart, G. W., Englund, P. T. 1990. Biosynthesis of glycosyl phosphatidylinositol membrane anchors. *J. Biol. Chem.* 265: 611–14
34. Eilam, Y., El-On, J., Spira, D. T. 1985. *Leishmania major:* excreted factor, calcium ions, and the survival of amastigotes. *Exp. Parasitol.* 59:161–68
35. Elhay, M., Kelleher, M., Bacic, A., McConville, M. J., Tolson, D. L., et al. 1990. Lipophosphoglycan expression and virulence in ricin-resistant variants of *Leishmania major*. *Mol. Biochem. Parasitol.* 40:255–68
36. Elhay, M. J., McConville, M. J., Handman, E. 1988. Immunochemical characterization of a glycoinositol-phospholipid membrane antigen of *Leishmania major*. *J. Immunol.* 141:1326–31
37. El-On, J., Bradley, D. J., Freeman, J. C. 1980. *Leishmania donovani:* action of excreted factor on hydrolytic enzyme activity of macrophages from mice with genetically different resistance to infection. *Exp. Parasitol.* 49:167–74
38. El-On, J., Schnur, L. F., Greenblatt, C. L. 1979. *Leishmania donovani:* physiochemical, immunological and biological characterization of excreted factor from promastigotes. *Exp. Parasitol.* 47:254–69
39. El-On, J., Zvillich, M., Sarov, I. 1990. *Leishmania major:* inhibition of the chemiluminescent response of human polymorphonuclear leukocytes by promastigotes and their excreted factors. *Parasite Immunol.* 12:285–95
40. Etges, R. J., Bouvier, J., Hoffman, R., Bordier, C. 1985. Evidence that the major surface protein of three *Leishmania* species are structurally related. *Mol. Biochem. Parasitol.* 14:141–49
41. Fan, X.-D., Goldberg, M., Bloom, B. R. 1988. Interferon-γ-induced transcriptional activation is mediated by protein kinase C. *Proc. Natl. Acad. Sci. USA* 85:5122–25
42. Farago, A., Nishizuka, Y. 1990. Protein kinase C in transmembrane signalling. *FEBS Lett.* 268:350–54
43. Franke, E. D., McGreevy, P. B., Katz, S. P., Sacks, D. L. 1985. Growth cycle-dependent generation of complement-resistant *Leishmania* promastigotes. *J. Immunol.* 134:2713–18
44. Frankenburg, S., Leibovici, V., Mansbach, N., Turco, S. J., Rosen, G. 1990. Effect of glycolipids of *Leishmania* parasites on human monocyte activity. Inhibition by lipophosphoglycan. *J. Immunol.* 145:4284–89
45. Glaser, T. A., Moody, S. F., Handman, E., Bacic, A., Spithill, T. W. 1991. An antigenically distinct lipophosphoglycan

of *Leishmania major*. *Mol. Biochem. Parasitol.* 45:337–44

46. Glew, R. H., Saha, A. K., Das, S., Remaley, A. T. 1988. Biochemistry of *Leishmania* parasites. *Microbiol. Rev.* 52:412–32
47. Goren, M. B., Brokl, O., Roller, P., Fales, H. M., Das, B. C. 1976. Sulfatides of *Mycobacterium tuberculosis:* the structure of the principal sulfatide (SL-1). *Biochemistry* 15:2728–35
48. Green, J. W. 1980. Oxidative reactions and degradations. In *The Carbohydrates,* ed. W. Pigman, 1B:1101–66. New York: Academic
49. Green, S. I., Crawford, R. M., Hockmeyer, J. T., Meltzer, M. S., Nacy, C. A. 1990. *Leishmania major* amastigotes initiate the L-arginine-dependent killing mechanism in IFN-γ-stimulated macrophages by induction of tumor necrosis factor-α. *J. Immunol.* 145:4290–97
50. Greenberg, M. E., Ziff, E. B. 1984. Stimulation of 3T3 cells induces transcription of the c-*fos* proto-oncogene. *Nature* 311:433–38
51. Greenblatt, C. L., Slutzky, G. M., de Ibarra, A. A. L., Snary, D. 1983. Monoclonal antibodies for serotyping of *Leishmania* strains. *J. Clin. Microbiol.* 18:191–93

51a. Greis, K. D., Turco, S. J., Thomas, J. R., McConville, M. J., Homans, S. W., Ferguson, M. A. J. 1992. Purification and characterization of an extracellular phosphoglycan from *Leishmania donovani*. *J. Biol. Chem.* In press

52. Grimaldi, G. Jr., Tesh, R. B., McMahon-Pratt, D. 1989. A review of the geographic distribution and epidemiology of leishmaniasis in the new world. *Am. J. Trop. Med. Hyg.* 41:687–725
53. Hamilton, T. A., Becton, D. L., Somers, S. D., Gray, P. W., Adams, D. O. 1985. Interferon-γ modulates protein kinase C activity in murine peritoneal macrophages. *J. Biol. Chem.* 260:1378–81
54. Handman, E., Greenblatt, C. L. 1977. Promotion of leishmanial infections in non-permissive host macrophages by conditioned medium. *Z. Parasitenkd.* 53:143–47
55. Handman, E., Goding, J. W. 1985. The *Leishmania* receptor for macrophages is a lipid-containing glycoconjugate. *EMBO J.* 4:329–36
56. Handman, E., Mitchell, G. F. 1985. Immunization with *Leishmania* receptors for macrophages protects mice against cutaneous leishmaniasis. *Proc. Natl. Acad. Sci. USA* 82:5910–14
57. Handman, E., Schnur, L. F., Spithill, T. W., Mitchell, G. F. 1986. Passive transfer of *Leishmania* lipopolysaccharide confers parasite survival in macrophages. *J. Immunol.* 137:3608–13
58. Hart, D. T., Opperdoes, F. R. 1984. The occurrence of glycosomes (microbodies) in the promastigote stage of four major *Leishmania* species. *Mol. Biochem. Parasitol.* 13:159–72
59. Herrmann, H., Gercken, G. 1980. Incorporation of octadecanol into the lipids of *Leishmania donovani*. *Lipids* 15:179–85
60. Hoffman, D. R., Hadju, J., Snyder, F. 1984. Cytotoxicity of platelet-activating factor and related alkyl-phospholipid analogs in human leukemia cells, polymorphonuclear neutrophils, and skin fibroblasts. *Blood* 63:545–52
61. Homans, S. W., Melhert, A., Turco, S. J. 1992. Solution structure of the lipophosphoglycan of *Leishmania donovani*. *Biochemistry*. 31:654–61
62. Houslay, M. D. 1991. Crosstalk: a pivotal role for protein kinase C in modulating relationships between signal transduction pathways. *Eur. J. Biochem.* 195:9–27

62a. Ilg, T., Etges, R., Overath, P., McConville, M. J., Thomas, J. R., et al. 1992. Structure of *Leishmania mexicana* lipophosphoglycan. *J. Biol. Chem.* In press

63. Ilg, T., Menz, B., Winter, G., Russell, D. G., Etges, R., et al. 1991. Monoclonal antibodies to *Leishmania mexicana* promastigote antigens. Secreted acid phosphatase and other proteins share epitopes with lipophosphoglycan. *J. Cell. Sci.* 99:175–80
64. Jaffe, C. L., Martinez, L. P., Sarfstein, R. 1989. *Leishmania tropica:* characterization of a lipophosphoglycan-like antigen recognized by species specific monoclonal antibodies. *Exp. Parasitol.* 70:12–24
65. Jaffe, C. L., Perez, L., Schnur, L. F. 1990. Lipophosphoglycan and secreted acid phosphatase of *Leishmania tropica* share species-specific epitopes. *Mol. Biochem. Parasitol.* 41:233–40
66. Jaffe, C. L., Shor, R., Trau, H., Passwell, J. H. 1990. Parasite antigens recognized by patients with cutaneous leishmaniasis. *Clin. Exp. Immunol.* 80: 77–82
67. Jardim, A., Tolson, D. L., Turco, S. J., Pearson, T. W., Olafson, R. W. 1991. The *Leishmania donovani* lipophosphoglycan T-lymphocyte reactive component is a tightly associated protein complex. *J. Immunol.* 147:3538–44
68. Joiner, K. A. 1988. Complement eva-

sion by bacteria and parasites. *Annu. Rev. Microbiol.* 42:201–30

69. Kaneshiro, E. S., Keka, J. A., Lester, R. L. 1986. Characterization of inositol lipids from *Leishmania donovani* promastigotes. Identification of inositol sphingo-phospholipids. *J. Lipid Res.* 27:1294–1303
70. Karp, C. L., Turco, S. J., Sacks, D. L. 1991. Lipophosphoglycan masks recognition of the *Leishmania donovani* promastigote surface by human kala-azar serum. *J. Immunol.* 147:680–84
71. Kemp, B. E., Theander, T. G., Handman, E., Hey, A. S., Kurtzhals, J. A. L., et al. 1991. Activation of human T lymphocytes by *Leishmania* lipophosphoglycan. *Scand. J. Immunol.* 33:219–24
72. King, D. L., Chang, Y. D., Turco, S. J. 1987. Cell surface lipophosphoglycan of *Leishmania donovani. Mol. Biochem. Parasitol.* 24:47–53
73. King, D. L., Turco, S. J. 1988. A ricin agglutinin-resistant clone of *Leishmania donovani. Mol. Biochem. Parasitol.* 28:285–94
74. Kruijer, W., Cooper, J. A., Hunter, T., Verma, I. 1984. Platelet-derived growth factor induces rapid but transient expression of the c-*fos* gene and protein. *Nature* 312:711–16
75. Laines, R. A., Hsieh, T. C. Y. 1987. Inositol-containing sphingolipids. *Methods Enzymol.* 138:186–95
76. Laskin, D. L., Gardner, C. R., Laskin, J. D. 1987. Induction of chemotaxis in mouse peritoneal macrophages by activators of protein kinase C. *J. Leuk. Biol.* 41:474–80
77. Lederkremer, R. M., Tanaka, C. T., Alves, M. J. M., Colli, W. 1977. Lipopeptidephosphoglycan from *Trypanosoma cruzi. Eur. J. Biochem.* 74:265–67
78. Louis, J., Milon, G. 1987. Immunobiology of experimental leishmaniasis. *Ann. Inst. Pasteur Immunol.* 138:737–95
79. McConville, M. J., Bacic, A. 1989. A family of glycoinositol phospholipids from *Leishmania major. J. Biol. Chem.* 264:757–66
80. McConville, M. J., Bacic, A. 1990. The glycoinositolphospholipids profiles of two *Leishmania major* strains that differ in lipophosphoglycan expression. *Mol. Biochem. Parasitol.* 38:57–68
81. McConville, M. J., Bacic, A., Mitchell, G. F., Handman, E. 1987. Lipophosphoglycan of *Leishmania major* that vaccinates against cutaneous leishmaniasis contains an alkylglycerophosphatidylinositol lipid anchor. *Proc. Natl. Acad. Sci. USA* 84:9841–45
82. McConville, M. J., Blackwell, J. M. 1991. Developmental changes in the glycosylated phosphatidylinositols of *Leishmania donovani. J. Biol. Chem.* 266:15170–79
83. McConville, M. J., Homans, S. W., Thomas-Oates, J. E., Dell, A., Bacic, A. 1990. Structures of the glycoinositolphospholipids from *Leishmania major:* a family of novel galactosylfuranose-containing glycolipids. *J. Biol. Chem.* 265:7385–94
84. McConville, M. J., Thomas-Oates, J. E., Ferguson, M. A. J., Homans, S. W. 1990. Structure of the lipophosphoglycan from *Leishmania major. J. Biol. Chem.* 265:19611–21
85. McNeely, T. B., Rosen, G., Londner, M. V., Turco, S. J. 1989. Inhibitory effects on protein kinase C by lipophosphoglycan and glycosyl-phosphatidylinositol antigens of *Leishmania. Biochem. J.* 259:601–4
86. McNeely, T. B., Tolson, D. L., Pearson, T. W., Turco, S. J. 1990. Characterization of *Leishmania donovani* variant clones using anti-lipophosphoglycan monoclonal antibodies. *Glycobiology* 1:63–69
87. McNeely, T. B., Turco, S. J. 1987. Inhibition of protein kinase C by the *Leishmania donovani* lipophosphoglycan. *Biochem. Biophys. Res. Commun.* 148:653–57
88. McNeely, T. B., Turco, S. J. 1990. Requirement of lipophosphoglycan for intracellular survival of *Leishmania donovani* within human monocytes. *J. Immunol.* 144:2745–50
89. Mendelzon, D. H., Previato, J. O., Parodi, A. J. 1986. Characterization of protein-linked oligosaccharides in trypanosomatid flagellates. *Mol. Biochem. Parasitol.* 18:355–67
90. Mendonça, S. C. F., Russell, D. G., Coutinho, S. G. 1991. Analysis of the human T cell responsiveness to purified antigens of *Leishmania:* lipophosphoglycan (LPG) and glycoprotein 63 (gp63). *Clin. Exp. Immunol.* 83:472–78
91. Menon, A., Mayor, S., Schwarz, R. T. 1990. Biosynthesis of glycosyl-phosphatidylinositol lipids in *Trypanosoma brucei:* involvement of mannosylphosphoryldolichol as the mannose donor. *EMBO J.* 9:4249–58
92. Mitchell, G. F., Handman, E. 1986. The glycoconjugate derived from a *Leishmania major* receptor for macrophages is a suppressogenic, disease-

promoting antigen in murine cutaneous leishmaniasis. *Parasite Immunol.* 8: 255–63

93. Mitchell, R. L., Zokas, L., Schreiber, R. D., Verma, I. M. 1985. Rapid induction of the expression of proto-oncogene fos during human monocytic differentiation. *Cell* 40:209–17
94. Moll, H., Mitchell, G. F., McConville, M. J., Handman, E. 1989. Evidence for T-cell recognition in mice of a purified lipophosphoglycan from *Leishmania major*. *Infect. Immunol.* 57:3349–56
95. Moody, S. F., Handman, E., Bacic, A. 1991. Structure and antigenicity of the lipophosphoglycan from *Leishmania major* amastigotes. *Glycobiology* 1:419–24
96. Mosser, D. M., Edelson, P. J. 1985. The mouse macrophage receptor for C3bi (CR3) is a major mechanism in the phagocytosis of *Leishmania* promastigotes. *J. Immunol.* 135:2785–89
97. Mosser, D. M., Edelson, P. J. 1987. The third component of complement (C3) is responsible for the intracellular survival of *Leishmania major*. *Nature* 327:329–31
98. Mosser, D. M., Wedgwood, J. F., Edelson, P. J. 1985. *Leishmania* amastigotes: resistance to complement-mediated lysis is not due to a failure to fix C3. *J. Immunol.* 134:4128–31
99. Müller, I., Pedrazzini, T., Farrell, J. P., Louis, J. 1989. T-cell responses and immunity to experimental infection with *Leishmania major*. *Annu. Rev. Immunol.* 7:561–78
100. Murray, H. W. 1981. Susceptibility of *Leishmania* to oxygen intermediates and killing by normal macrophages. *J. Exp. Med.* 153:1302–15
101. Nishizuka, Y. 1986. Studies and perspectives of protein kinase C. *Science* 233:305–12
102. Nishizuka, Y. 1988. The molecular heterogeneity of protein kinase C and its implication for cellular regulation. *Nature* 334:661–65
103. Olivier, M., Tanner, C. E. 1989. The effect of cyclosporin A in murine visceral leishmaniasis. *Trop. Med. Parasitol.* 40:32–38
104. Orlandi, P. A. Turco, S. J. 1987. Structure of the lipid moiety of the *Leishmania donovani* lipophosphoglycan. *J. Biol. Chem.* 262:10384–91
105. Pearson, R. D., Harcus, J. L., Roberts, D., Donowitz, G. R. 1983. Differential survival of *Leishmania donovani* amastigotes in human monocytes. *J. Immunol.* 131:1994–99
106. Pearson, R. D., Harcus, J. L., Symes, P. H., Romito, R., Donowitz, G. R. 1982. Failure of the phagocytic oxidative response to protect human monocyte-derived macrophages from infection by *Leishmania donovani*. *J. Immunol.* 129:1282–86
107. Pearson, R. D., Sousa, A. Q. 1985. *Leishmania* species. (Kala-azar, cutaneous, and mucocutaneous leishmaniasis). In *Principles and Practice of Infectious Diseases,* ed. G. L. Mandell, R. G. Douglas Jr., J. E. Bennett, pp. 1522–31. New York: Wiley
108. Pimenta, P. F. P., da Silva, R. P., Sacks, D. L., da Silva, P. 1989. Cell surface nanoanatomy of *Leishmania major* as revealed by fracture flip. A surface meshwork of 44 nm fusiform filaments identifies infective developmental stage promastigotes. *Eur. J. Cell Biol.* 48:180–90
109. Pimenta, P. F. P., Saraiva, E. M. B., Sacks, D. L. 1991. The comparative fine structure and surface glycoconjugate expression of three life stages of *Leishmania major*. *Exp. Parasitol.* 72: 191–204
110. Previato, J. O., Gorin, P. A. J., Mazurek, M., Xavier, M. T., Fournet, B., et al. 1990. Primary structure of the oligosaccharide chain of lipopeptidophosphoglycan of epimastigote forms of *Trypanosoma cruzi*. *J. Biol. Chem.* 265:2518–26
111. Puentes, S. M., da Silva, R. P., Sacks, D. L., Hammer, C. H., Joiner, K. A. 1990. Serum resistance of metacyclic stage *Leishmania* major promastigotes is due to release of C5b-9. *J. Immunol.* 145:4311–16
112. Puentes, S. M., Dwyer, D. M., Bates, P. A., Joiner, K. A. 1989. Binding and release of C3 from *Leishmania donovani* promastigotes during incubation in normal human serum. *J. Immunol.* 143: 3743–49
113. Puentes, S. M., Sacks, D. L., da Silva, R. P., Joiner, K. A. 1988. Complement binding by two developmental stages of *Leishmania major* promastigotes varying in expression of a surface lipophosphoglycan. *J. Exp. Med.* 167: 887–902
114. Radzioch, D., Bottazzi, B., Varesio, L. 1987. Augmentation of c-*fos* mRNA expression by activators of protein kinase C in fresh, terminally differentiated resting macrophages. *Mol. Cell. Biol.* 7: 595–99
115. Reiner, N. E. 1987. Parasite-accessory cell interactions in murine leishmaniasis.

I. Evasion and stimulus-dependent suppression of the macrophage interleukin I response by *Leishmania donovani*. *J. Immunol.* 138:1919–25
116. Reiner, N. E., Kazura, J. W. 1982. Oxidant-mediated damage of *Leishmania donovani* promastigotes. *Infect. Immunol.* 36:1023–27
117. Reiner, N. E., Ng, W., McMaster, W. R. 1987. Parasite-accessory cell interactions in murine leishmaniasis. II. *Leishmania donovani* suppresses macrophage expression of class I and II major histocompatibility complex gene products. *J. Immunol.* 138:1926–32
118. Rizvi, F. S., Ouaissi, M. A., Marty, B., Santoro, F., Capron, A. 1988. The major surface protein of *Leishmania* promastigotes is a fibronectin-like molecule. *Eur. J. Immunol.* 18:473–78
119. Rosen, G., Londner, M. V., Sevlever, D., Greenblatt, C. L. 1988. *Leishmania major:* glycolipid antigens recognized by immune human sera. *Mol. Biochem. Parasitol.* 27:93–100
120. Rosen, G., Pahlsson, P., Londner, M. V., Westerman, M. E., Nilsson, B. 1989. Structural analysis of glycosylphosphatidylinositol antigens of *Leishmania major*. *J. Biol. Chem.* 264: 10457–63
121. Russell, D. G. 1987. The macrophage-attachment glycoprotein gp63 is the predominant C3-acceptor site on *Leishmania mexicana* promastigotes. *Eur. J. Biochem.* 164:213–21
122. Russell, D. G., Alexander, J. 1988. Effective immunization against cutaneous leishmaniasis with defined membrane antigens reconstituted into liposomes. *J. Immunol.* 140:1274–79
123. Russell, D. G., Talamas-Rohana, P. 1989. *Leishmania* and the macrophage: a marriage of inconvenience. *Immunol. Today* 10:328–33
124. Russell, D. G., Wilhelm, H. 1986. The involvement of the major surface glycoprotein (gp63) of *Leishmania* promastigotes in attachment to macrophages. *J. Immunol.* 136:2613–20
125. Russell, D. G., Wright, S. D. 1988. Complement receptor type 3 (CR3) binds to an Arg-Gly-Asp-containing region of the major surface glycoprotein, gp63, of *Leishmania* promastigotes. *J. Exp. Med.* 168:279–92
126. Russo, D. M., Turco, S. J., Burns, J. M. Jr., Reed, S. G. 1992. Stimulation of human T lymphocytes by *Leishmania* lipophosphoglycan-associated proteins. *J. Immunol.* 148:202–7
127. Sacks, D. L. 1989. Metacyclogenesis in *Leishmania* promastigotes. *Exp. Parasitol.* 69:100–3
128. Sacks, D. L., Brodin, T., Turco, S. J. 1990. Developmental modification of the lipophosphoglycan from *Leishmania major* promastigotes during metacyclogenesis. *Mol. Biochem. Parasitol.* 42:225–34
129. Sacks, D. L., da Silva, R. P. 1987. The generation of infective stage *Leishmania major* promastigotes is associated with the cell-surface expression and release of a developmentally regulated glycolipid. *J. Immunol.* 139:3099–3106
130. Sacks, D. L., Hieny, S., Sher, A. 1985. Identification of cell surface carbohydrate and antigenic changes between noninfective and infective developmental stages of *Leishmania major* promastigotes. *J. Immunol.* 135:564–69
131. Sacks, D. L., Perkins, P. V. 1984. Identification of an infective stage of *Leishmania* promastigotes. *Science* 223: 1417–19
132. Sacks, D. L., Perkins, P. V. 1985. Development of infective stage *Leishmania* promastigotes within phlebotomine sandflies. *Am. J. Trop. Med. Hyg.* 34: 456–59
133. Schlein, Y., Schnur, L. F., Jacobson, R. L. 1990. Released glycoconjugate of indigenous *Leishmania major* enhances survival of a foreign *L. major* in *Phlebotomus papatasi*. *Trans. R. Soc. Trop. Med. Hyg.* 84:353–55
134. Schneider, P., Ferguson, M. A. J., McConville, M. J., Mehlert, A., Homans, S. W., et al. 1990. Structure of the glycosyl-phosphatidylinositol membrane anchor of the *Leishmania major* promastigote surface protease. *J. Biol. Chem.* 265:16955–64
135. Schnur, L. F., Zuckerman, A., Greenblatt, C. L. 1972. Leishmanial serotypes as distinguished by the gel diffusion of factors excreted *in vitro* and *in vivo*. *Isr. J. Med. Sci.* 8:932–42
136. Schutze, S., Nottrott, S., Pfizenmaier, K., Kronke, M. 1990. Tumor necrosis factor signal transduction. Cell-type-specific activation and translocation of protein kinase C. *J. Immunol.* 144: 2604–8
137. Sevlever, D., Pahlsson, P., Rosen, G., Nilsson, B., Londner, M. V. 1991. Structural analysis of a glycosylphosphatidylinositol glycolipid of *Leishmania donovani*. *Glycoconjug. J.* 8: 321–29
138. Sherry, B., Cerami, A. 1988. Cachectin/tumor necrosis factor exerts endocrine, paracrine, and autocrine control

of inflammatory responses. *J. Cell Biol.* 107:1269–77

139. Slutzky, G. M., El-On, J., Greenblatt, C. L. 1979. Leishmanial excreted factor: protein-bound and free forms from promastigote cultures of *Leishmania tropica* and *Leishmania donovani*. *Infect. Immunol.* 26:916–24
140. Snyderman, R., Pike, M. C. 1984. Chemoattractant receptors on phagocytic cells. *Annu. Rev. Immunol.* 2:257–81
141. Talamàs-Rohana, P., Wright, S. D., Lennartz, M. R., Russell, D. G. 1990. Lipophosphoglycan from *Leishmania mexicana* promastigotes binds to members of the CR3, pl5O,95, and LFA-1 family of leukocyte integrins. *J. Immunol.* 144:4817–24

141a. Thomas, J. R., McConville, M. J., Thomas-Oates, J. E., Homans, S. W., Ferguson, M. A. J., et al. 1992. Refined structure of the lipophosphoglycan of *Leishmania donovani*. *J. Biol. Chem.* In press

142. Tolson, D. L., Turco, S. J., Beecroft, R. P., Pearson, T. W. 1989. Immunochemical and cell surface arrangement of the *Leishmania donovani* lipophosphoglycan determined using monoclonal antibodies. *Mol. Biochem. Parasitol.* 35:109–18
143. Tolson, D. L., Turco, S. J., Pearson, T. W. 1990. Expression of a repeating phosphorylated disaccharide lipophosphoglycan epitope on the surface of macrophages infected with *Leishmania donovani*. *Infect. Immunol.* 58:3500–7
144. Towbin, H. Rosenfelder, G., Wieslander, J., Avila, J. L., Rojas, M., et al. 1987. Circulating antibodies to mouse laminin in Chagas disease, American cutaneous leishmaniasis, and normal individuals recognize terminal galactosyl(α1,3)galactose epitopes. *J. Exp. Med.* 166:419–32
145. Turco, S. J. 1990. The leishmanial lipophosphoglycan: a multifunctional molecule. *Exp. Parasitol.* 70:241–45
146. Turco, S. J., Hull, S. R., Orlandi, P. A. Jr., Shepherd, S. D., Homans, S. W., et al. 1987. Structure of the major carbohydrate fragment of the *Leishmania donovani* lipophosphoglycan. *Biochemistry* 26:6233–38
147. Turco, S. J., Orlandi, P. A. Jr., Homans, S. W., Ferguson, M. A. J., Dwek, R. A., et al. 1989. Structure of the phosphosaccharide-inositol core of the *Leishmania donovani* lipophosphoglycan. *J. Biol. Chem.* 264:6711–15
148. Turco, S. J., Sacks, D. L. 1991. Expression of a stage-specific lipophosphoglycan in *Leishmania major* amastigotes. *Mol. Biochem. Parasitol.* 45:91–100
149. Unanue, E. R., Allen, P. M. 1987. The basis for the immunoregulatory role of macrophages and other accessory cells. *Science* 236:551–57
150. Vachula, M., Holzer, T. J., Andersen, B. R. 1989. Suppression of monocyte oxidative response by phenolic glycolipid I of Mycobacterium leprae. *J. Immunol.* 142:1696–1701
151. Verghese, M. W., Smith, C. D., Charles, L. A., Jakoi, L., Snyderman, R. 1986. A guanine nucleotide regulatory protein controls polyphosphoinositide metabolism, Ca^{2+} mobilization, and cellular responses to chemoattractants in human monocytes. *J. Immunol.* 137:271–75
152. Wassef, M. K., Fioretti, T. B., Dwyer, D. M. 1985. Lipid analysis of isolated surface membranes of *Leishmania donovani* promastigotes. *Lipids* 20:108–15
153. Wilson, M. E., Hardin, K. K. 1988. The major concanavalin A-binding surface glycoprotein of *Leishmania donovani chagasi* promastigotes is involved in attachment to human macrophages. *J. Immunol.* 141:265–72
154. Wilson, M. E., Pearson, R. D. 1986. Evidence that *Leishmania donovani* utilizes a mannose receptor on human mononuclear phagocytes to establish intracellular parasitism. *J. Immunol.* 136:4681–88
155. Wilson, M. E., Pearson, R. D. 1988. Roles of CR3 and mannose receptors in the attachment and ingestion of *Leishmania donovani* by human mononuclear phagocytes. *Infect. Immunol.* 56:363–69
156. Wozencraft, A. O., Blackwell, J. M. 1987. Increased infectivity of stationary phase promastigotes of *Leishmania donovani:* correlation with enhanced C3 binding capacity and CR3 mediated attachment to host macrophages. *Immunology* 60:559–66
157. Wright, S. D., Silverstein, S. C. 1983. Receptors for C3b and C3bi promote phagocytosis but not the release of toxic oxygen from human phagocytes. *J. Exp. Med.* 158:2016–23

Annu. Rev. Microbiol. 1992. 46:95–116

REPLICATION CYCLE OF *BACILLUS SUBTILIS* HYDROXYMETHYLURACIL-CONTAINING PHAGES

P. P. Hoet, M. M. Coene, and C. G. Cocito

Microbiology and Genetics Unit, Institute of Cell Pathology, University of Louvain Medical School, Brussels 1200, Belgium

KEY WORDS: discontinuous synthesis on both DNA strands, recombinational replication, terminal repeats, abnormal base, bacteriophage 2C

CONTENTS

0066-4227/92/1001-0095$02.00

Abstract

The present review focuses on phage 2C, a member of a family of virulent phages that multiply in *Bacillus subtilis*. The best known members of this group are SPO1, Φe, H1, 2C, SP8, and SP82, the genomes of which are made of double-stranded DNA of about 150 kilobase pairs (kbp). The two DNA strands have different buoyant densities. Moreover, thymine (T) is completely replaced by hydroxymethyluracil (hmUra). Comparison of the phage DNAs has shown that both base substitutions and deletions have contributed to the evolution of their genomes. In addition, all of the hmUra-phage genomes contain colinear redundant ends, amounting to 10% of total bases. Two lines of evidence suggest that the redundant ends of 2C DNA, in spite of extensive homology, contain unique sequences.

Further studies focused on DNA replication during the lytic cycle. The semiconservative replication of the infecting viral genome is followed by extensive recombination. At the level of replication forks, viral DNA synthesis is discontinuous on both strands during the whole cycle. Deoxythymidinetriphosphate, required for viral DNA synthesis in permeabilized infected bacteria, was incorporated in small amounts into phage DNA. The putative primary origin of replication has been cloned and localized on the viral genome. Some viral promoters have been successfully cloned in *Escherichia coli*. These sequences, however, did not promote transcription in *B. subtilis*. The abnormal base might be required for promoter activity in the natural host.

INTRODUCTION

Phage 2C belongs to a group of large virulent *Bacillus subtilis* phages, whose genomes contain hydroxymethyluracil (hmUra) in place of thymine (T) (57). This phage, isolated from soil by Pène & Marmur (85), is virulent for the transformable strain of *B. subtilis* 168, and its DNA is infectious. The various isolated hmUra-containing phages seem to be closely related to each other (43). Other members of this group are SPO1, Φe, H1, SP8, and SP82.

Hydroxymethyluracil phages bear some morphological and biochemical resemblance to the hydroxymethylcytosine (hmC)-containing T even phages of *Escherichia coli*. However, unlike T4, virus 2C does not completely block the synthesis of host macromolecules (9, 10). These phages are immunologically related, because antiserum raised against 2C prevented *B. subtilis* infection by Φe, SP8, and SP82 (104). Because of the presence of the abnormal base, the biosynthesis of the viral genome can be selectively traced in infected cells, since host and phage DNA have different buoyant densities. In addition, after denaturation, the two strands of 2C DNA have different buoyant densities in CsCl gradients.

These properties make phage 2C a useful organism with which to study recombination, duplication, and repair of the viral genome; the relationship between structure and expression of viral genes; and viral-host interactions.

Biosynthesis of the abnormal base and sequential transcription of viral genes during the lytic cycle of SPO1, SP82, and Φe have been extensively studied (for reviews, see 34, 43, 71, 102). This article focuses on the structure, replication, and recombination of virus 2C DNA and on some aspects of host-virus interaction. We compare virus 2C with the other hmUra phages when relevant data are available.

PROPERTIES OF HYDROXYMETHYLURACIL-CONTAINING DNA AND BIOSYNTHESIS OF THE ABNORMAL BASE

Phage 2C, like the other hmUra phages, has a linear nonpermuted double-stranded DNA genome of 150 kilobase pairs (kbp) (43, 104). Native DNA has a G+C content of 38% (49). In most DNA molecules in nature, the relationship between base composition (% G+C), buoyant density, and melting point (T_m) is linear. A discrepancy between these parameters was reported for the DNA of hmUra phages (57). The buoyant density of native 2C DNA in CsCl was found to be 1.742 g/cm^3 (104), which corresponds to a content of 84% G+C, whereas the T_m (77.8°C) indicates a value of 17.5% G+C.

Similarly, 2C DNA was observed to deviate from the normal linear relationship between base composition and binding of fluorophoric adducts (25). The search for hexoses in the DNA of hmUra-containing phages was unsuccessful (2). The presence of hmUra was allegedly responsible for the abnormal properties of these DNAs. The abnormal base presumably modifies the tridimensional DNA structure. Likewise, cytosine methylation was reported to promote the conversion of a right-handed B helix into a left-handed Z helix, a shift that relies on both nucleotide sequence and superhelicity.

Another peculiar feature of hmUra DNA molecules is their bimodal behavior after denaturation. While denaturation of most DNA yields strands of equal density (monomodal DNA), those obtained from hmUra DNA have different densities (1.752 g/cm^3 and 1.762 g/cm^3 in the case of 2C DNA) (49). Several other unusual DNA bases were found in some *B. subtilis* phages, including uracil in phage PBS2, and 5-(4',5'-dihydroxypentyl)uracil linked to one or two molecules of glucose in phage SP15. In the latter case, only 41% of thymine is replaced by the abnormal base. This DNA displays a very low melting temperature (61.7°C) and a particularly high buoyant density (1.761 g/cm^3) (107).

Discrepancies occurred between the expected and observed number of restriction fragments when 2C DNA was submitted to restriction enzymes containing thymine in their recognition sequence (Table 1). *Eco*RI cleaves 2C DNA under standard conditions if the enzyme concentration is increased 10-fold. The restriction pattern is similar to that obtained under nonstandard (*Eco*RI*) conditions (33, 88, 110). According to Berkner & Folk (4), the kinetic properties of the enzyme, rather than the recognition of the specific sequence, are impaired by the presence of the abnormal base. The presence of hmUra may also explain the resistance of 2C DNA to restriction by *Pst*I (cloned 2C sequences harboring thymine instead of the abnormal base are cleaved by *Pst*I) (65).

However, some discrepancies between theoretical and experimental data were also observed for endonucleases without thymine in their recognition sequences, such as *Hpa*II, *Hae*III, and *Hha*I, which cleave a tetramer containing only guanine and cytosine. The abnormal base may thus interfere with the

Table 1 Number of restriction fragments produced by the action of different endonucleases on 2C DNA[a]

Restriction endonuclease	Recognition sequence (5' → 3')	Number of fragments[b]	
		Expected	Observed
*Mbo*I	G A^+ T C^o[c]	540	>100
*Sau*3AI	G A^o T C^+	540	>100
*Dpn*I	G A^M T C		0
*Bam*H1	G G A^o T C^+ C^o	23	0
*Bgl*II	A G A^o T C^+ T	50	12
*Hpa*II	C^+ C^+ G G	240	35
*Msp*I	C^+ C^o G G	240	35
*Sma*I	C C C^+ G G G	11	0
*Hae*III	G G C C	240	5
*Hha*I	G C G C	240	35
*Hin*dIII	A A G C T T	50	27
*Hpa*I	G T T A A C	50	0
*Eco*RI	G A A T T C	50	20[d]
*Pst*I	C T G C A G	23	0
*Sal*I	G T C G A C	23	4

[a] Data taken from refs. 14, 48.

[b] In parallel experiments on λ-DNA, the expected and the observed number of restriction fragments differed little. The expected number of restriction fragments was calculated by taking into account size and base composition of the genome and the enzyme recognition sequence.

[c] A^+ or C^+ indicates the inhibition of a restriction endonuclease by a N^6-methyladenine or 5-methylcytosine residue, respectively, within the recognition sequence. The symbols A^o or C^o show that digestion of DNA is not influenced by the presence of the respective modified bases within the sequence. A^M indicates that N^6-methyladenine within the recognition sequence is a prerequisite for the activity of *Dpn*I.

[d] 2C DNA was cleaved either in standard conditions, in the presence of a 10-fold excess of enzyme, or in the star conditions (88).

recognition mechanism of endonucleases, when flanking the recognized sequence, as suggested by the following data. If the nucleotide positions within the recognition sequence and its neighborhood are represented as follows:

-5-1-(G-G-C-C)-3-7-
-6-2-(C-C-G-G)-4-8-,

calculations of the base frequencies in 2C DNA indicate that out of the 250 expected -(G-G-C-C)- sequences, 38 sites have no hmUra in positions 1 to 4, and 6 sites have no abnormal base in positions 1 to 8. These numbers are indeed close to the observed values: 35 fragments for *Hpa*II and *Hha*I, and 5 for *Hae*III (Table 1).

The hmUra-containing phage H1 was not restricted upon infection of *B. subtilis* R cells, and its DNA was not cleaved by the restriction enzyme *Bsu*R, the isoschizomer of *Hae*III. When H1 DNA was cloned, and T replaced hmUra, the viral DNA was cleaved by *Bsu*R and one site was shown to have the sequence 5'-C-A-T-A-A-T-T-T-G-G-C-C-T-A-G-3'. Hydroxymethyluracil flanking the cleavage site affords protection because cleavage occurs when thymine replaces hmUra (6).

In addition to the presence of the abnormal base, resistance of 2C DNA to some restriction enzymes might result from a bias against certain sequences, as is the case with other phage DNAs (6, 52). Methylation of adenine or cytosine residues within specific sequences prevents endonucleases from recognizing overlapping sequences (7, 82). However, the use of groups of isoschizomers *Hpa*II/*Msp*I and *Mbo*I/*Sau*3AI/*Dpn*I (Table 1) excludes the occurrence of modified adenine and cytosine within the corresponding restriction sequences CCGG and GATC. The host, *B. subtilis,* is devoid of the *dam* and *dcm* enzymes, methylating respectively the sequences GATC and CC(A/T)GG (30). The DNAs of T even phages, containing glycosylated hmC, are resistant to most restriction enzymes, except *Taq*I, *Aha*III, and *Eco*RV. Unglycosylated hmC DNA, however, was cleaved partially by *Eco*RI and *Xba*I, and reproducible restriction patterns were obtained with mutant DNA in which most hydroxymethylcytosine was replaced by cytosine (64).

Virus-induced enzymes catalyze the biosynthesis of the nucleotide precursor bearing the abnormal base, as shown earlier for hmC in T even phages by Cohen & coworkers (15, 16, 60). Marmur & Greenspan (74) were the first to observe the synthesis of deoxycytidyl-deaminase in *B. subtilis* after infection with phage SP8: this synthesis was prevented by chloramphenicol. Roscoe & Tucker (93) found a similar enzyme after infection of *B. subtilis* with phage Φe. The product of this reaction, deoxyuridylic acid, was converted to 5-hydroxymethyldeoxyuridylic acid by a hydroxymethylase (92) (Figure 1).

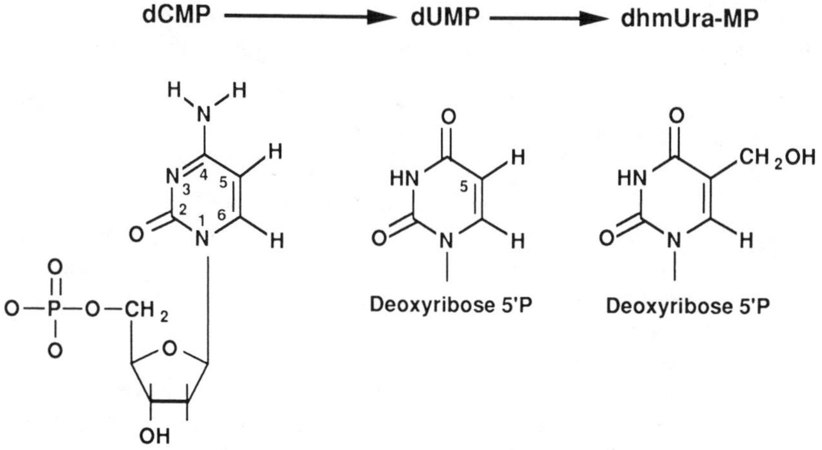

Figure 1 Biosynthetic pathway of hydroxymethyluracil (hmUra).

Three different mechanisms seem to be involved in the exclusion of thymine from viral DNA. (*a*) Deoxythymidylate synthetase activity (dUMP → dTMP) was lowered after infection of *B. subtilis* with phage Φe (94), an effect prevented by protein synthesis inhibitors. This result suggests the induction of either an inhibitor or a competitor of dTMP synthetase (42). (*b*) An additional mechanism is provided by the dTTP-dephosphorylating enzyme, which was induced after infection of *B. subtilis* with phages SP8 (56) and Φe (91). In fact, extracts of infected bacteria could hydrolyze both dUTP and dTTP, a single phage-dictated enzyme being responsible for both activities (31). (*c*) The induction of deoxythymidylate phosphatase (dTMP nucleotidylhydrolase) was observed in SP5-infected cells (3). However, similar enzymes were found in cells infected with phages whose DNA had either thymine (SP3) or uracil (PBS2). The occurrence of dTMP phosphatase activity is not restricted, therefore, to hmUra DNA phages (79).

Bacillus subtilis cells, made permeable after infection with phage 2C, incorporated deoxyribonucleoside triphosphates into DNA that had the buoyant density of phage DNA. Deoxy-hmUra-TP, which was not provided in the reaction mixture, presumably was incorporated from the endogenous pool, mimicking the conditions in infected cells (46). An unexpected requirement for dTTP was observed in this system. Indeed, labeled dTTP was incorporated in 2C DNA, as shown by CsCl isopycnic centrifugation. However, the amount of dTTP incorporated was 30-fold smaller than that of dATP, which points to the former as a minor precursor in the biosynthesis of viral DNA. Indeed, HPLC analysis revealed small amounts of thymine in 2C DNA hydrolysates (F. Hottat, unpublished results).

COMPARISON OF THE HYDROXYMETHYLURACIL PHAGES: PHYSICAL MAP AND REDUNDANT ENDS

Hydroxymethyluracil phages, because of the unique properties of their genome, are useful tools for a study of viral evolution. In the course of evolution, viral genomes may diverge as a result of base substitutions and of gene rearrangements such as deletions, insertions, and transpositions (17, 89, 101).

Analysis of the overall structure of hmUra-phage genomes by cross-hybridization showed more than 90% apparent homology between the DNA of 2C and related phages (104). Under more stringent hybridization conditions, however, a much lower apparent homology level was found (67). Sequence divergence can also be calculated from the number and sizes of restriction fragments unique to each DNA (77, 106). When this method was applied to available restriction patterns (*Eco*RI*, *Hin*dIII, *Hpa*II, and *Hha*I), 2.8–3.5% base substitution was found in pairwise comparison of hmUra DNAs. The percentage of inferred base substitutions increased to 6% when four viral DNAs were compared (48).

The occurrence of rearrangements of these phage genomes in the course of evolution was investigated by making an inventory of the restriction segments present in four hmUra genomes. Nucleotide stretches of 1.5 and 2.7 kbp in 2C DNA were absent from Φe DNA and SPO1 DNA, respectively (48): they accounted for 1–2% of the total genome, indicating that both base substitutions and deletions have contributed to the evolution of the genomes of this family of viruses.

The ability of specific restriction fragments to hybridize to both terminal fragments of 2C DNA suggested the occurrence of direct repeats of 12 kbp: this amount corresponds to about 1/11 of the total genome (14). When the map thus obtained was compared with maps of SPO1 (87) and SP82 (66), the overall organization of these viral genomes appeared very similar (Figure 2, *top*). Cross-hybridization patterns of the terminal redundancies, present in all members of this family of phages, indicate that the ends of the four hmUra-containing DNAs are colinear, highly homologous, and carry one *Hae*III cleavage site at different locations (14) (Figure 2, *bottom*). Such a shift, produced in the course of evolution, supposedly results from point mutations (one erasing a *Hae*III cleavage site in one place, and the other creating a new recognition sequence elsewhere).

The terminal redundancy might be involved not merely in replication but also in recombination and repair processes (61, 62). In addition, terminal repeats of both SPO1 and SP82 carry most of the genes transcribed early in infection (for review, see 102). Terminal repetitions observed in T phage genomes are much smaller than those in hmUra phages (170 bp for T7; 4 kbp for T4) (29, 64, 95).

GENETIC MAP AND CLONING OF GENOMIC FRAGMENTS

Thermosensitive 2C mutants were obtained by chemical mutagenesis. The relative order of mutated genes was deduced from their recombination frequencies, assuming a random occurrence of reciprocal cross-overs in pairwise infected *B. subtilis* (54). Data treatment by a computer program (58) has allowed the unambiguous location of 26 mutations clustered mostly at the right end of the genetic map (Figure 3).

The correspondence of the genetic and physical maps of phage 2C was

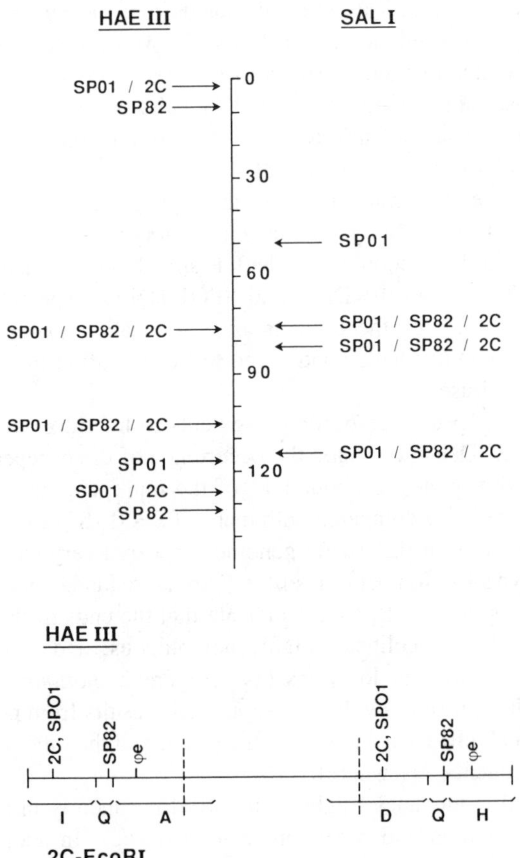

Figure 2 (*Top*) Restriction maps of the genomes (size in kbp) of three hmUra-containing phages (14, 67, 87). (*Bottom*) Physical map of the redundant ends of hmUra DNA molecules. The *Hae*III cleavage site in the terminal repetitive sequences of the chromosomes of phages 2C, SPO1, SP82, and Φe is indicated above the line. The *Eco*RI cleavage sites in the redundant ends of 2C genome are reported below the line.

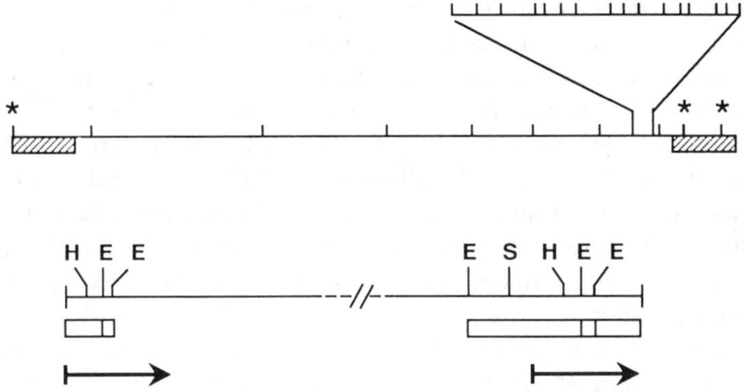

Figure 3 Features of 2C DNA terminal redundancies. (*Top*) The location of three unique thermosensitive mutations within the terminal redundancies is indicated by a star. Vertical lines indicate the location of 23 other thermosensitive mutations. Shaded boxes show the terminal redundancies. (*Bottom*) Location of cloned fragments on the 2C restriction map. Restriction sites: E, *Eco*RI; S, *Sal*I; H, *Hae*III. Boxes indicate the asymmetrical hybridization pattern of pHV33-2C recombinant plasmids with *Eco*RI restriction fragments of 2C DNA. The arrows indicate the terminal redundancies.

assessed by recombination of mutated genes with plasmid-borne wild-type sequences of known location. Fragments of 2C DNA cloned in pHV33 (32) were located using Southern hybridization on the available 2C restriction map. Marker rescue was observed in a limited number of combinations throughout the entire genome. The physical and genetic maps of phage 2C were thus aligned (58, 59). These studies revealed a peculiar feature of the terminal redundancies of 2C DNA, which contain single copies of some loci: two mutations within the right terminal redundancy, and one at the left end of the genome, did not have their equivalent counterparts in the opposite redundant end (Figure 3, *top*). Consequently, the two ends of the genome, in spite of an overall homology, contain unique sequences. Similar conclusions were drawn from the hybridization patterns of recombinant plasmids, some of which showed an unequal annealing with the two terminal repeats (Figure 3, *bottom*) (59).

A functional map of the genome was constructed by assigning the mutants to 11 different complementation groups of yet unknown functions.

DNA BIOSYNTHESIS: INITIATION AND ELONGATION

The replication origin of 2C DNA was identified through its ability to confer autonomous duplication to nucleotide sequences containing a suitable marker. The vectors used, which were unable to replicate in *B. subtilis*, included the chimaeric plasmids pSC540, pCP115, pHV32 (11, 78), and the *E. coli* pUC19-derived vector carrying a chloramphenicol resistance gene (Ph. Gillet, personal communication). The ability to replicate in *B. subtilis* was conferred upon those vectors by insertion of 2C DNA segments. Cloned viral segments hybridized to the redundant ends of the genome and to adjacent unique sequences. This observation is in agreement with density-shift experiments showing that the first round of replication of SPO1 DNA starts from origins located close to the terminal redundancies (35). Thence came the suggestion that these recombinant clones carried the primary replication origin(s) of the phage genome (65), which, although deprived of hmUra, could initiate plasmid replication.

The synthesis of phage DNA, traced by ^{3}H-uracil labeling of virus-infected cells, starts 10 min after infection and proceeds linearly for 30 min (8). These linear kinetics, confirmed by two other methods (26, 46), suggest a mechanism differing from a dichotomous replication of single genome units, entailing an exponential increase of viral DNA. These data are compatible with a rolling-circle type of replication or with progression of a replicative fork coupled with recombination, as shown in the next section.

Transcriptional and translational events occurring during the viral cycle were explored by addition of inhibitors at different moments of the lytic cycle (47). Rifampicin inhibited DNA synthesis during the first 10 min of the cycle and was ineffective afterwards. The requirement for host RNA-polymerase in DNA initiation is thus limited in time, while another initiation mechanism might be at work at later times. Similar observations with T4-infected *E. coli* indicated a secondary initiation from recombinational intermediates (76).

To isolate thermosensitive mutants, possibly affected in the initiation or the elongation of 2C DNA, the selective screening procedure described for SPO1 (102, 112) was applied. Of the 150 thermosensitive 2C mutants tested, all showed viral DNA synthesis at the restrictive temperature and half of them had a rate reduced, at the most, to 50% of that obtained with wild-type phage (58; Ph. Gillet, personal communication). This observation contrasts with SPO1 replication-deficient mutants, which could readily be isolated with some of them showing at the most 1% of wild-type DNA synthesis (112). We suggest that host functions or redundant phage pathways might be able to complement defective 2C DNA replicative mechanisms.

As to the molecular mechanism of DNA chain elongation, 2C DNA apparently duplicates according to a discontinuous model during the entire viral cycle, as shown with ^{3}H-uracil pulse-labeled DNA and with ultracentrifugal analysis of the denatured product in alkaline sucrose gradients (51). Okazaki pieces of about 1500 bp, the main species labeled in the early part of the cycle, were quite efficiently incorporated into full-size DNA molecules late in infection. This pattern corresponds to the scheme of discontinuous DNA duplication originally described by Okazaki et al (80) for the

T4-*E. coli* system. These DNA fragments may be intermediates of repair, rather than of true replication. Repair processes, involving shorter polynucleotide stretches, should not alter the buoyant density of molecules labeled after density shift. This possibility was ruled out, however, by experiments involving short ^{3}H-uracil pulses given to density-shifted infected bacteria. The radioactive precursor was incorporated into DNA segments that were lighter than the parental viral DNA. When the size of this DNA was reduced by sonication to an average size of 1500 bp, molecules with hybrid density (Okazaki fragments replicated semiconservatively) were produced.

Although DNA duplication is thought to be continuous for the leading strand and discontinuous for the lagging strand (1), a discontinuous duplication of both 2C DNA strands was suggested by the fact that short pulses of labeled precursors were almost exclusively incorporated into Okazaki fragments. Such a conclusion was supported by the finding that Okazaki fragments from virus-infected cells, upon denaturation and density gradient centrifugation, yielded two equivalent peaks, overlapping the H and L strands of reference 2C DNA and hybridizing with both viral strands. An alternative interpretation of these results would be a bidirectional synthesis of DNA involving a discontinuous replication of one strand in each direction (Figure 4). Indeed, in uninfected and phage T4-infected *E. coli,* DNA replication proceeded from a single origin in both directions; this scheme was claimed to be universal (28, 97). Further proof for a discontinuous replication of both DNA strands was obtained by demonstrating the ability of 2C DNA precursors to self-anneal. Indeed, after self-annealing, pulse-labeled Okazaki fragments, isolated from 2C-infected *B. subtilis* by alkaline sucrose gradient, withstood single-strand specific DNAase. Although a search for RNA primers was not done in the case of 2C, in other viral systems such as polyoma Okazaki fragments with RNA primers hybridized to both viral DNA strands (1).

REPLICATIONAL RECOMBINATION OF 2C DNA

The fate of an infecting viral genome depends on its structure, metabolism, and final packaging into the newly formed capsids. Depending on the virus-host systems, parental DNA is transferred to progeny to different extents ranging from no transfer in phage ΦX174 (98) to supposedly complete transfer in T even phages (44). Investigations on the fate of infecting phage DNA may yield information on the processes for replication, recombination, and repair of viral DNA.

When ^{32}P-labeled 2C phages infected unlabeled bacteria, 40% of the input radioactivity was found within progeny molecules, a transfer efficiency approaching that of T2 (44). Because the radioactivity of parental origin was

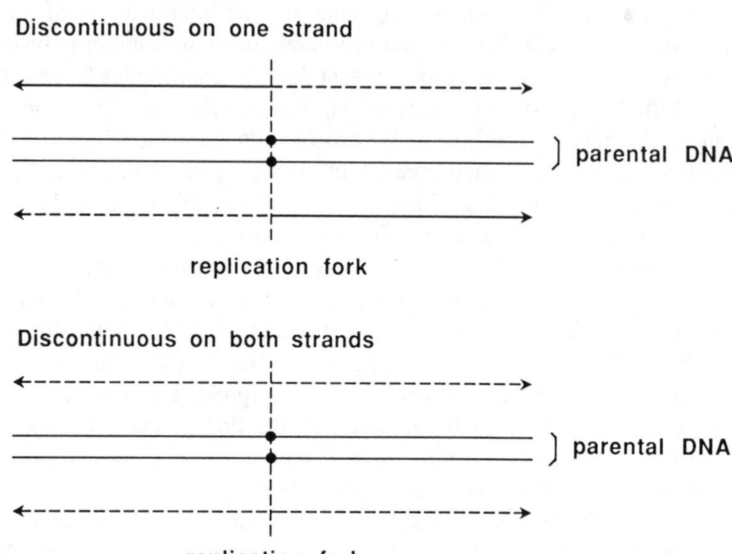

Figure 4 Two schemes of bidirectional DNA synthesis.

equally distributed among the two daughter strands, the transfer of parental 2C DNA to the offspring was symmetrical. A degradation of infecting DNA followed by random incorporation of the nucleotides was ruled out (49). Three possibilities ought to be considered: (*a*) the two parental strands are reencapsidated without being separated (conservative transfer); (*b*) intact strands of semiconservatively replicated DNA are found in progeny particles (single strand transfer); and (*c*) randomly produced segments of either parental or semiconservatively replicated DNA are transferred (dispersive transfer).

To approach this problem, unlabeled cells were infected with density- and radioactivity-labeled phages, and progeny 2C DNA, after controlled shear degradation, was submitted to molecular weight and density measurements, as schematically represented in Figure 5. The segments of parental 2C DNA that were transferred to progeny particles had an average size of 13 kbp. After sonication and denaturation of such DNA, a peak of heavy DNA corresponding to segments of parental origin was obtained. The overall interpretation of these data is as follows: 2C DNA, which replicates semiconservatively, undergoes extensive genetic recombination with newly formed viral DNA molecules in the vegetative pool. Parental DNA contributed to progeny virions single-stranded segments of DNA, presumably harboring clusters of genes. Never were intact parental molecules observed in progeny virions, thus excluding conservative transfer (50).

Recombination was monitored during the viral cycle by density-transfer

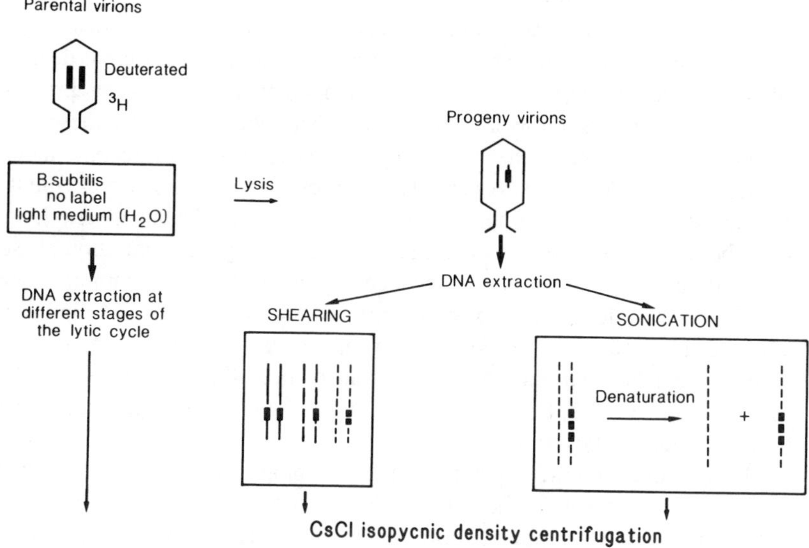

Figure 5 Scheme of experiments aiming to prove the replicational recombination of phage 2C DNA.

experiments (Figure 5). Some DNA possessing the original heavy buoyant density persisted during the entire lytic cycle: these were parental molecules that did not replicate nor recombine and were not found in progeny virions. Hybrid DNA (replicated, unrecombined molecules) appeared 13 min after infection. By 20 min, lighter DNA species appeared as well as hybrid molecules. At later times (35 min), true hybrid molecules had disappeared and were replaced by replicated-recombined molecules, showing that genetic recombination followed semiconservative duplication. Using similar techniques, Cregg & Stewart (21) observed only hybrid SPO1 molecules up to 24 min after infection.

Experiments were done in order to dissociate replication from recombination processes by adding transcriptional and translational inhibitors at different phases of the 2C replication cycle. This approach was based on the observation that rifampicin and chloramphenicol blocked DNA synthesis mainly during the early phase of infection (47). Double-labeling experiments revealed that genetic recombination between parental and progeny viral genomes was reduced by addition of chloramphenicol and rifampicin during the second part of the latent period, when viral DNA synthesis proceeded under these conditions, as confirmed by the accumulation of hybrid duplicational intermediates. Hence, recombination relies on the synthesis of virus-dictated enzymes, which apparently are synthesized later than those involved in viral DNA replication (45).

The occurrence of parent-to-parent recombination at the onset of the lytic cycle was explored by infecting the host with a mixture of different density-labeled phages in the presence of chloramphenicol (which prevents progeny DNA formation). Because no density shift was observed under these conditions, it was concluded that negligible parent-to-parent recombination occurs in the 2C-*B. subtilis* system. These observations are consistent with the absence of recombinational viral DNA molecules during the latent period, and with the involvement of virus-dictated enzymes during the maturation phase. This inference is supported by the occurrence of similar recombinational patterns in 2C-infected wild-type and *rec*$^-$ *B. subtilis* (45). It is tempting to attribute this lack of activity of cell enzymes to the presence of the unusual base on 2C DNA. On the other hand, transfection of *B. subtilis* with SP82 (37, 100) and Φe (72) DNA apparently involved parent-to-parent recombination mediated by host enzymes.

These studies on hmUra phages should be compared with those on T even phages. At the onset of the lytic cycle, T4 DNA replication depends on initiation from a few replication origins, involving RNA primers synthesized by unmodified *E. coli* RNA-polymerase. Replication at later times involves recombinational initiation of replication forks (63, 76), which generates concatemers prior to maturation. These data are consistent with our findings in the 2C–*B. subtilis* system.

REGULATION OF TRANSCRIPTION: CLONING OF PHAGE PROMOTERS

In 2C-infected *B. subtilis,* cellular and viral RNA species are both synthesized during the latent period, whereas preferential transcription of viral genes occurs at later times (8, 9). RNA labeled after infection hybridizes preferentially, but not exclusively with the heavy strand of 2C DNA (9).

The role of the abnormal base in promoter recognition has been mainly explored in SPO1, and investigators have relied largely on in vitro techniques (see section on selected data in SPO1 and SP82). In our studies, we cloned 2C restriction fragments in promoter-probe plasmids, searching for insertional activation of their silent gene. The promoter activity of cloned sequences, devoid of the abnormal base, could thus be estimated in vivo. In *E. coli,* promoter-carrying fragments were readily isolated by the use of several promoter-probe plasmids (pGR71, pKO1, pCED6, pTLXT-11, pAS3). Cross-hybridization showed that these fragments belonged to two or three classes (Ph. Gilot, personal commununication). Cloned sequences hybridized to the central part of the 2C genome. However, 2C promoters active in *E. coli* were inactive in *B. subtilis,* irrespective of the vectors used (pTL7O8, pTG4O2, pCED6). Note that: (*a*) the cloned sequences were intact, as indicated by the rescue of promoter activities when they were transfered back to *E. coli,* and (*b*) that *B. subtilis* chromosomal promoters were successfully cloned in the same vectors and expressed in the homologous host. The reason for the inactivity of the cloned viral promoters in *B. subtilis* remains unknown. It might result from the absence of the abnormal base. Nothing is known about the influence of the abnormal base on the helical structure and supercoiling of DNA, known to modulate *E. coli* RNA polymerase activity (70, 105).

SELECTED DATA ON SPO1 AND SP82

This section presents data on SPO1 and SP82, relevant to three topics studied in the 2C phages.

Genetic and Physical Maps

The presence in the genome of SPO1 and SP82 of distinct regions for DNA synthesis, head capsids, and morphogenesis of tail fibers has been observed (38, 54, 55, 81). Such an organization is similar to those of coliphages (75). Moreover, the dUMP-hydroxymethylase and deoxy-hmUra-kinase genes as well as 10 cistrons involved in hmUra DNA synthesis have been identified (55, 81). McAllister (73) describes an oriented penetration of SP82 genome into the host.

Two SPO1 mutants, yielding high recombination rates with many other mutants, were assigned to the terminal redundancy (21a). These were the only two mutations mapped in this region of the genome, which is known to contain at least 11 of the early transcribed genes of unknown function (87). Cell-free transcription/translation experiments revealed the presence of a middle promoter and of a middle gene within the terminal redundancy of SPO1 (5, 86).

Marker rescue experiments allowed the assignment of 26 out of 39 SPO1 mutations to specific restriction fragments (22, 23, 34, 87). Approximately 65% of genomic sequences were estimated to have been cloned. Yet four *Eco*RI fragments, three of which belong to the terminal redundancy, were not duplicated when cloned in *E. coli* or in *B. subtilis* (23): the expression of the genes carried by these fragments might be lethal for the host.

The function of several genes has been identified. Genes 28, 33, and 34 code for σ factors, determining the specificity of RNA-polymerase for middle and late viral promoters (for review, see 102). The product of gene 27 is required for late transcription and DNA replication (19). Gene 31 codes for phage SPO1 DNA polymerase (27), which was purified and characterized by

Yehle & Ganesan (111). This gene, as shown recently, contains a self-splicing group I intron (36) similar to those found in T4 (13, 99). Transcription factor 1 (TF1) is a DNA-binding protein that shows preference for hmUra-containing DNA. TF1 binds with high affinity to early promoters in the SPO1 genome (40) and is essential for viral multiplication (39, 41, 53, 96).

Terminal Redundancy

The existence of terminal repeats in SPO1 was first suggested when viral DNA molecules extracted from phage particles and from infected cells were compared. Two of the fragments produced by *Eco*RI* digestion of mature viral DNA were absent from digests of intracellular phage DNA. These fragments may be part of a terminal redundancy and participate in the formation of concatemeric structures. Watson (108) has proposed a scheme whereby linear T7 DNA molecules with repetitive terminal sequences are generated from concatemeric forms: this model accounts for the synthesis of the 5' ends of linear DNA molecules. Since DNA polymerases produce daughter molecules with protruding 3' ends, they could join in dimers if located within a terminal redundancy. Repetition of this process is expected to form higher order concatemers, which, at the time of maturation, are processed by staggered breaks in the two strands. The 5' protruding ends are templates for a duplication process that regenerates the terminal redundancy on each side of the break. Genetic data concerning a SPO1 mutant, located in the terminal redundancy, were interpreted in the light of the Watson model (21a).

Indeed, during the vegetative period of SPO1 infection in *B. subtilis,* long concatemeric forms of viral DNA equivalent to 20 genomic units accumulated: they were the precursors of mature viral DNA (69).

Transcription and Promoter Activity

At least six classes of mRNA have been identified during the lytic cycle of SPO1 and SP82 (102). At the onset of infection, early genes are transcribed by the bacterial RNA–polymerase holoenzyme. Four different viral proteins positively controlling SPO1 gene expression are subsequently synthesized. They are gp28, a σ factor recognizing middle promoters; gp33 and gp34, favoring the recognition of late promoters; and gp27, controlling late transcription and DNA replication (18–20). A similar pattern of development occurs in the SP82 lytic cycle. Such a cascade-type control, where the expression of one gene is required for the expression of subsequent genes, accounts for an irreversible regulatory mechanism akin to cellular differentiation (71, 103).

In both SPO1 and SP82, at least eleven early promoters were found within the terminally redundant segments. The early in vitro transcripts originating from these promoters overlapped and converged toward a bidirectional termination region. In vitro transcription initiation from these early promoters was specifically increased by the host protein delta, which is associated with RNA-polymerase during purification (109). Early transcripts of SP82 were shown to be processed by a *B. subtilis* endonuclease (83, 84).

Hydroxymethyluracil is required for in vitro recognition of middle promoters by the gp28-modified RNA polymerase, whereas early and late promoters were active regardless of the presence of the abnormal base in their sequence (22, 68). Moreover, hmUra replacement by thymine on a single DNA strand dramatically lowered transcription initiation at middle promoter sequences by the gp28-RNA polymerase complex (12). However, using different in vitro techniques, Romeo et al (90) showed that the gp28-RNA polymerase recognizes its cognate promoters in thymine-containing DNA. These studies are in agreement with in vivo experiments that show the activity of plasmid-borne SPO1 middle and late genes when thymine replaces hmUra (24).

CONCLUDING REMARKS AND PERSPECTIVES

Hdroxymethyluracil phages have some unique properties because of the presence of an unusual base in their genome. Distinct biophysical properties of hmUra DNA are a buoyant density different from that of the host genome and special interactions with given fluorophores. These properties, in addition to the bimodal distribution of the complementary DNA strands, have been exploited in the research work summarized in this review.

Several distinct modes of DNA replication seem to occur in 2C-infected cells. At the beginning of the lytic cycle, replication initiates from a region close to the ends of the genome. Autonomously replicating sequences, which were cloned from 2C DNA, hybridized to the terminal repeats and neighboring regions. They might correspond to the primary origins of replication, equivalent to those recognized in SPO1 by density-shift experiments. These sequences allow the autonomous replication of plasmids containing thymine. Small amounts of thymine polymerized in viral DNA (detected in permeabilized infected cells) might play a role in replication at the onset of the lytic cycle, before the abnormal base becomes available.

At later stages of the lytic cyle, 2C DNA is synthesized at a linear rate by semiconservative replication, followed by a recombinational replication. Only recombinant molecules are encapsidated during the maturation process: they presumably derive from the cleavage of long concatemeric structures similar to those described for SPO1. The involvement of the terminal redundant ends in this mechanism deserves further studies. Evidence suggests the presence of unique features within these terminal repeats, which contain overall homologous sequences.

A study of mutants with altered hmUra biosynthetic pathways might help to understand the in vivo role of this abnormal base. Thermosensitive 2C mutants, which synthesize viral DNA with reduced hmUra content, have been identified with a new fluorometric procedure (25; Ph. Gillet, work in progress). Previous work has related the occurrence of Φe mutants producing viable phages in which up to 20% of hmUra was replaced by thymine (102). Further work along these lines is required to identify the viral functions affected by the presence (or the absence) of the abnormal base.

ACKNOWLEDGMENTS

The authors thank C. Colson (University of Louvain, Belgium) and C. Stewart (Rice University, Houston, Texas) for critically reading the manuscript. P. Hoet is Research Director at the National Fund for Scientific Research (Belgium). Some of the research work on 2C has received the technical help of P. Rensonnet. The financial support of the National Fund for Medical Scientific Research (Belgium) is acknowledged.

Literature Cited

1. Alberts, B. M. 1987. Prokaryotic DNA replication mechanisms. *Philos. Trans. R. Soc. London Ser. B* 317:395–420
2. Alegria, A. H., Kahn, F. M. 1968. Attempts to establish whether glucose is attached to the deoxyribonucleic acid of certain bacteriophages infecting *Bacillus subtilis*. *Biochemistry* 7:1132–40
3. Aposhian, H. W., Tremblay, G. Y. 1966. Deoxythymidylate 5'-nucleotidase. Purification and properties of an enzyme found after infection of *B. subtilis* with phage SP5C. *J. Biol. Chem.* 241:5095–5101
4. Berkner, K. L., Folk, W. R. 1977. *Eco*RI cleavage and methylation of DNAs containing modified pyrimidine in the recognition sequence. *J. Biol. Chem.* 252:3185–93
5. Brennan, S. M., Geiduschek, E. P. 1983. Regions specifying transcriptional termination and pausing in the bacteriophage SPO1 terminal repeat. *Nucleic Acids Res.* 11:4157–75
6. Bron, S., Luxen, E., Venema, G. 1983. Resistance of bacteriophage H1 to restriction and modification by *Bacillus subtilis* R. *J. Virol.* 46:703–8
7. Brooks, J. E., Roberts, R. J. 1982. Modification profiles of bacterial genomes. *Nucleic Acids Res.* 10:913–14
8. Cocito, C. G. 1969. The action of virginiamycin on nucleic acid and protein synthesis in *B. subtilis* infected with bacteriophage 2C. *J. Gen. Microbiol.* 57: 195–206
9. Cocito, C. G. 1974. Origin and metabolic properties of the RNA species formed during the replication cycle of virus 2C. *J. Virol.* 14:1482–93
10. Cocito, C. G., Vanlinden, F. 1978. Polysomes and ribosome metabolism in virus 2C multiplication. *Biochimie* 60: 399–402
11. Chang, S., Cohen, S. N. 1979. High frequency transformation of *Bacillus subtilis* protoplasts by plasmid DNA. *Mol. Gen. Genet.* 168:111–15
12. Choy, H. A., Romeo, J. M., Geiduschek, E. P. 1986. Activity of a phage-modified RNA polymerase at hybrid promoters. Effects of substituting thymine for hydroxymethyluracil in a phage SPO1 middle promoter. *J. Mol. Biol.* 191:59–73
13. Chu, F. K., Maley, G. F., Maley, F., Belfort, M. 1984. An intervening sequence in the thymidylate synthase gene of bacteriophage T4. *Proc. Natl. Acad. Sci. USA* 81:3149–53
14. Coene, M. M., Hoet, P. P., Cocito, C. G. 1983. Physical map of virus 2C-DNA: evidence for the existence of large redundant ends. *Eur. J. Biochem.* 132: 69–75

15. Cohen, S. S. 1948. The synthesis of bacterial viruses. I. The synthesis of nucleic acid and protein in *E. coli* infected with T2R bacteriophage. *J. Biol. Chem.* 174:281–93
16. Cohen, S. S. 1968. *Virus-Induced Enzymes*. New York: Columbia Univ. Press
17. Cohen, S. S., Brevet, J., Cabello, F., Chang, A. C. Y., Chou, H., et al. 1978. Macro- and micro-evolution of bacterial plasmids. In *Microbiology,* ed. D. Schleissinger, pp. 217–20. Washington, DC: Am. Soc. Microbiol.
18. Costanzo, M., Brzustowicz, L., Hannett, N., Pero, J. 1984. Bacteriophage SPO1 genes 33 and 34. Location and primary structure of genes encoding regulatory subunits of *Bacillus subtilis* RNA polymerase. *J. Mol. Biol.* 180: 533–47
19. Costanzo, M., Hannett, N., Brzustowicz, L., Pero, J. 1983. Bacteriophage SPO1 gene 27: location and nucleotide sequence. *J. Virol.* 48:555–60
20. Costanzo, M., Pero, J. 1983. Structure of a *Bacillus subtilis* bacteriophage SPO1 gene encoding a RNA polymerase factor. *Proc. Natl. Acad. Sci. USA* 80: 1236–40
21. Cregg, J. M., Stewart, C. R. 1978. Timing of initiation of DNA replication in SPO1 infection of *Bacillus subtilis*. *Virology* 80:289–96
21a. Cregg, J. M., Stewart, C. R. 1978. Terminal redundancy of "high frequency of recombination" markers of *B. subtilis* phage SPO1. *Virology* 86:530–41
22. Curran, J. F., Stewart, C. R. 1982. Recombination and expression of a cloned fragment of the DNA of *Bacillus subtilis* bacteriophage SPO1. *Virology* 120:307–17
23. Curran, J. F., Stewart, C. R. 1985. Cloning and mapping of the SPO1 genome. *Virology* 142:78–97
24. Curran, J. F., Stewart, C. R. 1985. Transcription of *Bacillus subtilis* plasmid pBD64 and expression of bacteriophage SPO1 genes cloned therein. *Virology* 142:98–111
25. Daxhelet, G. A., Coene, M. M., Hoet, P. P., Cocito, C. G. 1989. Spectrofluorometry of dyes with DNAs of different base composition and conformation. *Anal. Biochem.* 179:401–3
26. Daxhelet, G. A., Kohnen, M., Coene, M. M., Hoet, P. P. 1990. Fluorometric in vivo determination of *B. subtilis* and phage 2C DNA. *Anal. Biochem.* 190: 116–19
27. De Antoni, G. L., Besso, N. E., Zanassi, G. E., Sarachu, A. N., Grau, O. 1985. Bacteriophage SPO1 DNA polymerase and the activity of viral gene 31. *Virology* 143:16–22
28. Delius, H., Howe, C., Kozinski, A. W. 1971. Structure of the replicating DNA from bacteriophage T4. *Proc. Natl. Acad. Sci. USA* 68:3049–53
29. Dreiseikelmann, B., Steger, U., Wackernagel, W. 1980. Length determination of the terminal redundant regions in the DNA of phage T7. *Mol. Gen. Genet.* 178:237–40
30. Dreiseikelmann, B., Wackernagel, W. 1981. Absence in *Bacillus subtilis* and *Staphylococcus aureus* of the sequence-specific deoxyribonucleic acid methylation that is conferred in *E. coli* K-12 by the *dam* and *dcm* enzymes. *J. Bacteriol.* 147:259–61
31. Dunham, L. F., Price, A. R. 1974. Deoxythymidine triphosphate-deoxyuridine triphosphate nucleotidohydrolase induced by *B. subtilis* bacteriophage Φe. *Biochemistry* 13:2667–72
32. Ehrlich, S. D. 1978. DNA cloning in *B. subtilis*. *Proc. Natl. Acad. Sci. USA* 75:1433–36
33. Gardner, R. C., Howarth, A. J., Messing, J., Shepherd, R. J. 1982. Cloning and sequencing of restriction fragments generated by EcoRI*. *DNA* 1: 109–15
34. Geiduschek, E. P., Ito, J. 1982. Regulatory mechanism in the development of lytic bacteriophages in *Bacillus subtilis*. In *The Molecular Biology of the Bacilli,* ed. D. Dubnau, 1:203–45. New York: Academic
35. Glassberg, J., Franck, M., Stewart, C. R. 1977. Multiple origins of replication for *B. subtilis* phage SPO1. *Virology* 78:433–41
36. Goodrich-Blair, H., Scarlato, V., Gott, J. M., Xu, M.-Q., Shub, D. A. 1990. A self-splicing group I intron in the DNA polymerase gene of *Bacillus subtilis* bacteriophage SPO1. *Cell* 63:417–24
37. Green, D. M. 1966. Physical and genetic characterization of sheared infective SP82 bacteriophage DNA. *J. Mol. Biol.* 22:15–22
38. Green, D. M., Laman, D. 1972. Organization of gene function in *B. subtilis* bacteriophage SP82G. *J. Virol.* 9:1033–46
39. Greene, J. R., Brennan, S. M., Andrew, D. J., Thompson, C. C., Richards, S. H., et al. 1984. Sequence of the bacteriophage SPO1 gene coding for transcription factor 1, a viral homologue of the bacterial type II DNA-binding proteins. *Proc. Natl. Acad. Sci. USA* 81:7031–35

40. Greene, J. R., Geiduschek, E. P. 1985. Site-specific DNA binding by the bacteriophage SPO1-encoded type II DNA-binding protein. *EMBO J.* 4:1345–49
41. Greene, J. R., Morrissey, L. M., Foster, L. M., Geiduschek, E. P. 1986. DNA binding by the bacteriophage SPO1-encoded type II DNA-binding protein, transcription factor 1. Formation of nested complexes at a selective binding site. *J. Biol. Chem.* 261:12820–27
42. Haslam, E. A., Roscoe, D. H., Tucker, R. G. 1967. Inhibition of thymidilate synthetase in bacteriophage infected *B. subtilis*. *Biochem. Biophys. Acta* 134: 312–26
43. Hemphill, H. E., Whiteley, H. R. 1975. Bacteriophages of *B. subtilis*. *Bacteriol. Rev.* 39:257–315
44. Hershey, A. D., Burgi, E. 1956. Genetic significance of the transfer of nucleic acid from parental to offspring phage. *Cold Spring Harbor Symp. Quant. Biol.* 21:91–101
45. Hoet, P. P., Cocito, C. G. 1982. Replication and recombination of the viral genome during lytic infection of *Bacillus subtilis* by phage 2C. In *Genetic Exchange,* ed. U. N. Streips, S. H. Goodgal, W. R. Guild, G. A. Wilson, pp. 275–82. New York: Dekker
46. Hoet, P. P., Coene, M. M., Cocito, C. G. 1978. Synthesis of phage 2C-DNA in permeabilized *B. subtilis*. *Mol. Gen. Genet.* 158:297–303
47. Hoet, P. P., Coene, M. M., Cocito, C. G. 1981. Action of inhibitors of macromolecule formation on duplication of *B. subtilis* phage 2C-DNA. *Biochem. Pharmacol.* 30:489–94
48. Hoet, P. P., Coene, M. M., Cocito, C. G. 1983. Comparison of the physical map and redundant ends of the chromosomes of viruses 2C, SPO1, SP82 and Φe. *Eur. J. Biochem.* 132:63–67
49. Hoet, P. P., Fraselle, G., Cocito, C. G. 1975. Transfer to progeny of both DNA strands of phage 2C. *Biochem. Biophys. Res. Commun.* 66:235–42
50. Hoet, P. P., Fraselle, G., Cocito, C. G. 1976. Recombinational type transfer of viral DNA during bacteriophage 2C replication in *B. subtilis*. *J. Virol.* 17:718–26
51. Hoet, P. P., Fraselle, G., Cocito, C. G. 1979. Discontinuous duplication of both strands of virus 2C DNA. *Mol. Gen. Genet.* 171:43–51
52. Ito, J., Roberts, R. J. 1979. Unusual base sequence arrangements in phage Φ29 DNA. *Gene* 5:1–7
53. Johnson, G. C., Geiduschek, E. P. 1972. Purification of the bacteriophage SPO1 transcription factor 1. *J. Biol. Chem.* 247:3571–78
54. Kahan, E. 1966. A genetic study of temperature-sensitive mutants of the *B. subtilis* phage SP82. *Virology* 30:650–60
55. Kahan, E. 1971. Early and late gene function in bacteriophage SP82. *Virology* 46:634–37
56. Kahan, F., Kahan, H., Riddle, B. 1964. Nucleotide metabolism in extracts of *B. subtilis* infected with phage SP8. *Fed. Proc.* 23:318
57. Kallen, R. G., Simon, M., Marmur, J. 1962. The occurrence of a new pyrimidine base replacing thymine in a bacteriophage DNA: 5-hydroxymethyluracil. *J. Mol. Biol.* 5:248–50
58. Kiss-Blümel, J. 1987. *Carte génétique et clonage de fragments génomiques du virus 2C de* Bacillus subtilis. PhD thesis. University of Louvain
59. Kiss-Blümel, J., Hoet, P. P. 1989. Marker-rescue of *Bacillus subtilis* phage 2C mutants by cloned viral DNA: unique features of the terminal redundancies of the phage genome. In *Advances in Genetic Transformation,* ed. L. O. Butler, pp. 243–51. Dorset, UK: Intercept
60. Kit, S., Dubbs, D. R. 1969. Enzyme induction by viruses. *Monogr. Virol.* 2: 1–96
61. Kornberg, A. 1980. *DNA Replication*. San Francisco: Freeman
62. Kornberg, A. 1982. *Supplement to DNA Replication*. San Francisco: Freeman
63. Kozinski, A. W. 1983. DNA metabolism. Origins of T4 DNA replication. See Ref. 75a, pp. 111–19
64. Kutter, E., Rüger, W. 1983. Structure, organization and manipulation of the genome. Map of the T4 genome and its transcription control sites. See Ref. 75a, pp. 277–90
65. Lannoy, N. N., Hoet, P. P., Cocito, C. G. 1985. Cloning of DNA segments of phage 2C, which allows autonomous plasmid replication in *Bacillus subtilis*. *Eur. J. Biochem.* 152:137–42
66. Lawrie, J. M., Whiteley, H. R. 1977. A physical map of bacteriophage SP82 DNA. *Gene* 2:233–50
67. Lawrie, J. M., Downard, J. S., Whiteley, H. R. 1978. *Bacillus subtilis* bacteriophage SP82, SPO1 and Φe: a comparison of DNAs and of peptides synthesized during infection. *J. Virol.* 27:725–37
68. Lee, G., Hannet, N. M., Korman, A., Pero, J. 1980. Transcription of cloned DNA from *B. subtilis* phage SPO1. *J. Mol. Biol.* 139:407–22
69. Levner, M. H., Cozzarelli, N. R. 1972.

Replication of viral DNA in SPO1-infected *Bacillus subtilis*. Replication intermediates. *Virology* 48:402–16

70. Liu, L. F., Wang, J. C. 1987. Supercoiling of the DNA template during transcription. *Proc. Natl. Acad. Sci. USA* 84:7024–27
71. Losick, R., Pero, J. 1981. Cascades of sigma factors. *Cell* 25:582–84
72. Loveday, K. S., Fox, S. M. 1978. The fate of bacteriophage Φe transfecting DNA. *Virology* 85:387–403
73. McAllister, W. T. 1970. Bacteriophage infection: which end of the SP82G genome goes in first? *J. Virol.* 5:194–98
74. Marmur, J., Greenspan, C. M. 1963. Transcription in vivo of DNA from bacteriophage SP8. *Science* 142:387–89
75. Mathews, C. K. 1971. *Bacteriophage Biochemistry*. New York: Van Nostrand Reinhold

75a. Mathews, C. K., Kutter, E. M., Mosig, G., Berget, P. B., eds. 1983. *Bacteriophage T4*. Washington: Am. Soc. Microbiol.

76. Mosig, G. 1983. Relationship of T4 DNA replication and recombination. See Ref. 75a, pp. 120–30
77. Nei, M., Li, W. H. 1979. Mathematical model for studying genetic variation in terms of restriction endonucleases. *Proc. Natl. Acad. Sci. USA* 76:5269–73
78. Niaudet, B., Ehrlich, S. D. 1979. In vitro genetic labeling of *Bacillus subtilis* cryptic plasmid pHV400. *Plasmid* 2:48–58
79. Nishihara, M., Crambach, A., Aposhian, H. V. 1967. The deoxycytidilate deaminase found in *Bacillus subtilis* infected with phage SP8*. *Biochemistry* 6:1877–86
80. Okazaki, R., Okazaki, R., Sakabe, K., Sugimoto, K., Kainuma, R., et al. 1968. In vivo mechanism of DNA chain growth. *Cold Spring Harbor Symp. Quant. Biol.* 33:129–43
81. Okubo, S. T., Yanagida, T., Fujita, D. J., Ohlsson-Wilhem, B. M. 1972. The genetics of bacteriophage SPO1. *Biken J.* 15:81–97
82. Padhy, R. N., Hottat, F. G., Coene, M. M., Hoet, P. P. 1988. Restriction analysis and quantitative estimation of methylated bases of filamentous and unicellular cyanobacterial DNAs. *J. Bacteriol.* 70:1934–39
83. Panganiban, A. T., Whiteley, H. R. 1983. Bacillus subtilis RNAase III cleavage sites in phage SP82 early mRNA. *Cell* 33:907–13
84. Panganiban, A. T., Whiteley, H. R. 1983. Purification and properties of a new *Bacillus subtilis* RNA processing enzyme. *J. Biol. Chem.* 258:12487–93
85. Pène, J. J., Marmur, J. 1964. Infectious DNA from a virulent bacteriophage active on transformable *B. subtilis*. *Fed. Proc.* 23:318
86. Perkus, M. E., Shub, D. A. 1985. Mapping the genes in the terminal redundancy of bacteriophage SPO1 with restriction endonucleases. *J. Virol.* 56:40–48
87. Pero, J., Hannett, N. M., Talkington, C. 1979. Restriction cleavage map of SPO1 DNA: general location of early, middle and late genes. *J. Virol.* 31:156–71
88. Poliski, B., Greene, P., Garfin, D. E., McCarthy, B. J., Goodman, H. M., et al. 1975. Specificity of substrate recognition by the EcoRI restriction endonuclease. *Proc. Natl. Acad. Sci. USA* 72:3310–14
89. Riley, M., Anilionis, A. 1978. Evolution of the bacterial genome. *Annu. Rev. Microbiol.* 32:519–60
90. Romeo, J. M., Greene, J. R., Richards, S. H., Geiduschek, E. P. 1986. The phage SPO1-specific RNA polymerase, $E.gp^{28}$, recognizes its cognate promoters in thymine-containing DNA. *Virology* 153:46–52
91. Roscoe, D. H. 1969. Thymidine triphosphate nucleotidohydrolase: a phage induced enzyme in *Bacillus subtilis*. *Virology* 38:520–26
92. Roscoe, D. H. 1969. Synthesis of DNA in phage-infected *Bacillus subtilis*. *Virology* 38:527–37
93. Roscoe, D. H., Tucker, R. G. 1964. The biosynthesis of a pyrimidine replacing thymine in bacteriophage DNA. *Biochem. Biophys. Res. Commun.* 16:106–10
94. Roscoe, D. H., Tucker, R. G. 1966. The biosynthesis of 5-hydroxymethyldeoxyuridylic acid in bacteriophage-infected *B. subtilis*. *Virology* 29:157–66
95. Rosenberg, A. H., Simon, M. N., Studier, F. W., Roberts, R. J. 1979. Survey and mapping of restriction endonuclease cleavage sites in bacteriophage T7 DNA. *J. Mol. Biol.* 135:907–15
96. Sayre, M. H., Geiduschek, E. P. 1988. TF1, the bacteriophage SPO1-encoded type II DNA-binding protein, is essential for viral multiplication. *J. Virol.* 62:3455–62
97. Schnös, M., Inman, R. B. 1970. Position of branch points in a replicating λ DNA. *J. Mol. Biol.* 51:61–73
98. Sinsheimer, R. L., Knippers, R., Komano, T. 1968. Stages in the replication of

bacteriophage ΦX174 DNA in vivo. *Cold Spring Harbor Symp. Quant. Biol.* 33:443–47
99. Sjöberg, B.-M., Hahne, S., Mathews, C. Z., Mathews, C. K., Rand, K. N., et al. 1986. The bacteriophage T4 gene for the small subunit of ribonucleotide reductase contains an intron. *EMBO J.* 5:2031–36
100. Spatz, H. C., Trautner, T. A. 1971. The role of recombination in transfection of *Bacillus subtilis. Mol. Gen. Genet.* 113:174–90
101. Starlinger, P. 1977. DNA rearrangements in procaryotes. *Annu. Rev. Genetics* 11:103–26
102. Stewart, C. 1988. Bacteriophage SPO1. In *The Bacteriophages,* ed. R. Calendar, 1:477–515. New York: Plenum
103. Stragier, P., Losick, R. 1990. Cascades of sigma factors revisited. *Mol. Microbiol.* 4:1801–6
104. Truffaut, N. B., Revet, B., Soulie, M. 1970. Etude comparative des DNA de phages 2C, SP8, SP82, Φe, SPO1 et SP50. *Eur. J. Biochem.* 15:391–400
105. Tsao, Y. P., Wu, H. Y., Liu, F. L. 1989. Transcription-driven supercoiling of DNA: direct biochemical evidence from in vitro studies. *Cell* 56:111–18
106. Upholt, W. B. 1977. Estimation of DNA sequence divergence from comparison of restriction endonuclease digests. *Nucleic Acids Res.* 4:1257–65
107. Warren, R. A. J. 1980. Modified bases in bacteriophage DNAs. *Annu. Rev. Microbiol.* 34:137–58
108. Watson, J. D. 1972. Origin of concatemeric T7 DNA. *Nat. New Biol.* 239:197–201
109. Williamson, V. M., Doi, R. H. 1978. Delta factor can displace sigma factor from *Bacillus subtilis* RNA polymerase holoenzyme and regulate its initiation activity. *Mol. Gen. Genet.* 161:135–41
110. Woodbury, C. P., Hagenbuckel, O., Von Hippel, P. H. 1980. DNA site recognition and reduced specificity of the EcoRI endonuclease. *J. Biol. Chem.* 255:11534–46
111. Yehle, C. O., Ganesan, A. T. 1973. Deoxyribonucleic acid synthesis in bacteriophage SPO1-infected *Bacillus subtilis.* II. Purification and catalytic properties of a deoxyribonucleic acid polymerase induced after infection. *J. Biol. Chem.* 248:7456–63
112. Glassberg, J., Slomirny, R. Stewart, C. R. 1977. Selective screening procedure for the isolation of heat- and cold-sensitive, DNA replication–deficient mutants of bacteriophage SPO1 and preliminary characterization of the mutants isolated. *J. Virol.* 21:54–60

Annu. Rev. Microbiol. 1992. 46:117–39

CONTROL OF CELL DENSITY AND PATTERN BY INTERCELLULAR SIGNALING IN *MYXOCOCCUS* DEVELOPMENT

Seung K. Kim and Dale Kaiser

Departments of Biochemistry and Developmental Biology, Stanford University School of Medicine, Stanford, California 94305

Adam Kuspa

Department of Biology, University of California at San Diego, La Jolla, California 92093

KEY WORDS: sporulation, signal transduction, morphogenesis, amino acid signaling, gene regulation

---

CONTENTS

0066-4227/92/1001-0117$02.00

Dedicated to Dr. Roland Thaxter, on the centennial of his seminal publication on the myxobacteria.

Abstract

Myxococcus xanthus cells feed, move, and develop cooperatively. Genetic, biochemical, and cell mosaic studies demonstrate that cells coordinate their multicellular behavior by transmission of intercellular signals. Starvation for amino acids at sufficiently high density on a solid surface initiates a series of events culminating in the formation of a multicellular structure called a fruiting body filled with dormant, environmentally resistant spores. This review discusses how myxobacteria use extracellular signals to sequentially check the density and arrangement of cells at different stages during development. For at least one early and one late developmental signal, cell density determines the efficiency of intercellular signaling. In turn, proper signaling insures that the appropriate cell density exists, thus controlling the progress of multicellular development in *M. xanthus*.

INTRODUCTION

Researchers have known for a long time that bacteria sense their surroundings in order to adapt to environmental change. Within colonies or in the dense bacterial populations that grow in surface films, other cells constitute a major part of that environment (10, 77). Also, bacteria clearly generate chemical and physical stimuli sensed by other bacteria, thereby establishing regulatory interactions between cells. If such an interaction performs an essential function, it might be revealed by dependence on cell density. Myxobacteria are notable for their cell interactions, which in fact are necessary for them to feed, move, sporulate, and construct multicellular fruiting bodies. This review aims to describe how *Myxococcus xanthus* monitors its relations to other cells in order to regulate its growth and development. Information about myxobacterial research may also be found in other review articles (13, 34, 35, 41, 79, 92) and a monograph (75).

Density-Dependent Growth

Myxobacteria are gram-negative gliding bacteria common in soil that feed with extracellular bacteriolytic, proteolytic, and other digestive enzymes (12). Rosenberg et al (76) measured the growth rate when the only source of carbon and nitrogen for *M. xanthus* cells in liquid culture was the polymeric substrate casein, so that proteolysis was the limiting first step. A twofold increase in growth rate was observed as the cell density rose above 10^4 cells/ml. When enzymatically prehydrolyzed casein was provided, the cells grew at the more rapid rate independent of cell density. At densities below 10^3 cells/ml, no colonies formed on agar with undigested casein. Evidently, extracellular digestion of protein is enhanced by cooperation between cells. The molecular basis of density-dependent growth is not yet clearly understood, but the higher local concentration of enzymes might increase the efficiency of hydrolysis. The growth advantage of cooperative feeding may have been the selective force behind the evolution of myxobacteria, which were multicellular at least one billion years ago (34, 66). In these terms, a swarm may be the unit of efficient cooperative feeding; a fruiting body may contain a number of spores sufficient for efficient dispersal and for rapid growth upon germination when nutrient is no longer limiting.

Control of Cell Movement

Myxobacteria move by gliding, which permits them to translocate on surfaces (6). Several hundred mutants have been isolated in *M. xanthus* that change the arrangement of cells in a gliding swarm (26). Observation of these mutants has revealed two sets of cell-pattern genes, each producing a characteristic cell distribution. Many single cells are evident in the right panel of Figure 1, where the active pattern genes belong to the set called "A" for adventurous. This pattern arises because single A-motile cells can move even when they are well separated from all other cells. Twenty-three A-motility genes are known (15, 25, 26, 87). Some A-motility genes are involved in the synthesis of cell surface lipopolysaccharides (15). Insertions of Tn*phoA* within A-motility genes results in secretion of alkaline phosphatase activity, suggesting that some A-motility gene products are exported to the cell membrane or periplasmic space (37, 58).

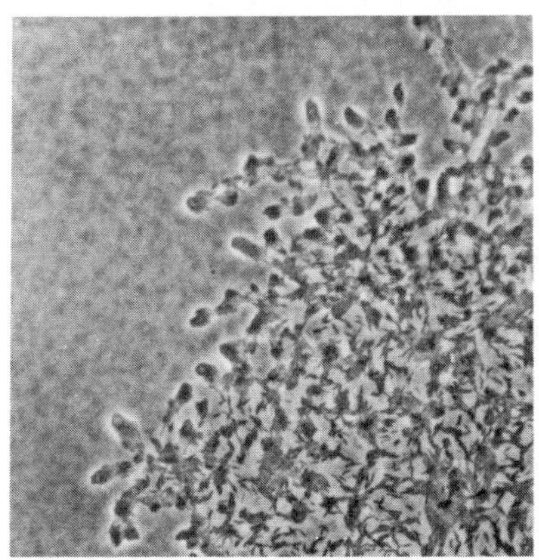

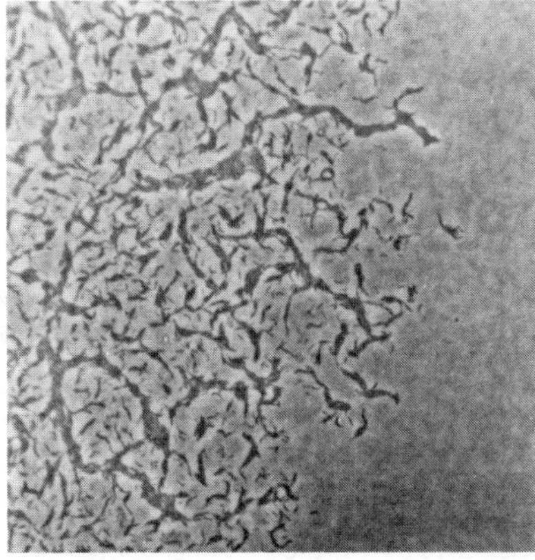

Figure 1 Two motility patterns: A, or adventurous, motility pattern (*right panel*) and S, or social, motility pattern (*left panel*). Reprinted, with permission, from Ref. 26.

The second set of pattern genes generates what is called social motility, because S-motile cells cluster in many groups of 10–20 cells with few isolated cells (left panel in Figure 1). S-motile cells move only if they are within a cell's length of another cell; consequently, the S pattern has almost no isolated cells (5, 36). At this time, 10 S-motility genes are known, but this gene system is not yet saturated genetically. One of these genes, *dsp,* is needed for production of an extracellular meshwork of fibrils (1; Behmlander & Dworkin, cited in 14), to which proteins bind (14, 32). The remaining nine are needed for the production of pili, which are located at one pole of *M. xanthus* cells (33). One of these nine, *tgl,* specifies a cell-surface lipoprotein (73).

Wild-type cells use both systems simultaneously. To obtain the two cell patterns shown in Figure 1, any gene of the A set is inactivated to give an A^-S^+ strain that has pure S-motility (left panel); likewise when any gene of the S set is inactivated to give an A^+S^- strain, pure A-motility remains (right panel) (26). More than 50 A^-S^- double mutants have been built with mutations in different A and S genes. All are nonmotile. Nonmotile cells may also arise following a single mutation of the *mglA* gene (89).

The A and the S cell arrangements are generated through interactions between cells. This is evident from the cell density dependence of the rate of movement of A- and of S-motile cells shown in Figure 2. In the figure, swarm expansion rate provides a global measure of cell movement. The ability of A-motile cells to move autonomously is reflected in a base-line rate of about 0.2 μm/min at the lowest cell density. A-motility increases rapidly as the cell density increases, while S-motility increases more gradually. S-motility starts from a zero base-line, reflecting the complete dependence on neighboring cells. S-motility continues to increase at intermediate cell densities where A-motility has saturated.

Wild-type (A^+S^+) cells show the 0.2 μm/min base-line characteristic of A-motility and its rapidly increasing response at low cell density. Wild-type also shows a slowly increasing response in the mid-density range, like S-motility. In these respects, A^+S^+ cells combine the A and S patterns. However, either A- or S-motility alone generates about 0.5 μm/min of movement (Figure 2). Wild-type cell movement reaches a maximum of 1.5 μm/min of movement, about 50% more than the sum of A and S, and the pattern includes a wider variety of cell clusters, both suggesting synergism between two active systems.

A second kind of density-dependent cell movement in *M. xanthus* is found with the stimulatable motility mutants. When certain pairs of nonmotile strains are mixed together on agar, one or both become temporarily motile and temporarily piliated (24–26, 33). Stimulation was observed only when cells of the two types touched each other, and when one strain carried a *cgl* mutation (in motility gene set A), or a *tgl* mutation (in motility gene set S), and the

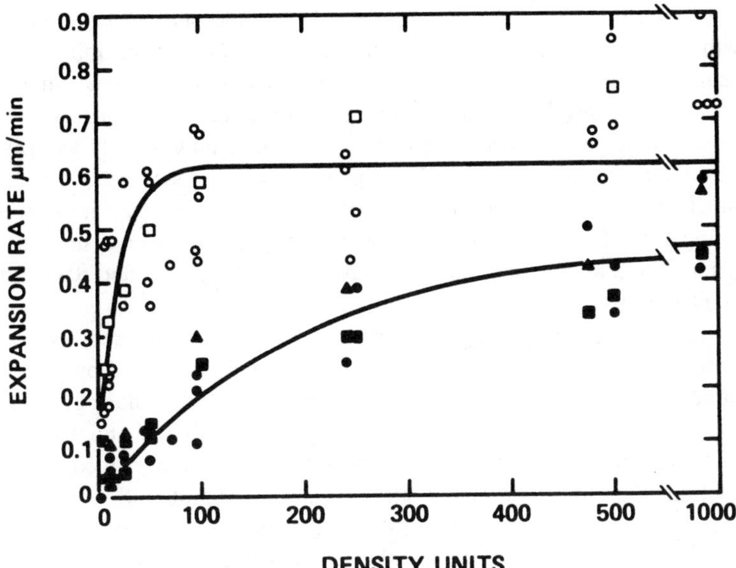

Figure 2 Rate of swarm expansion versus cell density for three A^-S^+ strains (*closed symbols*) and two A^+S^- strains (*open symbols*). The rate of swarm expansion was measured at cell densities ranging from 2.5 to 1000 density units. Each point represents the slope of a line plotting the linear increase in the radius of a circular spot of cells versus time. Modified, with permission, from Ref. 36.

other strain carried the corresponding cgl^+ or tgl^+ alleles (24). Stimulation is thus genetically constrained and requires close proximity of the two interacting cell types. Although the molecular basis of density-dependent motility stimulation is not yet understood, this phenomenon appears to play a role in the development of fruiting bodies (38, 47).

Six *frz* genes control the frequency of reversal of gliding direction (2, 3). The proteins encoded by all the *frz* genes except *frzB* are similar in sequence to those encoded by the chemotaxis (*che*) genes of *Escherichia coli* and *Salmonella* spp. (61). The *che* genes specify a protein phosphorylation cascade, and amino acids in the carboxy-terminal region of the *frzCD* protein are methylated and demethylated in response to starvation, as are the methylated chemotaxis proteins of enteric bacteria during adaptation (62, 63). An effect of cell-density on *frz* gene function has not yet been reported, but the *frz* sequences strongly suggest regulatory roles in cell movement.

Fruiting Body Morphogenesis and Development

M. xanthus builds fruiting bodies about 0.2 mm high that contain about 10^5 cells. Fruiting body development is induced when the available nutrient supply falls short. In addition, development requires a high cell density and a solid surface on which cells can glide (81). At the start of development, growth slows, and after 4 h the cells begin to congregate. Initially, the aggregates are asymmetric, but by 12 h thousands of cells have accumulated and the aggregates have become elliptical or circular mounds. Early in aggregation, ridge-like accumulations of cells move coordinately and rhythmically over the fruiting surface like ripples on a water surface (70, 83). At about 20 h, cells inside the mound progressively differentiate into spores, changing from 5 μm $\times$ 0.5 μm rods that have gram-negative structure to dormant spherical cells that have several coats and are heat and desiccation resistant (28, 91).

While the cells are aggregating and sporulating, new proteins are synthesized. More than 30 proteins, identified as bands separated by gel electrophoresis, and new cell-surface antigens become evident during development (16, 17, 27). Monoclonal antibodies to some of the antigens block development and these antigens may be involved in contact-mediated cell interactions (32). At 6 h post starvation, synthesis of protein S, a spore-coat protein, begins; later this protein self-assembles into the outermost spore coat (28–30, 92). Myxobacterial hemagglutinin (MBHA), a protein with lectin activity, starts to appear at 10 h (74). In addition, more than 29 promoters increase their activity according to a regular time schedule (50). These promotors can be monitored by an insertion of Tn*5 lac,* a transposable promoter probe that forms transcriptional fusions (48, 49). Figure 3 shows how the program combines differential gene expression and morphological change.

Cell-to-cell signals govern progress through this developmental program. Several signals were first implied by the characterization of four complementary classes of nonautonomous developmental mutants (22, 31, 57, 59, 81). These mutants were selected using cell-density dependence to find nonautonomous mutants, including those deficient in signal production. Mutants of the *asg, bsg, csg,* and *dsg* classes cannot sporulate when they are alone, but sporulation of the mutants can be rescued by codevelopment with wild-type cells (22). Each class of mutants modifies the program in a different way. Assay of β-galactosidase activity in the *lacZ* fusion strains shows that the developmental program of gene expression and morphogenesis is disrupted at different stages in *asg, bsg, csg,* and *dsg* mutants (as illustrated for *asg* and *csg* mutants in Figure 3) (43, 54, 59). These stages suggest a different time of action for each class of mutants (7, 19, 54, 59). Signal molecules for the *bsg* and *dsg* classes have not yet been identified, but the *bsgA* gene has been shown to specify an ATP-dependent protease (18), and *dsg,* a vital gene for *M. xanthus* (7, 8), shares 50% amino acid sequence identity with *E. coli* and *Bacillus sterothermophilus* translation initiation factor 3 (Y. Cheng, L. Kalman & D. Kaiser, unpublished observations).

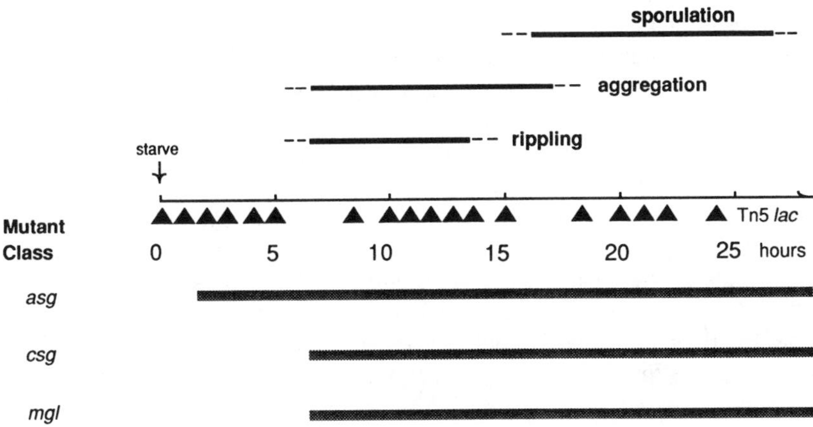

Figure 3 Morphologic events and patterns of gene expression during *M. xanthus* fruiting body development. Developmental time starting with starvation at 0 h is marked on the horizontal axis. Rippling and aggregation occur before sporulation. Broken lines denote that the exact timing of events depends on experimental conditions. Triangles denote gene expression measured by *lacZ* transcriptional fusions. Each class of developmental mutant arrests development at characteristic stages, as indicated by terminal morphology and by the set of *lacZ* transcriptional fusions expressed in the mutant. Developmental events or gene expression never achieved by each mutant class are denoted by the adjacent stippled bars.

Identification of the *asg* and *csg* signal molecules from wild-type cells has proven that these two classes are signal deficient. The sequence of steps in the developmental program can be determined by studying transcription in the signaling mutants. Figure 3 summarizes results of experiments showing which particular *lacZ* fusion strains are affected by *asg* and which by *csg*. If the *asg* and *csg* points of action are on the same regulatory sequence and if the transmission of A-signal starts earlier than the transmission of C-signal, then transcripts independent of A should be independent of C. As illustrated in Figure 3, all *asg*-independent *lacZ* transcriptional fusions are *csg* independent; all *csg*-dependent fusion strains are *asg* dependent; some *asg*-dependent strains are *csg* dependent and others are *csg* independent. These are the results expected if *asg* precedes *csg* on the same dependent developmental pathway, and they imply a series of sequential extracellular controls that monitor the progress of development (50).

A-FACTOR

A-Signaling Early in Development

Nearly half of the signaling mutants recovered by Hagen et al (22) belonged to the *asg* complementation class. Thirteen alleles were mapped to three genetic loci, *asgA*, *asgB*, and *asgC*, by means of linked transposon Tn*5* insertions

(53, 60). When starvation is severe, as in submerged culture development (51), *asgA* and *asgB* mutants produce 10^5-fold fewer spores than wild-type, and the *asgC* mutant produces 50-fold fewer spores than wild-type (53). The *asgA* and *asgB* cells do not aggregate at all, while *asgC* cells form small irregular aggregates that do not mature into fruiting bodies. Despite the severity of the mutant defects, addition of wild-type cells restores aggregation and the sporulation level of the mutants to near wild-type.

The *lacZ* fusions described above were used to find the time at which the *asg* mutations arrest development (54). The earliest affected fusion, Tn*5 lac* Ω4521, is expressed at 1–2 h by developing wild-type cells. Expression from the Ω4521 gene, as measured by β-galactosidase activity or accumulation of mRNA, rises more than 20-fold over vegetative levels in wild-type cells, but increases less than twofold in *asgA, B,* and *C* mutants (40, 54). Several *lacZ* fusions expressed earlier than Ω4521 are expressed normally in the *asg* mutants, indicating that the mutants begin to develop, but stop after about 2 h. Correctly timed expression of *asg*-dependent genes, including Ω4521, is restored in *asg* mutants when they are mixed with wild-type cells. Significantly, the expression of these genes is not rescued appreciably by mixing with other *asg* mutant cells (54, 60). The finding that expression of the earliest *asg*-dependent gene, along with later genes, is rescued with normal timing indicates that the rescued developmental program continues smoothly through the stage at which it would be arrested by the *asg* mutation in the absence of wild-type cells.

The rescue of Ω4521 gene expression provides a convenient assay for the A-signal molecules. It allowed the identification of a set of substances, collectively termed *A-factor,* that are released by wild-type cells and that rescue Ω4521 gene expression (54). A-factor is found in the medium of developing wild-type cells at the time (2 h) of the *asg* mutant arrest, and *asg* mutants release less than 5% of the wild-type levels of A-factor at this time (53, 54). A-factor from wild-type cells also restores the capacity of the *asgB* and *asgC* mutants to aggregate and sporulate, and restores the capacity of the *asgA* mutants to aggregate (54; A. Kuspa, unpublished observations).

Identification of A-Factor

The rescue of Ω4521 gene expression in an *asgB* mutant suspended in developmental buffer (the A-factor assay), was used as the primary assay for A-factor purification. A-factor is released from wild-type cells along with a complex mixture of small molecules, lipopolysaccharide, and protein (52). In this mixture, half of the A-factor activity is heat labile, and half is heat stable. The mixture from wild-type cells includes a particular set of amino acids at the time of A-factor release, and several pure amino acids have high A-factor–specific activity (55). Each of the high-specific-activity amino acids, added alone at concentrations of 100 to 500 μM, rescue Ω4521 gene expression to near-normal levels. Of the amino acids released by wild-type cells, proline, tyrosine, phenylalanine, tryptophan, leucine, isoleucine, and alanine have the highest activity and can account for about half of the heat-stable activity in crude A-factor. Extracellular proteolysis of developmentally released protein is observed in crude A-factor preparations in vitro, producing amino acids and peptides (56). Small peptides also have heat-stable A-factor activity in direct proportion to the activity of their constituent amino acids. Thus, heat-stable A-factor is a set of amino acids and a mixture of peptides containing these amino acids.

The finding that heat-stable A-factor is a mixture of amino acids and peptides is supported by the finding that heat-labile A-factor consists of at least two different proteases (69). Heat-labile A-factor proteins of 27 and 10 kilodaltons (kDa) have been identified by purification from crude A-factor preparations, and each has proteolytic activity with a distinct substrate specificity. These two proteins represent neither all of the heat-labile A-factor nor all the proteolytic activity in crude A-factor, so additional A-factor proteases probably exist. In fact, pronase, chymotrypsin, trypsin, and proteinase K have significant A-factor activity. Bovine trypsin has about half the A-factor specific activity as the highly purified 27-kDa myxobacterial protein. Inhibition of trypsin proteolytic activity by a specific trypsin inhibitor protein inhibited trypsin's A-factor activity, demonstrating that trypsin's proteolytic activity is required for its A-factor activity (69). Other tests were used to assess the ability of proteases and amino acids to serve as A-signal. Single amino acids or pure proteases restore normal aggregation to all of the *asg* mutants, and restore sporulation in the *asgB* and *asgC* mutants to near wild-type levels (56, 69). These assays confirm that amino acids and proteases can restore all the developmental defects of *asg* mutants.

The proteases with A-factor activity have different substrate specificities, suggesting that proteases do not signal by cleaving a specific bond but rather by producing extracellular amino acids. Because pronase hydrolyzes proteins to their constituent amino acids, its A-factor activity probably does not result from the activation of another protein by proteolytic cleavage. Since all (six) proteases tested have A-factor activity, but only particular amino acids have A-factor activity, it follows that these amino acids are the primary A-signal molecules missing in *asg* mutants, while the extracellular release of proteases and proteins by wild-type cells produces these amino acids. The amounts of amino acids released by wild-type cells are sufficient to account for the observed level of Ω4521 expression (52).

Below a threshold concentration, amino acids cannot rescue Ω4521 gene expression in the A-factor assay (56). For amino acids with high A-factor specific activity, this threshold concentration is 10 μM. This concentration may indicate the binding affinity of an amino acid receptor or may reflect the lower limit of an amino acid transporter under these conditions. Either way, a threshold suggests that cells use the low signaling levels of extracellular amino acids to, in effect, measure the cell density of the starving population. If cells are present at sufficient density to produce amino acids above the threshold concentration, they will continue to develop.

Several lines of evidence show that extracellular amino acids specify the minimum cell density required for the development of wild-type cells. First, single amino acids with high A-factor activity are able to elevate the expression level of the Ω4521 gene in developing wild-type cells (52), arguing against a simple nutritional effect on protein synthesis. Second, inhibition of (auto) proteolysis in crude A-factor inhibits A-factor activity to the same degree (52; A. Kuspa, unpublished observations). For example, 10 μM *o*-phenanthroline inhibits proteolysis and A-factor activity in crude A-factor by about 50%, and when added to developing wild-type cells, *o*-phenanthroline inhibits development completely and Ω4521 gene expression by 90% (52). This inhibition of Ω4521 gene expression is reversed by exogenous amino acids, suggesting that *o*-phenanthroline blocks development by inhibiting extracellular proteolysis.

Finally, when wild-type cells are diluted to the point that the calculated concentration of released extracellular amino acids falls short of threshold levels, Ω4521 gene expression decreases 80–90%, and aggregation and sporulation are reduced (56). Diluted wild-type (asg^+) cells behave like *asg* mutants. The expression of the Ω4521 gene by wild-type cells in dilute suspension can be restored by amino acids at the same concentrations needed to rescue the *asg* mutants. Proteases also rescue Ω4521 expression in diluted wild-type cells (56). These last results indicate that wild-type cells cannot develop below a critical cell density and that extracellular amino acids are monitored by the cells early in development to assess their density.

Cellular Response to A-Factor

Genes involved in the response to A-signal have been identified genetically. Mutants of an *asgB* strain carrying the Ω4521 *lacZ* fusion have been isolated that express Ω4521 in the absence of A-factor (40). Fifteen of the 17 mutants recovered map to one of two loci, *sasA* and *sasB* (for suppressor of *asg*). None of the *sas* mutations restore the capacity to produce A-factor nor do *sas* mutations appear in the Ω4521 gene itself; they are second-site suppressors of the *asgB* defect. Although the same *sasA* mutation restores Ω4521 expression in an *asgA*, *asgB*, or *asgC* mutant, neither *sasA* nor *sasB* mutations restore normal fruiting body development and sporulation. Supressing mutations in *sasA* and *sasB* were isolated at the same frequency as loss-of-function mutations in the *carR* locus (59) in the same experiment, and because *sasA* mutations are recessive, the *sas* alleles probably represent loss-of-function mutations. Since *sas* mutants render Ω4521 expression independent of A-factor, the *sasA*$^{+}$ and *sasB*$^{+}$ gene products apparently provide a function that represses Ω4521 gene expression, and A-factor appears to antagonize the repression of Ω4521 expression by *sasA* and *sasB*. In accord with these proposals, *sasA* mutations in an *asg*$^{+}$ background cause a developmental defect that reduces sporulation 10- to 100-fold and alters the expression of at least five developmental genes including Ω4521 (40). The Ω4521 gene expression displays normal developmental timing in *sasA-asgB* or *sasB-asgB* double mutants, indicating that another regulatory input independent of A-signaling also controls expression of the Ω4521 gene. Figure 4 shows a model for the action of amino acids as A-signal in the regulation of early *M. xanthus* development.

A-Signal Specificity

Beyond their service as substrates for polypeptide synthesis, extracellular amino acids are also used as a cell-cell signal in *M. xanthus* (Figure 4). This dual role is possible in early development for the following reasons. *M. xanthus* requires a threshold concentration of amino acids for growth; below the threshold, growth is arrested as fruiting body development is initiated. The threshold level of an equimolar mixture of the 20 common amino acids is approximately 10 mM for growth, while it is 0.05 mM for A-signaling (55, 56). Given these conditions, several other pieces of evidence highlight a signaling as opposed to a nutritional function. Individual amino acids that are not known to be converted to all other amino acids (tryptophan, for example), can nevertheless restore Ω4521 gene expression to an *asg* mutant. A-signaling levels of amino acids not only increase the production of β-galactosidase

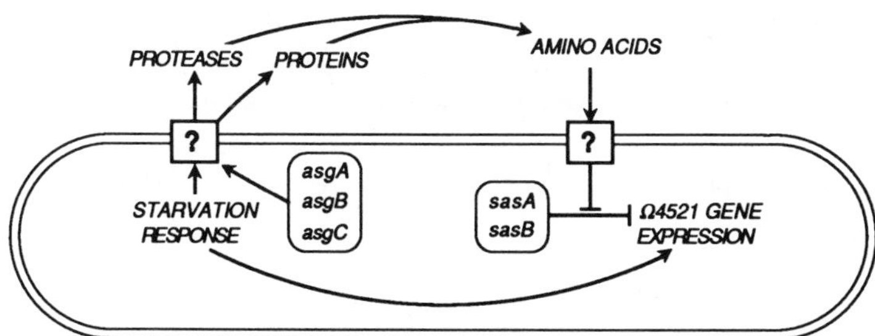

Figure 4 Model of A-factor signaling. A single *M. xanthus* cell illustrating the relationship between the functions of the various genes and signaling molecules is shown. Arrows indicate positive biochemical or genetic functions, and lines ending in bars indicate negative biochemical or genetic functions. The boxes indicate that the source of the proteins and proteases, and the mechanism of action of the amino acids, are not known. See text for details.

protein from the Ω4521 *lac* fusion, but also rescue the accumulation of Ω4521 RNA (40).

The use of amino acids for growth substrates and for developmental signaling raises a question of signal specificity. When amino acids are abundant in the extracellular environment during growth, what prevents them from prematurely signaling *asg*-dependent gene expression? Apparently, the answer is that for their expression, *asg*-dependent genes also depend on starvation (40). Thus, when amino acid levels are high enough to allow growth, Ω4521 is not expressed, and when these levels fall below the growth threshold, but remain above the signaling threshold, Ω4521 is expressed (55). Specificity for growth versus development may be provided temporally, perhaps through the action of the products of the *asg* genes or other *asg*-independent genes. During the first two hours of starvation, these early gene products may change the way the cells respond to low levels of extracellular amino acids. For instance, some of the *asg*-independent genes might encode parts of an amino acid receptor system, or they might encode new proteases. In fact, *asgA* has the amino acid sequence of a two-component-system sensor element (4; L. Plamann, personal communication).

The data suggest that *M. xanthus* monitors extracellular amino acids during the early development of starving cell populations to determine if conditions are appropriate for development. Only an appropriate amino acid level produced by cells at a particular cell density allows continued development of the population. Even when the nutrient level is below the normal growth threshold, if the cell density is too low for A-signaling, the cells begin to grow slowly. This was shown by LaRossa et al (57), who noticed that an *asg* mutant continued to grow on a minimal solid medium lacking phenylalanine, whereas wild-type cells (at high density) arrested growth and initiated development. Cell densities insufficient for development would be self-correcting in wild-type cells because growth may succeed in raising the cell density above the critical level for A-signaling. Evidence for intercellular signaling that specifies the cell density appropriate for continued development has been found in a wide variety of organisms, including bacteria, cellular slime molds, worms, and frogs (9, 20, 21, 72). Note added in proof: A density sensing factor in *Dictyostelium* has recently been identified (93).

C-FACTOR

Like the *asg* mutants, *csg* mutants were isolated as nonautonomous developmental mutants that fail to develop alone, but aggregate and sporulate when mixed with wild-type cells (22). However, *csg* mutants are blocked at a later developmental stage than *asg* mutants as revealed by patterns of morphogenesis and biochemical studies of developmental gene expression (Figure 3) (59). For instance, *csg* mutants produce A-factor (52). *M. xanthus csg* but not *asg* mutants synthesize developmental proteins S and H (57). All but three *lacZ* fusions tested absolutely require *asg* functions, whereas in *csg* mutants, production of β-galactosidase from *lacZ* fusions expressed before 6 h of development is normal but is reduced or abolished after 6 h (Figure 3) (43, 44, 59). These observations suggest C-signal functions on the same regulatory pathway as A-signal.

C-Factor Is a Short-Range Late Developmental Signal

Fruiting body morphogenesis is a starvation-induced multicellular process in which sporulation is delayed until the proper number of cells have assembled in the right way. Once cells in an undifferentiated swarm have detected amino acid limitation and complete A-signaling, they move coordinately to aggregation centers. There, motile rod-shaped cells synchronously differentiate to nonmotile dormant myxospores. How do cells verify that morphogenesis is complete so that differentiation can begin? Several lines of evidence indicate that the cell density-dependent intercellular transmission of C-factor, the protein product (42, 43) of the *csgA* gene (23, 80, 82), coordinates the later morphogenetic events of fruiting body formation.

Unlike wild-type cells, which aggregate into compact mounds by 12 h on a defined starvation agar medium, *csgA* mutants form diffuse mounds and ridges of cells only after 18 hours (84). Under the stringent starvation conditions of submerged culture (51), *csgA* cells fail to form detectable multicellular structures (Figure 3). Also, *csgA* mutants fail to ripple or to sporulate (84). Mixture with wild-type cells corrects all of the developmental defects of *csgA* mutants (59).

All existing members of the *csg* class result from mutation of a single genetic locus called *csgA* (80, 82). The DNA sequence indicates that *csgA* specifies a 17-kDa protein (23). The amino-terminal sequence of the *csgA* gene product resembles the signal sequence used by the enteric bacteria; otherwise no significant similarities to other protein sequences exist (23; S. K. Kim, unpublished observations). Purified polyclonal antibodies raised against a *lacZ-csgA* translational fusion protein produced in *E. coli* reveal a 17-kDa polypeptide in extracts made from *csgA*$^+$ cells but not from vegetative or *csgA*$^-$ cells (42, 85). Antibodies specific for the *csgA* gene product also block developmental aggregation and sporulation of wild-type cells (85).

Rescue of the developmental defects of mutants by mixture with wild-type cells provided a bioasssay for monitoring purification of the molecules responsible for rescue (43). A cell-associated protein with an apparent molecular weight of 17,000 was purified from extracts of nascent fruiting bodies (42). This protein, named C-factor, rescues *csgA* mutant aggregation, sporulation, and gene expression to wild-type levels when added at 1 nM. The size, amino acid sequence, pattern of expression, and the recognition of

C-factor by purified anti-*csgA* antibodies demonstrate that C-factor is the product of the *csgA* gene (42, 43, 85).

C-Signal Transmission and Response

The requirements for transmission of C-signal from cell to cell are different from the requirements for intercellular A-signal transmission. A-factor activity is recovered from supernatant washes of starved wild-type cells, in contrast to C-factor, which is found in the pelletable material following ultracentrifugation (42, 69). Separation of wild-type and *asg* strains by a porous membrane that is impermeable to cells does not prevent rescue of *asg* mutant development. When separated from wild-type cells by such a membrane, however, *csg* mutants do not develop (43). Thus A-factor functions as a diffusable signal that can act over relatively long distances, whereas C-factor appears to function over a short range and requires close contact of producing cells and responding cells.

One can also distinguish between transmission of A-signal and C-signal based on their requirements for cell motility. Expression of an A-dependent *lacZ* fusion such as Ω4521 is normal in a cell rendered nonmotile by mutation in the *mglA* gliding gene (47, 89), implying that A-factor transmission is independent of cell motility. However, expression of C-dependent *lacZ* fusions in nonmotile cells is reduced or abolished to the same degree as expression in *csgA* cells (Figure 3). Nonmotile cells, like motile *csgA* cells, fail to aggregate, ripple, or sporulate. These observations suggest that motility is necessary for proper C-signaling and that nonmotile cells fail to develop because they cannot complete C-signaling (47).

In confirmation of this hypothesis, C-factor purified from wild-type cells restores sporulation and gene expression to wild-type levels in *mglA* strains. Nonmotile cell aggregation is not restored because C-factor does not rescue the basic motility defect. However, mixture of intact wild-type cells with *mglA* cells does not rescue the developmental defects of *mglA* cells despite opportunities for cell contact in these mixtures ensured by high cell density. In addition, although wild-type concentrations of active C-factor can be purified from *mglA* cells, intact *mglA* cells cannot rescue adjacent *csgA* cells (44). These results indicate that C-signal transmission between cells during development requires the motility of both the donor and responder cells.

Why does normal intercellular transmission of C-signal depend on cell movement? One consequence of directed cell movement during fruiting body morphogenesis is to increase the local cell density by 10- to 50-fold. A roughly spherical nascent fruiting body with a 100-μm diameter contains about 100,000 rod-shaped cells organized in cohesive parallel arrays at a density of about 2×10^{11} cells per cm^3. At this density, cells are arranged end-to-end and side-by-side, much like close-packed bricks in a wall (51, 67).

These observations suggest that efficient C-signal transmission might require cell movement to create dense-packed mounds of cells at such high density.

In support of this hypothesis, Kroos et al (47) showed that a 10-fold increase in the density of nonmotile cells by sedimentation results in a 1000-fold increase in sporulation, to 1% of wild-type levels (Figure 5) (47). However, cells sedimented to a density of about 5×10^{10} cm^{-3}, the limits of this method, remain organized in very small randomly oriented groups of side-by-side cells. C-dependent gene expression in these sedimented cells does not increase above background levels (45).

To reconstruct the cellular organization found within a fruiting body, nonmotile cells were placed on a solid developmental surface that had been scored in one dimension with 5- to 10-μm aluminum oxide abrasive paper to create microscopic grooves. The aim was to test the idea that cell alignment, normally resulting from directed cell movement, is crucial for C-factor transmission. Nomarski optics reveal that the long axis of nonmotile cells oriented within grooves is parallel to the axis of the groove. Both C-dependent gene expression and sporulation are evident among cells so aligned. This treatment results in sporulation of nonmotile cells at 16% of wild-type levels (Figure 5). No increase in sporulation is detected following similar alignment of isogenic *csgA* cells (45).

How are the ordered cell arrays necessary for C-signaling stabilized, once formed? The cohesiveness of cells progressively increases during develop-

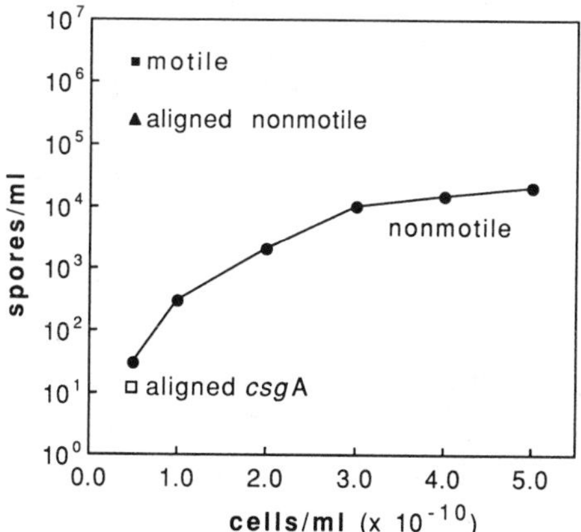

Figure 5 *M. xanthus* spore differentiation dependent on cell density and position. Heat-resistant, viable spores were quantitated. The sporulation defect of nonmotile cells is rescued to 1% of wild-type (motile) levels by a 10-fold increase in the density of cells induced to develop [data from Kroos et al (47)]. *Closed square,* motile, wild-type; *closed triangle,* aligned nonmotile cells in grooves; *closed circles,* unaligned nonmotile cells at increasing concentrations; *open square,* aligned *csgA*. Modified, with permission, from Ref. 45.

ment (78), and within a nascent fruiting body aligned cells appear to be more cohesive than unaligned cells (B. Sager & D. Kaiser, unpublished observations). One possible set of developmental cell surface molecules that may stabilize transient cell aggregates establishing the aligned state includes those specified by the S motility system. For example, *dsp* mutants, which are defective in the S motility system, are not cohesive, lack social motility, and fail to form ripples, fruiting mounds, or spores (78). Cohesion of *dsp* cells is rescued by a 1:1 mixture with wild-type cells. When mixed with developing wild-type cells, *dsp* mutants accumulate in fruiting mounds, but the *dsp* mutant sporulation defect is not rescued (the alignment of cells within a mound formed by these cell mixtures was not examined). Perhaps some motility functions are important for aggregate formation while others are necessary for appropriate alignment and subsequent sporulation. Developmental lectins such as MBHA comprise another type of surface molecule that might influence cell alignment. MBHA, which may comprise as much as 1–2% of the total cell protein during development (11), appears to accumulate at the poles of developing cells (65), and aggregation of cells that fail to produce this protein is delayed when magnesium levels are low (74).

The response of cells to C-factor involves modulating the activities of over 50 gene products whose functions include cell motility (7, 15, 26, 62), cohesion (78), aggregation (68), and cell morphology (28). Genes possibly involved in the response to C-signal have also been identified by genetic selection for cells with suppressors of the *csgA* defect (71). Second-site suppressor mutations have been mapped to six different loci and exhibit a variety of phenotypes. One group appears to bypass one or more intermediate steps in the developmental pathway as monitored by *lacZ* gene fusions, yet forms spore-filled fruiting bodies. Another group of suppressors restored all gene expression tested to levels seen in wild-type cells, yet failed to form fruiting bodies.

C-Factor Is a Density-Dependent Differentiation Signal

Early in development, following successful transmission of A-factor, C-factor stimulates aggregation of cells. The dependence of intercellular transmission on cell alignment may be less stringent at this stage, as cells are more randomly oriented and in less dense arrangements. Following aggregation, cells achieve a geometry that favors maximum cell packing, resulting in efficient C-signal transmission. As a result of C-factor signaling, cells then sporulate. Thus, cell alignment appears to function as a regulatory event during fruiting body morphogenesis.

How might cells in a nascent fruiting body sense as little as a fourfold increase in local cell density, the approximate fourfold difference between 5 $\times$ 10^{10} cells per ml (the concentration of nonmotile cells packed by centrifugation) and 2×10^{11} cells per ml (the concentration of cells aligned in a fruiting body)? One simple possibility is that increased cell density is measured by cells as an increase in local C-factor concentration. Transcription of *csgA* increases about three- to fourfold prior to the onset of sporulation, accounting for the fourfold increase in recoverable C-factor activity made by wild-type cells. Overall, the local C-factor concentration might increase by 12- to 16-fold, a product of increased *csgA* transcription and increased cell density (46).

How small a change in C-factor concentration can *M. xanthus* cells detect? Recently, we have found that, depending on its concentration, C-factor can elicit two distinct morphogenetic and transcriptional responses from *csgA* cells (Figure 6) (46). C-factor at 0.6 nM permits neither a detectable morphological response nor increased C-dependent transcription by *csgA*$^-$ cells. C-factor addition to a total of 0.8 nM rescues *csgA* cell aggregation and allows full expression of the C-dependent *lacZ* fusion Ω4499, normally expressed in aggregating cells. This amount of C-factor, however, fails to rescue *csgA* sporulation or expression of the C-dependent fusion Ω4435, which is primarily expressed in sporulating cells (59). However, addition of 1 nM purified C-factor leads to expression of both early and late C-dependent gene expression, aggregation, and sporulation. These data suggest that cells can distinguish and respond in a qualitatively different way to as little as a 25% change in C-factor concentration.

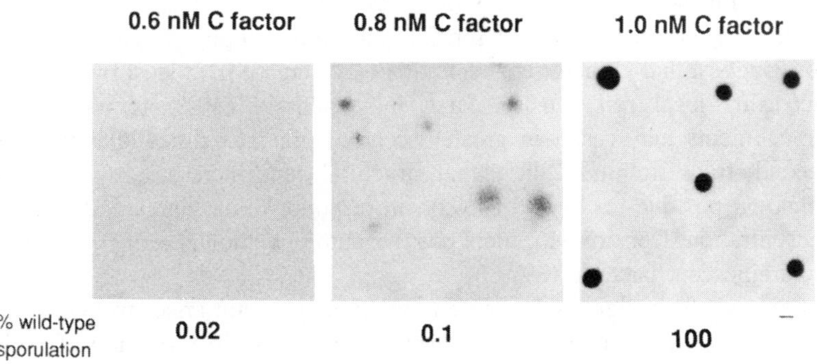

Figure 6 Aggregation and sporulation are initiated at different threshold concentrations of C-factor. Panel under the heading "0.6 nM C factor" shows that *csgA* cells receiving that amount of C-factor fail to aggregate. Panel under the heading "0.8 nM C factor" shows formation of *csgA* aggregates after 48 h. Panel under the heading "1.0 nM C factor" shows formation of darkened spore-filled fruiting bodies by *csgA* cells after 48 h. The levels of heat-resistant *csgA* spores formed after 6 days of the indicated treatments are presented below the panels. Bar equal to 100 μm.

The finding that C-factor can trigger sporulation and aggregation raises the possibility that the *csgA*-mediated interaction links these two developmental events. Previous studies of developmental mutants that either fail to aggregate but sporulate normally or fail to sporulate but aggregate normally led to the proposal that aggregation and sporulation result from two independent parallel pathways during fruiting body formation (64, 71). However, *asg, bsg, csg, dsg,* and *mglA* mutants fail to sporulate or aggregate, implying a pathway with functions common to both aggregation and sporulation. Yet during normal fruiting body morphogenesis, aggregation precedes sporulation. How is sporulation delayed until cells have aggregated?

C-factor is required to complete both aggregation and sporulation, and the temporal order of these developmental events may depend on the movement of cells into aligned patterns, creating an appropriately high cell density for efficient C-factor transmission. Midway in development, C-factor stimulates aggregation of cells. During aggregation, cells achieve a density that allows C-factor to trigger sporulation. This density dependence insures that cells normally sporulate only in groups.

C-Signal Amplification

An increase in the local C-factor signal concentration resulting from aggregate formation may be amplified by positive autoregulation of the C-factor promoter (46). A *csgA* mutant with a *csgA-lacZ* transcriptional fusion increases β-galactosidase production twofold when starved alone. When mixed with an equal number of wild-type cells or 1 nM purified C-factor, β-galactosidase production increases a total of fourfold during development. These observations suggest that C-factor concentration can increase in at least two ways by increasing local cell density. First, movement of cells into more dense arrangements may result in greater accumulation of extracellular proteins. Second, transcription of the signal structural gene increases, by an as yet unknown positive-feedback pathway, in response to an increase in C-factor concentration. Density-dependent positive autoregulation is well known in the bioluminescent bacteria (39, 86).

Evidence for a signal relay system involving C-factor arises from observations that C-factor induces cell movement, resulting in formation of multicellular aggregates, and that during aggregation C-factor positively regulates its own net synthesis. Kroos et al (47) hypothesized that such a positive feedback loop might produce a relayed signal that could manifest itself as the wave-like cell movements called rippling. The feedback loop would be broken by loss of motility or by *csgA* mutation, and both are needed for rippling.

Normally, streams of cells move directly into an aggregation center, but it is not yet clear how *M. xanthus* cells orient themselves in the proper direction.

Chemotaxis regulated by the *frz* genes toward a diffusable substance or toward molecules bound by extracellular matrix has been suggested. Alternatively, extracellular matrix conforming to local stress patterns in the gliding surface can orient the direction of myxobacteria, a pattern of movement called elasticotaxis (88). In either case, a small local focus of densely packed cells produced early in development by C-signaling could select the center of cell aggregation where a fruiting body is built.

CONCLUSION

Roland Thaxter (90) forsaw the experimental opportunity offered by the myxobacteria to probe basic developmental strategies 100 years ago. Density-dependent regulatory mechanisms are a general strategy used in the development of organisms ranging from bacteria to vertebrates. A- and C-signaling illustrate how such regulation operates in *M. xanthus*.

After starvation initiates development, *M. xanthus* uses A-factor to test the cell density. A-factor is a freely diffusable cell signal that does not require movement of the transmitting or receiving cells. If the A-signal is not transmitted because the cell density is too low, the cells use residual nutrient to grow and divide until the appropriate density is reached. Once the proper density has been achieved, the cells express A-dependent genes and proceed through the early stages of the developmental program.

C-factor functions later in development, first stimulating cells to aggregate into a cohesive organized mound, then triggering differentiation to spores. C-factor is well-suited for monitoring the quality of cell arrangements because it appears to remain tightly associated with the cell surface. Efficient C-signaling strictly requires that cells achieve the proper dense-packed geometry found within the nascent fruiting body. Cells cannot passively accumulate but must move into the proper signaling position. C-signaling allows cells to verify their positions, delays sporulation until the appropriate time, and helps insure that cells in the subsequent germinating swarm are motile.

The requirements for transmission of C-signal differ from those of A-signal in terms of cell density and arrangement. Nevertheless, A-signaling and C-signaling have common properties. Both apply tests of cell density. Both act to restrict further developmental choices: A-signal commits to early development as opposed to growth; C-signal commits to aggregation and, at a higher concentration, to sporulation. Identification and characterization of the cellular response elements necessary for transduction of information contained in A-factor and C-factor is an important future goal. Transmission of the A- and C-signals demonstrates how cell density–restricted intercellular signals can regulate development in both time and space. These signals may trap developmental states that are appropriate for progress in the program, creating the new biological order that is a hallmark of development.

ACKNOWLEDGMENTS

The authors thank their laboratory colleagues, past and present; L. Shimkets and L. Plamann for the communication of results prior to publication. A. K. also thanks W.F. Loomis for discussions that helped to form some of the ideas on A-signaling presented here. A. K. is supported by an American Cancer Society Postdoctoral Fellowship. S. K. is a student in the Medical Scientist Training Program. Research was supported by a grant from the NIH (GM23441) and from the NSF (DCB8903705) to D. K.

Literature Cited

1. Arnold, J. W., Shimkets, L. J. 1988. Cell surface properties correlated with cohesion in *Myxococcus xanthus*. *J. Bacteriol.* 170:5771–77
2. Blackhart, B. D., Zusman, D. R. 1985. Cloning and complementation analysis of the "Frizzy" genes of *Myxococcus xanthus*. *Mol. Gen. Genet.* 198:243–54
3. Blackhart, B. D., Zusman, D. R. 1986. Analysis of the products of the *Myxococcus xanthus frz* genes. *J. Bacteriol.* 166:673–78
4. Bourret, R. B., Borkovich, K. A., Simon, M. I. 1991. Signal transduction pathways involving protein phosphorylation in prokaryotes. *Annu. Rev. Biochem.* 60:401–41
5. Burchard, R. P. 1970. Gliding motility mutants of *Myxococcus xanthus*. *J. Bacteriol.* 104:940–47
6. Burchard, R. P. 1981. Gliding motility of prokaryotes: ultrastructure, physiology and genetics. *Annu. Rev. Microbiol.* 35:497–529
7. Cheng, Y., Kaiser, D. 1989. *dsg*, a gene required for cell-cell interactions early in *Myxococcus* development. *J. Bacteriol.* 171:3719–26
8. Cheng, Y., Kaiser, D. 1989. *dsg*, a gene required for *Myxococcus* development, is necessary for cell viability. *J. Bacteriol.* 171:3727–31
9. Clarke, M., Kayman, S. C., Riley, K. 1988. Density-dependent induction of discoidin-I synthesis in exponentially growing cells of *Dictyostelium discoideum*. *Differentiation* 34:79–87
10. Costerton, J. W., Lappin-Scott, H. M. 1989. Behavior of bacteria in biofilms. *ASM News* 55:650–58
11. Cumsky, M., Zusman, D. R. 1979. Myxobacterial hemagglutin, a development-specific lectin of *Myxococcus xanthus*. *Proc. Natl. Acad. Sci. USA*. 76:5505–9
12. Dworkin, M. 1973. Cell-cell interactions in the myxobacteria. In *Microbial Differentiation, Soc. Gen. Microbiol. Symp.*, ed. J. M. Ashworth, J. E. Smith, 23:125–42. Cambridge: Cambridge Univ. Press
13. Dworkin, M. 1985. The myxobacteria. In *Developmental Biology of the Bacteria*, ed. M. Dworkin, pp. 105–49. Menlo Park, CA: Benjamin/Cummings
14. Dworkin, M. 1991. Cell-cell interactions in myxobacteria. In *Microbial Cell-Cell Interactions*, ed. M. Dworkin, pp. 179–216. Washington, DC: Am. Soc. Microbiol.
15. Fink, J. S., Zissler, J. F. 1989. Defects in motility and development of *Myxococcus xanthus* lipopolysaccharide mutants. *J. Bacteriol.* 171:2042–48
16. Gill, J. S., Dworkin, M. 1986. Cell surface antigens during submerged development of *Myxococcus xanthus* examined with monoclonal antibodies. *J. Bacteriol.* 168:505–11
17. Gill, J. S., Jarvis, B. W., Dworkin, M. 1987. Inhibition of development in *Myxococcus xanthus* by monoclonal antibody 1604. *Proc. Natl. Acad. Sci. USA* 84:4505–8
18. Gill, R. E. 1992. *Proc. Natl. Acad. Sci. USA*. In press
19. Gill, R. E., Cull, M. G. 1986. Control of developmental gene expression by cell-to-cell interactions in *Myxococcus xanthus*. *J. Bacteriol.* 168:341–47
20. Grossman, A. D., Losick, R. 1988. Extracellular control of spore formation in *Bacillus subtilis*. *Proc. Natl. Acad. Sci. USA* 85:4369–73
21. Gurdon, J. B. 1988. A community effect in animal development. *Nature* 336:772–74
22. Hagen, D. C., Bretscher, A. P., Kaiser, D. 1978. Synergism between morphogenetic mutants of *Myxococcus xanthus*. *Dev. Biol.* 64:284–96

23. Hagen, T. J., Shimkets, L. J. 1990. Nucleotide sequence and transcriptional products of the *csg* locus of *Myxococcus xanthus*. *J. Bacteriol*. 172:15–23
24. Hodgkin, J., Kaiser, D. 1977. Cell to cell stimulation of movement in nonmotile mutants of *Myxococcus*. *Proc. Natl. Acad. Sci. USA* 74:2938–42
25. Hodgkin, J., Kaiser, D. 1979. Genetics of gliding motility in *Myxococcus xanthus* (Myxobacterales): genes controlling movements of single cells. *Mol. Gen. Genet*. 171:167–76
26. Hodgkin, J., Kaiser, D. 1979. Genetics of gliding motility in *Myxococcus xanthus* (Myxobacterales): two gene systems control movement. *Mol. Gen. Genet*. 171:177–91
26a. Hopwood, D. A., Chater, K. F., eds. 1989. *Genetics of Bacterial Diversity*. San Diego, CA: Academic
27. Inouye, M., Inouye, S., Zusman, D. R. 1979. Gene expression during development of *Myxococcus xanthus:* pattern of protein synthesis. *Dev. Biol*. 68:579–91
28. Inouye, M., Inouye, S., Zusman, D. R. 1979. Biosynthesis and self-assembly of protein S, a development specific protein of *Myxococcus xanthus*. *Proc. Natl. Acad. Sci. USA* 76:209–13
29. Inouye, S., Franceschini, T., Inouye, M. 1983. Structural similarities between the developmental-specific protein from a gram-negative bacterium, *Myxococcus xanthus,* and calmodulin. *Proc. Natl. Acad. Sci. USA* 80:6828–33
30. Inouye, S., Harada, W., Inouye, M. 1981. Development-specific protein S of *Myxococcus xanthus:* purification and characterization. *J. Bacteriol*. 148: 678–83
31. Janssen, G. R., Dworkin, M. 1985. Cell-cell interactions in developmental lysis of *Myxococcus xanthus*. *Dev. Biol*. 112:194–202
32. Jarvis, B. W., Dworkin, M. 1989. Role of *Myxococcus xanthus* cell surface antigen 1604 in development. *J. Bacteriol*. 171:4667–73
33. Kaiser, D. 1979. Social gliding is correlated with the presence of pili in *Myxococcus xanthus*. *Proc. Natl. Acad. Sci. USA* 76:5952–56
34. Kaiser, D. 1986. Control of multicellular development: *Dictyostelium* and *Myxococcus*. *Annu. Rev. Genet*. 20: 539–66
35. Kaiser, D. 1989. Multicellular development in myxobacteria. See Ref. 26a, pp. 243–66
36. Kaiser, D., Crosby, C. 1983. Cell movement and its coordination in swarms of *Myxococcus xanthus*. *Cell Motil*. 3:227–45
37. Kalos, M., Zissler, J. F. 1990. Transposon tagging of genes for cell-cell interactions in *Myxococcus xanthus*. *Proc. Natl. Acad. Sci. USA* 87:8316–20
38. Kalos, M., Zissler, J. F. 1990. Defects in contact-stimulated gliding during aggregation by *Myxococcus xanthus*. *J. Bacteriol*. 172:6476–93
39. Kaplan, H. B., Greenberg, E. P. 1987. Overproduction and purification of the *luxR* gene product: the transcriptional activation of the *Vibrio fischeri* luminescence system. *Proc. Natl. Acad. Sci. USA* 84:6639–43
40. Kaplan, H. B., Kuspa, A., Kaiser, D. 1991. Suppressors that permit A signal-independent developmental gene expression in *Myxococcus*. *J. Bacteriol*. 173: 1460–70
41. Kim, S. K. 1991. Intercellular signaling in *Myxococcus* development: the role of C factor. *Trends Genet*. 7:361–65
42. Kim, S. K., Kaiser, D. 1990. Purification and properties of C-factor, an intercellular signaling protein. *Proc. Natl. Acad. Sci. USA* 87:3635–39
43. Kim, S. K., Kaiser, D. 1990. C-factor: a cell-cell signaling protein required for fruiting body morphogenesis of M. xanthus. *Cell* 61:19–26
44. Kim, S. K., Kaiser, D. 1990. Cell motility is required for the transmission of C-factor, an intercellular signal which coordinates fruiting body morphogenesis of *Myxococcus xanthus*. *Genes Dev*. 4:896–905
45. Kim, S. K., Kaiser, D. 1990. Cell alignment required in differentiation of *Myxococcus xanthus*. *Science* 249:926–28
46. Kim, S. K., Kaiser, D. 1991. C-factor has distinct aggregation and sporulation thresholds during *Myxococcus* development. *J. Bacteriol*. 173:1722–28
47. Kroos, L., Hartzell, P., Stephens, K., Kaiser, D. 1988. A link between cell movement and gene expression argues that motility is required for cell-cell signaling during fruiting body formation. *Genes Dev*. 2:1677–85
48. Kroos, L., Kaiser, D. 1984. Construction of Tn*5 lac,* a transposon that fuses *lacZ* expression to exogenous promoters, and its introduction into *Myxococcus xanthus*. *Proc. Natl. Acad. Sci. USA* 81:5816–20
49. Kroos, L., Kaiser, D. 1987. Expression of many developmentally regulated genes in *Myxococcus* depends on a sequence of cell interactions. *Genes Dev*. 1:840–54
50. Kroos, L., Kuspa, A., Kaiser, D. 1986.

A global analysis of developmentally regulated genes in *Myxococcus xanthus*. *Dev. Biol.* 117:252–66
51. Kuner, J. M., Kaiser, D. 1982. Fruiting body morphogenesis in submerged cultures of *Myxococcus xanthus*. *J. Bacteriol.* 151:458–61
52. Kuspa, A. 1989. *Intercellular signalling in the regulation of early development in* Myxococcus xanthus. PhD thesis. Stanford Univ., Stanford, Calif.
53. Kuspa, A., Kaiser, D. 1989. Genes required for developmental signalling in *Myxococcus xanthus:* three *asg* loci. *J. Bacteriol.* 171:2762–72
54. Kuspa, A., Kroos, L., Kaiser, D. 1986. Intercellular signalling is required for developmental gene expression in *Myxococcus xanthus*. *Dev. Biol.* 117: 267–76
55. Kuspa, A., Plamann, L., Kaiser, D. 1992. Amino acids and peptides rescue the A-signal-defective developmental mutants of *M. xanthus*. *J. Bacteriol.* In press
56. Kuspa, A., Plamann, L., Kaiser, D. 1992. Evidence that A-signalling specifies the minimum cell density required for *Myxococcus xanthus* development. *J. Bacteriol.* In press
57. LaRossa, R., Kuner, J., Hagen, D., Manoil, C., Kaiser, D. 1983. Developmental cell interaction of *Myxococcus xanthus:* analysis of mutants. *J. Bacteriol.* 153:1394–1404
58. Manoil, C., Beckwith, J. 1986. A genetic approach to analyzing membrane protein topology. *Science* 233:1403–8
59. Martinez-Laborda, A., Murillo, F. J. 1989. Genic and allelic interactions in the carotenogenic response of *Myxococcus xanthus* to blue light. *Genetics* 122:481–90
60. Mayo, K. A., Kaiser, D. 1989. *asgB,* a gene required early for developmental signalling, aggregation, and sporulation of *Myxococcus xanthus*. *Mol. Gen. Genet.* 218:409–18
61. McBride, M. J., Weinberg, R. A., Zusman, D. R. 1989. "Frizzy" aggregation genes of the gliding bacterium *Myxococcus xanthus* show sequence similarities to the chemotaxis genes of enteric bacteria. *Proc. Natl. Acad. Sci. USA* 86:424–28
62. McCleary, W. R., McBride, M. J., Zusman, D. R. 1990. Developmental sensory transduction in *Myxococcus xanthus* involves methylation and demethylation of FrzCD. *J. Bacteriol.* 172:4877–87
63. McCleary, W. R., Zusman, D. R. 1990. FrzE of *Myxococcus xanthus* is homologous to both CheA and CheY of *Salmonella typhimurium*. *Proc. Natl. Acad. Sci. USA* 87:5898–5902
64. Morrison, C. E., Zusman, D. R. 1979. *Myxococcus xanthus* mutants with temperature-sensitive, stage-specific defects: evidence for independent pathways in development. *J. Bacteriol.* 140:1036–42
65. Nelson, D. R., Cumsky, M. G., Zusman, D. R. 1981. Localization of myxobacterial hemagglutinin in the periplasmic space and on the cell surface of *Myxococcus xanthus* during developmental aggregation. *J. Biol. Chem.* 256:12589–95
66. Ochman, H., Wilson, A. C. 1987. Evolution in bacteria: evidence for a universal substitution rate in cellular genomes. *J. Mol. Evol.* 26:74–86
67. O'Connor, K. A., Zusman, D. R. 1989. Patterns of cellular interactions during fruiting body formation in *Myxococcus xanthus*. *J. Bacteriol.* 171:6013–24
68. O'Connor, K. A., Zusman, D. R. 1990. Genetic analysis of the *tag* mutants of *Myxococcus xanthus* provides evidence for two developmental aggregation systems. *J. Bacteriol.* 172:3868–78
69. Plamann, L., Kuspa, A., Kaiser, D. 1992. Proteins that rescue A-signal-defective mutants of *Myxococcus xanthus*. *J. Bacteriol.* In press
70. Reichenbach, H. 1965. Rhythmische Vorgänge bei der Schwarmentfaltung von Myxobakterien. *Ber. Dtsch. Bot. Ges.* 78:102–5
71. Rhie, H. G., Shimkets, L. J. 1989. Developmental bypass suppression of *Myxococcus xanthus csgA* mutations. *J. Bacteriol.* 17:3268–76
72. Riddle, D. L. 1988. The dauer larvae. In *The Nematode* Caenorhabditis elegans, ed. W. B. Wood, pp. 398–400. Cold Spring Harbor, NY: Cold Spring Harbor Lab.
73. Rodriguez, J. 1991. PhD thesis. Stanford Univ., Stanford, Calif.
74. Romeo, J. M., Zusman, D. R. 1987. Cloning of the gene for myxobacterial hemagglutin and isolation and analysis of structural gene mutations. *J. Bacteriol.* 169:3801–8
75. Rosenberg, E., ed. 1984. *Myxobacteria: Development and Cell Interactions*. New York: Springer-Verlag
76. Rosenberg, E., Keller, K. H., Dworkin, M. 1977. Cell density–dependent growth of *Myxococcus xanthus* on casein. *J. Bacteriol.* 129:770–77
77. Shapiro, J. A. 1988. Bacteria as multicellular organisms. *Sci. Am.* 256:82–89
78. Shimkets, L. J. 1986. Role of cell cohe-

sion in *Myxococcus xanthus* fruiting-body formation. *J. Bacteriol.* 166:842–48

79. Shimkets, L. J. 1990. Social and developmental biology of the myxobacteria. *Microbiol. Rev.* 54:473–501
80. Shimkets, L. J., Asher, S. J. 1988. Use of recombination techniques to examine the structure of the *csg* locus of *Myxococcus xanthus*. *Mol.Gen. Genet.* 211: 63–71
81. Shimkets, L. J., Dworkin, M. 1981. Excreted adenosine is a cell density signal for the initiation of fruiting body formation in *Myxococcus xanthus*. *Dev. Biol.* 84:51–60
82. Shimkets, L. J., Gill, R. E., Kaiser, D. 1983. Developmental cell interactions in *Myxococcus xanthus* and the *spoC* locus. *Proc. Natl. Acad. Sci. USA* 80:1406–10
83. Shimkets, L. J., Kaiser, D. 1982. Induction of coordinated cell movement in *Myxococcus xanthus*. *J. Bacteriol.* 152: 451–61
84. Shimkets, L. J., Kaiser, D. 1982. Murein components rescue developmental sporulation of *Myxococcus xanthus*. *J. Bacteriol.* 152:462–70
85. Shimkets L. J., Rafiee, H. 1990. CsgA, an extracellular protein essential for *Myxococcus xanthus* development. *J. Bacteriol.* 172:5299–5306
86. Silverman, M., Martin, M., Engebrecht, J. 1989. Regulation of luminescence in marine bacteria. See Ref. 26a, pp. 71–86
87. Sodergren, E., Kaiser, D. 1983. Insertions of Tn*5* near genes that govern stimulatable cell motility in *Myxococcus*. *J. Mol. Biol.* 167:295–310
88. Stanier, R. Y. 1942. A note on elasticotaxis in myxobacteria. *J. Bacteriol.* 44:405–12
89. Stephens, K., Hartzell, P., Kaiser, D. 1989. Gliding motility of *Myxococcus xanthus: mgl* locus, RNA and predicted protein products. *J. Bacteriol.* 171:819–30
90. Thaxter, R. 1892. On the Myxobacteriaceae, a new order of Schizomycetes. *Bot. Gaz.* 17:389–406
91. White, D. 1981. Cell interactions and control of development in myxobacteria populations. *Int. Rev. Cytol.* 72:203–27
92. Zusman, D. R. 1984. Cell-cell interactions and development in *Myxococcus xanthus*. *Q. Rev. Biol.* 59: 119–38
93. Jain, R., Yuen, I. S., Taphouse, C. R., Gomer, R. H. 1992. A density-sensing factor controls development in *Dictyostelium*. *Genes Dev.* 6:390–400

Annu. Rev. Microbiol. 1992. 46:141–63

GENETICS OF RIBOSOMALLY SYNTHESIZED PEPTIDE ANTIBIOTICS

Roberto Kolter

Department of Microbiology and Molecular Genetics, Harvard Medical School, 200 Longwood Avenue, Boston, Massachusetts 02115

Felipe Moreno

Unidad de Genética Molecular, Hospital "Ramón y Cajal," Carretera de Colmenar, km. 9,100, Madrid 28034, Spain

KEY WORDS: lantibiotics, microcins, lactacins, lanthionine, peptide export

CONTENTS

0066-4227/92/1001-0141$02.00

Abstract

In recent years many peptide antibiotics have been shown to be ribosomally synthesized. Among these are many microcins, produced by diverse strains of gram-negative bacteria. While the structures and modes of action of these peptide antibiotics vary widely, many of them share several important features. Their synthesis is often induced by the cessation of growth. In addition, many of them undergo unusual posttranslational modifications to yield the mature molecule, which is often exported from the cell by a dedicated export apparatus. The genes involved in modification and export of the peptide antibiotics are generally found adjacent to the structural gene and are under the same regulation. The results supporting these conclusions are reviewed and discussed in this chapter.

INTRODUCTION

Antibiotics can be defined broadly as substances produced by one organism that have adverse effects on other organisms (21, 60). This operational definition, however, does not reflect the role of antibiotics in the natural environment, which remains poorly understood, although at least some of these substances confer a competitive advantage to the producer organism (81). Regardless of the ecological function of antibiotics, their impact on human health is obvious.

The ability of bacterial strains to produce antibiotic substances is widespread. In screens in which large numbers of natural isolates were tested for antibiotic production, a significant percentage of the strains yielded positive results (6, 119). Moreover, when the number of indicator microorganisms was increased, and the incubation conditions varied, the fraction of strains scored as antibiotic producers increased dramatically. Production of antibiotic substances by bacteria may be ubiquitous in nature (120). Given the large number of known antibiotics, the almost limitless variety of chemical structures exhibiting antibiotic properties is not surprising. Nor is it surprising that their mode of action can interfere with almost any of the essential functions of the living cell—transcription, translation, DNA replication, cell wall biosynthesis, or cell membrane integrity.

Many of the compounds classified as antibiotics are peptides or small proteins. These often contain amino acid residues not commonly found in proteins, immediately raising the question of how these peptides are synthesized. Are they products of an enzymatic pathway or do they represent ribosomally synthesized products that are posttranslationally modified? Indeed, both types of peptide antibiotics have been found. Enzymatically assembled peptide antibiotics include both linear and circular molecules, such as gramicidin, polymixin, and bacitracin (74). Nisin, subtilin, and microcin

B17 are examples of peptide antibiotics whose amino acid sequences are initially derived from RNA templates (8, 15, 18, 68). Early indications that some peptide antibiotics are synthesized from an RNA template included the detection of propeptides and the inhibition of antibiotic synthesis by protein synthesis inhibitors (62–64, 94). Final confirmation, as discussed below, came with the isolation and sequencing of the structural genes encoding the precursor peptides for these antibiotics.

This review looks at recent progress in the area of ribosomally synthesized peptide antibiotics, emphasizing genetic studies of this class of antibiotics. We present the genes responsible for peptide antibiotic production and the mechanisms by which the antibiotics kill sensitive cells, focusing on three peptide antibiotics produced by certain strains of *Escherichia coli,* microcin B17, microcin C7, and colicin V. Through classical and molecular genetic analyses of these we have gained important insights about their biology (40, 96, 110, 111). We also draw heavily from molecular genetic analyses of the more familiar peptide antibiotics from gram-positive bacteria, i.e. lantibiotics and nonlantibiotic heat-stable bacteriocins (66, 67, 72). The reader is referred to the many excellent reviews in the field of antibiotics for: (*a*) a more general treatment of the subject (61, 81), (*b*) a more biochemical treatment of peptide antibiotics (74), (*c*) a discussion of the larger sized bacteriocins/colicins (98, 99), and (*d*) other aspects of ribosomally synthesized peptide antibiotics (9, 66, 67, 72, 86). In addition, *The EMBO-FEMS-NATO Workshop on Bacteriocins, Microcins, and Lantibiotics* and *The First International Workshop on Lantibiotics* contain proceedings from recent international meetings on the subject (65, 67).

STATEMENT OF THE BIOLOGICAL PROBLEM

An organism that produces a peptide antibiotic must be able to: (*a*) synthesize the antibiotic, (*b*) export the antibiotic into the extracellular milieu, and (*c*) protect itself from the action of the antibiotic. Furthermore, the antibiotic, by definition, must interact with the sensitive cell and interfere with its growth or survival. This interaction may or may not require entry of the antibiotic into the cell. Figure 1 diagrams these general steps in antibiotic production and action. Synthesis of the antibiotic is presented as involving several steps: transcription, translation, posttranslational modification of the amino acid residues, and processing by cleavage of an N-terminal leader sequence.

Many peptide antibiotics contain unusual amino acid residues such as lanthionine and dehydroalanine (66). The function of these modified amino acids may be twofold. First, these small peptide molecules are less likely to achieve a stable conformation by the same means as much larger proteins. Yet their activity is generally extremely heat stable. Unusual cross links like the

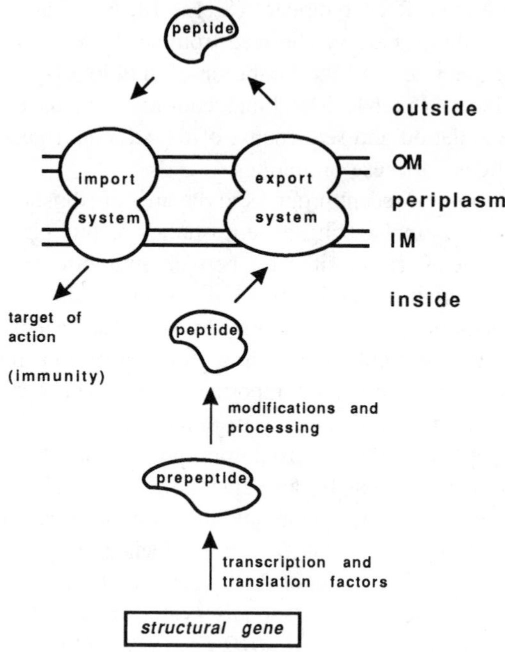

Figure 1 Diagram of the steps involved in peptide antibiotic production and action. Abbreviations are: OM, outer membrane; IM, inner membrane.

thioether bridge from lanthionine may allow the molecules to achieve and retain their proper folded structure despite their small size (47). Second, unusual side chains such as dehydroalanine may provide a wider repertoire of chemical reactivity that may be very important for the biological activity of these peptides (46).

SURVEY OF PEPTIDE ANTIBIOTICS

For each peptide antibiotic, specific strategies have evolved for its synthesis, export, and immunity. However, genetic and biochemical analyses of three broad classes of molecules have revealed common themes. The three classes of molecules are (*a*) microcins (from gram-negative bacteria), (*b*) lantibiotics (from gram-positive bacteria), and (*c*) nonlantibiotic heat-stable bacteriocins (from gram-positive bacteria). These represent a limited selection from the many known types of peptide antibiotics; for more complete surveys and descriptions, the reader should consult other sources (9, 66, 67, 72).

Microcins

The microcins comprise a family of antibiotic substances produced by diverse members of the *Enterobacteriaceae* (6, 9). They are distinguished from the majority of the colicins by their much lower molecular weight [less than 10 kilodaltons (kDa)] and because their synthesis is not induced by conditions that lead to induction of the SOS repair pathway. Rather, they are synthesized during stationary phase, similar to what is observed with most conventional antibiotics (81). The microcins were operationally defined as substances produced by gram-negative bacteria capable of passing through a cellophane membrane and inhibiting the growth of an indicator *E. coli* strain (6). Several substances known to be ribosomally synthesized fall within this group; however, this operational definition does not exclude nonpeptide microcins. Indeed, microcin A15 appears to be a derivative of methionine (1, 2).

MICROCIN C7 Microcin C7 (MccC7) is a heptapeptide containing modifications at both the N and C termini (see Figure 2) (35, 36). MccC7 blocks protein synthesis both in vivo (36) and in vitro (J. L. San Millán & F. Moreno, unpublished observation).

MICROCIN B17 Microcin B17 (MccB17) is a 43-residue peptide containing posttranslational modifications in several of its side chains (see Figure 2) (18). Although the exact nature of the modifications is unknown, the mature molecule may contain several covalently linked aromatic chromophores (135). Twenty-six of the 43 residues are glycine. MccB17 inhibits DNA replication and induces the SOS response (54, 55). The target of this antibiotic appears to be DNA gyrase (127).

COLICIN V Colicin V (ColV) kills sensitive cells by disrupting their membrane potential (134). The existence of ColV was first described in 1925; this description represents the first report of an antibiotic substance produced by *E. coli* (44). Although it is called a colicin, ColV appears to fit the description of a microcin (134). ColV has a molecular weight of only 6000 and its synthesis is not SOS-inducible (32, 56). Most colicins, by contrast, are proteins ranging in size from 25 to 80 kDa whose synthesis is SOS-inducible (98, 99). The structure of mature ColV is not known, with regard to either possible side-chain modifications or N-terminal end processing.

Lantibiotics

Many peptide antibiotics from diverse species of gram-positive bacteria have been identified and characterized structurally (66, 67, 72). Several contain sulfide rings that result from the unusual cross-linking residues lanthionine and methyllanthionine. Because of this particular structural motif, i.e. the

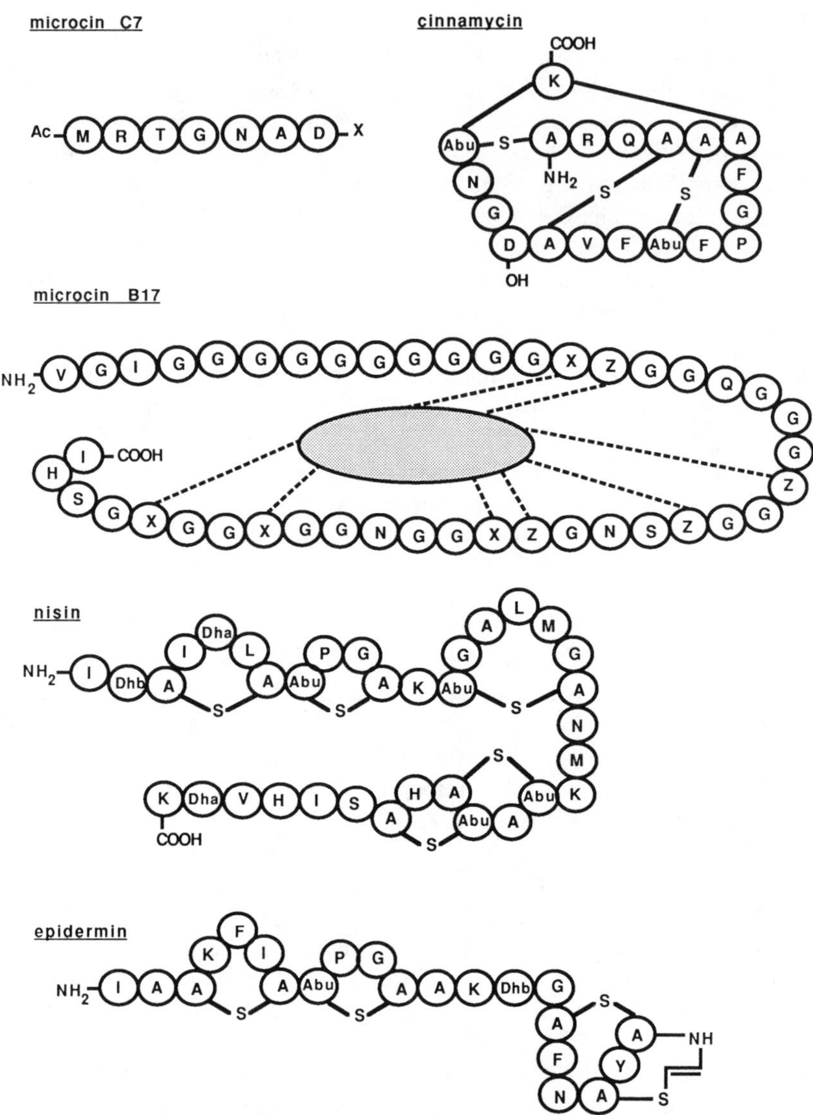

Figure 2 Schematic representation of the chemical structures of selected peptide antibiotics. For microcin C7, the N-terminal modification is an acetylation (Ac), and the C-terminal X designates an unknown modification. The amino acid designations within the circles follow the standard single-letter code. Unusual residues are designated as follows. For cinnamycin, nisin, and epidermin: Abu, aminobutyric acid; Dhb, dehydrobutyrine; and Dha, dehydroalanine. Thioether linkages are diagrammed as -S-; thus lanthionine is A-S-A and methyllanthionine is Abu-S-A. For microcin B17, X represents modified serine and Z represents modified cysteine. The exact structure of the modifications and the chromophores associated with microcin B17 are not known and are symbolized by the dashed lines and shaded oval.

presence of thioether linkages from lanthionine, the term *lantibiotics* has been given to this class of compounds (115). Lantibiotics can be divided into two subclasses, linear and circular. The linear class includes nisin (83, 102), subtilin (45), epidermin (4, 5), gallidermin (71), and Pep5 (109). The circular class includes cinnamycin (12) and ancovenin (131). For several lantibiotics, the basis of bactericidal activity is the formation of voltage-dependent pores in the cytoplasmic membrane of sensitive organisms (76, 108, 116). Some lantibiotics also activate cell-wall hydrolyzing enzymes and cause autolysis (13).

NISIN Nisin is perhaps the best known lantibiotic because of its importance as a food preservative (102). Produced by certain strains of *Lactococcus lactis,* nisin was first described in 1928 as a product of natural isolates found in milk (106). This 34-residue peptide contains five sulfide rings (Figure 2). In addition to lanthionine and methyllanthionine, nisin contains dehydroalanine and dehydrobutyrine (47). Some strains of *Bacillus subtilis* produce subtilin, a structurally similar lantibiotic (45). Subtilin contains 32 residues and conserves the number and positions of the thioether linkages present in nisin.

EPIDERMIN Several strains of *Staphylococcus epidermis* produce the 21-residue lantibiotic epidermin (5). Epidermin contains four sulfide rings: two result from lanthionine residues, one from a methyllanthionine residue, and the fourth from a very unusual amino acid, S(2-aminovinyl)-D-cysteine (see Figure 2). The efficacy of epidermin in inhibiting pathogenic bacteria such as *Propionibacterium acne* makes it a compound of potential therapeutic value in the topical treatment of acne (115). A closely related lantibiotic, gallidermin, differs from epidermin at only one residue (114). Another lantibiotic produced by *S. epidermis* is Pep5, which has a very different structure. It contains 34 residues and only 3 thioether linkages (70).

CINNAMYCIN Circular lantibiotics, among them cinnamycin, are produced by some *Streptomyces* strains (66, 67). The structure of cinnamycin is shown in Figure 2. Despite being rather small, only 19 residues, it contains 4 cross-links (92). Three of these are thioether linkages while the fourth is in the form of lysinoalanine. Cinnamycin inhibits phospholipase A_2, has inhibitory activity against Herpes simplex Type 1, and has immunopotentiating activities (69). Cinnamycin and related lantibiotics such as ancovenin may also act as inhibitors of the angiotensin-converting enzyme (131).

Nonlantibiotic Heat-Stable Bacteriocins

Many different peptide antibiotics that do not contain lanthionine have been identified from a variety of *Lactobacillus, Lactococcus, Pediococcus,* and

Enterococcus strains such as lactacins B (10, 11) and F (90, 91); lactocins S (87, 88) and 27 (122); sakacin A (112); pediocins A (17), PA-1 (43; S. T. Henderson, A. L. Chopko & P. D. van Wassenaar, submitted; P. A. Vanderberg, A. M. Ledeboer, C. M. Zoetmulder, M. J. Pucci & C. F. González, submitted), and AcH (101); diplococcin (19, 20); lactococcins A (59), B (123), and M (123–125); and AS48 (33, 34, 82). Most of these are hydrophobic peptides of differing sizes and amino acid composition and appear to kill sensitive cells by increasing membrane permeability.

GENES FOR PEPTIDE ANTIBIOTIC PRODUCTION AND ACTION

The Structural Gene

The structural genes encoding most of the peptide antibiotics mentioned above have been found on plasmids. Exceptions include the structural genes for gallidermin and subtilin (8, 114). In addition, the nisin gene has been found both on plasmids and on the chromosome and is part of a conjugative transposon (27).

Structural genes encoding several of the peptide antibiotics mentioned above have been cloned and sequenced. In the case of the microcins, the structural gene was first identified by analysis of mutants unable to produce microcin (39–41, 96, 110, 111). The structural genes for many lantibiotics and several other antibiotics from gram-positive bacteria, by contrast, were isolated using mixed oligonucleotide probes derived from known amino acid sequences of the antibiotics (8, 15, 59, 68, 70, 90, 91, 114, 115). Two important conclusions can be made regarding the coding information present in the structural-gene open reading frame viz a viz the known structure of the active peptide. First, in almost every case, the structural gene encodes an N-terminal leader sequence that is cleaved during maturation. Second, the specific amino acid residues that undergo posttranslational modifications can be inferred from the predicted sequence of the primary translation product.

The leader sequences, present only transiently in the prepeptide products, do not contain any of the characteristic features of typical signal sequences present at the N-termini of many exported proteins (128, 129). This observation suggests that even though these peptides are secreted to the medium, this secretion occurs in a signal sequence–independent fashion. Figure 3 presents the leader sequences of prepeptides for which both the sequence of the gene and the N-terminal sequence of the mature peptide are known (8, 15, 18, 59, 68–70, 90, 115, 124, 125; P. A. Vanderberg, A. M. Ledeboer, C. M. Zoetmulder, M. J. Pucci & C. F. González, submitted). The sequences are aligned at the site where cleavage occurs to yield the mature N termini.

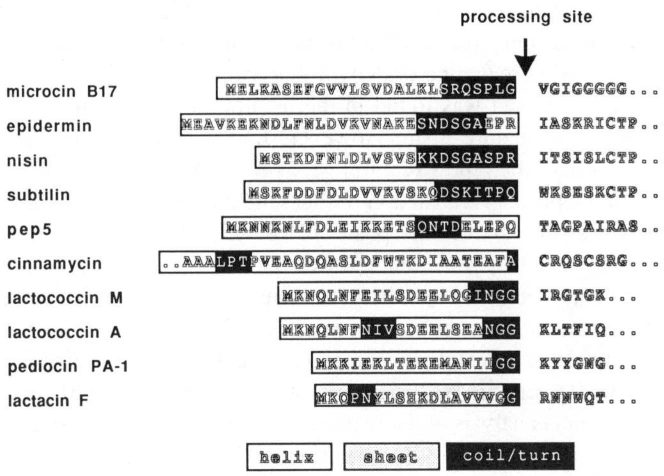

Figure 3 Leader sequences predicted for the prepeptide of several antibiotics.

Although there is no overall sequence conservation, the structure is generally conserved. Secondary structure–prediction programs indicate that these leader sequences are mostly helical. The helices, however, are broken at or before the processing sites. This common feature may make the sites accessible to the appropriate peptidases (66). For microcin B17 and most of the lantibiotics, the break in the helix partly results from a proline residue present two or three residues before the cleavage site. In the nonlantibiotics from gram-positive bacteria, the predicted break in helical structure results in part from two conserved glycine residues just prior to the cleavage site.

Several possible roles have been proposed for these leader sequences (66, 135). They may be responsible for keeping the peptide in a proper conformation for posttranslational modifications, serving as *cis*-acting chaperones. The leader sequence may also be important in the recognition of the peptide by the modification enzymes. Evidence indicates that the microcin B17 and Pep5 prepeptides are first modified and then processed, consistent with a role for the leader sequence in modification (133, 135). The leader sequences could also play a role in export through a dedicated translocation system; however, this does not seem to be the case for microcin B17, in which the N-terminal processing appears to occur prior to export, and the export machinery does not recognize the modified but unprocessed prepeptide (105, 135).

The sequence of the structural gene of several lantibiotics provided important evidence regarding the synthesis of dehydroalanine, lanthionine, and other unusual residues. A mechanism for the posttranslational synthesis of these residues was suggested more than 20 years ago (63, 64). Dehydroalanine and dehydrobutyrine were proposed to be dehydrated products of serine and threonine, respectively (see Figure 2). In addition, lanthionine and methyllanthionine were proposed to result from a linkage formed between cysteine and a dehydroalanine or dehydrobutyrine residue. Strong supportive evidence for this model was found in the incorporation of radioactive serine and threonine into mature nisin. However, formal proof of the mechanism was obtained only when the structural genes for lantibiotics were sequenced, showing that the primary translation product must contain serines, threonines, and cysteines at the appropriate positions (8, 15, 115).

When microcin B17 was first purified and its amino acid composition compared to that predicted from the gene sequence, discrepancies were found (18). Four serine and four cysteine residues were predicted from the DNA sequence but were not detected in the mature peptide. This finding led to the suggestion that microcin B17 might contain lanthionine (105). However, extensive chemical analysis of the peptide has shown that this residue does not occur in the active peptide (135). The exact structures of the posttranslational modifications found in microcin B17 are not yet known (see Figure 2), but they appear to result in cross-links and four heterocyclic chromophores (135). Thus to date, lantibiotics have been detected only from gram-positive organisms.

Other Genes Needed for Production

As indicated by the preceding discussion, a cell requires more than the structural gene to produce an extracellular antibiotic. Additional gene products are needed to mediate transcription regulation, posttranslational modifications and processing, extracellular export, and immunity against the lethal action of the antibiotic. To focus the presentation, we discuss five well-characterized genetic systems: microcin B17, microcin C7, colicin V, epidermin, and subtilin. Genes whose products function in one or more of these steps have been identified for all three types of antibiotics surveyed above. For example, the immunity determinants of Pep5 and lactococcin A appear to be located very close to the structural gene (59, 103, 123, 125). The finding that the Pep5 immunity determinant may involve the structural gene is particularly interesting (103).

Two approaches have been taken to identify additional genes required for normal production of peptide antibiotics. First, the power of *E. coli* genetics was exploited to obtain mutants defective in antibiotic production. The phenotypes of the resulting mutants were used to infer the functions of genes essential for synthesis and export of MccB17, MccC7, and ColV (26, 38–41, 96, 110, 111). Second, the DNA flanking the structural genes for subtilin and epidermin was directly sequenced. Subsequent homology searches and mutational analyses revealed open reading frames that are probably involved in the synthesis and export of these lantibiotics (7, 73, 113).

Figure 4 shows the clusters of genes involved in the production of MccB17 (37, 38), MccC7 (26, 96), ColV (40, 41), subtilin (73), and epidermin (7, 113). As evident from the figure, genes adjacent to the structural gene mediate several processes required for normal production of the peptide. Several unlinked genes have also been identified that play important roles in the normal production of these peptides.

MICROCIN B17 Seven contiguous genes, *mcbABCDEFG*, are involved in production of MccB17 (37, 38). The structural gene is a 69-codon open reading frame that codes for the MccB17 precursor (18). This precursor is posttranslationally modified by the action of the *mcbBCD* gene products to yield a pro-MccB17 that still contains the N-terminal leader sequence (135). The mechanism by which the McbBCD proteins mediate the posttranslational modifications remains unknown; the predicted amino acid sequences of these proteins do not show significant similarities to other proteins. The product of

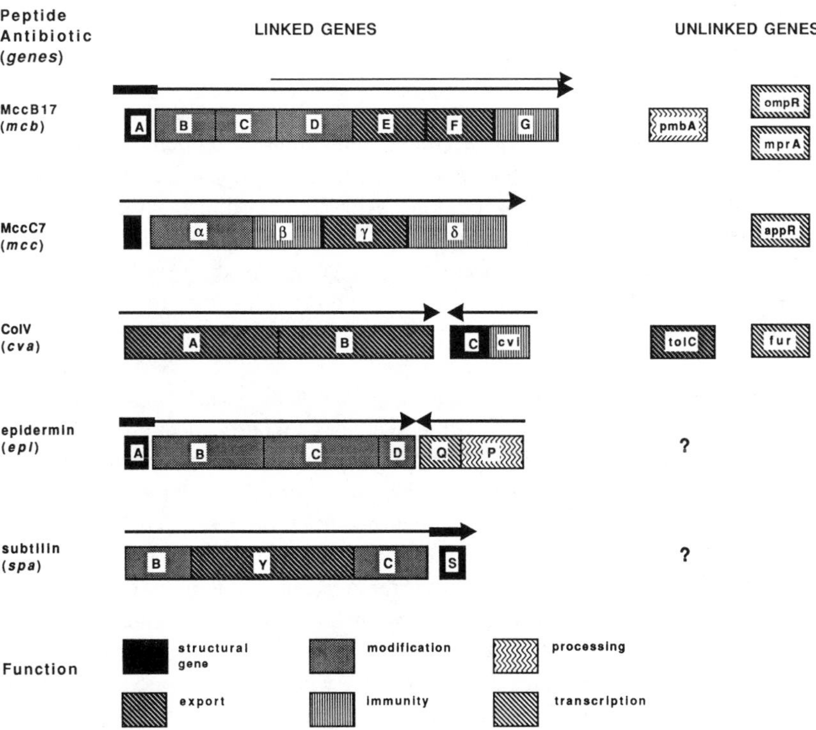

Figure 4 Linked and unlinked genes involved in the production of several peptide antibiotics. Arrows indicate transcriptional units. The thicker arrows indicate that those RNA molecules are known to be more abundant.

the unlinked gene *pmbA* is thought to process pro-MccB17 to the mature, active peptide (105). A complex of McbE and McbF constitutes a dedicated exporter for active MccB17 (37). McbE is predicted to span the membrane six times, while McbF contains sequences similar to other bacterial transport proteins and could be an ATP-binding protein (58). Interestingly, because these proteins can export the internal active microcin, they also confer a limited degree of immunity against the antibiotic. Full immunity, however, requires the product of the *mcbG* gene. The mechanism by which McbG confers immunity is not known. Finally, as discussed below, transcription of the *mcb* operon is stimulated by the product of the regulatory gene *ompR* (52, 53).

MICROCIN C7 The information required for MccC7 production and immunity is encoded in five kilobases of plasmid DNA from pMccC7 (96). This DNA has four regions genetically defined as α, β, γ, and δ, which are transcribed in the same orientation. Two of these regions, β and δ, confer immunity on the producing organism when independently cloned. Because MccC7 does not contain a leader sequence, no processing protein is required; both N- and C-terminal residues, however, are modified. The products required for these modifications and/or for the export of MccC7 might be encoded by the genes in regions α and γ. When the promoter located upstream of region α was sequenced, a seven–amino acid open reading frame encoding the sequence of MccC7 was identified (86; L. Díaz-Guerra & J. L. San Millán, unpublished observation). Although no direct evidence shows that this open reading frame codes for MccC7, its sequence and location argue compellingly for its role as the structural gene. As discussed below, the transcription of the *mcc* genes is under the control of the regulatory gene *appR* (26).

COLICIN V Four plasmid genes from pColV-K30, spanning 4.5-kilobases (kb) of DNA, are required for ColV synthesis, export, and immunity (40). These genes are arranged in two converging operons, whose transcription is induced during iron limitation under the control of the regulatory protein Fur (16, 39). The immunity gene, *cvi,* and the structural gene, *cvaC,* constitute one operon. The immunity protein is predicted to span the cytoplasmic membrane twice (29). This structure is reminiscent of immunity determinants of several other bacteriocins that act by damaging the cytoplasmic membrane (98, 99).

The other operon contains *cvaA* and *cvaB,* whose products are required for export of the antibiotic. The product of the unlinked, chromosomal gene *tolC* is also required to export ColV. The three protein components of the dedicated export system for ColV display a significant amount of amino acid sequence similarity with proteins that perform similar functions in the extracellular secretion of α-hemolysins (31, 130, 132), *Erwinia* and *Pseudomonas* proteases (25, 48), and *Bordetella* cyclolysin (42). These protein complexes are not only similar but are also functionally homologous because mutations in *cvaA* or *cvaB* or both can be complemented, albeit at a lowered efficiency, by the dedicated exporters of α-hemolysin and the *Erwinia* protease (30).

The CvaB protein is predicted to span the membrane six times and appears to contain an ATP-binding domain in its C-terminal cytoplasmic domain (41). The amino acid sequence of this ATP-binding domain is highly conserved among many proteins involved in export processes, including the mammalian MDR family of drug exporters (28); STE6, the a-factor exporter of *Saccharomyces cerevisiae* (77, 84); and CFTR, the chloride channel defective in patients with cystic fibrosis (104). A similar domain was found in a protein believed to export the lantibiotic subtilin (see below) (73).

The export signal of ColV, which is recognized by the CvaA/CvaB exporter, is located within the N-terminal 39 residues of the product of *cvaC* (41). These 39 residues do not show any features of a typical signal sequence for export (128, 129). When these CvaC residues are fused to an alkaline phosphatase moiety lacking its signal sequence, they can translocate the hybrid in a CvaA/CvaB-dependent fashion. To date, we have no direct evidence for an N-terminal processing step in the synthesis of ColV, but the ColV-PhoA fusions do show a reduction in apparent molecular weight, as judged in SDS-PAGE, after translocation across the membrane. If a processing step occurs in ColV maturation, then the exporters may carry out that function concomitant with export.

SUBTILIN By sequencing the DNA flanking the subtilin structural gene, Klein et al (73) detected three adjacent open reading frames. These genes, *spaBYC*, as shown by mutational analysis, are important for antibiotic production. Mutations in *spaB* and *spaC* prevented subtilin production, consistent with a role for their products in posttranslational modifications of subtilin. SpaY appears likely to be involved in the secretion of subtilin (73). Mutations in *spaY* result in reduced levels of extracellular subtilin, and cells harboring such mutations display aberrant morphology and very poor viability. The predicted amino acid sequence of SpaY indicates it belongs to the large family of exporters that also includes CvaB (29, 30, 41). The phenotype of cells with *spaY* mutations is reminiscent of the phenotype of cells unable to export microcin B17 (37).

EPIDERMIN When the 13.5-kb DNA fragment containing the epidermin structural gene (*epiA*) was sequenced, five adjacent open reading frames were found (7, 113). The open reading frames form two converging operons: *epiABCD* and *epiPQ*. The genes *epiBCD* may be involved in posttranslational modification of epidermin. The *epiBC* products show amino acid sequence similarity with the *spaBC* products found in the subtilin operon. The sequence similarity between *epiQ* and the regulatory gene *phoB* has been used as an argument for a role of EpiQ in the regulation of *epi* gene transcription. The sequence similarity between EpiP and the active sites of three serine proteases suggests that EpiP may be the protease that removes the N-terminal leader peptide of pre-epidermin.

Regulation of Antibiotic Biosynthesis

A common feature of most antibiotics is the timing of their production; they are made when exponential growth ceases and cultures enter the stationary phase (81). Some have argued that this timing of synthesis makes sense because antagonistic substances are needed only when growth conditions are unfavorable. Studies on the regulation of antibiotic biosynthesis are thus interesting not only from the perspective of antibiotics, but also because they provide insights into the regulatory mechanisms at work in the starved cell (117). The regulation of expression of the genes for the synthesis of microcins B17 and C7 is the best studied among the ribosomally synthesized peptide antibiotics, and the results are summarized here.

The transcriptional regulation of both the microcin B17 (*mcb*) and microcin C7 (*mcc*) genes has been studied using *lacZ* fusions. Even though both sets of genes are induced by the cessation of growth, the mechanisms by which this is accomplished are different.

Transcription of the *mcb* genes is stimulated by the regulatory protein OmpR (52, 53). This protein, in conjunction with EnvZ, mediates osmoregulation of the major porins, OmpC and OmpF, of *E. coli* (49). Surprisingly, expression of the *mcb* genes is not significantly affected by changes in the osmolarity of the medium, suggesting that the mechanism of activation by OmpR of the *mcb* promoter (P_{mcb}) differs from that at work on the *ompF* and *ompC* promoters (14). The sequences essential for OmpR-dependent activation of P_{mcb} lie between −166 and −54 relative to the start site of transcription. This position is consistent with the location of known OmpR binding sites in other promoters (85, 95). When these sequences are deleted from P_{mcb}, induction of transcription still occurs upon entry into stationary phase (14). OmpR appears to stimulate both the basal level seen during exponential growth phase and the induced level during stationary phase. Thus, while OmpR activates transcription from P_{mcb}, it does not mediate the stationary-phase induction of this promoter.

The sequences responsible for stationary phase induction of P_{mcb} are located between −54 and +10, directly overlapping with the sequences necessary for promoter activity (14). The sequence of this promoter resembles that of several other promoters induced during stationary phase. These have been named gearbox promoters (3). Their sequences do not contain the consensus -10 region typical of promoters recognized by the major sigma factor of *E. coli*, σ^{70} (50). Nonetheless, in a coupled in vitro transcription-translation system, σ^{70} was shown to be capable of initiating transcription from P_{mcb} (14). At present, we do not know what changes, if any, occur that allow the RNA polymerase containing σ^{70} to recognize the unusual -10 region of P_{mcb} during stationary phase. A negative regulatory gene affecting P_{mcb} transcription (*mprA*) has been identified whose product could act as a repressor during exponential growth (23, 24).

Transcription of the *mcc* genes is also induced by the cessation of growth. In this case, however, transcription depends on the presence of a functional *appR* product. This gene has been given several names: *appR* (121), *katF* (89), and *rpoS* (78, 79). It is a regulatory locus for several genes expressed maximally at the onset of stationary phase, and its sequence suggests that it may be a sigma factor (89). Interestingly, the inability to transcribe the *mcc* genes in the absence of AppR can be suppressed by insertions into *osmZ*, the gene encoding the histone-like protein HN-S (57; F. Moreno, unpublished observation).

Several of the peptide antibiotics discussed here share a common motif in the pattern of transcription of their structural genes (see Figure 3). The cluster of genes is transcribed as one (or two) operon(s). However, the transcript corresponding to the structural gene is much more abundant than that of the other genes (7, 8, 15, 38, 73). These small transcripts are also much more stable than most transcripts in bacteria, a feature that may allow for their continued translation even after the cells' transcriptional potential is reduced because of starvation. The case of microcin B17 also presents evidence for translational control. A lag of at least two hours between the appearance of the structural gene transcript and the appearance of its translation product was observed (135).

Recognition and Killing of Sensitive Cells

After the peptide antibiotic has been released into the medium, it must recognize sensitive cells and subsequently begin its lethal or inhibitory action on these cells. For peptides that kill gram-positive bacteria by increasing permeability across the cytoplasmic membrane, the recognition and killing activities may be coupled into one reaction—nonspecific insertion into a phospholipid bilayer. This is probably the case with most lantibiotics, including nisin, Pep5, and subtilin, which form voltage-dependent pores in membranes (75, 107, 116). Surprisingly, some antibiotics of gram-positive origin that kill by increasing membrane permeability (e.g. lactococcin A) have a very narrow spectrum of activity among gram-positive bacteria. When liposomes prepared with phospholipids of sensitive bacteria are treated with lactococcin A, no effect is seen, but membrane vesicles from the same bacterial species are affected by the antibiotic (123). These results have been interpreted to mean that incorporation into the membrane of the sensitive bacterium involves a specific interaction with a receptor protein.

Peptide antibiotics that kill gram-negative bacteria must first move across the outer membrane to reach their target of action. The study of the surface receptors that allow for this translocation has been greatly facilitated by the isolation of mutants resistant to the antibiotic.

For ColV, the only barrier between the outside of the cell and its target of action (the cytoplasmic membrane) is the outer membrane. Mutations in either of two loci are known to result in resistance to the action of ColV. Mutants lacking the outer membrane protein Cir are resistant to ColV (22). This membrane protein normally functions as a receptor for catechol siderophores (93). In addition, cells lacking TonB are resistant to ColV (100); TonB is an inner membrane protein required for translocation of many substances across the outer membrane (97). This protein is thought to transduce the energy available from the cytoplasmic membrane potential for transport processes across the outer membrane. Once in the periplasm, ColV presumably inserts into the cytoplasmic membrane and kills by disrupting the membrane potential (134).

Microcin B17 inhibits DNA replication and induces the SOS repair system of sensitive cells (54, 55). To reach its target of action, presumably DNA gyrase, microcin B17 must translocate across both outer and inner membranes. Mutations leading to resistance against extracellular microcin B17 have been isolated in two loci, *ompF* and *sbmA* (80). The *ompF* gene codes for an outer membrane porin that is most likely part of the receptor in the entry pathway of microcin B17. The product of *sbmA* is an inner membrane protein whose normal function in the cell remains unknown; the only known phenotype of mutations in this locus is resistance to microcin B17. SbmA seems to be required for translocation of microcin B17 from the periplasm into the cytoplasm since microcin B17 inhibits DNA replication in permeabilized cells lacking SbmA (O. Mayo & F. Moreno, unpublished results).

The identification of the target of action of microcin B17 was made possible by the characterization of mutants resistant to the lethal action of internal microcin B17. Producer cells containing a defective immunity mechanism were used to obtain such mutants (127). When microcin B17 producer cells carry a deletion of the immunity gene *mcbG,* they become conditionally lethal (37). They can grow in rich medium, where microcin production is reduced, but cannot grow in minimal glucose medium, where microcin production is substantially increased. Survivors able to produce microcin B17 and also able to grow in minimal medium were isolated. These survivors contained mutations in the *gyrB* gene, coding for the B subunit of DNA gyrase (127). Subsequent studies showed that microcin treatment leads to double-stranded DNA breaks in sensitive cells but not in resistant cells, which is analogous to what is seen after treatment with other antibiotics that inhibit DNA gyrase, such as oxolinic acid (118). Currently, microcin B17 is the only peptide antibiotic known to inhibit gyrase.

SUMMARY

At the outset, we presented a diagram indicating, in theory, the steps involved in the production and action of a peptide antibiotic. By way of summary, Figure 5 presents diagrams outlining the specific components involved in each step in the production and action of microcin B17 and colicin V.

This review has discussed recent advances in the genetic analysis of

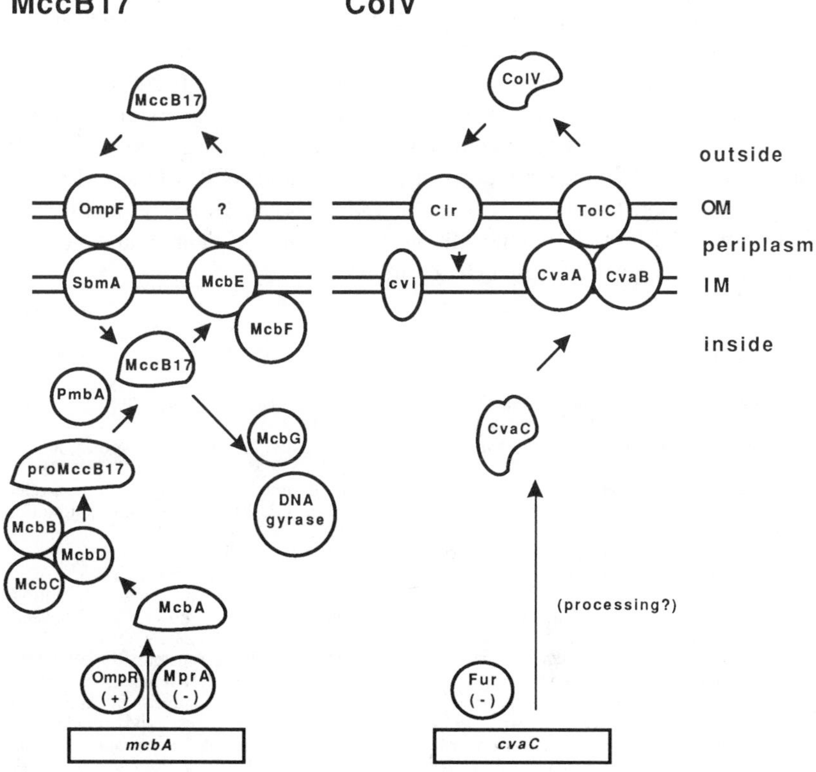

Figure 5 The steps in the production and action of microcin B17 (MccB17) and colicin V (Col V). OM, outer membrane; IM, inner membrane.

ribosomally synthesized peptide antibiotics. One goal has been to bring together two fields that had not been treated jointly before, microcins from gram-negative bacteria and lantibiotics from gram-positive bacteria, because many common strategies are utilized by both types of organisms in the production of these peptide antibiotics. The merged presentation of these two areas of research should help in providing a foundation for future scientific interactions among the investigators involved.

ACKNOWLEDGMENTS

We gratefully acknowledge our collaborators of recent years, the encouragement of Fernando Baquero, and the many colleagues that sent us published and unpublished material to help in the writing of this review. In particular, the contributions of Karl-Dieter Entian and Hans-Georg Sahl in the area of lantibiotics proved timely and invaluable. We also thank the participants of the EMBO-FEMS-NATO Workshop on Bacteriocins, Microcins, and Lantibiotics (September, 1991), who contributed with their discussions to the shaping of this review. We greatly appreciate the help of our coworkers Rachel Skvirsky, Devorah Sperling, Michael Fath, and Peter Yorgey in the critical reading of the manuscript. Finally, we are very grateful to the members of the Paul Ricard Foundation who hosted us in the Island of Bendor where most of this review was envisioned. Work in our laboratories on this subject is funded by grants from the Fondo de Investigaciones Sanitarias (FIS) and Comisión Interministerial de Ciencia y Tecnología (CICYT) to F. M. and from the National Institutes of Health (AI25944) and the Cystic Fibrosis Foundation (Z138) to R. K. In addition, R. K. is the recipient of an American Cancer Society Faculty Research Award.

Literature Cited

1. Aguilar, A., Baquero, F., Martínez, J. L., Asencio, C. 1983. Microcin 15N: a second antibiotic from *Escherichia coli* LP15. *J. Antibiot*. 36:325–27
2. Aguilar, A., Pérez-Díaz, J. C., Baquero, F., Asencio, C. 1982. Microcin 15m from *Escherichia coli:* mechanism of antibiotic action. *Antimicrob. Agents Chemother*. 21:381–86
3. Aldea, M., Garrido, T., Pla, J., Vicente, M. 1990. Division genes in *Escherichia coli* are expressed coordinately to cell septum requirements by gearbox promoters. *EMBO J*. 9:3787–94
4. Allagaier, H., Jung, G., Werner, R. G., Schneider, U., Zähner, H. 1985. Elucidation of the structure of epidermin, a ribosomally synthesized tetracyclic heterodetic polypeptide antibiotic. *Angew. Chem. Int. Ed. Engl.* 24:1051–53
5. Allagaier, H., Jung, G., Werner, R. G., Schneider, U., Zähner, H. 1986. Epidermin: sequencing of a heterodet tetracyclic 21-peptide amide antibiotic. *Eur. J. Biochem*. 160:9–22
6. Asensio, C. J., Pérez-Díaz, C., Martínez, M. C., Baquero, F. 1976. A new family of low molecular weight antibiotics from Enterobacteria. *Biochem. Biophys. Res. Commun*. 69:7–14
7. Augustin, J., Rosenstein, R., Wieland, B., Schneider, U., Schnell, N., et al. 1992. Genetic analysis of epidermin biosynthetic genes and epidermin-negative mutants of *Staphylococcus epidermidis*. *Eur. J. Biochem*. In press
8. Banerjee, S., Hansen, J. N. 1988. Struc-

ture and expression of a gene encoding the precursor of subtilin, a small protein antibiotic. *J. Biol. Chem.* 263:9508–14

9. Baquero, F., Moreno, F. 1984. The microcins. *FEMS Microbiol. Lett.* 23: 117–24
10. Barefoot, S. F., Klaenhammer, T. R. 1983. Detection and activity of lactacin B, a bacteriocin produced by *Lactobacillus acidophilus. Appl. Environ. Microbiol.* 45:1808–15
11. Barefoot, S. F., Klaenhammer, T. R. 1984. Purification and characterization of the *Lactobacillus acidophilus* bacteriocin lactacin B. *Antimicrob. Agents Chemother.* 26:328–34
12. Benedict, R. G., Dwonch, W., Shotwell, O. L., Pridham, T. G., Lindenfelser, L. A. 1952. Cinnamycin, an antibiotic from *Streptomyces cinnamoneus* nov. spec. *Antibiot. Chemother.* 2:591–94
13. Bierbaum, G., Sahl, H. G. 1987. Autolytic system of *Staphylococcus simulans* 22: influence of cationic peptides on activity of N-acetylmuramoyl-L-alanine amidase. *J. Bacteriol.* 169:5452–58
14. Bohannon, D. E., Connell, N., Keener, J., Tormo, A., Espinosa-Urgel, M., et al. 1991. Stationary-phase-inducible "gearbox" promoters: differential effects of *katF* mutations and the role of σ^{70}. *J. Bacteriol* 173:4482–92
15. Buchman, G. W., Banerjee, S., Hansen, J. N. 1988. Structure, expression, evolution of a gene encoding the precursor of nisin, a small protein antibiotic. *J. Biol Chem* 263:16260–66
16. Chehade, H., Braun, V. 1988. Iron-regulated synthesis and uptake of colicin V. *FEMS Microbiol. Lett.* 52:177–82
17. Daeschel, M. A., Klaenhammer, T. R. 1985. Association of a 13.6-megadalton plasmid in *Pediococcus pentosaceus* with bacteriocin activity. *Appl. Environ. Microbiol.* 50:1538–41
18. Davagnino, J., Herrero, M., Furlong, D., Moreno, F., Kolter, R. 1986. The DNA replication inhibitor microcin B17 is a forty-three amino-acid protein containing sixty percent glycine. *Proteins* 1:230–38
19. Davey, G. P. 1984. Plasmid associated with diplococcin production in *Streptococcus cremoris. Appl. Environ. Microbiol.* 48:895–96
20. Davey, G. P., Richardson, B. C. 1981. Purification and some properties of diplococcin from *Streptococcus cremoris* 346. *Appl. Environ. Microbiol.* 41:84–89
21. Davies, J. 1990. What are antibiotics? Archaic functions for modern activities. *Mol. Microbiol.* 4:1227–32
22. Davies, J. K., Reeves, P. 1975. Genetics of resistance to colicins in *Escherichia coli* K-12: cross-resistance among colicins of group B. *J. Bacteriol.* 123:96–101
23. del Castillo, I., Gómez, J. M., Moreno, F. 1990. *mprA*, an *Escherichia coli* gene that reduces growth-phase-dependent synthesis of microcins B17 and C7 and blocks osmoinduction of *proU* when cloned on a high-copy-number plasmid. *J. Bacteriol.* 172:437–45
24. del Castillo, I., González-Pastor, J. E., San Millán, J. L., Moreno, F. 1991. Nucleotide sequence of the *Escherichia coli* regulatory gene *mprA* and construction and characterization of *mprA*-deficient mutants. *J. Bacteriol.* 173: 3924–29
25. Delepelaire, P., Wandersman, C. 1990. Protein secretion in Gram-negative bacteria. *J. Biol. Chem.* 265:17118–25
26. Díaz-Guerra, L., Moreno, F., San Millán, J. L. 1989. *appR* gene product activates transcription of microcin C7 plasmid genes. *J. Bacteriol.* 171:2906–8
27. Dodd, H. M., Horn, N., Gasson, M. J. 1990. Analysis of the genetic determinant for the production of the peptide antibiotic nisin. *J. Gen. Microbiol.* 136:555–66
28. Endicott, J. A., Ling, V. 1989. The biochemistry of P-glycoprotein-mediated multidrug resistance. *Annu. Rev. Biochem.* 58:137–71
29. Fath, M. J., Skvirsky, R., Gilson, L., Mahanty, H. K., Kolter, R. 1992. See Ref. 65. In press
30. Fath, M. J., Skvirsky, R., Kolter, R. 1991. Functional complementation between bacterial MDR-like export systems: colicin V, α-hemolysin, *Erwinia* protease. *J. Bacteriol.* 173:7549–56
31. Felmlee, T., Pellet, S., Welch, R. A. 1985. Nucleotide sequence of an *Escherichia coli* chromosomal hemolysin. *J. Bacteriol.* 163:94–105
32. Frick, K. K., Quackenbush, R. L., Konisky, J. 1981. Cloning of immunity and structural genes for colicin V. *J. Bacteriol.* 148:498–507
33. Gálvez, A., Giménez-Gallego, G., Maqueda, M., Valdivia, E. 1989. Purification and amino acid composition of the peptide antibiotic AS-48 produced by *Streptococcus (Enterococcus) faecalis* subsp. *liquefaciens* S-48. *Antimicrob. Agents Chemother.* 33:437–41
34. Gálvez, A., Maqueda, M., Martínez-Bueno, M., Valdivia, E. 1991. Permeation of bacterial cells permeation of

cytoplasmic membrane and artificial membrane vesicles, channel formation on lipid bilayers by peptide antibiotic AS-48. *J. Bacteriol.* 173:886–92

35. García-Bustos, J. F., Pezzi, N., Asencio, C. 1984. Microcin C7: purification and properties. *Biochem. Biophys. Res. Commun.* 119:779–85
36. García-Bustos, J. F., Pezzi, N., Méndez, E. 1985. Structure and mode of action of microcin C7, an antibacterial peptide produced by *Escherichia coli*. *Antimicrob. Agents Chemother.* 27:791–97
37. Garrido, M. C., Herrero, M., Kolter, R., Moreno, F. 1988. The export of the DNA replication inhibitor microcin B17 provides immunity for the host cell. *EMBO J.* 7:1853–62
38. Genilloud, O., Moreno, F., Kolter, R. 1989. DNA sequence, products, transcriptional pattern of the genes involved in production of the DNA replication inhibitor microcin B17. *J. Bacteriol.* 171:1126–35
39. Gilson, L. 1990. *Signal-sequence-independent export of colicin V*. PhD thesis. Harvard Univ., Cambridge, Mass.
40. Gilson, L., Mahanty, H. K., Kolter, R. 1987. Four plasmid genes are required for colicin V synthesis, export, and immunity. *J. Bacteriol.* 169:2466–70
41. Gilson, L., Mahanty, H. K., Kolter, R. 1990. Genetic analysis of an MDR-like export system: the secretion of colicin V. *EMBO J.* 9:3875–84
42. Glaser, P., Sakamoto, H., Bellalou, J., Ullmann, A., Danchin, A. 1988. Secretion of cyclolysin, the calmodulin-sensitive adenylate cyclase-haemolysin bifunctional protein of *Bordetella pertussis*. *EMBO J.* 7:3997–4004
43. Gonzalez, C. F., Kunka, B. S. 1987. Plasmid-associated bacteriocin production and sucrose fermentation in *Pediococcus acidilactici*. *Appl. Environ. Microbiol.* 53:2534–38
44. Gratia, A. 1925. Sur un remarquable exemple d'antagonisme entre souches de colibacille. *C. R. Soc. Biol.* 93:1040–41
45. Gross, E., Kiltz, H. 1973. The number and nature of α,β-unsaturated amino acids in subtilin. *Biochem. Biophys. Res. Commun.* 50:559–65
46. Gross, E., Morell, J. L. 1967. The presence of dehydroalanine in the antibiotic nisin and its relationship to activity. *J. Am. Chem. Soc.* 89:2791–92
47. Gross, E., Morell, J. L. 1971. The structure of nisin. *J. Am. Chem. Soc.* 93:4634–35
48. Guzzo, J., Duong, F., Wandersman, C., Murgier, M., Lazdunski, A. 1991. The secretion genes of *Pseudomonas aeruginosa* alkaline protease are functionally related to those of *Erwinia chrysanthemi* proteases and *Escherichia coli* α-haemolysin. *Mol. Microbiol.* 5:447–53
49. Hall, M. N., Silhavy, T. J. 1981. The *ompB* locus and the regulation of the major outer membrane porin proteins of *Eshcherichia coli* K-12. *J. Mol. Biol.* 146:23–43
50. Harley, C. B., Reynolds, R. P. 1987. Analysis of the *E. coli* promoter sequences. *Nucleic Acids Res.* 15:2343–61
51. Deleted in proof
52. Hernández-Chico, C., Herrero, M., Rejas, M., San Millán, J. L., Moreno, F. 1982. Gene *ompR* and regulation of microcin B17 and colicin E2 synthesis. *J. Bacteriol.* 152:897–900
53. Hernández-Chico, C., San Millán, J. L., Kolter, R., Moreno, F. 1986. Growth phase and OmpR regulation of transcription of the microcin B17 genes. *J. Bacteriol.* 167:1058–65
54. Herrero, M., Kolter, R., Moreno, F. 1986. Effects of microcin B17 on microcin B17-immune cells. *J. Gen. Microbiol.* 132:403–10
55. Herrero, M., Moreno, F. 1986. Microcin B17 blocks DNA replication and induces the SOS system in *Escherichia coli*. *J. Gen. Microbiol.* 132:393–402
56. Herschman, H. R., Helinski, D. R. 1967. Comparative study of the events associated with colicin induction. *J. Bacteriol.* 94:691–99
57. Higgins, C. F., Dorman, C. J., Stirling, D. A., Waddell, L., Booth, I. R., et al. 1988. A physiological role for DNA supercoiling in the osmotic regulation of gene expression in *S. typhimurium* and *E. coli*. *Cell* 52:569–84
58. Higgins, C. F., Hiles, I. D., Salmond, G. P. C., Gill, D. R., Evans, J. A. D. I. J., et al. 1986. A family of related ATP-binding subunits coupled to many distinct biological processes in bacteria. *Nature* 323:448–50
59. Holo, H., Nilssen, Ø., Nes, I. F. 1991. Lactococcin A, a new bacteriocin from *Lactococcus lactis* subsp. *cremoris*: isolation and characterization of the protein and its gene. *J. Bacteriol.* 173: 3879–87
60. Hopwood, D. A. 1989. Antibiotics: opportunities for genetic manipulation. *Philos. Trans. R. Soc. London (Biol.)* 324:549–62
61. Hopwood, D. A., Sherman, D. H. 1990. Molecular genetics of polyketides and its comparison to fatty acid biosynthesis. *Annu. Rev. Genet.* 24:37–66

62. Hurst, A., Paterson, G. M. 1971. Observations on the conversion of an inactive precursor protein to the antibiotic nisin. *Can. J. Microbiol.* 17:1379–84
63. Ingram, L. 1970. A ribosomal mechanism for synthesis of peptides related to nisin. *Biochim. Biophys. Acta* 224:263–65
64. Ingram, L. C. 1969. Synthesis of the antibiotic nisin: formation of lanthionine and beta-methyl-lanthionine. *Biochim. Biophys. Acta* 184:216–19
65. James, R., Lazdunski, C., Pattus, F., eds. 1992. *Bacteriocins, Microcins, and Lantibiotics.* Heidelberg: Springer/NATO Series H—Cell Biology
66. Jung, G. 1991. Lantibiotics—ribosomally synthesized biologically active polypeptides containing sulfide bridges and α-β-didehydroamino acids. *Angew. Chem. Int. Ed.* 30:1051–68
67. Jung, G., Sahl, H. G., eds. 1991. *Nisin and Novel Lantibiotics: Proceedings of the First International Workshop on Lantibiotics, April 15–18, 1991.* Leiden: ESCOM Sci. Publishers B. V.
68. Kaletta, C., Entian, K.-D. 1989. Nisin, a peptide antibiotic: cloning and sequencing of the *nisA* gene and posttranslational processing of its peptide product. *J. Bacteriol.* 171:1597–1601
69. Kaletta, C., Entian, K.-D., Jung, G. 1991. Prepeptide sequence of cinnamycin Ro09–0198: the first structural gene of a duramycin-type lantibiotic. *Eur. J. Biochem.* 199:411–15
70. Kaletta, C., Entian, K.-D., Kellner, R., Jung, G., Reis, M., Sahl, H.-G. 1989. Pep5, a new lantibiotic: structural gene isolation and prepeptide sequence. *Arch. Microbiol.* 152:16–19
71. Kellner, R., Jung, G., Hörner, T., Schnell, N., Entian, K.-D., Götz, F. 1988. Gallidermin: a new lanthionine containing polypeptide antibiotic. *Eur. J. Biochem.* 177:53–59
72. Klaenhammer, T. R. 1988. Bacteriocins of lactic acid bacteria. *Biochimie* 70:337–49
73. Klein, C., Kaletta, C., Schnell, N., Entian, K.-D. 1992. Analysis of genes involved in the biosynthesis of lantibiotic subtilin. *Appl. Environ. Microbiol.* 58:132–42
74. Kleinkauf, H., von Döhren, H. 1987. Biosynthesis of peptide antiobiotics. *Annu. Rev. Microbiol.* 41:259–89
75. Kordel, M., Benz, R., Sahl, H.-G. 1988. Mode of action of the staphylococcinlike peptide Pep5: voltage-dependent depolarization of bacterial and artificial membranes. *J. Bacteriol.* 170:84–88
76. Kordel, M., Sahl, H.-G. 1986. Susceptibility of bacterial, eukaryotic and artificial membranes to the disruptive action of the cationic peptides Pep5 and nisin. *FEMS Microbiol. Lett.* 34:139–44
77. Kuchler, K., Sterne, R. E., Thorner, J. 1989. *Saccharomyces cerevisiae* STE6 gene product: a novel pathway for protein export in eukaryotic cells. *EMBO J.* 8:3973–84
78. Lange, R., Hengge-Aronis, R. 1991. Growth phase-regulated expression of *bolA* and morphology of stationary phase *Escherichia coli* cells is controlled by the novel sigma factor σ^S (*rpoS*). *J. Bacteriol.* 173:4474–81
79. Lange, R., Hengge-Aronis, R. 1991. Identification of a central regulator of stationary phase gene expression in *Escherichia coli. Mol. Microbiol.* 5:49–59
80. Laviña, M., Pugsley, A. P., Moreno, F. 1986. Identification, mapping, cloning and characterization of a gene (*sbmA*) required for microcin B17 action on *Escherichia coli. J. Gen. Microbiol.* 132:1685–93
81. Martin, J. F., Demain, A. L. 1980. Control of antibiotic synthesis. *Microbiol. Rev.* 44:230–51
82. Martínez-Bueno, M., Gálvez, A., Valdivia, E., Maqueda, M. 1990. A transferable plasmid associated with AS-48 production in *Enterococcus faecalis. J. Bacteriol.* 172:2817–18
83. Mattick, A. T. R., Hirsch, A. 1944. A powerful inhibitory substance produced by group N streptococci. *Nature* 154: 551
84. McGrath, J. P., Varshavsky, A. 1989. The yeast STE6 gene encodes a homologue of the mammalian multidrug resistance P-glycoprotein. *Nature* 340: 400–4
85. Mizuno, T., Mizushima, S. 1986. Characterization by deletion and localized mutagenesis in vitro of the promoter region of the *Escherichia coli ompC* gene and importance of the upstream DNA domain in positive regulation by the OmpR protein. *J. Bacteriol.* 168:86–95
86. Moreno, F., San Millán, J. L., Hernández-Chico, C., Kolter, R. 1992. Genetics of microcin biosynthesis. In *Genetics of Antibiotic Biosynthesis,* ed. L. Vining. Boston: Heinemann. In press
87. Mørtvedt, C. I., Nes, I. F. 1990. Plasmid-associated bacteriocin production by a *Lactobacillus sake* strain. *J. Gen. Microbiol.* 136:1601–7
88. Mørtvedt, C. I., Nissen-Meyer, J., Slet-

ten, K., Nes, I. 1991. Purification and amino acid sequence of lactocin S, a bacteriocin produced by *Lactobacillus sake* L45. *Appl. Environ. Microbiol.* 57:1829–34

89. Mulvey, M. R., Loewen, P. C. 1989. Nucleotide sequence of *katF* of *Escherichia coli* suggests KatF protein is a novel σ transcription factor. *Nucleic Acids Res.* 17:9979–91
90. Muriana, P. M., Klaenhammer, T. R. 1991. Cloning, phenotype expression, and DNA sequence of the gene for lactacin F, an antimicrobial peptide produced by *Lactobacillus* spp. *J. Bacteriol.* 173: 1779–88
91. Muriana, P. M., Klaenhammer, T. R. 1991. Purification and partial characterization of lactacin F, a bacteriocin produced by *Lactobacillus acidophilus* 11088. *Appl. Environ. Microbiol.* 57: 114–21
92. Naruse, N., Tenmyo, O., Tomita, K., Konishi, M., Miyaki, T., Kawaguchi, H. 1989. Lanthiopeptin, a new peptide antibiotic: production, isolation and properties of lanthiopeptin. *J. Antibiot.* 42:837–45
93. Nikaido, H., Rosenberg, E. Y. 1990. Cir and Fiu proteins in the outer membrane of *Escherichia coli* catalyze transport of monomeric catechols: study with β-lactam antibiotics containing catechol and analogous groups. *J. Bacteriol.* 172:1361–67
94. Nishio, C., Komura, S., Kurahashi, K. 1983. Peptide antibiotic subtilin is synthesized via precursor proteins. *Biochem. Biophys. Res. Commun.* 116: 751–58
95. Norioka, S., Ramakrishnan, G., Ikenaka, K., Inouye, M. 1986. Interaction of a transcriptional activator, OmpR, with reciprocally osmoregulated genes, *ompF* and *ompC*, of *Escherichia coli*. *J. Biol. Chem.* 261:17113–19
96. Novoa, M. A., Díaz-Guerra, L., San Millán, J. L., Moreno, F. 1986. Cloning and mapping of the genetic determinants for microcin C7 production and immunity. *J. Bacteriol.* 168:1384–91
97. Postle, K. 1990. TonB and the gram-negative dilemma. *Mol. Microbiol.* 4:2019–25
98. Pugsley, A. P. 1984. The ins and outs of colicins. Part I: production and translocation across membranes. *Microbiol. Sci.* 1:168–75
99. Pugsley, A. P. 1984. The ins and outs of colicins. Part II: lethal action, immunity and ecological implications. *Microbiol. Sci.* 1:203–5
100. Pugsley, A. P., Schwartz, M. 1985. Export and secretion of proteins by bacteria. *FEMS Microbiol. Rev.* 32:3–38
101. Ray, S. K., Kim W. J., Ray, B. 1989. Conjugal transfer of a plasmid encoding bacteriocin production and immunity in *Pediococcus acidilactici* H. *J. Appl. Bacteriol.* 66:393–99
102. Rayman, K., Hurst, A. 1984. Nisin: properties, biosynthesis and fermentation. In *Biotechnology of Industrial Antibiotics*, ed. E. J. Vandamme, pp. 607–28. New York: Dekker
103. Reis, M., Sahl, H. G. 1991. Genetic analysis of the producer self-protection mechanism ("immunity") against Pep5. See Ref. 67, pp. 320–31
104. Riordan, J. R., Rommens, J. M., Kerem, B.-S., Alon, N., Rozmahel, R., et al. 1989. Identification of the cystic fibrosis gene: cloning and characterization of complementary DNA. *Science* 245:1066–72
105. Rodríguez-Sainz, M. C., Hernández-Chico, C., Moreno, F. 1990. Molecular characterization of *pmbA*, an *Escherichia coli* chromosomal gene required for the production of the antibiotic peptide MccB17. *Mol. Microbiol.* 4:1921–32
106. Rogers, L. A. 1928. The inhibiting effect of *Streptococcus lactis* on *Lactobacillus bulgaricus*. *J. Bacteriol.* 16: 321–25
107. Ruhr, E., Sahl, H.-G. 1985. Mode of action of the peptide antibiotic nisin and influence on the membrane potential of whole cells and on cytoplasmic and artificial membrane vesicles. *Antimicrob. Agents Chemother.* 27:841–45
108. Sahl, H. G. 1985. Influence of staphylococcinlike peptide Pep5 on membrane potential of bacterial cells and cytoplasmic vesicles. *J. Bacteriol.* 162:833–36
109. Sahl, H. G., Brandis, H. 1981. Production, purification and chemical properties of an antistaphylococcal agent produced by *Staphylococcus epidermidis*. *J. Gen. Microbiol.* 127:377–84
110. San Millán, J. L., Hernández-Chico, C., Pereda, P., Moreno, F. 1985. Cloning and mapping of the genetic determinants for microcin B17 production and immunity. *J. Bacteriol.* 163:275–81
111. San Millán, J. L., Kolter, R., Moreno, F. 1985. Plasmid genes involved in microcin B17 production. *J. Bacteriol.* 163:1016–20
112. Schillinger, U., Lücke, F.-K. 1989. Antibacterial activity of *Lactobacillus sake* isolated from meat. *Appl. Environ. Microbiol.* 55:1901–6
113. Schnell, N., Engelke, G., Augustin, J.,

Rosenstein, R., Ungermann, V., et al. 1992. Analysis of genes involved in the biosynthesis of lantibiotic epidermin. *Eur. J. Biochem.* 204:57–68

114. Schnell, N., Entian, K.-D., Götz, F., Hörner, T., Kellner, R., Jung, G. 1989. Structural gene isolation and prepeptide sequence of gallidermin, a new lanthionine containing antibiotic. *FEMS Microbiol. Lett.* 58:263–68
115. Schnell, N., Entian, K.-D., Schneider, U., Götz, F., Zähner, H., et al. 1988. Prepeptide sequence of epidermin, a ribosomally synthesized antibiotic with four sulphide-rings. *Nature* 333:276–78
116. Schüller, F., Benz, R., Sahl, H. G. 1989. The peptide antibiotic subtilin acts by formation of voltage-dependent multi-state pores in bacterial and artificial membranes. *Eur. J. Biochem.* 182:181–86
117. Siegele, D. A., Kolter, R. 1992. Life after log. *J. Bacteriol.* 174:345–48
118. Snyder, M., Drlica, K. 1979. DNA gyrase on the bacterial chromosome: DNA cleavage induced by oxolinic acid. *J. Mol. Biol.* 131:287–302
119. Tagg, J. R. 1976. Bacteriocins of gram-positive bacteria. *Bacteriol. Rev.* 40: 722–56
120. Tagg, J. R. 1992. BLIS production in the genus streptococcus. See Ref. 65, pp. 405–8
121. Touati, E., Dassa, E., Boquet, P. L. 1986. Pleiotropic mutations in *appR* reduce pH 2.5 acid phosphatase expression and restore succinate utilization in CRP-deficient strains of *Escherichia coli*. *Mol. Gen. Genet.* 202:257–64
122. Upreti, G. C., Hinsdill, R. D. 1975. Production and mode of action of lactocin 27: bacteriocin from a homofermentative *Lactobacillus*. *Antimicrob. Agents Chemother.* 7:139–45
123. van Belkum, M. J. 1991. *Lactococcal bacteriocins: genetics and mode of action*. PhD thesis. Rijksuniv. Groningen, Groningen, The Netherlands
124. van Belkum, M. J., Hayema, B. J., Geis, A., Kok, J., Venema, G. 1989. Cloning of two bacteriocin genes from a lactococcus bacteriocin plasmid. *Appl. Environ. Microbiol.* 55:1187–91
125. van Belkum, M. J., Hayema, B. J., Jeeninga, R. E., Kok, J., Venema, G. 1991. Organization and nucleotide sequences of two lactococcal bacteriocin operons. *Appl. Environ. Microbiol.* 57:492–98
126. Deleted in proof
127. Vizán, J. L., Hernández-Chico, C., del Castillo, I., Moreno, F. 1991. The peptide antibiotic microcin B17 induces double-strand cleavage of DNA mediated by *E. coli* DNA gyrase. *EMBO J.* 10:467–76
128. von Heijne, G. 1986. A new method for predicting signal sequence cleavage sites. *Nucleic Acids Res.* 14:4683–90
129. von Heijne, G. 1988. Transcending the impenetrable: how proteins come to terms with membranes. *Biochem. Biophys. Acta* 947:307–33
130. Wagner, W., Vogel, M., Goebel, W. 1983. Transport of hemolysin across the outer membrane of *Escherichia coli* requires two functions. *J. Bacteriol.* 154:200–10
131. Wakamiya, T., Ueki, Y., Shiba, T., Kido, Y., Motoki, Y. 1985. The structure of ancovenin, a new peptide inhibitor of angiotensin I converting enzyme. *Tetrahedron Lett.* 26:665–68
132. Wandersman, C., Delepelaire, P. 1990. TolC, an *Escherichia coli* outer membrane protein required for hemolysin secretion. *Proc. Natl. Acad. Sci. USA* 87:4776–80
133. Weil, H.-P. Beck-Sicklinger, A. G., Metzger, J., Stevanovic, S., Jung, G., et al. 1990. Biosynthesis of the lantibiotic Pep5. *Eur. J. Biochem.* 194:217–23
134. Yang, C. C., Konisky, J. 1984. Colicin V-treated *Escherichia coli* does not generate membrane potential. *J. Bacteriol.* 158:757–59
135. Yorgey, P., Lee J., Kolter, R. 1992. The structure and maturation pathway of microcin B17. See Ref. 65, pp. 19–32

Annu. Rev. Microbiol. 1992. 46:165–91

MOLECULAR BIOLOGY OF METHANOGENS

John N. Reeve

Department of Microbiology, The Ohio State University, Columbus, Ohio 43210

KEY WORDS: *Archaea (Archaebacteria)*, thermophily, biotechnology, gene cloning, gene regulation

---

CONTENTS

0066-4227/92/1001-0165$02.00

Abstract

Methanogens are a very diverse group of the *Archaea* (*Archaebacteria*). Their genomic DNAs range from 26 to 68 mol% G+C; they exhibit all known prokaryotic morphologies and inhabit anaerobic environments as varied as the human gut and deep-sea volcanic vents. They are, nevertheless, unified by their ability to gain energy by reducing CO, CO_2, formate, methanol, methylamines, or acetate to methane. Methanogen genes are reviewed and analyzed in terms of their organization, structure, and expression and are compared with their bacterial (eubacterial) and eukaryal (eukaryotic) counterparts. Many methanogens are thermophiles, and some are hyperthermophiles. The influence of these extreme environments on their macromolecular structures is also addressed. Methanogens are oxygen-sensitive, fastidious anaerobes, and therefore their experimental manipulation in research laboratories has been very limited. The majority of the information currently available describing their molecular biology has been gained by gene cloning. With improvements in anaerobic handling procedures, this is beginning to change, and several experimentally tractable regulated systems of gene expression in methanogens are discussed.

Anaerobic biodegradation terminating in methane biogenesis is an established, economically very important biotechnology used world-wide both to reduce waste and to generate fuel-grade biogas. The substantial progress made over the past decade, reviewed here, in understanding the molecular biology of methanogens should now provide a data base for considering genetic approaches to improving this process.

INTRODUCTION

Approximately a decade ago (131), "Genetics with Methanogens" began. This article reports the progress made since that date, not only in methanogen genetics but also in gaining knowledge of their gene structure. Because methanogens are also *Archaea* (7, 79, 179–181), studies of their molecular biology are almost always discussed from two points of view. First, the results are evaluated for their intrinsic novelty and importance; then they are usually also compared to the established norms in *Bacteria* and *Eukarya*. Many methanogens are also thermophiles and some are hyperthermophiles, (7, 27, 67, 79, 91), so methanogen publications frequently also discuss the results obtained in terms of their relevance to thermophily. This review addresses these aspects to the extent that they directly relate to the molecular biology of methanogens, but this is not a review of *Archaea*-ology (22, 34, 188) nor of thermophily. Details of the biochemistry of methanogenesis are also included, where necessary, to introduce the structure and regulation of expression of genes whose products play a direct role in this process, but reviewing methanogenesis in detail is also beyond the scope of this article. Several recent reviews of the phylogenetics, physiology, biochemistry, genetics, and applications of methanogens are available (7, 12, 22, 33, 49, 79, 90, 117, 133, 171, 188).

GENOME SIZE AND STRUCTURE

Denaturation and renaturation experiments showed that the genome of *Methanobacterium thermoautotrophicum* is typically prokaryotic and somewhat smaller than the genome of *Escherichia coli* (110). Using the same procedures, Klein & Schnorr (88) obtained similar results for the genomes of *Methanobrevibacter arboriphilicus, Methanosarcina barkeri,* and *Methanococcus voltae,* and a physical map was recently constructed for the genome of *M. voltae* that confirms these conclusions (146). This genome is a circular double-stranded DNA molecule ~1.9 Mbp in length, or ~45% the size of the *E. coli* genome. Southern hybridization experiments placed genes at positions all around this molecule and gave no evidence for many repetitive DNA sequences. The majority of this archaeal genome therefore appears to be unique sequences, presumably coding regions and intergenic regulatory sequences, as found in bacterial genomes. The fortuitous underrepresentation of the sequence 5'GATC in the genome of *M. voltae* (73) facilitated the construction of *Bam*HI, *Bgl*II, *Bcl*I, and *Pvu*I physical maps. This sequence is, however, present at much higher frequencies in genomic DNAs from other methanogens (P. T. Hamilton & J. N. Reeve, unpublished results). Methanogen genomic DNAs range overall from 26 to 68 mol% G+C (79), although intergenic regions are frequently more A+T rich than the average value for the genome (2, 22). Several restriction enzymes have been isolated from methanogens, and some have been synthesized in *E. coli* by expression of the cloned encoding genes for development as commercial products (100, 137, 160).

Small, basic DNA-binding proteins have been isolated and sequenced from several methanogens (28, 29, 93, 116, 135). Presumably these proteins contribute to the architecture of the methanogen genome in vivo, although this hypothesis has yet to be investigated directly. The genes that encode the subunits of the DNA binding proteins HMf (135) and HMt (R. Tabassum, K. M. Sandman & J. N. Reeve, unpublished results) in *Methanothermus fervidus* and *M. thermoautotrophicum* strain ΔH, respectively, have been cloned and sequenced. The amino acid sequences deduced for HMf and HMt are >80% identical to each other and >30% identical to a consensus sequence for eukaryal core histones (135). HMf and HMt bind to DNA in vitro, forming nucleosome-like structures in which the DNA molecule is wrapped in a positive toroidal supercoil (115, 116); it is not negatively supercoiled, as in eukaryal nucleosomes. The overall superhelicity of a methanogen genome in vivo must, however, reflect not only the effects of DNA-binding proteins, but also the activities of enzymes such as DNA polymerase (186), topoisomerases (20), and RNA polymerase (RNAP) (162) and the contributions of physical parameters, such as the internal salt concentration and temperature (116). In this regard, hyperthermophilic methanogens contain an unusual topoisomerase, designated reverse gyrase (20), and very high intracellular concentrations of an unusual salt, potassium 2'3'(cyclic)-diphosphoglycerate (64, 98). These novel components and DNA binding proteins, such as HMf, may have evolved to protect the genomes of these hyperthermophiles from heat denaturation (116).

EXTRACHROMOSOMAL ELEMENTS

Cryptic plasmids have been isolated from several methanococcal species, *Methanosarcina acetivorans, Methanolobus vulcani, Methanobacterium thermoformicicum,* and *Methanobacterium thermoautotrophicum* strain Marburg (22, 104, 119, 152, 177, 183, 187). Restriction maps have been constructed, and the entire 4439-bp DNA sequence determined for pME2001 from *M. thermoautotrophicum* strain Marburg (18). Although pME2001 contains several open reading frames (ORFs) and is transcribed in vivo (106), no recognizable phenotype is associated with its presence. A *M. thermoautotrophicum* strain Marburg isolate, cured of pME2001, is fully viable. Several derivatives of pME2001 have been constructed for use as cloning vehicles and shuttle vectors, which carry selectable markers plus origins of replication from *Bacteria* and *Eukarya* (105, 107).

Viruses (archaeophages?) have been isolated that productively infect *Methanobrevibacter* species and *M. thermoautotrophicum* strain Marburg (22, 103). Despite the fact that it has a burst size of only 5, phage ψM1, which grows on *M. thermoautotrophicum* strain Marburg, has been studied in some detail (80, 103). Phage particles have a flexible, probably noncontractile tail attached to a polyhedral head. ψM1 particles can package ~30 kb of circularly permuted and terminally redundant phage DNA or concatamers of pME2001 or random fragments of chromosomal DNA. In the latter case, they have been used in generalized transduction to transfer genes between stains of *M. thermoautotrophicum* strain Marburg. A virus-like particle (VLP) has been identified that accumulates in the medium during growth of cultures of *M. voltae* A3 (184). The DNA molecule inside these VLPs, designated pURB600, is circular and 23 kb in length. Identical DNA molecules are present, as plasmids, in the cytoplasm of *M. voltae* A3 cells, and one copy of pURB600 is integrated into the genome. Although infectivity has not been demonstrated, this element has many properties suggesting that it is, or was, a temperate phage.

The only methanogen-derived, mobile genetic element so far identified, ISM1 from *Methanobrevibacter smithii,* has features typical of prokaryotic insertion sequences (57). ISM1 is 1183 bp in length and has 29-bp inverted repeat sequences at its termini. It is present in ~10 copies per genome and duplicates 8 bp at the sites of its insertion. One ORF occupies ~87% of the length of ISM1 and probably encodes the ISM1 transposase. Introduction of a selectable gene, possibly the puromycin resistance gene (*pac*) from *Streptomyces alboniger* (50, 124) or the pseudomonic acid–resistance gene (*ileS*) from a mutant of *M. thermoautotrophicum* (78), into ISM1 without interrupting this ORF might generate very useful transposons.

GENE ORGANIZATION AND EXPRESSION SIGNALS

The recognition of the *Archaea* as a separate biological domain (179–181) promised novelties in their molecular biology, but so far, very few major differences have been forthcoming (22, 34). The organization and mechanisms of gene expression in methanogens resemble the well-established bacterial patterns. Many genes are located in tightly linked clusters, conventionally designated as operons, even though operator sequences per se have not been identified. As in *Bacteria,* these operons are usually transcribed from one or more upstream promoters into polycistronic RNAs. Although introns are present in some tRNA and rRNA genes in halophilic and sulfur-dependent *Archaea* (32, 81, 86), introns have not yet been discovered in methanogens. Some methanogen RNAs do have poly-A$^+$ tails, but these average only 12 bases in length, typical of bacterial poly-A$^+$ sequences (23). Protein-encoding genes employ the standard genetic code, and codon usage patterns reflect both the overall base composition of the methanogen genome and the level of gene expression (30, 87, 170). Codons GTG and TTG are used relatively frequently as translation-initiation codons (17, 21, 38, 44, 88a, 108, 148, 157). Immediately preceding most ORFs is a sequence that, when transcribed, is complementary to the 3' terminal sequence of the methanogen's 16S rRNA. These transcribed regions are presumed to be ribosome-binding sites (RBS) (22, 34). Highly expressed genes have RBSs with more bases complementary to the 16S rRNA sequence than do RBSs upstream of housekeeping genes. As in *Bacteria,* these are considered to be strong RBSs that direct more frequent translation initiation (22, 30, 87). As the 3' terminal sequences of 16S rRNAs are very similar in methanogens and in *Bacteria,* it is assumed, but not proven, that methanogen RBSs can function correctly in bacterial cells. Several sequenced methanogen genes encode proteins that appear to be synthesized initially as precursors, with amino-terminal signal (leader) peptides (21, 35, 84). These signal sequences are not present in the mature proteins and must therefore be removed by maturation systems analogous to the signal peptidase systems of *Bacteria*

(123). As expression of a *secY*-related gene from *Methanococcus vannielii* (3, 96) in *E. coli* suppresses the defective phenotype of an *E. coli secY* mutant (5), some events during membrane insertion and secretion of proteins may be conserved in these two prokaryotes. However, not all of the putative signal peptides identified in methanogens (35, 84) resemble their bacterial counterparts (123).

One conspicuous difference between *Archaea* and *Bacteria* is the subunit composition of their DNA-dependent RNA polymerases (RNAP). Both prokaryotic groups do appear to have only one RNAP, and not three separate enzymes as in *Eukarya*. Archaeal RNAPs are, however, much more complex, containing 8–10 different polypeptide subunits (10, 22, 34, 88a 136, 138, 162, 188), and show no semblance of the standard $\alpha_2\beta\beta'\sigma$ subunit pattern found consistently in bacterial RNAPs. Although immunological cross reactivities and deduced primary amino acid sequences (10, 136, 138, 188) indicate that all RNAPs have a common ancestor, contemporary archaeal RNAPs clearly have structures more similar to eukaryal RNAPs than to bacterial RNAPs (88a, 188). This observation is reflected in the sequences recognized and used as promoters by archaeal RNAPs (2, 22, 24, 25, 34, 51, 61, 163–165, 172, 188). The major element determining transcription initiation by methanogen RNAPs is a TATA box with the consensus sequence 5'TTTA(T/A)ATA. This eukaryal-like element was identified initially on the basis of its conservation upstream of many genes from a wide range of methanogens and was designated *boxA*. A second conserved element, designated *boxB,* with the consensus sequence 5'ATGC, was also identified approximately 25 bp downstream from the TATA box. This region is now known to be the site of transcription initiation. Studies using in vitro transcription systems derived from several different methanogens (47, 48, 51, 61, 89, 161) clearly show that the TATA-box is required for RNAP binding and that transcription is initiated 22–27 bp downstream of the TATA box, at a purine-pyrimidine dinucleotide, ideally within a *boxB* sequence (61). The TATA box appears to be conserved as the primary element determining transcription initiation of both protein- and stable RNA–encoding genes in a wide range of methanogens (22, 48, 51, 61). It is, at least in part, also conserved as a major promoter element in nonmethanogenic *Archaea* (34, 188). Methanogen RNAPs resemble bacterial RNAPs in that they bind directly to promoter-containing DNA sequences (24, 25, 163, 165); however as in *Eukarya,* transcription factors are also required for specific promoter recognition (48).

Transcription termination sites, identified downstream of methanogen genes, appear to conform to one of two motifs (22, 34, 113, 188). In some cases, transcription is terminated following an inverted repeat sequence that, by analogy with ρ-independent terminators in *Bacteria,* probably forms a stem-loop structure in the transcript to direct termination (113). The second transcription terminator appears to be an oligo-T sequence, and in some cases, several tandemly arranged oligo-T sequences reside immediately downstream of methanogen genes. To date, only oligo-T terminators have been found in the hyperthermophile *M. fervidus* (21, 54, 135, 169), suggesting that the spontaneous formation of stem-loop structures in transcripts may not be very effective as a regulatory mechanism in cells growing at >80°C.

STRUCTURE AND ANALYSIS OF CLONED METHANOGEN GENES

Amino Acid and Purine Biosynthetic Pathways

The first methanogen genes cloned and sequenced were obtained by complementation of auxotrophic mutations in *E. coli* (31, 131, 182). Complementing the same mutation with DNA cloned from different methanogens has enabled researchers to isolate and compare the organization and primary sequences of essentially the same gene from different methanogens (31, 58, 112). Methanogen genes isolated using this procedure have been given the designation of the complemented *E. coli* gene. To date these are *hisA, hisI, argG, proC, trpBA,* and *purEK* (8, 31, 57, 58, 108, 112, 143, 168, 182). In some cases, DNA sequencing has revealed additional genes in the same biosynthetic pathway flanking the methanogen gene that provided the initial selection. These additional genes have also been given the designations of their *E. coli* counterparts (108).

Three methanococcal *hisA* genes have been cloned and sequenced that are clearly related to each other (31, 168). Their flanking regions have, however, undergone different genomic rearrangements and do not appear to encode histidine biosynthetic enzymes. A *hisI* gene, which apparently is also within an operon, has been cloned and sequenced from *M. vannielii* (8), but it is not closely linked to the *M. vannielii hisA* gene. Transcription of the *M. vannielii hisA* gene has been investigated in both *M. vannielii* and in *E. coli*. Transcription is initiated at an appropriate distance downstream from a TATA box in *M. vannielii,* but at a different site in *E. coli* (25). Intergenic regions in methanogens are often very A+T rich (2, 22), and therefore sequences that conform to the consensus sequence for *E. coli* promoters occur frequently. Such a sequence presumably functions fortuitously in *E. coli* to direct the transcription of the *M. vannielii hisA* gene. Transcription of the *hisA* gene in *M. voltae* is stimulated by the presence of aminotriazole, an inhibitor of histidine biosynthesis (147).

Related *argG* genes, which encode argininosuccinate synthetase, have been cloned and sequenced from *M. vannielii* and *Methanosarcina barkeri* (111, 112). These genes are flanked by unrelated sequences. Upstream of the *M.*

barkeri argG gene is an ORF that has been designated *carB* as it appears to encode carbamyl phosphate synthetase, an enzyme also involved in arginine biosynthesis. A tandem duplication has been found in all bacterial and eukaryal *carB* genes, and this duplication is also present in the *M. barkeri carB* (G. P. Schofield, personal communication). The duplication must therefore have been present in the ancestral sequence that gave rise to the *carB* genes now found in all three biological domains. The 389-bp intergenic region separating the *carB* and *argG* genes in *M. barkeri* contains six directly repeated copies of a 14-bp sequence, three directly repeated copies of a 29-bp sequence, and a 9-bp inverted repeat, indicating a region of complex regulation. The *M. barkeri argG* gene also complements mutations in the *argA* gene of *Bacillus subtilis* (111).

DNA sequencing has revealed that clusters of tryptophan biosynthetic genes, cloned by *trpBA* complementation, are *trpEGCFBAD* from *M. thermoautotrophicum* strain Marburg (108) and *trpDFBA* from *M. voltae* (143). In *M. thermoautotrophicum,* these *trp* genes form a single operon in which the *trpE* and *trpG* genes overlap by 2 bp and the remainder are separated by intergenic regions ranging from 5 to 56 bp in length. Downstream from the TATA box in the *M. thermoautotrophicum trp* promoter region is a region of dyad symmetry reminiscent of the *E. coli trp* operator sequence. In some *Bacteria,* tryptophan biosynthesis is subject to feedback regulation directed by tryptophan binding to a conserved site in anthranilate synthetase, the product of the *trpE* gene. This site is also conserved in the *M. thermoautotrophicum trpE* gene product.

Genes have been cloned from *M. thermoautotrophicum* strain ΔH and *Methanobrevibacter smithii* that complement mutations in the *purE* locus of *E. coli* (57, 58). This locus was subsequently shown to contain two genes, *purE* encoding 5'phosphoribosyl-5-aminoimidazole carboxylase and *purK* encoding a CO_2-binding activity (166). As the methanogen *purE* gene products complement mutations in both the *E. coli purE* and *purK* genes, they must embody both activities. The CO_2 binding function (*purK* complementation) has been located, by Tn*5* mutagenesis, in the carboxyl terminal region of the *M. smithii purE* gene product (57, 166).

A glutamine synthetase (GS)–encoding gene (*glnA*) was cloned from *M. voltae* by using a bacterial *glnA* gene as a heterologous hybridization probe (126). The methanogen *glnA* gene appears to form a monocistronic transcriptional unit. The *boxA* and *boxB* promoter elements upstream of the *glnA* gene are separated by a palindromic sequence similar to a palindrome found in the regulatory region upstream of the *Methanococcus thermolithotrophicus nifH1* gene (150). In enterobacteria, GS activity is regulated by adenylation of a conserved tyrosine, but this residue is replaced by a phenylalanine in the *M. voltae* GS, arguing against conservation of this regulatory mechanism.

Methanogenesis-Related Functions

Methanogens employ hydrogenase, formate dehydrogenase, carbon monoxide dehydrogenase, and secondary alcohol dehydrogenase activities to obtain reducing equivalents for methanogenesis from molecular hydrogen, formate, acetate, and secondary alcohols, respectively. Methyl groups, either produced by the reduction of CO, CO_2, or formate or obtained directly from methanol, methylamines, or acetate, are then reduced by methyl reductase in all methanogens to generate methane (7, 12, 79, 133, 139).

Most methanogens contain at least two different hydrogenases, the cofactor F_{420}–reducing hydrogenase (FRH) and the methyl-viologen–reducing hydrogenase (MVH) (129). Genes encoding these enzymes have been cloned and sequenced from *M. thermoautotrophicum* strain ΔH (1, 130). The α, β, and γ subunits of FRH are encoded by the *frhA, frhB,* and *frhG* genes, arranged *frhADGB* within a single transcriptional unit. The function of the *frhD* gene product is unknown, but based on its sequence, its product is related to the product of the *hydD* gene located within a hydrogenase-encoding gene cluster in *E. coli* (1). The α, γ, and δ subunits of MVH are encoded by the *mvhA, mvhG,* and *mvhD* genes, respectively, which are also located within a single transcriptional unit, arranged *mvhDGAB* (130). The *mvhB* gene encodes a protein that has been designated as a polyferredoxin because it is predicted to contain six tandemly arranged ferredoxin-like domains. The precise function of the polyferredoxin is currently unknown, but it seems likely to be involved in electron transport, most probably accepting electrons from MVH. The polyferredoxin has been isolated from extracts of *M. thermoautotrophicum* strain ΔH and demonstrated in *M. thermoautotrophicum* strain Marburg, and in *M. fervidus* by Western blotting, using antibodies raised against the product of a *lacZ-mvhB* gene fusion synthesized in *E. coli* (V. J. Steigerwald, T. D. Pihl & J. N. Reeve, unpublished results). Polyferredoxin-encoding *mvhB* genes, linked to MVH-encoding genes, have also been cloned and sequenced from *M. fervidus* (154) and *M. voltae* (56). These polyferredoxins are also predicted to contain six ferredoxin-like domains. The intergenic region upstream of the MVH operon has been sequenced from three different strains of *M. thermoautotrophicum*. A sequence, over 100 bp in length, is highly conserved in all three upstream regions and includes the TATA-box promoter element (129). Comparisons of amino acid sequences have identified cysteine and histidine residues conserved in the methanogen FRH and MVH and in several bacterial [Ni,Fe]-hydrogenases. These residues are likely to be involved in metal cofactor binding. Their conservation indicates a common evolutionary ancestry for all these prokaryotic [Ni,Fe]-hydrogenases (1, 129, 130, 154).

The genes *fdhA* and *fdhB,* which encode the two subunits of formate dehydrogenase (FDH) in *Methanobacterium formicicum,* have been cloned and sequenced (141). They overlap by 1 bp and are cotranscribed in the order *fdhAB* as part of a 12-kb transcript. FDH contains a molybdopterin cofactor, and starvation for molybdenum increases transcription of these *fdhAB* genes. Similar sequences upstream of the FDH-encoding genes in *M. formicicum* and in *E. coli* suggest a conserved regulatory function (122). In methanogens, both FDH and FRH catalyze hydride transfer to cofactor F_{420}. Conserved amino acid sequences identified in the β-subunits of both FDH and FRH are therefore strong candidates for the sites of this common activity (1).

Carbon monoxide dehydrogenase (CODH) catalyzes both CO reduction and C-C bond cleavage during catabolism of acetate to methane (52a). Synthesis of CODH in *Methanosarcina* species is stimulated by the presence of acetate and inhibited by the energetically more productive methanogenic substrates, methanol and methylamines (13, 159). In *M. thermophila,* acetate increases transcription of the gene (*cdhA*) that encodes the large subunit of CODH (K. R. Sowers, T. T. Thai & R. P. Gunsalus, personal communication). In *Methanothrix soehngenii,* the large and small subunit-encoding genes of CODH, *cdhA* and *cdhB,* respectively, are separated by only 19 bp and are cotranscribed in the order *cdhAB* (39). These genes have been expressed in *E. coli,* but the polypeptides synthesized lacked catalytic activity. In *M. soehngenii,* the substrate for CODH activity, acetyl-coenzyme A, is generated from acetate by acetyl coenzyme A synthetase (ACS). The gene (*acs*) that encodes this enzyme has also been cloned and sequenced (38). It forms a monocistronic transcriptional unit that, when expressed in *E. coli,* directs the synthesis of an active enzyme.

In the final step in methanogenesis, a methyl moiety attached to coenzyme M (CH_3-S-CoM) is reduced to methane. This reaction is catalyzed by methyl coenzyme M reductase (MCR), an enzyme that appears to be conserved in all methanogens. The genes *mcrA, mcrB,* and *mcrG* that encode the α, β, and γ subunits of MCR have been cloned and sequenced from *M. vannielii, M. voltae, M. barkeri, M. thermoautotrophicum* strain Marburg, and *M. fervidus* (16, 17, 30, 87, 169, 170) and cloned and partially sequenced from *Methanopyrus kandleri* (J. R. Palmer, C. J. Daniels & J. N. Reeve, unpublished results). In every case, they are arranged within a single transcriptional unit, *mcrBDCGA,* designated the *mcr* operon (87, 170). The functions of the *mcrD* and *mcrC* gene products remain unknown, although an association of *gpmcrD* with MCR has been demonstrated in extracts of *M. vannielii* (140). As all *mcr* operons appear to have evolved from a common ancestor and because MCR retains a common function, primary sequence comparisons should provide valid phylogenetic data. They should also be useful in correlating amino acid substitutions with mesophilic, thermophilic, and hyperthermophilic growth (87, 170). Two MCR isoenzymes (MCRI and MCRII) were recently isolated and characterized from *M. thermoautotrophicum* strain Marburg (132). Southern blot analyses corroborated this discovery, revealing two related but clearly separate *mcr* operons in genomic DNA from this methanogen (63). Two *mcr* operons were also detected in genomic DNA from *M. thermoautotrophicum* strain ΔH, *Methanobacterium wolfei,* and *M. fervidus,* but only one *mcr* operon was apparent in genomic DNAs from *M. vannielii, M. voltae, M. thermolithotrophicus,* and *M. kandleri* (63; A. N. Hennigan, J. R. Palmer, C. J. Daniels & J. N. Reeve, unpublished results). In *M. thermoautotrophicum* strain Marburg, MCRII is synthesized predominantly during exponential growth and MCRI during less active growth and in stationary phase. The *mcr* operon first cloned and sequenced from *M. thermoautotrophicum* encodes MCRI (16, 132). The MCRII operon has now been cloned and is located immediately downstream of the *mvh* operon (130).

One of the reactions on the reductive pathway from CO_2 to CH_4 transfers a formyl group from the cofactor methanofuran to a second cofactor, tetrahydromethanopterin (12, 133, 139, 171). The gene (*ftr*) that encodes formylmethanofuran:tetrahydromethanopterin formyl transferase, the enzyme that catalyzes this reaction, has been cloned and sequenced from *M. thermoautotrophicum* strain ΔH and functionally expressed in *E. coli* (37). It appears to be the promoter distal gene in an operon, but the other genes within this unit have not been identified. Several of the other enzymes involved in the reductive pathway from CO_2 to CH_4 have also been purified (101, 139). Cloning of their encoding genes should now be possible, especially as aminoterminal amino acid sequences are already available (101).

ATPases

The genes *atpA* and *atpB* that encode the α and β subunits, respectively, of a membrane-bound ATPase in *M. barkeri* have been cloned and sequenced (69). They are separated by only 2 bp and are cotranscribed in the order *atpAB*. They appear to have evolved from the same ancestral sequence, following a gene duplication event, and to have the same common ancestor as bacterial F_1 H^+-ATPases and eukaryal vacuolar H^+-ATPases. In contrast, the sequence of a gene that encodes a vanadate-sensitive ATPase in *M. voltae* bears no obvious relation to any previously sequenced gene (35, 36). This *M. voltae* ATPase appears to be synthesized as a precursor with an aminoterminal signal peptide containing 12 amino acid residues that are not present in the mature protein.

Cofactor Biosynthesis

S-adenosyl-L-methionine:uroporphyrinogenIII methyltransferase (SUMT) catalyzes the synthesis of precorrin II, a precursor of the MCR cofactor F_{430}. The SUMT-encoding gene (*cofA*) cloned and sequenced from *Methanobacterium ivanovii* contains sequences also found in SUMT-encoding genes sequenced from several different *Bacteria* (15).

Intermediary Metabolism

GLYCERALDEHYE-3-PHOSPHATE DEHYDROGENASE Genes (*gap*) encoding glyceraldehye-3-phosphate dehydrogenase (GAPDH) have been cloned and sequenced from *Methanobrevibacter bryantii, Methanobacterium formicicum,* and *Methanothermus fervidus* (43, 45, 65). The sequences of the two GAPDHs from the mesophiles *M. bryantii* and *M. formicicum* are ~90% identical to each other but only ~70% identical to the sequence of the GAPDH from the hyperthermophile, *M. fervidus*. The *M. fervidus gap* gene has been expressed in *E. coli,* and the GAPDH synthesized was found to be active and to have retained its inherent heat resistance (46). Mosaic *gap* genes containing regions of the *M. fervidus gap* gene fused to regions of the *M. bryantii gap* gene have also been constructed and expressed in *E. coli* (14). The products of these gene fusions retain most of the heat resistance of the *M. fervidus* enzyme when they contain a short amino acid sequence from the carboxyl terminal region of the *M. fervidus* protein (14).

L-MALATE DEHYDROGENASE Determining the sequence of the L-malate dehydrogenase (MDH)-encoding gene (*mdh*) cloned from *M. fervidus* (66) revealed only very limited similarity to bacterial and eukaryal MDHs. Bacterial and eukaryal MDHs have amino acid sequences more similar to the sequences of their L-lactate dehydrogenases (LDH) than to the MDH from *M. fervidus*. Therefore, MDH- and LDH-encoding genes in *Bacteria* and *Eukarya* apparently diverged from each other after the progenitor of this methanogen MDH-encoding sequence separated from the common ancestral line.

3-PHOSPHOGLYCERATE KINASE The 3-phosphoglycerate kinase (PGK)-encoding genes cloned and sequenced from *M. bryantii* and *M. fervidus* encode polypeptides with amino acid sequences that are 61% identical to each other and 32–36% identical to the sequences of bacterial and eukaryal PGKs (44). Amino acid substitutions, which may confer increased resistance to heat denaturation, have been identified and compared in the *M. fervidus* PGK, MDH, GAPDH, and MCR sequences (43–45, 66, 169).

Superoxide Dismutase

Although methanogens evolved in an anoxic world and are strict anaerobes, a superoxide dismutase (SOD)-encoding gene (*sod*) has been cloned and sequenced from *M. thermoautotrophicum* strain Marburg and expressed in *E. coli* (157). The 205–amino acid sequence encoded by this *sod* gene most closely resembles the sequences of Mn-SODs, but atomic absorption spectroscopy has established that the enzyme is a Fe-SOD (158). The intergenic region upstream of the *sod* gene contains a TATA box and an inverted repeat sequence similar to an inverted repeat sequence found, at the same location, upstream of a SOD-encoding gene in *Halobacterium halobium* (156). Within the coding sequence of the *M. thermoautotrophicum sod* gene, immediately preceding the translation-termination codon, is an inverted repeat sequence that may be a transcription terminator.

Nitrogen Fixation

Diazotrophic growth occurs in several methanogens (9, 114). The *Anabaena nifH* gene hybridized to sequences in genomic DNAs from 14 different methanogens, and the *nifDK* region from *Klebsiella pneumoniae* hybridized to genomic DNAs from 4 methanogens (125, 144, 145). Two different *nifH* genes (*nifH1* and *nifH2*) have been cloned and sequenced from *M. ivanovii, M. thermolithotrophicus,* and *M. barkeri* (142, 148, 150), and one *nifH* gene has been cloned from *M. voltae* (149), even though this methanogen does not exhibit diazotrophic growth. The products of these methanogen *nifH* genes, which all appear to be the Fe-protein components of MoFe nitrogenases, have amino acid sequences that are only 47–55% conserved. In contrast, the sequences of all *gpnifH*s from the bacterial domain are 67–97% conserved. This extreme divergence of methanogen *nifH* genes indicates that nitrogen fixation is likely to be a very ancient property of methanogens (145).

The methanogen *nifH1* genes reside within conserved gene clusters (142, 144–150). In *M. thermolithotrophicus,* the arrangement is *nifH*-ORF105-ORF128-*nifD-nifK* in which *nifD* and *nifK* overlap by 8 bp. They form two transcriptional units, *nifH1*-ORF105-ORF128 and *nifD-nifK,* which are transcribed only under diazotrophic growth conditions. The genes downstream of *nifH1,* designated ORF105-ORF128 in *M. thermolithotrophicus,* ORF105-ORF122 in *M. barkeri,* and ORF105-ORF123 in *M. ivanovii,* are all related and appear to have diverged following an ancestral gene duplication event. These ORFs all encode proteins with sequences ~50% identical to the sequences of the *glnB*-encoded P_{II}-proteins that regulate Fe-protein (*gpnifH*) activity in *Bacteria* (142).

The *nifH2* in *M. thermolithotrophicus* also appears to be within a polycistronic transcriptional unit arranged ORF162-*nifH2*-ORF102, but transcription of this unit has not been detected so far in vivo, even under diazotrophic growth conditions (148).

Surface-Layer Glycoproteins

Methanothermus fervidus and *Methanothermus sociabilis* are closely related hyperthermophiles that grow at temperatures >80°C. They have surface layers (S-layers) composed of glycoprotein arrays, and the genes (*slgA*) encoding the protein components of these glycoproteins have been cloned and sequenced (21). These two *slgA* genes are almost identical, encoding proteins containing 593 amino acid residues that differ at only 3 positions. Both proteins appear to be synthesized with amino-terminal signal peptides, containing 22 amino acid residues, that must be removed during translocation of the proteins to the cell surface. The secondary structures and protease-cleavage sites predicted for these methanogen signal peptides conform well to the consensus structures for such elements in *Bacteria* (123). The carbohydrate moiety, which contributes ~17% of the mass of these glycoproteins, is composed of mannose, 3-0-methyl glucose, galactose, N-acetyl glucosamine, and N-acetyl galactosamine (60).

Flagellins

Archaeal flagella are thinner than bacterial flagella and may contain several different flagellin subunits (85, 151). Four adjacent and related flagellin genes, which form two transcriptional units, *flaA* and *flaB1-flaB2-flaB3,* in the genome of *M. voltae,* have been cloned and sequenced (84). The encoded flagellins appear to be synthesized as precursors with signal peptides that contain only 11 or 12 amino acid residues and that do not resemble bacterial signal peptides (123). This precursor structure appears to be conserved in flagellins in the nonmethanogenic, halophilic *Archaea* (85). The sequences of the *M. voltae* flagellins contain potential sites for glycosylation, although glycosylation per se has not been demonstrated (84).

COMPONENTS OF THE TRANSLATION MACHINERY

Ribosome Structure

The *Archaea* were identified initially as a separate and coherent phylogenetic group on the basis of their 16S rRNA sequences (7, 79, 179–181). Subsequent studies of 5S rRNAs, 23S rRNAs, and ribosomal proteins (r-proteins) have consolidated the archaeal domain. Archaeal ribosomes are intermediate in size between bacterial and eukaryal ribosomes (3). They contain bacterial-sized rRNAs but have additional r-proteins, some of which are related to r-proteins previously found only in eukaryal ribosomes (3, 4, 6, 96, 155). Antibiotics that inhibit translation by modifying or binding to specific targets in bacterial and/or eukaryal ribosomes have been used to probe for the presence of these targets in archaeal ribosomes. Mosaic patterns of bacterial and eukaryal sensitivities to antibiotics have been obtained (22, 34, 42, 55, 59, 68, 72, 90, 117, 124, 128).

Elongation Factors and Ribosomal Proteins

Diphtheria toxin inhibited translation in an in vitro system derived from *Methanococcus vannielii* (95), indicating the presence of elongation factors (EF) related to eukaryal EFs. This result was confirmed by cloning and sequencing the *M. vannielii* genes encoding EF-2 and EF-1α (3, 70, 94–96).

These genes are, however, located within an operon, arranged ORF1-ORF2-S12-S7-EF-2-EF-1α-S10, which is very similar to the *E. coli* streptomycin operon except for the presence of the additional ORFs and the gene encoding the r-protein S10. ORF1 encodes a protein with an amino acid sequence 37% identical to that of the rat r-protein L30 (3, 96). Approximately 30 kb of genomic DNA separate this operon from the *M. vannielii* equivalents of the *E. coli* S10 and spectinomycin operons. The S10-like operon in *M. vannielii* has genes arranged L22-S3-L29, which is the same gene order as in the *E. coli* S10 operon, but lacks several r-protein-encoding genes, including the S10 gene. Immediately downstream from this operon is the *M. vannielii* equivalent of the *E. coli* spectinomycin operon that contains genes arranged ORFa-ORFb-S17-L14-L24-ORFc-L5-S14-S8-L6-ORFd-ORFe-L18-S5-L30-L15 (3, 4). This is the same gene order as in the *E. coli* operon but also contains the S17 gene and ORFs c, d, and e, which encode r-proteins related to mouse L32, rat L19, and yeast S6. A promoter upstream of ORFa directs the synthesis of an 0.8-kb transcript that terminates following ORFb and an 8-kb transcript that terminates downstream of L15 (4). A second promoter, located upstream of the S17 gene, directs the synthesis of a 7-kb transcript that also terminates downstream of L15. An additional example of the similar but different arrangement of r-protein genes is the L1-L10-L12 transcriptional unit found in *M. vannielii* (6). In *E. coli,* these genes are arranged L11-L1-L10-L12-β-β'. They form two transcriptional units, L11-L1 and L10-L12, plus the β-β' genes that encode these subunits of *E. coli* RNAP (127). The leader sequences of both the methanogen and the bacterial L1–containing transcripts can form secondary structures that resemble the sites on the 23S rRNAs in their ribosomes where the L1 protein binds. In *E. coli,* excess L1 protein binds to this region of its own mRNA and inhibits translation, providing a feed-back mechanism that regulates the level of synthesis of both the L1 and L11 proteins. The conservation of this structure in the *M. vannielii* L1-L11-L12 transcript suggests that this autogenous regulatory system for r-protein synthesis may also exist in this methanogen (6). The *M. vannielii* gene encoding the equivalent of the β' subunit of *E. coli* RNAP has also been located. It is separated from its L10 and L12 genes and is located, together with the gene (*rpoH*) encoding subunit H of its RNAP (88a), immediately upstream of the *M. vannielii* equivalent of the streptomycin operon (3).

Ribosomal RNAs and 7S RNA

Most methanogen rRNA-encoding genes are arranged, as in *Bacteria,* in 16S-23S-5S operons that appear to be cotranscribed into primary transcripts that must then be cleaved and processed to generate the mature rRNAs. The sequences between the rRNA coding regions can be arranged into secondary structures very similar to RNaseIII recognition and cleavage sites in *E. coli.*

There are two such rRNA operons in the genomes of *M. thermoautotrophicum* (121, 178), *M. formicicum* (97), *M. fervidus* (53), and *M. soehngenii* (40, 41); one in *M. voltae* (175); and four in *M. vannielii* (74–77). A $tRNA^{Ala}$ gene resides in the spacer region between the 16S and 23S rRNA-encoding sequences in most of these operons (75). In *M. thermoautotrophicum* and *M. fervidus,* 7S RNA- and $tRNA^{Ser}$-encoding genes are located immediately upstream of one of their two rRNA operons, forming a 7SRNA-$tRNA^{Ser}$-16SrRNA-$tRNA^{Ala}$-23SrRNA-5SrRNA cluster of genes that may be a single transcriptional unit (53, 121). Large amounts of a 7S RNA molecule are present in all archaeal species so far investigated (99); however, this molecule's function(s) remains unknown. The primary sequences and predicted secondary structures for archaeal 7S RNAs are similar to the 7S RNA components of eukaryal signal-recognition particles and to the small cytoplasmic scRNA of *Bacillus subtilis* and the 4·5S RNA of *E. coli* (53). Genes encoding 7S RNAs have also been cloned and sequenced from *M. voltae* (83) and *Methanosarcina acetivorans* (82), but linkage to rRNA operons was not reported.

M. vannielii and *M. voltae* contain one and two additional 5S rRNA-encoding genes, respectively, clustered with tRNA genes (173, 175). These 5S rRNAs have sequences that are more similar to each other than to the sequences of the 5S rRNAs encoded in their rRNA operons. The sequence of the *Methanopyrus kandleri* 16S rRNA indicates that this hyperthermophile belongs to an entirely separate phylogenetic lineage (27, 67, 91). Preliminary evidence also indicates that this methanogen has only one 16S and one 23S rRNA-encoding gene and that, unlike the situation in all other methanogens, these rRNA genes are not closely linked to each other nor to tRNA-encoding genes (J. R. Palmer, C. J. Daniels & J. N. Reeve, unpublished results).

tRNAs and Amino Acyl-tRNA Synthetase

Many tRNA genes have been cloned and sequenced from several different methanogens (22). Similar clusters of tRNA genes have been found in *M. vannielii, M. voltae,* and *M. fervidus* (54, 173–175). The secondary structures predicted for the same tRNAs, from mesophilic and thermophilic methanogens, indicate that tRNAs from thermophilic methanogens have more base pairs and preferentially employ G+C base pairs (54). The four tRNA genes cloned from *Methanopyrus kandleri* encode tRNAs with double-stranded regions formed almost exclusively by G+C pairs (J. R. Palmer, C. J. Daniels & J. N. Reeve, unpublished results). These *M. kandleri* tRNA genes also encode the 3'-terminal CCA sequence found in the mature tRNAs. With few exceptions (173, 174), this is not usually the case for methanogen tRNA genes. Although introns have been discovered in tRNA genes from halophilic *Archaea* (32) and in tRNA and rRNA genes from the sulfur-dependent thermophilic *Archaea* (81, 86), they have not yet been reported in methanogen genes.

Pseudomonic acid is a competitive inhibitor of isoleucyl-tRNA synthetases. The gene (*ileS*) encoding this synthetase has been cloned and sequenced from both the wild-type and a spontaneously pseudomonic acid–resistant mutant of *M. thermoautotrophicum* strain Marburg (78). The primary sequence of the methanogen isoleucyl-tRNA synthetase conforms well to the consensus sequence for class I isoleucyl-tRNA synthetases derived from bacterial and eukaryal sequences (26). The mutation that confers resistance to pseudomonic acid is a single base change in the *ileS* gene. A glycine residue, at position 590 of the 1045–amino acid sequence, is replaced by an aspartic acid residue, resulting in an enzyme with a much reduced affinity for pseudomonic acid (78). Both the wild-type and mutated *ileS* genes have been expressed in *E. coli,* but the mutated gene did not provide *E. coli* with a substantial increase in pseudomonic acid resistance. The *ileS* gene in *M. thermoautotrophicum* appears to be cotranscribed with an upstream ORF of unknown function and with a downstream gene (*purL*) predicted to encode formylglycineaminidine ribonucleotide synthase II (78).

REGULATED SYSTEMS OF GENE EXPRESSION

Experimental manipulations with growing cultures of methanogens are technically difficult. Simple procedures, such as changing the growth medium without interrupting growth, are not trivial tasks when the growth substrate is a pressurized, strictly anaerobic mixture of gases. Some methanogens are even sensitive to light (120). As a consequence, very few studies have been designed specifically to investigate gene regulation in growing methanogens. Different enzyme profiles have been reported in cells of the same methanogen grown under different conditions (13, 118, 153, 159), indicating regulated gene expression, but rarely have the effects of changes in growth conditions on gene expression in growing cells been investigated. This will change as handling procedures continue to improve and tractable systems for study are identified. In this regard, already documented is that substrate availability regulates CODH synthesis (13, 159), fixed nitrogen regulates *nif* gene expression (150), growth phase regulates *mcr* expression in methanogens with two *mcr* operons (132), and molybdenum and nickel availabilities regulate formate dehydrogenase (102, 153) and hydrogenase synthases (52, 129), respectively. As described in this review, the regulated genes in all these systems are already in hand and regulatory studies can therefore be undertaken. It has also been shown that *M. voltae* exhibits a classic heat shock response (62), that *M. thermolithotrophicus* synthesizes additional proteins in response to increased hydrostatic pressure (71), and that the synthases of secondary alcohol dehydrogenases and methyl transferases are substrate dependent (118, 167, 176). Although the structural genes regulated in these systems have not yet been cloned, these are clearly experimentally tractable systems that could also be developed to study gene regulation in methanogens. A heat-shock *dnaK*-homologous gene has been cloned and sequenced from *Methanosarcina mazei* S6 (101a).

METHANOGEN GENETICS

Methanogens are resistant to most of the antibiotics routinely used as selective agents in bacterial genetics (22, 117, 128). In addition, some of the few effective antibiotics are not stable for sufficient lengths of time, at elevated temperatures, to be useful with the thermophilic methanogens widely used for biochemical studies (133, 139). Despite these problems, inhibitor-based selections have been developed and resistant mutants isolated (22, 55, 59, 68, 72, 78, 90, 92, 107, 109, 117, 124, 128). Techniques to enrich for auxotrophs, analogous to the penicillin-enrichment procedure used with bacteria, have also been developed that employ bacitracin with *Methanobacterium* species (59, 72) and base analogs with methanococci (92, 109).

Generalized transduction of chromosomal DNA between wild-type and auxotrophic mutants of *M. thermoautotrophicum* strain Marburg has been obtained using phage ψM1 (80, 103). Chromosomal transformation has also been demonstrated for this methanogen by mixing donor DNA directly with cells grown on the surface of a gellan gum–solidified medium (185). Chromosomal transformation of *Methanococcus voltae* PS and *Methanococcus maripaludis* has been achieved by incubating donor DNA with suspensions of recipient cells (11, 50, 109, 134). Electroporation increases the frequency of this chromosomal transformation of *M. voltae* PS (109). Transformation of a methanogen resulting in a stably replicating extrachromosomal element has yet to be reported. Genes conferring resistance to inhibitors have been combined with methanogen-derived plasmids and origins of replication (50, 105, 107, 124), but transformation with these constructs has not yet resulted in an autonomously replicating, resistance-conferring plasmid in a methanogen. A very promising observation is that a *Streptomyces alboniger pac* gene, which encodes puromycin acetyl transferase, confers puromycin resistance on *M. voltae* and *M. maripaludis* when introduced into these methanococci by transformation (50, 134). Resistance is obtained by integrating the *pac* gene into the methanogen's genome via homologous recombination, using flanking methanogen sequences. This procedure can be used to direct the insertion of the *puc* gene into specific chromosomal locations.

Substantial progress has been made during the past decade in studying genes from methanogens. Transformation systems are now needed to return these genes to their native environments to analyze the consequences of in vitro manipulations. Methanogens play such very important roles in the environment and in the economies of the world (33, 49) that their continued study is assured. Undoubtedly, investigations of their molecular biology will continue, perhaps masquerading as biotechnology (90), and will contribute substantially during the next decade to our understanding of the roles of methanogens in anaerobic biodegradation, waste treatment, biogas production, and global warming.

ACKNOWLEDGMENTS

Funding for methanogen research in the author's laboratory is provided by the Basic Biology Program of the Department of Energy and the Biological Sciences Division of the Office of Naval Research. Their firm support, during the fledgling stages of this science, has been very important.

Literature Cited

1. Alex, L. A., Reeve, J. N., Orme-Johnson, W. H., Walsh, C. T. 1990. Cloning sequence determination, and expression of the genes encoding the subunits of the nickel-containing 8-hydroxy-5-deazaflavin reducing hydrogenase from *M. thermoautotrophicum* strain ΔH. *Biochemistry* 24:7237–44
2. Allmansberger, R., Knaub, S., Klein, A. 1988. Conserved elements in the transcription initiation regions preceding highly expressed structural genes of methanogenic archaebacteria. *Nucleic Acids Res.* 16:7419–35
3. Auer, J., Lechner K., Böck, A. 1989. Gene organization and structure of two transcriptional units from *Methanococcus* coding for ribosomal proteins and elongation factors. *Can. J. Microbiol.* 35:200–4
4. Auer, J., Spicker, G., Böck, A. 1989. Organization and structure of the *Methanococcus* transcriptional unit homologous to the *Escherichia coli* 'spectinomycin operon'. *J. Mol. Biol.* 209:21–36
5. Auer, J., Spicker, G., Böck, A. 1991. Presence of a gene in the archaebacterium *Methanococcus vannielii* homologous to *secY* of eubacteria. *Biochimie (Paris)* 73:683–88
6. Baier, G., Piendl, W., Redl, B., Stöffler, G. 1990. Structure, organization and evolution of the L1 equivalent ribosomal protein gene of the archaebacterium *Methanococcus vannielii*. *Nucleic Acids Res.* 18:719–24
7. Balch, W. E., Fox, G. E., Magrum, L. J., Woese, C. R., Wolfe, R. S. 1979. Methanogens: reevaluation of a unique biological group. *Microbiol. Rev.* 43: 260–96
8. Beckler, G. S., Reeve, J. N. 1986. Conservation of primary structure in the *his*I gene of the archaebacterium *Methanococcus vannielii,* the eubacterium *Escherichia coli* and the eucaryote *Saccharomyces cerevisiae*. *Mol. Gen. Genet.* 204:133–40
9. Belay, N., Sparling, R., Daniels, L. 1984. Dinitrogen fixation by a thermophilic methanogenic bacterium. *Nature* 312:286–88
10. Berghöfer, B., Kröckel, L., Körtner, C., Truss, M., Schallenberg, J., et al. 1988. Relatedness of archaebacterial RNA polymerase core subunits to their eubacterial and eukaryotic equivalents. *Nucleic Acids Res.* 16:8113–28
11. Bertani, G., Baresi, L. 1987. Genetic transformation in the methanogen *Methanococcus voltae*. *J. Bacteriol.* 169:2730–38
12. Bhatnagar, L., Jain, M. K., Zeikus, J. G. 1991. Methanogenic bacteria. In *Variations in Autotrophic Life,* pp. 251–70. Orlando, FL: Academic
13. Bhatnagar, L., Krzycki, J. A., Zeikus, J. G. 1987. Analysis of hydrogen metabolism in *Methanosarcina barkeri,* regulation of hydrogenase and role of CO-dehydrogenase in H_2 production. *FEMS Microbiol. Lett.* 41:337–43
14. Biro, J., Fabry, S., Dietmaier, W., Bogedain, C., Hensel, R. 1990. Engineering thermostability in archae-

bacterial glyceraldehyde-3-phosphate dehydrogenase. *FEBS Lett.* 275:130–34
15. Blanche, F., Robin, C., Couder, M., Faucher, D., Cauchois, L., et al. 1991. Purification, characterization, and molecular cloning of 5-adenosyl-L-methionine: uroporphyrinogen III methyltransferase from *Methanobacterium ivanovii. J. Bacteriol.* 173:4637–45
16. Bokranz, M., Bäumner, G., Allmansberger, R., Ankel-Fuchs, D., Klein, A. 1988. Cloning and characterization of the methyl coenzyme M reductase genes from *Methanobacterium thermoautotrophicum. J. Bacteriol.* 170:568–77
17. Bokranz, M., Klein, A. 1987. Nucleotide sequence of the methyl coenzyme M reductase gene cluster from *Methanosarcina barkeri. Nucleic Acids Res.* 15:4350–51
18. Bokranz, M., Klein, A., Meile, L. 1990. Complete nucleotide sequence of plasmid pME2001 of *M. thermoautotrophicum* (Marburg). *Nucleic Acids Res.* 18:363
19. Deleted in proof
20. Bouthier de la Tour, C., Portemer, C., Nadal, M., Stetter, K. O., Forterre, P. 1990. Reverse gyrase, a hallmark of the hyperthermophilic archaebacteria. *J. Bacteriol.* 172:6303–8
21. Bröckl, G., Behr, M., Fabry, S., Hensel, R., Kaudewitz, H., et al. 1991. Analysis and nucleotide sequence of the genes encoding the surface-layer glycoproteins of the hyperthermophilic methanogens *Methanothermus fervidus* and *Methanothermus sociabilis. Eur. J. Biochem.* 199:147–52
22. Brown, J. W., Daniels, C. J., Reeve, J. N. 1989. Gene structure, organization and expression in archaebacteria. *C. R. C. Crit. Rev. Microbiol.* 16:287–338
23. Brown, J. W., Reeve, J. N. 1985. Polyadenylated, noncapped RNA from the archaebacterium *Methanococcus vannielii. J. Bacteriol.* 162:909–17
24. Brown, J. W., Reeve, J. N. 1989. Transcription initiation and a RNA polymerase binding site upstream of the *purE* gene of the archaebacterium *Methanobacterium thermoautotrophicum* strain ΔH. *FEMS Microbiol. Lett.* 60:131–36
25. Brown, J. W., Thomm, M., Beckler, G. S., Frey, G., Stetter, K. O., et al. 1988. An archaebacterial RNA polymerase binding site and transcription initiation of the *hisA* gene in *Methanococcus vannielii. Nucleic Acids Res.* 16:315–49
26. Burbaum, J. J., Schimmel, P. 1991. Structural relationships and the classification of aminoacyl-tRNA synthetases. *J. Biol. Chem.* 266:16965–68
27. Burggraf, S., Stetter, K. O., Rouvière, P., Woese, C. R. 1991. *Methanopyrus kandleri,* an archaeal methanogen unrelated to all other known methanogens. *Syst. Appl. Microbiol.* 14:346–51
28. Chartier, F., Laine, B., Bélaîche, D., Sautière, P. 1989. Primary structure of the chromosomal proteins MC1a, MC1b, and MC1c from the archaebacterium *Methanothrix soehngenii. J. Biol. Chem.* 264:17006–15
29. Chartier F., Laine B., Sautière, P., Touzel, J.-P., Albagnac, G. 1985. Characterization of the chromosomal protein HMb isolated from *Methanosarcina barkeri. FEBS Lett.* 183:119–23
30. Cram, D. S., Sherf, B. A., Libby, R. T., Mattaliano, R. J., Ramachandran, K. L., Reeve, J. N. 1987. Structure and expression of the genes *mcrBDCGA,* which encode the subunts of component C of methyl coenzyme M reductase in *Methanococcus vannielii. Proc. Natl. Acad. Sci. USA* 84:3992–96
31. Cue, D., Beckler, G. S., Reeve, J. N., Konisky, J. 1985. Structure and sequence divergence of two archaebacterial genes. *Proc. Natl. Acad. Sci. USA* 82:2407–11
32. Daniels, C. J., Gupta, R., Doolittle, W. F. 1985. Transcription and excision of a large intron in the $tRNA^{trp}$ gene of an archaebacterium, *Halobacterium volcanii. J. Biol. Chem.* 260:3132–34
33. Daniels, L. 1984. Biological methanogenesis: physiological and practical aspects. *Trends Biotechnol.* 2:91–98
34. Dennis, P. P. 1986. Molecular biology of archaebacteria. *J. Bacteriol.* 168: 471–78
35. Dharmavaram, R., Gillevet, P., Konisky, J. 1991. Nucleotide sequence of the gene encoding the vanadate-sensitive membrane-associated ATPase of *Methanococcus voltae. J. Bacteriol.* 173: 2131–33
36. Dharmavaram, R., Konisky, J. 1989. Characterization of a P-type ATPase of the archaebacterium *Methanococcus voltae. J. Biol. Chem.* 264:14085–89
37. DiMarco, A. A., Sment, K. A., Konisky, J., Wolfe, R. S. 1990. The formyl methanofuran: tetrahydromethanopterin formyl transferase from *M. thermoautotrophicum* strain ΔH. Nucleotide sequence and functional expression of the cloned gene. *J. Biol. Chem.* 265:472–76
38. Eggen, R. I. L., Geerling, A. C. M.,

Boshoven, A. B. P., DeVos, W. M. 1991. Cloning, sequence analysis, and functional expression of the acetyl coenzyme A synthetase gene from *Methanothrix soehngenii* in *Escherichia coli. J. Bacteriol.* 173:6383–89

39. Eggen, R. I. L., Geerling, A. C. M., Jetten, M. S. M., DeVos, W. M. 1991. Cloning, expression and sequence analysis of the genes for carbon monoxide dehydrogenase of *Methanothrix soehngenii. J. Biol. Chem.* 266:6883–87
40. Eggen, R., Harmsen, H., de Vos, W. M. 1990. Organization of a ribosomal RNA gene cluster from the archaebacterium *Methanothrix soehngenii. Nucleic Acids Res.* 18:1306
41. Eggen, R., Harmsen, H., Geerling, A., de Vos, W. M. 1989. Nucleotide sequence of a 16S rRNA encoding gene from the archaebacterium *Methanothrix soehngenii. Nucleic Acids Res.* 17:9469
42. Elhardt, D., Böck, A. 1982. An in vitro polypeptide synthesizing system from methanogenic bacteria: sensitivity to antibiotics. *Mol. Gen. Genet.* 188:128–34
43. Fabry, S., Hensel, R. 1988. Primary structure of the glyceraldehyde-3-phosphate dehydrogenase deduced from the nucleotide sequence of the thermophilic archaebacterium *Methanothermus fervidus. Gene* 64:189–97
44. Fabry, S., Heppner, P., Dietmaier, W., Hensel, R. 1990. Cloning and sequencing the gene encoding 3-phosphoglycerate kinase from mesophilic *Methanobacterium bryantii* and thermophilic *Methanothermus fervidus. Gene* 91: 19–25
45. Fabry, S., Lang, J., Niermann, T., Vingron, M., Hensel, R. 1989. Nucleotide sequence of the glyceraldehyde-3-phosphate dehydrogenase gene from the mesophilic methanogenic archaebacteria *Methanobacterium bryantii* and *Methanobacterium formicicum. Eur. J. Biochem.* 179:405–13
46. Fabry, S., Lehmacher, A., Bode, W., Hensel, R. 1988. Expression of the glyceraldehyde-3-phosphate dehydrogenase gene from the extremely thermophilic archaebacterium *Methanothermus fervidus* in *E. coli. FEBS Lett.* 237:213–17
47. Frey, G., Thomm, M., Brüdigam, B., Gohl, H. P., Hausner, W. 1991. An archaebacterium cell-free transcription system. The expression of tRNA genes from *Methanococcus vannielii* is mediated by a transcription factor. *Nucleic Acids Res.* 18:1361–67
48. Frey, G., Thomm, M., Gohl, H. P., Brüdigam, B., Hausner, W. 1990. An archaebacterial cell-free transcription system. The expression of tRNA genes from *Methanococcus vannielii* is mediated by a transcription factor. *Nucleic Acids Res.* 18:1361–67
49. Garcia, J. L. 1990. Taxonomy and ecology of methanogens. *FEMS Microbiol. Rev.* 87:297–308
50. Gernhardt, P., Possot, O., Foglino, M., Sibold, L., Klein, A. 1990. Construction of an integration vector for use in the archaebacterium *Methanococcus voltae* and expression of a eubacterial resistance gene. *Mol. Gen. Genet.* 221: 273–79
51. Gohl, H. P., Hausner, W., Thomm, M. 1992. Cell-free transcription of the *nif*H1 gene of *Methanococcus thermolithotrophicus* indicates that promoters of archaeal *nif* genes share basic features with the methanogen consensus promoter. *Mol. Gen. Genet.* 231:286–95
52. Graf, E. G., Thauer, R. K. 1981. Hydrogenase from *Methanobacterium thermoautotrophicum,* a nickel containing enzyme. *FEBS Lett.* 136:165–69

52a. Grahame, D. A. 1991. Catalysis of acetyl-CoA cleavage and tetrahydrosarcinapterin methylation by a carbon monoxide dehydrogenase–corrinoid enzyme complex. *J. Biol. Chem.* 266: 227–33

53. Haas, E. S., Brown, J. W., Daniels, C. J., Reeve, J. N. 1990. Genes encoding the 7S RNA and $tRNA^{Ser}$ are linked to one of the two rRNA operons in the genome of the extremely thermophilic archaebacterium *Methanothermus fervidus. Gene* 90:51–59
54. Haas, E. S., Daniels, C. J., Reeve, J. N. 1989. Genes encoding 5S rRNA and tRNAs in the extremely thermophilic archaebacterium *Methanothermus fervidus. Gene* 77:253–63
55. Haas, E. S., Hook, L. A., Reeve, J. N. 1986. Antibiotic resistance caused by permeability changes of the archaebacterium *Methanococcus vannielii. FEMS Microbiol. Lett.* 33:185–88
56. Halboth, S., Klein, A. 1991. *Methanococcus voltae* harbors two gene groups each homologous to (NiFe) and (NiFeSe) hydrogenases which reduce cofactor F_{420} or only electron accepting dyes. *Gene Bank Accession No. X61204*
57. Hamilton, P. T., Reeve, J. N. 1985. Structure of genes and an insertion element in the methane producing archaebacterium *Methanobrevibacter smithii. Mol. Gen. Genet.* 200:47–59

58. Hamilton, P. T., Reeve, J. N 1985. Sequence divergence of an archaebacterial gene cloned from a mesophilic and a thermophilic methanogen. *J. Mol. Evol.* 22:351–60
59. Harris, J. E., Pinn, P. A. 1985. Bacitracin-resistant mutants of a mesophilic *Methanobacterium species*. *Arch. Microbiol.* 143:151–53
60. Hartmann, E., König, H. 1989. Uridine and dolichyl diphosphate activated oligosaccharides are intermediates in the biosynthesis of the S-layer glycoprotein of *Methanothermus fervidus*. *Arch. Microbiol.* 151:274–81
61. Hausner, W., Frey, G., Thomm, M. 1991. Control regions of an archaeal gene: a TATA box and an initiator element promote cell-free transcription of the $tRNA^{Val}$ gene of *Methanococcus vannielii*. *J. Mol. Biol.* 222:495–508
62. Hebert, A. M., Kropinski, A. M., Jarrel, K. F. 1991. Heat shock response of the archaebacterium *Methanococcus voltae*. *J. Bacteriol.* 173:3224–27
63. Hennigan, A. N., Stroup, D., Palmer, J. P., Reeve, J. N. 1991. Identification and quantitation of the transcript and products of the methyl coenzyme M reductase operon in *Methanococcus vannielii*. *Abstr. Annu. Meeting Am. Soc. Microbiol.* I118
64. Hensel, R., König, H. 1988. Thermoadaptation of methanogenic bacteria by intracellular ion concentration. *FEMS Microbiol. Lett.* 49:75–79
65. Hensel, R., Zwickel, P., Fabry, S., Lang, J., Palm, P. 1989. Sequence comparison of glyceraldehyde-3-phosphate dehydrogenases from the three urkingdoms: evolutionary implications. *Can. J. Microbiol.* 35:81–85
66. Honka, E., Fabry, S., Niermann, T., Palm, P., Hensel, R. 1990. Properties and primary structure of the L-malate dehydrogenase from the extremely thermophilic archaebacterium *Methanothermus fervidus*. *Eur. J. Biochem.* 188:623–32
67. Huber R., Kurr, M., Jannasch, H. W., Stetter, K. O. 1989. A novel group of abyssal methanogenic archaebacteria (Methanopyrus) growing at 110°C. *Nature* 342:833–34
68. Hummel, H., Böck, A. 1985. Mutations in *Methanobacterium formicicum* conferring resistance to anti-80S ribosome-targeted antibiotics. *Mol. Gen. Genet.* 198:529–33
69. Inatomi, K.-I., Eya, S., Maeda, M., Futai, M. 1989. Amino acid sequence of the α and β subunits of *Methanosarcina barkeri* ATPase deduced from cloned genes. *J. Biol. Chem.* 264:10954–59
70. Iwake, N., Kuma, K., Hasegawa, M., Osawa, S., Miyata, T. 1989. Evolutionary relationships of archaebacteria, eubacteria, and eukaryotes inferred from phylogenetic trees of duplicated genes. *Proc. Natl. Acad. Sci. USA* 86:9355–59
71. Jaenicke, R., Bernhardt, G., Lüdemann, H.-D., Stetter, K. O. 1988. Pressure-induced alterations in the protein pattern of the thermophilic archaebacterium *Methanococcus thermolithotrophicus*. *Appl. Environ. Microbiol.* 54:2375–80
72. Jain, M. K., Zeikus, J. G. 1987. Methods for isolation of auxotrophic mutants of *Methanobacterium ivanovii* and initial characterization of acetate auxotrophs. *Appl. Environ. Microbiol.* 53:1387–90
73. Jarrell, K. F., Julseth, C., Pearson, B., Kuzio, J. 1987. Paucity of the *Sau*3AI recognition sequence (GATC) in the genome of *Methanococcus voltae*. *Mol. Gen. Genet.* 208:191–94
74. Jarsch, M., Altenbuchner, J., Böck, A. 1983. Physical organization of the genes for ribosomal RNA in *Methanococcus vannielii*. *Mol. Gen. Genet.* 189:41–47
75. Jarsch, M., Böck, A. 1983. DNA sequence of the 16SrRNA/23SrRNA intercistronic spacer of two rDNA operons of the archaebacterium *Methanococcus vannielii*. *Nucleic Acids Res.* 11:7537–44
76. Jarsch, M., Böck, A. 1985. Sequence of the 16S ribosomal RNA gene from *Methanococcus vannielii:* evolutionary implications. *Syst. Appl. Microbiol.* 6: 54–59
77. Jarsch, M., Böck, A. 1985. Sequence of the 23S rRNA gene from the archaebacterium *Methanococcus vannielii:* evolutionary and functional implications. *Mol. Gen. Genet.* 200:305–12
78. Jenal, U., Rechsteiner, T., Tan, P.-Y., Bühlmann, E., Meile, L., et al. 1991. Isoleucyl-tRNA synthetase of *Methanobacterium thermoautotrophicum* Marburg. Cloning of the gene, nucleotide sequence, and localization of a base change conferring resistance to pseudomonic acid. *J. Biol. Chem.* 266:10570–77
79. Jones, W. J., Nagle, D. P. Jr., Whitman, W. B. 1987. Methanogens and the diversity of the archaebacteria. *Microbiol. Rev.* 51:135–77
80. Jordan, M., Meile, L., Leisinger, T. 1989. Organization of *Methanobacterium thermoautotrophicum* bacterio-

phage ψM1 DNA. *Mol. Gen. Genet.* 220:161–64

81. Kaine, B. P. 1987. Intron-containing tRNA genes of *Sulfolobus solfataricus*. *J. Mol. Evol.* 25:248–54
82. Kaine, B. P. 1990. Structure of the archaebacterial 7S RNA molecule. *Mol. Gen. Genet.* 221:315–21
83. Kaine, B. P., Merkel, V. L. 1989. Isolation and characterization of the 7S RNA gene from *Methanococcus voltae*. *J. Bacteriol.* 171:4261–66
84. Kalmokoff, M. L., Jarrell, K. F. 1991. Cloning and sequencing of a multigene family encoding the flagellins of *Methanococcus voltae*. *J. Bacteriol.* 173: 7113–25
85. Kalmokoff, M. L., Karnauchow, T. M., Jarrell, K. F. 1990. Conserved N-terminal sequences in the flagellins of archaebacteria. *Biochem. Biophys. Res. Commun.* 167:154–60
86. Kjems, J., Garrett, R. A. 1985. An intron in the 23S ribosomal RNA gene of the archaebacterium *Desulfurococcus mobilis*. *Nature (London)* 318:675–77
87. Klein, A., Allmansberger, R., Bokranz, M., Knaub, S., Müller, B., et al. 1988. Comparative analysis of genes encoding methyl coenzyme M reductase in methanogenic bacteria. *Mol. Gen. Genet.* 213:409–20
88. Klein, A., Schnorr, M. 1984. Genome complexity of methanogenic bacteria. *J. Bacteriol.* 158:628–31

88a. Klenk, H.-P., Palm, P., Lottspeich, F., Zillig, W. 1992. Component H of the DNA-dependent RNA polymerases of *Archaea* is homologous to a subunit shared by three eukaryal nuclear RNA polymerases. *Proc. Natl. Acad. Sci. USA* 89:407–10

89. Knaub, S., Klein, A. 1990. Specific transcription of cloned *Methanobacterium thermoautotrophicum* transcription units by homologous RNA polymerase in vitro. *Nucleic Acids Res.* 18:1441–46
90. Konisky, J. 1989. Methanogens for biotechnology: application of genetics and molecular biology. *Trends Biotechnol.* 7:88–92
91. Kurr, M., Huber, R., König, H., Jannasch, H. W., Fricke, H., et al. 1991. *Methanopyrus kandleri,* gen. and sp. nov. represents a novel group of hyperthermophilic methanogens growing at 110°C. *Arch. Microbiol.* 156:239–47
92. Ladapo, J., Whitman, W. B. 1990. Method for isolation of auxotrophs in the methanogenic archaebacteria: role of the acetyl-CoA pathway of autotrophic CO_2 fixation in *Methanococcus maripaludis*. *Proc. Natl. Acad. Sci. USA* 87:5598–5602
93. Laine, B., Culard, F., Maurizot, J.-C., Sautière, P. 1991. The chromosomal protein MC1 from the archaebacterium *Methanosarcina* sp. CHT155 induces DNA bending and supercoiling. *Nucleic Acids Res.* 19:3041–45
94. Lechner, K., Böck, A. 1987. Cloning and nucleotide sequence of the gene for an archaebacterial protein synthesis factor Tu. *Mol. Gen. Genet.* 208:523–28
95. Lechner, K., Heller, G., Böck, A. 1988. Gene for the diphtheria toxin-susceptible elongation factor 2 from *Methanococcus vannielii*. *Nucleic Acids Res.* 16:7817–26
96. Lechner, K., Heller, G., Böck, A. 1989. Organization and nucleotide sequence of a transcriptional unit of *Methanococcus vannielii* comprising genes for protein synthesis elongation factors and ribosomal proteins. *J. Mol. Evol.* 29:20–27
97. Lechner, K., Wich, G., Böck, A. 1985. The nucleotide sequence of the 16S rRNA gene and flanking regions from *Methanobacterium formicicum*. *Syst. Appl. Microbiol.* 6:157–63
98. Lehmacher, A., Vogt, A.-B., Hensel, R. 1990. Biosynthesis of cyclic 2,3-diphosphoglycerate. Isolation and characterization of 2-phosphoglycerate and cyclic 2,3-diphosphoglycerate synthetase from *Methanothermus fervidus*. *FEBS Lett.* 272:94–98
99. Luehrsen, K. R., Nicholson, D. E. Jr., Fox, G. E. 1985. Widespread distribution of a 7S RNA in archaebacteria. *Curr. Microbiol.* 12:69–72
100. Lunnon, K. D., Morgan, R. D., Timan, C. J., Krzycki, J. A., Reeve, J. N., et al. 1989. Characterization and cloning of *Mwo*I (GCN_7GC), a new type-II restriction-modification system from *Methanobacterium wolfei*. *Gene* 77:11–19
101. Ma, K., Linder, D., Stetter, K. O., Thauer, R. K. 1991. Purification and properties of N^5, N^{10}-methylene tetrahydromethanopterin reductase (Coenzyme F_{420}-dependent) from the extreme thermophile *Methanopyrus kandleri*. *Arch. Microbiol.* 155:593–600

101a. Macario, A. J. V., Dugan, C. B., Conway de Macario, E. 1991. A *dnaK* homolog in the archaebacterium *Methanosarcina mazei* S6. *Gene* 108:133–37

102. May, H. D., Patel, P. S., Ferry, J. G. 1988. Effect of molybdenum and tungsten on synthesis and composition of for-

mate dehydrogenase in *Methanobacterium formicicum*. *J. Bacteriol.* 170:3384–89

103. Meile, L., Jenal, U., Studer, D., Jordan, M., Leisinger, T. 1989. Characterization of ψM1, a virulent phage of *Methanobacterium thermoautotrophicum* Marburg. *Arch. Microbiol.* 152:105–10
104. Meile, L., Keiner, A., Leisinger, T. 1983. A plasmid in the archaebacterium *Methanobacterium thermoautotrophicum*. *Mol. Gen. Genet.* 191:480–84
105. Meile, L., Leisinger, T., Reeve, J. N. 1985. Cloning of DNA sequences from *Methanococcus vannielii* capable of autonomous replication in yeast. *Arch. Microbiol.* 143:253–55
106. Meile, A., Madon, J., Leisinger, T. 1988. Identification of a transcript and its promoter region on the archaebacterial plasmid pME2001. *J. Bacteriol.* 170:478–81
107. Meile, L., Reeve, J. N. 1985. Potential shuttle vectors based on the methanogen plasmid pME2001. *BioTechnology* 3: 69–72
108. Meile, L., Stettler, R., Banholzer, R., Kotik, M., Leisinger, T. 1991. Tryptophan gene cluster of *Methanobacterium thermoautotrophicum* Marburg: molecular cloning and nucleotide sequence of a putative *trpEGCFBAD* operon. *J. Bacteriol.* 173:5017–23
109. Micheletti, P. A., Sment, K. A., Konisky, J. 1991. Isolation of a coenzyme M auxotrophic mutant and transformation by electroporation of *Methanococcus voltae*. *J. Bacteriol.* 173:3414–18
110. Mitchell, R. M., Loeblich, L. A., Klotz, L. C., Loeblich, A. R. III, 1979. DNA organization in *Methanobacterium thermoautotrophicum*. *Science* 204: 1082–84
111. Morris, C. J., Reeve, J. N. 1984. Functional expression of an archaebacterial gene from the methanogen *Methanosarcina barkeri* in *Escherichia coli* and *Bacillus subtilis*. In *Microbial Growth on C1 Compounds,* ed. R. L. Crawford, R. S. Hanson, pp. 205–9. Washington, DC: Am. Soc. Microbiol.
112. Morris, C. J., Reeve, J. N. 1988. Conservation of structure in the human gene encoding argininosuccinate synthetase and the *argG* genes of the archaebacteria *Methanosarcina barkeri* MS and *Methanococcus vannielii*. *J. Bacteriol.* 170: 3125–30
113. Müller, B., Allmansberger, R., Klein, A. 1985. Termination of a transcription unit comprising highly expressed genes in the archaebacterium *Methanococcus voltae*. *Nucleic Acids Res.* 13:6439–45
114. Murray, P. A., Zinder, S. H. 1984. Nitrogen fixation by a methanogenic archaebacterium. *Nature* 312:284–86
115. Musgrave D. R., Sandman, K. M., Reeve, J. N. 1991. DNA binding by the archaeal histone HMf results in positive supercoiling. *Proc. Natl. Acad. Sci. USA* 88:10397–10401
116. Musgrave, D. R., Sandman, K. M., Stroup, D., Reeve, J. N. 1992. DNA binding proteins and genome topology in thermophilic procaryotes. In *Biocatalysis Near or Above 100°C,* ed. M. W. W. Adams, R. M. Kelly. Washington, DC: Am. Chem. Soc. In press
117. Nagle, D. P. Jr. 1989. Development of genetic systems in methanogenic archaebacteria. *Dev. Ind. Microbiol.* 30:43–51
118. Nauman, E., Fahlbusch, K., Gottschalk, G. 1984. Presence of a trimethylamine: HS-coenzyme M methyl transferase in *Methanosarcina barkeri*. *Arch. Microbiol.* 138:79–83
119. Nölling, J., Frijlink, M., DeVos, W. M. 1991. Isolation and characterization of plasmids from different strains of *Methanobacterium thermoformicicum*. *J. Gen. Microbiol.* 137:1981–86
120. Olson, K. D., McMahon, C. W., Wolfe, R. S. 1991. Light sensitivity of methanogenic archaebacteria. *Appl. Environ. Microbiol.* 57:2683–86
121. Østergaard, L., Larsen, N., Leffers, H., Kjems, J., Garrett, R. 1987. A ribosomal RNA operon and its flanking region from the archaebacterium *Methanobacterium thermoautotrophicum* Marburg strain: transcription signals, RNA structure and evolutionary implications. *Syst. Appl. Microbiol.* 9:199–209
122. Patel, P. S., Ferry, J. G. 1988. Characterization of the upstream regions of the formate dehydrogenase operon of *Methanobacterium formicicum*. *J. Bacteriol.* 170:3390–95
123. Pollitt, S., Inouye, M. 1987. Structure and functions of the signal peptide. In *Bacterial Outer Membranes as Model Systems,* ed. M. Inouye, pp. 117–39. New York: Wiley
124. Possot, O., Gernhardt, P., Klein, A., Sibold, L. 1988. Analysis of drug resistance in the archaebacterium *Methanococcus voltae* with respect to potential use in genetic engineering. *Appl. Environ. Microbiol.* 54:734–40
125. Possot, O., Henry, M., Sibold, L. 1986. Distribution of DNA sequences

homologous to *nifH* in archaebacteria. *FEMS Microbiol. Lett.* 34:173–77
126. Possot, O., Sibold, L., Aubert, J.-P. 1989. Nucleotide sequence and expression of the glutamine synthetase structural gene *glnA* of the archaebacterium *Methanococcus voltae*. *Res. Microbiol.* 140:355–71
127. Post, L. E., Strycharz, G. D., Nomura, M., Lewis, H., Dennis, P. P. 1979. Nucleotide sequence of the ribosomal protein gene cluster adjacent to the gene for RNA polymerase subunit β in *Escherichia coli*. *Proc. Natl. Acad. Sci. USA* 76:1693–1701
128. Rechsteiner, T., Kiener, A., Leisinger, T. 1986. Mutants of *Methanobacterium thermoautotrophicum*. *Syst. Appl. Microbiol.* 7:1–4
129. Reeve, J. N., Beckler, G. S. 1990. Conservation of primary structure in prokaryotic hydrogenases. *FEMS Microbiol. Rev.* 87:419–24
130. Reeve, J. N., Beckler, G. S., Cram, D. S., Hamilton, P. T., Brown, J. W., et al. 1989. A hydrogenase-linked gene in *Methanobacterium thermoautotrophicum* strain ΔH encodes a polyferredoxin. *Proc. Natl. Acad. Sci. USA* 86:3031–35
131. Reeve, J. N., Trun, N. J., Hamilton, P. T. 1982. Beginning genetics with methanogens. In *Genetic Engineering of Microorganisms for Chemicals,* ed. A. Hollaender, R. D. DeMoss, S. Kaplan, J. Konisky, D. Savage, R. S. Wolfe, pp. 233–44. New York: Plenum
132. Rospert, S., Linder, D., Ellermann, J., Thauer, R. K. 1990. Two genetically distinct methyl-coenzyme M reductases in *Methanobacterium thermoautotrophicum* strains Marburg and ΔH. *Eur. J. Biochem.* 194:871–77
133. Rouvière, P. E., Wolfe, R. S. 1988. Novel biochemistry of methanogenesis. *J. Biol. Chem.* 263:7913–16
134. Sandbeck, K. A., Leigh, J. A. 1991. Recovery of an integration shuttle vector from tandem repeats in *Methanococcus maripaludis*. *Appl. Environ. Microbiol.* 57:2762–63
135. Sandman, K., Krzycki, J. A., Dobrinski, B., Lurz, R., Reeve, J. N. 1990. HMf, a DNA-binding protein isolated from the hyperthermophilic archaeon *Methanothermus fervidus,* is most closely related to histones. *Proc. Natl. Acad. Sci. USA* 87:5788–91
136. Schallenberg, J., Moes, M., Truss, M., Reiser, W., Thomm, M., et al. 1988. Cloning and physical mapping of RNA polymerase genes from *Methanobacterium thermoautotrophicum* and comparison of homologies and gene orders with those of RNA polymerase genes from other methanogenic archaebacteria. *J. Bacteriol.* 170:2247–53
137. Schmidt, K., Thomm, M., Laminet, A., Laue, F. G., Kessler, C., et al. 1984. Three new restriction endonucleases *Mae*I, *Mae*II and *Mae*III from *Methanococcus aeolicus*. *Nucleic Acids Res.* 12:2619–28
138. Schnabel, R., Thomm, M., Gerardy-Schann, R., Zillig, W., Stetter, K. O. 1983. Structural homology between different archaebacterial DNA-dependent RNA polymerases analyzed by immunological comparison of their components. *EMBO J.* 2:751–55
139. Schwörer, B., Thauer, R. K. 1991. Activities of formyl methanofuran dehydrogenase, methylenetetrahydromethanopterin dehydrogenase, methylenetetrahydromethanopterin reductase, and heterodisulfide reductase in methanogenic bacteria. *Arch. Microbiol.* 155:459–65
140. Sherf, B. A., Reeve, J. N. 1990. Identification of the *mcrD* gene product and its association with component C of methyl coenzyme M reductase in *Methanococcus vannielii*. *J. Bacteriol.* 172: 1828–33
141. Shuber, A. P., Orr, E. C., Recny, M. A., Schendel, P. F., May, H. D., et al. 1986. Cloning, expression, and nucleotide sequence of the formate dehydrogenase genes from *Methanobacterium formicicum*. *J. Biol. Chem.* 261:12942–47
142. Sibold, L., Henriquet, M., Possot, O., Aubert, J.-P. 1991. Nucleotide sequence of *nifH* regions from *Methanobacterium ivanovii* and *Methanosarcina barkeri* 227 and characterization of *glnB*-like genes. *Res. Microbiol.* 142:5–12
143. Sibold, L., Henriquet, M. 1988. Cloning of *trp* genes from the archaebacterium *Methanococcus voltae,* nucleotide sequence of the *trpBA* genes. *Mol. Gen. Genet.* 214:439–50
144. Sibold, L., Pariot, D., Bhatnagar, L., Henriquet, M., Aubert, J.-P. 1985. Hybridization of DNA from methanogenic bacteria with nitrogenase structural genes. *Mol. Gen. Genet.* 200:40–46
145. Sibold, L., Souillard, N. 1988. Genetic analysis of nitrogen fixation in methanogenic archaebacteria. In *Nitrogen Fixation: Hundred Years After,* ed. H. Bothe, F. J. de Bruijn, W. E. Newton, pp. 705–10. Stuttgart: Gustav Fischer Verlag

146. Sitzmann, J., Klein, A. 1991. Physical and genetic map of the *Methanococcus voltae* genome. *Mol. Microbiol.* 5:505–13
147. Sment, K. A., Konisky, J. 1986. Regulated expression of the *Methanococcus voltae hisA* gene. *Syst. Appl. Microbiol.* 7:90–94
148. Souillard, N., Maget, M., Possot, O., Sibold, L. 1988. Nucleotide sequence of regions homologous to *nifH* (nitrogenase Fe protein) from the nitrogen-fixing archaebacteria *Methanococcus thermolithotrophicus* and *Methanobacterium ivanovii*. Evolutionary implications. *J. Mol. Evol.* 27:65–77
149. Souillard, N., Sibold, L. 1986. Primary structure and expression of a gene homologous to *nifH* (nitrogenase Fe proteins) from the archaebacterium *Methanococcus voltae*. *Mol. Gen. Genet.* 203:21–28
150. Souillard, N., Sibold, L. 1989. Primary structure, functional organization and expression of nitrogenase structural genes of the thermophilic archaebacterium *Methanococcus thermolithotrophicus*. *Mol. Microbiol.* 3:541–55
151. Southam, G., Kalmokoff, M. L., Jarrell, K. F., Koral, S. F., Beveridge, T. J. 1990. Isolation, characterization, and cellular insertion of the flagella from two strains of the archaebacterium *Methanospirillum hungatei*. *J. Bacteriol.* 172:3221–28
152. Sowers, K. R., Gunsalus, R. P. 1988. Plasmid DNA from the acetotrophic methanogen *Methanosarcina acetivorans*. *J. Bacteriol.* 170:4979–82
153. Sparling, R., Daniels, L. 1990. Regulation of formate dehydrogenase activity in *Methanococcus thermolithotrophicus*. *J. Bacteriol.* 172:1464–69
154. Steigerwald, V. J., Beckler, G. S., Reeve, J. N. 1990. Conservation of hydrogenase and polyferredoxin structures in the hyperthermophilic archaebacterium *Methanothermus fervidus*. *J. Bacteriol.* 172:4715–18
155. Strobel, O., Köpke, A. K. E., Kamp, R. M., Böck, A., Wittmann-Liebold, B. 1988. Primary sequence of the archaebacterial *Methanococcus vannielii* ribosomal protein L12. Amino-acid sequence determination, oligonucleotide hybridization and sequencing of the gene. *J. Biol. Chem.* 263:6538–46
156. Takao, M., Kobayashi, T., Oikawa, A., Yasui, A. 1989. Tandem arrangement of photolyase and superoxide dismutase genes in *Halobacterium halobium*. *J. Bacteriol.* 171:6323–29
157. Takao, M., Oikawa, A., Yasui, A. 1990. Characterization of a superoxide dismutase gene from the archaebacterium *Methanobacterium thermoautotrophicum*. *Arch. Biochem. Biophys.* 283:210–16
158. Takao, M., Yasui, A., Oikawa, A. 1991. Unique characteristics of superoxide dismutase of a strictly anaerobic archaebacterium *Methanobacterium thermoautotrophicum*. *J. Biol. Chem.* 266:14151–54
159. Terlesky, K. C., Nelson, J. K., Ferry, J. G. 1986. Isolation of an enzyme complex with carbon monoxide dehydrogenase activity containing corrinoid and nickel from acetate-grown *Methanosarcina thermophila*. *J. Bacteriol.* 168:1053–58
160. Thomm, M., Frey, G., Belton, B. J., Lane, F., Kessler, C., et al. 1988. *Mvn*I: a restriction enzyme in the archaebacterium *Methanococcus vannielii*. *FEMS Microbiol. Lett.* 52:229–34
161. Thomm, M., Frey, G., Hausner, W., Brüdigam 1990. An archaebacterial in vitro transcription system. In *Microbiology and Biochemistry of Strict Anaerobes Involved in Interspecies Transfer,* ed. J.-P. Belaich, M. Bruschi, J. L. Garcia, pp. 305–12. New York: Plenum
162. Thomm, M., Madon, J., Stetter, K. O. 1986. DNA dependent RNA polymerases of the three orders of methanogens. *Biol. Chem. Hoppe-Seyler* 367:473–81
163. Thomm, M., Sherf, B. A., Reeve, J. N. 1988. RNA polymerase-binding and transcription initiation sites upstream of the methyl reductase operon of *Methanococcus vannielii*. *J. Bacteriol.* 170:1958–61
164. Thomm, M., Wich, G. 1988. An archaebacterial promoter element for stable RNA genes with homology to the TATA box of higher eukaryotes. *Nucleic Acids Res.* 16:151–63
165. Thomm, M., Wich, G., Brown, J. W., Frey, G., Sherf, B. A., et al. 1989. An archaebacterial promoter sequence assigned by RNA polymerase binding experiments. *Can. J. Microbiol.* 35:30–35
166. Tiedeman, A. A., Keyhani, J., Kamholz, J., Daum, M. A. III, Gots, J. S., et al. 1989. Nucleotide sequence analysis of the *purEK* operon encoding 5'phosphoribosyl-5-aminoimidazole carboxylase of *Escherichia coli* K-12. *J. Bacteriol.* 171:205–12
167. van der Meijden, P., Heythuysen, H. J., Pouwels, A., Houwen, F., van der

Drift, C., et al. 1983. Methyl transferases involved in methanol conversion by *Methanosarcina barkeri*. *Arch. Microbiol.* 134:238–42
168. Weil, C. F., Beckler, G. S., Reeve, J. N. 1987. Structure and organization of the *hisA* gene of the thermophilic archaebacterium *Methanococcus thermolithotrophicus*. *J. Bacteriol.* 169: 4857–60
169. Weil, C. F., Cram, D. S., Sherf, B. A., Reeve, J. N. 1988. Structure and comparative analysis of the genes encoding component C of methyl coenzyme M reductase in the extremely thermophilic archaebacterium *Methanothermus fervidus*. *J. Bacteriol.* 170:4718–26
170. Weil, C. F., Sherf, B. A., Reeve, J. N. 1989. A comparison of the methyl reductase genes and gene products. *Can. J. Microbiol.* 35:101–8
171. Whitman, W. B. 1985. Methanogenic bacteria. In *The Bacteria,* C. R. Woese, R. S. Wolfe, 8:3–84. Orlando, FL: Academic
172. Wich, G., Hummel, H., Jarsch, M., Bar, U., Böck, A. 1986. Transcription signals for stable RNA genes in *Methanococcus*. *Nucleic Acids Res.* 14:2459–79
173. Wich, G., Jarsch, M., Böck, A. 1984. Apparent operon for a 5S ribosomal RNA gene and for tRNA genes in the archaebacterium *Methanococcus vannielii*. *Mol. Gen. Genet.* 196:146–51
174. Wich, G., Sibold, L., Böck, A. 1986. Genes for tRNAs and their putative expression signals in *Methanococcus*. *Syst. Appl. Microbiol.* 7:18–25
175. Wich, G., Sibold, L., Böck, A. 1987. Divergent evolution of 5S rRNA genes in Methanococcus. *Z. Naturforsch.* 42c:373–80
176. Widdel, F., Wolfe, R. S. 1989. Expression of secondary alcohol dehydrogenases in methanogenic bacteria and purification of the F_{420}-specific enzyme from *Methanogenium thermophilum* strain TC1. *Arch. Microbiol.* 152:322–28
177. Wilharm, T., Thomm, M., Stetter, K. O. 1986. Genetic analysis of plasmid PMP1 from *Methanolobus vulcani*. *Syst. Appl. Microbiol.* 7:401
178. Willekens, P., Huysmans, E., Vandenberghe, A., deWachter, R. 1986. Archaebacterial 5S ribosomal RNA: nucleotide sequence in two methanogen species, secondary structure models, and molecular evolution. *Syst. Appl. Microbiol.* 7:151–59
179. Woese, C. R. 1987. Bacterial evolution. *Microbiol. Rev.* 51:221–71
180. Woese, C. R., Kandler, O., Wheelis, M. L. 1990. Towards a natural system of organisms: proposal for the domains Archaea, Bacteria and Eucarya. *Proc. Natl. Acad. Sci. USA* 87:4576–79
181. Woese, C. R., Magrum, L. J., Fox, G. E. 1978. Archaebacteria. *J. Mol. Evol.* 11:245–52
182. Wood, A. G., Redborg, A. H., Cue, D. R., Whitman, W. B., Konisky, J. 1983. Complementation of *argG* and *hisA* mutations of *Escherichia coli* by DNA cloned from the archaebacterium *Methanococcus voltae*. *J. Bacteriol.* 156:19–29
183. Wood, A. G., Whitman, W. B., Konisky, J. 1985. A newly isolated marine methanogen harbors a small cryptic plasmid. *Arch. Microbiol.* 142:259–61
184. Wood, A. G., Whitman, W. B., Konisky, J. 1989. Isolation and characterization of an archaebacterial virus like particle from *Methanococcus voltae*. *J. Bacteriol.* 171:93–98
185. Worrell, V. E., Nagle, D. P., McCarthy, D., Eisenbraun, A. 1988. Genetic transformation system in the archaebacterium *Methanobacterium thermoautotrophicum* Marburg. *J. Bacteriol.* 170: 653–56
186. Zabel, H.-P., Fischer, H., Holler, E., Winter, J. 1985. In vivo and in vitro evidence for eucaryotic α-type DNA-polymerases in methanogens. Purification of the DNA-polymerase of *Methanococcus vannielii*. *Syst. Appl. Microbiol.* 6:111–18
187. Zhao, H., Wood, A. G., Widdel, F., Bryant, M. P. 1988. An extremely thermophilic *Methanococcus* from a deep sea hydrothermal vent and its plasmid. *Arch. Microbiol.* 150:178–83
188. Zillig, W., Palm, P., Reiter, W.-D., Gropp, F., Puhler, G., et al. 1988. Comparative evaluation of gene expression in archaebacteria. *Eur. J. Biochem.* 173:473–82

Annu. Rev. Microbiol. 1992. 46:193–218

THE BIOLOGY AND GENETICS OF THE GENUS *RHODOCOCCUS*

William R. Finnerty

Finnerty Enterprises, Inc., Athens, Georgia 30605

KEY WORDS: taxonomy, physiology and biochemistry, industrial and environmental applications, pathogenicity, biological response modifiers

---

CONTENTS

Abstract

The genus *Rhodococcus* is a unique taxon consisting of microorganisms that exhibit broad metabolic diversity, particularly to hydrophobic compounds

0066-4227/92/1001-0193$02.00

such as hydrocarbons, chlorinated phenolics, steroids, lignin, coal, and petroleum. Advances in chemical, numerical, and molecular systematic methods have contributed greatly to the circumspection of the rhodococci, including the development of diagnostic fluorogenic probes for improved biochemical profiling and identification. Bioprocessing systems employing various *Rhodococcus* strains are operational for industrial and environmental applications. Such applications include production of acrylic acid and acrylamide, steroid conversions, and bioremediation of chlorinated hydrocarbons and phenolics. Progress on the genetic systems of the rhodococci is rather limited, although a number of plasmids, cloning vectors, and DNA transfer systems have been reported recently, such that progress should be rapid. Certain members of the genus *Rhodococcus* are known pathogens for humans, animals, and plants. Recent trends indicate that rhodococci of animal origin are opportunistic human pathogens, indicating the need for a greatly improved recognition and understanding of the virulence factors associated with the genus *Rhodococcus*.

INTRODUCTION

Members of the genus *Rhodococcus* are common throughout nature; some members show pathogenicity for humans, animals, and plants. The taxonomic positions of *Rhodococcus* species have undergone incisive changes over the past several years, and today remain in a state of flux as false names are corrected, existing strains are sorted out into newly recognized genera (e.g. *Gordona* and *Tsukamurella*), and new species are identified. The techniques of modern biology are just beginning to emerge in the study of the rhodococci, enabling the application of sophisticated genetic and molecular biology analyses of the rhodococci. This article reviews advances that have occurred over the past 10 years in identification and classification, physiology and biochemistry, genetics and molecular biology, pathogenicity, and industrial and environmental applications of the rhodococci. These topics were reviewed in 1985 in the context of a comprehensive overview of the nocardioform actinomycetes, including the rhodococci (105), so this article concentrates primarily on those developments occurring since then and only on the genus *Rhodococcus*.

GENERAL BIOLOGY

Identification and Classification

The classification of mycolic acid–containing nocardioform actinomycetes has undergone frequent revisions, the direct result of improvements in chemical, numerical, and molecular systematic methods (18, 56, 184). The extensive and sophisticated studies of Goodfellow and colleagues have contributed significantly to the circumspection of nocardioform actinomycetes, including the rhodococci (52, 56). Mycolate-containing nocardioform actinomycetes form a distinct suprageneric group, encompassing the genera *Corynebacterium, Gordona, Mycobacterium, Nocardia, Rhodococcus,* and *Tsukamurella* (53).

The genus *Rhodococcus* was reintroduced in 1974 (169) and redefined by Goodfellow & Alderson in 1977 (53, 55). Introduction of the genera *Gordona* (158) and *Tsukamurella* (28) redefined microorganisms previously considered as rhodococci, leaving the genus *Rhodococcus* as a homogeneous taxon consisting of 13 species: *R. aichiensis, R. chlorophenolicus, R. coprophilus, R. equi, R. erythropolis, R. fascians, R. globerulus, R. luteus, R. marinonascens, R. maris, R. rhodnii, R. rhodochrous,* and *R. ruber*. The genus *Gordona* is considered to contain organisms previously classified as *R. bronchialis, R. rubropertinctus, R. sputi,* and *R. terrae* (57). The genus *Tsukamurella* contains organisms previously classified as *Corynebacterium paurometabolum* and *R. aurantiacus*; the latter now forms one species, *Tsukamurella paurometabola* (57). A second species, *T. wratislaviensis,* was recently described (57).

The genus *Rhodococcus* is defined on the basis of cell wall composition and is restricted to nocardioform actinomycetes with the following characteristics: (*a*) the peptidoglycan consists of N-acetylglucosamine, N-glycolylmuramic acid, D- and L-alanine, D-glutamic acid, and *meso*-diaminopimelic acid; (*b*) arabinose and galactose comprise the cell wall carbohydrates; (*c*) phospholipids consist of cardiolipin, phosphatidylethanolamine, phosphatidylinositol, and phosphatidylinositol mannosides; (*d*) normal saturated and unsaturated fatty acids are accompanied by tuberculostearic acid; (*e*) mycolic acids have 34–52 carbon atoms; and (*f*) the dihydrogenated menaquinones have eight isoprene units. Table 1 compares major characteristics of nocardioform actinomycetes.

The rhodococci are aerobic, gram-positive, nonmotile nocardioform actinomycetes with a growth cycle ranging from cocci or short rods to more complex growth phases that form filaments with short projections, elementary branching, or, in some species, extremely branched hyphae. The rhodococci are partially acid-fast with an oxidative type of metabolism and grow well in standard laboratory media. Some strains require thiamine when grown on chemically defined media.

Rapid and reliable methods in the differentiation of nocardioform actinomycetes are certainly needed. As new genera emerge within the actinomycete family line and investigators recognize their importance in medicine, industry, and bioremediation, the correct identification and classification of these organisms becomes essential. To this end, enzyme tests have been developed for improved biochemical profiling.

Table 1 Characteristics of nocardioform actinomycetes

	Corynebacterium	*Gordona*	*Mycobacterium*	*Nocardia*	*Rhodococcus*	*Tsukamurella*
Mycelia	−		−, +	+	+	−
Tuberculo stearic acid	−	+	−, +	+	+	+
Phosphatidyl ethanolamine	−	+	+	+	+	+
Menaquinones	−8 (H_2)	−9 (H_2)	−9 (H_2)	−8 (H_4)	−8 (H_2)	−9
Mycolic acids	22–38C	48–66C	60–90C	46–60	32–66	62–74
Mole % G+C	51–59	63–69	62–70	64–72	67–73	67–68

Fluorogenic probes have been used to classify and identify clinically significant actinomycetes and mycolic acid–containing bacteria belonging to the genera *Gordona, Rhodococcus,* and *Tsukamurella* (54, 56, 57). These studies show that fluorogenic probes prepared from 4-methylumbelliferone and 7-amino-4-methylcoumarin provide a rapid and reliable means of detecting enzyme activities in small samples of intact microorganisms. These fluorogenic substrates measure endopeptidases, exopeptidases, glycosidases, and hydrolysis of inorganic and organic esters. Enzyme tests coupled to nutritional and antibiotic tolerance profiles offer a battery of potentially useful diagnostic tests for the identification of nocardioform actinomycetes to the species level. The development of well-characterized type species will provide the basis for the further development of tests that can be weighted for diagnostic purposes.

PHYSIOLOGY AND BIOCHEMISTRY

The rhodococci exhibit an unusual armamentarium of novel enzymatic capabilities for the transformation and degradation of diverse classes of substrates, and some transformations result in useful commercial processes. Unfortunately, basic physiological studies are few and generally emphasize a product or process of industrial importance or problems involving the pathogenicity of specific members of the genus *Rhodococcus*. A comprehensive review addresses the biology of the nonstreptomycetes (rare actinomycetes), including the rhodococci (105).

Intermediary Metabolism

Glycolysis occurs through the Embden-Meyerhof-Parnas pathway in *R. erythropolis* and through the Entner-Doudoroff pathway in *R. opacus* (105). Analyses of enzyme activities in *R. erythropolis* showed that isocitrate dehydrogenase and 6-phosphogluconate dehydrogenase were constitutive, whereas malic dehydrogenase, mannose dehydrogenase, and glucose-6-phosphate dehydrogenase activities were inducible. Glucose utilization by *R. erythropolis* AN-13 is increased by the addition of acetate and amino acids (8), presumably through stimulation of tricarboxylic acid (TCA) cycle enzymes.

A formaldehyde dehydrogenase of *R. erythropolis* is reportedly a trimeric subunit structure (44,000 daltons per subunit) and is induced by growth on substrates containing methyl groups such as methylamine and 3,4-dimethoxybenzoate (42). An inducible phenylalanine dehydrogenase derived from *R. maris* K-18 has been characterized as an unstable 74,000 dalton dimer that is stabilized by either 0.25 M malonate or 0.5 M glutarate (119).

Hydrocarbon Metabolism

The metabolism of gaseous and liquid hydrocarbons by rhodococci is not uncommon. *R. rhodochrous* metabolizes acetylene to acetaldehyde, ethanol, acetate, and CO_2 (51). Propane-grown *Rhodococcus* species convert ethylene, prop-1-ene and but-1-ene to the corresponding 1,2-epoxy alkanes (182). The oxidation of propane by *R. rhodochrous* PNKB1 occurs at either the terminal or penultimate carbon atom of propane (181). Oxidation of carbon-1 yields propionate, while oxidation of carbon-2 gives acetone and acetate. A NAD^+-linked primary and secondary alcohol dehydrogenase has been purified from propane-grown *R. rhodochrous* PNKB1 as a dimeric enzyme of 86,000 daltons (13). The substrate range for this bifunctional enzyme consists of C_2-C_8 primary and secondary alcohols, glycerol, 1,2- and 1,3-dihydroxypropane, cyclohexanol, and phenylethyl alcohol. The oxidation reaction is optimal at pH 10 with reduction of ketones occurring at pH 6.5 in the presence of NADH. The methylmalonyl-coenzyme A (CoA) pathway has been reported in propane-grown *R. rhodochrous* (111).

The conversion of alkanes and esterified fatty acid to *cis*-unsaturated alkenes and unsaturated fatty acids is catalyzed by *Rhodococcus* species KSM-B-MT66 (165). Alkanes (C_{13}–C_{14}) are desaturated at carbon-9 at yields of 18 g/liter per 3 days. Isopropyl fatty acids are desaturated at carbon-6 at yields of 50 g/liter per 3 days. Free fatty acids do not serve as substrates in the oxidative desaturation reaction.

A hydrophobic polysaccharide has been characterized from a hexadecane-grown *Rhodococcus* species (129). This polysaccharide consists of the repeating unit glucuronic acid, glucose, galactose, and rhamnose in molar ratios of 1:1:1:2 with one acetate residue per repeating unit. The polysaccharide appears to contribute to the hydrophobicity of the cell surface. The amounts of polysaccharide increased when cells were grown at 20°C.

The oxidation of long-chain alcohols by *R. equi* was studied in organic media (164, 174). Tetradecane-grown cells were prepared as either acetone powders or lyophilized cells, suspended in isooctane and supplemented with either primary alcohols or tetradecan-2-ol as substrates. Primary alcohols (C_4–C_{14}) were oxidized to acids and tetradecan-2-ol was oxidized to tetradecan-2-one at efficiencies of 80–100%.

A novel reaction catalyzed by *R. rhodochrous* IFO 3338 quantitatively converts 2-nitro-1-phenyl-prop-1-ene to 2-nitro-1-phenylpropane (145). This is an unusual hydrogenation reaction for an aerobic microorganism. *Rhodococcus* species BPM 1613 grows on the isoprenoid hydrocarbon phytane, norpristane and farnesane converting each substrate to the corresponding primary alcohol and fatty acid (126). This pristane-utilizing isolate oxidizes pristane to pristanol, pristanic acid and pristylpristinate.

Surface-Active Lipids

Several reports have appeared over the last decade describing the production of surface-active glycolipids by alkane-grown rhodococci (19, 100, 140, 155–157). A soil isolate identified as *R. rhodochrous* utilized dodecane as a sole source of carbon and energy (157). Characterization of the cellular lipids indicated steroids, monoglyceride, phosphatidylcholine, and glycolipid as the major cell-associated lipids. *T. paurometabola* (*R. aurantiacus*) synthesized four different glycolipid species when grown on n-alkanes (139). The glycolipids were cell-associated and extracellular, exhibiting surface tension values of 26–30 mN/m. Wagner and associates have studied the synthesis of trehalose-2,2',3,4-tetraester by *R. erythropolis* as a function of nitrogen limitation (91, 143). Over-production of this glycolipid occurred under conditions of nitrogen limitation followed by a temperature and pH shift, with C_{10}, C_{14}, and C_{15} alkanes as the sole source of carbon and energy. These authors reported minimal interfacial tension values of less than 1 mN/m at a critical micelle concentration of 15 mg/liter. A study of the physiology of glycolipid synthesis by *Rhodococcus* species H13-A grown at the expense of either hexadecane or hexadecanol showed that surface-active glycolipids accumulate in the growth medium during the stationary growth phase in response to limiting ammonium ion concentration (155). Cultures of H13-A grown on triglycerides, fatty acids, organic acids, or carbohydrates did not produce extracellular glycolipid.

Rhodococcus species H13-A contains three indigenous plasmids (pMVS100, pMVS200, pMVS300); neither pMVS200 nor pMVS300 are involved in glycolipid synthesis or hexadecane dissimilation (155). The extracellular glycolipids synthesized by hexadecane-grown *Rhodococcus* species H13-A are anionic, consisting of trehalose, C_{10} to C_{22} saturated and unsaturated fatty acids, C_{35} to C_{40} mycolic acids, hexanedioic acid and decanedioic acid, 10-methyl hexadecanoic acid, and 10-methyl octadecanoic acid (156). The major glycolipid species has been identified as 2,3,4,6,2',3',4',6'-octaacyltrehalose plus minor di-, tetra-, and hexaacyltrehalose derivatives. The purified extracellular glycolipids exhibited a critical micelle concentration of 1.5 gm/liter and a minimal interfacial tension value of 0.02 mN/m. In the presence of the cosurfactant, pentan-1-ol, minimal interfacial tension values in the range of 0.00005 mN/m against decane are possible.

Hexadecane-grown *R. erythropolis* produces the extracellular accumulation of 2,2',3,4,-di-O-succinyl-di-O-alkanoyl-α, α'-trehalose, with both mono- and disuccinyl trehalose lipids lacking mycolic acids (172, 173). The glycolipid fatty acids were hexadecanoic and tetradecanoic acids with extracellular glycolipid yields of 40 g/liter in 10 days.

A novel glycolipid, trehalose-2,3,6-mycolate, has been isolated from a psychrophilic *T. paurometabola* (*R. aurantiacus*) (167). The mycolic acids were polyunsaturated with a carbon number ranging from C_{62} to C_{74}. The glycolipids appear to consist of four molecular species with C_{62} to C_{74} polyenoic mycolic acids.

Microbial flocculants produced by rhodococci have been reported (102, 103). An extracellular flocculant was produced by *R. erythropolis* grown on 1% glucose, urea, and yeast extract under conditions of low aeration, pH 8.5–9.5 at 30°C. This extracellular product consisting of protein and carbohydrate flocculates inanimate particles such as kaolin and bacteria such as *Escherichia coli*.

Enzymatic Hydrolysis of Complex Lipids

The rhodococci reportedly exhibit unique and novel complex-lipid hydrolyzing enzyme activities not previously described in prokaryotes. Ito & Yamagata first observed the presence of specific endoglycosidase and endoglycoceramidase activities in several actinomycetes, including rhodococci (81). Three distinct molecular species of endoglycoceramidase with differing substrate specificities have been purified from the spent culture broth of a mutant strain of a *Rhodococcus* species (82). Type I enzyme was specific for ganglio-, lacto- and globo-type glycosphingolipids yielding oligosaccharide and ceramide. Type II enzyme was specific for ganglio- and lacto-type glycosphingolipids yielding oligosaccharide and ceramides, and the enzyme was inactive in the hydrolysis of globo- and gala-type glycosphingolipids, phospholipids, cerebrosides, sulfatides, glycoglycerolipids, or sphingomyelins. Type III enzyme was specific only to gala-type glycosphingolipids.

An aryl acylamidase has been studied in *R. erythropolis* NCIB 12273 that hydrolyzes N-acyl primary aromatic amines to primary amines and fatty acids (176). The enzyme is not subject to either carbon or nitrogen repression and exhibits a pH optimum of 8.0.

Carotenoid pigments from *R. rhodochrous* RNMS 1 include two monocylic carotenoids, each with a tertiary hydroxyl group, a cartenoid monoglycoside, and a carotenoid monoglycoside fatty acid ester; the latter two are both novel derivatives (163).

Steroid Oxidations

Researchers have become increasingly interested in the stereo- and regiospecific transformation of steroids by microorganisms, primarily because of the ability of microorganisms to catalyze such specific biotransformations. The rhodococci exhibit such specific enzymic capability for the 9α-hydroxylation of δ5-3 β-hydroxysteroids (35), the synthesis of 20-carboxypregna-1,4-dien-3-one from β-sitosterol (78, 79), conversions of squalene to an α,β-unsaturated ketone (150), and the conversion of cholesterol to 4-cholesten-3-one (1–4, 14, 84, 85, 178, 179). A methylperhydroindanedione reductase has been identified in *R. equi* that reduces the 5-ketogroup of steroids to a 5α-hydroxyl group (117).

Environmental Applications

The rhodococci exhibit a broad substrate diversity for the degradation of phenols (70, 159), aromatic acids (24, 46, 130), halogenated phenols (5, 9–12, 29, 46, 61–65, 83), halogenated alkanes (146), substituted benzenes (60), anilines (7), and quinolines (148, 149). The isolation and identification of a new chlorophenol-degrading microorganism, *Rhodococcus chlorophenolicus,* stimulated the development of bioremediation processes for the removal of chlorinated phenolics from soil (12). An exhaustive study detailed the degradation of polychlorinated phenols by *R. chlorophenolicus* (9). This bacterium efficiently degraded a series of chlorinated phenols ranging from dichloro to pentachloro compounds. The total dechlorination of tetrachlorohydroquinone by cell-free extracts prepared from pentachlorophenol-induced *R. chlorophenolicus* reductively dechlorinated tetrachlorohydroquinone to 1,2,4-trihydroxybenzene through the intermediates dichlorotrihydroxybenzene and monochlorotrihydroxybenzene. The reaction involves total dechlorination prior to ring cleavage rather than ring cleavage and then dechlorination, as described for other microorganisms degrading mono- and dichlorinated phenols.

Noninduced rhodococcal strains with no known history of exposure to chlorinated aromatic compounds hydroxylated mono-, di-, and trichlorophenols at the ortho position, forming chlorinated catechols (65). The ability of these rhodococci to hydroxylate chlorophenols correlated directly with their ability to grow at the expense of unsubstituted phenols as the sole source of carbon and energy. Several rhodococci hydroxylated trichlorophenols to trichlorocatechols and then O-methylated this product to chloroquiacol and chloroveratrole. Because the rhodococci are widely distributed in soil and sludge, these results indicate a role for the rhodococci in the biotransformation of chlorinated phenols in the environment. *R. chlorophenolicus* has been employed for the bioremediation of chlorophenol-containing ground water under simulated conditions (175). Polyurethane-immobilized cells reduced the influent concentration of chlorophenol 1000- to 10,000-fold over 4 months. This biofilter reduced chlorophenol in the absence of supplementary carbon sources and in process streams containing low levels of nitrogen and phosphate nutrients. *R. chlorophenolicus* was inoculated into various soil types containing pentachlorophenol in the range 30–600 mg pentachlorophenol/kg soil (118). Inoculation of 10^5–10^8 cells/g soil initiated pentachlorophenol degradation with 75% removal in highly polluted soils and 50% removal in lightly polluted soil. The degradation of substituted aromatic ring compounds by the rhodococci appears to be through the 3-oxo-adipate pathway, although variations and modifications exist. *R. rhodochrous* and other nocardioform actinomycetes contain a novel enzyme that catalyzes the highly specific isomerization of 4-methyl-enelactone to the 3-methyl-enelactone, thereby allowing further metabolism (23). This 4-methylmucono lactone methyl isomerase has been purified and characterized as a 75,000-dalton tetramer in *R. rhodochrous* N75 (25).

A three-member consortium consisting of *Rhodococcus, Pseudomonas,* and a *Flavobacterium* was established to study the degradation of alicyclic hydrocarbons (108). The three organisms were isolated from an oil refinery site and tested individually and as a consortium for their ability to grow at the expense of a wide range of alicyclic hydrocarbons. Individually, the microorganisms did not utilize alicyclic hydrocarbons as a growth substrate. However, the consortium could grow on most of the test alicyclic hydrocarbons, suggesting a metabolic interdependence in metabolizing these growth substrates.

A soil isolate identified as *R. rhodochrous* reportedly catalyzes a highly specific carbon-sulfur bond cleavage reaction (89). Organic sulfur-containing heterocycles such as dibenzothiophene are converted to 2-hydroxybiphenyl and sulfate. The application of this microorganism for the removal of organic sulfur from high-sulfur coal and oil is under study (90). A *Rhodococcus* species has been isolated from activated sludge that grows at the expense of thiophene-2-carboxylic acid and 5-methylthiophene-2-carboxylic acid as sole sources of carbon and energy (88). Thiophene, methylthiophene, dibenzothiophene, dimethylsulfide, and pyrrole-2-carboxylic acid were not suitable growth substrates.

Industrial Applications

The commercial exploitation of the rhodococci appears to be rather limited, as evidenced by patent activity since 1973 (Table 2). Rhodococci do have, however, commercial applications, most notably in the production of acrylic acid and acrylamide in Japan and the use of specific rhodococci for steroid modifications. Compared to considerable industrially focused research and development activity on methylotrophic microorganisms for commercial exploitation, which apparently has not resulted in a major commercial success to date (107), the limited commercial exploitation of the rhodococci appears to represent an overwhelming success. Rhodococci will probably attract increased commercial interest in the areas of bioremediation and chemical production in future years, based on their ability to catalyze the oxidation and metabolism of diverse and unusual substrates.

A stable L-phenylalanine dehydrogenase isolated from *Rhodococcus* species M4 is induced by growth on L-phenylalanine (77). Enzyme prepara-

Table 2 Patents issued since 1973 concerning *Rhodococcus* species

Patent	Title	Year
FR 7333613	Acrylic acid synthesis using bacteria	1973
US 4000081	Acrylamide production by bacteria	1977
JP 129190	Acrylamide and methacrylamide production using microorganisms	1979
JP 7946887	Acrylamide from acrylonitrile using microorganisms	1979
JP 8111799	Fermentative production of 9α-hydroxysteroids	1981
JP 162193	Preparation of amides using microorganisms	1986
US 4582804	Microbiological synthesis of hydroxy fatty acids and keto fatty acids	1986
US 4584270	Process for preparing optically active 4-amino-3-hydroxy butyric acid	1986
JP 1199588	Production of unsaturated fatty acid or hydrocarbons by *Rhodococcus*	1989
JP 1229097	Decomposition of cholesterol in fat and oil by *Rhodococcus equi*	1989
US 4952500	Cloning systems for *Rhodococcus* and related bacteria	1990

tions have been used for the continuous production of L-phenylalanine from phenylpyruvate, with 95% conversion efficiencies in a membrane-bound enzyme bioreactor.

The conversion of the L-(+)-isomer to the optically active D-(−)-isomer in a racemic mixture of pantoyl lactone is reportedly a simple one-step quantitative conversion by *R. erythropolis* IFO 12540 (153). The D-(−)-isomer is an important intermediate in the synthesis of the D-(+)-pantothenic acid. This stereo-specific oxidation-reduction reaction yields a 94% enantiomeric excess of D-(−)-pantoyl lactone from a racemic mixture.

Nitrilases and nitrile hydratases are enzyme activities present in various rhodococci that transform aliphatic and aromatic nitriles to acids and amides, respectively (15, 16, 27, 45, 76, 87, 92–94, 96–99, 113–115, 122, 123, 177). These enzymes are highly stable and function under mild reaction conditions, have broad substrate ranges, and exhibit conversion efficiencies approaching 100%. Nitrilase is used for the production of nicotinic acid (113), *p*-amino benzoic acid (95), pyrazinoic acid (96), acrylic acid, and methacryclic acid (125).

Nitrile hydratase produces acrylamide (123), isonicotinamide (114), pyrazinamide (114), and nicotinamide (114, 123). The discovery of ϵ-caprolactam as a new powerful inducer of *R. rhodochrous* J1 nitrilase was reported recently (124) as was the hyperinduction of an aliphatic nitrilase in *R. rhodochrous* K22 (99).

GENETICS AND MOLECULAR BIOLOGY

Genetic information among the rhodococci is extremely limited. Genetic linkage maps have been published for *R. rhodochrous, R. erythropolis,* and *R. canicruria* using natural mating and recombination (22). The linear map of *R. erythropolis* contains 65 genetic traits, including three mating genes.

A lysogenic actinophage, ϕEC, has been physically mapped using restriction analysis (21). This actinophage is transferred either as a prophage or plasmid between other strains of the *R. erythropolis* mating complex and to *R. globerula*. Also, *R. calcarea* plus ϕEC DNA can be transfected with *R. erythropolis* protoplasts. A generalized transducing phage (Q4) for *R. erythropolis* has been described that efficiently transduced several unlinked genes (30). Actinophages (ϕA and ϕB) derived from *R. australis* CSIR-A201 and *R. equi* strain CSIR-A655 liberated spontaneously two new actinophages (ϕA and B) (71). Although the actinophages have similar morphologies, restriction analysis indicated that they are quite dissimilar. The actinophages ϕA and ϕB also infect *R. rubropertinctus* ATCC 14352.

Native plasmids that confer cadmium resistance have been described in *R. fascians* (36). *R. opacus* (39, 141), *Rhodococcus* species H13-A (154), and a *Rhodococcus* isolate (34) have plasmids that confer resistance to arsenate, arsenite, cadmium, and chloramphenicol.

Rhodococcal cloning vectors have been constructed by subcloning the origins of replication from indigenous plasmids present in specific rhodococci (Table 3). An *E. coli–Rhodococcus* shuttle vector, pMVS301, was constructed from pMVS300 derived from *Rhodococcus* sp. H13-A by cloning a 3.8-kb restriction fragment of pMVS300 into pIJ30, which contained ampicillin and thiostrepton resistance determinants (154). This recombinant plasmid, pMVS301, was transformed into *Rhodococcus* sp. AS-50 by the polyethylene glycol–assisted transformation of *Rhodococcus* protoplasts and by selection for thiostrepton resistant transformants. Optimization of the transformation procedure resulted in frequencies of 10^5 transformants per μg of DNA. The plasmid host range extends to *R. erythropolis, R. globulerus,* and *R. equi* but not *R. rhodochrous*. A partial restriction map demonstrated 14 unique restriction sites in pMVS301, some of which may be useful for molecular cloning in *Rhodococcus* spp. and other actinomycetes. This is the first report of transformation and of heterologous gene expression in *Rhodococcus* species. Studies of whether the origin of replication of the *Rhodococcus* sp. H13-A indigenous plasmids pMVS100 and pMVS200 relates to the pMVS301 origin of replication indicated no relatedness. Hybridization studies on the purified plasmids were conducted under stringent conditions (W. R. Finnerty, unpublished results).

The transformation of *R. erythropolis* and *R. equi* with the arsenate-resistant plasmid pD37 also employed the protoplast technique (33). When pD37 was isolated for use in transformation in different organisms, the transformation efficiency was reduced 10-fold. This same reduction in transformation efficiency was noted in *Rhodococcus* sp. H13-A (154). When pMVS301 was isolated from *E. coli* (pMVS301) and used to transform

Table 3 *Rhodococcus* plasmids and cloning vectors

Plasmid	Size (kb)	Resistance markers	Replicon	Cloning sites	Comments
pD188	138	Cd			Transferred within *R. fascians* strains by conjugation.
pDA37	14.6	Ap, As, Q4 immunity		*Bgl*II	*E. coli–Rhodococcus* positive selection shuttle vector.
pMVS301	10.1	Ap, Thiostrepton	pBR322	14 sites	A 3.8-kb fragment unique to *Rhodococcus* pMVS300 cloned into pIJ30.
pRF28	10.5	Cm	pRF2	*Stu*I, *Xba*I	Origin of replication of 160-kb indigenous plasmid from *R. fascians*.
pRF30	13.2	Ap, Cm	pRF2, pUC13	*Stu*I	pRF28 was cloned into *Xba*I site of pUC13.
pRF37	10.8	Ap, Bl	pRF2, pUC13	*Hin*dIII	A 5.1-kb fragment of pRF28 was cloned into pUC13 plus a 3-kb Bl^R gene from *Streptomyces*.

Rhodococcus sp. AS-50, 500-fold lower transformation frequencies resulted. These separate results indicate a restriction-modification system that can be overcome by using plasmid DNA isolated from the *Rhodococcus* species used for transformation.

A 138-kb cadmium-resistant plasmid (pD188) was isolated from *R. fascians*. The conjugal transfer of pD188 between *R. fascians* strains was described, which represents the first report of conjugation in *Rhodococcus* (36). Desomer and colleagues (37) also constructed a series of *E. coli–Rhodococcus* shuttle vectors containing the origin of replication of an indigenous plasmid from *R. fascians*. This plasmid DNA was introduced in *R. fascians* by electroporation, yielding 10^5 transformants per μg of DNA.

The nitrile hydratase gene from *Rhodococcus* species N-774 has been cloned and sequenced, and its expression has been studied in *E. coli* (80). Nitrile hydratase is expressed in *E. coli* as an insoluble protein precipitate when cloned downstream of the *lacPO* promoter. The purification, cloning, and primary structure determination of an enantiomer-selective amidase from *Brevibacterium* species R312 has been reported (116). This *Brevibacterium* and the enzyme amidase appear similar, if not identical, to *Rhodococcus* species R-N774 and nitrile hydratase (69).

The amidase (*amdA*) structural gene is located 73 base pairs upstream of the nitrile hydratase structural gene, suggesting both genes are part of the same operon.

Two genes encoding pigment production in *Rhodococcus* species ATCC 21145 have been cloned and expressed in *E. coli* (67, 72). Expression of the two unidentified pigment genes in *E. coli* produced pink or blue colonies.

The nucleotide sequence of the 5S ribosomal RNA derived from *R. erythropolis* is related more to *Micrococcus luteus* than to *Mycobacterium tuberculosis* (135). The nucleotide sequence of the 5S ribosomal RNA derived from *R. fascians* is similar to almost all actinomycetes (109).

A new type II restriction endonuclease (*Rsp*XI) was purified from a *Rhodococcus* isolate (168). *Rsp*XI recognizes the sequence 5'TCATGA3' and cleaves between T and C, giving a 4-base 5' overhang. The number of fragments generated by digestion with *Rsp*XI on the following DNAs are: pUC 19, three; pBR322, four; M13mp19, one; SV40, two; λ, nine; adenovirus 2, four; ϕ174, three.

PATHOGENICITY

Human

An apparent increased incidence of rhodococcal infections has developed over the past decade. Whether these reports are the result of improved diagnostic methods for recognizing and identifying gram-positive bacteria in the clinical environment, an enhanced awareness of opportunistic gram-positive pathogens by medical personnel, or an actual increased incidence is uncertain. Regardless, a significant number of reports are appearing that describe infectious lesions caused by various rhodococci previously regarded as nonpathogenic in humans. These infections are appearing in both nonimmuno-compromised and immuno-compromised hosts. Table 4 summarizes the types of rhodococcal infections currently recognized in nonimmuno-compromised hosts.

The situation in human immunodeficiency virus (HIV)–infected or acquired immuno deficiency syndrome (AIDS) patients indicates opportunistic rhodococci are present and are a contributory infectious agent in lobar pneumonias of this group of people (41, 44, 47, 68, 86, 101, 137, 147, 180). Differential diagnosis of cavitating lobar pneumonias should probably be expanded to include unusual gram-positive bacteria.

Animals

R. equi has been recognized for decades as a important veterinary pathogen causing broncho pneumonia, enteritis, lymphadenitis, abortion, and other

Table 4 Types of human infections caused by *Rhodococcus* species

Microorganism	Lesion	Reference
Gordona (Rhodococcus) bronchialis	Bypass surgery	142
Tsukamurella paurometabola (Gordona aurantiacus)	Meningitis	138
Rhodococcus equi	Endophthalmitis	38
Rhodococcus equi	Superficial wound	120
Tsukamurella paurometabola (Gordona aurantiacus)	Abscess and necrotizing tenosynvitis	170
Rhodococcus species	Skin and lymphadenitis	112
Rhodococcus species	Actinomycetoma	152
Rhodococcus species	Pulmonary malacoplakia	26
Rhodococcus equi	Pulmonary infections	41, 110
Rhodococcus equi	Endophthalmitis	74
Rhodococcus equi	Brain abscess	132
Rhodococcus equi	Pneumonia	6
Rhodococcus species	Pulmonary infections	133
Rhodococcus equi	Sarcoidosis	73
Rhodococcus erythropolis	Peritonitis	20
Rhodococcus rhodochrous	Corneal ulcer	58
Rhodococcus equi	Osteomyelitis	131
Rhodococcus species	Pulmonary infection	66

diseases in animals (40, 43, 50, 134, 136, 171). The chronic nature of these infections suggest that foals in particular are exposed to the infectious organism weeks before the onset and diagnosis of the disease. Antibiotic treatment at this stage of the disease is largely ineffective.

Seven capsular serotypes are recognized in *R. equi* strains. Serotypes 1 and 2 represent the majority of isolates found in infected animals. The capsular polysaccharide of *R. equi* serotype I consists of D-glucose, D-mannose, and D-glucuronic acid in molar proportions of 1:1:2 (106). This acidic high molecular weight polysaccharide is a repeating tetrasaccharide with an O-acetyl substituent and an acetal-linked pyruvic acid moiety. The high molecular weight acidic polysaccharide of *R. equi* serotype 2 consists of the repeating tetrasaccharide D-glucose, D-mannose, D-glucuronic acid, and 3-0-[-(S)-1-carboxyethyl]-L-rhamnose in equimolar ratios (151). The development of *R. equi* vaccines has been of interest because of the high rates of mortality from chronic bronchitis in foals. Limited success has been achieved to date in the control and prevention of *R. equi* pneumonia using crude whole cell vaccines, even though the efficacy of such vaccines is still uncertain.

R. equi and other atypical mycobacteria have been isolated from the lymph nodes of healthy animals (162). The question addressing virulence factor(s) associated with *R. equi* remains unanswered. The mycolic-containing glycolipids of *R. equi* have been introduced into mice as a possible virulence factor

(59). The injection of purified glycolipids into mice caused granulomas and liver damage. Glycolipids containing mycolic acids C_{44}–C_{46} were more toxic than those glycolipids containing C_{34}–C_{36} mycolates. In addition, the presence of the 15- to 17-kilodalton (kDa) antigens of *R. equi* suggests a role for these major cell-associated proteins as virulence factors (160). These proteins were present in all clinical isolates examined of virulent *R. equi* strains and were absent in all avirulent strains.

A 85-kb plasmid was recently isolated from virulent strains of *R. fascians* that also possessed the 15- to 17-kDa antigens (161). All strains examined possessed the 15- to 17-kDa antigens and the 85-kb plasmid and were virulent in mice. Virulent strains cured of the 85-kb plasmid lacked the 15- to 17-kDa antigen and were significantly less lethal in mice. These findings suggest that the 85-kb plasmid is required for virulence and expression of the 15- to 17-kDa antigens of *R. fascians.*

Fifty-four strains of *R. equi* isolated from different clinical environments were screened for plasmid content (166). Plasmids were present in 49 of the 54 isolates. An 80-kb plasmid was isolated from 21 of 22 isolates from horses and 20 of 28 isolates from pigs, as was a 105-kb plasmid from 7 of 28 isolates from pigs. Restriction analysis of the purified 80-kb plasmid DNA indicated that the plasmids of *R. equi* isolated from equine and porcine hosts as well as the 105-kb plasmid were related.

Plants

R. fascians is a plant pathogen causing fasciation disease of dicotyledons and monocotyledons that is characterized by the loss of apical dominance and outgrowth of lateral buds. Several *R. fascians* strains harbor plasmids (see Genetics section). The bacteria are epiphytic, living on the surface of host tissue where they produce cytokinins. *R. fascians* strains vary greatly in their degree of pathogenicity. Virulent strains excrete larger amounts of cytokinins than weak or avirulent strains. The major cytokinin produced by a virulent strain of *R. fascians* was identified as N6-(δ-2-isopentyl)adenine (121). This virulent strain harbored three plasmids, whereas avirulent strains lacked plasmids. Virulent and avirulent strains of *R. fascians* reportedly harbor a 112-kb plasmid, which leads to questions about the relationship between plasmids and virulence (104).

Virulent strains of *R. fascians* can utilize agmatine and proline as a sole nitrogen source. The growth of *R. fascians* in complex media at 37°C results in the loss of virulence as well as the ability of the resulting avirulent strain to utilize agmatine or proline as the nitrogen source (144). Virulent strains retained their virulence characteristics for pea seedlings when grown in minimal medium at 37°C with ammonium ion as the nitrogen source.

Rhodococcal Biological-Response Modifiers

The immunotherapy of malignant tumors is an area of active interest for the treatment of various malignancies. Interferons, tumor necrosis factor, and cytokines have been induced by priming with whole cells of *Propionibacter acne,* BCG, and lipopolysaccharide (127). Specific components derived from rhodococcal strains have been studied for their immuno-modifying activities.

A β-ketomycolic acid-containing glycolipid derived from *Gordona (Rhodococcus) terrae* 70012 caused prominent granulomas in the lungs and spleen of mice following intravenous injection (128, 183). Multiple intravenous injections of the glycolipid and trehalose-6,6'-dimycolate showed antitumor activity against sarcoma-180 accompanied by granulomatous changes and growth suppression in mice. The multiple injection of trehalose-6,6'-dimycolate and glucose-6-monomycolate derived from *Nocardia rubra* (*Rhodococcus* sp.) (C_{36}–C_{48} mycolates) exhibited antitumor effects against sarcoma-180. Glycolipids containing mannose or fructose did not exhibit significant tumor inhibition or granuloma formation (127). The glucose-6-monomycolate derived from *G. terrae* showed inductive activity for tumor necrosis factor when coadministered with *E. coli* lipopolysaccharide (127).

A purified arabinogalactan-peptidoglycan complex derived from *R. lentifragmentus* suppresses the growth of the syngeneic fibrosarcoma Meth A cells implanted in BALB/c mice (75). The arabinogalacton-peptidoglycan complex has immunoadjuvant activity in stimulating circulating antibody formation, resulting in the induction of cell-mediated immunity and antitumor activity in experimental animals.

The glycolipid trehalose-2,3,6'-trimycolate derived from *T. paurometabola (G. aurantiacus)* induced the formation of tumor necrosis factor and interferons in mice (49). This unsymmetrical glycolipid also showed a protective effect against *Yersinia enterocolitica* infection in mice (48).

Fusidic acid is a antibiotic that inhibits gram-positive bacteria. Resistance to fusidic acid in *R. rhodococcus* results from an inducible extracellular inactivating enzyme (31). Rifampicin inactivation by *R. rhodococcus* results from an induced rifampicin-inactivating enzyme (32). Prokaryotes generally acquire resistance to rifampicin through the mutational change in the target moiety, the DNA-dependent RNA polymerase, with alteration of the RNA polymerase β-subunit.

A unique symbiotic relationship exists between *Rhodnius prolixus* and *R. rhodnii* (17). *Rhodnius prolixus* is an important vector of Chagas disease in Central and South America. This insect vector carries *R. rhodnii* as a symbiont in its gut. Researchers have approached the possibility of developing a vaccine against infection by *Rhodnius prolixus* through the inhibition or suppression of growth of the symbiont, *R. rhodnii*. *Rhodnius prolixus* was fed the blood of rabbits immunized against *R. rhodnii*. The growth of *R. rhodnii* was reduced in the host gut, and the host showed prolonged molting times, incomplete development, and malformed limbs. These symptoms of retarded development in the host were interpreted as indications of symbiont deficiency. The biochemical basis for this symbiotic relationship is unclear, but it does suggest that *R. rhodnii* plays a significant role in the normal development of *Rhodnius prolixus*.

CONCLUDING REMARKS

The genus *Rhodococcus* represents a taxon encompassing microorganisms with a unique and facile armamentarium of enzymatic capabilities for catalyzing the biotransformation of diverse organic compounds. A broader substrate diversity will be realized as new representatives of this taxon emerge. The rhodococci and their products may well prove useful for applications such as lignites, high-sulfur coals, and oil updgrading, for the reduction of heavy oil viscosity, and possibly for the enhancement of oil recovery technologies. Although commercial exploitation of the rhodococci is currently limited, commercial development of the rhodococci for bioremediation may in the future be rapid. The use of rhodococcal-focused bioremediation strategies such as degradation of chlorophenolics appears to represent promising technologies for environmental management systems.

Genetic and molecular biological studies on any organism require the availability of a few basic techniques. Systems must be available for the transfer of DNA, and methods for identifying and isolating mutants, and cloning vector systems, must be readily available. Knowledge of the fundamental molecular properties of the rhodococci will assist genetic manipulation. Because little information is available, these are areas in need of further research. Molecular approaches for the study of the rhodococci, although currently quite limited, are becoming available; future advances should occur rapidly, expanding our understanding of the molecular biology of this genus.

The apparent increased incidence of rhodococcal infections may pose an important problem to the medical community. Many infections are being caused by organisms previously considered nonpathogenic to humans and animals. Such infectious threats are most probably opportunistic, but much more information is needed on the virulence factors of the rhodococci and on the mechanism(s) involved in the initiation of invasive threats to humans and animals. The genus *Rhodococcus* stands as a complex group of microorganisms direly in need of focused research on the basic physiology and biochemistry of the rhodococci, the genetics and molecular biology of the constituent microorganisms, expanded research initiatives into the pathogenicity of the microorganisms, and applied research programs for their potential application to diverse environmental problems and economically feasible industrial processes.

Literature Cited

1. Ahmad, S., Johri, B. N. 1991. A cholesterol degrading bacteria: isolation, characterization and bioconversion. *Indian J. Exp. Biol.* 29:76–77
2. Aihara, H., Watanabe, K., Nakamura, R. 1986. Characterization and production of cholesterol oxidases in three Rhodococcus strains. *J. Appl. Bacteriol.* 61:269–74
3. Aihara, H., Watanabe, K., Nakamura, R. 1988. Degradation of cholesterol in egg yolk by *Rhodococcus-equi* no. 23. *J. Food Sci.* 53:659–60
4. Aihara, H., Watanabe, K., Nakamura, R. 1988. Degradation of cholesterol in hen's egg yolk and its low density lipoprotein by extracellular enzymes of *Rhodococcus-equi* no. 23. *Lebensm. Wiss. Technol.* 21:342–45
5. Allard, A. S., Remberger, M., Neilson, A. H. 1987. Bacterial O-methylation of halogen-substituted phenols. *Appl. Environ. Microbiol.* 53:839–45
6. Allen, V. D., Niec, A., Kerem, E., Greenberg, M., Gold, R. 1989. *Rhodococcus equi* pneumonia in a child with leukemia. *Pediatr. Infect. Dis. J.* 8:656–58
7. Aoki, K., Uemori, T., Shinke, R., Nishira, H. 1985. Further characterization of bacterial production of anthranilic acid from aniline. *Agric. Biol. Chem.* 49:1151–58
8. Aoki, K., Uemori, T., Shinke, R., Nishira, H. 1988. Promotion of glucose utilization by organic acids and amino acids in aniline-assimilating *Rhodococcus erythropolis* AN-13. *Agric. Biol. Chem.* 52:113–20
9. Apajalahti, J. H. A., Salkinoja-Salonen, M. S. 1986. Degradation of polychlorinated phenols by *Rhodococcus chlorophenolicus*. *Appl. Microbiol. Biotechnol.* 25:62–67
10. Apajalahti, J. H. A., Salkinoja-Salonen, M. S. 1987. Dechlorination and parahydroxylation of polychlorinated phenols by *Rhodococcus chlorophenolicus*. *J. Bacteriol.* 169:675–81
11. Apajalahti, J. H. A., Salkinoja-Salonen, M. S. 1987. Complete dechlorination of tetrachlorohydroquinone by cell extracts of pentachlorophenol-induced *Rhodococcus chlorophenolicus*. *J. Bacteriol.* 169:5125–30
12. Apajalahti, J. H. A., Karpanoja, P., Salkinoja-Salonen, M. S. 1986. *Rhodococcus chlorophenolicus* new species: a chlorophenol-mineralizing actinomycete. *Int. J. Syst. Bacteriol.* 36:246–51
13. Ashraf, W., Murrell, J. C. 1990. Purification and characterization of a NAD-dependent secondary alcohol dehydrogenase from propane-grown *Rhodococcus rhodochrous*. *Arch. Microbiol.* 153:163–68
14. Bar, R. 1988. Ultrasound enhanced bioprocesses: cholesterol oxidation by *Rhodococcus erythropolis*. *Biotechnol. Bioeng.* 32:655–63
15. Bengis-Garber, C., Gutman, A. L. 1988. Bacteria in organic synthesis: selective conversion of 1,3-dicyanobenzene into 3-cyanobenzoic acid. *Tetrahedron Lett.* 29:2589–90
16. Bengis-Garber, C., Gutman, A. L. 1989. Selective hydrolysis of dinitriles into cyano-carboxylic acids by *Rhodococcus rhodochrous* N. C. I. B. 11216. *Appl. Microbiol. Biotechnol.* 32:11–16
17. Ben-Yakir, D. 1987. Growth retardation of *Rhodnius prolixus* symbionts by immunizing host against *Nocardia* (*Rhodococcus*) *rhodnii*. *J. Insect Physiol.* 33: 379–83
18. Bowden, G. H., Goodfellow, M. 1990. The actinomycetes: *Actinomyces, Nocardia* and related genera. In *Topley and Wilson's Principles of Bacteriology, Virology and Immunity,* ed. M. T. Parker, B. I. Duerden, 2:32–53. London: Edward Arnold. 496 pp.
19. Breda, M., Ioneda, T. 1988. Production of glycolipids by *Rhodococcus rhodochrous* grown on galactose. *Rev. Microbiol.* 19:202–6
20. Brown, E., Hendler, E. 1989. *Rhodococcus peritonitis* in a patient treated with peritoneal dialysis. *Am. J. Kidney Dis.* 14:417–18
21. Brownell, G. H., Saba, J. A., Denniston-Thompson, K., Enquist, L. W. 1982. The development of a *Rhodococcus*-actinophage cloning system. *Dev. Ind. Microbiol.* 23:287–98
22. Brownell, G. H., Denniston, K. 1984.

Genetics of the nocardio form bacteria. In *The Biology of the Actinomycetes,* ed. M. Goodfellow, M. Mordarski, S. T. Williams, pp. 201–8. New York: Academic. 465 pp.

23. Bruce, N. C., Cain, R. B. 1988. Betamethylmuconolactone, a key intermediate in the dissimilation of methylaromatic compounds by a modified 3-oxoadipate pathway evolved in nocardioform actinomycetes. *FEMS Microbiol. Lett.* 50:233–40
24. Bruce, N., Cain, B. 1990. Hydroaromatic metabolism in *Rhodococcus rhodochrous:* purification and characterisation of its NAD-dependent quinate dehydrogenase. *Arch. Microbiol.* 154: 179–86
25. Bruce, N. C., Cain, R. B., Pieper, D. H., Engesser, K. H. 1989. Purification and characterization of 4-methylmuconolactone methyl-isomerase, a novel enzyme of the modified 3-oxoadipate pathway in nocardioform actinomycetes. *Biochem. J.* 262:303–12
26. Byard, R. W., Thorner, P. S., Edwards, V., Greenberg, M. 1990. Pulmonary malacoplakia in a child. *Pediatr. Pathol.* 10:417–24
27. Cohen, M. A., Sawden, J., Turner, N. 1990. Selective hydrolysis of nitriles under mild conditions by an enzyme. *Tetrahedron Lett.* 31:7223–26
28. Collins, M. D., Smida, J., Dorsch, M., Stackebrandt, E. 1988. *Tsukamurella* gen. nov. harboring *Corynebacterium paurometabolum* and *Rhodococcus aurantiacus. Int. J. Syst. Bacteriol.* 88:385–91
29. Cook, A. M., Hutter, R. 1986. Ring dechlorination of deethylsimazine by hydrolases from a *Rhodococcus corallinus. FEMS Microbiol. Lett.* 34:335–38
30. Dabbs, E. R. 1987. A generalized transducing bacteriophage for *Rhodococcus erythropolis. Mol. Gen. Genet.* 206:116–20
31. Dabbs, E. R. 1987. Fusidic acid resistance in *Rhodococcus erythropolis* due to an inducible extracellular inactivating enzyme. *FEMS Microbiol. Lett.* 40: 135–38
32. Dabbs, E. R. 1987. Rifampicin inactivation by *Rhodococcus* and *Mycobacterium* species. *FEMS Microbiol. Lett.* 44:395–400
33. Dabbs, E. R., Gowan, B., Andersen, S. J. 1990. Nocardioform arsenic resistance plasmids and construction of *Rhodococcus* cloning vectors. *Plasmid* 23:242–47
34. Dabbs, E. R., Sole, G. J. 1988. Plasmid-borne resistance to arsenate, arsenite, cadmium, and chloramphenicol in a *Rhodococcus* species. *Mol. Gen. Genet.* 211:148–54
35. Datcheva, V. K., Voishvillo, N. E., Kamernitskii, A. V., Vlahov, R. J., Reshetova, I. G. 1989. Synthesis of 9 alpha-hydroxysteroids by a *Rhodococcus* species. *Steroid* 54:271–86
36. Desomer, J., Dhaese, P., Van Montagu, M. 1988. Conjugative transfer of cadmium resistance plasmids in *Rhodococcus fascians* strains. *J. Bacteriol.* 170: 2401–5
37. Desomer, J., Dhaese, P., Van Montagu, M. 1990. Transformation of *Rhodococcus fascians* by high-voltage electroporation and development of *Rhodococcus fascians* cloning vectors. *Appl. Environ. Microbiol.* 56:2818–25
38. Ebersole, L., Paturzo, J. L. 1988. Endophthalmitis caused by *Rhodococcus equi* prescott serotype 4. *J. Clin. Microbiol.* 26:1221–22
39. Ecker, C., Reh, M., Schlegel, H. G. 1986. Enzymes of the autotrophic pathway in mating partners and transconjugants of *Nocardia opaca* 1B and *Rhodococcus erythropolis. Arch. Microbiol.* 145:280–86
40. Edwards, J. F., Simpson, R. B. 1988. Nocardioform actinomycete (*Rhodococcus rubropertinctus*)-induced abortion in a mare. *Vet. Pathol.* 25:529–30
41. Egawa, T., Hara, H., Kawase, I., Masuno, T., Asari, S., et al. 1990. Human pulmonary infection with *Corynebacterium equi. Eur. Respir. J.* 3:240–42
42. Eggeling, L., Sahm, H. 1985. The formaldehyde dehydrogenase of *Rhodococcus erythropolis,* a trimeric enzyme requiring a cofactor and active with alcohols. *Eur. J. Biochem.* 150:129–34
43. Elliott, G., Lawson, G. H. K., Mackenzie, C. P. 1986. *Rhodococcus equi* infection in cats. *Vet. Rec.* 118:693–94
44. Emmons, W., Reichwein, B., Winslow, D. L. 1991. *Rhodococcus-equi* infection in the patient with AIDS: literature review and report of an unusual case. *Rev. Infect. Dis.* 13:91–96
45. Endo, T., Watanabe, I. 1989. Nitrile hydratase of *Rhodococcus sp.* N-774. Purification and amino acid sequences. *FEBS Lett.* 243:61–64
46. Engesser, K. H., Cain, R. B., Knackmuss, H. J. 1988. Bacterial metabolism of side chain fluorinated aromatics: cometabolism of 3-trifluoromethyl (TFM)-benzoate by *Pseudomonas putida* (arvilla) mt-2 and *Rhodococcus rubropertinctus* N657. *Arch. Microbiol.* 149:188–97
47. Fierer, J., Wolf, P., Seed, L., Gay, T.,

Noonan, K., et al. 1987. Nonpulmonary *Rhodococcus equi* infections in patients with acquired immune deficiency syndrome (AIDS). *J. Clin. Pathol.* 40:556–58

48. Fujita, T., Kawata, Y., Kitamura, F., Sugimoto, N., Yano, I., et al. 1988. Protective effect of glycolipids containing mycolic acid derived from *Gordona aurantiaca* against *Yersinia enterocolitica*. *J. Jpn. Allerg. Infect. Dis.* 62:26–31
49. Fujita, T., Sugimoto, N., Yokoi, F., Ohtsubo, Y., Ikutoh, M., et al. 1990. Induction of interferons (IFNs) and tumor necrosis factor (TNF) in mice by a novel glycolipid trehalose 2,3,6'-trimycolate form *Rhodococcus aurantiacus* (*Gordona aurantiaca*). *Microbiol. Immunol.* 34:523–32
50. Garg, D. N., Kapoor, P. K. 1986. Isolation and characterization of *Rhodococcus* (*Corynebacterium*) *equi* from cows with mastitis. *Indian J. Comp. Microbiol. Immunol. Infect. Dis.* 7:91–95
51. Germon, J. C., Knowles, R. 1988. Metabolism of acetylene and acetaldehyde of *Rhodococcus rhodochrous*. *Can. J. Microbiol.* 34:242–48
52. Goodfellow, M. 1987. The taxonomic status of *Rhodococcus equi*. *Vet. Microbiol.* 14:205–9
53. Goodfellow, M. 1989. Suprageneric classification of actinomycetes. In *Bergey's Manual of Systematic Bacteriology*, ed. S. T. Williams, M. E. Sharpe, J. G. Holt, 4:2333–39. Baltimore: Williams & Wilkins. 2785 pp.
54. Goodfellow, M. 1991. Identification of some mycolic acid containing actinomycetes using fluorogenic probes based on 7-amino-4-methylcoumarin and 4-methylumbelliferone. *Actinomycetal* 5: 21–27
55. Goodfellow, M., Alderson, G. 1977. The actinomycete genus *Rhodococcus:* a home for the 'rhodochrous' complex. *J. Gen. Microbiol.* 100:99–122
56. Goodfellow, M., Thomas, E. G., Ward, A. C., James, A. L. 1990. Classification and identification of rhodococci. *Zentralbl. Bakteriol.* 274:299–315
57. Goodfellow, M., Zakrzewska-Czerwinska, J., Thomas, E. G., Mordarski, M., Ward, A. C., James, A. L. 1991. Polyphasic taxonomic study of the genera *Gordona* and *Tsukamurella* including the description of *Tsukamurella wratislaviensis* species nov. *Zentralbl. Bakteriol.* 275:162–78
58. Gopaul, D., Ellis, C., Maki, A. Jr., Joseph, M. G. 1988. Isolation of *Rhodococcus rhodochrous* from a chronic corneal ulcer. *Diag. Microbiol. Infect. Dis.* 10:185–90
59. Gotah, K., Mitsuyama, M., Imaizumi, S., Kawanana, I., Yano, J. 1991. Mycolic-acid containing glycolipid as a possible virulence factor of *Rhodococcus equi* for mice. *Microbiol. Immunol.* 35:175–85
60. Goulding, C., Gillen, C. J., Bolton, E. 1988. Biodegradation of substituted benzenes. *J. Appl. Bacteriol.* 65:1–5
61. Haggblom, M. M., Apajalahti, J. H. A., Salkinoja-Salonen, M. S. 1988. O-methylation of chlorinated para-hydroquinones by *Rhodococcus chlorophenolicus*. *Appl. Environ. Microbiol.* 54:1818–24
62. Haggblom, M. M., Apajalahti, J. H. A., Salkinoja-Salonen, M. S. 1988. Hydroxylation and dechlorination of chlorinated guaiacols and syringols by *Rhodococcus chlorophenolicus*. *Appl. Environ. Microbiol.* 54:683–87
63. Haggblom, M. M., Janke, D., Middeldorp, P. J. M., Salkinoja-Salonen, M. S. 1989. O Methylation of chlorinated phenols in the genus *Rhodococcus*. *Arch. Microbiol.* 152:6–9
64. Haggblom, M. M., Janke, D., Salkinoja-Salonen, M. S. 1989. Hydroxylation and dechlorination of tetrachlorohydroquinone by *Rhodococcus* species strain CP-2 cell extracts. *Appl. Environ. Microbiol.* 55:516–19
65. Haggblom, M. M., Janke, D., Salkinoja-Salonen, M. S. 1989. Transformation of chlorinated phenolic compounds in the genus *Rhodococcus chlorophenolicus*. *Microbiol. Ecol.* 18:147–59
66. Hart, D. H., Peel, M. M., Andrew, J. H., Burdon, J. G. 1988. Lung infection caused by *Rhodococcus*. *Aust. N. Z. J. Med.* 18:790–91
67. Hart, S., Kirby, R., Woods, D. R. 1990. Structure of a *Rhodococcus* gene encoding pigment production in *Escherichia coli*. *J. Gen. Microbiol.* 136:1357–63
68. Harvey, R. L., Sunstrum, J. C. 1991. *Rhodococcus equi* infection in patients with and without human immunodeficiency virus infection. *Rev. Infect. Dis.* 13:139–45
69. Hashimoto, Y., Nishiyama, M., Ikehata, O., Horinouchi, S., Beppu, T. 1991. Cloning and characterization of an amidase gene from *Rhodococcus* species N-774 and its expression in *Escherichia coli*. *Biochim. Biophys. Acta* 1088:225–33
70. Hensel, J., Straube, G. 1990. Kinetic studies of phenol degradation by *Rhodococcus* sp. P1. II. Continuous cultiva-

tion. *Antonie van Leeuwenhoek J. Microbiol. Ser.* 57:33–36
71. Hiddema, R., Curran, M. D., Ferreira, N. P., Coetzee, J. N., Lecatsas, G. 1985. Characterization of phages derived from strains of *Rhodococcus australis* and *R. equi*. *Intervirology.* 23:109–11
72. Hill, R., Hart, S., Illing, N., Kirby, R., Woods, D. R. 1989. Cloning and expression of *Rhodococcus* genes encoding pigment production in *Escherichia coli. J. Gen. Microbiol.* 135:1507–13
73. Hillerdal, G., Riesenfeldt-Orn, I., Pedersen, A., Ivanicova, E. 1988. Infection with *Rhodococcus equi* in a patient with sarcoidosis treated with corticosteroids. *Scand. J. Infect. Dis.* 20:673–77
74. Hillman, D., Garretson, B., Fiscella, R. 1989. *Rhodococcus equi* endophthalmitis. Case report. *Arch. Ophthalmol.* 107:20
75. Hirai, O., Fujitsu, T., Mori, J., Kikuchi, H., Koda, S., et al. 1987. Antitumor activity of purified arabinogalactan-peptidoglycan complex of the cell wall skeleton of *Rhodococcus lentifragmentus. J. Gen. Microbiol.* 133:369–73
76. Hjort, C. M., Godtfredsen, S. E., Emborg, C. 1990. Isolation and characterization of a nitrile hydratase from a *Rhodococcus* species. *J. Chem. Technol. Biotechnol.* 48:217–26
77. Hummel, W., Schuette, H., Schmidt, E., Wandrey, C., Kula, M. R. 1987. Isolation of L-phenylalanine dehydrogenase from *Rhodococcus* sp M4 and its application for the production of L-phenylalanine. *Appl. Microbiol. Biotechnol.* 26:409–16
78. Iida, M., Murohisa, T., Yoneyama, A., Iizuka, H. 1985. Microbiological production of 20 carboxypregna-1,4-dien-3-one from beta-sitosterol with high yield by a *Rhodococcus* species. *J. Ferment. Technol.* 63:559–62
79. Iida, M., Tsuyuki, K. I., Kitazawa, S., Iizuka, H. 1987. Production of 20 carboxypregna-1,4-dien-3-one from sterols by mutants of *Rhodococcus* species. *J. Ferment. Technol.* 65:525–30
80. Ikehata, O., Nishiyama, M., Horinouchi, S., Beppu, T. 1989. Primary structure of nitrile hydratase deduced from the nucleotide sequence of a *Rhodococcus* species and its expression in *Escherichia coli. Eur. J. Biochem.* 181:563–70
81. Ito, M., Yamagata, T. 1986. Glycosphingolipid-degrading enzymes of actinomycetes. *J. Biol. Chem.* 261: 14278–82
82. Ito, M., Yamagata, T. 1989. Purification and characterization of glycosphingolipid-specific endoglycosidases (endoglycoceramidases) from a mutant strain of *Rhodococcus* species. Evidence for three molecular species of endoglycoceramidase with different specificities. *J. Biol. Chem.* 264:9510–19
83. Janke, D., Ihn, W. 1989. Cometabolic turnover of aniline phenol and some of their monochlorinated derivatives by the *Rhodococcus* mutant strain AM 144. *Arch. Microbiol.* 152:347–52
84. Johnson, T. L., Somkuti, G. A. 1990. Properties of cholesterol dissimilation by *Rhodococcus equi. J. Food Prot.* 53: 332–35
85. Johnson, T. L., Somkuti, G. A. 1991. Isolation of cholesterol oxidases from *Rhodococcus equi* ATCC 33706. *Biotechnol. Appl. Biochem.* 13:196–204
86. Jones, M. R., Neale, T. J., Say, P. J., Horne, J. G. 1989. *Rhodococcus equi:* an emerging opportunistic pathogen? *Aust. N. Z. J. Med.* 19:103–7
87. Kakeya, H., Sakai, N., Sugai, T., Ohta, H. 1991. Microbial hydrolysis as a potent method for the preparation of optically active nitriles, amides and carboxylic acids. *Tetrahedron Lett.* 32:1343–46
88. Kanagawa, T., Kelly, D. P. 1987. Degradation of substituted thiophenes by bacteria isolated from activated sludge. *Microbiol. Ecol.* 13:47–58
89. Kilbane, J. J. 1989. Desulfurization of coal: the microbial solution. *Trends Biotechnol.* 7:97–100
90. Kilbane, J. J. 1990. Sulfur-specific microbial metabolism of organic compounds. *Resour. Conserv. Recycl.* 3:69–79
91. Kim, J. S., Powalla, M., Lang, S., Wagner, F., Luensdorf, H., et al. 1990. Microbial glycolipid production under nitrogen limitation and resting cell conditions. *J. Biotechnol.* 13:257–66
92. Klempier, N., De Raadt, A., Faber, K., Griengl, H. 1991. Selective transformation of nitriles into amides and carboxylic acids by an immobilized nitrilase. *Tetrahedron Lett.* 32:341–44
93. Kobayashi, M., Nagasawa, T., Yamada, H. 1988. Regiospecific hydrolysis of dinitrile compounds by nitrilase from *Rhodococcus rhodochrous* J1. *Appl. Microbiol. Biotechnol.* 29:231–33
94. Kobayashi, M., Nagasawa, T., Yamada, H. 1989. Nitrilase of *Rhodococcus rhodochrous* J1. Purification and characterization. *Eur. J. Biochem.* 182:349–56
95. Kobayashi, M., Nagasawa, T., Yanaka,

N., Yamada, H. 1989. Nitrilase-catalyzed production of *p*-aminobenzoic acid from *p*-aminobenzonitrile with *Rhodococcus rhodochrous* J1. *Biotechnol. Lett.* 11:27–30

96. Kobayashi, M., Yanaka, N., Nagasawa, T., Yamada, H. 1990. Nitrilase-catalyzed production of pyrazinoic acid, an antimycobacterial agent from cyanopyrazine by resting cells of *Rhodococcus rhodochrous* J1. *J. Antibiot.* 43:1316–20
97. Kobayashi, M., Yanaka, N., Nagasawa, T., Yamada, H. 1990. Monohydrolysis of an aliphatic dinitrile compound by nitrilase from *Rhodococcus rhodochrous* K22. *Tetrahedron* 46:5587–90
98. Kobayashi, M., Yanaka, N., Nagasawa, T., Yamada, H. 1990. Purification and characterization of a novel nitrilase of *Rhodococcus rhodochrous* K22 that acts on aliphatic nitriles. *J. Bacteriol.* 172:4807–15
99. Kobayashi, M., Yanaka, N., Nagasawa, T., Yamada, H. 1991. Hyperinduction of an aliphatic nitrilase by *Rhodococcus rhodochrous* K22. *FEMS Microbiol. Lett.* 77:121–24
100. Kretschmer, A., Bock, H., Wagner, F. 1982. Chemical and physical characterization of interfacial-active lipids for *Rhodococcus erythropolis* grown on n-alkanes. *Appl. Environ. Microbiol.* 44:864–70
101. Kunke, P. J. 1987. Serious infection in an AIDS patient due to *Rhodococcus equi*. *Clin. Microbiol. Newsl.* 9:163–64
102. Kurane, R., Takeda, K., Suzuki, T. 1986. Screening for and characteristics of microbial flocculants. *Agric. Biol. Chem.* 50:2301–8
103. Kurane, R., Toeda, K., Takeda, K., Suzuki, T. 1986. Culture conditions for production of microbial flocculant by *Rhodococcus erythropolis*. *Agric. Biol. Chem.* 50:2309–14
104. Lawson, E. N., Gantotti, B. V., Starr, M. P. 1982. A 78-megadalton plasmid occurs in avirulent as well as virulent strains of *Corynebacterium fascians*. *Curr. Microbiol.* 7:327–32
105. Lechevalier, M. P., Lechevalier, H. 1985. Biology of actinomycetes not belonging to genus *Streptomyces*. In *Biology of Industrial Microorganisms,* ed. A. L. Demain, N. A. Solomon, pp. 315–58. Menlo Park, CA: Benjamin/Cummings. 573 pp.
106. Leitch, R. A., Richards, J. C. 1990. Structural analysis of the specific capsular polysaccharide of *Rhodococcus equi* serotype 1. *Biochem. Cell Biol.* 68:778–89
107. Lidstrom, M. E., Stirling, D. I. 1990. Methylotrophs: genetic and commercial applications. *Annu. Rev. Microbiol.* 44:27–58
108. Lloyd-Jones, G., Trudgill, P. W. 1989. The degradation of alicyclic hydrocarbons by a microbial consortium. *Int. Biodeterior.* 25:197–206
109. Luehrsen, K., Woese, C. R., Wolters, J., Stackebrandt, E. 1989. Nucleotide sequence of 5S ribosomal RNA of *Rhodococcus fascians*. *Nucleic Acids Res.* 17:5378
110. MacGregor, J. H., Samuelson, W. M., Sane, D. C., Godwin, J. D. 1986. Opportunistic lung infection caused by *Rhodococcus equi*. *Radiology* 160:83–84
111. MacMichael, G. J., Brown, L. R. 1987. Role of carbon dioxide in catabolism of propane by *Nocardia paraffinicum* and *Rhodococcus rhodochrous*. *Appl. Environ. Microbiol.* 53:65–69
112. Martin, T., Hogan, D. J., Murphy, F., Natyshak, I., Ewan, E. P. 1991. *Rhodococcus* infection of the skin with lymphadenitis in a nonimmunocompromised girl. *J. Am. Acad. Dermatol.* 24:328–32
113. Mathew, C. D., Nagasawa, T., Kobayashi, M., Yamada, H. 1988. Nitrilase-catalyzed production of nicotinic acid from 3-cyanopyridine in *Rhodococcus rhodochrous* J1. *Appl. Environ. Microbiol.* 54:1030–32
114. Mauger, J., Nagasawa, T., Yamada, H. 1988. Nitrile hydratase-catalyzed production of isonicotinamide picolinamide and pyrazinamide from 4-cyanopyridine, 2-cyanopyridine and cyanopyrazine in *Rhodococcus rhodochrous* J1. *J. Biotechnol.* 8:87–95
115. Mauger, J., Nagasawa, T., Yamada, H. 1989. Synthesis of various aromatic amide derivatives using nitrile hydratase of *Rhodococcus rhodochrous* J1. *Tetrahedron* 45:1347–54
116. Mayaux, J. F., Cerebelaud, E., Soubrier, F., Faucher, D., Petre, D. 1990. Purification, cloning, and primary structure of an enantiomer-selective amidase from *Brevibacterium* species strain R312: structural evidence for genetic coupling with nitrile hydratase. *J. Bacteriol.* 172:6764–73
117. Miclo, A., Germain, P. 1990. Catabolism of methylperhydroindanedione propionate by *Rhodococcus equi:* evidence of a MEPHIP-reductase activity. *Appl. Microbiol. Biotechnol.* 32:594–99
118. Middeldorp, P. J. M., Briglia, M., Salkinoja-Salonen, M. S. 1990. Biodegradation of pentachlorophenol in natural

soil by inoculated *Rhodococcus chlorophenolicus*. *Microbiol. Ecol.* 20:123–39
119. Misono, H., Yonezawa, J., Nagata, S., Nagasaki, S. 1989. Purification and characterization of a dimeric phenylalanine dehydrogenase from *Rhodococcus maris* K-18. *J. Bacteriol.* 171:30–36
120. Muller, F., Schaal, K. P., von Graevenitz, A., von Moos, L., Woolcock, J. B., et al. 1988. Characterization of *Rhodococcus equi*-like bacterium isolated from a wound infection in a noncompromised host. *J. Clin. Microbiol.* 26:618–20
121. Murai, N., Skoog, F., Doyle, M. E., Hanson, R. S. 1980. Relationship between cytokinin production, presence of plasmids and fasciation caused by strains of *Corynebacterium fascians*. *Proc. Natl. Acad. Sci. USA* 77:619–23
122. Nagasawa, T., Kobayashi, M., Yamada, H. 1988. Optimum culture conditions for the production of benzonitrilase by *Rhodococcus rhodochrous* J1. *Arch. Microbiol.* 150:89–94
123. Nagasawa, T., Mathew, C. D., Mauger, J., Yamada, H. 1988. Nitrile hydratase-catalyzed production of nicotinamide from 3-cyanopyridine in *Rhodococcus rhodochrous* J1. *Appl. Environ. Microbiol.* 54:1766–69
124. Nagasawa, T., Nakamura, T., Yamada, H. 1990. Epsilon caprolactam, a new powerful inducer for the formation of *Rhodococcus rhodochrous* J1 nitrilase. *Arch. Microbiol.* 155:13–17
125. Nagasawa, T., Nakamura, T., Yamada, H. 1990. Production of acrylic acid and methacrylic acid using *Rhodococcus rhodochrous* J1 nitrilase. *Appl. Microbiol. Biotechnol.* 34:322–24
126. Nakajima, K., Sato, A., Takahara, Y., Iida, T. 1985. Microbial oxidation of isoprenoid alkanes phytane, norpristane and farnesane. *Agric. Biol. Chem.* 49: 1993–2002
127. Natsuhara, Y., Oka, S., Kaneda, K., Kata, Y., Yano, I. 1990. Parallel antitumor, granuloma-forming and tumor-necrosis-factor-priming activities of mycoloyl glycolipids from *Nocardia rubra* that differ in carbohydrate moiety: structure-activity relationships. *Cancer Immunol. Immunother.* 31:99–106
128. Natsuhara, Y., Yoshinaga, J., Shogaki, T., Sumi-Nishikawa, Y., Kurano, S., et al. 1990. Granuloma-forming activity and antitumor activity of newly isolated mycoloyl glycolipid from *Rhodococcus terrae* 70012 (Rt. GM-2). *Microbiol. Immunol.* 34:45–53
129. Neu, T. R., Poralla, K. 1988. An amphiphilic polysaccharide from an adhesive *Rhodococcus* strain. *FEMS Microbiol. Lett.* 49:389–92
130. Ninnekar, H. Z., Pujar, B. G. 1985. Degradation of dimethylterephthalate by a *Rhodococcus* species. *Indian J. Biochem. Biophys.* 22:232–35
131. Novak, R. M., Polisky, E. L., Janda, W. M., Libertin, C. R. 1988. Osteomyelitis caused by *Rhodococcus equi* in a renal transplant recipient. *Infection* 16:186–88
132. Obna, W. G., Scannell, K. A., Jacobs, R., Greco, C., Rosenblum, M. L. 1991. A case of *Rhodococcus equi* brain abscess. *Surg. Neurol.* 35:321–24
133. Osoagbaka, O. U. 1989. Evidence for the pathogenic role of *Rhodococcus* species in pulmonary diseases. *J. Appl. Bacteriol.* 66:497–506
134. Oxenford, C. J., Ratcliffe, R. C., Ramsay, G. C. 1987. *Rhodococcus equi* infection in a cat. *Aust. Vet. J.* 64:121
135. Park, Y. H., Hori, H., Suzuki, K., Osawa, S., Kamagata, K. 1987. Nucleotide sequence of 5S ribosomal RNA from *Rhodococcus erythropolis*. *Nucleic Acids Res.* 15:365
136. Perdrizet, J. A., Scott, D. W. 1987. Cellulitis and subcutaneous abscesses caused by *Rhodococcus equi* infection in a foal. *J. Am. Vet. Med. Assoc.* 190: 1559–61
137. Prescott, J. F. 1991. *Rhodococcus equi:* an animal and human pathogen. *Clin. Microbiol. Rev.* 4:20–34
138. Prinz, G., Ban, E., Fekete, S., Szabo, Z. 1985. Meningitis caused by *Gordona aurantiaca* (*Rhodococcus aurantiacus*). *J. Clin. Microbiol.* 22:472–74
139. Ramsay, B., McCarthy, J., Guerra-Santos, L., Kappeli, O., Fiechter, A., et al. 1988. Biosurfactant productions and diauxic growth of *Rhodococcus aurantiacus* when using N-alkanes as the carbon source. *Can. J. Microbiol.* 34:1209–12
140. Rapp, P., Bock, H., Wray, V., Wagner, F. 1979. Formation, isolation and characterization of trehalose dimycolates from *Rhodococcus erythropolis* grown on n-alkanes. *J. Gen. Microbiol.* 115: 491–503
141. Reh, M., Schlegel, H. G. 1981. Physiological studies on auxotrophic mutants of *Nocardia opaca*. *J. Gen. Microbiol.* 126:327–36
142. Richet, H. M., Craven, P. C., Brown, J. M. Lasker, B. A., Cox, C. D., et al. 1991. A cluster of *Rhodococcus bronchialis* sternal-wound infections after coronary-artery bypass surgery. *New Engl. J. Med.* 324:104–9

143. Ristau, E., Wagner, F. 1983. Formation of novel anionic trehalose tetra esters from *Rhodococcus erythropolis* under growth limiting conditions. *Biotechnol. Lett.* 5:95–100
144. Sabart, P. R., Gakovich, D., Hanson, R. S. 1986. Avirulent isolates of *Corynebacterium fascians* that are unable to utilize agmatine and proline. *Appl. Environ. Microbiol.* 52:33–36
145. Sakai, K., Nakazawa, A., Kondo, K., Ohta, H. 1985. Microbial hydrogenation of nitroolefins. *Agric. Biol. Chem.* 49:2231–2336
146. Sallis, P. J., Armfield, S. J., Bull, A. T., Hardman, D. J. 1990. Isolation and characterization of a haloalkane halidohydrolase from *Rhodococcus erythropolis* Y2. *J. Gen. Microbiol.* 136:115–20
147. Scannell, K. A., Portoni, E. J., Finkle, H. I., Rice, M. 1990. Pulmonary malacoplakia and *Rhodococcus equi* infection in a patient with AIDS. *Chest* 97:1000–1
148. Schwarz, G., Bauder, R., Speer, M., Rommel, T. O., Lingens, F. 1989. Microbial metabolism of quinoline and related compounds. II.+ Degradation of quinoline by *Pseudomonas fluorescens* 3, *Pseudomonas putida* 86 and *Rhodococcus* species B1. *Biol. Chem. Hoppe-Seyler* 370:1183–89
149. Schwarz, G., Senghas, E., Erben, A., Schaefer, B., Lingens, F., Hoeke, H. 1988. Microbial metabolism of quinoline and related compounds. I. Isolation and characterization of quinoline-degrading bacteria. *Syst. Appl. Microbiol.* 10:185–90
150. Setchell, C. H., Bonner, J. F., Wright, S. J., Caunt, P., Baker, P. B. 1985. Microbial transformation of squalene: formation of a novel ketone from squalene by a *Rhodococcus* species. *Appl. Microbiol. Biotechnol.* 21:255–57
151. Severn, W. B., Richards, J. C. 1990. Structural analysis of the specific capsular polysaccharide of *Rhodococcus equi. Carbohydr. Res.* 206:311–32
152. Severo, L. C., Petrillo, V. F., Coutinho, L. M. B. 1987. Actinomycetoma caused by *Rhodococcus* species. *Mycopathologia.* 98:129–31
153. Shimizu, S., Hattori, S., Hata, H., Yamada, H. 1987. One-step microbial conversion of a racemic mixture of pantoyl lactone to optically active D-(−)-pantoyl lactone. *Appl. Environ. Microbiol.* 53:519–22
154. Singer, M. E., Finnerty, W. R. 1988. Construction of an *Escherichia coli-Rhodococcus* shuttle vector and plasmid transformation in *Rhodococcus* species. *J. Bacteriol.* 170:638–45
155. Singer, M. E. V., Finnerty, W. R. 1990. Physiology of biosurfactant synthesis by *Rhodococcus* species H13-A. *Can. J. Microbiol.* 36:741–45
156. Singer, M. E. V., Finnerty, W. R., Tunelid, A. 1990. Physical and chemical properties of a biosurfactant synthesized by *Rhodococcus* species H13-A. *Can. J. Microbiol.* 36:746–50
157. Sorkhoh, N. A., Ghannoum, M. A., Ibrahim, A. S., Stretton, R. J., Radwan, S. S. 1990. Sterols and diacylglycerophosphocholines in the lipids of the hydrocarbon-utilizing prokaryote *Rhodococcus rhodochrous. J. Appl. Bacteriol.* 69:856–63
158. Stackebrandt, E., Smida, J., Collins, M. D. 1988. Evidence of phylogenetic heterogeneity within the genus *Rhodococcus:* renewal of the genus *Gordona* (*Tsukamurella*). *J. Gen. Appl. Microbiol.* 34:341–48
159. Straube, G., Hensel, J., Niedan, C., Straube, E. 1990. Kinetic studies of phenol degradation by *Rhodococcus* sp. P1. I. Batch cultivation. *Antonie van Leeuwenhoek J. Microbiol. Ser.* 57:29–32
160. Takai, S., Koike, K., Ohbushi, S., Izumi, C., Tsubaki, S. 1991. Identification of 15-kilodalton to 17-kilodalton antigens associated with virulent *Rhodococcus-equi. J. Clin. Microbiol.* 29:439–43
161. Takai, S., Sekizaki, T., Ozawa, T., Sugawara, T., Watanabe, Y., Tsubaki, S. 1991. Association between a large plasmid and 15- to 17-kilodalton antigens in virulent *Rhodococcus equi. Infect. Immun.* 59:4056–60
162. Takai, S., Takeuchi, T., Tsubaki, S. 1986. Isolation of *Rhodococcus* (*Corynebacterium*) *equi* and atypical mycobacteria from the lymph nodes of healthy pigs. *Jpn. J. Vet. Sci.* 48:445–48
163. Takaichi, S., Ishidsu, J. I., Seki, T., Fukada, S. 1990. Carotenoid pigments from *Rhodococcus rhodochrous* RNMS1: two monocyclic carotenoids, a carotenoid monoglycoside and carotenoid glycoside monoesters. *Agric. Biol. Chem.* 54:1931–38
164. Takazawa, Y., Sato, S., Takahashi, J. 1984. Microbial oxidation of tetradecanols and related substances in organic media. *Agric. Biol. Chem.* 48:2489–95
165. Takeuchi, K., Koike, K., Ito, S. 1990. Production of *cis*-unsaturated hydrocarbons by a strain of *Rhodococcus* in repeated batch culture with a phase-inversion hollow-fiber system. *J. Biotechnol.* 14: 179–86

166. Tkachuk-Saad, O., Prescott, J. 1991. *Rhodococcus equi* plasmids: isolation and characterization. *J. Clin. Microbiol.* 29:2696–2700
167. Tomiyasu, I., Yoshinaga, J., Kurano, F., Kata, Y., Kaneda, K., et. al. 1986. Occurrence of a novel glycolipid, trehalose-2,3,6-trimycolate in a psychrophilic acid-fast bacterium *Rhodococcus aurantiacus* (*Gordona aurantiacus*). *FEMS Lett.* 203:239–42
168. Tsui, W. C., Elgar, G., Merrill, C., Maunders, M. 1988. RspX I: a new restriction endonuclease with a recognition sequence of 5'TCATGA3'. *Nucleic Acids Res.* 16:4178
169. Tsukamura, M. 1974. A further numerical taxonomic study of the rhodococcus group. *Jpn. J. Microbiol.* 18:37–44
170. Tsukamura, M., Hikosaka, K., Nishimura, K., Hara, S. 1988. Severe progressive abscesses and necrotizing tenosynovitis caused by *Rhodococcus aurantiacus*. *J. Clin. Microbiol.* 26: 201–5
171. Tsukamura, M., Komatsuzaki, C., Sakai, R., Kaneda, K., Kudo, T., et al. 1988. Mesenteric lymphadenitis of swine caused by *Rhodococcus sputi*. *J. Clin. Microbiol.* 26:155–57
172. Uchida, Y., Misawa, S., Nakahara, T., Tabuchi, T. 1989. Factors affecting the production of succinoyl trehalose lipids by *Rhodococcus erythropolis* SD-74 grown on n alkanes. *Agric. Biol. Chem.* 53:765–70
173. Uchida, Y., Tsuchiya, R., Chino, M., Hirano, J., Tabuchi, T. 1989. Extracellular accumulation of mono and disuccinoyltrehalose lipids by a strain of *Rhodococcus erythropolis* grown on n-alkanes. *Agric. Biol. Chem.* 53:757–64
174. Ueda, M., Mukataka, S., Sato, S., Takahashi, J. 1986. Conditions for the microbial oxidation of various higher alcohols in isooctane. *Agric. Biol. Chem.* 50:1533–38
175. Valo, R., Haggblom, M. M., Salkinoja-Salonen, M. S. 1990. Bioremediation of chlorophenol-containing simulated ground water by immobilized bacteria. *Water Res.* 24:253–58
176. Vaughan, P. A., Hall, G. F., Best, D. J. 1990. Aryl acylamidase from *Rhodococcus erythropolis* NCIB 12273. *Appl. Microbiol. Biotechnol.* 34:42–46
177. Watanabe, I., Satoh, Y., Enomoto, K., Seki, S., Sakashita, K. 1987. Optimal conditions for cultivation of *Rhodococcus* sp. N-744 and for conversion of acrylonitrile to acrylamide by resting cells. *Agric. Biol. Chem.* 51:3201–6
178. Watanabe, K., Aihara, H., Nakagawa, Y., Nakamura, R., Sasaki, T. 1989. Properties of the purified extracellular cholesterol oxidase from *Rhodococcus equi* no. 23. *J. Agric. Food Chem.* 37:1178–82
179. Watanabe, K., Aihara, H., Nakamura, R. 1989. Degradation of cholesterol in lard by the extracellular and cell-bound enzymes from *Rhodococcus equi* no. 23. *Lebensm. Wiss. Technol.* 22:98–99
180. Weingarten, J. S., Huang, D. Y., Jackman, J. D. Jr. 1988. *Rhodococcus equi* pneumonia. An unusual early manifestation of the acquired immunodeficiency syndrome (AIDS). *Chest* 94:195–96
181. Woods, N. R., Murrell, J. C. 1989. The metabolism of propane in *Rhodococcus rhodochrous* PNKb1. *J. Gen. Microbiol.* 135:2335–44
182. Woods, N. R., Murrell, J. C. 1990. Epoxidation of gaseous alkenes by a *Rhodococcus* species. *Biotechnol. Lett.* 12:409–14
183. Yano, I., Tomiyasu, I., Kaneda, K., Kato, Y., Sumi, Y., et al. 1987. Isolation of mycolic acid–containing glycolipids in *Nocardia rubra* and their granuloma forming activity in mice. *J. Pharmacobiodyn.* 10:113–23
184. Zakrzewska-Czerwinska, J., Mordarski, M., Goodfellow, M. 1988. DNA base composition and homology values in the classification of some *Rhodococcus* species. *J. Gen. Microbiol.* 134:2807–13

Annu. Rev. Microbiol. 1992. 46:219–52

BIODIVERSITY AS A SOURCE OF INNOVATION IN BIOTECHNOLOGY

Alan T. Bull

Biological Laboratory, University of Kent, Canterbury, Kent CT2 7NJ, United Kingdom

Michael Goodfellow

Department of Microbiology, University of Newcastle upon Tyne, Newcastle upon Tyne, NE2 4HH, United Kingdom

J. Howard Slater

School of Pure and Applied Biology, University of Wales, Cardiff, CF1 3TL, Wales, United Kingdom

KEY WORDS: microbial diversity, genetic diversity, isolation and enrichment, search and discovery, taxonomic data bases

---

CONTENTS

0066-4227/92/1001-0219$02.00

Abstract

The object of this article is to draw attention to the significance of microbial diversity as a major resource for biotechnological products and processes. The topic is approached from two complementary standpoints. First, an attempt is made to assess the extent of biodiversity, particularly microbial diversity. In this context, the application of the modern techniques of molecular biology is enabling the detection of hitherto completely unknown groups of microbes and, also, is revealing the extent of genetic diversity within microbial taxa. The case is made for the establishment of sound microbial taxonomies both on the basis of satisfying fundamental scientific needs, and for designing effective isolation strategies. The impact of an ecological approach to search and discovery of novel organisms and properties also is emphasized and illustrated. Second, the question of screening a collection of appropriate microorganisms for the desired attributes is considered. The focus here is placed on modern intelligent or targeted screening, and on the power of molecular biology to extend the range of screening options.

Discussions of microbial ecology or diversity only rarely touch upon questions of gene pool conservation. The point made here is that loss of biodiversity should be as ominous for microbiologists and biotechnologists as it is to conservationists. The article concludes with thoughts on some means of conserving microbial diversity.

INTRODUCTION

This review presents the case for the utilization of natural biodiversity, in Edward Wilson's words (123), our planet's greatest but least developed resource for biotechnological innovation. This discussion is timely for several reasons. Biotechnology in recent times has been dominated by spectacular developments in recombinant DNA technology, and without question, these have provided unique opportunities for commercialization. However, this option may be inappropriate or even proscribed for many biotechnological products and processes—hence the strong incentive to utilize natural organisms for the production of novel or replacement antibiotics and biopharmaceutins, to meet growing consumer demands for natural products, and for agriculture and clean-up operations in the context of an increasingly environmentally alert society. The conservation and managed exploitation of biodiversity is one of the most urgent problems confronting humankind and rightly has become an issue of international concern.

The most useful definition of biodiversity is that given by the International Union for Conservation of Nature and Natural Resources: biodiversity encompasses all life forms, ecosystems, and ecological processes, and acknowledges the hierarchy at genetic, taxon, and ecosystem levels (71). The focus of this review is microbial diversity, but paradoxically, despite the acknowledged commercial value of microorganisms, our knowledge of their diversity and many of their key roles in sustaining global life support systems is extraordinarily meager.

The starting point for biotechnological developments is the search for and discovery of exploitable biological phenomena; the effective planning and optimization of search and discovery programs is just as crucial as any other phase of the bioprocess operation. Search and discovery can be regarded as a sequence of unit stages (12) starting with the assembly of appropriate biological material, then moving on to screening for a desired property, selection of the best option from the subset of screened organisms, and culminating in the development and acceptance of the final product or process. Screening implies the bringing together of genetic variation and screening criteria: the former is found among natural biodiversity and/or through the production of mutations and recombinations, while the latter depend only on the ingenuity of the investigator. In the past, biological material has been assembled haphazardly and microorganisms isolated and manipulated using easy and ubiquitous methods (28). Continuing to use such an approach, without eliminating recurring organisms at an early stage of the operation, clearly compromises the efficiency of the screen. Similarly, screens have tended to be random and empirical. Therefore, a major intention of this review is to show how the adoption of innovative methods and thinking is transforming our ability to detect novelty at the organism and biochemical levels, and to discover specified activities through rational target-directed screening. Although we refer to the paucity of knowledge of microbial diversity, and the enormous potential for biotechnological innovation that its eventual inventory will release, one must not underestimate the diversity provided by known but neglected groups of microorganisms. The basidiomycete fungi, for example, which number about 30,000 species, have not been the subject of extensive screening, yet recent surveys reveal novel metabolites with antibiotic, antiviral, phytotoxic, and cytostatic activities (2). Reichenbach's group (91) pointed out the value of investigating myxobacteria in view of the wide spectrum of novel secondary metabolites they synthesize and because of the interesting biological activities they possess.

THE EXTENT OF MICROBIAL DIVERSITY

Facts and Estimates

The current inventory of the world's biodiversity is very incomplete (Table 1) and that of viruses, microorganisms, and invertebrates is especially deficient. Hawksworth's (51) recent analysis of fungal diversity is a model of how to make informed estimates of the total number of species in a particular taxonomic group. His estimate of 1.5 million fungal species is based principally on a ratio of vascular plants to fungi of about 1:6, and although six times greater than any previous estimate, it is thought to be conservative. Among the reasons for proposing a yet higher figure are: (*a*) the probable underestimate of total numbers of vascular plants; (*b*) the vascular plants:fungi ratio is rising with time (decadal rates of increase in numbers of fungi in different habitats range from about 20 to 50%); (*c*) plants:fungi ratios tend to reflect the more intensive mycological studies made in the north temperate regions but may be higher in the tropics; and (*d*) no separate estimate was made for fungi associated with the enormous insect diversity (Table 1). Moreover, as poorly explored regions and habitats are examined, the fungal inventory grows; some examples include marine ecosystems, cryptoendolithic communities (57), cryoconite holes on glaciers (116), and mycorrhizal and mycophyllal associations (17).

Attempts to estimate total numbers of bacteria, archaea, and viruses are even more problematical because of difficulties such as detection in and recovery from the environment, our very incomplete knowledge of obligate microbial associations [e.g. the case of *Symbiobacterium thermophilum* (108)], and the problem of species concept in these groups. Take the case of mycoplasmas, for example. These prokaryotes have various obligate associations with eukaryotic organisms (commensals, mutualists, pathogens), frequently have fastidious nutritional requirements or are nonculturable, and appear to have remarkable diversity. Evidence indicates that one genus, *Spiroplasma,* although only discovered in 1972, may be the largest genus on Earth (117). *Spiroplasma* species are principally associated with insects, and the overall rate of new species isolations from such sources of 6% annually (48) indicates incredible species richness. Similarly, marine ecosystems likely support a luxuriant microbial diversity. At the level of animal phyla, marine biodiversity is greater than that of any other environment (46) (28 phyla, 13 of which are endemic). Unfortunately, knowledge of marine bacterial diversity

Table 1 Known and estimated numbers of biological species[a]

Group	Known species	Estimated total species	Percentage of known species
Viruses	5,000	130,000	4
Bacteria	4,760	40,000	12
Fungi	69,000	1,500,000	5
Algae	40,000	60,000	67
Bryophytes	17,000	25,000	68
Gymnosperms	750		
Angiosperms	250,000	270,000	93
Protozoa	30,800	100,000	31
Porifera	5,000		
Cnidaria	9,000		
Nematodes	15,000	500,000	3
Crustacea	38,000		
Insects	800,000	6–10,000,000	8–13
Other arthropods/minor invertebrates	132,460		
Molluscs	50,000		
Echinoderms	6,100		
Amphibians	4,184		
Reptiles	6,380		
Fishes	19,000	21,000	90
Birds	9,198		~100
Mammals	4,170		~100

[a] Data from DiCastri & Younes (23), McNeely et al (71), Bull & Hardman (14), Hawksworth (51), Select Committee on Science and Technology (96).

is slight, but the recent discovery of picoplankton, and the extent of bacteria-invertebrate associations, gives credence to speculations about diversity.

The diversity of microorganisms, therefore, appears in large measure to reflect obligate or facultative associations with higher organisms and to be determined by the spatio-temporal diversity of their hosts or associates (117). A potential paradox arises in this context: how much of this great diversity is functionally redundant within an ecosystem? Several authors (51, 87) have argued that such redundancy imparts resilience to ecosystems such that the loss of certain species is buffered by the presence of others with similar functions. Such bootstrapping of ecosystems contrasts with those in which the loss of keystone species leads to major disturbance of the ecosystem. While such concepts derive from macroecology, they are relevant to microbial diversity and to the conservation of microbial gene pools.

Modern Methods for Revealing Microbial Diversity

The perception of microbial diversity is being radically altered by recombinant DNA techniques such as DNA-DNA hybridization, nucleic acid fingerprinting and methods of assessing the outcome of DNA probing, and, perhaps most importantly at present, rRNA sequencing (80, 85, 104, 105).

RNA-BASED ANALYSIS The use of probes based on 5S, 16S, and 23S rRNA together with the isolation, separation, and sequencing of RNA molecules has led to the establishment of new phylogenetic relationships. Thus, analysis of 16S rRNA has radically altered the classification of microorganisms into three domains, the *Bacteria, Archaea,* and *Eukarya* (125). By choosing the appropriate probe for a particular rRNA sequence, one can analyze members of communities at domain-, kingdom-, phylum-, genus-, or species-specific levels of resolution (106).

Recently, Amann et al (1) used the design of 16S rRNA sequence probes that were conjugated to fluorescent dyes and characteristic of *Fibrobacter* species, in one case the subspecies *F. succinogenes* subsp. *succinogenes,* to locate and identify individual bacteria in the mouse caecum. This investigation was possible because of the realization that bacteria are permeable to short oligonucleotide probes (34) that can be correlated with individual cells in the caecum contents by fluorescence microscopy. Such an approach enables the researcher to directly monitor in situ the effectiveness of experimental protocols designed to culture specific microorganisms. Any increase in the abundance of the specific strains or groups indicates that the treatment is positive for the selective growth of the targeted organism. Variations in media design can be evaluated, and the approach might be valuable in the process of medium improvement, especially in the context of designing media for previously nonculturable microorganisms. As the range of probes expands, the use of the technique will eventually help to resolve the number of genuinely unknown microorganisms present in a particular habitat, however complex it may be.

En passant we would make brief reference to the physiological condition known as the viable but nonculturable state. This state has been recognized for many years and is described for a variety of bacteria. Its implication for public-health microbiology, the release of organisms into the environment, and the recovery of novel organisms and properties for biotechnology is very evident. Yet little is known about conditions that trigger this physiological state or lead to resuscitation. Recent work on nonculturable *Vibrio vulnificus* suggests ways of approaching this enigma. Cells of *V. vulnificus* exposed to a temperature downshift to 5°C rapidly became unculturable (79), but their complete resuscitation was possible by imposing a temperature upshift on the population for 2–3 days (76). These authors proposed that the nonculturable response is linked to the starvation response because when *V. vulnificus* was prestarved at room temperature before being subjected to the downshift, the nonculturable response was not expressed. Synthesis of starvation stress proteins may be responsible for repressing the viable but nonculturable state. The wider implication of stress proteins for the entry into the nonculturable state, and means of resuscitating such populations, are problems meriting major investigation.

DNA-BASED ANALYSIS DNA fingerprinting by restriction fragment length polymorphism (RFLP) analysis is another accepted technique for evaluating relationships between organisms, especially if they are closely related. The utility of the technique is illustrated below for the analysis of a collection of strains of *Lactobacillus plantarum*.

DNA probes have been used to search for particular microorganisms within the total DNA pool of an ecosystem. The DNA has to be carefully separated from materials such as soil particles and humic acids and purified. Holben et al (53) detected *Bradyrhizobium japonicum* selectively at densities as low as 4.3×10^3 organisms per gram dry soil. Earlier attempts to use DNA probing enabled different strains of *Rhizobium trifolii* isolated from root nodules to be identified unequivocally using DNA colony hybridization techniques (52). Specific probes have been used to search for organisms in natural communities that contained the mercury resistance operon (4), while Sayler et al (94) used DNA colony hybridization to look for bacteria competent in toluene and naphthalene degradation. Yates et al (126) also used this method to detect organisms in biomining operations. Many other such studies have been made illustrating the power of the method for revealing the diversity and distribution of particular organisms and particular properties (14).

Detection of Previously Unknown Organisms

David Ward and his colleagues (114) have examined a thermophilic microbial community from Octopus Spring in Yellowstone National Park, USA, by evaluating the sequences of cDNA molecules synthesized from 16S rRNA using a primer complementary to the universally conserved region of rRNA. The rRNA was extracted from a cyanobacterial mat that was thought to have been well characterized by conventional analyses (113, 124) as a result of the isolation and growth of *Synechococcus lividus, Chloroflexus aurantiacus,* and several heterotrophic bacteria. Earlier studies had indicated that this community may have contained other noncultured microorganisms (112). Analysis of the cDNA revealed eight sequence groups; none were identical to any of the microorganisms isolated previously from Octopus Spring, nor to any other microbe characteristically associated with hot springs. The greatest similarity was with rRNA from *C. aurantiacus* (94.7%), but this finding served to show that the original identification was not completely accurate at this level of resolution. Ward et al (114) concluded that all eight sequence groups originated from bacterial, and not archaeal, members of the microbial community that had not been isolated and cultured in the laboratory. Detailed evaluation of the sequences and comparison with appropriate data bases showed two of the sequences could be assigned to known taxa with reasonable certainty, namely green nonsulphur bacteria and spirochaetes. There was reasonable correlation with the cyanobacterium *S. lividus,* but the percentage of sequence similarity was higher when compared with another cyanobacterium, *Anacystis nidulans,* thus raising doubts that *S. lividus* was the sole or dominant cyanobacterium. More importantly, the authors concluded that "other sequence types bear too little similarity to known eubacterial phyla to presently enable further assignments; some of these may represent novel phyla" (114).

In context with the theme of this review, the most significant feature is that the conclusions reached by Ward and his colleagues related to a microbial community believed to be relatively simple and well studied over many years (85). If such uncertainties arise in this case, even greater uncertainty almost certainly must surround the microbial composition of more complex and less well (if at all) studied habitats.

The phylogenetic (16S rRNA) analysis of bacterioplankton from the Sargasso Sea (33) also is revealing. Giovannoni concluded that one of the bacterial groups corresponded to oxygenic phototrophs (cyanobacteria, prochlorophytes, eukaryotic chloroplasts) but that none of the sequences matched known cultivated marine cyanobacteria. These observations suggest that the physiological, biochemical, and genetic conclusions drawn from all past laboratory studies were of limited value because they are not drawn from the dominant members of marine bacterioplankton communities. Furthermore, the rRNA sequence analysis identified a completely novel microbial taxon present as a major component of the Sargasso oligotrophic bacterioplankton. In this case, comparison with more than 200 known 16S rRNA bacterial sequences suggested that the closest similarity (82%) was with *Agrobacterium tumefaciens* and showed that the group was obviously distinct from the major heterotrophic group that is normally cultured. Subsequently, Britschgi & Giovannoni (9) identified two new gene classes among the Sargasso bacterioplankton indicative of unknown species of α- and γ-proteobacteria.

Giovannoni (34) also noted that the results of the analyses demonstrated a substantial genetic variability even within members of the same group, suggesting that closely related species coexist within stable microbial communities. This phenomenon seems to minimize the importance of gene transfer events (at least as measured by rRNA genes), which in turn suggests that it is even more important to identify unknown microorganisms and to develop methods by which they can be cultured. Microbiologists searching for novel exploitable properties cannot rely on the eventual transfer of desired characteristics into culturable microorganisms.

The analysis of conserved genes such as rRNAs as a means of detecting microbial diversity, especially among nonculturable organisms, has been greatly facilitated by application of the polymerase chain reaction (PCR). However, certain risks may be attached to PCR gene amplification from mixed microbial populations. Stackebrandt's group (67) recently demonstrated these potential risks. The 16S rRNA genes of an apparent monoculture of a psychrophilic, strict barophilic bacterium (WHB46) were examined using PCR amplification and by cloning into phage M13. The experiment revealed two different 16S rDNA types with a 90% homology, indicating that WHB46 consisted of two closely related species shown to be members of the γ-subdivision of Proteobacteria. Analysis of one other clone showed it to be a 16S rDNA hybrid molecule, thereby indicating a risk feature of the technology, i.e. that it may suggest the existence of organisms that do not exist in the sample.

Genetic Diversity

The new techniques are already making major impacts on our understanding of microbial diversity within apparently well defined groups. For example, Stahl and his colleagues have been interested in the obligately anaerobic bacteria and eukaryotes in the rumen. This is a very complex habitat containing representatives of all the domains and organisms present at high densities (105). Classical taxonomy determined that one of the more important cellulolytic rumen bacteria was *Bacteroides succinogenes*. However, this "species" is a cluster of genetically distinct bacteria that have been reclassified as two species of *Fibrobacter* (73). The two species, *F. succinogenes* and *F. intestinalis,* are 90% similar, and hence are less closely related than *Escherichia coli* and *Proteus vulgaris* (93% similarity) (106). Clearly, these two species must be physiologically closely related, accounting for the earlier grouping as a single species. To date, the genetic diversity at this level has not been reflected in any defined physiological or biochemical differences that may account for the obvious evolutionary divergence seen within the 16S rRNA sequences. It seems improbable, however, that these two species did not evolve separately in response to unique and distinct environmental pressures.

Similarly, we have examined the difference between 11 strains of *Lactobacillus plantarum* obtained from a culture collection compared with one novel strain isolated from a silage fermentation (49). All the strains uniformly conformed to expected characteristics for this species and were further characterized by pairwise comparisons of conserved restriction fragment length polymorphisms after genomic DNA probing with randomly generated probes constructed from one of the strains. Analysis of the DNA fingerprints produced two clear clusters, A and B, of the strains screened (Figure 1). Clusters A and B showed greater than 90% DNA sequence divergence, suggesting that this species needs some redefinition at the generic level. Furthermore, at the 95% sequence divergence level, cluster B could be further divided into subcluster B1 containing five of the strains assayed and subcluster B2 with four strains. Again, the physical differences between these groupings are not clear. Strain BTLS1 produced lactic acid at high rates. All members of the A cluster may possess similar attributes, distinguishing them from the B cluster. If this is the case, then screening programs to isolate novel strains with high rates of acid production might be well advised to focus attention on those new isolates that are related to A-cluster strains. This approach indicates one way in which screening programs may be made more focused or directed, concentrating only on those strains that are more likely to possess the required characteristic.

Restriction fragment length polymorphism analysis has been used in an attempt to correlate taxonomic divergence with geographical isolation for the nitrogen-fixing microbe *Rhizobium galegae* (58). This species of rhizobia nodulates two species of goat's rue, *Galega officinalis* and *G. orientalis*. Probing with DNA coding for common nodulating (*nod*) and nitrogen-fixing genes (*nif*) showed that different strains of *R. galegae* were tightly host-plant specific with little geographical correlation. However, if the same strains were analyzed using DNA probes made from genes not associated with the specific attribute of nitrogen fixation, then very different groupings were

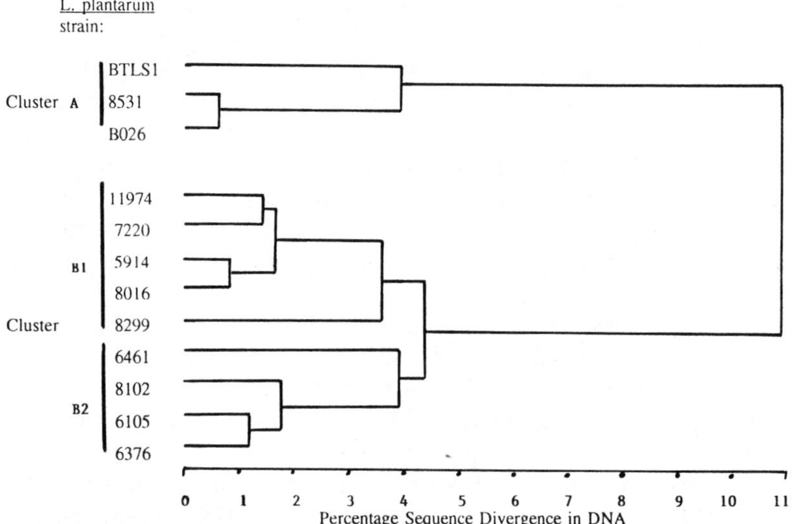

Figure 1 Dendrogram illustrating relationships between 12 strains of *Lactobacillus plantarum* determined on the basis of percent sequence divergence.

observed. Four strains of *R. galegae* isolated from *G. officinalis* were more closely related to rhizobial strains isolated from *G. orientalis*. This observation may indicate that, because the *nod* and *nif* genes are encoded on plasmids that may be transmissible, plasmid transfer has blurred the host specificity at the level of genes not directly involved in nitrogen fixation. Indeed, in the absence of geographical correlations, the DNA RFLP patterns observed are apparently the direct consequence of a recent transfer of the symbiotic sequences within two geographically distinct populations of bacteria.

Similarly, RFLP analysis of genomic DNA from a range of *Chlamydia psittaci* (23 isolates) isolated from numerous avian and mammalian hosts and of DNA from *Chlamydia trachomatis* (3 isolates) from mammalian sources revealed genetic diversity (29). Despite earlier studies that showed 100% DNA-DNA homology between all strains of *C. psittaci,* the 23 strains could be grouped into four related clusters with less than 30% DNA homology between strains of the different groups. Some regions of the genome were highly conserved, which probably indicates that these groups had a common ancestor, but nevertheless, the differences were substantial and correlated extremely well on the basis of host specificity. The single *C. trachomatis* strain isolated from mouse showed less than 50% DNA homology with either of the two strains isolated from humans. A case can be made here for a reclassification.

The DNA molecular diversity within bacterial strains isolated from the same geographical site can be extremely variable, raising the question of the usefulness of comparing strains of the same species on a wide geographical range. For example, Masters et al (69) isolated 19 new strains of *Deinococcus radiopugnans,* 18 of which were obtained from three 2-g soil samples taken from adjacent sites close to a Nottingham (UK) lake. By conventional criteria, these strains were identical, but RFLP analysis by three different probes distinguished 17 different RFLP groupings with no overlap of RFLP types between the groups. This study showed great genome heterogeneity within *D. radiopugnans* taken from essentially the same sampling site. Most microbiologists would have made the assumption that "if multiple isolates cultured from a single natural sample appear identical by classical methods, it is tacitly assumed that these isolates are monoclonal" (69). No explanation has yet been proposed to explain the observed diversity, but it does serve as a caution for accepting a single isolate as typical of the complete population of that species.

In other instances, existing classifications, essentially based on nutritional and physiological characteristics, have been shown to be fundamentally consistent with DNA analyses (19, 59). Such was the case of the gram-negative nonspore forming sulphate-reducing bacteria (20). However, more recent analysis has indicated that the genus *Desulfovibrio* ought to be subdivided into five separate species (21) on the basis of rRNA sequence analysis and acceptance of the levels of relatedness normally indicative of genera (56). This study does, however, introduce an interesting dilemma, as Devereux et al (21) observed: "It is not appropriate to reclassify these species solely on the basis of rRNA sequence data and in the absence of distinguishing phenotypic features." In each case, the new genera would all exhibit the key characteristics of the present *Desulfovibrio,* namely: incomplete oxidation of lactate to acetate; utilization of hydrogen, formate, and ethanol; and an inability to grow on fatty acids. This reinforces the view previously expressed that because of the nutritional bias in isolation and identification procedures, major distinguishing factors, which would lead to a fuller appreciation of microbial diversity, are being missed and neglected. On that basis, it seems valid to presage, at least, a reclassification. It also raises the question as to the wisdom of using rRNA sequence data alone as a basis for classification.

THE NEED FOR GOOD TAXONOMIES

"Systematics, the study of biological diversity, is sometimes portrayed as the mere classification of organisms, but in fact its range and challenges are among the greatest in biology" (122). Omura (82) strongly advocated taxonomic studies on strains shown to be producers of new compounds because other members or relatives of that taxon also may yield novel exploitable properties. This broad theme is the subject of this section. The discovery of new natural products generally occurs either with the introduction of mass screening systems or when novel organisms are examined using existing screens. Some groups of microorganisms are known to be better sources of new products than others, which influences procedures for the selective isolation of biological material. The attention given to actinomycetes and fungi in such work reflects their diverse metabolic potential; the actinomycetes are used as an example here because of their unrivaled capacity to produce new commercially significant bioactive compounds, notably antibiotics, antitumor agents, enzymes, enzyme inhibitors, and immunomodifiers (43). Actinomycetes produce approximately two thirds of the 6,000 or so known antibiotics.

The search for industrially significant actinomycetes was initially empirical (41, 119). Innumerable aerobic heterotrophic strains that grow well at neutral pH on conventional isolation media have been isolated and screened in the hope that something valuable will turn up. This blunderbuss approach to selective isolation continues to yield a steady flow of isolates with the ability to produce new metabolites, but it increasingly leads to the costly rediscovery of known compounds. Obviously therefore, the effectiveness of search and discovery programs could be improved by the early elimination of previously studied organisms. It is rarely appreciated that the choice of microorganisms for industrial screening programs, especially those with a low throughput, is primarily a problem of distinguishing between known organisms and of recognizing novel ones. This problem is partly historical, because the need to classify and identify organisms has rarely been seen as an important task in industrial biotechnology. We should, therefore, not be surprised that current procedures for the selection of actinomycetes for screening owe as much to custom and practice as to any underlying rationale. For practical purposes, novel microorganisms can be regarded either as those that cannot be assigned to validly described taxa or as known organisms that have not been examined using in-house screening systems. These definitions of novelty presuppose the availability of reliable taxonomies and accurate identification procedures.

The developing microbial systematics can also be used to help meet several challenges facing microbial biotechnologists by:

1. generating high quality data bases that can be used to improve existing systems for microbial classification and identification and to formulate new media for the selective isolation of rare and novel microbes;
2. allowing rapid detection, circumscription, and identification of novel and target organisms on primary isolation plates, and the recognition of colonies that have arisen from identical propagules;
3. enabling ecological approaches to selective isolation;
4. providing comprehensive descriptions of patent strains; and
5. determining the extent of prokaryotic diversity, including determining the geographical distribution of industrially significant microorganisms, and in so doing redress the current tendency in microbial ecology to focus on function rather than promote an understanding of the roles of specific microbial taxa in microbial ecosystems.

The full extent of this exciting prospectus can only be appreciated within the context of the new microbial systematics.

The New Systematics

The new dawn in microbial systematics can be traced back to the introduction and application of new taxonomic concepts and techniques in the late 1950s and early 1960s (38). The most significant developments arose from the application of numerical, molecular, and chemical taxonomic procedures.

NUMERICAL TAXONOMY Computer-assisted classification or numerical taxonomy, the grouping by numerical methods of taxonomic units into taxa based on shared characters, was first applied to microorganisms in the late 1950s (39). The primary objective was to sort individual bacterial strains into homogeneous groups (taxospecies) using large sets of phenetic data, and to use the quantitative results generated for the numerically circumscribed groups to devise improved identification schemes. Well-planned and executed numerical taxonomic surveys have contributed to the development of sound bacterial taxonomies, and to a lesser extent to fungal classification (8). Indeed, the current classification of most bacterial genera has benefited from a reassessment of their pre-1960 classification by numerical taxonomic studies; good examples include the aerobic, endospore-forming bacilli (89), actinoplanetes (42), and the streptomycetes (60).

The numerical taxonomic methodology has also been used to assign environmental isolates to new taxospecies (42), though the full potential of this approach in charting microbial diversity has still to be realized. Such studies on culturable microorganisms will be particularly valuable in detecting bacterial diversity at and below the genus level and in determining the range and type of nucleic acid probes and chemical markers for detecting these bacteria in situ. Microbiologists have also been slow to appreciate the value of good quality phenetic data bases in designing computer probability matrices for the identification of unknown organisms (121) and for formulating new media selective for specific fractions of complex microbial communities that occur in natural habitats (41, 55, 119).

MOLECULAR SYSTEMATICS The renewed interest in microbial taxonomy owes much to the advances in molecular biology (103). Nucleic acid hybridization and sequencing techniques have attained prominence in bacterial systematics mainly because of their impact in unraveling relationships at different levels in the taxonomic hierarchy (see Modern Methods for Revealing Microbial Diversity). Nevertheless, DNA base composition studies still provide a useful and routine way of distinguishing between phenetically similar but genetically unrelated strains. DNA base composition values now form part of the minimum standards for the description of genera and species (66). It is also now generally accepted that microorganisms showing DNA base composition values differing by more than 5% guanine (G) plus cytosine (C) should not be assigned to the same species, and those with more than 10% differences in G-C range should not be classified in the same genus.

Nucleic acid hybridization methods provide a more exacting way of establishing relationships between many bacteria species, though it is not possible to define borderline values for the delineation of genera, despite earlier statements to this effect (100). Genomic species encompass strains with approximately 70% or greater DNA relatedness and with 5°C or less ΔT_m (115), although the exact level below which organisms are considered to belong to different species varies. Ideally, genomic species and taxospecies defined using nucleic acid hybridization should correspond to those defined in numerical taxonomy. DNA-rRNA hybridization is a useful tool for establishing relationships at intra- and intergeneric levels given appropriate reference strains.

In the context of rRNA sequencing techniques, the validity of taxonomic conclusions should be tested by comparative sequence analysis of other highly conserved genes, such as those coding for elongation factors and the β-subunit of ATPase (95).

CHEMOSYSTEMATICS Chemosystematics is a rapidly evolving area in which information derived from chemical analyses of whole organisms or cell fractions is used for classification and identification (40, 44). Chemical characters are increasingly being used to describe, separate, and identify microorganisms, notably actinomycetes (36). Increasingly sophisticated chemical procedures are now available to determine lipid, cell-wall amino acid and sugar, and whole-organism protein composition. Also, pilot experiments show that rapid typing of industrially significant actinomycetes can be achieved using Curie-point pyrolysis mass spectrometry (93).

Taxonomic Databases and Selective Isolation Programs

By combining the needs of pharmaceutical screening programs and advances in microbial classification and identification, totally new selective media can be formulated to promote the growth of rare, uncommon, and novel microorganisms, especially actinomycetes (41, 78, 110, 111, 119, 120). In particular, numerical taxonomic data bases have been examined for nutrient components, physiological requirements, and resistance markers that can be used to formulate media selective for markers of industrially significant clusters or taxospecies. The availability of improved diagnostic methods ensures the selection of suitable target organisms for screening (Figure 2). This taxonomic approach to selective isolation has been used to expand the range of actinomycete diversity screened and thereby reduces the likelihood of encountering previously described metabolites. In other words, the quality of biological material tested is greatly improved, thereby raising the prospect of a much higher hit rate than with a random population.

The cycle outlined in Figure 2 can be entered at any point. For example, organism(s) found to be active in high throughput screens can be identified then characterized for properties that can be added to the appropriate data base, which can then be searched to find relevant selective agents. If the medium design has been successful, a number of new organisms closely related to the original culture(s) will be available. These will be tested in the screen, thereby completing the cycle (Figure 3). In many cases, the target organisms will produce similar secondary metabolite profiles, but some may have technical advantages over the original isolate(s). The advantage could be in terms of the rate of production of the desired metabolite, suitability for genetic manipulation, the production of a modified and more desirable analog of the original molecule, or robustness in fermentation conditions.

FORMULATION OF NEW SELECTIVE MEDIA The discovery that diagnostic sensitivity test agar was selective for diverse *Nocardia asteroides* strains (83)

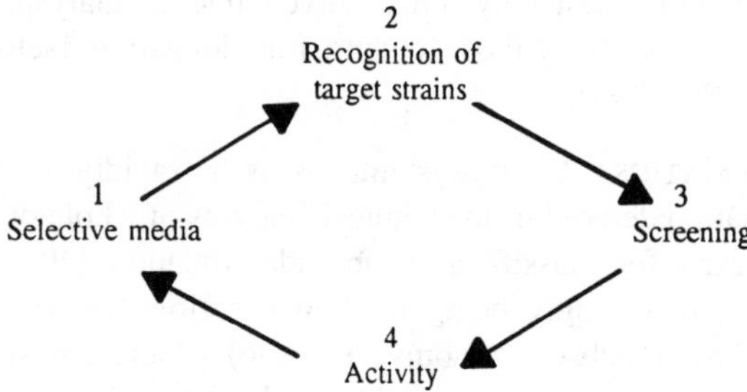

Figure 2 Use of information in taxonomic data base to formulate media selective for members of target taxospeices.

was based on antibiotic sensitivity test data and the product of an earlier numerical taxonomic survey. A logical development from this work was the visual scanning of percentage positive frequency tables, derived from numerical taxonomic studies, to highlight antibiotics that could form the basis of selective media. Several such media were subsequently designed for the isolation of *Actinomadura, Saccharomonospora, Saccharopolyspora,* and *Thermomonospora* strains (41).

Another development has involved tailoring selective media to the nutritional requirements of the target organisms. Williams and his colleagues (111, 119, 120) used particular combinations of carbohydrate and amino acid, with and without antibacterial antibiotics, to favor the outgrowth of members of uncommon streptomycete species known to be promising sources of antibiotically active metabolites, or to inhibit the growth of *Streptomyces albidoflavus,* which appears to be a common species in soils. The selective agents were chosen by examining a streptomycete data base (118) using a computer program [DIACHAR (102)] that ranks the most diagnostic characters for individual clusters or taxospecies. Several successful and similarly designed media have been used to isolate acidophilic streptomycetes, rhodococci, and members of the family Pseudonocardiaceae (41, 116).

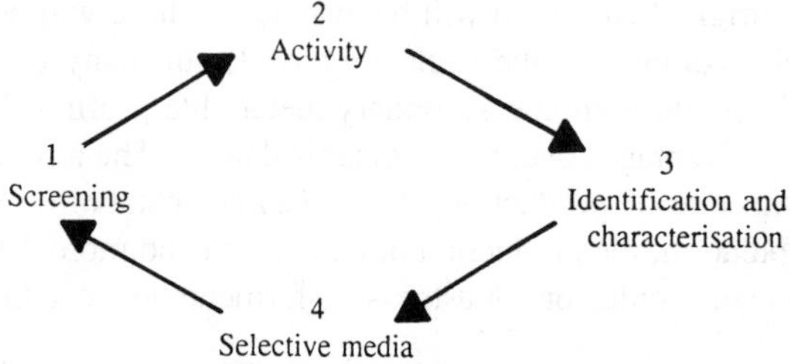

Figure 3 Search and discovery of additional active organisms following identification and the formulation of new selective media.

Instead of incorporating antibiotics into selective isolation media, Huck et al (55) used stepwise discriminant analysis to identify nutritional and physiological factors that increased the competitiveness of antibiotic-producing actinomycetes on agar plates. The strategy involved examining the effects of many factors on colony radial growth rates of soil isolates, which included not only antibiotic-producing and nonproducing actinomycetes but also eubacteria and fungi. Bacteria and fungi often are troublesome on isolation plates because they out-compete actinomycetes. Discriminant analysis was done in two rounds: first to find selective pressures resulting in enhanced competitiveness of actinomycetes, and second to find conditions that selectively encouraged antibiotic-producing actinomycetes. The analysis ranks each factor according to its relative importance as a discriminant function, and thus prioritized, the factors are used to design the selective media. In this case, an artificial intelligence program prioritized the factors to make the final choice of selective conditions. Media were designed that greatly increased the ratio of actinomycetes to eubacteria over that on commonly used isolation media (selective characters: 1% proline and 0.1% humic acid as sources of C and N, growth on nitrate as N-source, pH 7.7–8.0). Similarly, media were designed that allowed antibiotic-producing actinomycetes to be differentiated from nonproducers on agar plates (selective characters: 1% proline, 0.1% humic acid, asparagine as N-source, vitamin supplementation). Huck and her colleagues highlight the use of taxometric methods for designing media as a means of increasing the efficacy of isolating target groups of microorganisms from different environments and at different seasons, under both of which conditions the proportions of organisms present might vary greatly.

EVALUATION OF SELECTIVE MEDIA It is essential to evaluate the effectiveness of computer-generated media and where necessary to alter combinations of selective agents to enhance the isolation of target organisms. Vickers et al (111) showed that it is possible to increase the number of particular neutrotolerant streptomycete species and decrease others using computer-designed media and a computer probability matrix for the identification of unknown streptomycetes. It was noted that some species increased or decreased in a way that was not predicted given information in the data base. However, one should remember that isolation plates carry a variety of interacting bacterial colonies so that modifications to media can influence the growth of several species, which can in turn stimulate or discourage the growth of others. Consequently, the selectivity of isolation media will be raised or lowered by the mix of species in the inoculum placed upon it and able to grow. This very unpredictability can be turned to good effect when selected pressures generated by a medium allow the growth of rare and novel species from environmental samples.

Recognition of Target and Novel Organisms

The sure and definite identification of microorganisms is difficult to achieve without resorting to the use of specialized techniques and expertise. The lack of universally applicable diagnostic techniques compounds the problem because highly specialized procedures are often needed for the identification of particular groups. In the actinomycetes, identification at the genus level and above can be achieved using a combination of morphological and chemical approaches. In recent years, the well-established use of wall chemotypes and whole-organism sugar patterns has been supplemented by additional chemical information drawn from the analysis of fatty acids, including mycolic acids, menaquinones, and polar lipids (Table 2). In contrast, species identification, especially in large genera like *Streptomyces,* remains a difficult and time-consuming process. Enzyme tests based on the use of fluorogenic substrates may break this logjam (37). The exclusion of previously isolated organisms, or the recognition of colonies on primary isolation plates that have arisen from identical environmental propagules is more a question of pattern recognition than identification. Such analyses have to be done quickly and economically with high sample throughput.

Table 2 Some approaches to the recognition of target and novel actinomycetes

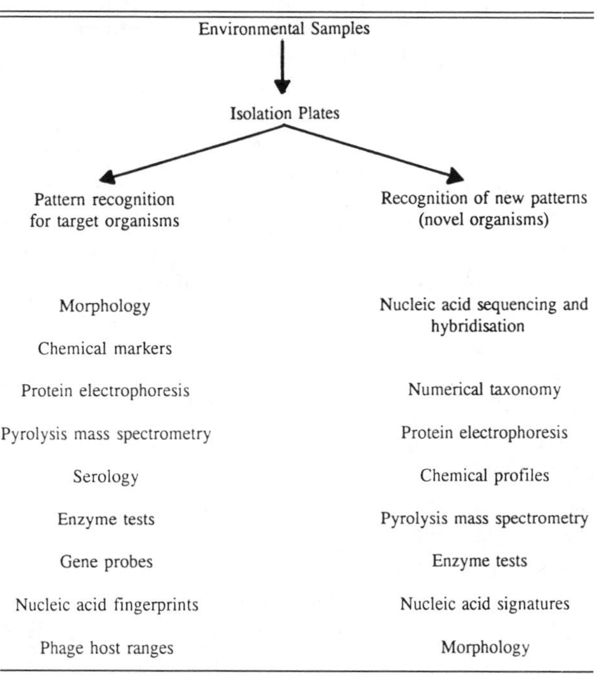

Environmental Samples

↓

Isolation Plates

Pattern recognition for target organisms	Recognition of new patterns (novel organisms)
Morphology	Nucleic acid sequencing and hybridisation
Chemical markers	
Protein electrophoresis	Numerical taxonomy
Pyrolysis mass spectrometry	Protein electrophoresis
Serology	Chemical profiles
Enzyme tests	Pyrolysis mass spectrometry
Gene probes	Enzyme tests
Nucleic acid fingerprints	Nucleic acid signatures
Phage host ranges	Morphology

Pyrolysis techniques have advantages over other chemical procedures as they are rapid, highly reproducible, and do not require sample preparation. Curie-point pyrolysis mass spectrometry has been successfully used to identify target actinomycetes directly from isolation plates and to separate streptomycetes known to belong to different genomic species (93). This technique can also be used to eliminate identical strains from screening programs and as a first step in the detection and circumscription of novel actinomycetes. The ability to analyze small amounts of biological material with minimum sample preparations to obtain, in minutes, fingerprint data that can be used for identification and typing is unparalleled by other methods, including nucleic acid and fingerprinting methods. Table 2 lists suggested protocols for the recognition of target and novel actinomycetes.

ASSEMBLING THE BIOLOGICAL MATERIAL

The sources of microorganisms for biotechnological innovation are culture collections and isolations from nature. As computerized-sequence data bases grow and become more comprehensive in their holdings, they also could become a screening source, not for organisms per se but for properties of interest.

Sampling, Isolation, and Enrichment

Because of the great uncertainty surrounding the magnitude of microbial diversity, it will be prudent to sample from the widest range of pristine and disturbed environments, and to pay attention to microorganisms associated with plants, insects, and other invertebrates. We have already opined that the discovery of novel compounds and properties will depend largely on devising strategies for the isolation of rare or novel microorganisms. Thus, the choice of environment for sampling will not be irrelevant.

The act of sampling the environment frequently is based on experience and intuition. In our experience, the following are useful guides in this endeavor:

1. Be aware of both the macro- and microhabitats within the ecosystem being examined. Examples include the zones of different fungal colonization within decaying wood, often distinguished by visible barrier lines, and microhabitats in soil such as rhizoplane, rhizosphere, soil crumbs, cryptoendolithic microsites, and humus. Special treatments, such as mild surfactant exposure, will often be required to recover organisms from such microhabitats.
2. Collect several or multiple samples from a given site.

3. Sample a given site during different seasons.
4. Analyze environmental samples as soon as possible after their collection in the field. Differential loss of microbial viability in stored samples is common.
5. Incubate isolation cultures for a prolonged time in order to detect very slowly growing organisms or organisms that develop only after modification of the environment by other strains.
6. Relate the objective of the search program to the physico-chemical characteristics of the habitat and to prevailing, or dynamic, climatic factors. Populations will have adapted to such environmental conditions.
7. Be aware that, in very broad terms, those habitats that exert weak selective pressures (so-called neutral environments) probably will have the greatest biodiversity, and vice versa.
8. Pretreatment or baiting of a habitat prior to sampling may be an effective way of encouraging competitive growth of microorganisms with the desired properties; examples include dry heating, treatment with surfactants, surface sterilization, flotation of spores, treatment with polyvalent bacteriophages, and placement of capillary tubes, membrane filters, and practice golf balls.
9. Recognize that while exotic habitats may harbor the greatest microbial diversity, novel organisms are also found in commonplace habitats close to one's own laboratory or home.
10. Be prepared to follow one's own instinct and not to be unduly influenced by conventional wisdom.

Once samples have been obtained, the next stage involves isolation of appropriate microorganisms before putting them through screening procedures. Two strategies can be used: direct isolation and enrichment isolation. It is sometimes argued that direct isolation (onto agar plates or into liquid media) may elicit recovery of the maximum diversity of microorganisms from a sample, but in the absence of totally nonselective media and isolation conditions, this proposition is debatable. Usually in search and discovery programs, selective isolation conditions are imposed either negatively to repress the establishment of unwanted types or positively to promote the establishment of desired types; the latter procedure defines enrichment isolation culture. Often, organisms with sought properties fail to grow when plated directly onto agar media and need to pass through a preliminary phase of enrichment culture.

The design of enrichment isolation media often is given rather scant attention. The protocol recommended by Hutter (55a) for eliciting the greatest gene expression of microorganisms is readily adapted for the purposes of enrichment isolation. Thus, the enrichment medium can be formulated such that different nutrients can be made growth-rate limiting; the growth-limiting nutrient can be presented in different chemical forms, and catabolite repression prevented. Enrichment media should be formulated such that they are balanced with respect to trace elements and have buffering capacity, just as fermentation media. Similar thought must be given to aeration; the effects of different nutrient limitations, for example, on enrichment may be completely masked if the culture is or rapidly becomes oxygen limited (15).

A second major consideration when designing an enrichment protocol is the choice of culture system to be used. There are four basic types distinguished by whether they are closed or open in operation or represent microcosms or models (86). A microcosm can be defined as a laboratory system that attempts to simulate, as far as possible, conditions prevailing in the whole or in a part of a habitat under investigation (e.g. alkaline soil microcosm, aquatic microcosm with terrestrial interface). A laboratory model system operates under conditions that are functionally similar to part of an environment (e.g. simple, multiple, and counter-flow chemostats). Clearly a single-enrichment system will not be appropriate for all situations; we review briefly some advantages and limitations of the principal techniques.

Closed or batch-type enrichments are characterized by temporal changes in the environment and hence in the selection conditions, circumstances that may make it difficult to isolate organisms of interest because they appear only transiently in the microbial succession. Also, under batch conditions, selection may largely be based on maximum specific growth rate because of the substrate excess regime; organisms so selected tend to have high growth rates and low substrate specificities. There also is a higher risk of selecting for auxotrophes (86). Chemostat enrichments are characterized by substrate-limited growth (the nature of the growth-limiting substrate can be chosen at will), and selection can be made on the basis of specific growth rate and substrate affinity, thereby enabling a wider range of physiological types to be enriched. Enrichments can be made under an enormously wide range of steady-state conditions but also under defined transient-state conditions. The investigator can also enrich cultures under conditions of cell recycle (13); this option may be useful if enrichment conditions are toxic or when organisms with very high substrate affinities are required. Chemostat enrichments usually select for mixed microbial populations. The latter are often the cultures of choice for environmental biotechnology, as testified by many studies; in our laboratories, we have enriched mixed cultures for degrading toxic halogenated xenobiotics (97, 101), mycotoxins (5), and for removing toxic metals (25). For certain purposes, it may be propitious to enrich in the presence of inert particles or macrosurfaces, as in cases where the organism(s) will be operating under fluidized or fixed-bed process conditions, e.g. the fixed-bed biotreatment process for pentachlorophenol (88).

The Ecological Approach to the Discovery of Novel Taxa and Activities

The rationale guiding the choice of organisms for passing through a screening program is very much project determined or prompted by the interests (or idiosyncrasies) of the individual investigator. Commercial targets, experience, lateral thinking, and, most importantly, insatiable curiosity [for example, consider the discovery of alkalophilic microorganisms (54)] can influence the route taken. A knowledge of ecology is invaluable in this process. On the one hand, one may have a biotechnological objective in mind from the outset and decide to focus on isolations of so-called high probability taxa. For example, as already mentioned, actinomycetes are producers of antibiotics par excellence; therefore, one could concentrate on isolating novel members of this group. An understanding of actinomycete ecology will greatly aid this task as the very high decadal increases in newly described streptomycetes and nonstreptomycetes attest (41). On the other hand, biotechnological objectives may be secondary to basic ecological studies. Then it becomes a matter for others to screen any biological novelty that results from such research. For instance, recent reports from Breznak's group on the gut ecology of wood-feeding and soil-feeding termites describe new genera (*Acetonema*) and species (*Clostridium mayombei*) of acetogenic bacteria (61, 62)—recall the estimated diversity of insects!

The differentiation of various microorganisms is regulated by steroid, peptide, and other microbial hormones (signal molecules), and the view that cell-regulatory mechanisms in higher vertebrates evolved from primitive origins is gaining support. A strategy employed at Xenova (77) is to screen fungi, in particular, for biopharmaceutins based on the above premise. Again, a knowledge of fungal ecology can aid the effectiveness of isolating and collecting unusual species. A somewhat less defined thinking surrounds current interest in hyperthermophilic microorganisms: their ecology is so extreme that something of biotechnological consequence is certain to be discovered. The choice of possible topics to discuss under this heading is enormous, so we intend to illustrate the current thrust of the ecological strategy by brief reference to alkalophiles and hyperthermophiles.

Considerable interest has developed in alkalophiles and halophiles in recent years (45, 54), and certain of their properties have been exploited in biotechnology, e.g. cyclodextrinase. The extent of the taxonomic novelty appearing is well illustrated by the discovery of *Flexistipes sinusarabici* (27). This gram-negative bacterium was isolated from Red Sea Atlantis II Deep hot brines, an extreme environment of high salinity (26%), temperature (up to 64°C), and heavy-metal concentrations, and anoxia. *F. sinusarabici* is the first bacterium to be isolated from this environment. It was initially described as a new genus but subsequent work (68) has placed it in a new bacterial phylum. 16S rRNA sequence analysis revealed that *F. sinusarabici* had no specific relationship to any known eubacterial phylum, but sequence analysis of the elongation factor Tu gene indicated its affinity with the domain *Bacteria*.

A search for alkalophiles possessing the ability to accumulate gallium illustrates the targeted approach. The requirement was for gallium bioaccumulation under conditions of high pH with the objective of recovering gallium from Bayer liquor resulting from the processing of bauxite. Knowing that Ga(III) has a very similar atomic size to Fe(III), Gascoyne et al (31) isolated alkalophilic bacteria that could produce gallium-sequestering siderophores under highly alkaline conditions. Enrichments were made under conditions of low available iron [to induce siderophore production and to trick the bacteria into taking up Ga(III)] and high pH. Several bacteria with this capability were isolated (32). In a simultaneous study of the growth efficiency of alkalophiles, the investigators postulated that siderophore production might be growth limiting at pH 10.5 because all alkalophilic *Bacillus* species examined have very high cytochrome contents under such conditions (47).

Within a decade of the discovery of hyperthermophiles in submarine thermal vent systems, a large number of archaeal genera and one bacterial genus have been isolated from marine and continental geothermal environments (107). The pace of discovery and initial exploration of physiology and biochemistry has been dramatic. The most spectacular properties to date have been ascribed to species of *Pyrococcus,* an organism that has obvious biotechnological potential. *Pyrococcus furiosus* and *P. woesei* can be grown on starch under a hydrogen–carbon dioxide atmosphere in the presence of sulfur, but in complex media under a nitrogen–carbon dioxide atmosphere, elemental sulfur is not required. These organisms produce several amylolytic enzymes with temperature optima of 100°C or higher and impressive catalytic half-lives. The α-amylase of *P. woesei* has now been purified and characterized (63). The enzyme is metal-independent and is active over the range 40–130°C. Activity can still be measured after autoclaving the enzyme for 5 h at 2 bar and 120°C. *P. furiosus* produces an equally remarkable serine protease that has a half-life of 33 h at 98°C (7). Two genera of hyperthermophilic methanogens are known that have maximum growth temperatures in the range 86–97°C. A third genus, *Methanopyrus,* was recently reported (64). *Methanopyrus kandleri* was isolated from marine hydrothermal systems; it can grow at temperatures up to 110°C and is an obligate chemolithoautotroph. Obligately anaerobic carboxydotrophic bacteria that produce carbon dioxide from volcanic gases have been isolated from similar marine hydrotherms (109).

SCREENING STRATEGIES

Screening defines the deployment of selective procedures for detecting chemicals, enzymes, or any other biological product or activity of interest.

Recent reviews of this topic (12, 28, 72, 77, 82) consistently highlight the following basic features of screening operations:

1. An imperative to focus on new detection methods and novel organisms in order to increase the chance of detecting new activities; examples are the discovery of the novel nonpeptide cholecystokinin antagonist, asperlicin, by use of a new receptor ligand screen, and the discovery of the broad-spectrum antifungal antibiotic pradimicin A by investigating a novel actinomycete, *Actinomadura hibisea* (6).
2. Means of recognizing and eliminating as early as possible previously discovered properties of known organisms (the latter has been emphasized in an earlier section). The former may comprise an automated bioassay in conjunction with gas chromatography–mass spectrometry (GC-MS), for example.
3. Assay procedures that are specific, sensitive, and robust such that they give reliable information.
4. A requirement for interdisciplinary operations in order to cope satisfactorily with the isolation and selection of organisms, cultivation, assay systems, and chemical characterization.

The early screening strategies tended to be labor-intensive, empirical, and capable only of low hit rates; the current emphasis is on targeted screening in which specific detection methods are founded on good scientific understanding and which produce higher hit rates. Although such screens have been developed largely in the search for novel antibiotics and biopharmaceutins, modulators of the immune system and of oncogene function, the same intelligent thinking is being adopted for other product types and processes. This section examines selected cases of targeted screening particularly to reveal the ingenuity and outcome of screens based upon mode of action, enzyme inhibitions, and receptorology.

Targeted Screening

MODE-OF-ACTION SCREENS These screens were developed during the search for new antibacterial antibiotics as a consequence of detailed knowledge of the chemistry and biosynthesis of bacterial walls. Thus, when researchers discovered that the penicillin target was transpeptidase, screens were developed to search for inhibitors of D-ala-D-ala carboxypeptidase and transpeptidase. Similarly, β-lactamase-based screens sought specific inhibitors of this enzyme, which came to prominence during investigations of penicillin resistance. Such screening strategies revealed major new classes of β-lactam antibiotics such as cephamycins, carbapenems, and thienamycins.

Considerable sophistication has been brought to this field. For example, carbapenems are sensitive to renal dehydropeptidase and thus lose their potency in vivo. Streptomycete metabolites have been discovered that mimic the activity of cilastatin as dehydropeptidase inhibitors (50) so that coadministration with the antibiotic protects the antibiotic potency. A second example of interest is glycopeptide antibiotics, of which vancomycin is a member. These antibiotics exhibit specific binding to acyl-D-ala-D-ala heptapeptide intermediates of cell-wall synthesis. This fact has been exploited in effective tripeptide (e.g. diacetyl-L-lys-D-ala-D-ala) reversal screening for novel glycopeptide antibiotics. Filter discs loaded with such a mimetic tripeptide together with a test fermentation broth sample are placed onto agar plates seeded with a sensitive test bacterium. Glycopeptide antibiotics in the broth are revealed by the antagonistic reaction of the tripeptide in comparison to discs lacking the tripeptide. Several new antibiotics were discovered by this route, among them aridicins, kibdelins (from *Kibdelosporangium* gen. nov.) and synmonicins (from *Synnemomyces* gen. nov.) (90, 98, 99).

ENZYME INHIBITORS Enzyme inhibitors have also been developed successfully in the search for pharmacologically active compounds [biopharmaceutins (72)]. The range of prospective targets is extremely wide, and microbial metabolites that specifically inhibit key enzymes in the following pathologies have been reported: hypercholesterolaemia, hypertension, gastric ulcers, inflammation, muscular dystrophy, benign prostate hyperplasia, systemic lupus erythemosus. Commercial development of anticholesterolaemic drugs via this route has been very successful (mevinolin from *Aspergillus,* monacolin from *Monascus,* compactin from *Penicillium*) and are now on the market. These drugs were discovered in screens against 3-hydroxy-3-methyl glutaryl coenzyme A reductase, which catalyzes mevalonic acid synthesis. Mevinolin and monacalin are polyketides and very potent inhibitors of the reductase (nanomolar range).

Protease inhibitors have been proposed as a possible new class of therapeutic drugs for retroviral disease (24). The target here is the virally encoded protease involved in posttranslational processing of the *gag* and *pol* gene products.

RECEPTOROLOGY Screening for receptor-ligand binding (agonists, antagonists) is complementary to enzyme inhibitor screening. It is a fast-growing activity and already is established practice in several pharmaceutical companies. Receptor binding may elicit much more effective drugs than enzyme inhibitors; for example, interference with 5-hydroxytryptamine action in the central nervous system (CNS) is more direct and likely to be more effective than searching for inhibitors of the key enzyme tryptophan hydroxylase. The gastrointestinal hormone cholecystokinin (CCK) controls various digestive functions such as pancreatic secretion and gall bladder contraction. As a result of receptor screening, a fungal metabolite, asperlicin (from *Aspergillus alliaceus*), was discovered (18) that had greatly increased affinity for CCK receptors compared with the available drug. However, the native compound lacked sufficient potency and oral bioavailability. Subsequently, analogs were prepared for use as CCK and gastrin receptor antagonists that retained receptor affinities and also had satisfactory oral bioavailability and duration-of-action properties (26). The avermectins antagonize γ-amino butyric acid receptors and have become established veterinary drugs. They block GABA-neurotransmission and are very effective against parasitic invertebrates. The avermectin analog, ivermectin, is being considered for the treatment and control of human parasites (16). To date it has been approved for the treatment of onchocerciasis.

The field of targeted screening is truly enormous and it is impossible to do it justice in a review such as this. The excellent discussions of Nisbet & Porter (77), Monaghan & Tkacz (72), and Franco & Coutinho (28) are recommended.

The Impact of Molecular Biology

Genetics and molecular biology have had a substantial effect on screening and opened up totally new opportunities for biotechnological innovations. Our checklist is not exhaustive but illustrates the range of impact.

1. Provision of test organisms that have increased sensitivities, or resistances, to known agents, e.g. antibiotic super-sensitive strains in the search for new β-lactams.
2. Production of enzymes for inhibitor screens by cloning and overexpression.
3. Cloning of receptors, thereby providing opportunities for receptor screens either as receptor binding or as a functional assay that reflects the effect of binding.
4. Provision of reporter gene assays, e.g. reporter gene constructs with viral *trans*-activating factors to search for metabolites that disrupt virus replication (28).
5. The development of molecular probes for in situ or laboratory-based screening.
6. The development of assay procedures such as ELISA.

Know-How and Serendipity

Targeted search and discovery programs notwithstanding, one should be aware of the contributions that experimental know-how and accidental findings (serendipity) make to biotechnology.

Know-how can reveal itself as: a knowledge of where and how to find organisms of interest; the knowlege that methanogenic archaea are light sensitive and the use of this fact to prevent their growth and thereby enrich for other strict anaerobes in environmental samples (81); an appreciation that the isolation of novel organisms sometimes may be expedited by enriching for mixed microbial communities [e.g. *Desulfomonile* gen. nov. in a 3-chlorobenzoate degrading community (22)]; a knowledge of how to elicit secondary-metabolite synthesis; and so on, ad infinitum. We suspect that serendipitous events are usually not reported as such in the scientific literature, but the point is clearly illustrated by the discovery of magnetotactic bacteria, and the chance findings of *Pectinatus* gen. nov. in stale beer (65) or *Actinopolyspora* gen. nov. in contaminated medium intended for halophile cultivation (35), or the ability of white rot fungi to transform certain xenobiotic chemicals. Thus, prepared minds are likely to be rewarded.

CONSERVATION AND MANAGEMENT OF MICROBIAL GENE POOLS

The resource that biodiversity presents for biotechnological innovation is, at present, incalculable. However, the prospect of the loss of biodiversity should be as alarming and unacceptable to microbiologists and those involved in biotechnology as it is to conservationists. Traditionally, the microbiological community has not identified itself with broad conservation issues, although the UNEP-Unesco Microbial Resource Centers attempt to utilize microbial gene pools for sustainable development. The realization that biotechnology has a major biodiversity perspective has taken place only recently (10), but interest and concern is now growing quickly.

Is Microbial Diversity Being Lost?

Precise, or even approximate, measures of the loss of biodiversity are elusive (70) and will remain so until we know how much exists. At the beginning of this review, we showed how flimsy knowledge is of microbial diversity. Consequently, statements about loss of microbial diversity are at best anecdotal (92) or otherwise only intelligent guesses. However, given the association of microorganisms with animal and plant species and with different types of environments, the easier estimates of the loss and degradation of the latter reveal the implications for microbial diversity. Overall losses of tropical forest biodiversity are occurring at approximately 1.8% per year; those of wetlands, freshwater, marine, and other wildlands are less easy to quantify but are substantial and, in particular cases, catastrophic (84). Economic valuations of environmental functions and biodiversity are now starting to be made (3).

In Situ and Ex Situ Conservation

This review has argued that a crucial requirement for biotechnological innovation is knowledge of microbial diversity and the means for its discovery and retrieval from the environment. Consequently, we argue that conservation in situ is essential for the protection of microbial genetic diversity. The figures presented in Table 1 clearly show that ex situ culture collections will never be able to hold more than a small fraction of this genetic diversity. In discussing plant germplasm banks, Gamez (30) commented that while the banks represent an important and valuable strategy, their limitations are numerous. For example, gene banks halt evolutionary processes. Similar arguments might be made with respect to microbial culture collections. The question then arises: which genes, species, and habitats should be conservation priorities? Criteria and methods have been proposed for determining priorities, but the attention is on macro-, not microorganisms. With our present knowledge, therefore, conservation priorities for microorganisms must apparently piggyback on those established by macroecology.

Megadiversity Regions

We can achieve a partial, interim resolution of this question of priorities for the conservation of microbial gene pools by emphasizing the so-called megadiversity regions of the planet. These are regions that have high species diversities and high endemism; they include Amazonia, Madagascar, and Indonesia. Moreover, we can identify within such regions hot spots of disproportionately high biodiversity (74) typified by tropical rain forests and marine environments such as coral-reef ecosystems. One such action that seeks to address the issue of microbial diversity for biotechnology innovation is the UK-Indonesia cooperative program with which the authors are associated (11).

CONCLUSIONS

An Action Plan for Microbial Diversity

In the past two years, several recommendations have been made regarding biodiversity and conservation. The U.S. National Science Board (75) recommended the following actions in order to provide acceptable knowledge about the world's biota: (*a*) produce a complete global inventory, (*b*) focus on representative and threatened ecosystems, (*c*) implement conservation, (*d*) develop comprehensive data bases, (*e*) develop human resources, i.e. taxonomists; and (*f*) promote international programs. Similarly, McNeely (70), in preparation for a Biodiversity Conservation Strategy, has proposed the "Five-I Approach": investigate how natural ecosystems function; information—best available to make informed decisions; incentives as an aid to conserve biodiversity; integration—a cross-sectoral approach to conservation; and international support for conserving biodiversity. Finally, there is the prospect of an action plan directed at microbial diversity per se. Microbial Diversity 21 is a draft action statement emanating from an IUBS-IUMS workshop held in Amsterdam in September 1991. Among the actions proposed are: (*a*) establish a major international initiative, a Decade of Microbial Diversity; (*b*) inventory all known species; (*c*) develop standards for sampling microbial communities; (*d*) list habitats meriting conservation; (*e*) encourage conservation of environmental samples from threatened habitats by long-term cryopreservation; (*f*) develop techniques for isolation, culture, and long-term preservation of microorganisms; (*g*) convene a conference on the species concept in microbial groups; and (*h*) provide inputs to the selection of nature reserves and other sites to be protected in the long-term. We commend such considerations to the international community of microbiologists.

Literature Cited

1. Amann, R., Krumholz, L., Stahl, D. A. 1990. Fluorescent-oligonucleotide probing of whole cells for determinative, phylogenetic and environmental studies in microbiology. *J. Bacteriol.* 172:762–70
2. Anke, T. 1989. Basidiomycetes: a source for new bioactive secondary metabolites. In *Bioactive Metabolites from Micro-organisms,* ed. M. E. Bushell, U. Graffe, pp. 51–66. Amsterdam: Elsevier
3. Aylward, B., Barbier, E. B. 1992. Valuing environmental functions in developing countries. *Biodivers. Conserv.* 1:34–50
4. Barkay, T., Fouts, D. L., Olson, B. H. 1985. Preparation of a DNA gene probe for detection of mercury resistance genes in Gram-negative bacterial communities. *Appl. Environ. Microbiol.* 49:686–92
5. Beeton, S., Bull, A. T. 1989. Biotransformation and detoxification of T-2 toxin by soil and freshwater bacteria. *Appl. Environ. Microbiol.* 55:190–97
6. Birnbaum, J. 1989. *Elements of a successful natural products discovery program.* Pres. at Princeton Drug Res. Symp. on Microbial Metabolites as Sources of New Drugs, Princeton
7. Blumenthals, H., Robinson, A. S., Kelly, R. M. 1990. Characterisation of sodium dodecyl sulfate-resistant proteolytic activity in the hyperthermophilic archaebacterium *Pyrococcus furiosus. Appl. Environ. Microbiol.* 56:1992–98
8. Bridge, P. D., Hawksworth, D. L., Kozakiewicz, Z., Onions, A. H. S., Pateson, R. R. M., et al. 1989. A reappraisal of the terverticillate penicillia using biochemical, physiological and morphological features. 1. Numerical taxonomy. *J. Gen. Microbiol.* 135:2941–66
9. Britschgi, T. B., Giovannoni, S. J. 1991. Phylogenetic analysis of a natural marine bacterioplankton population by rRNA gene cloning and sequencing. *Appl. Environ. Microbiol.* 57:1707–13
10. Bull, A. T. 1987. The biotechnologies of the nineties. *Proc. European Congr. Biotechnology, 4th, Amsterdam, 1987,* pp. 189–202. Amsterdam: Elsevier
11. Bull, A. T. 1991. Biotechnology and biodiversity. In *The Biodiversity of Microorganisms and Invertebrates: Its role in Sustainable Agriculture,* ed. D. L. Hawksworth, pp. 203–19. Wallingford: CAB International
12. Bull, A. T. 1992. Isolation and screening of industrially important organisms. In *Recent Advances in Industrial Applications of Biotechnology,* ed. F. Vardar-Sukan, S. S. Sukan, pp. 1–17. Dordrecht: Kluwer Academic
13. Bull, A. T. 1992. Innovations in fermentation technology: requirements and options. *Aust. J. Microbiol.* In press
14. Bull, A. T., Hardman, D. J. 1991. Microbial diversity. *Curr. Opin. Biotechnol.* 2:421–28
15. Bushell, M. E. 1989. *Promotion of secondary metabolism.* Pres. at Princeton Drug Res. Symp. on Microbial Metabolites as Sources of New Drugs, Princeton

16. Campbell, W. C. 1991. Ivermectin as an antiparasitic agent for use in humans. *Annu. Rev. Microbiol.* 45:445–74
17. Carroll, G. 1988. Fungal endophytes in stems and leaves: from latent pathogen to mutualistic symbiont. *Ecology* 69:2–9
18. Chang, R. S. L., Lotti, V. J., Monaghan, R. L., Birnbaum, J., Stapely, E. O., et al. 1985. A potent nonpeptide cholecystokinin selective for peripheral tissues isolated from *Aspergillus alliaceus*. *Science* 230:177–85
19. Clark-Curtiss, J. E., Walsh, G. P. 1989. Conservation of genomic sequences among isolates of *Mycobacterium leprae*. *J. Bacteriol.* 171:4844–51
20. Devereux, R., Delany, M., Widdel, F., Stahl, D. A. 1989. Natural relationships among sulfate-reducing eubacteria. *J. Bacteriol.* 171:6689–95
21. Devereux, R., He, S.-H., Doyle, C. L., Orkland, S., Stahl, D. A., et al. 1990. Diversity and origin of *Desulfovibrio* species: phylogenetic definition of a family. *J. Bacteriol.* 172:3609–19
22. DeWeerd, K. A., Mandelco, L., Tanner, R. S., Woese, C. R., Suflita, J. M. 1990. *Desulfomonile tiedjei* gen. nov. and sp. nov., a novel anaerobic, dehalogenating, sulfate-reducing bacterium. *Arch. Microbiol.* 154:23–30
23. DiCastri, F., Younes, T. 1990. Ecosystem function of biological diversity. *Biol. Int. Spec. Issue* 22:1–20
24. Dreyer, G. B. 1989. *Inhibitors of HIV-1 protease: design enzymology, and antiviral activity*. Pres. at Princeton Drug Res. Symp. on Microbial Metabolites as Sources of New Drugs, Princeton
25. Dunn, G. M., Bull, A. T. 1983. Bioaccumulation of copper by a defined community of bacteria. *Eur. J. Appl. Microbiol. Biotechnol.* 17:30–34
26. Evans, B. 1989. *Asperlicin—discovery and exploitation*. Pres. at Princeton Drug Res. Symp. on Microbial Metabolites as Sources of New Drugs, Princeton
27. Fiala, G., Woese, C. R., Langworthy, T. A., Stetter, K. O. 1990. *Flexistipes sinusarabici,* a novel genus and species of eubacteria occurring in the Atlantis II Deep Brines of the Red Sea. *Arch. Microbiol.* 154:120–26
28. Franco, C. M. M., Coutinho, L. E. L. 1991. Detection of novel secondary metabolites. *Crit. Rev. Biotechnol.* 11: 193–276
29. Fukushi, H., Hirai, K. 1989. Genetic diversity of avian and mammalian *Chlamydia psittaci* strains and relation to host origin. *J. Bacteriol.* 171:2850–55
30. Gamez, R. 1989. Threatened habitats and germplasm preservation: a Central American perspective. See Ref. 62a, pp. 477–92
31. Gascoyne, D. J., Connor, J. A., Bull, A. T. 1991. Isolation of bacteria producing siderophores under alkaline conditions. *Appl. Microbiol. Biotechnol.* 36:130–35
32. Gascoyne, D. J., Connor, J. A., Bull, A. T. 1991. Capacity of siderophore-producing alkalophilic bacteria to accumulate iron, gallium and aluminium. *Appl. Microbiol. Biotechnol.* 36:136–41
33. Giovannoni, S. J., Britschgi, T. B., Moyer, C. L., Field, K. G. 1990. Genetic diversity in Sargasso Sea bacterioplankton. *Nature* 345:60–63
34. Giovannoni, S. J., DeLong, E. F., Olsen, G. J., Pace, N. R. 1988. Phylogenetic group-specific oligooxynucleotide probes for identification of single microbial cells. *J. Bacteriol.* 170:720–26
35. Gochnauer, M. B., Leppard, G. G., Komeratat, P., Kates, M., Novitsky, T., Kushner, D. J. 1975. Isolation and characterisation of *Actinopolyspora halophila* gen. et sp. nov., an extremely halophilic actinomycete. *Can. J. Microbiol.* 21:1500–1511
36. Goodfellow, M. 1989. Suprageneric classification of actinomycetes. In *Bergey's Manual of Systematic Bacteriology*, ed. S. T. Williams, M. E. Sharpe, J. G. Holt, 4:2333–39. Baltimore: Williams & Wilkins
37. Goodfellow, M. 1991. Identification of some mycolic acid containing actinomycetes using fluorogenic probes based on 7-amino-4-methylcoumarin and 4-methylumbelliferone. *Actinomycetology* 5: 21–27
38. Goodfellow, M., Board, R. G., eds. 1980. *Microbiological Classification and Identification*. London: Academic Press. 408 pp
39. Goodfellow, M., Jones, D., Priest, F. G., eds. 1985. *Computer-Assisted Bacterial Systematics*. London: Academic. 443 pp.
40. Goodfellow, M., Minnikin, D. E., eds. 1985. *Chemical Methods in Bacterial Systematics*. London: Academic. 410 pp.
41. Goodfellow, M., O'Donnell, A. G. 1989. Search and discovery of industrially-significant actinomycetes. In *Microbial Products: New Approaches,* ed. S. Baumberg, I. S. Hunter, P. M. Rhodes, pp. 343–83. Cambridge: Cambridge Univ. Press. 383 pp.
42. Goodfellow, M., Stanton, L. J., Simpson, K. E., Minnikin, D. E. 1990. Nu-

merical and chemical classification of *Actinoplanes* and some related actinomycetes. *J. Gen. Microbiol.* 136:19–36

43. Goodfellow, M., Williams, S. T., Mordarski, M., eds. 1988. *Actinomycetes in Biotechnology*. London: Academic. 501 pp.
44. Gottschalk, G., ed. 1985. *Methods in Microbiology,* Vol. 18. London: Academic. 376 pp.
45. Grant, W. D., Mwatha, W. E., Jones, B. E. 1990. Alkaliphiles: ecology, diversity and applications. *FEMS Microbiol. Rev.* 75:255–70
46. Grassle, J. F., Lasserre, P., McIntyre, A. D., Ray, G. C. 1990. Marine biodiversity and ecosystem function. *Biol. Int. Spec. Issue* 23:1–16
47. Guffanti, A. A., Hicks, D. B. 1991. Molar growth yields and bioenergetic parameters of extremely halophilic *Bacillus* species in batch cultures, and growth in a chemostat at pH 10.5. *J. Gen. Microbiol.* 127:2375–79
48. Hackett, K. J., Clark, T. B. 1989. The ecology of spiroplasmas. In the *Mycoplasmas,* ed. R. F. Whitcomb, J. G. Tully, 5:113–99. New York: Academic
49. Hall, B. G. D., Slater, J. H. 1986. Determination of identity between two organisms. *International Patent Application W086/02101*. pp. 1–70
50. Hashimoto, S., Mural, H., Ezaki, M., Morikawa, N., Hatanaka, H., et al. 1990. Studies on new dehydrogenase inhibitors I. Taxonomy, fermentation, isolation and physico-chemical properties. *J. Antibiot.* 43:29–35
51. Hawksworth, D. L. 1991. The fungal dimension of biodiversity: magnitude, significance, and conservation. *Mycol. Res.* 95:641–55
52. Hogson, A. L. M., Roberts, W. P. 1983. DNA colony hybridisation to identify *Rhizobium* strains. *J. Gen. Microbiol.* 129:207–12
53. Holben, W. E., Jansson, J. K., Chelm, B. K., Tiedje, J. M. 1988. DNA probe method for the detection of specific microorganisms in the soil bacterial community. *Appl. Environ. Microbiol.* 54:701–11
54. Horikoshi, K. 1991. *Microorganisms in Alkaline Environments*. Tokyo: Kodansha. 275 pp.
55. Huck, T. A., Porter, N., Bushell, M. E. 1991. Positive selection of antibiotic-producing soil isolates. *J. Gen. Microbiol.* 137:2321–29

55a. Hutter, R. 1982. Design of culture media capable of provoking wide gene expression. In *Bioactive Microbial Products: Search and Discovery,* ed. J. D. Bu'Lock, L. J. Nisbet, D. J. Winstanley, pp. 37–50. London: Academic

56. Johnson, J. L. 1984. Nucleic acids in bacterial classification. In *Bergey's Manual of Systematic Bacteriology,* ed. N. R. Krieg, J. G. Holt, 1:8–11. Baltimore: Williams & Wilkins
57. Johnston, C. G., Vestal, J. R. 1991. Photosynthetic carbon incorporation and turnover in Antarctic cryptoendolithic microbial communities: are they the slowest-growing communities on earth? *Appl. Environ. Microbiol.* 57:2308–11
58. Kaijalainen, S., Lindstrom, K. 1989. Restriction fragment length polymorphism analysis of *Rhizobium galegae* strains. *J. Bacteriol.* 171:5561–66
59. Kakoyiannis, C., Winter, P. J., Marshall, R. B. 1984. Identification of *Campylobacter coli* isolates from animals and humans by bacterial restriction endonuclease DNA analysis. *Appl. Environ. Microbiol.* 46:545–49
60. Kämpfer, P., Kroppenstedt, R. M., Dott, W. 1991. A numerical classification of the genera *Streptomyces* and *Streptoverticillium* using miniaturized physiological tests. *J. Gen. Microbiol.* 137:1831–91
61. Kane, M. D., Brauman, A., Breznak, J. A. 1991. *Clostridium mayombei* sp. nov., an H_2/CO_2 acetogenic bacterium from the gut of the African soil-feeding termite, *Cubitermes speciosus. Arch. Microbiol.* 156:99–104
62. Kane, M. D., Breznak, J. A. 1991. *Acetonema longum* gen. nov. sp. nov. an H_2/CO_2 acetogenic bacterium from the termite, *Pterotermes occidentis. Arch. Microbiol.* 156:91–98

62a. Knutson, L., Stoner, A. K., eds. 1989. *Biotic Diversity and Germplasm Preservation, Global Imperatives*. Amsterdam: Kluwer Academic

63. Koch, R., Spreinat, A., Lemke, K., Antranikian, G. 1991. Purification and properties of a hyperthermoactive α-amylase from the archaeabacterium *Pyrococcus woesei. Arch. Microbiol.* 155:572–78
64. Kurr, M., Huber, R., Konig, H., Jannasch, M. W., Fricke, M., et al. 1991. *Methanopyrus kandleri* gen. and sp. nov. represents a novel group of hyperthermophilic methanogens, growing at 110°C. *Arch. Microbiol.* 156:239–47
65. Lee, S. Y., Mabee, M. S., Jangaard, N. O. 1978. Pectinatus, a new genus of the family Bacterioidaceae. *Int. J. Syst. Bacteriol.* 28:582–94
66. Levy-Frebault, V. V., Portaels, F.

1992. Proposal for recommended minimal standards for the genus *Mycobacterium* and for newly described slowly growing *Mycobacterium* species. *Int. J. Syst. Bacteriol.* 42: In press

67. Liesack, W., Weyland, H., Stackebrandt, E. 1991. Potential risks of gene amplification by PCR as determination by 16S rDNA analysis of a mixed-culture of strict basophilic bacteria. *Microb. Ecol.* 21:191–98
68. Ludwig, W., Wallner, G., Tesch, A., Klink, F. 1991. A novel eubacteria phylum: comparative nucleotide sequence analysis of a *tuf*-gene of *Flexistipes sinusarabici*. *FEMS Microb. Lett.* 78:139–44
69. Masters, C. I., Murray, R. G. E., Moseley, B. E. B., Minton, K. W. 1991. DNA polymorphisms in new isolates of *Deinococcus radiopugnans*. *J. Gen. Microbiol.* 137:1459–69
70. McNealy, G. A. 1992. The sinking Ark: pollution and the worldwide loss of biodiversity. *Biodivers. Conserv.* 1:2–18
71. McNeely, G. A., Miller, K. R., Reid, W. V., Mittermeier, R. A., Werner, T. R. 1990. *Conserving the World's Biological Diversity*. Gland: Int. Union for Conservation of Nature and Natural Resources
72. Monaghan, R. L., Tkacz, J. S. 1990. Bioactive microbial products: focus upon mechanism of action. *Annu. Rev. Microbiol.* 44:271–301
73. Montgomery, L., Flesher, B., Stahl, D. A. 1988. Transfer of *Bacteriodes succinogenes* (Hungate) to *Fibrobacter* gen. nov. as *Fibrobacter succinogenes* comb. nov. and description of *Fibrobacter intestinalis* sp. nov. *Int. J. Syst. Bacteriol.* 38:430–35
74. Myers, N. 1988. Threatened biotas: "hotspots" in tropical forests. *Environmentalist* 8:1–20
75. National Science Board. 1989. *Loss of Biological Diversity: a Global Crisis Requiring International Solutions*. Washington, DC: Natl. Sci. Found.
76. Nilsson, L., Oliver, J. D., Kjelleberg, S. 1991. Resuscitation of *Vibrio vulnificus* from the viable but nonculturable state. *J. Bacteriol.* 173:5054–59
77. Nisbet, L. J., Porter, N. 1989. The impact of pharmacology and molecular biology on the exploitation of microbial products. In *Bioactive Microbial Products,* ed. S. M. Baumberg, I. S. Hunter, P. M. Rhodes, pp. 309–42. Cambridge: Cambridge Univ. Press
78. Nolan, R. D. 1988. Exploitation of taxonomic databases in selective isolation programmes. In *Prospects in Systematics,* ed. D. L. Hawksworth, pp. 357–662. Oxford: Clarendon. 454 pp.
79. Oliver, J. D., Nilsson, L., Kjelleberg, S. 1992. Formation of nonculturable *Vibrio vulnificus* cells and its relationship to the starvation state. *Appl. Environ. Microbiol.* 57:2640–44
80. Olsen, G. J., Lane, D. J., Giovannoni, S. J., Pace, N. R., Stahl, D. A. 1986. Microbial ecology and evolution: a ribosomal RNA approach. *Annu. Rev. Microbiol.* 40:337–65
81. Olson, K. D., McMahon, C. W., Wolfe, R. S. 1991. Light sensitivity of methanogenic archaebacteria. *Appl. Environ. Microbiol.* 57:2683–86
82. Omura, S. 1986. Philosophy of new drug discovery. *Microbiol. Rev.* 50: 259–79
83. Orchard, V. A., Goodfellow, M. 1980. Numerical classification of some named strains of *Nocardia asteroides* and some related isolates from soil. *J. Gen. Microbiol.* 118:295–312
84. Overseas Development Administration. 1991. *Biological Diversity and Developing Countries. Issues and Options*. London: ODA
85. Pace, N. R., Stahl, D. A., Lane, D. J., Olsen, G. J. 1986. The analysis of natural microbial populations by ribosomal RNA sequences. *Adv. Microb. Ecol.* 9:1–55
86. Parkes, R. J. 1982. Methods for enriching, isolating and analysing microbial communities in laboratory systems. In *Microbial Interactions and Communities,* ed. A. T. Bull, J. H. Slater, pp. 45–102. London: Academic
87. Perry, D. A., Amaranthus, M. P., Borchers, J. G., Borchers, S. L., Brainerd, R. E. 1989. Bootstrapping in ecosystems. *BioScience* 39:230–37
88. Pignatello, J. J., Martinson, M. M., Steiert, J. G., Carlson, R. E., Crawford, R. L. 1983. Biodegradation and photolysis of pentachlorophenol in artificial freshwater streams. *Appl. Environ. Microbiol.* 46:1024–31
89. Priest, F. G., Goodfellow, M., Todd, C. 1988. A numerical classification of the genus *Bacillus*. *J. Gen. Microbiol.* 134: 1847–82
90. Rao, V. A., Ravishanker, D., Sadhukhan, A. K., Ahmed, S. M., Goel, A. K., et al. 1986. Synmonicins: a novel antibiotic complex produced by *Synemomyces mamnoorii* gen. et sp. nov. *Abstr. Intersci. Conf. Antimicrob. Agents Chemother. 26th, New Orleans*
91. Reichenbach, H., Gerth, K., Irschick, H., Kunze, B., Hofle, G. 1988. Myxo-

bacteria: a source of new antibiotics. *TIBTECH* 6:115–21

92. Rifai, M. A. 1989. Astounding fungal phenomena as manifestations between tropical plants and microorganisms. In *Interactions Between Plants and Microorganisms,* ed. G. Lim, K. Katsya, pp. 1–8. Singapore: Natl. Univ. Singapore
93. Sanglier, J.-J., Whitehead, D., Saddler, G. S., Ferguson, E. V., Goodfellow, M. 1992. Pyrolysis mass spectrometry as a method for the classification, identification and selection of actinomycetes. *Gene*. In press
94. Sayler, G. S., Shields, M. S., Tedford, E. T., Breen, A., Hooper, S. W., et al. 1985. Application of DNA-DNA colony hybridization to the detection of catabolic genotypes in environmental samples. *Appl. Environ. Microbiol.* 49:1295–1303
95. Schleifer, K. H., Ludwig, W. 1990. Phylogenetic relationships among bacteria. In *Hierarchy of Life,* ed. B. Fernholm, K. Bremer, H. Jornwall, pp. 103–17. Amsterdam: Elsevier Science
96. Select Committee on Science and Technology. 1991. *Systematic Biology Research, Written evidence. House of Lords HL Paper 41*. London: Her Majesty's Stationery Office
97. Senior, E., Bull, A. T., Slater, J. H. 1976. Enzyme evolution in a microbial community growing on the herbicide Dalapon. *Nature* 263:476–79
98. Shearer, M. C., Actor, P., Bowie, P. A., Grappel, S. F., Nash, C. H., et al. 1985. Aridicins, novel glycopeptide antibiotics I. Taxonomy, production and biological activity. *J. Antibiot.* 38:555–63
99. Shearer, M. C., Giovenella, A. J., Grappel, S. F., Hedde, R. D., Metha, R. J., et al. 1986. Kibdelins, novel glycopeptide antibiotics I. Discovery, production and biological evaluation. *J. Antibiot.* 39:1386–93
100. Schleifer, K. H., Stackebrandt, E. 1983. Molecular systematics of prokaryotes. *Annu. Rev. Microbiol.* 37:143–87
101. Slater, J. H., Bull, A. T. 1982. Environmental microbiology: biodegradation. *Philos. Trans. R. Soc. London Ser. B* 297:575–97
102. Sneath, P. H. A. 1980. BASIC program for the most diagnostic properties of groups from an identification matrix of percent positive characters. *Comp. Geosci.* 6:21–26
103. Stackebrandt, E., Goodfellow, M., eds. 1991. *Nucleic Acid Techniques in Bacterial Systematics*. Chichester: Wiley. 329 pp.
104. Stahl, D. A. 1988. Phylogenetically based studies of microbial ecosystem perturbation. In *Biotechnology for Crop Protection,* ed. P. A. Hedin, J. J. Menn, R. M. Hollingworth, pp. 373–90. Washington DC: Am. Chem. Soc.
105. Stahl, D. A., Flesher, B., Mansfield, H. R., Montgomery, L. 1988. Use of phylogenetically based hybridisation probes for studies of ruminal microbial ecology. *Appl. Environ. Microbiol.* 54:1079–84
106. Stahl, D. A., Amann, R. 1991. Development and application of nucleic acid probes. In *Nucleic Acid Techniques in Bacterial Systematics,* ed. E. Stackebrandt, M. Goodfellow, pp. 205–48. Chichester: Wiley
107. Stetter, K. O., Fiala, G., Huber, G., Huber, R., Sagerer, A. 1990. Hyperthermophilic microorganisms. *FEMS Microbiol. Rev.* 75:117–24
108. Suzuki, S., Horinouchi, S., Beppu, T. 1988. Growth of a tryptophanase-producing thermophile, *Symbiobacterium thermophilum* gen. nov., sp. nov., is dependent on co-culture with a *Bacillus* sp. *J. Gen. Microbiol.* 134:2353–62
109. Svetlichny, V. A., Sokolora, T. G., Gerhardt, M., Kostrikina, N. A., Zavarzin, G. A. 1991. Anaerobic extremely thermophilic carboxydotrophic bacteria in hydrotherms of Kuril Islands. *Microb. Ecol.* 21:1–10
110. Thomas, E. G. 1991. *Numerical classification and selective isolation of* Rhodococcus *and related actinomycetes*. MS thesis. Univ. Newcastle-upon-Tyne
111. Vickers, J. C., Williams, S. T., Ross, G. W. 1984. A taxonomic approach to selective isolation of streptomycetes from soil. In *Biological, Biochemical and Biomedical Aspects of Actinomycetes,* ed. L. Ortiz-Ortiz, L. F. Bojalil, V. Yakoleff, pp. 553-61. Orlando: Academic
112. Ward, D. M., Brassell, S. C., Eglington, G. 1985. Archaebacterial lipids in hot-spring microbial mats. *Nature* 318:656–59
113. Ward, D. M., Tayne, T. A., Anderson, K. L., Bateson, M. M. 1987. Community structure and interactions among community members in hot spring cyanobacterial mats. In *Ecology of Microbial Communites,* ed. M. Fletcher, T. R. G. Gray, T. G. Jones, pp. 179–210. Cambridge: Cambridge Univ. Press
114. Ward, D. M., Weller, R., Bateson, M. M. 1990. 16S RNA sequences reveal

numerous uncultured microorganisms in a natural community. *Nature* 345:63–65
115. Wayne, L. G., Brenner, D. J., Colwell, R. R., Grimont, P. A. D., Kandler, O., et al. 1987. Report of the ad hoc committee on reconciliation of approaches to bacterial systematics. *Int. J. Syst. Bacteriol.* 37:463–64
116. Wharton, R. A., McKay, C. P., Simmons, G. M., Parker, B. C. 1985. Cryoconite holes on glaciers. *Bioscience* 35:499–503
117. Whitcomb, R. F., Hackett, K. J. 1989. Why are there so many species of mollicutes. An essay on prokaryotic diversity. See Ref. 62a, pp. 205–40
118. Williams, S. T., Goodfellow, M., Alderson, G., Wellington, E. M. H., Sneath, P. H. A., Sackin, M. J. 1983. Numerical classification of *Streptomyces* and related genera. *J. Gen. Microbiol.* 129:1743–1813
119. Williams, S. T., Goodfellow, M., Vickers, J. C. 1984. New microbes from old habitats? In *The Microbes 1984: Prokaryotes and Eukaryotes,* ed. D. P. Kelly, N. G. Carr, pp. 219–56. Cambridge: Cambridge Univ. Press
120. Williams, S. T., Vickers, J. C. 1988. Detection of actinomycetes in natural habitats—problems and perspectives. In *Biology of Actinomycetes '88,* ed. Y. Okami, T. Beppu, H. Ogawara, pp. 265–70. Tokyo: Jpn. Sci. Soc.
121. Williams, S. T., Vickers, J. C., Goodfellow, M. 1985. Application of new theoretical concepts to the identification of streptomycetes. See Ref. 39, pp. 289–306
122. Wilson, E. O. 1985. Time to revive systematics. *Science* 230:1227–29
123. Wilson, E. O., ed. 1988. *Biodiversity.* Washington, DC: Natl. Acad.
124. Woese, C. R. 1987. Bacterial evolution. *Microbiol. Rev.* 51:221–71
125. Woese, C. R., Kandler, O., Wheelis, M. L. 1990. Towards a natural system of organisms: proposal for the domains Archaea, Bacteria, and Eucarya. *Proc. Natl. Acad. Sci. USA* 87:4576–79
126. Yates, J. R., Lobos, J. H., Holmes, D. S. 1986. The use of genetic probes to detect microorganisms in biomining operations. *J. Ind. Microbiol.* 1:129–35

Annu. Rev. Microbiol. 1992. 46:253–76

THE STRUCTURE AND REPLICATION OF HEPATITIS DELTA VIRUS

John M. Taylor

Fox Chase Cancer Center, Philadelphia, Pennsylvania 19111

KEY WORDS: RNA-directed RNA synthesis, RNA editing, RNA processing, virus assembly

---

CONTENTS

0066-4227/92/1001-0253$02.00

Abstract

Hepatitis delta virus exists in nature as a satellite of hepatitis B virus. This review emphasizes studies during the past few years that have clarified much about this satellite relationship. Many unique and intriguing features have been assigned to delta and its replication. In addition, certain unresolved questions are emphasized, and consideration is even given to the application of delta as a vector.

INTRODUCTION AND OBJECTIVES

Hepatitis delta virus (HDV) is a novel infectious agent quite unlike anything that has been described among the infectious agents of animals. Its definition does not actually lie within that of a virus (31); it is a subviral agent that replicates only with help from another virus. In nature, this helper virus is human hepatitis B virus (HBV) (84). With this helper, HDV can replicate most efficiently in hepatocytes and can greatly increase the severity of liver damage caused by an HBV infection. As such, HDV is a significant human pathogen.

HDV has been the subject of other recent reviews (8, 44, 98–100) and so, as with any review, the intention here is to avoid redundancy. The two objectives are, first, to bring up to date our current understanding of the molecular biology of HDV, and second, to make clear what might be the most profitable areas for future research.

DISCOVERY, NATURAL INFECTIONS, AND EPIDEMIOLOGY

HDV was discovered primarily because a group of investigators were trying to understand why certain patients with HBV infections had a more serious disease. Rizzetto and coworkers demonstrated that these patients had a novel antigen in liver cells, which they called the delta antigen (83). Subsequently, experimental transmission into chimpanzees (84), and even later, transmission into woodchucks (79) showed that the delta antigen was part of the novel infectious agent that we now know as HDV.

From the start, investigators noted that experimental infection of primates with HDV depended upon the presence of HBV. Similarly, with woodchucks, HDV was always accompanied by woodchuck hepatitis virus (WHV), a hepadnavirus that is very similar to HBV (35). The research showed that these hepadnaviruses provided their own particular envelope proteins, and that these proteins were required for HDV assembly. It is because of this dependence on a helper virus that we refer to HDV as subviral (31).

Both HDV and its helper hepadnavirus are released from the infected hepatocytes and accumulate in the blood stream. For an infected animal, there can be as many as 10^{12} virions per milliliter of blood (80).

Natural infections with HDV seem to be transmitted in the same manner as those caused by HBV. That is, the virus present in the blood of an infected individual is spread parenterally. This includes sexual transmission, use of contaminated needles, or accidental transfusions of contaminated blood or blood products (8, 44, 50, 81).

The epidemiology of delta infections might be expected to mirror the geographic distribution of HBV, but this is not the case; there are areas with high levels of HBV but relatively little if any HDV (44). Particularly puzzling is that the outcome of HDV infections seems to vary around the world. For example in certain geographic areas, infections with HDV can cause significant liver damage, with as much as a 20% chance of liver collapse and death (8, 15). In contrast, in a specific region of Greece, the levels of HDV associated with HBV are high, and yet HDV-associated liver damage is apparently no more than that caused by HBV alone (45).

TREATMENT AND PREVENTION

Prevention is much easier than treatment. Because HBV provides the envelope for HDV, vaccination against HBV envelope should also work against HDV infection. It does (44, 50), but the primary basis for such success might be indirect, i.e. against the HBV upon which HDV depends for its spread.

A high risk of fulminant hepatitis and death can be associated with HDV infections (15, 81). This is true for coinfections, in which both HBV and HDV are received at the same time, and to a greater extent for superinfections, in which an HDV infection is superimposed on a prior chronic HBV infection. Survivors of coinfections are most likely to resolve the infection, whereas survivors of superinfections are most likely to be chronically infected with HDV (8, 50). World-wide, there are currently about 25 million chronic HDV carriers (44). Such chronicity, coupled with chronicity for HBV, usually will have an increased association of liver damage. Thus, with chronic HBV infections (2), the risk of developing liver cancer is greatly increased.

Treatment of infected patients with high doses of recombinant interferon can suppress HDV replication, and there is a good chance of this leading to a cure (28, 50). Such treatment can even cure the HBV infection for certain patients that have only recently become chronically infected. However, interferon treatment of chronic HDV patients has been reported to be harmful and even fatal (86).

RELATIONSHIP TO HELPER HEPADNAVIRUSES

As mentioned above, the evidence is clear that the life cycle of HDV needs help from a hepadnavirus. The data indicate that this help does not involve genome replication but seems to be limited to aspects of packaging. The situation is more complicated, however, in that the replication of HDV can interfere with the replication of the helper hepadnavirus.

Interference

At one point in a HDV superinfection, the level of HDV replication reaches a dramatic peak. The evidence indicates that the level of replication of the helper hepadnavirus is dramatically depressed at this time, and if the HDV infection is then resolved, the hepadnavirus may return to its earlier level (22, 94). This transient interference can be detected in terms of virus particles detected in serum and liver of the infected animal. The nature of this apparent interference mechanism has not yet been clarified. Attempts have been made, however, to reproduce the phenomenon in cultured cells. Again, an interference has been detected (40, 68–70, 110), but it is not yet clear whether the underlying basis for the phenomenon is the same in the two situations.

Packaging

As mentioned earlier, HDV certainly depends upon the use of hepadnavirus envelope proteins during virion assembly. This is discussed in some detail in a subsequent section. What is relevant here is that because HDV particle assembly typically yields at least 1000-fold more particles than a HBV infection, part of the HDV-induced suppression of HBV replication may be at the level of particle assembly. As implied in the previous section, however, we could readily imagine other levels at which the interference could be achieved.

SYSTEMS FOR THE STUDY OF HDV GENOME REPLICATION

Prior to a consideration of topics such as the structure of the virions or the RNA that acts as the genome for HDV, a review of the sources of information such as the different experimental systems that allow HDV genome replication is important.

Animals

In addition to natural HDV infection of humans, experimental transmission can be achieved in primates (17, 73, 82, 84) and woodchucks (79). These experimental animals provide valuable models for study of normal human infections. A model in smaller experimental animals would be much more convenient, and several labs have tried to establish HDV infections in ducks, using duck hepatitis B virus as the helper. However, such studies have not been reproducibly successful (69; J. M. Taylor, L. Coates, unpublished observations).

In recent years, the increased application of liver transplantation procedures in humans with terminal liver failure has created two new possibilities for studying HDV. The first is that the removed liver can become a source of human hepatocytes (see following section) for further study. The second is that if the host was virus positive, it may be possible to study the de novo infection of the transplanted donor liver. Some interesting findings have already come from such efforts (41, 72, 76).

Hepatocytes

Currently, we have only one type of cultured hepatocyte for studies of HDV replication: the primary hepatocyte culture. Such cultures can be difficult to obtain; for example, one first has to perfuse the liver with collagenase to help release the hepatocytes, and then the hepatocytes require special conditions to maintain their viability and differentiated state. They may also need as a substrate a layer of feeder cells or some form of collagen matrix. Nevertheless, such cultures have been set up and found to be susceptible to HDV infection. Replication of HDV has been demonstrated in primary hepatocyte cultures from the woodchuck (25, 26, 102) and the chimpanzee (93, 95). And, in the latter case, even the release of progeny virions was recently demonstrated (C. Sureau, personal communication).

Several laboratories have devised methods to conditionally transform hepatocytes into permanent cell lines without the loss of the differentiated state as a second and possibly more convenient source of hepatocytes. Such procedures reportedly maintain susceptibility to two other hepatitis viruses, but data for HDV have not yet been presented.

DNA-Transfected Cells

The cDNA cloning of the HDV RNA genome has made it possible to test the ability of such a cDNA, when appropriately inserted into a eukaryotic expression vector and then transfected into an animal cell line, to initiate replication of the HDV genome (24, 61, 69). To date, this transfection approach has worked in every animal cell tested, independent of the species or tissue of origin (61). For example, mouse fibroblasts readily support such replication (18). However, such studies only demonstrate replication of the genome and do not typically aim for the production of new virus particles. Nevertheless, these studies establish three facts regarding the replication of the genome. (*a*) Cells derived from liver tissue are not required under these conditions. (*b*) The hepadnavirus is not needed for genome replication. (*c*) The delta antigen, which, as discussed later, is the only known protein of HDV, is essential.

Now if the proteins of a hepadnavirus are made available in such transfections, for example by cotransfection, the release of virus particles, which in some cases have been tested and proven to be infectious (87), occurs (16, 87, 110); this is discussed in a subsequent section.

DNA-Transfected Animals

Alternatively, an animal already chronically infected with a hepadnavirus can provide the hepadnavirus proteins. Thus, when a cDNA clone of HDV inserted into an appropriate eukaryotic expression vector was transfected into the liver of a chimpanzee chronically infected with HBV, that animal went on to develop a full-blown HDV infection (94). A similar experiment has been done with a woodchuck chronically infected with WHV (A. Ponzetto, personal communication).

RNA-Transfected Cells

The expression vector constructs used in the cDNA transfections of animal cells produced multimeric transcripts of HDV genomic or even antigenomic RNA. However, when such experiments were modified to directly transfect with genomic RNA, as synthesized in vitro, genome replication did not occur. Yet as Glenn et al (37) first showed, if the recipient cells were modified to express stably the delta antigen, then genome replication was initiated. In initial experiments, a special liposome-mediated delivery system was used to deliver many molecules of HDV genomic RNA (37). In subsequent experiments using lipofection, a less efficient delivery procedure, induction of HDV genome replication, still occurred (18, 38).

Such RNA transfections have been successful using in vitro–synthesized RNA that is either genomic in polarity or the complement, the so-called antigenome (23). Also, the RNA can be no larger than the unit genome size and still work. Even when this RNA is flanked by non-HDV RNA sequences, correct replication still occurs; that is, examination of the replicating molecules reveals that the non-HDV sequences are precisely removed (37). Such removal may be via some kind of repair process, but this intriguing phenomenon remains to be clarified.

In Vitro Studies

A cell-free system that allows one to study the replication of the HDV genome in more detail would be very valuable. However, in attempting such studies, caution must be exercised. Symons stressed this point in relation to in vitro replication of certain similar agents of plants (97). He pointed out that replication as achieved outside the cell must be tested in relation to what happens inside infected cells; apparently many RNA polymerases are promiscuous in vitro when offered rod-like RNAs as a template for transcription. When it comes to HDV RNA synthesis, the mammalian RNA polymerase II seems guilty of in vivo promiscuity; nevertheless, we must still establish that any results obtained in vitro are relevant to the in vivo situation (97).

ISOLATED NUCLEI When nuclei are isolated from cells that are replicating HDV RNA, one can demonstrate the synthesis of HDV RNA. This observation has provided good evidence that an α-amanitin–sensitive RNA polymerase, like that of the host RNA polymerase II, is needed for HDV genome replication (68, 69). Other important results should emerge from the application of this system. To date, no one has reported similar success in adding HDV RNA to nuclear extracts from uninfected cells. It may be that this will be achieved by the coaddition of appropriate forms of the delta antigen.

PURIFIED POLYMERASES If it can be shown that a single RNA polymerase is sufficient for RNA-directed RNA synthesis, then such studies with purified polymerase will be very informative. Clearly, we need to know in detail the nature of the HDV RNA that allows it to be recognized by what is apparently the host RNA polymerase II, and which therefore makes it a template for RNA-directed RNA synthesis.

VIRION STRUCTURE

As mentioned above, the virions of HDV might be similar to their helper hepadnaviruses at least in terms of the presence of the same envelope proteins. HBV makes three different species of envelope proteins. These so-called surface antigen proteins come in three sizes, designated as *L, M,* and *S,* in order of decreasing size. They have common central and C-terminal domains. They differ in terms of the length of their N-terminal extensions. In HBV, the proportions of the three proteins are about 1:1:4, respectively (14, 35, 49, 103). In contrast, for purified HDV, the proportions are about 1:5:95 (6, 7, 9, 36). As described below, HDV particles containing the viral genome can be constructed that are 0:0:100. Such results have significant implications for identification of the HDV cell-surface receptor, because the HBV evidence is that the N-terminal domain unique to *L* is essential for receptor binding in the liver (75).

HBV infections produce not only an infectious particle 42 nm in diameter but also a relatively large excess of empty particles. These empty particles include spheres and filaments 22 nm in diameter. The spherical particles, like HDV, contain relatively low amounts of *L* and *M*. Thus, HDV assembly may exploit the assembly of what might otherwise be strictly empty particles.

Empty HBV particles and HDV differ significantly. Characterization of naturally occurring HDV particles indicates that the particles are about 38 nm in diameter. Such measurements are based either on indirect filtration procedures, direct electron microscopy, or both (48). Also, the latter studies indicate that the particles are somewhat heterogeneous in size, with a range of 30–40 nm.

The delta antigen is inside the HDV particle. This is the same protein that was initially discovered in the nuclei of liver biopsies, along with a small single-stranded RNA. Early studies did not detect a nucleocapsid structure (7). One study did claim to immunoprecipitate HDV RNA by means of an antibody directed against the protein (66). Recently, we obtained confirmatory evidence that the RNA and protein exist in a ribonucleoprotein, and also, with UV-cross-linking studies, have begun to answer questions about these particles (W.-S. Ryu & J. M. Taylor, unpublished observations). In new electron microscope studies, we have obtained evidence for a capsid structure within some of the virions (87).

GENOME AND INTRACELLULAR RNAS

HDV particles contain a single species of RNA called the genome. Inside an infected liver, one can find significant amounts of two other species (Figure 1). First, an RNA species, called the antigenome, is a full-length complement of the genome. Second, another species of complementary RNA is less than full length and is the mRNA for the delta antigen (23, 51). These three HDV RNAs are discussed in more detail in the following sections. [Additional, but less abundant HDV RNA species can also be found in the liver. These correspond to dimers and even trimers of the genomic and antigenomic RNA (23).]

Structure of Genomic RNA

In 1986, three laboratories presented evidence that the genomic RNA of HDV was unique because it was smaller than that of any animal virus (around 1700 nucleotides) and has a circular conformation (23, 58, 107). Moreover, both electron microscopy (58) and predictions based on the nucleotide sequence (107) indicated that the RNA can fold into an unbranched rod-like structure.

A later calculation deduced that the negative free energy for such folding was 840 kcal/mol (62), with approximately 70% of the nucleotides involved in Watson-Crick pairings.

Sequence of Genomic RNA

Several full (21, 29, 55, 59, 62, 71, 107) and partial sequences (58, 88, 89) of HDV have now been reported. These reports show that certain domains are more conserved than others and have been designated as the coding region for the delta antigen and the sequences needed for the self-cleavage of genomic and antigenomic RNA.

Another aspect of variation seems to be specific to HDV: the mutation that removes the termination codon for the 195–amino acid open reading frame (ORF) for the delta antigen, leading to the synthesis of a 214–amino acid ORF. This variation results from a RNA-editing process and seems to have profound significance in the life cycle of HDV.

Many of the unique structural properties of HDV are shared with the viroids. Because the viroids are several times smaller than HDV and do not code for any protein, it has been tempting to divide HDV into a viroid-like domain primarily responsible for *cis*-acting sequences involved in genome replication and a protein-encoding domain (10, 11, 21). UV cross-linking

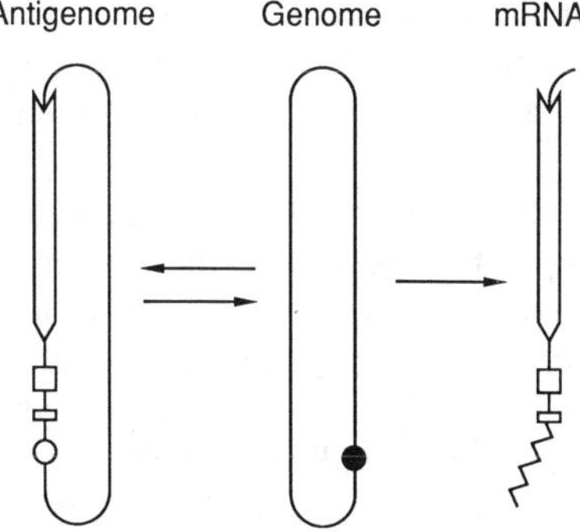

Figure 1 A representation of the three major RNAs of HDV genome replication. The central species is the genomic RNA, as found inside virions. It is also found in the nucleus of infected cells along with its exact complement, the antigenome, as indicated at the left side. Also present in an infected cell, but in the cytoplasm, is a species of polyadenylated RNA. This RNA, as indicated at the right side, is less than full length and is complementary to the genome. The diagram attempts to show that the genome and antigenome are both circular and are able to fold into an unbranched rod-like structure. Small closed and open circles indicate the sites of self-cleavage on the genome and antigenome, respectively. The open arrow indicates the open reading frame for the delta antigen. And the open square and rectangle indicate the polyadenylation signal, AAUAAA, and polyadenylation site, respectively. This figure is reproduced with permission from *Cell* (101).

data in the so-called viroid-like domain have supported such a division (10). Nevertheless, the concept may be too simplistic; certainly the initiation of antigenomic RNA synthesis occurs at what might be called the top of the rod (Figure 1), which is at the end opposite to the putative viroid-like domain (51).

Self-Cleavage and Self-Ligation

HDV and the RNAs of certain pathogenic agents of plants are strongly analogous. Prior findings of self-cleavage and self-ligation in plant agents (54, 96) led to the search for corresponding phenomena in HDV.

A self-cleavage site was first found on the antigenomic RNA of HDV (90). This observation was subsequently confirmed and extended to identify a similar site on the genomic RNA (63, 111–113). There are apparently no other sites. They are related in terms of the rod-like structure. As can be seen in Figure 1, the site on the genome has a sequence that is about 70% complementary on the opposite side of the rod; the latter in turn has a site 100% complementary on the antigenome—the antigenomic self-cleavage site. This domain is just downstream of the mRNA for the delta antigen. As explained later in terms of a model of HDV genome replication, there is a very good reason for such a location.

Early estimates by Kuo et al (63) of the minimum amount of sequence needed to achieve self-cleavage in vitro have been revised (3, 77, 78, 85). Attempts have been made to define the precise folding needed for such self-cleavage. Such experiments have already been done for other self-cleaving RNAs (96), and the results for HDV are clearly different (13, 78, 85).

Self-cleaving reactions share a kind of unity over and above the observation that they all depend upon heating RNA in the presence of divalent metal ions. Investigators have known for many years that such treatments can lead to RNA cleavage, but such cleavage is typically not site-specific (33). What the so-called self-cleaving RNAs do is provide preferred locations at which the metal ion will bind, thereby providing a specific site for self-cleavage. This unifying concept is based on the initial work on specific cleavage of tRNA by lead ions (57).

The HDV self-cleavage sequence can be divided into two parts: one that acts as the enzyme and one that acts as the substrate (13). Such separations have been reported for other self-cleaving RNAs (96, 104).

Soon after HDV was first shown to self-cleave, further experiments were carried out to attempt the reverse reaction, self-ligation. As reviewed elsewhere (91), there were three good reasons to test this, and the outcome was successful. The reaction depended upon the use of the rod-like structure as a kind of backbone to bring the two sides of the cleavage site into apposition.

The reaction was surprisingly efficient; more than 40% ligation occurred in vitro, even at 4°C, as a trimolecular reaction (91). Thus, in vivo, as this appears to be a concentration-independent monomolecular reaction, the efficiency could be close to 100%. In a later section, this reaction is incorporated into a model of HDV genome replication.

mRNA

Because HDV genome replication takes place predominantly in the nucleus (39, 102), a cytoplasmic mRNA must also be present for the translation of the delta antigen. Chen et al (23) initially detected a candidate RNA of antigenomic polyadenylated RNA, which was 800 bases in length and came from approximately the correct location on the antigenomic sequence. This RNA species has since been confirmed for transfected cells, been proven to be cytoplasmic, and been characterized in more detail (24, 51, 56). As represented in Figure 1, its 5'-end is about five bases from the top of the rod, and the site of polyadenylation has also been precisely mapped (51). The synthesis of this RNA depends upon the presence, just upstream of the polyadenylation site, of a consensus polyadenylation signal, AAUAAA. This signal is the same as that found for typical cellular mRNAs (5), and without it, the HDV genome cannot replicate, unless the delta antigen is provided *in trans,* such as by transcription from another plasmid (51).

Because there are two different forms of antigenomic HDV RNA, the full-length circular species and the mRNA, some kind of mechanism must regulate the synthesis of one form relative to the other. In Figure 1, the two forms are indicated as arising from alternative and separate pathways. Such a separation does apply for certain negative-strand RNA viruses (109), but according to findings of a recent study (53), it does not occur in HDV.

The three important conclusions are as follows: (*a*) Although the HDV self-cleavage domain is just downstream of the polyadenylation site, it is not needed for polyadenylation. (*b*) As part of the typical polyadenylation process for cellular mRNAs, a specific RNA cleavage is generated: the 5'-fragment becomes the acceptor for the polyadenylation, while the 3'-fragment is apparently of no further use and is promptly destroyed. Delta is different. The HDV self-cleavage reaction, which can occur just downstream of where the polyadenylation cleavage occurs, somehow stabilizes this downstream fragment. [Such stabilization might not be achieved if the self-cleavage domain were not located as it is, just downstream of the processing site for polyadenylation (51).] This stabilization is proposed to be very important to the HDV life cycle because it means that a single initiation of antigenomic RNA transcription can lead to the generation of at least two stable RNAs. (*c*) The third finding concerns the ultimate method by which the polyadenylation can be suppressed; unless it is suppressed, there can never be a full-length antigenomic RNA. When a nascent antigenomic HDV RNA can fold into the rod-like structure (which the normal HDV mRNA cannot do) and if the delta antigen is present (and as mentioned earlier, it can bind specifically to that rod-like structure), then polyadenylation is suppressed. Apparently, this suppression occurs at or before the cleavage reaction that occurs as part of the polyadenylation process. These three findings are incorporated into a model of HDV genome replication (Figure 2).

RELATIONSHIP TO PLANT AGENTS

An early hypothesis regarding the nature of HDV was that it could be like one of the pathogenic RNA agents of plants (31). Such agents include what are known as viroids, virusoids, satellite RNAs, and satellite viruses (31). The finding that the genome structure for HDV is circular and rod-shaped dramatically confirmed this hypothesis (23, 58, 107). The analogy was extended further by the findings that HDV could self-cleave (90) and self-ligate (91). What then are the limits of the analogy?

The plant agents are considered to replicate by what are referred to as rolling-circle models (12, 96). There are two types, depending upon whether circular RNAs occur for both genome and antigenome or just for the genome. Some of the concepts do apply to HDV, but the model for HDV (Figure 2) is more complex.

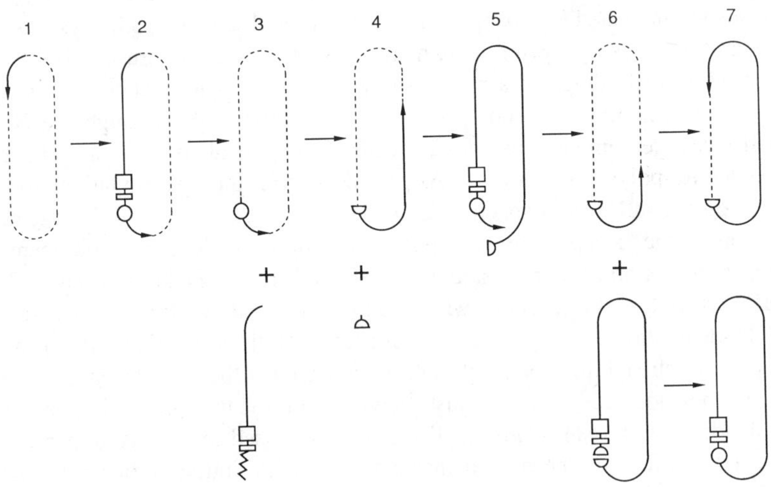

Figure 2 A model of the transcription from the HDV genome of mRNA and antigenomic RNA. Explanation is given in the text. This figure is reproduced with permission from *Journal of Virology* (53).

Other aspects of the analogy remain in question; of these, the most interesting is the nature of the RNA polymerase used for HDV genome replication. Evidence indicates the involvement of RNA polymerase II in the plant viroids. The data may not be rigorous, and further confusion arises from the additional claim for coinvolvement of a pol I activity (92). In HDV, pol II appears to be the enzyme involved, but again the data may not be adequate as yet (69).

Is it possible to say which of the plant pathogenic RNA HDV most resembles? Using a computer-based phylogeny, Elena et al (34) recently concluded that HDV is most like a virusoid at the nucleotide sequence level. However, among the plant agents only viroids are able to replicate their genomes by redirecting a host polymerase, i.e. without the help of a polymerase provided by another virus. Thus, on the one hand, HDV might be considered viroid-like; yet on the other, since HDV has an ORF for an essential protein, it is like a plant satellite virus. A subsequent section considers some of the other models that have been proposed for the origin of HDV.

DELTA ANTIGEN

The delta antigen is the only protein made by HDV. In a previous section, we considered the mRNA. In this section, we consider some of the many interesting aspects of this protein.

Predictions Based on Primary Sequence

This protein actually exists as two related species of 195 and 214 amino acids, referred to as the S (small) and L (large) forms, respectively. From direct nucleotide sequencing of viral genomes, Wang et al (107) first showed that the S and L correspond to a specific sequence microheterogeneity; on some genomes, the amber termination codon of the S is specifically altered so as to allow the synthesis of the L.

This interpretation has been supported by expression of this region both in yeast and bacteria (108). In fact, in *Escherchia coli* that contain an amber suppressor, the coding frame for the 195–amino acid sequence opens up, as expected, to allow the synthesis of the 214–amino acid species. The expression has been confirmed in animal cells as well (4, 17, 20, 108).

Other predictions come simply from an examination of the predicted amino acid sequences of the S and L forms. Some of these predictions have already been put to experimental tests. First, the protein is predicted to be highly basic. It should have a net positive charge of 12 in the pH range of 6–9 (62, 71). Such a basic nature would be consistent with RNA binding ability and has been confirmed. Second, in two regions on the predicted sequence, the structure is not only likely to be α-helical, but also to have what has been called a leucine-zipper or coiled-coil motif (64–66). Such regions have in some cases been invoked to explain protein-protein interactions. Clearly, such interactions occur between the two forms of the delta antigen. Third, a sequence has been noted on the predicted sequence that is very much like a nuclear localization signal of the glucocorticoid receptor protein (20). In this case, while the delta antigen does accumulate in the nucleus (61, 64, 68–70), there is evidence against the predicted site actually being responsible (64).

Small-to-Large Transition

As mentioned above, naturally occurring HDV genomes seem to encode either the S or the L form of the delta antigen. However, transfection experiments using HDV cDNA clones to initiate genome replication made clear that an unusual phenomenon was occurring. Specifically, when a cDNA clone was used that encoded only the S form, not only the S protein but also what looked like the L protein appeared in the transfected cell during genome replication. Such results were obtained in transfected cells (20), a transfected chimpanzee (94), and most recently, a transfected woodchuck (A. Ponzetto, personal communication).

Polymerase chain reaction (PCR) procedures allowed investigators to test the actual HDV genomes that were replicating in such situations. And, sure enough, the same nucleotide microheterogeneity in the amber termination codon was reappearing (67). As many as 40% of the genomes could be shown to be so changed. It was then found that only the S form would support genome replication; the L form would not. Moreover, the L form could act as a dominant negative mutation to suppress replication supported by the S form (20).

The change in the amber termination codon was from UAG to UGG. Evidence was assembled to implicate a known RNA modifying activity in the change. The prime suspect was an activity, also known as unwindase, that can act on double-stranded RNA and convert an adenosine to inosine (1, 105). If this occurred on antigenomic HDV RNA, it would lead, as a consequence of subsequent RNA-directed RNA synthesis, to the equivalent of replacing adenosine with guanosine. Regardless of several lines of circumstantial evidence, unwindase is apparently innocent.

More recent studies both by H. Zheng, T.-B. Fu, D. Lazinski, and myself (unpublished observations) and by J. L. Casey, K. F. Bergmann, T. Brown & J. L. Gerin (personal communication) have used a related PCR approach to study the nucleotide change. Both studies agree that the substrate for the change is not the antigenomic RNA, as implied above, but actually the genomic RNA. Thus, a uridine is somehow modified to something that in a PCR assay behaves as cystidine. Moreover, the substrate specificity for the change requires that the uridine not be presented as part of a RNA folded into the rod-like structure. Zheng's group has shown that nuclear extracts from uninfected cells can bring about this nucleotide modification. This result establishes that the change arises not by misincorporation during RNA-directed RNA synthesis but by a form of RNA editing and is now leading to studies to determine the exact nature of the modification. Currently, important questions remain as to what this enzyme normally does and how it recognizes the HDV RNA site with such specificity.

The above discussion only addresses the questions relating to how the HDV RNA target is changed. Equally important is the question, why it is changed? That is, what is the relevance of the change in the life cycle of HDV? As mentioned earlier, one part of the answer might be that only the S form will support genome replication. The appearance of the L form, as indicated by cotransfection studies, could lead to a suppression of such replication. The other part of the answer was recently obtained by Chang et al (16) and confirmed by Ryu et al (87). They found that the L form is necessary for virus assembly. The S form per se is not sufficient.

Dimeric Interactions

As mentioned above, the delta antigen has a sequence by which dimers might be formed. Lin et al (66) using cross-linking of delta protein, as translated in vitro from cRNA, have seen some dimers. And in other studies, they used delta fusion proteins to demonstrate more convincingly a dimerization reaction inside cells (64). D. Lazinski and I (unpublished observations) have also obtained evidence for antigen dimers, and even tetramers, inside the cell, by means of a glutaraldehyde cross-linking procedure (115). Also, Ryu et al (87), as part of virus-assembly studies discussed below, convincingly showed that inside transfected cells the two forms of the delta antigen are associated. Such dimer formation by the delta antigen is just one of several ways in which the delta antigen resembles the core antigen of the hepadnavirus.

RNA and Rod Binding

The delta antigen is predicted to be highly basic. Consistent with this is its behavior as a RNA-binding protein. Two laboratories have used in vitro reconstitution experiments to show that the L form of the delta antigen can be used in assays that reveal HDV specificity in this binding (17, 66, 70). Lin et al (66) showed that only the middle one-third of the delta antigen was necessary for this binding. Since then, a third study made use of both the S and L forms in an assay involving an additional renaturation step and obtained binding specificity for only certain HDV RNA sequences (19). Only those RNAs that could fold into the rod-like structure could yield this specific binding. Such binding was achieved with regions of the genome and antigenome, but always required a rod-like structure. As discussed in a previous section, such an interpretation has been incorporated into a model of regulation of polyadenylation of antigenomic HDV RNA.

The above studies refer to the reconstitution of RNA binding. However, obtaining data for such binding in vivo has been more difficult. Certainly in transfected cells the antigen localizes primarily to the nucleolus (17, 61); this might represent a localization based on nonspecific interactions with nascent ribosomal RNA species. Other studies have been done with virions.

After disrupting virions with a nonionic detergent, Bonino et al (7) did not obtain any evidence of a ribonucleoprotein structure (RNP). However, Chang et al (17) did have some success and used an antibody specific for the delta antigen to immunoprecipitate HDV genomic RNA from disrupted virions, thereby implying that at least some of the RNA in virions were bound to antigen. Recently, W.-S. Ryu and I (unpublished observations) used cross-linking with UV, prior to virion disruption, and thereby obtained confirmatory evidence for a RNP. An important problem at this time is to determine the exact number of delta antigen molecules bound per genomic RNA. Certainly we also need to know whether any special motifs are within the rod-like structure that are recognized in such binding.

VIRION ASSEMBLY

After it was shown that HDV genome replication could occur in the absence of any HBV sequences or functions, it became clear that the only role of the HBV was in virus entry and virus assembly. Recently, HDV assembly has been studied by means of cotransfecting cells with constructs that not only initiate HDV genome replication but also lead to the synthesis of specific hepadnavirus proteins. This section discusses such assembly experiments at three levels: the hepadnavirus proteins, the delta proteins, and the delta RNA.

In terms of the hepadnavirus proteins, the cotransfection studies (16, 87, 110) have confirmed what we already knew from the earlier experimental-animal studies (80, 84): both HBV and WHV can provide the surface proteins needed for HDV assembly (16, 87, 110). HBV and WHV each synthesize three related forms of surface antigens (35); as reviewed earlier, these are *L*, *M*, and *S*. The cotransfection studies with *S* alone show virus particle assembly; *L* and *M* are not needed (16, 87). It is not yet known whether such particles, when injected into a woodchuck chronically infected with WHV, lead to HDV genome replication. If they do, then maybe HDV uses a receptor different from that used by hepadnaviruses (75).

The next level of assembly concerns the involvement of the delta antigens. Evidence clearly indicates that of the two forms of the delta antigen, the L form is essential for virus assembly (16). The S form, per se, will not function. Thus, somehow the *L* form can interact with a hepadnavirus surface antigen, presumably *S* alone, so as to become incorporated into the released particles. Such delta antigen interactions must normally also include interactions with HDV genomic RNA. However, even in the absence of genomic HDV RNA, particles are still released (87, 106). It is not yet known whether such particles contain cellular RNAs instead. What has become clear concerns the *S* form of the delta antigen, which as mentioned above, cannot by itself support assembly; if *S* and *L* are present in the same transfected cell, then not only is *L* packaged but some of the *S* is as well (87). In other words, important interactions can take place between *S* and *L* at or prior to release from the cell.

A MODEL OF THE LIFE CYCLE

The typical picture of the life cycle of a virus in relation to the cell it infects requires an understanding of at least seven steps: adsorption, penetration, uncoating, transcription, translation, virus assembly, and release. With HDV, we clearly have a long way to go. The ligand on the virus used in virus adsorption has not yet been identified. If it proves to be different from that used by the helper hepadnavirus, then the receptor on the host hepatocyte will most likely also be different. The hepadnavirus may provide a valuable but limited analogy to help us understand penetration and uncoating. Genome replication in the nucleus (39, 102) and cytoplasmic translation of mRNA (51) are aspects that have become accessible, allowing progress to be made with virus assembly and release (16, 87, 110), where again, since hepadnavirus proteins are involved, some valuable analogies can be made. For example, the hepadnavirus core protein and the delta antigen share several functional similarities, including dimerization, RNA-binding, nuclear localization, and capsid-forming ability (35).

Recently, a model for HDV genome replication was presented (53) that incorporates much of the knowledge as presented in previous sections of this review. This model, as reproduced in Figure 2, specifically addresses the synthesis of antigenomic RNA, indicated by a solid line, from a genomic RNA template, indicated by a broken line. Transcription, presumably via the host RNA polymerase II, is considered to begin close to the top of the rod, at the location deduced to be the 5' end of the mRNA (steps 1 and 2). This nascent transcript is subsequently processed via the host polyadenylation apparatus to yield a mRNA (step 3), which acts in the cytoplasm as a template for delta antigen synthesis. Then as described earlier, the downstream transcript is stabilized via the self-cleavage reaction (step 4) and can continue on to become even greater than full-length antigenomic RNA (step 5). This downstream species is different from the original transcript in two important ways: first, it can fold into the rod structure, and second, if the delta antigen is present in sufficient amounts, the downstream species can be bound by this antigen. As previously shown, these two properties lead to the suppression of events that normally lead to polyadenylation. Thus, the downstream transcript undergoes only one kind of processing, namely a second self-cleavage (step 6). The fragment so produced will promptly fold into the rod structure and is expected to be rapidly converted into a circular RNA by means of the self-ligation reaction (step 7). A similar reaction scheme is probably involved for the synthesis of new genomic RNA from this antigenomic RNA; it will, however, be less complicated because a competing polyadenylation reaction is no longer involved.

APPLICATIONS

The information gained so far on HDV and how it replicates has opened up several possible applications. One simple and straightforward application is the isolation and characterization of the self-cleavage domains in order to make *cis*- and *trans*-acting ribozymes. Such ribozyme domains were first found for the hammerhead ribozyme of certain plant agents (47, 104). This finding has been extended to the hairpin-type ribozymes (27, 46), and now comparable studies have been reported for HDV (13).

A quite different application of HDV is to use it as a vector for the delivery of novel RNA sequences. HDV has certain unique advantages in this respect; the high copy number and the nuclear location of the HDV genome are the major advantages. To date, little more than a feasibility study has been reported for such an application (52). About 30 nucleotides of additional sequence could be inserted at the bottom of the rod-like structure of the RNA genome. This sequence not only did not interfere with the replication of the genome, it was also maintained as a stable sequence, at least in transfected cells. Moreover, when this insert was chosen to be a ribozyme specific for the cleavage *in trans* of another RNA, that RNA was apparently cleaved in vivo as a consequence of the replication of the modified HDV (52). The outlook then is that such inserts on the HDV genome may be designed with particular biological activities. For example, a modified HDV might be used to interfere with the expression of cellular genes or even the genes of other viruses. It has even been proposed that a modified HDV genome might be used as part of a strategy to interfere with chronic HBV infections (52).

ORIGIN

Several attempts have been made to determine the origin of HDV (11, 101). This is particularly difficult because HDV is clearly unlike any other known infectious agent of animals. However, as discussed in a previous section, many of the features of HDV can be found among certain infectious agents of plants.

The circular conformation of the HDV RNA is a property shared with the plant viroids and virusoids. Diener has emphasized the advantages of such circularity for primitive replication systems; he sees it as a potential relic of precellular evolution (30). Some of these plant agents may have arisen as "escaped introns," and this analogy has been extended to HDV (32, 43). Such claims are based on small amounts of consensus sequences shared among certain introns, but not on a relationship to any one particular intron. Also some of the plant agents show a patch of sequence homology to 7S L RNA, the RNA component of the signal recognition particle (42); it was thus quite intriguing when a similar patch of homology was found on the HDV genome (74, 114).

A quite different approach to determining the origin of HDV has been to examine its relatedness to its normal helper virus, HBV. One group has asserted that HDV contains an ORF that overlaps with a part of the delta antigen but is out of frame with it and that shows some relationship with part of the predicted terminal protein of HBV (60).

FUTURE DIRECTIONS

A previous section considers the potential applications of HDV. In terms of the nonapplied or pure HDV research, this review has attempted to point out new areas of investigation. Examples include the determination of the various interactions that the two species of the delta antigen make with each other, with HDV RNA, and with the surface proteins of the helper hepadnavirus. Such information will need to be put into the context of the full crystal structure of the delta antigen. Other important questions involve the mechanism of entry of the HDV—the receptor-ligand interaction will need to be clarified and compared, and possibly contrasted, with that of the helper hepadnavirus. Another significant question is the pathogenesis of HDV. How does HDV replication lead to liver damage? Are there really strains with little or no pathogenicity? Or, could we modify the genome to make such a strain?

The outlook for future HDV research shows many opportunities for interesting work. Already delta has been a gold mine of interesting surprises. The study has encompassed more than just a simple extrapolation from plant agents. And the possibility remains that HDV is but the prototype of a class of successful infectious agents, parasites of parasites, that may be involved in other diseases of animals.

ACKNOWLEDGMENTS

I was supported by grants CA-06927, RR-05539, and AI-26522 from the National Institutes for Health, grant MV-7M from the American Cancer Society, and by an appropriation from the Commonwealth of Pennsylvania. Valuable critical readings were given by Bill Mason, Wang-Shick Ryu, and Hans Netter. Also, in an attempt to keep the manuscript current, citations have been made of as yet unpublished studies from this and other laboratories.

Literature Cited

1. Bass, B. L., Weintraub, H. 1989. Biased hypermutation of viral RNA genomes could be due to unwinding/modification of double-stranded RNA. *Cell* 56:331
2. Beasley, R. P. 1988. Hepatitis B virus: the major etiology of hepatocellular carcinoma. In *Accomplishments in Cancer Research: 1987 Prize Year,* ed. J. G. Fortner, J. E. Rhoads, pp. 80–106. Philadelphia: Lippincott
3. Belinsky, M., Dinter-Gottlieb, G. 1991. Characterizing the self-cleavage of a 135 nucleotide ribozyme from genomic hepatitis delta virus. See Ref. 35a, pp. 265–74
4. Bergmann, K. F., Gerin, J. L. 1986. Antigen of hepatitis delta virus in the liver and serum of humans and animals. *J. Infect. Dis.* 154:702–6
5. Bernstein, P., Ross, J. 1989. Poly(A), poly(A) binding protein and the regulation of mRNA stability. *Trends Biochem. Sci.* 14:373–77
6. Bonino, F., Heermann, K. H., Rizzetto, M., Gerlich, W. H. 1986. Hepatitis delta virus: protein composition of delta antigen and its hepatitis B virus-derived envelope. *J. Virol.* 58:945–50
7. Bonino, F., Hoyer, W., Shih, J. W.-K., Rizzetto, M., Purcell, R., et al. 1984. Delta hepatitis agent: structural and antigenic properties of the delta-associated particles. *Infect. Immunol.* 43:1000–5
8. Bonino, F., Negro, F., Brunetto, M. R., Verme, G. 1989. Hepatitis delta infection. *Prog. Liver Dis.* 9:485–96
9. Bonino, F., Rizzetto, M., Gerlich, W. H. 1986. Hepatitis delta virus protein composition of delta antigen and its hepatitis B virus–derived envelope. *J. Virol.* 58:945–50
10. Branch, A. D., Benenfeld, B. J., Baroudy, B. M., Wells, F. V., Gerin, J. L., et al. 1989. An ultraviolet-sensitive RNA structural element in a viroid-like domain of the hepatitis delta virus. *Science* 243:649–51
11. Branch, A. D., Levine, B. J., Robertson, H. D. 1990. The brotherhood of circular RNA pathogens: viroids, circular satellites, and the delta agent. *Sem. Virol.* 1:143–52
12. Branch, A. D., Robertson, H. D. 1984. A replication cycle for viroids and small infectious RNAs. *Science* 223:450–55
13. Branch, A. D., Robertson, H. D. 1991. Efficient *trans* cleavage and a common structural motif for the ribozymes of the human hepatitis δ agent. *Proc. Natl. Acad. Sci. USA* 88:10163–67
14. Bruss, V., Ganem, D. 1991. The role of envelope proteins in hepatitis B virus assembly. *Proc. Natl. Acad. Sci. USA* 88:1059–63
15. Buetrago, B., Popper, H., Hadler, S. C., Thung, S. N., Gerber, M. A., et al. 1986. Specific histologic features of Santa Marta hepatitis: a severe form of hepatitis 5–virus infection in northern south America. *Hepatology* 6:1285–91
16. Chang, F.-L., Chen, P.-J., Tu, S.-J., Chiu, M.-N., Wang, C.-J., et al. 1991. The large form hepatitis δ antigen is crucial for the assembly of hepatitis δ virus. *Proc. Natl. Acad. Sci. USA* 88:8490–94
17. Chang, M.-F., Baker, S. C., Soe, L. H., Kamahora, T., Keck, J. G., et al. 1988. Human hepatitis delta antigen is a nuclear phosphoprotein with RNA binding activity. *J. Virol.* 62:2403–10
18. Chao, M. 1991. *Replication of hepatitis delta virus*. PhD thesis. Univ. Penn., Philadelphia
19. Chao, M., Hsieh, S.-Y., Taylor, J. 1991. The antigen of hepatitis delta virus: examination of in vitro RNA binding specificity. *J. Virol.* 65:4057–62
20. Chao, M., Hsieh, S.-Y., Taylor, J. 1990. Role of two forms of the hepatitis delta virus antigen: evidence for a mech-

anism of self-limiting genome replication. *J. Virol.* 64:5066–69

21. Chao, Y.-C., Chang, M.-F., Gust, I. D., Lai, M. M. C. 1990. Sequence conservation and divergence of hepatitis 5 virus RNA. *Virology* 178:384–92
22. Chen, P.-J., Chen, D.-S., Chen, C.-R., Chen, Y.-Y., Chen, H.-M. H., et al. 1988. Infection in asymtomatic carriers of hepatitis B surface antigen: low prevalence of δ activity and effective suppression of hepatitis B virus replication. *Hepatology* 8:1121–24
23. Chen, P.-J., Kalpana, G., Goldberg, J., Mason, W., Werner, B., et al. 1986. Structure and replication of the genome of hepatitis delta virus. *Proc. Natl. Acad. Sci. USA* 83:8774–78
24. Chen, P.-J., Kuo, M. Y.-P., Chen, M.-L., Tu, S.-J., Chiu, M.-N., et al. 1990. Continuous expression of the hepatitis δ virus genome in Hep G2 hepatoblastoma cells transfected with cloned viral DNA. *Proc. Natl. Acad. Sci. USA* 87:5253–57
25. Choi, S.-S., Rasshofer, R., Roggendorf, M. 1988. Propagation of woodchuck hepatitis delta virus in primary woodchuck hepatocytes. *Virology* 167: 451–57
26. Choi, S.-S., Rasshofer, R., Roggendorf, M. 1989. Inhibition of hepatitis delta virus RNA replication in primary woodchuck hepatocytes. *Antivir. Res.* 7:213–22
27. Chowrira, B. M., Berzal-Herranz, A., Burke, J. M. 1991. Novel guanosine requirement for catalysis by the hairpin ribozyme. *Nature (London)* 354:320–22
28. Davis, G. L., Balart, L. A., Schiff, E. R., Lindsay, K., Bodenheimer, H. C., et al. 1989. Treatment of chronic hepatitis C with recombinant interferon alpha: a multicenter randomized, controlled trial. *New Eng. J. Med.* 321:1501–10
29. Deny, P., Zignego, A. L., Rascalou, N., Ponzetto, A., Tiollais, P., et al. 1991. Nucleotide sequence analysis of three different hepatitis delta viruses isolated from a woodchuck and humans. *J. Gen. Virol.* 72:735–39
30. Diener, T. O. 1989. Circular RNAs: relics of precellular evolution? *Proc. Natl. Acad. Sci. USA* 86:9370–74
31. Diener, T. O., Prusiner, S. B. 1985. The recognition of subviral pathogens. See Ref. 71a, pp. 3–20
32. Dinter-Gottlieb, G. 1986. Viroids and virusoids are related to group I introns. *Proc. Natl. Acad. Sci. USA* 83:6250–54
33. Eichhorn, G. L., Tarien, E., Butzow, J. J. 1971. Interaction of metal ions with nucleic acids and related compounds. XVI. Specific cleavage effects in depolymerization of ribonucleic acids by zinc(II) ions. *Biochemistry* 10:2014–19
34. Elena, S. F., Dopazo, J., Flores, R., Diener, T. O., Moya, A. 1991. Phylogeny of viroids, viroidlike satellite RNAs, and the viroidlike domain of hepatitis δ virus RNA. *Proc. Natl. Acad. Sci. USA* 88:5631–34
35. Ganem, D. 1991. Assembly of hepadnviral virions and subviral particles. *Curr. Top. Microbiol. Immunol.* 168: 61–83

35a. Gerin, J. L., Purcell, R. H., Rizetto, M., eds. 1991. *The Hepatitis Delta Virus*. New York: Wiley-Liss

36. Gerlich, W. H., Heermann, K. H., Ponzetto, A., Crivelli, O., Bonino, F. 1987. Proteins of hepatitis delta virus. See Ref. 83a, pp. 97–103
37. Glenn, J. S., Taylor, J. M., White, J. M. 1990. In vitro–synthesized hepatitis delta virus RNA initiates genome replication in cultured cells. *J. Virol.* 64:3104–7
38. Glenn, J. S., White, J. M. 1991. *trans*-Dominant inhibition of human hepatitis delta virus genome replication. *J. Virol.* 65:2357–61
39. Gowans, E. J., Baroudy, B. M., Negro, F., Ponzetto, A., Purcell, R., et al. 1988. Evidence for the replication of hepatitis delta virus RNA in nuclei after in vivo infection. *Virology* 167:274–78
40. Gowans, E. J., Macnaughton, T. B., Jilbert, A. R., Burrell, C. J. 1991. Cell culture model systems to study HDV expression, replication, and pathogenesis. See Ref. 35a, pp. 299–308
41. Grendele, M., Gridelli, B. B., Colledan, M., Rossi, G., Doglia, M., et al. 1991. Good news about HBV-HDV reinfection in liver transplant patients. See Ref. 49a, pp. 495–96
42. Haas, B., Klanner, A., Ramm, K., Sanger, H. L. 1989. The 7S RNA from tomato leaf tissue resembles a signal recognition particle RNA and exhibits a remarkable sequence complementarity to viroids. *EMBO J.* 7:4063–74
43. Hadidi, A. 1986. Relationships of viroids and certain other plant pathogenic nucleic acids to group I and II introns. *Plant Mol. Biol.* 7:129–42
44. Hadler, S. C., Fields, H. A. 1991. Hepatitis delta virus. In *Textbook of Human Virology*, ed. R. B. Belshe, pp. 749–66. St. Louis: Mosby Year Book. 2nd ed.
45. Hadzyiannis, S. J., Hatzakis, A., Papiannou, C., Anastassakos, C. 1987. Endemic hepatitis delta virus infection in a Greek community. See Ref. 83a, pp. 181–202

46. Hampel, A., Tritz, R., Hicks, M., Cruz, P. 1990. "Hairpin" catalytic RNA model: evidence for helices and sequence requirement for substrate RNA. *Nucleic Acids Res.* 18:299–304
47. Hazeloff, J., Gerlach, W. L. 1988. Simple RNA enzymes with new and highly specific endoribonuclease activities. *Nature (London)* 334:585–91
48. He, L.-F., Ford, E., Purcell, R. H., London, W. T., Phillips, J., et al. 1989. The size of the hepatitis delta agent. *J. Med. Virol.* 27:31–33
49. Heermann, K. H., Goldmann, U., Schwartz, W., Seyffarth, T., Baumgarten, H., et al. 1984. Large surface proteins of hepatitis B virus containing the pre-S sequence. *J. Virol.* 52:396–92
49a. Hollinger, F. B., Lemmon, S. M., Margolis, H. S., eds. 1991. *Viral Hepatitis and Liver Disease*. Baltimore: Williams & Wilkins
50. Hoofnagle, J. H. 1989. Type D (Delta) hepatitis. *J. Am. Med. Assoc.* 261:1321–25
51. Hsieh, S.-Y., Chao, M., Coates, L., Taylor, J. 1990. Hepatitis delta virus genome replication: a polyadenylated mRNA for delta antigen. *J. Virol.* 64: 3192–98
52. Hsieh, S.-Y., Taylor, J. 1992. Delta virus as a vector for the delivery of biologically-active RNAs: possible ribozyme specific for chronic hepatitis B virus infection. In *Innovations in Antiviral Development and the Detection of Virus Infections,* ed. T. M. Block, D. L. Jungkind, R. L. Crowell, M. R. Dennison, L. Walsh. New York: Plenum. In press
53. Hsieh, S.-Y., Taylor, J. M. 1991. Regulation of polyadenylation of hepatitis delta virus antigenomic RNA. *J. Virol.* 65:6438–46
54. Hutchins, C. J., Rathjen, P. D., Forster, A. C., Symons, R. H. 1986. Self-cleavage of plus and minus RNA transcripts of avocado sunblotch viroid. *Nucleic Acids Res.* 14:3627–40
55. Imazeki, F., Omata, M., Ohto, M. 1991. Complete nucleotide sequence of hepatitis delta virus RNA in Japan. *Nucleic Acids Res.* 19:5439
56. Jilbert, A. R., Gowans, E. J., Macnaughton, T. B., Burrell, C. J. 1991. Characterization of a subgenomic HDV-specific poly(A)+ RNA species isolated from human hepatoma cells supporting HDV RNA replication. See Ref. 35a, pp. 309–14
57. Klug, A. 1979. The assembly of tobacco mosaic virus: structure and specificity. *Harvey Lect.* 74:141–72
58. Kos, A., Dijkema, R., Arnberg, A. C., van der Meide, P. H. Schellekens, H. 1986. The hepatitis delta virus possesses a circular RNA. *Nature (London)* 322:558–60
59. Kos, T., Molijn, A., van Doorn, L.-J., van Belkum, A., Dubbeld, M., et al. 1991. Hepatitis delta virus cDNA sequence from an acutely HBV-infected chimpanzee: sequence conservation in experimental animals. *J. Med. Virol.* 34:268–79
60. Khudyakov, Y. E., Makhov, A. M. 1990. Amino acid sequence similarity between the terminal protein of hepatitis B virus and predicted hepatitis delta virus gene product. *FEBS Lett.* 262:345–48
61. Kuo, M. Y.-P., Chao, M., Taylor, J. 1989. Initiation of replication of the human hepatitis delta virus genome from cloned DNA: role of the delta antigen. *J. Virol.* 63:1945–50
62. Kuo, M. Y.-P., Goldberg, J., Coates, L., Mason, W., Gerin, J., et al. 1988. Molecular cloning of hepatitis delta virus RNA from an infected woodchuck liver: sequence structure and applications. *J. Virol.* 62:1855–61
63. Kuo, M. Y.-P., Sharmeen, L., Dinter-Gottlieb, G., Taylor, J. 1988. Characterization of self-cleaving RNA sequences on the genome and antigenome of human hepatitis delta virus. *J. Virol.* 62:4439–44
64. Lai, M. M. C. , Chao, Y.-C. , Chang, M.-F. , Lin, J.-H., Gust, I. 1991. Functional studies of hepatitis delta antigen and delta virus RNA. See Ref. 35a, pp. 283–92
65. Landschulz, W. H., Johnson, P. F., McKnight, S. L. 1989. The DNA binding domain of the rat liver nuclear protein C/EBP is bipartite. *Science* 243: 1681–88
66. Lin. J.-H., Chang, M.-F., Baker, S. C., Govindarajan, S., Lai, M. M. C. 1990. Characterization of hepatitis delta antigen: specific binding to hepatitis delta virus RNA. *J. Virol.* 64:4051–58
67. Luo, G., Chao, M., Hsieh, S.-Y., Sureau, C., Nishikura, K., et al. 1990. A specific base transition occurs on replicating hepatitis delta virus RNA. *J. Virol.* 64:1021–27
68. Macnaughton, T. B., Gowans, E. J., Jilbert, A. R., Burrell, C. J. 1990. Hepatitis delta virus RNA, protein synthesis and associated toxicity in stably transfected cell line. *Virology* 177:692–98
69. Macnaughton, T. B., Gowans, E. J., Qiao, M., Burrell, C. J. 1991. Simul-

taneous expression from cDNA of markers of hepatitis delta virus and hepadnavirus infection in continuous culture. See Ref. 49a, pp. 465–68

70. Macnaughton, T. B., Gowans, E. J., Reinboth, B., Jilbert, A. R., Burrell, C. J. 1990. Stable expression of hepatitis delta virus antigen in a eukaryotic cell line. *J. Gen. Virol.* 71:1339–45

71. Makino, S., Chang, M.-F., Shieh, C.-K., Kamahora, T., Vannier, D. V., et al. 1987. Molecular cloning and sequencing of a human hepatitis delta virus RNA. *Nature (London)* 329:343–46

71a. Marmorosch, K., McKelvey, J. J. Jr., eds. 1985. *Subviral Pathogens of Plants and Animals: Viroids and Prions*. Florida: Academic

72. Mason, W. S., Taylor, J. M. 1991. Liver transplantation: a model for the transmission of hepatitis delta virus. *Gasteroenterology* 101:1741–43

73. Negro, F., Bergmann, K. F., Baroudy, B. M., Satterfield, W. C., Popper, H., et al. 1988. Chronic hepatitis D virus (HDV) infection in hepatitis B virus carrier chimpanzees experimentally superinfected with HDV. *J. Infect. Dis.* 158:151–59

74. Negro, F., Gerin, J. L., Purcell, R. H., Miller, R. H. 1989. Basis of hepatitis delta virus disease. *Nature* 341:111

75. Neurath, A. R., Kent, S. B. H., Strick, N., Parker, K. 1986. Identification and chemical synthesis of a host cell receptor binding site on hepatitis B virus. *Cell* 46:429–36

76. Ottobrelli, A., Marzano, A., Smedile, A., Recchia, S., Salizzoni, M., et al. 1991. Patterns of hepatitis delta virus reinfection and disease in liver transplantation. *Gasteroenterology* 101: 1649–55

77. Perrotta, A. T., Been, M. D. 1990. The self-cleaving domain from the genomic RNA of hepatitis delta virus: sequence requirements and the effects of denaturant. *Nucleic Acids Res.* 18:6821–27

78. Perrotta, A. T., Been, M. D. 1991. A pseudoknot-like structure required for efficient self-cleavage of hepatitis delta virus RNA. *Nature (London)* 350:434–36

79. Ponzetto, A., Cote, P. J., Popper, H., Boyer, B. H., London, W. T., et al. 1984. Transmission of hepatitis B–associated delta agent to the eastern woodchuck. *Proc. Natl. Acad. Sci. USA* 81:2208–11

80. Ponzetto, A., Negro, F., Popper, H., Bonino, F., Engle, R., et al. 1988. Serial passage of hepatitis delta virus infection in chronic hepatitis B virus carrier chimpanzees. *Hepatology* 8: 1655–61

81. Popper, H., Thung, S. N., Gerber, M. A., Hadler, S. C., De Monzon, M., et al. 1983. Histological studies of severe delta agent infection in Venezuelan Indians. *Hepatology* 3:906–12

82. Purcell, R. H., Satterfield, W. C., Bergmann, K. F., Smedile, A., Ponzetto, A., et al. 1987. Experimental hepatitis delta virus infection in the chimpanzee. See Ref. 83a, pp. 27–36

83. Rizzetto, M., Canese, M. G., Arico, J., Crivelli, O., Bonino, F., et al. 1977. Immunofluorescence detection of a new antigen-antibody system (delta-anti-delta) associated to the hepatitis B virus in the liver and in the serum of HBsAg carriers. *Gut* 18:997–93

83a. Rizetto, M., Gerin, J. L., Purcell, R. H., eds. 1987. *The Hepatitis Delta Virus and Its Infection*. New York: Liss

84. Rizzetto, M., Hoyer, B., Canese, M. G., Shih, J. W.-K., Purcell, R. H., et al. 1980. Delta agent: association of delta antigen with heptatitis B surface antigen and RNA in serum of delta-infected chimpanzees. *Proc. Natl. Acad. Sci. USA* 77:6124–28

85. Rosenstein, S. P., Been, M. D. 1991. Evidence that genomic and antigenomic RNA self-cleaving elements from hepatitis delta virus have similar secondary structures. *Nucleic Acids Res.* 19:5409–16

86. Rossetti, F., Pontini, F., Crivellaro, C., Cadrobbi, P., Guido, M., et al. 1991. Interferon therapy of chronic hepatitis delta virus in patients cured of pediatric malignancies: possible harmful effect. *Liver* 11:255–59

87. Ryu, W.-S., Bayer, M., Taylor, J. M. 1992. Assembly of hepatitis delta virus particles. *J. Virol.* 66:2310–15

88. Saldanha, J. A., Thomas, H. C., Monjardino, J. P. 1987. Cloning and characterization of a delta virus cDNA sequence derived from a human source. *J. Med. Virol.* 22:323–31

89. Saldanha, J. A., Thomas, H. C., Monjardino, J. P. 1990. Cloning and sequencing of RNA of hepatitis delta virus isolated from human serum. *J. Gen. Virol.* 71:1603–6

90. Sharmeen, L., Kuo, M. Y.-P., Dinter-Gottlieb, G., Taylor, J. 1988. The antigenomic RNA of human hepatitis delta virus can undergo self-cleavage. *J. Virol.* 62:2674–79

91. Sharmeen, L., Kuo, M. Y.-P., Taylor, J. 1989. Self-ligating RNA sequences on

the antigenome of human hepatitis delta virus. *J. Virol.* 63:1428–30

92. Spiesmacher, E., Muhlbach, H.-P., Tabler, M., Sanger, H. L. 1985. Synthesis of (+) and (−) RNA molecules of potato spindle tuber viroid (PSTV) in isolated nuclei and its impairment by transcription inhibitors. *Biosci. Rep.* 5:251–65
93. Sureau, C., Jacob, J. R., Eichberg, J. W., Lanford, R. E. 1991. Tissue culture system for infection with human hepatitis delta virus. *J. Virol.* 65:3443–50
94. Sureau, C., Taylor, J., Chao, M., Eichberg, J. E., Lanford, R. E. 1989. Cloned hepatitis delta virus complementary DNA is infectious in the chimpanzee. *J. Virol.* 63:4292–97
95. Sureau, C., Taylor, J., Chao, M., Eichberg, J. W., Lanford, R. E. 1991. In vitro tissue culture system for the study of HDV infectivity. See Ref. 35a, p. 464
96. Symons, R. H. 1989. Self-cleavage of RNA in the replication of small pathogens of plants and animals. *TIBS* 14:445–50
97. Symons, R., Haseloff, J., Visvader, J. E., Keese, P., Murphy, P. J., et al. 1985. On the mechanism of replication of viroids, virusoids, and satellite RNAs. See Ref. 71a, pp. 235–63
98. Taylor, J. 1990. Structure and replication of hepatitis delta virus. *Sem. Virol.* 1:135–41
99. Taylor, J. 1990. Hepatitis delta virus: cis and trans functions needed for replication. *Cell* 61:371–73
100. Taylor, J. M. 1991. Human hepatitis delta virus. *Curr. Top. Microbiol. Immunol.* 168:141–66
101. Taylor, J., Chao, M., Kuo, M., Sharmeen, L., Hsieh, S.-Y. 1990. Human hepatitis delta virus: unique or not unique. In *New Aspects of Positive-Strand Viruses,* ed. M. Brinton, pp. 20–24. Washington: Am. Sci. Microbiol.
102. Taylor, J., Mason, W., Summers, J., Goldberg, J., Aldrich, C., et al. 1987. Replication of human hepatitis delta virus in primary cultures of woodchuck hepatocytes. *J. Virol.* 61:2891–95
103. Ueda, K., Tsurimoto, T., Matsubara, K. 1991. Three envelope proteins of hepatitis B virus: large S, middle S, and major S proteins needed for the formation of Dane particles. *J. Virol.* 65:352 1–29
104. Uhlenbeck, O. C. 1987. A small catalytic oligoribonucleotide. *Nature (London)* 328:596–600
105. Wagner, R. W., Nishikura, K. 1988. Cell cycle expression of RNA duplex unwindase activity in cells. *Mol. Cell Biol.* 8:770–77
106. Wang, C.-J., Chen, P.-J., Wu, T.-C., Patel, D., Chen, D.-S. 1991. Small-form hepatitis B surface antigen is sufficient to help in the assembly of hepatitis delta virus-like particles. *J. Virol.* 65:6630–36
107. Wang, K.-S., Choo, Q.-L., Weiner, A. J., Ou, J.-H., Najarian, C., et al. 1986. Structure, sequence and expression of the hepatitis delta viral genome. *Nature (London)* 323:508–13
108. Weiner, A. J., Choo, Q.-L., Wang, K.-S., Govindarajan, S., Redeker, A. G., et al. 1988. A single antigenomic open reading frame of the hepatitis delta virus encodes the epitope(s) of both hepatitis delta antigen polypeptides p24$^{\delta}$ and p27$^{\delta}$. *J. Virol.* 62:594–99
109. Wertz, G. W., Howard, M. B., Davis, N., Patton, J. 1987. The switch from transcription to replication of a negative-strand virus. *Cold Spring Harbor Symp. Quant. Biol.* 52:367–71
110. Wu, J.-C., Chen, P.-J., Kuo, M. Y.-P., Lee, S.-D., Chen, D.-S., et al. 1991. Production of hepatitis delta virus and suppression of helper hepatitis B virus in a human hepatoma cell line. *J. Virol.* 65:1099–1104
111. Wu, H.-N., Lai, M. M. C. 1989. Reversible cleavage and ligation of hepatitis delta virus RNA. *Science* 243:652–54
112. Wu, H.-N., Lai, M. M. C. 1990. RNA conformational requirements for self-cleavage of hepatitis delta virus RNA. *Mol. Cell Biol.* 10:5575–79
113. Wu, H.-N., Lin, Y.-J., Lin, F.-P., Makino, S., Chang, M.-F., et al. 1989. Human hepatitis δ virus RNA subfragments contain an autocleavage activity. *Proc. Natl. Acad. Sci. USA* 86: 1831–35
114. Young, B., Hicke, B. 1990. Delta virus as a cleaver. *Nature (London)* 343:28
115. Zapp, M. L., Hope, T. J., Parslow, T. G., Green, M. R. 1991. Oligomerization and RNA binding domains of the human type 1 immunodeficiency virus Rev protein: a dual function for an arginine-rich binding motif. *Proc. Natl. Acad. Sci. USA* 88:7734–38

Annu. Rev. Microbiol. 1992. 46:277–305

THE ELECTRON-TRANSPORT PROTEINS OF HYDROXYLATING BACTERIAL DIOXYGENASES

Jeremy R. Mason and Richard Cammack

Divisions of Biosphere Sciences and Biomolecular Sciences, King's College London, Campden Hill Road, London W8 7AH, United Kingdom

KEY WORDS: flavoprotein, ferredoxin, iron-sulfur protein, Rieske-type [2Fe-2S] cluster, nonheme iron

CONTENTS

Abstract

The degradation of aromatic compounds by aerobic bacteria frequently begins with the dihydroxylation of the substrate by nonheme iron-containing dioxygenases. These enzymes consist of two or three soluble proteins that interact to form an electron-transport chain that transfers electrons from reduced nucleotides (NADH) via flavin and [2Fe-2S] redox centers to a terminal dioxygenase. The dioxygenases may be classified in terms of the number of constituent components and the nature of the redox centers. Class I consists of two-component enzymes in which the first protein is a reductase

0066-4227/92/1001-0277$02.00

containing both a flavin and a [2Fe-2S] redox center and the second component is the oxygenase; Class II consists of three-component enzymes in which the flavin and [2Fe-2S] redox centers of the reductase are on a separate flavoprotein and ferredoxin, respectively; and Class III consists of three-component enzymes in which the reductase contains both a flavin and [2Fe-2S] redox center but also requires a second [2Fe-2S] center on a ferredoxin for electron transfer to the terminal oxygenase. Further subdivision is based on the the type of flavin (FMN or FAD) in the reductase, the coordination of the [2Fe-2S] center in the ferredoxin, and the number of terminal oxygenase subunits.

From the deduced amino acid sequence of several dioxygenases the ligands involved in the coordination of the nucleotides, iron-sulfur centers, and mononuclear nonheme iron active site are proposed. On the basis of their spectroscopic properties and unusually high redox potentials, the [2Fe-2S] clusters of the ferredoxins and terminal oxygenases have been assigned to the class of Rieske-type iron-sulfur proteins. The iron atoms in the Rieske iron-sulfur cluster are coordinated to the protein by two histidine nitrogens and two cysteine sulfurs.

INTRODUCTION

The benzene nucleus is one of the most abundant chemical structures in the biosphere, originating from both natural and anthropogenic sources. The biodegradation of compounds containing this structure is carried out almost exclusively by microorganisms (26), and with few exceptions, the initial step is the introduction of two hydroxyl groups into the benzene ring. This action is a prerequisite for further fission and catabolism of the aromatic compound. The enzymes that perform this specialized function of catalyzing the insertion of molecular oxygen into the organic substrate are termed *oxygenases*.

This review describes the properties and structure of a class of dioxygenases that convert aromatic rings to *cis*-diols. The enzymes generally comprise several protein subunits, an iron center, iron-sulfur clusters, and a flavin, arranged in an electron-transfer chain. The structures of these components vary in several ways. The structural evidence derives from very limited X-ray crystallographic results, extensive amino-acid sequence data and comparison with other electron-transfer proteins, and the application of spectroscopic methods.

Oxygenases

The most extensively investigated oxygenase enzymes are those isolated from *Pseudomonas* species (54). They activate molecular oxygen and incorporate it directly into the chemical structure of the organic molecule. Although the reactions catalyzed by oxygenases are highly exergonic, the free energy released is not conserved by the formation of ATP. Thermodynamically, the effect of the oxygenase reaction is to destabilize the aromatic growth substrate and render the catabolism of the aromatic compounds irreversible (22).

The various oxygenases differ in structure, mechanism, and in coenzyme requirements. They can, however, be classified into two groups: the monooxygenases, which incorporate one atom of oxygen into one molecule of substrate, and the dioxygenases, which add both atoms of oxygen into the substrate. In monooxygenases, the other atom of oxygen is reduced to water so that these enzymes function as part oxygenase and part oxidase; hence, they are also termed *mixed-function oxidases* (73). The distinction between dioxygenase and monooxygenase activities is, however, not always absolute. For example, both toluene and naphthalene dioxygenases can catalyze monooxygenase reactions as in the oxidation of indene and indan to 1-indenol and 1-indanol, respectively (112), while 4-methoxy benzoate monooxygenase dihydroxylates the vinylic side chain of 4-methoxystyrene with both atoms derived from molecular oxygen (114, 115).

Dioxygenases are involved at several stages in the pathways for the catabolism of aromatic compounds and may be considered to fall into two groups:

1. Dioxygenases involved in ring hydroxylation. These all require reduced cofactors, either NADH or NADPH, in addition to oxygen. They dihydroxylate aromatic substrates to produce *cis*-diols, e.g. toluene dioxygenase:

$$\text{toluene} + O_2 + \text{NADH} + H^+ \longrightarrow \textit{cis}\text{-toluene dihydrodiol} + \text{NAD}^+$$

toluene — *cis*-toluene dihydrodiol

2. Dioxygenases involved in ring fission. These have no cofactor requirement and cleave the benzene ring of hydroxylated (di- or tri-) aromatic substrates, e.g. catechol 1,2-dioxygenase:

$$\text{catechol} + O_2 \longrightarrow \textit{cis, cis}\text{-muconic acid}$$

catechol — *cis, cis*-muconic acid

This review deals with the former group of dioxygenases, which are essential for the activation of aromatic compounds.

STRUCTURE OF RING-HYDROXYLATING DIOXYGENASES

Gibson and coworkers (40, 41) were the first to show that the dioxygenase system of *Pseudomonas putida* oxidized benzene to generate *cis*-1,2-dihydroxy-cyclohexa-3,5-diene. Several aromatic ring-hydroxylating dioxygenases have since been isolated and characterized (2, 30, 49, 50, 101, 102, 120, 123). The enzymes are nonheme dioxygenases that oxidize the aromatic substrates to give *cis*-dihydrodiols or *cis*-diol carboxylic acids. They require, in addition to molecular oxygen, cofactors, including reduced pyridine nucleotides and Fe(II), for the reactions they catalyze.

In the ring-hydroxylating oxygenases so far investigated, the preferred reductant is NADH. All are soluble, multicomponent enzymatic systems comprising the same types of redox components (Table 1). The details of the electron-transfer chains have not been elucidated in all cases; however, a typical arrangement is a short electron-transfer chain:

NADH → Flavin	→ Fe-S_{Fd} → Fe-S_R →	Fe → O_2
Reductase site	Electron-transfer chain	Oxygenase site

Here, Fe-S_{Fd} and Fe-S_R represent ferredoxin-type and Rieske-type [2Fe-2S] iron-sulfur clusters respectively, as defined below. Fe represents the iron-binding site proposed to be the terminal oxygenase center.

The distribution of the redox components, and the number and size of protein subunits, depend on the type of the ring-hydroxylating dioxygenases being considered (Table 1). All of the terminal dioxygenase proteins of this class consist of an iron-sulfur protein with a Rieske-type [2Fe-2S] cluster and an iron site, but the reductase chain, which transfers reducing equivalents from NADH to the terminal dioxygenase, may consist of either one or two separate proteins.

Reductase System

Table 1 shows that the reductase chains of different ring-hydroxylating dioxygenases occur in three different arrangements, depending on the number of subunits and the disposition of iron-sulfur clusters. They may be further subdivided on the basis of the types of iron-sulfur clusters and flavin present:

1. Two-component dioxygenases, in which the flavin and iron-sulfur cluster are combined in the same protein. Examples include benzoate 1,2-dioxygenase (EC 1.13.99.2) (119), phthalate dioxygenase (EC 1.14.12.7) (5), 4-chlorophenylacetate-3,4-dioxygenase (71, 95), and 4-sulfo-

Table 1 Multicomponent hydroxylating oxygenases

Enzyme System	Components		Prosthetic group(s)	Gene designation	Class[a]	References
Phthalate dioxygenase	Reductase	34,000	FMN, [2Fe-2S]		IA	3, 5
(*P. cepacia*)	Oxygenase	$\alpha(48{,}000)_4$	$4[2Fe\text{-}2S]_R$, 4Fe			3, 5
4-chlorophenylacetate 3,4-dioxygenase	Reductase				IA	71
(*P.* sp. CBS)	Oxygenase	$\alpha(46{,}000)_3$	3[2Fe-2S]			71, 72
4-sulfobenzoate 3,4-dioxygenase	Reductase				IA	69
(*Comamonas testosteroni* T-2)	Oxygenase	$\alpha(50{,}000)_2$				69
Benzoate dioxygenase	Reductase	37,500	FAD, 2Fe-2S	*benC* (*xylZ*)	IB	75, 79, 122
(*P. arvilla* C-1, *Acinetobacter calcoaceticus*)	Oxygenase	$\alpha(50{,}000)_3$ $\beta(20{,}000)_3$	3[2Fe-2S], 3Fe	*benAB* (*xylXY*)		79, 122
4-Methoxybenzoate O-demethylase	Reductase	42,000	FMN, [2Fe-2S]		IB	9, 91
(*P. putida* DSM No. 1868)	Putidamonooxin	$\alpha(41{,}000)_3$	[2Fe-2S], Fe			8, 9
Benzene, 1,2-dioxygenase	$Reductase_{BED}$	$\alpha(42{,}000)_2$	FAD	*bedA* (*bnzC*)	IIB	2, 38, 59, 105
(*P. putida* ML2 NCIB 12190)	$Ferredoxin_{BED}$	11,860	$[2Fe\text{-}2S]_R$	*bedB* (*bnzB*)		2, 38, 59, 105
	ISP_{BED}	$\alpha(54{,}500)_2$ $\beta(23{,}500)_2$	$2[2Fe\text{-}2S]_R$, Fe	*bedC1C2* (*bnzAB*)		2, 38, 59, 105, 126
Pyrazon dioxygenase	Reductase	67,000	FAD		IIA	93
(*Pseudomonas*)	Ferredoxin	12,000	[2Fe-2S]			93
	Oxygenase	180,000	[2Fe-2S]			93
Toluene dioxygenase	$Reductase_{TOL}$	46,000	FAD	*todA*	IIB	43, 101, 111
(*P. putida* F1)	$Ferredoxin_{TOL}$	11,900	$[2Fe\text{-}2S]_R$	*todB*		43, 101
	ISP_{TOL}	$\alpha(52{,}500)_2$ $\beta(20{,}800)_2$	$2[2Fe\text{-}2S]_R$, Fe	*todC1C2*		43, 93, 101
Naphthalene dioxygenase	$Reductase_{NAP}$	36,300	FAD, [2Fe-2S]		III	31, 50
(*P. putida* NCIB 9816)	$Ferredoxin_{NAP}$	15,300	[2Fe-2S]	*ndoA*		31, 49, 67
	ISP_{NAP}	$\alpha(55{,}000)_2$ $\beta(20{,}000)_2$	2[2Fe-2S], 2Fe	*ndoBC*		29, 31, 67

[a] The classes of dioxygenases are as defined by Batie et al (4).

benzoate-3,4-dioxygenase (69). These are Class I dioxygenases in the classification of Batie et al (4) and are subdivided into Class IA, containing FMN, and IB, containing FAD.

2. Three-component dioxygenases, in which the electron-transfer chain comprises a flavoprotein and a separate iron-sulfur protein (ferredoxin). Class IIA comprises pyrazon dioxygenase (EC 1.14.12), which has a plant-type ferredoxin as judged by its electron paramagnetic resonance (EPR) spectrum (93). Class IIB includes benzene dioxygenase (EC 1.14.12.3) (2, 39) and toluene dioxygenase (101, 102) which, as discussed later, appear to have Rieske-type iron-sulfur clusters in their ferredoxins.
3. Three-component dioxygenases, in which the electron-transfer chain comprises both an iron-sulfur flavoprotein and a ferredoxin. The sole representative, so far, of Class III is naphthalene dioxygenase (49, 50).

The differences between the dioxygenases may be related to function; the products of the three-component dioxygenases are all *cis*-dihydrodiols, whereas those of the two-component dioxygenases are either *cis*-diol carboxylic acids or catechols. In the latter case (e.g. 4-sulfobenzoate-3,4-dioxygenase), a hypothetical dihydrodiol intermediate represents a highly unstable configuration that spontaneously loses a group (sulfite or chloride) with the concomitant energetically favorable rearomatization.

RELATED SYSTEMS The electron-transfer chains of the aromatic dioxygenases may be considered in the wider context of mixed-function oxidases. This review refers to other oxygenases that have similar electron-transfer chains from NAD(P)H.

For example, 4-methoxybenzoate monooxygenase from *P. putida* (DSM No. 1868) is a closely related system that has a similar reductase chain. It does not catalyze dioxygenation of the aromatic ring (9), but as already noted, it has dioxygenase activity towards certain substrates (115). It comprises a nonheme iron-containing terminal oxygenase, which Bernhardt and coworkers (8, 109) named putidamonooxin, and a reductase that contains FMN and a [2Fe-2S] cluster. The vanillate demethylase of *Pseudomonas* strain ATCC 19151 catalyzes a similar monooxygenase reaction and consists of two similar components, the sequences of which were recently determined (16).

Another similar oxygenase is the cytochrome P-450 monooxygenase, in which the terminal oxygenase component is heme (83). All P-450 systems incorporate a pyridine nucleotide-linked flavoprotein, and some, but not all, involve ferredoxins with [2Fe-2S] clusters. They are typified by the well-studied camphor monooxygenase of *P. putida* (strain CIB), which comprises a flavoprotein (putidaredoxin reductase), a ferredoxin (trivial name putidaredoxin), and the terminal monooxygenase P-450$_{CAM}$ (28, 98). Overall, more is known about the mechanism of oxygen activation in the P-450 monooxygenases than in any other hydroxylase.

Finally, the third similar oxygenase is methane monooxygenase, in which the terminal oxygenase component is a dimeric nonheme iron cluster, probably with μ-oxo bridging ligands (34, 85, 118). This has a NADH-linked reductase comprising an iron-sulfur flavoprotein (70, 86).

We now consider the various components of the aromatic ring-hydroxylating dioxygenases, and what is known about their composition, catalytic activity and structure.

SIMPLE FLAVOPROTEINS In the three-component dioxygenase systems, the flavoprotein, the first component of the electron-transfer chain, acts as an oxidoreductase catalyzing the transfer of two electrons from NADH. The flavoprotein consists of a single polypeptide with relative molecular mass (M_r) ranging from 42 to 67 kilodaltons (kDa) (Table 2). The protein is reportedly monomeric for toluene and pyrazon dioxygenases (93, 101). It contains FAD as the only prosthetic group, which is readily lost during purification (50, 101). These flavoproteins belong to a large family of FAD-containing pyridine nucleotide oxidoreductases. They vary in size, substrate specificities, and preference for NADH or NADPH (51).

The reductase components of the benzene dioxygenase and toluene dioxygenase contain FAD as their prosthetic groups (2, 101). The flavoproteins show absorption spectra typical of FAD with maxima at around 375 and 450 nm. They function as oxidoreductases by transferring electrons from NADH to the ferredoxin components. Both enzymes are specific for NADH. By contrast, the flavoprotein of naphthalene dioxygenase (50) is significantly reduced when NADPH is substituted for NADH as the coenzyme.

The probable cofactor binding sites in the reductase proteins may be inferred from their amino acid sequences and comparison with other flavoproteins. NADH-dependent reductases of dioxygenases may be compared with the FMN-containing putidaredoxin reductase, which performs an analogous function in the cytochrome P-450–dependent camphor hydroxylase (63, 91). Other NADH-dependent flavoproteins found in *Pseudomonas* sp. include dihydrolipoamide dehydrogenase [*P. fluorescens* (7)] and the flavin-containing monooxygenases, *p*-hydroxyphenylacetate hydroxylase (87) and salicylate hydroxylase (61). Examples of NADPH-dependent flavoproteins include mercuric reductase [*P. aeruginosa* (15)], glutathione reductase [*P. aeruginosa* (N. L. Brown, cited in 96)], and *p*-hydroxybenzoate hydroxylase [*P. fluorescens* (56)]. (Note that the above-mentioned flavoprotein monooxygenases contain no metal centers.)

The proteins are each expected to contain two binding sites for ADP moieties, one in FAD and the other in NAD. For convenience, these sites are referred to as the FAD-binding and NAD-binding sites. Common to the

Table 2 Properties of the reductase component of aromatic dioxygenases

Dioxygenase	Component	Prosthetic group(s)	Absorption λ_{max} (nm)	EPR g_x, g_y, g_z	References
Benzene	A_2 (Ferredoxin $_{BED}$ reductase)	FAD	385, 425, 455		2
Toluene	$Ferredoxin_{TOL}$ reductase	FAD	372, 488, 475		101
Pyrazon	A2	FAD	450, 475		93
Naphthalene	$Ferredoxin_{NAP}$ reductase	FAD, [2Fe-2S]	278, 340, 420, 460, 540[a]		50
Benzoate	Reductase	FAD, [2Fe-2S]	273, 340, 402, 467		119
Phthalate	Reductase	FMN, [2Fe-2S]	330, 420, 462, 495[a], 530[a]	1.900, 1.949, 2.008, 2.041	4, 24
4-Chlorophenylacetate 3,4-dioxygenase	B	FMN, [2Fe-2S]	336, 394, 458	1.90, 1.94, 2.004, 2.03	71
4-Sulfobenzoate	B	FMN, [2Fe-2S]	~330, ~400, 465		69

[a] Shoulder.

FAD-binding (ADP-binding) site of the flavoprotein oxidoreductases is the highly conserved sequence

Gly-X-Gly-X-X-Gly-X-X-X-Ala-X-X-X-X-X-X-Gly,

where X represents any amino acid, situated near the amino terminus of the protein (7, 51, 99). The Gly-X-Gly-X-X-Gly sequence constitutes a tight turn between the first strand of a β-sheet and the succeeding α-helix (117). The NAD-binding (ADP-binding) domain of many enzymes has also been reported to be a $\beta\alpha\beta$-fold centered around a highly conserved Gly-X-Gly-X-X-Gly sequence (7, 96). A comparison of the primary structure of reductase$_{BED}$ and reductase$_{TOL}$ with the amino acid sequence of other flavoprotein oxidoreductases reveals two such regions as shown in Figure 1. The first sequence of Gly-X-Gly-X-X-Gly-X-X-X-Gly-X-X-X-X-X-X-Gly, which is conserved in the dioxygenase flavoproteins examined, is located close to the amino terminus of the protein and is most probably the FAD-binding site of the reductase component of the dioxygenase.

Some 140 residues along the protein from the first conserved Gly, one finds another dinucleotide-binding consensus sequence, although with significant differences (Figure 2). First, the conserved alanine residue in the dioxygenase reductase is replaced by a glycine in the other oxidoreductases. Second, a glutamate residue is always seven residues farther along the protein from the last conserved glycine. This negatively charged residue at the carboxy end of

Enzyme	Amino acid sequence
Reductase$_{BED}$	[1]MANHVAIIGNGVAGFTTAQALRAEGYEGRISLIGEEQHLP
Reductase$_{TOL}$	[1]MATHVAIIGNGVGGFTTAQALRAEGFEGRISLIGDEPHLP
LipoamideDH	[2]QKFDVVVIGAGPGGYVAAIRAAQLGLKTACIEKYIGKEGK
PutidaredoxinR	[3]ANDNVVIVGTGLAGVEVAFGLRASGWEGNIRLVGDAWVIP
ρ-OH benzoateH	[2]MKTQVAIIGAGPSGLLLGQLLHKAGIDNVILERQTPDYVL
Consensus sequence	G G G A G G
	....
Secondary structure	ββββββTαααααααααααααTTββββββ

Figure 1 Alignment of the putative FAD-binding sites of reductase$_{BED}$ and reductase$_{TOL}$ with those of other *Pseudomonas* oxidoreductases. The numbers indicate the position of the amino acid in the protein sequence. In the secondary structure, α, β, and T indicate α-helix, β-sheet, and turn, respectively. The sequences presented are: reductase$_{TOL}$ (127); lipoamide dehydrogenase (7); putidaredoxin reductase (90), and *p*-hydroxybenzoate hydroxylase (113).

the second β-strand is involved in binding the 2'-OH of the ribose of NADH (96). In NADPH-dependent oxidoreductases, this residue is replaced by an uncharged amino acid, presumably to accommodate the 2'-phosphate group on the coenzyme. The other difference in the NADP-binding domain is the replacement of the third glycine residue of the conserved trio by alanine, with another alanine four residues farther along the helix. The substitution of alanine for glycine is thought to modify the local polypeptide conformation associated with the difference in coenzyme specificity (96). These differences in the amino acid sequence could explain why NADPH cannot replace NADH as the coenzyme in the benzene and toluene dioxygenase systems (2, 101). The other salient feature of the two conserved regions in all these enzymes is their similar distance from each other, which could indicate similar overall folding of the polypeptide.

IRON-SULFUR FLAVOPROTEINS In contrast to the two-component reductase chains in which the iron-sulfur cluster resides on a separate polypeptide, dioxygenases have been isolated in which this redox center is arranged on the same polypeptide as the flavoprotein. Table 2 lists the properties of these iron-sulfur flavoprotein reductases. There are two classes, depending on whether they contain FAD or FMN. They each contain a [2Fe-2S] cluster.

Although they fulfill the same role as the two-component electron-transfer

Enzyme	Amino acid sequence
Reductase$_{BED}$	^{141}TPNTRLLIVGGGLIGCEVATTARKLGLSVTILEAGDELLV
Reductase$_{TOL}$	^{141}TSATRLLIVGGGLIGVEVATTARKLGLSVTILEAGDELLV
Lipoamide dehydrogenase *P. fluorescens*	^{178}AVPKKLGVIGAGVIGLELGSVWARLGAEVTVLEALDKFLP
A. vinelandii	^{178}NVPGKLGVIGAGVIGLELGSVWARLGAEVTVLEAMDKFLP
E. coli	^{178}EVPERLLVMGGGIIGLEMGTVYHALGSQIDVVEMFDQVIP
Consensus sequence	G G G A G E G
Secondary structure	.... ββββββTαααααααααααααTTββββββ

Figure 2 Alignment of the putative NAD-binding sites of reductase$_{BED}$ and reductase$_{TOL}$ with those of other oxidoreductases. The numbers indicate the position of the amino acid in the protein sequence. In the secondary structure, α, β, and T indicate α-helix, β-sheet, and turn, respectively. The sequences presented are: reductase$_{BED}$ (59), reductase$_{TOL}$ (127), and lipoamide dehydrogenase [*P. fluorescens* (7), *Azotobacter vinelandii* (116), *E. coli* (99)].

systems, these "true" reductases show little homology with the former. Thus, the reductase component of the benzoate 1,2-dioxygenase, although showing a high degree of immuno-cross-reactivity with the toluate dioxygenase, showed no such homology with the components of the benzene or naphthalene dioxygenases (75).

The recently determined structure of phthalate dioxygenase reductase (4) shows that the protein comprises three domains. The N-terminal domain, containing FMN, has the same structural topology as the plant ferredoxin: NADP reductase (62), consisting of a six-stranded β-barrel. The intermediate NAD-binding domain has the typical Rossman α/β fold of nucleotide-binding proteins. The C-terminal domain resembles the plant-type ferredoxins from cyanobacteria (35, 92, 108). The sequences of the iron-sulfur flavoproteins of the benzoate and toluate dioxygenases (79) show homologies with the plant-type ferredoxins, as do the reductases of vanillate demethylase (16), xylene monooxygenase (103), and methane monooxygenase (97).

The iron-sulfur flavoproteins can transfer electrons directly from NADH to an artificial electron acceptor. Thus, the reductase components of benzoate 1,2-dioxygenase, phthalate dioxygenase, and 4-chlorophenylacetate 3,4-dioxygenase reduce cytochrome *c* directly in the presence of NADH (5, 95, 119, 123). These proteins are monomeric with M_r of 34–38 kDa and contain a [2Fe-2S] cluster in addition to the flavin prosthetic group. Consequently, the absorption spectra of the oxidized proteins show absorption maxima at 340 and 460 nm. The EPR spectra of the reduced proteins show typical features of plant-type [2Fe-2S] clusters, together with a signal around $g = 2.004$ from the flavosemiquinone radical (5, 95).

Yamaguchi & Fujisawa (119) studied the iron-sulfur flavoprotein (designated NADH-cytochrome *c* reductase) of the benzoate 1,2-dioxygenase of *P. arvilla*. They succeeded in removing the [2Fe-2S] cluster by treatment with *p*-chloromercuriphenyl-sulfonate and in reconstituting it by incubation with iron, sulfide, and mercaptoethanol (121). This technique has been successfully applied to ferredoxins (e.g. 107a), but it is unusual to achieve reconstitution with an enzyme of this complexity. On removal of the [2Fe-2S] cluster, the protein retained some ability to transfer electrons from NADH to artificial acceptors such as ferricyanide (60%), but not to the dioxygenase protein (less than 1%). Full activity was restored when the [2Fe-2S] cluster was replaced.

Ferredoxin$_{NAP}$ reductase resembles the reductases described above in that it can catalyze the reduction of cytochrome *c* independently. The absorption spectrum is also similar to the spectra of the iron-sulfur flavoproteins with absorption maxima at 340, 420, and 460 nm, and a broad shoulder at 540 nm. Unlike these proteins, however, ferredoxin$_{NAP}$ reductase requires an additional protein, ferredoxin$_{NAP}$ to transfer electrons to the terminal dioxygenase component (49).

FERREDOXINS The intermediate electron-transfer carrier in the aromatic dioxygenases is a ferredoxin containing an iron-sulfur cluster of the [2Fe-2S] type. Ferredoxins are small proteins that contain iron-sulfur clusters, whose sole function is to transfer electrons (17).

Several different types of [2Fe-2S] clusters are known in proteins. The largest class consists of the plant-type ferredoxins, of which the first known example was spinach ferredoxin, found in chloroplast membranes. Similar proteins have been found in cyanobacteria, and the structures of three of these have been determined (35, 92, 108). Ferredoxins with similar properties are involved in electron transfer to certain cytochrome P-450–dependent monooxygenases; these include putidaredoxin from *P. putida* camphor monooxygenase, and adrenodoxin of the steroid monooxygenase of mammalian adrenal mitochondria. NMR studies indicate that the structures of these proteins have a similar fold (45, 82). All of these are small hydrophilic proteins containing one [2Fe-2S] cluster per molecule, serving as one-electron carriers. The [2Fe-2S] clusters contain two atoms of iron, bridged by two labile sulfide atoms, and coordinated to the protein by four sulfide ligands contributed by four cysteine residues. They are reddish in color; the visible absorption principally results from sulfur-to-Fe(III) charge-transfer transitions (81). Upon reduction, one of the iron ions is reduced to Fe(II), and the visible absorption decreases to about half.

^{57}Fe Mössbauer spectroscopy enables the investigator to distinguish the individual iron atoms in the iron-sulfur clusters. In reduced ferredoxins, the spectra of Fe(III) and Fe(II) ions can be clearly identified as quadrupole doublets with chemical shifts characteristic of their valence states (20, 27, 78). The electronic structure of the [2Fe-2S] clusters is such that the spins on the iron ions are antiferromagnetically coupled. As a result, the oxidized protein has a ground state with zero spin, and there is no electron paramagnetic resonance (EPR) spectrum. Upon reduction, one of the iron ions becomes high-spin Fe(II), and the net spin of the coupled Fe(III) and Fe(II) ions is S = 1/2. Thus, unlike most iron compounds, the [2Fe-2S] clusters give an EPR signal in the reduced state, when measured at low temperatures, which has an average g-factor of less than 2 (44). In the plant-type ferredoxins g_{av} is around 1.96.

Plant-type ferredoxin

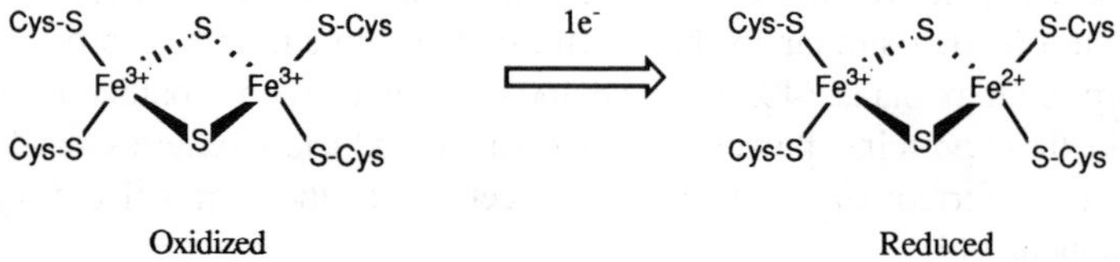

Cys-S S S-Cys
Fe^{3+} Fe^{3+}
Cys-S S S-Cys
Oxidized

$1e^-$ ⟹

Cys-S S S-Cys
Fe^{3+} Fe^{2+}
Cys-S S S-Cys
Reduced

Another class of proteins that contain [2Fe-2S] clusters comprises the Rieske proteins. They are named after J. S. Rieske, who described a protein isolated from the cytochrome bc_1 complex of mitochondria (89). Similar proteins are associated with the b_6-f complex of the thylakoid membrane of chloroplasts (58, 88) and the plasma membrane of some aerobic and photosynthetic bacteria (32, 36); for a recent review on microbial *bc* complexes, see the report by Trumpower (106). Among the distinguishing properties of the Rieske iron-sulfur clusters are red-shifted visible spectra with maximum absorbance about two-thirds those of the plant-type ferredoxins and relatively positive midpoint redox potentials. In the EPR spectra of the reduced proteins, g_{av} is lower, around 1.91, with a characteristic sharp derivative peak at $g \simeq 2.02$ and a broad trough around $g \simeq 1.80$. The plant-type and Rieske-type [2Fe-2S] clusters can be distinguished using several different spectroscopic methods, including optical absorption, EPR, Mössbauer, resonance Raman, and X-ray absorption. As explained below, there is evidence that two of the ligands to the Rieske iron-sulfur cluster are the nitrogens of histidine residues (33, 48).

The ferredoxins of the ring-hydroxylase dioxygenase systems are small acidic proteins of M_r ranging from 12 to 15 kDa, containing 2 gram atoms of iron and acid-labile sulfide per mole (Table 3). The ferredoxins are one-electron acceptors (25, 102). In addition to transferring electrons from the flavoprotein to the terminal oxygenase component, the ferredoxin catalyzes a flavoprotein-dependent reduction of other electron acceptors such as cytochrome *c* (2, 102).

Gibson and coworkers found that analogous electron-transfer proteins, such as spinach ferredoxin, putidaredoxin, adrenodoxin, and the 2[4Fe-4S] ferredoxin from *Clostridium pasteurianum,* cannot replace ferredoxin$_{TOL}$ in either cytochrome *c* reduction or toluene oxidation (43, 102). Similarly, ferredoxin$_{TOL}$ could not substitute for ferredoxin$_{NAP}$ in the naphthalene dioxygenase system (49). Thus, the ferredoxins are apparently rather specific for the dioxygenase system, in which they act as electron-transfer proteins. This specificity is reflected in the primary structure of the proteins.

The ferredoxins of the different ring-hydroxylating dioxygenases are similar in size: ferredoxin$_{NAP}$ comprises 104 amino acids, while ferredoxin$_{BED}$ and ferredoxin$_{TOL}$ have three amino acids more. Ferredoxin$_{BED}$ and ferredoxin$_{TOL}$ are almost identical in sequence, but ferredoxin$_{NAP}$ reportedly shares only 34% homology with ferredoxin$_{BED}$ in amino acid composition (67). In all other [2Fe-2S] ferredoxins [with the exception of a [2Fe-2S] protein of *C. pasteurianum,* which appears to represent yet another family of iron-sulfur proteins (74)], the cysteine residues involved in binding the cluster are arranged as

Cys-X-X-X-Cys-X-X-Cys–29 amino acids–Cys

Table 3 Properties of the ferredoxin component of aromatic dioxygenases

Dioxygenase	Component	Prosthetic group	Absorption λ_{max} (nm)	EPR g_x, g_y, g_z	Redox potential (mV)	References
Benzene	B (Ferredoxin$_{BED}$)	[2Fe-2S]	280, 320, 456	1.834, 1.890, 2.026	−155	25, 77
Toluene	Ferredoxin$_{TOL}$	[2Fe-2S]	277, 327, 460	1.81, 1.86, 2.01	−109	102
Naphthalene	Ferredoxin$_{NAP}$	[2Fe-2S]	280, 325, 460			49
Pyrazon	B	[2Fe-2S]	411, 453	1.94, 2.02		93

where X denotes any amino acid (125). The arrangement of the cysteines in the dioxygenase ferredoxins is quite different, however. Figure 3 shows an alignment of the conserved regions in the four sequences determined so far. Because the ferredoxin serves as an intermediate electron carrier in these dioxygenases, it is conceivable that some of the conserved regions in the proteins represent domains involved in recognition of the flavoprotein and terminal oxygenase components. However, examination of the conserved regions shows that they include three cysteine and two histidine residues, suggesting that these include ligands involved in binding to the [2Fe-2S] cluster, in which case the ligation is different from that in the plant-type ferredoxins.

The EPR spectrum of the ferredoxin of the pyrazon dioxygenase resembles those of the P-450 monooxygenases such as putidaredoxin with *g*-factors of 2.02 and 1.94 (93), indicating that it is a typical [2Fe-2S] cluster with cysteine ligands. By contrast, the fact that ferredoxins of the toluene and benzene dioxygenases have unusual [2Fe-2S] clusters is shown by their spectroscopic

Ferredoxin homology

```
    45       53        63        77        87
----CTHG-  -SdGYLeG- -ECpLH-  -KALCAP- -IKtyPvKiE-          ND
    43       51        61        77        85
----CTHG-  -SeGYLdG- -ECtLH-  -KALPAC- -IKvyPiKiE-          BED1
    41       49        59        75        83
----CTHG-  -SdGYLdG- -ECtLH-  -KALPAC- -IKvfPiKvE-          BED2
    43       51        61        77        85
----CTHG-  -SdGYLdG- -ECtLH-  -KALPAC- -IKvfPiKvE-          TOD
```

Terminal oxygenase (ISP) α subunit homology

```
    81                 101
----CRHRG-             -CSYHGW-                             ND
    96                 116
----CRHRG-             -CSYHGW-                             BED1
    96                 116
----CRHRG-             -CSYHGW-                             TOD
```

Rieske iron-sulfur protein homology

```
   108                 126
----CTHLGCV-           -CPCHGS-                             CYPETC
   159                 178
----CTHLGCV-           -CPCHGS-                             SCRIP1
   174                 193
----CTHLGCV-           -CPCHGS-                             NCUCR
```

Figure 3 Conserved cysteines and histidines in the putative cluster-binding sequences of Rieske-type [2Fe-2S] clusters in ferredoxins, dioxygenases, and Rieske proteins. Ferredoxins: ND, naphthalene dioxygenase from *P. putida* 9816 (67); BED1, benzene dioxygenase from *P. putida* ML2 (77); BED2, benzene dioxygenase from *P. putida* BE81 (59); TOD, toluene dioxygenase from *P. putida* F1 (127); CYPETC, Rieske iron-sulfur protein (PetC) from the cyanobacterium *Nostoc* PCC 7906; SCRIP1, Rieske iron-sulfur protein 1 from *Saccharomyces cerevisiae* (6); NCUCR, ubiquinol-cytochrome *c* reductase from *Neurospora crassa* (55).

and redox properties (Table 3). Evidently, they have features that differ from the plant and other bacterial [2Fe-2S] ferredoxins. These include the absence of the typical doublet peak in the 410- to 463-nm region of the absorption spectrum exhibited by other well-studied [2Fe-2S] ferredoxins (81). The molar absorbance at λ_{max} = 460 nm is only about 70% of the value for plant-type ferredoxins at λ_{max} = 415 nm (102), and is nearly equivalent to those for Rieske iron-sulfur proteins (33). The midpoint redox potentials of the ferredoxins of benzene (ferredoxin$_{BED}$) and toluene dioxygenase (ferredoxin$_{TOL}$) are −155 mV and −105 mV, respectively (39, 102). This is significantly more positive than spinach ferredoxin [−420 mV (104)], adrenodoxin [−270 mV (57)], or putidaredoxin [−239 mV (46)], but more negative than the Rieske clusters, which range from −140 to +350 mV (19, 65). Furthermore, the EPR spectra of reduced ferredoxin$_{BED}$ and ferredoxin$_{TOL}$ have average *g*-factors of 1.92 and 1.84, respectively. These may be compared with g_{av} = 1.96 for most plant-type ferredoxins (11), and g_{av} = 1.91 for the Rieske-type proteins (32, 48). These properties appear more similar to those of the Rieske-type iron-sulfur proteins (32, 33) than the plant-type [2Fe-2S] ferredoxins.

The Catalytic Terminal Oxygenase Component

The terminal oxygenase, also an iron-sulfur protein, is the catalytic component of the dioxygenase enzyme. For hydroxylation of the aromatic substrate, both O_2 and Fe^{2+} must be present. Thus, in addition to a substrate binding site, the terminal dioxygenase component also contains an iron-binding site and a Rieske-type [2Fe-2S] cluster (5, 39, 73).

The terminal dioxygenase components are large oligomeric proteins of M_r 150–200 kDa (Table 4). The majority of them have two dissimilar subunits, α and β of about 50 and 20 kDa, respectively. They are organized either as $\alpha_2\beta_2$ for benzene, toluene, and naphthalene dioxygenases (42, 93, 126) or $\alpha_3\beta_3$ for benzoate dioxygenase (122). The exceptions are the terminal dioxygenase components of 4-sulfobenzoate 3,4-dioxygenase (69), 4-chlorophenylacetate 3,4-dioxygenase (72), and phthalate dioxygenase (5), which comprise two, three, and four identical monomers, respectively. All the terminal dioxygenase components have [2Fe-2S] clusters of the Rieske-type with one [2Fe-2S] cluster per $\alpha\beta$ dimer or α monomer. For the terminal dioxygenase components with an $\alpha\beta$ dimeric configuration, research has indicated that the α subunit contains the [2Fe-2S] cluster [benzoate 1,2-dioxygenase (122)], while the β subunit has been implicated as the subunit involved with substrate recognition [toluate 1,2-dioxygenase (53)].

IRON CENTER OF THE TERMINAL OXYGENASE Several aromatic dioxygenases require Fe(II) for activity. The iron is presumed to bind to a specific site in the protein, where it may be oxidized to Fe(III). The iron-

Table 4 Properties of the oxygenase component of aromatic dioxygenases

Dioxygenase	Component	Prosthetic group	Absorption λ_{max} (nm)	EPR g_x, g_y, g_z	Redox potential (mV)	References
Benzene	A_1 (ISP_{BED})	2[2Fe-2S]	326, 450, 550[a]	1.754, 1.917, 2.018	−115	25, 38, 126
Toluene	ISP_{TOL}	2[2Fe-2S]	326, 450, 550[a]			100
Naphthalene	ISP_{NAP}	2[2Fe-2S]	334, 462, 566[a]			29
Pyrazon	A_1	[2Fe-2S]	445, 545[a]	1.79, 1.91, 2.02		93
Benzoate	Oxygenase	3[2Fe-2S]	325, 464, 565[a]			120, 122
Phthalate	Oxygenase	4[2Fe-2S]	325, 460, 560[a]	1.73, 1.91, 2.01		5
4-Chlorophenylacetate	A	3[2Fe-2S]	325, 458, 564			72
4-Sulfobenzoate	A	2[2Fe-2S]	327, 467, 560			69

[a] Shoulder.

binding site is the least well understood component, so any discussion of its structure and function will necessarily be speculative. A center capable of single-electron transfer is required for efficient activation of the O_2, which has a triplet ground state. Although all the terminal dioxygenases contain Rieske-type clusters, there is no precedent for a [2Fe-2S] cluster serving as an oxygen-binding site; the sulfide is prone to oxidative damage. Analogy with other types of oxygenases indicates that oxygen activation occurs at the iron site. Several classes of oxygenase involve an iron site without an NADH-linked reductase chain. An example is protocatechuate dioxygenase from *P. aeruginosa;* the structure of this molecule has been determined using X-ray crystallography (80). The protein ligands to the iron were found to be two tyrosines and two histidines. Spectroscopy experiments revealed that tyrosine is a ligand in other nonheme iron-containing dioxygenases, including 4-hydroxyphenylpyruvate dioxygenase (14). If the iron represents the site at which O_2 reacts, it is presumed to be close to the substrate-binding site.

Amino acid sequences of the terminal oxidases of benzene, toluene, benzoate, toluate, and naphthalene dioxygenases all contain, in addition to the characteristic Rieske sequence near the N terminus (Figure 4), two conserved histidines and two conserved tyrosines near the middle of the sequence (59, 67, 79, 127). This site represents a likely binding domain for the iron center. Interestingly, the iron-containing monooxygenases for alkanes (64) and xylene (103) also have a homologous region rich in histidines and tyrosines, which is dissimilar from those of the dioxygenases. The iron-binding site presumably dictates the type of oxygenation reaction that takes place.

One should be able to observe this reaction using EPR, though for some types of ligand field, the reaction may become too broad to be detected. No EPR signals due to high-spin Fe(III) have yet been reported for the aromatic dioxygenases. The comparable iron site in 4-methoxybenzoate monooxygenase has been examined using EPR and Mössbauer spectroscopy (12, 13, 109). In the oxidized protein, the iron center is expected to be high-spin Fe(III). A distribution of a complex series of EPR signals was around $g = 6.0$ and 4.3, corresponding to high-spin Fe(III) in a tetragonal and rhombic symmetry, respectively. The $g = 6.0$ signals shifted upon addition of substrates; the $g = 4.3$ did not. A comparison of enzyme reoxidized by oxygen with enzyme oxidized anaerobically demonstrated that these signals do not correspond to Fe(III)·O_2·substrate complexes.

The mechanism of the dioxygenase reaction remains to be elucidated. A proposal by Twilfer et al (110) envisaged the activation of oxygen at the iron center by successive electron transfers from the reductase:

$$\text{Fe (III)} \xrightarrow{1\,e^-} \text{Fe (II)} \xrightarrow{O_2} \text{Fe (II)} \cdot O_2 \rightarrow \text{Fe (III)} \cdot O_2^- \xrightarrow{e^-} \text{Fe (II)} \cdot O_2^- \rightarrow \text{Fe (III)} \cdot O_2^{2-}$$

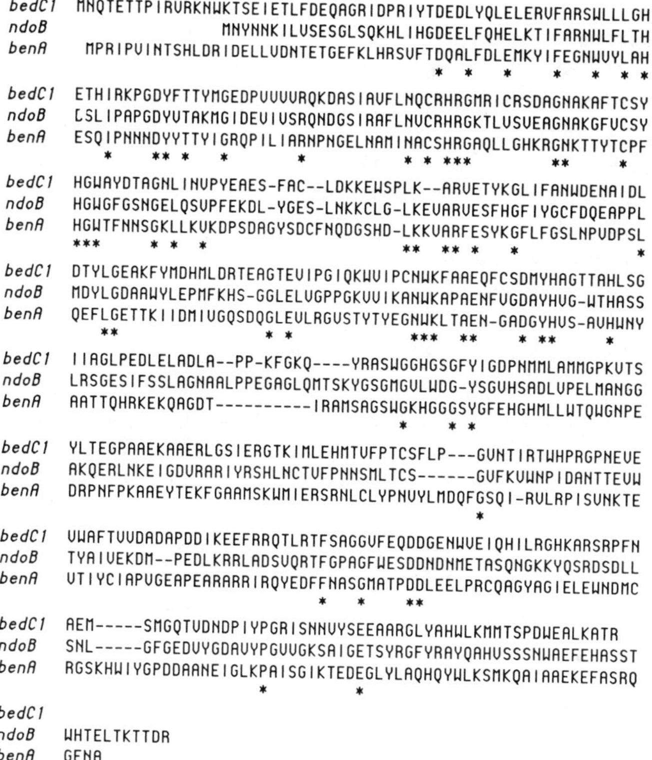

Figure 4 Comparison of the amino acid homology of the ISP α subunits from the Class IIB benzene (*bedC1*) (59), Class III, naphthalene (*ndoB*) (67), and Class IB benzoate (*benA*) (79) dioxygenases. The asterisks indicate conserved amino acids.

followed by reaction of the peroxo complex with the substrate. Twilfer et al (110) obtained a series of nitrosyl analogs of the oxo-derivatives by using nitric oxide. Since NO is an odd-electron species, it can convert Fe(II), which is normally undetectable by EPR, to an EPR-detectable species. EPR signals were observed for high-spin (S = 3/2) Fe(II)·NO (or Fe(III)·NO^-) species, with g-factors $g_\perp = 4$, $g_{||} = 2$. The g-factors of signals changed upon binding of several substrates (110).

COORDINATION OF THE IRON-SULFUR CLUSTERS The [2Fe-2S] clusters of the Rieske proteins of electron-transfer chains, and of the aromatic dioxygenases, appear to have similar chemical structures, as judged by the similarities in amino-acid sequences and spectroscopic properties (Figure 3). On this assumption, the results of the Rieske-protein studies, such as that from *Thermus thermophilus* (32), have been extrapolated to those of the dioxygenases.

The sequences for the Rieske iron-sulfur protein from various organisms have been determined by amino acid sequencing [beef heart (94)] or deduced from their DNA sequence [spinach cytochrome b_6-f complex (94), *Neurospora crassa* (55), *Saccharomyces cerevisiae* (6), *Rhodobacter capsulatus* (36), and *Paracoccus denitrificans* (68)]. Analysis of these amino acid sequences reveals that the cysteine residues are conserved in two regions (94). From a comparison of these conserved regions with those observed in the terminal oxygenase and the ferredoxin components of the dioxygenases, shown in Figure 3, the following consensus sequence emerged:

Cys-X-His–15 to 17 amino acids–Cys-X-X-His

(where X is any amino acid). Because all the above-mentioned proteins contain an iron-sulfur cluster that has Rieske-type properties, we propose that this consensus sequence is the motif involved in binding to the Rieske-type [2Fe-2S] cluster.

In the conserved regions of sequences of the Rieske proteins, each histidine residue is flanked by a single glycine residue either next to it or one residue away. Glycine residues occur in the turns of protein secondary structure (51), and the [2Fe-2S] cluster may be situated in a cleft of the protein. One of the two conserved regions in these Rieske proteins is found in a hydrophobic part of the protein and the other in a more amphipathic environment [hydrophobicity plots (94)].

Mössbauer spectra of the dioxygenase protein of benzene dioxygenase from *P. putida* ML2 (37) indicated the presence of [2Fe-2S] clusters similar to those observed in the plant-type ferredoxins. The main difference was the isomer shift of one of the iron sites, which was greater than in the ferredoxins. This effect is similar to that observed in Rieske-type proteins such as that from *Thermus thermophilus* (32). The Mössbauer parameters were compared with those of the monooxygenase ferredoxin putidaredoxin, which has a [2Fe-2S] cluster with all-sulfur ligation (78). In the reduced state, the Fe(III) site of the dioxygenase Fe-S clusters was similar to the corresponding ion in the ferredoxin, but the Fe(II) sites were different, as shown by the electric-field gradient and hyperfine tensors. This difference between the two types of protein might be explained by different distortions of the cluster geometry, or alternatively by their having different ligands. The latter interpretation considered it likely that the Fe(II) iron site had nonsulfur ligands, because in the Rieske cluster, this iron atom had a larger Mössbauer isomer shift than that in putidaredoxin (78).

Electron-spin relaxation processes are characteristic of the iron-sulfur clusters. In EPR spectra, the relaxation rate affects the temperature range over which the spectrum can be determined; in Mössbauer spectra it determines the highest temperature at which hyperfine structure is observed. It enables the reduced [2Fe-2S] clusters to be distinguished, for example, from the more rapidly relaxing [4Fe-4S] clusters. The electron-spin relaxation rate of the [2Fe-2S] clusters in the dioxygenases and the Rieske proteins is slower than in plant-type ferredoxins, but faster than in hydroxylase ferredoxins such as putidaredoxin and adrenodoxin. Bertrand et al (10) interpreted such effects in terms of differences in geometry of the [2Fe-2S] cluster.

Cline et al (23), by using electron nuclear double resonance (ENDOR) and electron spin-echo envelope modulation (ESEEM) spectroscopy, observed hyperfine interactions in the [2Fe-2S] clusters in the respiratory Rieske iron-sulfur protein of *Thermus thermophilus* and in the phthalate dioxygenase of *Pseudomonas cepacia*. All of the spectra showed resonances attributed to protons that were weakly coupled to the iron-sulfur clusters. Similar proton interactions have been found in all iron-sulfur proteins. However, in addition, unusually strong hyperfine resonances were observed, which were assigned to the ^{14}N of two histidines.

In a recent series of measurements of the phthalate dioxygenase of *P. cepacia,* Gurbiel et al (48) established the nature of the nitrogen-containing ligands to the [2Fe-2S] Rieske-type cluster. Cells of a histidine auxotroph were grown on ^{15}N-labeled histidine and ^{14}N background (using natural-abundance NH_4^+ as nitrogen source), and on ^{14}N histidine in a ^{15}N background. In spectra recorded at Q-band microwave frequency (35 GHz), the ENDOR lines due to nitrogen were clearly distinguished from those due to protons. The pairs of nitrogen hyperfine lines were assigned as resulting from two nitrogens with similar coupling constants. All of the ENDOR lines assigned to nitrogen behaved as expected for ^{15}N (nuclear spin $I = 1/2$) in the ^{15}N-histidine protein, and for ^{14}N (nuclear spin $I = 1$) in the ^{14}N-histidine protein. These results demonstrate the presence of two histidines in the first coordination sphere of iron. Furthermore, specific ^{15}N labeling showed that the δ-nitrogens of the imidazole ring bind the iron (47). Based on these results, the following structure is proposed as a model of the Rieske iron-sulfur cluster:

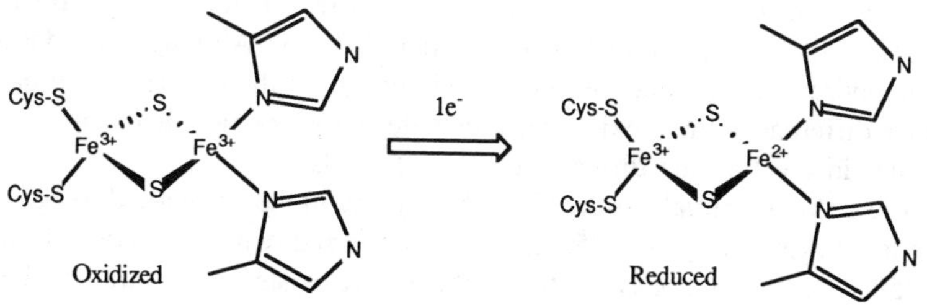

Kuila et al (66) determined the resonance Raman spectra of the Rieske protein of *T. thermophilus* and the phthalate dioxygenase of *P. cepacia*. Resonance Raman spectroscopy examines the vibrational modes of the chromophoric metal center. For the plant-type ferredoxin [2Fe-2S] cluster, the modes are as expected for a centrosymmetric cluster. For the dioxygenase cluster, the spectrum showed additional vibrational modes, indicating a major perturbation from centrosymmetry. The spectrum was analyzed in terms of the above model, with two sulfur and two nitrogen ligands in various relative positions. Of the three possible arrangements, the above, with both nitrogens on one iron atom, was considered to be the most likely.

X-ray absorption spectroscopy has been applied to the Rieske protein of heart mitochondria (84) and the terminal dioxygenase component of the phthalate dioxygenase of *P. cepacia* (107). In order to observe the Rieske cluster in the dioxygenase protein, without interference from the Fe center, the latter was eliminated by depletion of iron and replacement with cobalt or zinc. Extended X-ray absorption fine structure (EXAFS) spectra of an atom such as an iron site depend critically on distances of nearby atoms, particularly heavy atoms. They depend less on the number of atoms and the bond angles. Based on the above model, Tsang et al (107) estimated the following distances for the oxidized cluster: Fe-S (bridging), 0.22 nm; Fe-S (terminal), 0.231 nm; Fe-Fe, 0.268 nm. The Fe-Fe distance expanded slightly on reduction. Unfortunately, it was not possible to estimate the number or distances of any nitrogen ligands.

FACTORS INFLUENCING THE MIDPOINT REDOX POTENTIALS OF THE IRON-SULFUR CLUSTERS Apparently, two, or possibly three, classes of proteins contain Rieske-type [2Fe-2S] clusters. The first are the membrane-bound proteins involved in respiration and photosynthesis, which contain a single [2Fe-2S] cluster and are characterized by having an unusually high redox potential that varies with pH. The protein from *T. thermophilus* that has similar properties reportedly contains two [2Fe-2S] clusters per molecule (32) and might therefore represent another class of proteins; the sequence is not yet determined. The other group includes the NADH-dependent dioxygenase components and the 4-methoxybenzoate monooxygenase. By contrast, the negative redox potential of the [2Fe-2S] cluster in the dioxygenase components does not appear to be pH dependent (33, 66). Although the coordination environment is spectroscopically similar in all three classes of proteins, other differences must exist that modulate the properties of the iron-sulfur cluster in meeting the different functional needs.

The redox potentials of the [2Fe-2S] clusters in the aromatic dioxygenases are less negative, by 100–250 mV, than the ferredoxins of photosynthesis or P-450 systems, but not as positive as the Rieske clusters of respiratory chains. These differences should have a functional significance. In photosynthesis, a strong reductant is required to reduce NADP, while in complexes of the bc_1 type, the Rieske clusters must be reducible by quinones. The dioxygenase systems apparently need not have as negative a potential as that in the P-450_{CAM} system. This may be an advantage, as a higher-potential ferredoxin would be less likely to react with O_2 to generate superoxide.

Most iron-sulfur clusters are strong reducing agents. Their midpoint potentials are generally more negative than those of simple iron centers. This may be the consequence of the net negative charge on the clusters, which results from the charges on the cysteine ligands and bridging sulfides. However, in addition to the Rieske-type iron-sulfur proteins, other types of iron-sulfur clusters have less negative midpoint potentials, including the high-potential iron-sulfur proteins (HiPIPs), in which the cluster can take up a higher oxidation state with fewer electrons, and the [3Fe-4S] clusters. In addition, the midpoint potentials of iron-sulfur clusters of a particular type differ considerably depending on their protein environment (19).

The most important factors determining the midpoint potential are probably the electrostatic charge on the cluster and the polarity of the environment (see 76). Other more subtle factors include hydrogen bonds from the sulfides to polypeptide amides (1, 21) and the bond lengths and bond angles of the iron-sulfur cluster (81). The ligands to the Fe atoms are expected to play a major role (107). In the plant-type ferredoxins, the ligands to the [2Fe-2S] cluster are all cysteine sulfurs. The replacement of thiolate by carboxylate ligands for [4Fe-4S] clusters resulted in substantial positive shifts of the redox potential [100 mV per substituted ligand (60)]. Probably, substitutions of thiolate by nitrogen-containing residues will have a similar effect on the redox potential. The formal charge on a nitrogen ligand is 0, compared with -1 for a Cys^- ligand. The above structure would have a formal charge of 0 in the oxidized form compared to -2 for the classical $[2Fe\text{-}2S]^{2+}(CysS^-)_4$ structure. A single electron would therefore be more easily stabilized on the Rieske cluster (65). The electrostatic effect of charged amino acids around the clusters could also influence the midpoint potential. These might include histidines that are close to the cluster but not ligands to it. Notably, the lower-potential photosynthetic ferredoxins have no histidine residues in the vicinity of the cysteine residues. In this context, the surrounding amino acid residues of the conserved regions could be targets for substitution studies. Using site-directed mutagenesis, one can replace these residues systematically and examine the effects on the redox potential and its function in transferring electrons.

CONCLUDING REMARKS

The aromatic ring-hydroxylating dioxygenases and related oxygenases are built up of protein domains and redox centers. Several of these resemble other electron-transfer systems such as those of respiration and photosynthesis. However, the aromatic dioxygenases show considerable diversity (Table 1) and appear to have multiple origins. The iron-sulfur flavoprotein might have evolved by gene fusion between the separate ferredoxin and flavoprotein (18), as found in the benzene dioxygenase. However, the iron-sulfur component of the flavoproteins is related to the plant-type ferredoxins (52), while the ferredoxin components are more closely related to the Rieske proteins (Figure 3). The diversity of the electron-transfer chains in the dioxygenases so far discovered indicates that further permutations of electron carriers will almost certainly be found in others.

ACKNOWLEDGMENTS

We thank Dr. D. P. Ballou for providing manuscripts prior to publication and Dr. A. Cooke for helpful discussions. The work described from this laboratory was supported by Shell Research Ltd. (JRM) and the Science and Engineering Research Council (RC).

Literature Cited

1. Adman, E. T. 1979. A comparison of the structures of electron transfer proteins. *Biochim. Biophys. Acta* 549:107–44
2. Axcell, B. C., Geary, P. J. 1975. Purification and some properties of a soluble benzene-oxidizing system from a strain of *Pseudomonas*. *Biochem. J.* 146:173–83
3. Batie, C. J., Ballou, D. P. 1990. Phthalate dioxygenase. *Methods Enzymol.* 188:61–70
4. Batie, C. J., Ballou, D. P., Correll, C. J. 1991. Phthalate dioxygenase reductase and related flavin-iron-sulfur containing electron transferases. In *Chemistry and Biochemistry of Flavoenzymes*, ed. F. Müller, pp. 544–54. Boca Raton: CRC Press
5. Batie, C. J., LaHaie, E., Ballou, D. P. 1987. Purification and characterization of phthalate oxygenase and phthalate oxygenase reductase from *Pseudomonas cepacia*. *J. Biol. Chem.* 262:1510–18
6. Beckmann, J. D., Ljungdahl, P. O., Lopez, J. L., Trumpower, B. L. 1987. Isolation and characterization of the nuclear gene encoding the Rieske iron-sulfur protein (RIP1) from *Saccharomyces cerevisiae*. *J. Biol. Chem.* 262:8901–9
7. Benen, J. A. E., Van Berkel, W. J. H., Van Dongen, W. M. A. M., Müller, F., De Kok, A. 1989. Molecular cloning and sequence determination of the *lpd* gene encoding lipoamide dehydrogenase from *Pseudomonas fluorescens*. *J. Gen. Microbiol.* 135:1787–97
8. Bernhardt, F.-H., Heyniann, E., Traylor, P. S. 1978. Chemical and spectral properties of putidamonooxin, the iron-containing and acid-labile-sulfur-containing monooxygenase from *Pseudomonas putida*. *Eur. J. Biochem.* 92:209–23
9. Bernhardt, F.-H., Pachowsky, H., Staudinger, H. 1975. A 4-Methoxybenzoate O-demethylase from *Pseudomonas putida*. A new type of monooxygenase. *Eur. J. Biochem.* 57:241–56
10. Bertrand, P., Gayda, J.-P., Fee, J. A., Kuila, D., Cammack, R. 1987. Comparison of the spin-lattice relaxation properties of the two classes of [2Fe-2S] clusters in proteins. *Biochim. Biophys. Acta* 916:24–28
11. Bertrand, P., Guigliarelli, B., Gayda, J. P., Beardwood, P., Gibson, J. 1985. A ligand-field model to describe a new class of 2Fe-2S clusters in proteins and their synthetic analogues. *Biochim. Biophys. Acta* 831:261–66
12. Bill, E., Bernhardt, F.-H., Trautwein, A. X. 1981. Mössbauer studies on the active Fe. . .[2Fe-2S] site of putidamonooxin, its electron transport and dioxygen activation mechanism. *Eur. J. Biochem.* 121:39–46
13. Bill, E., Bernhardt, F.-H., Trautwein, A. X., Winkler, H. 1985. Mössbauer investigation of the cofactor iron of puti-

damonooxin. *Eur. J. Biochem.* 147: 177–82

14. Bradley, F. C., Lindstedt, S., Lipscomb, J. D., Que, L., Roe, L., Rundgren, M. 1986. 4-Hydroxyphenylpyruvate dioxygenase is an iron tyrosinate protein. *J. Biol. Chem.* 261:11693–96
15. Brown, N. L., Ford, S. J., Pridmore, R. D., Fritzinger, D. C. 1983. Nucleotide sequence of a gene from the *Pseudomonas* transposon Tn*501* encoding mercuric reductase. *Biochemistry* 22:4089–95
16. Brunel, F., Davison, J. 1988. Cloning and sequencing of *Pseudomonas* genes encoding vanillate demethylase. *J. Bacteriol.* 171:4924–30
17. Bruschi, M., Guerlesquin, F. 1988. Structure, function and evolution of bacterial ferredoxins. *FEMS Microbiol. Rev.* 54:155–76
18. Cammack, R. 1983. Evolution and diversity in the iron-sulphur proteins. *Chem. Script.* 21:87–95
19. Cammack, R. 1984. Midpoint potentials of iron-sulphur proteins—a survey. In *Charge and Field Effects in Biosystems,* ed. M. J. Allen, P. N. R. Usherwood, pp. 41–51. Tunbridge Wells: Abacus
20. Cammack, R., Rao, K. K., Hall, D. O., Johnson, C. E. 1971. Mössbauer studies of adrenodoxin. The mechanism of electron transfer in a hydroxylase iron-sulphur protein. *Biochem. J.* 125:849–56
21. Carter, C. W., Jr. 1977. New stereochemical analogies between iron-sulfur electron transfer proteins. *J. Biol. Chem.* 252:7802–11
22. Clarke, P. H., Ornston, L. N. 1975. Metabolic pathways and regulation. In *Genetics and Biochemistry of* Pseudomonas, ed. P. H. Clarke, M. H. Richmond, pp. 191–261. London: Wiley
23. Cline, J. F., Hoffman, B. M., Mims, W. B., LaHaie, E., Ballou, D. P., Fee, J. A. 1985. Evidence for N coordination to Fe in the (2Fe-2S) clusters of Thermus Rieske protein and phthalate dioxygenase from *Pseudomonas. J. Biol. Chem.* 260:3251–54
24. Correll, C. C., Batie, C. J., Ballou, D. P., Ludwig, M. L. 1985. Crystallographic characterization of phthalate oxygenase reductase, an iron-sulphur flavoprotein from *Pseudomonas cepacia. J. Biol. Chem.* 260:14633–36
25. Crutcher, S. E., Geary, P. J. 1979. Properties of the iron-sulphur proteins of the benzene dioxygenase system from *Pseudomonas putida. Biochem. J.* 177:393–400
26. Dagley, S. 1986. Biochemistry of aromatic hydrocarbon degradation in pseudomonads. In *The Bacteria,* ed. J. R. Sokatch, 10:527–55. London: Academic
27. Dunham, W. R., Bearden, A. J., Salmeen, I. T., Palmer, G., Sands, R. H., et al. 1971. The two-iron ferredoxins in spinach, parsley, pig adrenal cortex, *Azotobacter vinelandii* and *Clostridium pasteurianum:* studies by magnetic field Mössbauer spectroscopy. *Biochim. Biophys. Acta* 253:134–41
28. Dus, K. M. 1984. Camphor hydroxylase of *Pseudomonas putida:* vestiges of sequence homology in cytochrome P-450_{CAM}, putidaredoxin, and related proteins. *Proc. Natl. Acad. Sci. USA* 81:1664–68
29. Ensley, B. D., Gibson, D. T. 1983. Naphthalene dioxygenase: purification and properties of a terminal oxygenase component. *J. Bacteriol.* 155:505–11
30. Ensley, B. D., Gibson, D. T., Laborde, A. L. 1982. Oxidation of naphthalene by a multicomponent enzyme system from *Pseudomonas* sp. strain NCIB 9816. *J. Bacteriol.* 149:948–54
31. Ensley, B. D., Haigler, B. E. 1990. Naphthalene dioxygenase from *Pseudomonas* NCIB 9816. *Methods Enzymol.* 188:46–52
32. Fee, J. A., Findling, K. L., Yoshida, T., Hille, R., Tarr, G. E., et al. 1984. Purification and characterization of the Rieske iron-sulfur protein from *Thermus thermophilus*. Evidence for a [2Fe-2S] cluster having non-cysteine ligands. *J. Biol. Chem.* 259:124–33
33. Fee, J. A., Kuila, D., Mather, M. W., Yoshida, T. 1986. Respiratory proteins from extremely thermophilic bacteria. *Biochim. Biophys. Acta* 853:153
34. Fox, B. C., Surerus, K. K., Munck, E., Lipscomb, J. D. 1988. Evidence for μ-oxo-bridged binuclear iron cluster in the hydroxylase component of methane monooxygenase. Mössbauer and EPR studies. *J. Biol. Chem.* 263:10553–56
35. Fukuyama, K., Hase, T., Matsumoto, S., Tsukihara, T., Katsube, Y., et al. 1980. Structure of *S. platensis* [2Fe-2S] ferredoxin and evolution of the chloroplast-type ferredoxins. *Nature* 286:522–24
36. Gabellini, N., Sebald, W. 1986. Nucleotide sequence and transcription of the *fbc* operon from *Rhodopseudomonas sphaeroides*. Evaluation of the deduced amino acid sequences of the FeS protein, cytochrome *b* and cytochrome c_1. Eur. J. Biochem. 154:569–79
37. Geary, P. J., Dickson, D. P. E. 1981. Mössbauer spectroscopic studies of the terminal dioxygenase from *Pseudomonas putida. Biochem. J.* 195:199–203

38. Geary, P. J., Mason, J. R., Joannou, C. L. 1990. Benzene dioxygenase from *Pseudomonas putida* (NCIB 12190). *Methods Enzymol.* 188:52–60
39. Geary, P. J., Saboowalla, F., Patil, D. S., Cammack, R. 1984. An investigation of the iron-sulphur proteins of benzene dioxygenase from *Pseudomonas putida* by electron-spin-resonance spectroscopy. *Biochem. J.* 217:667–73
40. Gibson, D. T., Cardini, G. E., Maseles, F. C., Kallio, R. E. 1970. Incorporation of oxygen-18 into benzene by *Pseudomonas putida*. *Biochemistry* 9:1631–35
41. Gibson, D. T., Koch, J. R., Kallio, R. E. 1968. Oxidative degradation of aromatic hydrocarbons by microorganisms. I. Enzymatic formation of catechol from benzene. *Biochemistry* 7:2653–62
42. Gibson, D. T., Subramanian, V. 1984. Microbial degradation of aromatic hydrocarbons. In *Microbial Degradation of Organic Compounds,* ed. D. T. Gibson, pp. 181–252. New York: Marcel Dekker
43. Gibson, D. T., Yeh, W.-K., Liu, T.-N., Subramanian, V. 1982. Toluene dioxygenase: a multicomponent enzyme system from *Pseudomonas putida*. In *Oxygenases and Oxygen Metabolism,* ed. M. Nozaki, S. Yamamoto, Y. Ishimura, M. J. Coon, L. Ernster, R. W. Estabrook, pp. 51–62. London: Academic
44. Gibson, J. F., Hall, D. O., Thornley, J. H. M., Whatley, F. R. 1966. The iron complex in spinach ferredoxin. *Proc. Natl. Acad. Sci. USA* 56:987–90
45. Greenfield, N. J., Wu, X. H., Jordan, F. 1989. Proton magnetic resonance spectra of adrenodoxin—features of the aromatic region. *Biochim. Biophys. Acta* 995:246–54
46. Gunsalus, I. C., Lipscomb, J. D. 1973. Structure and reactions of a microbial monooxygenase: the role of putidaredoxin. See Ref. 69a, pp. 151–71
47. Gurbiel, R. J., Ohnishi, T., Robertson, D. E., Daldal, F., Hoffman, B. M. 1991. Q-band ENDOR spectra of the Rieske protein from *Rhodobactor capsulatus* ubiquinol cytochrome-*c* oxidoreductase show 2 histidines coordinated to the [2Fe-2S] cluster. *Biochemistry* 30:11579–84
48. Gurbiel, R. J., Batie, C. J., Sivaraja, M., True, A. E., Fee, J. A., et al. 1989. Electron nuclear double resonance spectroscopy of N-15-enriched phthalate dioxygenase from *Pseudomonas cepacia* proves that 2 histidines are coordinated to the [2Fe-2S] Rieske-type clusters. *Biochemistry* 28:4861–71
49. Haigler, B. E., Gibson, D. T. 1990. Purification and properties of ferredoxin$_{NAP}$, a component of naphthalene dioxygenase from *Pseudomonas* sp. strain NCIB 9816. *J. Bacteriol.* 172:465–68
50. Haigler, B. E., Gibson, D. T. 1990. Purification and properties of NADH-ferredoxin$_{NAP}$ reductase, a component of naphthalene dioxygenase from *Pseudomonas* sp. strain NCIB-9816. *J. Bacteriol.* 172:457–64
51. Hanukoglu, I., Gutfinger, T. 1989. cDNA Sequence of adrenodoxin reductase—identification of NADP-binding sites in oxidoreductases. *Eur. J. Biochem.* 180:479–84
52. Harayama, S., Polissi, A., Rekik, M. 1991. Divergent evolution of chloroplast-type ferredoxins. *FEBS Lett.* 285:85–88
53. Harayama, S., Rekik, M., Timmis, K. N. 1986. Genetic analysis of a relaxed substrate specificity aromatic ring dioxygenase, toluate 1,2-dioxygenase, encoded by TOL plasmid pWWO of *Pseudomonas putida*. *Mol. Gen. Genet.* 202:226–34
54. Harayama, S., Timmis, K. N. 1989. Catabolism of aromatic hydrocarbons by *Pseudomonas*. In *Genetics of Bacterial Diversity,* ed. D. A. Hopwood, K. F. Chater, pp. 151–74. London: Academic
55. Harnisch, U., Weiss, H., Sebald, W. 1985. The primary structure of the iron-sulphur subunit of ubiquinol-cytochrome *c* reductase from *Neurospora,* determined by cDNA and gene sequencing. *Eur. J. Biochem.* 149:95–99
56. Howell, L. G., Spector, T., Massey, V. 1972. Purification and properties of p-hydroxybenzoate hydroxylase from *Pseudomonas fluorescens*. *J. Biol. Chem.* 247:4340–50
57. Huang, J. J., Kimura, T. 1973. Studies on adrenal steroid hydroxylases. Oxidation-reduction properties of adrenal iron-sulfur protein (adrenodoxin). *Biochemistry* 12:406–9
58. Hurt, E., Hauska, G. 1981. A cytochrome *f/b6* complex of five polypeptides with plastoquinol-plastocyanin-oxidoreductase activity from spinach chloroplasts. *Eur. J. Biochem.* 117:591–99
59. Irie, S., Doi, S., Yorifuji, T., Takagi, M., Yano, K. 1987. Nucleotide sequencing and characterization of the genes encoding benzene oxidation enzymes of *Pseudomonas putida*. *J. Bacteriol.* 169:5174–79
60. Johnson, R. W., Holm, R. H. 1978. Reaction chemistry of the iron-sulphur protein site analogues $[Fe4S4(SR)_4]^{2-}$. Sequential thiolate ligand substitution reactions with electrophiles. *J. Am. Chem. Soc.* 100:5338–44

61. Kamin, H., White-Stevens, R. H., Presswood, R. P. 1978. Salicylate hydroxylase. *Methods Enzymol.* 53:527–43
62. Karplus, P. A., Daniels, M. J., Herriott, J. R. 1991. Atomic structure of ferredoxin-NADP+ reductase—prototype for a structurally novel flavoenzyme family. *Science* 251:60–66
63. Koga, H., Yamaguchi, E., Matsunaga, K., Aramaki, H., Horiuchi, T. 1989. Cloning and nucleotide sequences of NADH-putidaredoxin reductase gene (Cama) and putidaredoxin gene (Camb) involved in cytochrome-P-450_{CAM} hydroxylase of *Pseudomonas putida*. *J. Biochem.* 106:831–36
64. Kok, M., Oldenhuis, R., Van der Linden, M. P. G., Raatjes, P., Kingma, J., et al. 1989. The *Pseudomonas oleovorans* alkane hydroxylase gene. Sequence and expression. *J. Biol. Chem.* 264: 5435–41
65. Kuila, D., Fee, J. A. 1986. Evidence for a redox linked ionizable group associated with the [2Fe-2S] cluster of *Thermus* Rieske protein. *J. Biol. Chem.* 261:2768–71
66. Kuila, D., Fee, J. A., Schoonover, J. R., Woodruff, W. H., Batie, C. J., Ballou, D. P. 1987. Resonance Raman spectra of the [2Fe-2S] clusters of the Rieske protein from *Thermus* and phthalate dioxygenase. *J. Am. Chem. Soc.* 109:1559–61
67. Kurkela, S., Lehvaslaiho, H., Palva, E. T., Teeri, T. H. 1988. Cloning, nucleotide sequence and characterization of genes encoding naphthalene dioxygenase of *Pseudomonas putida* strain NCIB 9816. *Gene* 73:355–62
68. Kurowski, B., Ludwig, B. 1987. The genes of the *Paracoccus denitrificans* bc_1 complex. Nucleotide sequence and homologies between bacterial and mitochondrial subunits. *J. Biol. Chem.* 262:13805–11
69. Locher, H. H., Leisinger, T., Cook, A. M. 1991. 4-Sulphobenzoate 3,4-dioxygenase. Purification and properties of a desulphonative two-component enzyme system from *Comamonas testosteroni* T-2. *Biochem. J.* 274:833–42
69a. Lovenberg, A., ed. 1973. *Iron-Sulphur Proteins*. New York: Academic
70. Lund, J., Woodland, M. P., Dalton, H. 1985. Electron transfer reactions in the soluble methane monooxygenase of *Methylcoccus capsulatus*. *Eur. J. Biochem.* 147:297–305
71. Markus, A., Klages, U., Krauss, S., Lingens, F. 1984. Oxidation and dehalogenation of 4-chlorophenylacetate by a two-component enzyme system from *Pseudomonas* sp. strain CBS3. *J. Bacteriol.* 160:618–21
72. Markus, A., Krekel, D., Lingens, F. 1986. Purification and some properties of component A of the 4-chlorophenylacetate 3,4-dioxygenase from *Pseudomonas* species strain CBS. *J. Biol. Chem.* 261:12883–88
73. Mason, J. R. 1988. Oxygenase catalyzed hydroxylation of aromatic compounds: simple chemistry by complex enzymes. *Int. Ind. Biotechnol.* 8:19–24
74. Meyer, J., Bruschi, M., Bonicel, J. J., Bovier-Lapierre, G. E. 1986. Amino acid sequence of [2Fe-2S] ferredoxin from *Clostridium pasteurianum*. *Biochemistry* 25:6054–61
75. Moodie, F. D. L., Woodland, M. P., Mason, J. R. 1990. The reductase component of the chromosomally encoded benzoate dioxygenase from *Pseudomonas putida* C-1 is immunologically homologous with a product of the plasmid encoded *xylD* gene (toluate dioxygenase) from *Pseudomonas putida* mt-2. *FEMS Microbiol. Lett.* 71:163–67
76. Moore, G. D., Pettigrew, G. W., Rogers, N. K. 1986. Factors influencing redox potentials of electron transfer proteins. *Proc. Natl. Acad. Sci. USA* 83:4998–5000
77. Morrice, M., Geary, P., Cammack, R., Harris, A., Beg, F., Aitken, A. 1988. Primary structure of protein B from *Pseudomonas putida,* a member of a new class of 2Fe-2S ferredoxins. *FEBS Lett.* 231:336–40
78. Münck, E., Debrunner, P. G., Tsibris, J. C. M., Gunsalus, I. C. 1972. Mössbauer parameters of putidaredoxin and its selenium analogue. *Biochemistry* 11:855–63
79. Neidle, E. L., Hartnett, C., Ornston, L. N., Bairoch, A., Rekik, M., Harayama, S. 1991. Nucleotide sequences of the *Acinetobacter calcoaceticus benABC* genes for benzoate 1,2-dioxygenase reveal evolutionary relationships among multicomponent oxygenases. *J. Bacteriol.* 173:5385–95
80. Ohlendorf, D. H., Lipscomb, J. D., Weber, P. C. 1988. Structure and assembly of protocatechuate 3, 4-dioxygenase. *Nature* 336:403
81. Palmer, G. 1973. Current insights into the active center of spinach ferredoxin and other iron-sulphur proteins. See Ref. 69a, pp. 285–325
82. Pochapsky, T. C., Ye, X. M. 1991. ^{1}H identification of a β-sheet structure and description of folding topology in putidaredoxin. *Biochemistry* 30:3850–56
83. Poulos, T. L., Finzel, B. C., Gunsalus,

I. C., Wagner, G. C., Kraut, J. 1986. The 2.6 Å crystal structure of *Pseudomonas putida* cytochrome P-450. *J. Biol. Chem.* 260:16122–30

84. Powers, L., Schagger, H., VonJagow, G., Smith, J., Chance, B., Ohnishi, T. 1989. EXAFS studies of the isolated bovine heart Rieske $[2Fe\text{-}2S]^{1+(1+,\ 2+)}$ cluster. *Biochim. Biophys. Acta* 975:293–98
85. Prince, R. C., George, G. N., Savas, J. C., Cramer, S. P., Patel, R. N. 1988. Spectroscopic properties of the hydroxylase of methane monooxygenase. *Biochim. Biophys. Acta* 952:220–29
86. Prince, R. C., Patel, R. N. 1986. Redox properties of the flavoprotein of methane monooxygenase. *FEBS Lett.* 203:127–31
87. Raju, S. G., Kamath, A. V., Vaidyanathan, C. S. 1988. Purification and properties of 4-hydroxyphenylacetic acid 3-hydroxylase from *Pseudomonas putida*. *Biochem. Biophys. Res. Commun.* 154:537–43
88. Riedel, A., Rutherford, A. W., Hauska, G., Muller, A., Nitschke, W. 1991. Chloroplast Rieske protein—EPR study on its spectral characteristics, relaxation and orientation properties. *J. Biol. Chem.* 266:17838–44
89. Rieske, J. S., Hansen, R. E., Zaugg, W. S. 1964. Studies on the electron transfer system. LVIII. Properties of a new oxidation-reduction component of the respiratory chain as studied by electron paramagnetic resonance spectroscopy. *J. Biol. Chem.* 239:3017–21
90. Romeo, C., Moriwaki, N., Yasunobu, K. T., Gunsalus, I. C., Koga, H. 1987. Identification of the coding region for the putidaredoxin reductase gene from the plasmid of *Pseudomonas putida*. *J. Protein Chem.* 6:253–61
91. Roome, P. W. J., Philley, J. C., Peterson, J. A. 1983. Purification and properties of putidaredoxin reductase. *J. Biol. Chem.* 258:2593–98
92. Rypniewski, W. R., Breiter, D. R., Benning, M. M., Wesenberg, C., Oh, B. H., et al. 1991. Crystallization and structure determination to 2.5-Å resolution of the oxidized [Fe2-S2] ferredoxin isolated from *Anabaena*-7120. *Biochemistry* 30:4126–31
93. Sauber, K., Frohner, C., Rosenberg, G., Eberspacher, J., Lingens, F. 1977. Purification and properties of pyrazon dioxygenase from pyrazon-degrading bacteria. *Eur. J. Biochem.* 74:89–97
94. Schagger, H., Borchart, U., Machleidt, W., Link, T. A., Von Jagow, G. 1987. Isolation and amino acid sequence of the "Rieske" iron sulfur protein of beef heart ubiquinol: cytochrome *c* reductase. *FEBS Lett.* 219:161–68
95. Schweizer, D., Markus, A., Seez, M., Ruf, H. H., Lingens, F. 1987. Purification and some properties of component B of the 4-chlorophenylacetate 3,4-dioxygenase from *Pseudomonas* species strain CBS3. *J. Biol. Chem.* 262:9340–46
96. Scrutton, N. S., Berry, A., Perham, R. N. 1990. Redesign of the coenzyme specificity of a dehydrogenase by protein engineering. *Nature* 343:38–43
97. Stainthorpe, A. C., Lees, V., Salmond, G. P. C., Dalton, H., Murrell, J. C. 1990. The methane monooxygenase gene cluster of *Methylococcus capsulatus* (Bath). *Gene* 91:27–34
98. Stayton, P. S., Poulos, T. L., Sligar, S. G. 1989. Putidaredoxin competitively inhibits cytochrome b_5–cytochrome-P-450_{CAM} association—a proposed molecular model for a cytochrome-P-450_{CAM} electron-transfer complex. *Biochemistry* 28:8201–5
99. Stephens, P. E., Lewis, H. M., Darlison, M. G., Guest, J. R. 1983. Nucleotide sequence of the lipoamide dehydrogenase gene of *Escherichia coli* K12. *Eur. J. Biochem.* 135:519–27
100. Subramanian, V., Liu, T.-N., Yeh, W. K., Gibson, D. T. 1979. Toluene dioxygenase: purification of an iron-sulphur protein by affinity chromatography. *Biochem. Biophys. Res. Commun.* 91:1131–39
101. Subramanian, V., Liu, T.-N., Yeh, W. K., Narro, M., Gibson, D. T. 1981. Purification and properties of NADH-ferredoxin$_{TOL}$ reductase. A component of toluene dioxygenase from *Pseudomonas putida*. *J. Biol. Chem.* 256:2723–30
102. Subramanian, V., Liu, T.-N., Yeh, W. K., Serdar, C. M., Wackett, L. P., Gibson, D. T. 1985. Purification and properties of ferredoxin$_{TOL}$. A component of toluene dioxygenase from *Pseudomonas putida* F1. *J. Biol. Chem.* 260:2355–63
103. Suzuki, M., Hayakawa, T., Shaw, J. P., Rekik, M., Harayama, S. 1991. Primary structure of xylene monooxygenase: similarities to and differences from the alkane hydroxylation system. *J. Bacteriol.* 173:1690–95
104. Tagawa, K., Arnon, D. I. 1968. Oxidation-reduction potentials and stoichiometry of electron transfer in ferredoxins. *Biochim. Biophys. Acta* 153: 602–13
105. Tan, H.-M., Mason, J. R. 1990. Cloning and expression of the plasmid-encoded benzene dioxygenase genes

from *Pseudomonas putida* ML2. *FEMS Microbiol. Lett.* 72:259–64

106. Trumpower, B. L. 1990. Cytochrome bc_1 complexes of microorganisms. *Microb. Rev* 54:101–29
107. Tsang, H. T., Batie, C. J., Ballou, D. P., Penner-Hahn, J. E. 1989. X-ray absorption spectroscopy of the [2Fe-2S] Rieske cluster in *Pseudomonas cepacia* phthalate dioxygenase—determination of core dimensions and iron ligation. *Biochemistry* 28:7233–40

107a. Tsibris, J. C. M., Tsai, R. L., Gunsalus, I. C., Orme-Johnson, W. H., Hansen, R. E., Beinert, H. 1968. The number of iron atoms in the paramagnetic centre ($g = 1.94$) of reduced putidaredoxin, a nonheme iron protein. *Proc. Natl. Acad. Sci. USA* 59:959–65

108. Tsukihara, T., Fukuyama, K., Mizushima, M., Harioka, T., Kusunoki, M., et al. 1990. Structure of the [2Fe-2S] Ferredoxin-I from the blue-green alga *Aphanothece sacrum* at 2.2 Å resolution. *J. Mol. Biol.* 216:399–410
109. Twilfer, H., Bernhardt, F.-H., Gersonde, K. 1981. An electron-spin-resonance study on the redox-active centers of the 4-methoxybenzoate monooxygenase from *Pseudomonas putida*. *Eur. J. Biochem.* 119:595–602
110. Twilfer, H., Bernhardt, F. H., Gersonde, K. 1985. Dioxygen-activating iron center in putidamonooxin. Electron spin resonance investigation of the nitrosylated putidamonooxin. *Eur. J. Biochem.* 147:171–76
111. Wackett, L. P. 1990. Toluene dioxygenase from *Pseudamonas putida* F1. *Methods Enzymol.* 188:39–46
112. Wackett, L. P., Kwart, L. D., Gibson, D. T. 1988. Benzylic monooxygenation catalyzed by toluene dioxygenase from *Pseudomonas putida*. *Biochemistry* 27: 1360–67
113. Weijer, W. J., Hofsteenge, J., Vereijken, J. M., Jekel, P. A., Beintema, J. J. 1982. Primary structure of p-hydroxybenzoate hydroxylase from *Pseudomonas fluorescens*. *Biochim. Biophys. Acta* 704:385–88
114. Wende, P., Bernhardt, F.-H., Pfleger, K. 1989 Substrate-modulated reactions of putidamonooxin—the nature of the active oxygen species formed and its reaction mechanism. *Eur. J. Biochem.* 181:189–97
115. Wende, P., Pfleger, K., Bernhardt, F.-H. 1982. Dioxygen activation by putidamonooxin: substrate-modulated reaction of activated dioxygen. *Biochem. Biophys. Res. Commun.* 104:527–32
116. Westphal, A. H., DeKok, A. 1988. Lipoamide dehydrogenase from *Azotobacter vinelandii;* molecular cloning, organization and sequence analysis of the gene. *Eur. J. Biochem.* 172:299–305
117. Wierenga, R. K., Maeyer, M. C. H., Hol, W. G. J. 1985. Interaction of pyrophosphate moieties with α-helixes in dinucleotide binding proteins. *Biochemistry* 24:1346–57
118. Woodland, M. P., Patil, D. S., Cammack, R., Dalton, H. 1986. ESR studies of protein A of the soluble methane moooxygenase from *Methylococcus capsulatus* (Bath). *Biochim. Biophys. Acta* 873:237–42
119. Yamaguchi, M., Fujisawa, H. 1978. Characterization of NADH-cytochrome *c* reductase, a component of benzoate 1,2-dioxygenase system from *Pseudomonas arvilla* C-1. *J. Biol. Chem.* 253:8848–53
120. Yamaguchi, M., Fujisawa, H. 1980. Purification and characterization of an oxygenase component in benzoate 1,2-dioxygenase system from *Pseudomonas arvilla* C-1. *J. Biol. Chem* 255:5058–63
121. Yamaguchi, M., Fujisawa, H. 1981. Reconstitution of iron-sulfur cluster of NADH-cytochrome *c* reductase, a component of benzoate 1,2-dioxygenase system from *Pseudomonas arvilla* C-1. *J. Biol. Chem.* 256:6783–87
122. Yamaguchi, M., Fujisawa, H. 1982. Subunit structure of oxygenase component in benzoate 1,2-dioxygenase system from *Pseudomonas arvilla* C-1. *J. Biol. Chem.* 257:12497–12502
123. Yamaguchi, M., Yamauchi, T., Fujisawa, H. 1975. Studies on mechanism of double hydroxylation. I. Evidence for participation of NADH-cytochrome *c* reductase in the reaction of benzoate 1,2-dioxygenase (benzoate hydroxylase). *Biochem. Biophys. Res. Commun.* 67: 264–71
124. Deleted in proof
125. Yasunobu, K. T., Tanaka, M. 1973. The types, distribution in nature, structure-function, and evolutionary data of the iron-sulphur proteins. See Ref. 69a, pp. 27–130
126. Zamanian, M., Mason, J. R. 1987. Benzene dioxygenase in *Pseudomonas putida*. Subunit composition and immuno-cross-reactivity with other aromatic dioxygenases. *Biochem. J.* 244:611–16
127. Zylstra, G. J., Gibson, D. T. 1989. Toluene degradation by *Pseudomonas putida*-F1—nucleotide sequence of the *todC1C2BADE* genes and their expression in *Escherichia coli*. *J. Biol. Chem.* 264:14940–46

Annu. Rev. Microbiol. 1992. 46:307–46

EXOPOLYSACCHARIDES IN PLANT-BACTERIAL INTERACTIONS

John A. Leigh

Department of Microbiology, University of Washington, Seattle, Washington 98195

David L. Coplin

Department of Plant Pathology, The Ohio State University, Columbus, Ohio 43210-1087

KEY WORDS: *Rhizobium,* plant pathogen, root nodule, capsular polysaccharide, extracellular polysaccharide

---

CONTENTS

Abstract

Rhizobial plant symbionts and bacterial plant pathogens produce exopolysaccharides that often play essential roles in the plant interaction. Many

0066-4227/92/1001-0307$02.00

of these exopolysaccharides are acidic heteropolysaccharides that have repeating subunit structures with carbohydrate and noncarbohydrate substituents, while others are homopolysaccharides such as alginate, levan, cellulose, and glucan. While the homopolysaccharides are synthesized by mechanisms that vary with the particular polysaccharide, the heteropolysaccharides as a rule are synthesized by subunit assembly from nucleotide diphosphate-sugar precursors on a membrane-bound lipid carrier followed by polymerization and secretion. Many mutants in exopolysaccharide synthesis have been isolated, and in several cases this has led to the identification of genes that function in particular steps of biosynthesis, as well as in regulation of exopolysaccharide biosynthesis. The genetic regulation of exopolysaccharide synthesis in many plant pathogens is complex, perhaps reflecting the various niches, free living and *in planta,* in which exopolysaccharides function. In some cases, exopolysaccharide synthesis is regulated coordinately with other virulence factors, and in other cases separately. Regulatory genes that have homology to the two-component sensor and transcriptional effector systems are a common motif. In *Rhizobium* species, exopolysaccharide synthesis is regulated by transcriptional as well as posttranslational mechanisms. Exopolysaccharides function differently in the root-nodule symbiosis versus plant pathogenesis. Specific *Rhizobium* exopolysaccharide structures promote nodule development and invasion in legumes that form indeterminate nodules. In plant pathogenesis, less specific mechanisms of pathogenesis occur: exopolysaccharides cause wilting by blocking xylem vessels, are partly responsible for water-soaked lesions, and may also aid in invasion, growth, and survival in plant tissues.

INTRODUCTION

Bacterial exopolysaccharides provide important model systems for the study of molecular assembly and secretion, gene regulation, cell-cell interactions, symbiosis, and pathogenesis. Exopolysaccharides (EPSs) are carbohydrate polymers that are secreted by a wide variety of bacteria. They can remain associated with the cell wall to form a bound capsule layer, or they can be released into the cell's milieu as extracellular slime. EPSs enable free-living bacteria to adhere to and colonize solid surfaces, where nutrients accumulate (47). Capsules coat the cell and may protect it from desiccation and other environmental stresses, or, owing to the capsules' anionic nature, may help to hold minerals and nutrients near the bacterial cell (167, 184). EPSs also play important roles in pathogenesis and symbiosis. In human pathogens, they can function as adherence factors or prevent opsonization and ingestion by macrophages. EPSs are also very important in plant pathogenesis and in the *Rhizobium*-legume root nodule symbiosis. Our purpose here is to describe the structures, biosynthesis, genetics, and biological functions of EPSs that are produced by rhizobial symbionts and bacterial plant pathogens. Although we discuss a broad group of microorganisms, this review focuses on EPSs in *Rhizobium meliloti* and the wilt-inducing pathogens *Erwinia amylovora, Erwinia stewartii, Pseudomonas solanacearum,* and *Xanthomonas campestris* pv. *campestris* as model systems. Several minireviews have appeared recently (42, 81, 148).

CHEMICAL STRUCTURE OF EPS

EPSs can be either homopolymers or heteropolymers and may carry a variety of noncarbohydrate substituents. In the plant interactive bacteria, the most common homopolymers are the periplasmic β-1,2 glucans found in the *Rhizobiaceae* (90), cellulose in *Agrobacterium* (122) and *Rhizobium* (156) species, levan in *Erwinia* (11) and *Pseudomonas* (70) species, and some alginates in fluorescent *Pseudomonas* sp. (70). Levan is a β-2,6 linked polyfructan that is synthesized from sucrose by a single enzyme, levansucrase. Alginate is a copolymer of *O*-acetylated β-1,4 linked D-mannuronic acid and its C-5 epimer L-guluronic acid. The heteropolysaccharides are acidic polymers composed of linear arrangements of repeating units containing neutral sugars and uronic acids as well as noncarbohydrate substituents such as acetate, pyruvate, hydroxybutyrate, and succinate (167, 184). Figure 1 shows the repeat-unit structures of representative heteropolysaccharide EPSs.

EPSs Produced by Rhizobia

Rhizobium meliloti strain SU47 produces succinoglycan, also known as EPSa or EPS I (4, 102) (Figure 1). Strains of *Agrobacterium* and *Alcaligenes* also produce succinoglycan (89). Interestingly, a strain of *Agrobacterium radiobacter* produces an EPS with a heptadecasaccharide repeat unit that resembles two consecutive succinoglycan repeat units except for a glucose β(1-2) linked to every second glucose that is β(1-3) linked to galactose and for the positioning of pyruvate on the penultimate instead of the terminal glucose of each side chain (132). *R. meliloti* strain SU47 also can, under certain conditions (76, 195, 197), produce a different EPS termed EPSb or EPS II (88, 119) (Figure 1). *R. meliloti* strain YE-2 can produce both EPSs simultaneously, EPSb in the high-molecular-weight fraction and succinoglycan in the low-molecular-weight fraction (192). Still other strains of *R. meliloti* produce EPSs that differ from both succinoglycan and EPSb (5, 191). The *Rhizobium* sp. strain NGR234 EPS (63) (Figure 1) has a repeat unit structure that is identical to succinoglycan in its reducing terminal half.

Rhizobium leguminosarum consists of three biovars that nodulate different plant hosts. The EPSs of many strains of *R. leguminosarum* have the same

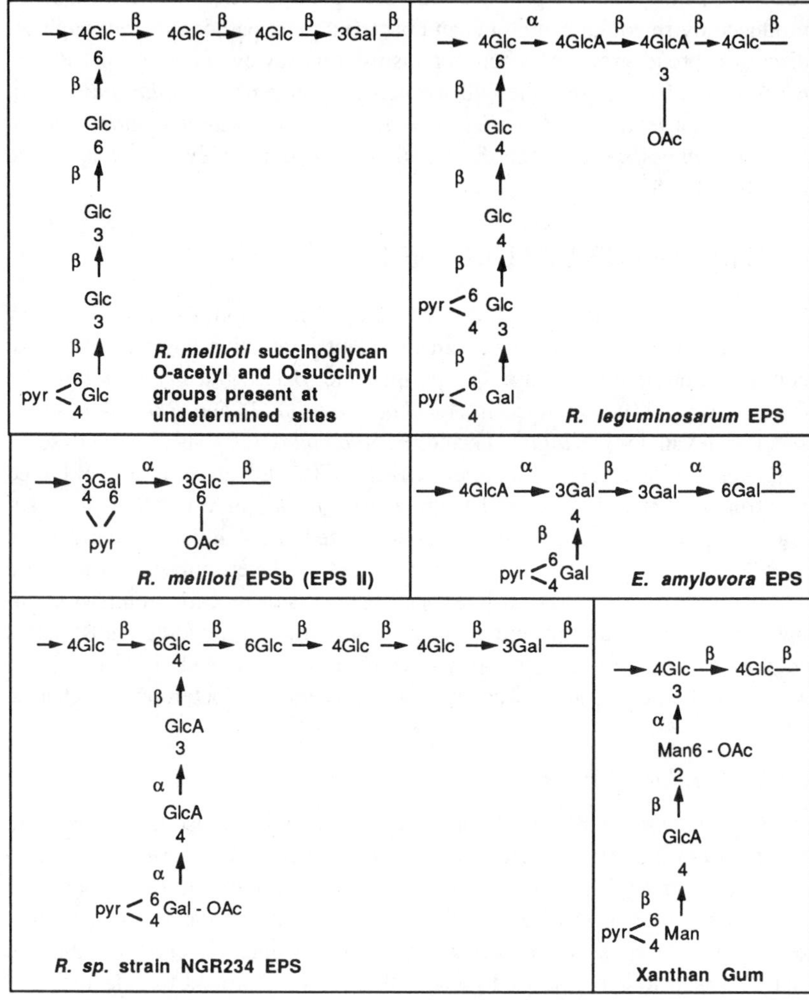

Figure 1 Repeat-unit structures of some representative heteropolysaccharide EPSs. Abbreviations are: Glc, glucose; Gal, galactose; Man, mannose; GlcA, glucuronic acid; pyr, pyruvate; OAc, *O*-acetyl.

basic structure (30, 91, 123, 150) (Figure 1) regardless of biovar. Besides those noncarbohydrate substituents shown, additional *O*-acetyl and 3-hydroxybutanoyl substituents are found. The amount and pattern of noncarbohydrate substitution may vary slightly (123), and whether this variation is correlated with biovar and hence host specificity has been controversial (30, 131, 135, 136; G. Orgambide, S. Philip-Hollingsworth, L. Cargill & F.

Dazzo, submitted). At least, the variation is not universal among all strains of *R. leguminosarum,* so EPS structure cannot be a primary determinant of host specificity. The primary determinants of host specificity are now known and are decribed elsewhere in this volume (53). Some strains of *R. leguminosarum* bv. *phaseoli* have EPS structures that have longer side chains than that shown in Figure 1 and have only one pyruvate (123). Some strains may also differ in other ways (75).

At least several species of *Rhizobium* produce EPS of a given repeat unit structure in two major molecular weight ranges, a very high-molecular-weight (HMW) polymer, and a low-molecular-weight (LMW) oligomer, both of which are found in culture supernatants (6, 62, 116, 193). LMW succinoglycan from *Rhizobium meliloti* can be further fractionated into repeat unit monomers, trimers, and tetramers, and each of these multiplicity classes may have various degrees of anionic character (9a). The biosynthetic relationships among these size classes of EPS are not clear. Some could be derived from others (e.g. LMW from HMW by hydrolysis), or they could be produced by independent processes after the repeat unit–synthesis stage. The distinctions between various forms of a given EPS become important in the consideration of their biological activities, as described below.

EPSs Produced by Plant Pathogens

ERWINIA AND OTHER ENTERIC BACTERIA Bacteria in the *Enterobacteriaceae* synthesize a wide range of capsular polysaccharides (CPS) (100, 167). *Escherichia coli* strains, alone, produce over 70 different types of capsules. A given strain, however, makes only one CPS. The CPSs made by erwinias are typical of those found in other enterics. These capsules fall into two classes distinguished on the basis of their: (*a*) acidic component, (*b*) charge density, (*c*) substitution with lipid and attachment to lipopolysaccharide (LPS) at the reducing end, and (*d*) expression at low temperature. Colanic acid (M-antigen) and the K30 CPS are typical of group I capsules, which are complex heteropolysaccharides containing hexuronic sugars that are synthesized only at low temperatures (below 30°C). They are found in a limited range of *E. coli* O serotypes and thought to be attached to lipid A. Group II CPSs are usually lower in molecular weight, substituted with phosphatidic acid at the reducing end, and may contain as major components hexuronic acids, N-acetyl neuraminic acid (NeuAc), or 2-keto-3-deoxy-octulosonic acid (KDO). They are associated with many *E. coli* O serotypes and are not produced below 20°C. The capsules produced by the plant pathogens *E. amylovora* and *E. stewartii* can be classified as group I capsular polysaccharides with respect to their composition, production at low temperatures, and molecular genetics (see below). The EPSs produced by soft-rotting erwinias have not been characterized, but the abundance of serotypes among strains of *Erwinia carotovora* and *Erwinia chrysanthemi* suggests that this group may produce both group I and group II polysaccharides.

E. amylovora produces a large [50 to 150 megadaltons (MDa)] capsular polysaccharide composed of galactose, glucuronic acid, and pyruvate. This EPS was originally named amylovorin (77), but more recently it has been called amylovoran to be more consistent with the naming of other polysaccharides. Figure 1 shows the structure of amylovoran proposed by Smith et al (157). The capsular EPS of *E. stewartii* is also very large (45 mdal), viscous, and highly charged (50, 78). Preliminary compositional analysis, ^{1}H and ^{13}C NMR, and methylation analysis of alditol acetate derivatives indicate that the polysaccharide from strain SS104 has a repeating unit of seven monosaccharides and contains glucose, galactose, and glucuronic acid in a ratio of 4:2:1 (46). Methylation studies show that the polysaccharide has a β(1-6) linked backbone of glucose and galactose with 1-4 and 1-3 branch points on one of the galactose moieties. One side chain contains glucose and the other side chain consists of a β-linked glucuronic acid with a terminal glucose. The order of glucose and galactose in the backbone has not yet been determined, nor has the orientation of linkages in the side chains. This is the only EPS that has been found in cultures of this strain.

PSEUDOMONAS AND *XANTHOMONAS* Most pseudomonads and xanthomonads produce loose slime layers but rarely form bound capsules. Pathovars of *Pseudomonas syringae,* which constitute the majority of the plant-pathogenic *Pseudomonas* sp., produce alginates and levans. On the other hand, the EPS produced by *P. solanacearum* is quite different from that produced by other plant-pathogenic bacteria and consists of a highly complex mixture of polysaccharides. Orgambide et al (133) fractionated the EPS from strain GMI 1000 into two main components: an acidic heteropolysaccharide and a mainly noncarbohydrate subfraction each comprising about 40% of the total EPS. The latter component contained mannose and glucose (4:1), but these sugars constituted only 10% of the total weight of the polymer. A mixture of glucans and a rhamnose-rich polyoside were also present as minor components. The acidic EPS is a linear polymer of N-acetylgalactosamine and two unusual sugars, 2-N-acetyl-2-deoxy-L-galacturonic acid and 2-N-acetyl-4-N-(3-hydroxybutanoyl)-2,4,6-trideoxy-D-glucose. These sugars were previously known to occur only in the LPS O-antigens of certain *Pseudomonas aeruginosa* strains. Other studies (1, 56, 66) reported rhamnose and glucose as significant components of unfractionated EPS, so the types of EPS produced under different environmental conditions by different strains may vary considerably.

The EPS produced by all *X. campestris* pathovars, known commercially as xanthan gum, is similar in chemical structure to the class 1 capsular polysaccharides produced by enterics. The polymer consists of a cellulose backbone with trisaccharide side chains (101) as shown in Figure 1.

BIOSYNTHESIS OF EPS

Biosynthesis of Levan and Alginate

Levan is the simplest EPS produced by plant pathogens. In *E. amylovora,* it is synthesized extracellularly from sucrose by one enzyme, levansucrase (83). Other simple polymers are glucans and alginates. The pathways used to produce these polysaccharides involve relatively few enzymes beyond those needed for the synthesis of nucleotide sugar precursors. Alginate biosynthesis has not been studied in plant pathogenic pseudomonads, but the pathway they use may be similar to that found in other rRNA homology group I *Pseudomonas* sp. In *P. aeruginosa,* alginate is first synthesized as a homopolymer of mannuronic acid and then is variably acetylated and epimerized (reviewed in 57). The first three steps in the synthesis of alginate lead to the synthesis of GDP-mannose. The enzymes phosphomannose isomerase (PMI), phosphomannomutase (PMM), and GDP-mannose pyrophosphorylase (GMP) are also involved in general carbohydrate metabolism and are found in many bacteria. GDP-mannose dehydrogenase (GMD), which catalyses the oxidation of GDP-mannose to GDP-mannuronic acid, is specific to this pathway and represents the primary means for channeling intermediates into alginate. Little is known about how alginate is polymerized, acetylated, and transported to the cell surface, or how random mannuronic acid residues are epimerized to guluronic acid. The synthesis of alginate differs from that of other microbial polysaccharides in that the final epimerization step takes place after export of the polymer.

Biosynthesis of Heteropolysaccharides

The structure and synthesis of the acidic heteropolysaccharide capsules produced by enteric bacteria and xanthomonads, as well as EPSs produced by rhizobia, are more complex. Most of what is known about the synthesis of heteropolysaccharide EPSs comes from studies of cell-free enzyme systems from *Klebsiella aerogenes* (176), *Xanthomonas campestris* (15, 98, 99, 182), and *R. meliloti* (172) and by analogy to similar studies on LPS synthesis in *Salmonella typhimurium* and group II capsule production in *E. coli* (176, 183). The biosynthesis of these polysaccharides has been reviewed (21, 100, 165–167, 176) and involves four stages: (*a*) synthesis of nucleotide sugar diphosphate intermediates, (*b*) stepwise assembly of the repeating oligosaccharide subunit of the polymer by transfer of monosaccharides from the corresponding nucleotide to the carrier lipid (C-55 undecaprenyl phosphate) located in the cell membrane, (*c*) addition of decorations, such as pyruvate, acetate, succinate, hydroxybutyrate, and sulfate, and (*d*) transfer of the growing polysaccharide chain from its carrier lipid to the new subunit. Very little is known about the actual mechanism of polymerization, but it is believed to occur on the inner face of the cytoplasmic membrane, after which the EPS is transported to the cell surface through Bayer adhesion zones between the outer and inner membranes (114). Thus, EPS synthesis requires enzymes for production of each nucleotide sugar precursor, separate transferases for each monosaccharide in the subunit, one or more polymerases, and proteins involved in export of the polysaccharide. Pathways for synthesis of nucleotide sugars and carrier lipid also provide key intermediates for peptidoglycan and LPS biosynthesis.

One well-characterized instance of EPS biosynthesis is that of xanthum gum. Ielpi et al (98, 99) proposed a pathway (Figure 2) based on studies of the order of nucleotide sugar incorporation into EPS using a cell-free system

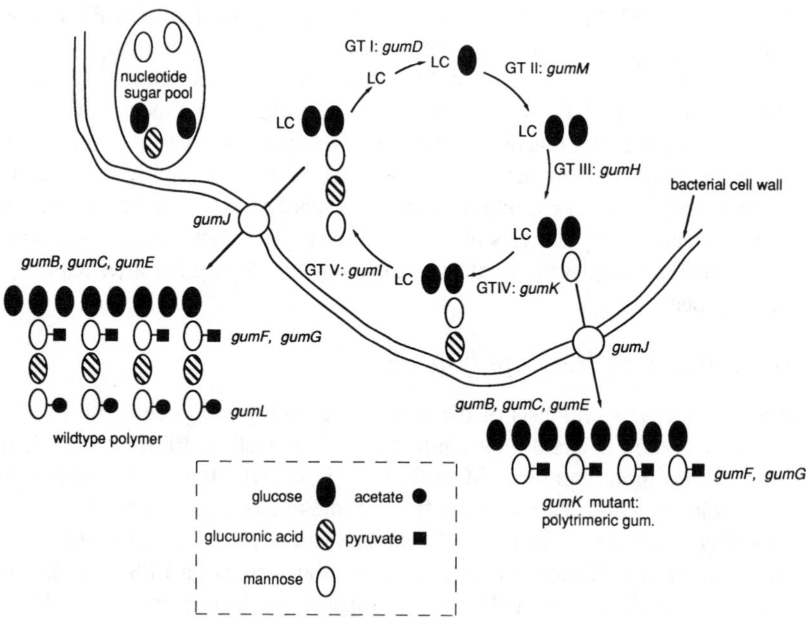

Figure 2 Simplified model of xanthan gum biosynthesis (42). Lipid-linked oligosaccharides are assembled and polymerized intracellularly from nucleotide sugars and then extruded from the cell. Abbreviations are: GT, glycosyltransferase; LC, lipid carrier. [Reproduced from Coplin & Cook (42) with permission from APS Press.]

consisting of ethylenediaminetetraacetic acid (EDTA)-treated cells. The oligosaccharide subunit is assembled upon an isoprenoid lipid carrier in the following order: Glu, Glu, Man, GlcA, and Man. The additions are accomplished by glycosyltransferases I through V, respectively. Pyruvate and acetate are added to the oligosaccharide by the corresponding ketolase and acetylase, and the polysaccharide chain is then polymerized. The xanthan molecule is thought to be added to the lipid-linked repeating unit by a tail-to-head polymerization. Finally, the carrier lipid is recycled by a dephosphorylation step and the polymer is released into the milieu as free slime.

GENETICS AND REGULATION OF EPS BIOSYNTHESIS

EPS Synthesis Genes and Their Functions

Molecular characterization of EPS genes has been aided greatly by the finding that, without exception, the genes for the biosynthetic enzymes are clustered. This type of physical arrangement is common among catabolic genes, but also extends to such idiosyncratic functions as symbiosis and pathogenicity. One may speculate that basic sets of EPS genes evolved together in gram-negative bacteria, moved by horizontal gene-exchange to new species, and then diverged to produce different polysaccharides.

ERWINIA Studies on the synthesis of class I capsular polysaccharides in *E. coli* K12 have provided a very useful model for investigating the genetics and regulation of EPS synthesis in *Erwinia* sp. and for this reason are briefly described here. The *cps* genes determine colanic acid synthesis in *E. coli*. These genes comprise six complementation groups that have been genetically mapped to two regions of the chromosome (175): *cpsA–E* are located at 44 min (adjacent to the *rfb* cluster and near *his*), and *cpsF* is at 90 min on the *E. coli* map. Only the function of *cpsB* is presently known. This gene encodes a mannose-1-phosphate guanyltransferase, which is part of a pathway for synthesis of GDP-fucose, a precursor of colanic acid (161).

As in *E. coli,* most of the *cps* genes for capsular polysaccharide biosynthesis in *E. stewartii* reside in a large cluster near *his* (65). This cluster contains at least five complementation groups, designated *cpsA* through *cpsE* (44). The *galE* gene (UDP-galactose-4-epimerase), which provides UDP-sugar precursors for EPS synthesis, is immediately adjacent to this cluster and not part of the *gal* operon. This arrangement is also found in *Vibrio cholerae* (94) and *R. meliloti* (27a, 29). Random Tn*5* mutagenesis has identified two other unlinked loci, *cpsF* and *cpsG* (D. Coplin, unpublished data). Other than *galE,* the enzymatic functions of the *cps* genes are not known. Tn*phoA* mutagenesis recently showed that *cpsB* and *cpsC* encode exported or membrane proteins (D. Coplin, unpublished data), which suggests they could be associated with the cytoplasmic membrane during synthesis of the repeating unit or involved with export of the polymer.

A similar *ams* gene cluster for the biosynthesis of amylovoran has been found in *E. amylovora* (F. Bernhard, D. Coplin & K. Geider, in preparation). The amylovoran gene cluster is ~6 kb in size and contains five complementation groups, *amsA–E*. In reciprocal complementation tests, four of the *E. amylovora ams* groups were functionally equivalent to *E. stewartii cps* groups; a mutation mapping near *cpsB* corresponded to *amsB, cpsC* to *amsA,* and *cpsD* to *amsC–E*. Mucoidy and full virulence were restored to the *E. stewartii* mutants by *ams* clones, but the cloned *cps* genes did not restore virulence to the *E. amylovora* mutants. Analysis of the EPS produced by the *ams/cps* merodiploids revealed that either mixed polymers were synthesized or alterations in EPS structure had resulted. The proportion of galactose in EPS from *E. stewartii cpsB, cpsC,* and *cpsD* mutants complemented with the *ams* genes was greatly increased, whereas the amount of galactose in total EPS from an *E. amylovora amsE* mutant complemented with the *cps* cluster was much less. Mucoidy was also restored to *E. stewartii cpsD* and *E. amylovora amsE* mutants by a clone of the *R. meliloti exoA* gene, which is believed to encode a glucosyltransferase (L. Reuber, personal communication). These results suggest that the gene products CpsB–CpsD and AmsA–AmsE are related glycosyltransferase activities, which are not specific to the synthesis of a particular polysaccharide. However, the biosynthetic cluster in the two species appears to be organized somewhat differently; the *cpsD* homologues in *E. amylovora* form three transcriptional units rather than one. *E. stewartii cpsA* and *galE* mutants were not complemented by the *ams* clone, so either these genes are absent in this cluster or the homologous enzymes are not functionally equivalent.

XANTHOMONAS CAMPESTRIS Studies on the xanthan gum of *X. campestris* pv. *campestris* provide a good model for EPS synthesis because this is one of the few systems in which both the biosynthetic genes and their enzymic products are known (13–15, 182). The *gum* genes for xanthan synthesis are physically separate from those for synthesis of nucleotide sugars. They have been cloned and are organized in a contiguous 16-kb cluster (9, 86, 182). When transferred to *Pseudomonas fluorescens* and *Pseudomonas stutzeri,* the *gum* cluster directed synthesis of a small amount of xanthan gum (13), indicating that it contains all of the genes necessary for xanthan synthesis, provided nucleotide sugar precursors are available. M. Capage and collaborators at Synergen, Inc. (personal communication) sequenced the entire region coding for synthesis of EPS from strain NRRL-B1459, and the protein products were determined (182). Twelve open reading frames (ORFs) were present and have been designated *gumB* through *gumM*. Sequence analysis and transcription mapping has revealed only one promoter region and no known internal termination signals (M. Capage, personal communication). This, together with polarity data, suggests that the region is one very large operon. Hötte et al (92) have characterized a second large gene cluster required for xanthan production, which contains 12 complementation groups. The functions of these genes are unknown, but mutations in this cluster affect other surface properties of the bacteria, so they may specify synthesis of nucleotide sugars.

The Synergen, Inc. group constructed nonpolar mutations in each open reading frame of the *gum* cluster and then characterized the altered polysaccharides and/or lipid-linked oligosaccharide intermediates produced by each mutant in a cell-free system or in vivo. This enabled them to assign functions to most of the genes of the *gum* cluster in the following manner (Figure 2): *gumD* (transferase I) mutants did not incorporate any label; *gumM* (transferase II) mutants charged the carrier lipid with only Glu; *gumH* (transferase III) mutants produced cellobiose in the lipid fraction; *gumK* (transferase IV) mutants incorporated label into both soluble and lipid fractions and produced a polytrimeric gum; and *gumI* (transferase V) mutants produced a less viscous polytetrameric gum, which lacked the terminal mannose. Transferase I, II, and III mutants did not incorporate UDP-GlcA or GDP-Man, and transferase IV mutants did not incorporate UDP-GlcA. Polymerase mutants (*gumB, gumC,* and *gumE*) incorporated all three nucleotide sugars but only into the lipid fraction. The *gumF* and *gumG* mutants were defective in acetylation and *gumL* mutants in pyruvylation. Tait & Sutherland (169) and Whitfield et al (185) have reported similar mutants producing altered xanthans. The *gumJ* mutants are blocked in a postpolymerization step in xanthan production that may be polymer export.

PSEUDOMONAS SOLANACEARUM Transposon mutagenesis has identified two major gene clusters in *P. solanacearum* (56, 159, 188). The cluster discovered by T. Denny and associates appears to encode the biosynthetic genes, whereas the second cluster investigated in L. Sequeira's laboratory may determine the synthesis of nucleotide sugars. Denny et al (56) obtained two classes of EPS-impaired mutants on rich media; class I mutants produce about 95% less EPS on both rich and minimal media and are greatly reduced in virulence, whereas class II mutants are impaired only on rich medium, produce normal amounts of EPS on minimal culture media and *in planta,* and retain virulence. The two classes of mutations have been mapped to adjacent loci (regions I and II, respectively) of the chromosome (54). Region I contains at least two genes and spans at least 9 kb. It is located 7 kb upstream from region II, which is about 2.5 kb. Two of the EPS^- Tn*5* mutants isolated by Xu et al (188) map near region I and may extend it slightly (104).

Cook & Sequeira (41) identified a second gene cluster, unlinked to regions I and II, that is required for normal EPS production and virulence. The genes in this cluster were originally designated *eps,* but they were later renamed *ops* (outer membrane polysaccharide) after it was discovered that they also affect synthesis of LPS (105). The *ops* region is 6.5 kb in size and contains seven complementation groups (41, 105). Mutations in *opsA–C* and *opsG* yield strains that produce no visible EPS, whereas *opsD* and *opsF* strains have an intermediate colony morphology. Tn*3* insertions in *opsE* reduce virulence, but do not affect colony morphology. Analysis of EPS by gas-liquid chromatography of alditol acetate derivatives revealed that all of these mutants are affected in EPS production, even those with apparently wild-type colony morphologies. Preliminary sequence analysis of *opsG* shows that it encodes two polypeptides, which have greater than 70% identity to the rhamnose synthetase of *Salmonella typhimurium* (C. C. Kao & L. Sequeira, unpublished data). These results suggest that *ops* genes may be involved in production of nucleotide sugars that are used for both LPS and EPS synthesis. Alternatively, some *ops* genes may encode transferases or transport proteins that are common to the synthesis of these polysaccharides.

ALGINATE SYNTHESIS IN PSEUDOMONADS The genes for alginate biosynthesis in *P. aeruginosa* are clustered at 34 min on the chromosome (for reviews see 57, 72). The cluster contains *algD* encoding GDP-mannose dehydrogenase (GMD), *algA* encoding a bifunctional enzyme with phosphomannose isomerase (PMI) and GDP-mannose pyrophosphorylase (GMP) activities, *algG* involved in the epimerization process, and seven other genes of unknown function. A phosphomannomutase (PMM) activity has been cloned, but it is not known if it encodes a biosynthetic or regulatory function. The *alg* genes are regulated positively by *algR* and a histone-like protein and negatively by several *muc* loci. In a study to determine the distribution of alginate genes in *Pseudomonas* and related genera, Fialho et al (72) used cloned *algA, pmm, algD* and *algR* genes as hybridization probes. Sequences hybridizing with all of the probes were found in all rRNA homology group I *Pseudomonas* species tested, which included the plant pathogens *P. chicorii, P. marginalis,* and several pathovars of *P. syringae*. These results suggest that the pathway for alginate biosynthesis and its regulation in phytopathogenic, fluorescent pseudomonads may be similar to that in *P. aeruginosa*.

RHIZOBIUM MELILOTI In *R. meliloti* SU47, two clusters of genes, one for succinoglycan synthesis (121, 147) and one for EPSb (EPS II) synthesis (76, 197), are found on the indigenous megaplasmid pRmeSU47b (pSymb) (73, 97). The succinoglycan synthesis genes are called *exo,* while EPSb synthesis genes are *exp* or *muc*. The *exo* cluster has been characterized by genetic complementation and sequencing (27a, 121, 139, 147) and consists of at least 12 genetic complementation groups (likely transcription units) within a 22 kb region (Figure 3). In addition to genes in the *exo* cluster, several loci outside of the cluster affect succinoglycan synthesis. Less is known about the *exp* cluster, which contains at least seven complementation groups (76).

Mutations in eight complementation groups of the *exo* cluster (*exoP, M, A, L, T, YF, Q,* and *B*), and in one chromosomal locus (*exoC*) completely abolish succinoglycan production; colonies are not stained with the succinoglycan-binding dye Calcofluor, and neither high-molecular-weight nor low-molecular-weight EPS can be detected in liquid culture supernatants (J. Leigh and colleagues, unpublished data). The *exoB* gene encodes the *R. meliloti* counterpart of *galE* (27a, 29), while the *exoC* locus is necessary for phosphoglucomutase activity (177). The *exoB* and *exoC* mutants are deficient in several surface carbohydrates as expected for mutants that lack precursor synthesis (76, 116, 197). ExoA seems to function as a glucosyltransferase that adds the first glucose residue following galactose to the succinoglycan repeat unit (L. Reuber, personal communication). None of the genes in the *exo* cluster appears to be a transcriptional regulator, because mutations that span the entire region have no effect on the expression of *exoY-lacZ* fusions (H. Zhan & J. Leigh, unpublished data).

Several *R. meliloti exo* genes complement and hybridize with *Rhizobium* sp. strain NGR234 *exo* genes (194). Interspecies complementation experiments showed that NGR234 *exoC* corresponds to *R. meliloti exoB* and is

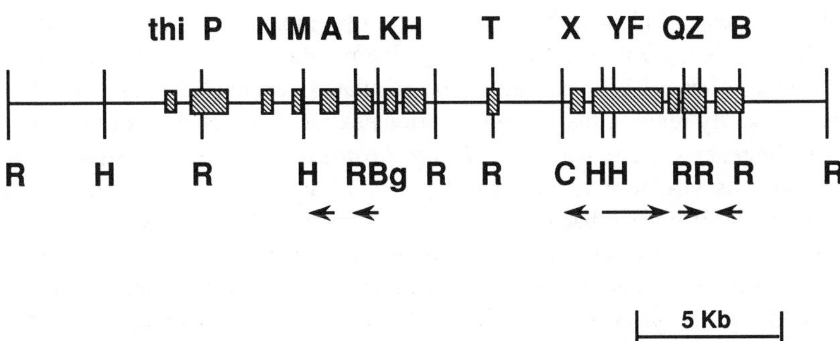

Figure 3 Genetic map of the *R. meliloti* SU47 *exo* gene cluster. The map is derived from Long et al (121), with modifications (27a, 126, 145, 147, 194, 196). For additional details of the *exoX, F, Z,* and *B* regions, see elsewhere (27a, 126). Shaded boxes represent the minimum boundaries of the *exo* loci designated above the line. Thi indicates a thiamine biosynthetic locus (73). Vertical lines indicate sites of cutting by restriction enzymes designated below the line. Abbreviations are: R, *Eco*RI; H, *Hin*dIII; Bg, *Bgl*II; and C, *Cla*I. Not all of the sites for the latter two enzymes have been determined. Horizontal arrows indicate direction of transcription.

therefore a *galE* homologue. NGR234 *exoD* is an operon that includes genes corresponding to *R. meliloti exoM, A,* and *L.* These similarities in gene function may reflect similarities in structure between the *R. meliloti* and *Rhizobium* sp. strain NGR234 EPSs (Figure 1), or similar mechanisms of polymerization and secretion. In addition, a similarity in gene organization (194) may reflect a common evolutionary origin of the two *exo* regions.

Two classes of *R. meliloti exo* mutants have been reported that secrete a structurally altered form of succinoglycan. The *exoH* mutants map to a locus within the *exo* cluster and secrete succinoglycan that lacks the succinyl substituent (117). Succinoglycan produced by *exoH* mutants is also greatly decreased in the LMW fraction (116). The causal relationship between succinylation and molecular weight is unclear. The other mutation that reportedly causes secretion of chemically altered succinoglycan, a nonpyruvylated succinoglycan, is in a chromosomal locus (126). These mutants have been useful in the analysis of structure-function relationships in nodulation (see below).

Mutations at several genetic loci produce decreased levels of succinoglycan but do not abolish it entirely. The *exoN, exoK* (121), and *exoZ* (27a) mutants map to the *exo* cluster. The *exoG* and *exoJ* mutants (121) are alleles of *exoY* and *exoX* (145) and may affect regulation of succinoglycan synthesis (see below). The *exoD* (143, 144) and *exoZ* (34) genes are outside of the *exo* cluster and may affect EPS synthesis indirectly, perhaps by altering membrane properties.

Regulatory Genes and Their Modes of Action

REGULATION OF EPS SYNTHESIS IN PLANT PATHOGENS The regulation of EPS synthesis in plant pathogenic bacteria is complex and involves multiple systems utilizing both positive and negative regulation. In *X. campestris* and *P. solanacearum,* EPS can either be regulated separately or as part of a pathogenicity regulon that includes various extracellular enzymes implicated in pathogenesis. *E. stewartii* and *E. amylovora* do not produce extracellular enzymes as virulence factors and regulate EPS by mechanisms similar to those in other enteric bacteria. A common feature of all the systems is the participation of two-component regulators consisting of an environmental sensor protein located in the cytoplasmic membrane and an effector protein located in the cytoplasm (2, 152, 162). These systems can increase EPS production to meet specific needs of the bacterium during pathogenesis or to enhance its survival in the environment. Global regulatory mechanisms also adjust EPS synthesis according to the general physiological status of the cell, turning it up under nutritional stress or down when carbon and energy must be diverted to growth and repair processes.

Erwinia Our understanding of EPS regulation in *Erwinia* sp. derives from a model proposed for colanic acid synthesis in *E. coli* (79). In this model

(Figure 4), at least two activator proteins, RcsA and RcsB, exert control over the *cps* loci. In culture, RcsA availability normally limits EPS synthesis, because RcsA is rapidly degraded by the Lon protease (173). RcsA is stabilized in *lon* mutants, and the bacterial colonies are mucoid. Similarly, overproduction of RcsA results in mucoidy of lon^{+} hosts. A two-component regulatory system consisting of an effector, RcsB, and a sensor, RcsC, provides additional control of EPS synthesis. The sequences of RcsB and RcsC share high identity with a large class of two-component prokaryotic regulators (163), so RcsC is probably a phosphorylase/kinase that activates RcsB. Complex dominance relationships among the *rcs* genes suggest that transcription of the *cps* genes is most likely initiated by an effector dimer. This might be either a phosphorylated RcsB dimer or a RcsA:RcsB dimer, but not a RcsA dimer. Because overexpression of RcsB will suppress the phenotype of *rcsA* mutations, in the presence of the appropriate environmental stimulus, activated RcsB apparently can stimulate *cps* transcription by itself. Gottesman & Stout (79) hypothesize that RcsA may therefore be an accessory transcription factor that interacts with RcsB to form a temperature-sensitive complex, thereby protecting RcsA from degradation by Lon. Moreover, the

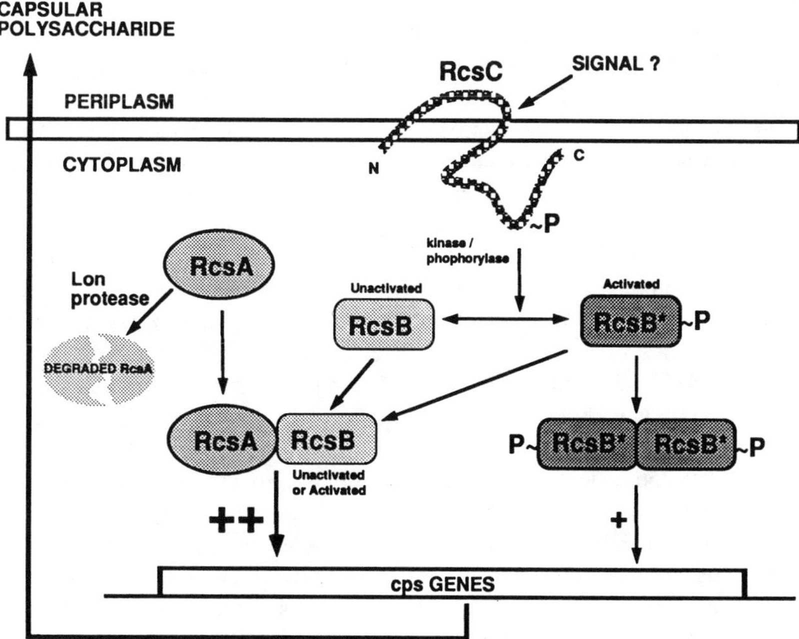

Figure 4 Proposed model for the regulation of capsular polysaccharides in *E. coli* and *Erwinia* sp. Based on the ideas of Gottesman & Stout (79).

RcsA:RcsB complex may be an alternative mechanism for activating the *cps* genes in the absence of RcsC-mediated phosphorylation of RcsB. Additional regulation of colanic acid synthesis may lie in the regulation of *rcsA* and *rcsB* transcription. The *rcsB* promoter depends on the alternate σ factor, RpoN (163), suggesting that it in turn may require an effector. The *rcsD* locus, which is near *trp,* was recently identified as a negative regulator of *cpsB* that appears to modify signal perception by RcsC (T. Kloptowksi, personal communication). The *rfa1* mutation that affects synthesis of the LPS core polysaccharide also results in mucoidy and increased *cpsB-lacZ* expression in a RcsC-dependent manner (C. T. Parker, A. W. Kloser, C. A. Schnaitman, M. A. Stein, S. Gottesman & B. W. Gibson, submitted).

The *rcsA* gene is conserved among the *Enterobacteriaceae* and has also been identified in *Klebsiella pneumoniae* (3) and *Erwinia* sp. In wild-type *E. coli* strains, *rcsA* appears to control the synthesis of several different group I capsular polysaccharides (107). Torres-Cabassa et al (174) showed that the *cpsA-cpsD* genes in *E. stewartii* are regulated by a *rcsA* gene that is functionally equivalent to *rcsA* in *E. coli*. Each cloned *E. stewartii* and *E. coli rcsA* gene complements *rcsA* mutants and activates *cps* genes in the other species. Similar interspecific complementation experiments have identified a *rcsA* gene in *E. amylovora* (12, 35, 40). *E. amylovora rcsA* mutants cannot make either amylovoran or levan (12) and do not express an *ams-lacZ* fusion (J. C. Chang & K. Geider, personal communication). Not surprisingly, the nucleotide sequences of the *Erwinia rcsA* genes (12, 137) have high identity with *rcsA* genes from *E. coli* (164) and *K. pneumoniae* (3). Comparison of all four *rcsA* sequences revealed predicted amino acid homologies ranging from 55% between *E. amylovora* and *K. pneumoniae* to 82% between *E. stewartii* and *E. amylovora* (137). Stout et al (164) conducted homology searches with the *E. coli rcsA* sequence and found that RcsA belongs to a large family of regulatory proteins that includes LuxR from *Vibrio fischeri* (87). The highest degree of homology between RcsA and the other activators was located in a region near the C-terminal end of the protein. This region is highly conserved in the proteins of the LuxR family and contains a helix-turn-helix DNA-binding motif. Interestingly, RcsB also belongs to this family and has significant homology with RcsA, suggesting that they may have similar DNA-binding properties.

The two-component, RcsB-RcsC system is also present in *E. stewartii* (45). RcsB is required for EPS synthesis and positively regulates *cps-lacZ* gene fusions in *E. stewartii*. Like RcsA, it is interchangeable with its *E. coli* homologue in reciprocal, interspecific complementation tests using *rcsB* clones and mutants. Nucleotide sequence analysis of the *rcsB* region from *E. stewartii* revealed an ORF that has 90.7% amino acid identity with *E. coli* RcsB. In addition, partial sequencing of the region adjacent to *rcsB* revealed high identity to *E. coli rcsC*. In *E. coli*, null mutations in *rcsC* lack an apparent phenotype in culture, but a recessive point mutation, *rcsC137*, results in overexpression of CPS due to accumulation of phosphorylated RcsB. The cloned *rcsB-rcsC* region from *E. stewartii* could suppress an *E. coli rcsC137* mutation *in trans*, indicating that the *E. stewartii rcsB–C* genes are both structural and functional homologues of *E. coli rcsB–C* and are similarly linked on the chromosome. The *rcsB* and *rcsC* genes have not been characterized in *E. amylovora*, but a preliminary report suggests that an *rcsB* homologue may be present in this species (124).

The model for regulation of EPS synthesis in *Erwinia* sp. and *E. coli* offers interesting possibilities for control of EPS synthesis by both physiological and environmental factors, because it features two pathways for activation of capsule synthesis that meet in a common regulatory element, RcsB. Capsule synthesis can be turned up either by an increase in the stability of RcsA or by RcsC-mediated phosphorylation of RcsB. Because null mutations in *rcsC* do not have a recognizable phenotype in *E. stewartii* or *E. coli* (45, 163), RcsA-mediated activation of capsule synthesis appears to be dominant in culture. Whether RcsC-mediated activation is increasingly important *in planta* or in the environment and what stimuli RcsC senses will be interesting to learn. Given their different life styles, *Erwinia* and *E. coli* may differ in this aspect of the *rcs* regulatory circuit.

Xanthomonas In *X. campestris* pv. *campestris*, EPS can be coordinately regulated with other pathogenicity factors. At least two positive-acting, two-component regulators and a balancing negative regulatory system control xanthan gum synthesis in a wild-type strain. Daniels et al (48) isolated a nonpathogenic mutant that was defective in production of extracellular protease and polygalacturonic acid lyase and that was severely depressed in synthesis of EPS, amylase, and endoglucanase. Cloning and genetic analysis of the corresponding locus revealed a cluster of seven positive regulatory genes (*rpfA-F*, regulation of pathogenicity factors) that have coordinate effects on all four degradative enzymes, EPS, and pathogenicity (171). The sequence of *rpfC* has strong homology with both the sensor and effector proteins of the same prokaryotic two-component regulators as do RcsC and RcsB. The N-terminal end is strongly homologous with conserved domains of the sensor proteins RcsC, EnvZ, and PhoR, and the C-terminal region contains homology to the effectors NtrC, OmpR, CheY, and PhoB.

The Rpf system appears to be balanced, in turn, by a separate, parallel negative regulatory function. Tang et al (170) cloned a locus that, when present in a multicopy plasmid, coordinately represses extracellular enzyme and polysaccharide synthesis. Disruption of this locus in the chromosome produced a pathogenic strain with elevated levels of degradative enzymes and

EPS. At the level of global regulation, a catabolite activation protein (Clp) also influences extracellular enzyme production and EPS synthesis, and Clp$^-$ mutants produced decreased amounts of an EPS with altered physical properties (51). Although Clp will suppress *crp* mutants of *E. coli,* it appears to regulate phytopathogenicity rather than carbon source catabolism in *X. campestris* pv. *campestris*.

An additional two-component system, which more specifically regulates EPS synthesis, was identified in a *X. campestris* pv. *campestris* library by means of oligonucleotide probes to conserved regions of known regulators (134). This gene pair also has extensive homology with both the sensor and effector proteins of known two-component systems. Mutants of the effector gene, constructed by marker-exchange, were affected in EPS production and resistance to salt and chloramphenicol.

Pseudomonas solanacearum Spontaneous, weakly virulent EPS-deficient strains of *P. solanacearum* are easily obtained following prolonged growth in stationary broth cultures (109). In addition to their obvious EPS deficiency, these strains are affected in other characteristics including decreased production of endoglucanase and increased polygalacturonase synthesis, indole-3-acetic acid production, pigmentation, and motility (27, 125). Brumbley & Denny (26) have termed this pleotrophic change phenotype conversion (PC). For a long time it did not appear likely that these strains could arise by a simple mutation, because reversion to wild-type has never been observed. Now, more than one type of genetic rearrangement apparently can lead to PC, and several factors may be involved. Most spontaneous PC mutants in strain AW, however, have mutations in a single positive regulatory gene, *phcA* (26). Inactivation of *phcA* results in the complete PC phenotype and decreased transcription of *eps* regions I and II and the gene for endoglucanase (95). The PhcA protein is about 38 kilodaltons (kDa) in size and has N-terminal homology to NahR, a transcriptional activator in *Pseudomonas putida* and a member of the LysR family (25).

Preliminary results from the laboratories of T. Denny and M. Schell (personal communications) show that *phcA* is part of a much more complex regulatory circuit involving an endogenous, volatile inducer; a second activator protein, XpsR; and two putative membrane-associated proteins, VsrA and VsrB. Mutant strain AW1-83 is PC-like, but when grown on split-plates with a wild-type strain, it produces normal levels of EPS and endoglucanase (38). This effect is PhcA dependent. PhcA, in turn, may act through *xpsR,* a positive regulatory gene linked to region II. The XpsR protein could be the direct transcriptional activator of the *eps* genes in region I, because increased expression of *xpsR* will override a *phcA* mutation. Tn*phoA* mutagenesis has identified two additional regulatory loci, *vsrA* and *vsrB,* that produce membrane or periplasmic proteins. VsrB may stimulate *eps* expression directly, but the effect of VsrA also seems to involve an interaction with XpsR. In contrast to the *eps* genes in regions I and II, the *ops* genes, which are required for both EPS and LPS synthesis, are expressed in a constitutive fashion (D. Cook & L. Sequeira, personal communication).

REGULATION OF EPS SYNTHESIS IN *RHIZOBIUM* EPS synthesis in *Rhizobium* spp. is regulated at both the transcriptional and posttranslational levels. Although the best-characterized transcriptional regulator acts negatively, it may be too early to stipulate that the dominant mode of regulation is negative.

Rhizobium meliloti In *R. meliloti,* a negative regulatory locus, *exoR,* affects succinoglycan synthesis (64, 146). This gene is chromosomal and unlinked to the *exo* cluster on pSymb. The negative regulatory nature of *exoR* is supported by the existence of several independent insertion mutations, all of which cause increased EPS synthesis, and by the recessive nature of the mutations. Alkaline phosphatase activities of *exo-phoA* fusions indicate that *exoR* regulates the transcription or translation of most of the genes of the *exo* cluster with the exception of the *galE* equivalent *exoB* (64, 147). The *exo* mRNA levels increase in *exoR* mutants as expected if transcriptional regulation is occurring. The *exoR* gene has been sequenced but lacks homology to known transcriptional regulators. Apparently, *exoR* mutants are insensitive to NH_4^+ inhibition of succinoglycan synthesis. ExoR could be involved directly in the response to nitrogen, or *exoR* mutants could bypass the nitrogen-sensitive step by stimulating EPS synthesis by another mechanism.

Another succinoglycan regulatory gene, *exoS,* resembles *exoR* in its lack of genetic linkage to the *exo* cluster and the effect of a mutation on succinoglycan synthesis and *exo* gene expression (64, 147). Only one mutant of *exoS* has been isolated, and the locus has not yet been thoroughly characterized.

Relatively little is known of the genetic regulation of EPSb synthesis in *R. meliloti*. Two chromosomal mutations, *mucR* (197) [probably similar to *rexA* (139)] and *expR* (76) cause increased EPSb synthesis. The *expR* mutations, at least, markedly affect EPSb-related gene expression. The *mucR* mutation was caused by transposon Tn*5* insertion and, if null, would indicate that the *mucR* locus negatively regulates EPSb synthesis. In addition, the *mucR* locus may regulate succinoglycan synthesis positively, since very little succinoglycan is produced in *mucR* mutant backgrounds (197). Besides *mucR* and *expR* mutations, EPSb synthesis also results from the presence of cosmids containing the *muc* or *exp* cluster of EPSb synthesis genes in an otherwise wild-type background (76, 197). Whether regulatory genes exist in this cluster has not been reported.

The exoX (psiA)–exoY (pssA) *posttranslational regulatory system of* Rhizobium *sp.* A novel regulatory mechanism, mediated by the *exoX* and *exoY* genes, is conserved among several species of *Rhizobium*. The *exoX* gene is found in *Rhizobium* sp. strain NGR234 (80) and *R. meliloti* (145, 196) and encodes a small protein with hydrophobic as well as hydrophilic domains. Multicopy *exoX* inhibits EPS synthesis, and *exoX* mutants overproduce EPS. ExoX appears to inhibit EPS synthesis posttranslationally, because the gene dosage of *exoX* does not alter the expression of translational fusions to other *exo* genes (145, 196). In *R. meliloti, exoX* affects succinoglycan synthesis but not EPSb synthesis (145, 195).

The inhibitory effect of *exoX*$^+$ is counterbalanced in a copy number–dependent manner by *exoY*$^+$, which stimulates EPS synthesis. Homologues of *exoY* have been found in *Rhizobium* sp. strain NGR234 (80), *R. meliloti* (145, 196), *R. leguminosarum* (16, 17, 115), *Xanthomonas campestris* (16, 145), and even *Streptococcus agalactiae* (C. Rubens, personal communication). In *Rhizobium* sp. strain NGR234, *exoY* does not affect the transcription of *exoX* (B. Rolfe, personal communication), and a posttranslational interaction of ExoX with ExoY has been proposed (81). In this model, the two proteins form a regulatory complex; when sufficient ExoX is bound to ExoY, the biosynthetic function of the latter is inhibited. This complexing could occur as part of the membrane-bound complex that synthesizes EPS, because both ExoX and ExoY have hydrophobic domains. In another type of model (145), the two proteins interact biosynthetically without necessarily complexing physically; for example, ExoY may function in the initiation of EPS synthesis by facilitating attachment of the first sugar residues to the lipid carrier, and ExoX may have the reverse effect by removing EPS subunits. As a variant on this model, ExoX could compete with ExoY by charging the lipid carrier for synthesis of a different polysaccharide. However, ExoX does not seem to be required for normal LPS, EPSb, or periplasmic glucan synthesis (82). All of these models are consistent with the observation that the ExoY homolog in *Xanthomonas campestris,* GumD, is required for the addition of the first sugar residue to the lipid carrier (145).

The same mode of regulation is found in *R. leguminosarum,* where the *psi-pss* system resembles the *exoX-exoY* system. *R. leguminosarum psiA* (18, 19) is a symbiotic plasmid-borne gene that resembles *exoX* in size, hydrophobicity pattern, and sequence similarity in a short central region, and has a similar posttranslational effect on EPS synthesis (115). Similarly, *pssA* (*pss2*) (16, 17, 115) is homologous to *exoY* and counterbalances the effect of *psiA* in a posttranslational manner. Both PsiA and PssA are cell-surface associated (115). The *pssA* gene may affect membrane targeting of a protein encoded upstream of *psiA,* though not PsiA itself, as evidenced by alkaline phosphatase activities arising from translational fusions to *phoA* (115).

Effects of nod *regulatory genes on EPS synthesis* Certain genes that affect the expression of nodule formation (*nod*) genes (see 53) also affect the level of EPS synthesis. In *R. meliloti, syrM* and *syrA* lie in the *nod* region, and plasmid-borne $syrM^+$ and $syrA^+$ together enhance EPS production (127). Plasmid-borne $syrM^+$ also works in concert with $nodD3^+$ to enhance *nod* gene expression. However, mutants in either gene are symbiotically normal on alfalfa. The *syrM-syrA* system could have evolved to assure that sufficient EPS is produced at the critical early stages of nodulation when EPS is needed. A contrasting situation was found in *Bradyrhizobium japonicum* (7), in which EPS synthesis is not symbiotically essential; plasmid-borne $nodD2^+$ appeared to inhibit EPS synthesis.

FUNCTIONS OF EPS IN SYMBIOSIS AND PATHOGENESIS

The functions of EPS in the root nodule symbiosis and plant pathogenesis may be quite different. In symbiosis, a specific signaling role appears likely, although nonspecific functions are not ruled out. In pathogenesis, little evidence suggests specificity, and the role of EPSs may be purely a function of the same physical properties that are important for survival in the environment: hygroscopy, viscosity, and charge.

Functions of EPS in Symbiosis

REVIEW OF DEVELOPMENTAL EVENTS AND SIGNALING IN THE *RHIZOBIUM*-LEGUME SYMBIOSIS Root nodulation consists of the initiation of nodule formation, subsequent nodule development, nodule invasion, and bacteroid maturation (120). After inoculation of the legume root with rhizobia, meristematic activity in the root cortex marks the beginning of nodule formation, while root hair curling may represent the first step in invasion. Invasion occurs via a plant cell wall–derived tubular infection thread that forms at the root-hair curl and grows into the root. Once inside the nodule cells, bacteria differentiate into bacteroids, their nitrogen-fixing forms. The *Rhizobium*-legume interaction involves considerable host-symbiont specificity; a given species or biovar of *Rhizobium* typically nodulates a restricted set of legumes, and a given legume is susceptible to only one or a few species or biovars of *Rhizobium*.

Separate sets of signals seem to control the initiation of nodule formation and subsequent nodule development, including invasion. Nodule formation is stimulated by modified oligosaccharides of N-acetyl glucosamine that require the *nod* gene products for their synthesis [see review in this volume (53)]. These nodule-formation signals are primary determinants of host specificity. The pea lectin can also determine specificity for *R. leguminosarum* bv. *viciae*

(58), but it is not clear what carbohydrate component of the *R. leguminosarum* bv. *viciae* cell surface is recognized. Legume lectins in many cases bind specifically to the surfaces of appropriate biovars of *Rhizobium* (85), and this may enhance the efficiency of nodulation by binding compatible bacterial cells to the root. However, the binding of lectins to EPS per se cannot in general be a primary determinant of host specificity because EPS mutants still form nodules on appropriate host plants (see below).

In contrast to nodule formation, later steps in nodule development involve a multitude of signals that include EPSs. With the exception of *syrM* and sometimes *nodD* (above), *nod* genes do not seem to affect the quantity, biochemical properties, or biological activity of EPS in *R. meliloti* (9a, 30, 131), nor do *exo* genes seem to affect *nod* gene function or the nodule-formation process (112). How EPSs function in nodule development is the subject of the following sections.

EVIDENCE FOR EPS FUNCTION The following discussion pertains to heteropolysaccharide EPSs; cellulose and periplasmic glucan are discussed subsequently. Genetic evidence indicates that *Rhizobium* EPSs are required for nodule development in plants that form indeterminant (continually elongating) nodules, but not in plants that form determinate (round) nodules. Thus, nodules formed by *exo* mutants of the appropriate species or biovars are defective on alfalfa (118, 121), clover (33), pea (16), vetch (17), and *Leucaena leucocephala* (37, 93), but are normal on bean (16), soybean (110, 113), and *Lotus* (93) as well as several tropical legumes of the determinate nodule type (37). The relationship between nodule type and the importance of EPS is underscored by instances in which the same strain or species of *Rhizobium* can nodulate plants of both nodule types. Diebold & Noel (60) constructed different biovars of *R. leguminosarum* containing identical Exo$^-$ mutations and showed that the mutations blocked nodule development on clover and pea but not on bean. Chen et al (37) found that Exo$^-$ mutants of *Rhizobium* sp. strain NGR234 formed defective nodules on *Leucaena* but normal nodules on several tropical legumes of the determinate nodule type. Hotter & Scott (93) found that Exo$^-$ mutants of *Rhizobium loti* formed defective nodules on *Leucaena* but normal nodules on *Lotus*. The basis for the difference in EPS requirement between the two nodule-types of plants is not clear. Stacey et al (158) pointed out that indeterminate nodules have broader infection threads than determinate nodules, and EPS could form an important component of the infection-thread matrix (fibrillar interior of the infection thread) in broad infection threads. However, evidence discussed below suggests that this could not be the only function of EPS in nodule invasion. Another important difference is that bacteria in indeterminant nodules spread into new cells by continued infection-thread penetration, while in determinate nodules, bacteria spread primarily by the division of cells already containing bacteria. Therefore, the difference in the EPS requirement may reflect the function of EPS in infection-thread penetration as described below.

DEVELOPMENTAL STAGE OF EPS FUNCTION The developmental abnormality observed with *R. meliloti exo* mutants on alfalfa is manifested at several morphological (117, 118, 189) and molecular levels. Some of these abnormalities may be direct consequences of EPS deficiency, while others may represent later consequences of an earlier block in the developmental pathway: root-hair curling is delayed; infection threads form but do not penetrate the nodule; structures are observed around bacteria and aborted infection threads that suggest plant defense responses (139); the nodules that form are fully differentiated but are not elongate and lack a discrete, persistent meristem (189); once nodules have formed, only 2 nodulins (nodule-specific plant-gene products) are found, compared with approximately 18 in normal nodules (59, 117, 129). These abnormalities are consistent for all *exo* mutants tested, including the *exoH* mutant that produces nonsuccinylated succinoglycan. Although all of these defects are consequences of EPS deficiency, it is not known which are the most immediate consequences and therefore mark a step in nodulation that may occur as a direct response to EPS. Yang et al (189) pointed out that the lack of a discrete persistent meristem cannot result solely from the lack of invasion because certain empty nodules that form in the absence of bacteria do have discrete persistent meristems.

Although EPS and nodule formation signals are generated independently, they may work cooperatively during invasion of alfalfa nodules. In coinoculation experiments, *nod* mutants together with *exo* mutants invaded nodules only when cell-to-cell contacts were allowed (106). This observation was interpreted to mean that the coinoculants invaded in close association with one another.

Although the defect in nodule development is generally consistent for all *exo* mutants of a given species on a given plant, it varies with the particular *Rhizobium*-plant pair. Thus, in contrast to *R. meliloti* and alfalfa, *R. leguminosarum* Exo$^-$ mutants fail to form any visible nodules on peas (16, 60); *Rhizobium* sp. strain NGR234 *exo* mutants form undifferentiated calluses on *Leucaena* (37); and *R. leguminosarum* bv. *trifolii exo* mutants form nodules on clover that do not elongate but seem to proceed weakly to the bacterial release stage (33, 60). The role of EPS may be entirely different in each *Rhizobium*-plant pair. However, it is equally likely that a similar mechanism operates in each case and that the affected stage in nodule development varies.

Although no evidence for specific regulation of EPS synthesis by the plant has been obtained, it does appear that improper regulation or overproduction of EPS synthesis during invasion can be deleterious. *R. meliloti exoR* mutants that overproduce succinoglycan give rise to normal nodules, but only pseudorevertants that produce normal levels of succinoglycan reach the interior of the nodule (64). The *exoS* and *exoX* mutants, which overproduce succinoglycan but less markedly than *exoR* mutants, form normal nodules without any apparent genetic alteration (64, 145, 196). Similar to *exoR* in *R. meliloti, psiA* mutants in *R. leguminosarum* bv. *phaseoli* are defective in invasion of bean (115).

Once nodule development has proceeded past the point of invasion, EPS production does not appear to be necessary in alfalfa (106, 196). Indeed, *exo* gene expression seems to decrease after invasion. Although isolated bacteroids exhibited *exo* gene expression (108), in situ measurements showed that repression of *exo* gene expression occurs after invasion (147). Similarly, *pssA* gene expression was not detected in bean nodules induced by *R. leguminosarum* bv. *phaseoli* (115).

EPS STRUCTURAL SPECIFICITY AND MECHANISM OF ACTION A variety of possible functions for rhizobial EPSs have been considered. Nonspecific mechanisms that do not depend on the particular chemical structure of the EPS could include a role in the attachment of the bacterial cell to the plant root or a contribution to the morphology of nodule-invasion structures. As a capsule, EPS could also function nonspecifically in avoiding a plant-defense response. It could cover another surface determinant that would otherwise induce plant defenses, or it could protect the bacterial cell from them. Models entailing EPS structural specificity could include signaling that leads to a plant cellular response involved in some critical stage in nodule development, activation of a critical enzyme such as a polysaccharase that may be involved in root hair-curling or infection-thread growth, or active suppression of plant-defense responses.

Experimental results have not supported nonspecific mechanisms. In the *R. meliloti*-alfalfa system, no significant defect in attachment has been observed with *R. meliloti exo* mutants (L. A. Lagares, personal communication; C. C. Lee & J. Leigh, unpublished data). A passive role for EPS in the avoidance of plant defenses seems unlikely as well, at least as an essential invasion function. When *exo* mutants are coinoculated onto alfalfa with exo$^+$ (*nod*$^-$) strains, both partners invade nodules (106, 111, 126). If *exo* mutants induced a defense response due to an unmasking of another determinant, this would occur after coinoculation as well, preventing invasion. If *exo* mutants were susceptible to a defense response due to lack of a protective coating, they would not invade after coinoculation. A specific role for EPS as an active suppressor of plant defenses remains possible.

Early indications that EPS structure is important in nodule development came from studies with *Rhizobium* mutants that produce structurally altered EPSs. *R. meliloti exoH* mutants, which produce nonsuccinylated succinoglycan and very little LMW succinoglycan, and another *R. meliloti* mutant that produces nonpyruvylated succinoglycan, fail to invade alfalfa nodules (117, 126). Perhaps some of the most useful studies dealing with EPS structure and function have been ones in which isolated EPSs have had marked biological activity. *Rhizobium* sp. strain NGR234 EPS restored nodule development to *Leucaena* plants inoculated with *Rhizobium* sp. strain NGR234 *exo* mutants, and *R. leguminosarum* bv. *trifolii* EPS restored nodule development to clover plants inoculated with *R. leguminosarum* bv. *trifolii exo* mutants (62). EPS from the heterologous species that normally do not interact with the particular plant had no effect. Similarly, the addition of the *R. meliloti* EPS succinoglycan restored nodule invasion of alfalfa by *R. meliloti exo* mutants (9a, 176a), while EPSs from a variety of other species had no effect. In this case, only LMW, not HMW, succinoglycan was active. Further fractionation of LMW succinoglycan yielded a single active fraction that consisted of a tetramer of the succinoglycan repeat unit that contained a high degree of noncarbohydrate substitution (9a). LMW succinoglycan was active at a concentration as low as 1 μM. These studies demonstrated three important points: (*a*) that the chemical structure of EPS is important in nodule development, (*b*) that the EPS need not be produced by the same bacterial cell that invades, and (*c*) that EPS is active at a low concentration. A specific signaling role for *Rhizobium* EPSs in nodule development seems likely.

Despite the importance of EPS structure, certain polysaccharides of different structure can also be active in place of a particular EPS. Genetic studies showed that the production of the *R. meliloti* second EPS, EPSb (see Figure 1), was sufficient to allow *R. meliloti* to invade alfalfa in the absence of succinoglycan production (76, 197). LPS from a particular strain of *R. meliloti* can also substitute (140, 186). However, redundancy of mechanism is quite possible, and alfalfa may retain the ability to recognize several different structures at a particular stage in invasion. Also, the three alternate carbohydrates, succinoglycan, EPSb, and LPS may have similar three-dimensional structures. Succinoglycan and EPSb could have similar *O*-6-acetylglucose-β(1-3)galactose motifs (76, 148). Genetic evidence has also indicated that the *R. meliloti* EPS succinoglycan can substitute for the *Rhizobium* sp. strain NGR234 EPS in the invasion of *Leucaena* (82). However, the chemical structures of NGR234 EPS and succinoglycan are very similar (Figure 1), and both may satisfy a specific recognition requirement for *Leucaena*.

Specific mechanisms of EPS action do not necessarily exclude nonspecific mechanisms. For example, perhaps HMW EPS influences the efficiency of invasion by providing a morphological component of the infection-thread interior. Also, EPS-deficient mutants can negatively affect plants, causing chlorosis on bean (130).

OTHER *RHIZOBIUM* SURFACE FEATURES THAT FUNCTION IN NODULE DEVELOPMENT In addition to heteropolysaccharide EPSs, several other bacterial factors are essential for nodule invasion and development. In the nodulation of alfalfa by *R. meliloti,* genes required for the synthesis of a periplasmic cyclic β-(1-2)-glucan are required for nodule invasion at a stage similar to the *exo* gene requirement (67, 68). However, the periplasmic glucan may not be directly required for normal nodule development, because pseudorevertants that nodulate normally still fail to produce the glucan (68).

Genetic evidence suggests that just as EPS is important in early stages in indeterminate nodules, LPS may be important in early stages in determinate nodules (31, 32, 128, 158). LPS mutants of indeterminate nodulators are in some cases symbiotically defective as well, but the defect occurs at a later stage in the nodulation process than it does in the case of EPS mutants (24, 52, 138). In the *R. meliloti*-alfalfa symbiosis, LPS does not appear to be important at any stage (39), except as a replacement for EPS in certain cases (see above).

Cellulose fibrils produced by *R. leguminosarum* appear to play a role in attachment but are not essential for successful nodulation (156).

Functions of EPS in Plant Pathogenesis

CROWN GALL TUMORIGENESIS The *Agrobacterium tumefaciens* EPS, which is identical in structure to the *R. meliloti* EPS succinoglycan, is apparently unnecessary for the formation of crown gall tumors. A variety of *A. tumefaciens exo* mutants corresponded by genetic complementation to *R. meliloti exo* mutants and were deficient in EPS production, but were fully virulent (28). The only exceptions were pleiotropic *exoC* mutants defective in precursor synthesis for a variety of carbohydrate polymers (103, 177).

Periplasmic glucan appears to be essential in the virulence of *A. tumefaciens*. The *chvB* and *chvA* mutants are deficient in the synthesis and export of periplasmic cyclic β-(1-2)-glucan, fail to attach effectively to plant surfaces, and are avirulent (141). Cellulose may also function in attachment but is not required for virulence (122).

WILTS, CANKERS, AND LEAF BLIGHTS Bacterial EPS plays two major roles in causing disease symptoms on plants: it can cause wilting by blocking xylem vessels and it is partly responsible for the water-soaking symptom typical of bacterial leaf spots. In addition, EPS may function in plant-bacteria interactions by aiding the movement of bacteria through plant tissues, promoting their growth in intercellular spaces, and helping them to avoid plant defenses. The plant pathogens featured in this review are representative of different types of plant diseases in which EPS plays an important role in determining the pathogen's ability to colonize plant tissue and cause symptoms. *P. solanacearum* and certain pathovars of *X. campestris* are wilt-inducing pathogens that grow in the xylem vessels of their hosts, where the EPS they produce interferes with water transport. Bacteria in other pathovars of *X. campestris* and *P. syringae* are necrogens that cause lesions on leaves. The slime that they produce fills the intercellular spaces of the leaves and maintains a moist environment in which the bacteria can grow. This action makes the lesions appear water-soaked and also causes droplets of ooze to form on leaf surfaces. *E. stewartii* and *E. amylovora* can be considered both necrogenic and wilt-inducing pathogens. *E. stewartii* causes Stewart's wilt of corn, and *E. amylovora* causes fireblight on rosaceous hosts, such as apple and pear. Both are xylem-inhabiting bacteria, but each can affect other plant parts. *E. amylovora* can grow in cortical tissues and produce large cankers, and *E. stewartii* can grow in the intercellular spaces of corn leaves and produce water-soaked lesions. Although they infect completely different hosts, these two *Erwinia* species are very closely related and have similar mechanisms of pathogenicity. Infection requires *hrp*-like genes and the production of EPS (43–45, 149, 187), but not toxins or degradative enzymes (23, 149).

Plant pathologists have long noted the correlation between mucoid colony-types and high virulence in different bacterial pathogens, but genetically defined mutants have been used only recently to determine the role of EPS in pathogenicity. EPS-mutants of *E. stewartii* (22, 44, 65), *E. amylovora* (8, 10, 160, 198), *X. campestris* (9, 142, 168, 185), and *P. solanacearum* (27, 41, 54, 56, 96, 159) have reduced virulence. However, in a few instances, certain nonmucoid strains retained partial to full virulence (9, 20, 44, 142, 188). These mutant studies have focused primarily on acidic heteropolysaccharides, and the roles of alginate and levan remain unclear. Both polymers appear to be produced in plants (71, 84), but a mutant study has only been done with *E. amylovora*. Levan-deficient mutants of *E. amylovora* spread less rapidly in host tissue, although levan does not appear to be required for pathogenicity (74).

Controversy around the role of EPS in wilt diseases has emerged from the variability observed in the pathogenicity of EPS^- mutants. We are now beginning to understand that this apparent variability results in part from the nature of the particular EPS mutants involved. There is no situation where a mutant that was completely EPS deficient remained fully virulent under all conditions. In the cases of EPS^- mutants that were still quite virulent, many turned out to be EPS impaired and still produce enough slime to cause delayed wilting. The problems encountered in such studies are that colony-type is simply not a reliable method of determining EPS production, and the results of pathogenicity tests depend greatly on how the assay is done. For example, region II mutants of *P. solanacearum,* which appear nonmucoid on rich media, are fully virulent because they still produce EPS *in planta* (54), and the fully virulent EPS$^-$ mutant of *P. solanacearum* reported by Xu et al (188) really produces 20% of the normal amount of EPS and exhibits decreased virulence at low inoculum concentrations (104). Likewise, *X. campestris* pv. *campestris* xanthan mutants can be virulent when inoculated by infiltration or wounding, but are weakly virulent when introduced through hydrathodes, their natural point of entry (49). On the other hand, the decreased virulence of an EPS mutant does not always result solely from the mutant's loss of EPS, because some have pleotrophic mutations that affect other pathogenicity factors or cell growth. For example, *P. solanacearum phcA* mutants and *X. campestris rpf* mutants are also impaired in the synthesis of other extracellular pathogenicity factors, such as endoglucanase or protease (26, 55, 171). Likewise, *P. solanacearum ops* mutants are defective in LPS synthesis, which may in turn affect many outer membrane functions (105). Still other EPS mutants, such as *cpsE* strains of *E. stewartii* (44), are probably avirulent because they grow poorly under conditions that favor EPS production. Poor growth is frequently associated with certain EPS mutations and may result from reduced availability of lipid carrier for cell-wall synthesis because the lipid carrier becomes complexed with incomplete EPS subunits and is not recycled.

EPS has been implicated as a mechanism for vascular occlusion (36, 178) and symptom expression by wilt-inducing bacteria. However, whether wilting ability is purely a function of vascular plugging by any large, viscous molecule or it is a specific function of the EPS structure is still a subject for debate. Van Alfen (178) points out that plugging the pit membranes between vessels is more likely to cause wilting than blocking the vessels themselves with slime, bacterial cells, plant gums, or tyloses. This observation explains how very small amounts of large EPS molecules, amounts much smaller than could be reasonably expected to plug the vessels, can cause plant cuttings to wilt. Pit membranes normally serve to reduce the spread of air embolisms in the xylem, but they can also act as molecular filters. As water moves within the vascular system of a plant it must pass through the very small capillaries in these membranes. In studies on the mechanism of wilt induction by *Clavibacter michiganense* subsp. *insidiosum* in alfalfa plants, Van Alfen's laboratory (179, 180) has shown that the pit membrane pores of alfalfa are not the same size throughout the plant; their effective sizes are larger in stems than in leaf traces. Thus, a smaller molecule is required to travel from a site of infection in the stem to the leaf traces and cause plugging there. Moreover, an EPS molecule must be just the right size to get caught in a pore; smaller or larger molecules will not cause blockage. *C. michiganense* subsp. *insidiosum* produces three different sizes of EPS molecules, each the appropriate size to plug one of the three known capillary pore sizes in its host. This finding explains how resistance to water flow can occur at sites distal to the location of bacteria. One can speculate that *C. michiganense* subsp. *insidiosum* EPSs have evolved so that they are the optimal sizes for blocking pit membranes, but it is not apparent in any wilt disease what selective advantage wilting per se confers on the bacterium that causes it.

In vascular and canker diseases, EPS may aid in the spread of the bacteria though plant tissue. Hydrostatic pressure created by EPS in blocked xylem vessels can cause them to rupture and release bacteria into adjacent vessels or the surrounding stem cortex and pith. Certain *cps* mutants of *E. stewartii* lack the ability to cause systemic infections even though they can produce local lesions (44), and Schouten (154) has suggested that the swelling pressure of amylovoran and levan enables *E. amylovora* to colonize host cortical tissue. Another mechanism by which EPS may aid in systemic movement is to prevent immobilization of bacteria by plant products. Xylem vessels of apple trees contain a protein agglutinin that retards passage of acapsular mutants of *E. amylovora* (151), and corn contains a similar agglutinin that binds nonmucoid strains of *E. stewartii* (22). In both cases, wild-type EPS-producing strains are not agglutinated. Young & Sequeira (190) also observed that EPS^- mutants of *P. solanacearum* are rapidly agglutinated by plant cell-wall fragments, and that EPS prevents this process.

Bacteria grow best in leaves that are water congested. In fact, a relative humidity of near 100% in the intercellular spaces is required for infection to occur. Native EPS is mostly water, and its ability to absorb moisture from the atmosphere and modify the immediate environment of a bacterium probably helps bacterial pathogens to infect and colonize plants. After infection, the bacterial slime formed in the intercellular spaces can hold the water and nutrients released from damaged cells and thereby contribute to both water-soaking symptoms and maintenance of favorable conditions for bacterial multiplication (44, 153). For example, EPS^- mutants of *E. stewartii* form smaller lesions that turn necrotic very soon after infection (44). Studies with purified EPS preparations further support this role for EPS, even though infiltrating plants with concentrations of EPS as high as those that occur during infections is difficult. When xanthan gum is infiltrated into cotton leaves, it alone can cause the appearance of water-soaking for as long as the plants are kept at high relative humidity (M. Zachowsky & K. Rudolph, personal communication). Furthermore, if pathogenic *X. campestris* pv. *malvacearum* cells are infiltrated along with the xanthan preparations, symptoms appear sooner and the bacteria grow much faster than without added EPS

(153). When these experiments are done with incompatible races of *X. campestris* pv. *malvacearum,* the purified EPS prevents a hypersensitive response. Thus, an added advantage of EPS production may be to block induced resistance responses by preventing recognition of potential bacterial cell-surface ligands or extracellular products by plant cell-wall receptors.

Interestingly, mutant studies with *E. carotovora* and *E. chrysanthemi* have never implicated EPS as a virulence factor in soft rot diseases. Perhaps these pathogens are far less subtle parasites, and the rapid maceration and liquefaction of plant tissues that they cause eliminates any need for EPS in retaining water and nutrients and shielding the bacteria against induced host responses.

Very little is known about how the structure of EPS influences pathogenicity. With regard to wilt-induction, anything that effects the size, charge, or viscosity of EPS may be important. We know the most about the structure of xanthan gum, but the defined *gum* mutants of *X. campestris* used for biochemical studies at Synergen have not been examined for changes in pathogenicity. Ramirez et al (142), however, examined a similar range of chemically induced *gum* mutants of NRRL-B1459 for a relationship between gum quality and virulence on *Brassica oleracea*. By inoculating plants via the hydrathodes, they found a positive correlation between lesion area and the final viscosity of the culture, the viscosifying capacity of the polymer, and the amount of acetyl substituents in the gum. Likewise, physical modifications of amylovoran (155) indicate that charge and viscosity are more important in producing wilt than molecular weight. Direct effects of EPS on plant cells have been reported (reviewed in 178), but still no good evidence indicates that EPS can function as a toxin. Although purified EPS from *P. syringae* pathovars and *X. campestris* pv. *malvacearum* (69) causes electrolyte leakage and *C. michiganense michiganense* EPS interferes with tomato callus development (181), the cause of these effects is unknown. Unlike symbionts, the EPSs of plant pathogens do not appear to have a specific signaling role in plant-bacteria interactions. Moreover, the fact that many *X. campestris* and *P. syringae* pathovars produce essentially the same xanthans or alginates suggests that EPS structure is not involved in determining host specificity.

CONCLUDING REMARKS

Virtually all plant-associated bacteria produce EPS and, as discussed above, EPSs are probably required nodulation factors for many symbionts and required virulence factors for all pathogens that cause blights, cankers, and wilts. Aside from their importance in plant interactions, the bacterial EPSs discussed in this review are almost certainly important in other niches as well. EPSs may influence survival of bacteria in soil, rhizospheres, or water and aid in insect transmission. Phytobacteriologists may therefore be egocentric in thinking that these polymers are made primarily for plant-bacteria interactions. The same physical properties of EPS that are important *in planta* are also of value for survival outside of hosts. The complex regulation of EPS synthesis reflects this dual role. Specific chemical signals from plants that modulate EPS synthesis have not been detected. Instead, EPS synthesis is stimulated in media that are poor in nutrients, such as nitrogen or phosphate, and peaks in the early stationary phase of growth, as is found generally in bacteria. The need to modulate EPS synthesis in a variety of niches may help to explain the multiplicity of genetic regulatory mechanisms that has become evident, particularly in the plant pathogens. We are just beginning to understand the nature of the signals and the regulatory responses that govern EPS synthesis to suit these niches. In plant pathogenesis and root nodule symbiosis, nature has found new uses for a nearly ubiquitous class of substances, the exopolysaccharides.

ACKNOWLEDGMENTS

We thank G. Walker, L. Reuber, P. Albersheim, F. Dazzo, A. Johnston, G. Stacey, J. Handelsman, C. Kado, D. Keister, C. Rubens, M. Capage, V. Stout, T. Denny, C. Kao, L. Sequeira, S. Gottesman, K. Geider, and K. Rudolph for providing unpublished data and helpful discussions. Support was provided by NIH grant GM39785 to J. L. and by state and federal funds to the Ohio Agricultural Research and Development Center, The Ohio State University to D. C.

Literature Cited

1. Akiyama, Y., Eda, S., Nishikawaji, S., Tanaka, H., Fujimori, T., Kato, K., Ohnishi, A. 1986. Extracellular polysaccharide produced by a virulent strain (U-7) of *Pseudomonas solanacearum*. *Agric. Biol. Chem.* 50:747–51
2. Albright, L. M., Huala, E., Ausubel, F. M. 1988. Prokaryotic signal transduction mediated by sensor and regulator protein pairs. *Annu. Rev. Genet.* 23: 311–36
3. Allen, P., Hart, C. A., Saunders, J. R. 1987. Isolation from *Klebsiella* and characterization of two *rcs* genes that activate colanic acid capsular biosynthesis in *Escherichia coli*. *J. Gen. Microbiol.* 133:331–40
4. Aman, P., McNeil, M., Franzen, L., Darvill, A. G., Albersheim, P. 1981. Structural elucidation, using HPLC-MS and GLC-MS, of the acidic polysaccharide secreted by *R. meliloti* strain 1021. *Carbohydr. Res.* 95:263–82
5. Amemura, A., Hisamatsu, M., Ghai, S. K., Harada, T. 1981. Structural studies on a new polysaccharide, containing D-riburonic acid, from *Rhizobium meliloti* IFO 13336. *Carbohydr. Res.* 91:59–65
6. Amemura, A., Hisamatsu, M., Mitani, H., Harada, T. 1983. Cyclic (1–2)-β-D-glucan and the octasaccharide repeating units of extracellular acidic polysaccharides produced by *Rhizobium*. *Carbohydr. Res.* 114:277–85
7. Applebaum, E. R., Thompson, D. V., Idler, K., Chartrain, N. 1988. *Rhizobium japonicum* USDA 191 has two *nodD* genes that differ in primary structure and function. *J. Bacteriol.* 170:12–20
8. Ayers, A. R., Ayers, S. B., Goodman, R. N. 1979. Extracellular polysaccharide of *Erwina amylovora:* a correlation with virulence. *Appl. Environ. Microbiol.* 38:659–66
9. Barrère, G. C., Barber, C. E., Daniels, M. J. 1986. Molecular cloning of genes involved in the production of the ex-

tracellular polysaccharide xanthan by *Xanthomonas campestris* pv. *campestris*. *Int. J. Biol. Macromol.* 8:372–74

9a. Battisti, L., Lara, J. C., Leigh, J. A. 1992. A specific oligosaccharide form of the *Rhizobium meliloti* exopolysaccharide promotes nodule invasion in alfalfa. *Proc. Natl. Acad. Sci. USA* In press

10. Bennett, R. A., Billing, E. 1978. Capsulation and virulence in *Erwina amylovora*. *Ann. Appl. Biol.* 89:44–45

11. Bennett, R. A., Billing. E. 1980. Origin of the polysaccharide ooze from plants infected with *Erwina amylovora*. *J. Gen. Microbiol.* 116:341–49

12. Bernhard, F., Poetter, K., Geider, K., Coplin, D. L. 1990. The *rcsA* gene from *Erwina amylovora:* identification, nucleotide sequence, and regulation of exopolysaccharide biosynthesis. *Mol. Plant-Microbe Interact.* 3:429–37

13. Betlach, M. R., Campbell, D. S., Capage, M. A., Doherty, D. H., Gold, M., et al. 1989. *Recombinant DNA mediated biosynthesis of xanthan gum in denitrifying pseudomonads under anaerobic conditions*. Pres. Annual AIChE Meeting., Nov. 5–10, 1989

14. Betlach, M. R., Capage, M. A., Doherty, D. H., Hassler, R. A., Henderson, N. M., et al. 1988. *Molecular biology of xanthan gum biosynthesis in* Xanthomonas campestris. Pres. 1988 Am. Chem. Soc. Meeting, Sept., 1988

15. Betlach, M. R., Capage, M. A., Doherty, D. H., Hassler, R. A., Henderson, N. M., et al. 1987. Genetically engineered polymers: manipulation of xanthan biosynthesis. In *Progress in Biotechnology 3. Industrial Polysaccharides: Genetic Engineering, Structure/Property Relations and Applications,* ed. M. Yalpani, pp. 35–50. Amsterdam: Elsevier

16. Borthakur, D., Barbur, C. E., Lamb, J. W., Daniels, M. J., Downie, J. A., Johnston, A. W. B. 1986. A mutation that blocks exopolysaccharide synthesis prevents nodulation of peas by *Rhizobium leguminosarum* but not of beans by *R. phaseoli* and is corrected by cloned DNA from *Rhizobium* or the phytopathogen *Xanthomonas*. *Mol. Gen. Genet.* 203:320–23

17. Borthakur, D., Barbur, R. F., Latchford, J. W., Rossen, L., Johnston, A. W. B. 1988. Analysis of *pss* genes of *Rhizobium leguminosarum* required for exopolysaccharide synthesis and nodulation of peas: their primary structure and their interaction with *psi* and other nodulation genes. *Mol. Gen. Genet.* 213:155–62

18. Borthakur, D., Downie, J. A., Johnston, A. W. B., Lamb, J. W. 1985. *psi,* a plasmid-linked *Rhizobium phaseoli* gene that inhibits exopolysaccharide production and which is required for symbiotic nitrogen fixation. *Mol. Gen. Genet.* 200:278–82

19. Borthakur, D., Johnston, A. W. B. 1987. Sequence of *psi,* a gene on the symbiotic plasmid of *Rhizobium phaseoli* which inhibits exopolysaccharide synthesis and nodulation and demonstration that its transcription is inhibited by *psr,* another gene on the symbiotic plasmid. *Mol. Gen. Genet.* 207:149–54

20. Boucher, C. A., Barberis, P. A., Trigalet, A., Demery, D. A. 1985. Transposon mutagenesis of *Pseudomonas solanacearum:* isolation of Tn*5*-induced avirulent mutants. *J. Gen. Microbiol.* 131:2449–57

21. Boulnois, G. J., Jann, K. 1989. Bacterial polysaccharide capsule synthesis, export and evolution of structural diversity. *Mol. Microbiol.* 3:1819–23

22. Bradshaw-Rouse, J. J., Whatley, M. A., Coplin, D. L., Woods, A., Sequeira, L., Kelman, A. 1981. Agglutination of strains of *Erwina stewartii* with a corn agglutinin: correlation with extracellular polysaccharide production and pathogenicity. *Appl. Environ. Microbiol.* 42:344–50

23. Braun, E. J. 1990. Colonization of resistant and susceptible maize plants by *Erwina stewartii* strains differing in exopolysaccharide production. *Physiol. Mol. Plant Pathol.* 36:363–79

24. Brink, B. A., Miller, J., Carlson, R. W., Noel, K. D. 1990. Expression of *Rhizobium leguminosarum* CFN42 genes for lipopolysaccharide in strains derived from different *R. leguminosarum* soil isolates. *J. Bacteriol.* 172:548–55

25. Brumbley, S. M., Carney, B. F., Denny, T. P. 1991. Cloning and characterization of *phcA,* a regulatory gene of *Pseudomonas solanacearum*. *Phytopathology* 81:1145 (Abstr.)

26. Brumbley, S. M., Denny, T. P. 1990. Cloning of wild-type *Pseudomonas solanacearum phcA,* a gene that when mutated alters expression of multiple traits that contribute to virulence. *J. Bacteriol.* 172:5677–85

27. Buddenhagen, I., Kelman, A. 1964. Biological and physiological aspects of bacterial wilt caused by *Pseudomonas*

solanacearum. *Annu. Rev. Phytopathol.* 2:203–30

27a. Buendia, A. M., Enenkel, B., Köplin, R., Niehaus, K., Arnold, W., Pühler, A. 1991. The *Rhizobium meliloti exoZ*/*exoB* fragment of megaplasmid 2: *exoB* functions as a UDP-glucose 4-epimerase and ExoZ shows homology to NodX of *Rhizobium leguminosarum* biovar *viciae* strain TOM. *Molecular Microbiol.* 5: 1519–30

28. Cangelosi, G. A., Hung, L., Puvanesarajah, V., Stacey, G., Ozga, D. A., et al. 1987. Common loci for *Agrobacterium tumefaciens* and *Rhizobium meliloti* exopolysaccharide synthesis and their roles in plant interactions. *J. Bacteriol.* 169:2086–91

29. Canter Cremers, H. C. J., Batley, M., Redmond, J. W., Eydems, L., Breedveld, M. W., et al. 1990. *Rhizobium leguminosarum exoB* mutants are deficient in the synthesis of UDP-glucose 4'-epimerase. *J. Biol. Chem.* 265: 21122–27

30. Canter Cremers, H. C. J, Batley, M., Redmond, J. W., Wijfjes, A. H. M., Lugtenberg, B. J. J., Wijffelman, C. A. 1991. Distribution of *O*-acetyl groups in the exopolysaccharide synthesized by *Rhizobium leguminosarum* strains is not determined by the Sym plasmid. *J. Biol. Chem.* 266:9551–64

31. Carlson, R. W., Kalembasa, S., Turowski, D., Pachori, P., Noel, K. D. 1987. Characterization of the lipopolysaccharide from a *Rhizobium phaseoli* mutant that is defective in infection thread development. *J. Bacteriol.* 169: 4923–28

32. Cava, J. R., Elias, P. M., Turowski, D. A., Noel, K. D. 1989. *Rhizobium leguminosarum* CFN42 genetic regions encoding lipopolysaccharide structures essential for complete nodule development on bean plants. *J. Bacteriol.* 171:8–15

33. Chakravorty, A. K., Zurkowski, W., Shine, J., Rolfe, B. G. 1982. Symbiotic nitrogen fixation: molecular cloning *Rhizobium* genes involved in exopolysaccharide synthesis and effective nodulation. *J. Mol. Appl. Genet.* 1:585–96

34. Charles, T. C., Newcomb, W., Finan, T. M. 1991. *ndvF,* a novel locus on megaplasmid pRmeSU47b (pEXO) of *Rhizobium meliloti,* is required for normal nodule development. *J. Bacteriol.* 173:3981–92

35. Chatterjee, A., Chun, W., Chatterjee, A. K. 1990. Isolation and characterization of an *rcsA*-like gene of *Erwina amylovora* that activates extracellular polysaccharide production in *Erwina* species, *Escherichia coli,* and *Salmonella typhimurium*. *Mol. Plant-Microbe Interact.* 3:144–48

36. Chatterjee, A. K., Vidaver, A. K. 1986. *Advances in Plant Pathology,* Vol. 4, *Genetics of Pathogenicity Factors: Application to Phytopathogenic Bacteria,* ed. D. S. Ingram, P. H. Williams. Orlando: Academic. 224 pp.

37. Chen, H., Batley, M., Redmond, J., Rolfe, B. G. 1985. Alteration of the effective nodulation properties of a fast-growing broad host range *Rhizobium* due to changes in exopolysaccharide synthesis. *J. Plant Physiol.* 120:331–49

38. Clough, S. J., Denny, T. P. 1991. Regulation of virulence in *Pseudomonas solanacearum* by an endogenous volatile compound. *Phytopathology* 81:1144 (Abstr.)

39. Clover, R. H., Kieber, J., Signer, E. R. 1989. Lipopolysaccharide mutants of *Rhizobium meliloti* are not defective in symbiosis. *J. Bacteriol.* 171:3961–67

40. Coleman, M., Pearce, R., Hitchin, E., Busfield, F., Mansfield, J. W., Roberts, I. S. 1990. Molecular cloning, expression and nucleotide sequence of the *rcsA* gene of *Erwina amylovora,* encoding a positive regulator of capsule expression: evidence for a family of related capsule activator proteins. *J. Gen. Microbiol.* 136:1799–1806

41. Cook, D., Sequeira, L. 1991. Genetic and biochemical characterization of a *Pseudomonas solanacearum* gene cluster required for extracellular polysaccharide production and for virulence. *J. Bacteriol.* 173:1654–62

42. Coplin, D. L., Cook, D. 1990. Molecular genetics of extracellular polysaccharide biosynthesis in vascular phytopathogenic bacteria. *Mol. Plant-Microbe Interact.* 3:271–79

43. Coplin, D. L., Frederick, R. D., Majerczak, D. R., Tuttle, L. D. 1992. Characterization of a gene cluster that specifies pathogenicity in *Erwina stewartii*. *Mol. Plant-Microbe Interact.* 5:81–88

44. Coplin, D. L., Majerczak, D. R. 1990. Extracellular polysaccharide genes in *Erwina stewartii:* directed mutagenesis and complementation analysis. *Mol. Plant-Microbe Interact.* 3:286–92

45. Coplin, D. L., Majerczak, D. R., Poetter, K. 1992. Genetics of extracellular polysaccharide biosynthesis in *Erwinia stewartii*. In *Biotechnology and Plant*

Protection: Bacterial Pathogenesis and Disease Resistance. Boston: Butterworth-Heineman. In press

46. Costa, J. B. 1991. *Structural studies of some viscous, acidic bacterial exopolysaccharides*. PhD thesis. The Ohio State Univ. Columbus, Ohio
47. Costerton, J. W., Cheng, K. J., Geesey, G. G., Ladd, T. I., Nickel, J. C., et al. 1987. Bacterial biofilms in nature and disease. *Annu. Rev. Microbiol.* 41:435–64
48. Daniels, M. J., Barber, C. E., Turner, P. C., Cleary, W. G., Sawczyc, M. K. 1984. Isolation of mutants of *Xanthomonas campestris* pv. *campestris* showing altered pathogenicity. *J. Gen. Microbiol.* 130:2447–54
49. Daniels, M. J., Osbourne, A. E., Tang, J. L. 1989. Regulation in *Xanthomonas*-plant interactions. In *Signal Molecules in Plants and Plant-Microbe Interactions. NATO ASI Series,* ed. B. J. J. Lugtenberg. H36:189–96. Berlin: Springer-Verlag
50. Darus, A. 1980. The glycosyltransferase complex isolated from *Erwina stewartii*. PhD dissertation, Univ. of Missouri, Columbia, Missouri
51. de Crécy-Lagard, V., Glaser, P., Lejeune, P., Sismeiro, O., Barber, C. E., et al. 1990. A *Xanthomonas campestris* pv. *campestris* protein similar to catabolite activation factor is involved in regulation of phytopathogenicity. *J. Bacteriol.* 172:5877–83
52. de Maagd, R. A., Rao, A. S., Mulders, I. H. M., Goosen–de Roo, L., van Loosdrecht, M. C. M., et al. 1989. Isolation and characterization of mutants of *Rhizobium leguminosarum* bv. *viciae* 248 with altered lipopolysaccharides: possible role of surface charge or hydrophobicity in bacterial release from the infection thread. *J. Bacteriol.* 171:1143–50
53. Dénarié, J., Debellé, F., Rosenberg, C. 1992. Signaling and host range variation in nodulation. *Annu. Rev. Microbiol.* 46:497–531
54. Denny, T. P., Baek, S. R. 1991. Genetic evidence that extracellular polysaccharide is a virulence factor of *Pseudomonas solanacearum*. *Mol. Plant-Microbe Interact.* 2:198–206
55. Denny, T. P., Carney, B. F., Schell, M. A. 1990. Inactivation of multiple virulence genes reduces the ability of *Pseudomonas solanacearum* to cause wilt symptoms. *Mol. Plant-Microbe Interact.* 5:293–300
56. Denny, T. P., Makini, F. W., Brumbley, S. M. 1988. Characterization of *Pseudomonas solanacearum* Tn*5* mutants deficient in extracellular polysaccharide. *Mol. Plant-Microbe Interact.* 1:215–23
57. Deretic, V., Mohr, C. D., Martin, D. W. 1991. Mucoid *Pseudomonas aeruginosa* in cystic fibrosis: signal transduction and histone-like elements in the regulation of bacterial virulence. *Mol. Microbiol.* 5:1577–83
58. Diaz, C. L., Melchers, L. S., Hooykaas, P. J. J., Lugtenberg, B. J. J., Kijne, J. W. 1989. Root lectin as a determinant of host-plant specificity in the *Rhizobium*-legume symbiosis. *Nature* 338:579–81
59. Dickstein, T., Bisseling, T., Reinhold, V. N., Ausubel, F. 1988. Expression of nodule-specific genes in alfalfa root nodules blocked at an early stage of development. *Genes Dev.* 2:677–87
60. Diebold, R., Noel, K. D. 1989. *Rhizobium leguminosarum* exopolysaccharide mutants: biochemical and genetic analyses and symbiotic behavior on three hosts. *J. Bacteriol.* 171:4821–30
61. Deleted in proof
62. Djordjevic, S. P., Chen, H., Batley, M., Redmond, J. W., Rolfe, B. G. 1987. Nitrogen fixation ability of exopolysaccharide synthesis mutants of *Rhizobium* sp. NGR234 and *Rhizobium trifolii* is restored by the addition of homologous exopolysaccharides. *J. Bacteriol.* 169:53–60
63. Djordjevic, S. P., Rolfe, B. G., Batley, M., Redmond, J. W. 1986. The structure of the exopolysaccharide from *Rhizobium* sp. strain ANU280 (NGR234). *Carbohydr. Res.* 148:87–99
64. Doherty, D., Leigh, J. A., Glazebrook, J., Walker, G. C. 1988. *Rhizobium meliloti* mutants that overproduce the *R. meliloti* acidic Calcofluor-binding exopolysaccharide. *J. Bacteriol.* 170: 4249–56
65. Dolph, P. J., Majerczak, D. R., Coplin, D. L. 1988. Characterization of a gene cluster for exopolysaccharide biosynthesis and virulence in *Erwina stewartii*. *J. Bacteriol.* 170:865–71
66. Drigues, P., Demery-Lafforgue, D., Trigalet, A., Dupin, P., Samain, D., Asselineau, J. 1985. Comparative studies of lipopolysaccharide and exopolysaccharide from a virulent strain of *Pseudomonas solanacearum* and from three avirulent mutants. *J. Bacteriol.* 162:504–9
67. Dylan, T., Ielpi, L., Stanfield, S., Kashyap, L., Douglas, C., et al. 1986.

Rhizobium meliloti genes required for nodule development are related to chromosomal virulence genes in *Agrobacterium tumefaciens. Proc. Natl. Acad. Sci. USA* 83:4403–7

68. Dylan, T., Nagpal, P., Helinski, D. R., Ditta, G. S. 1990. Symbiotic pseudorevertants of *Rhizobium meliloti ndv* mutants. *J. Bacteriol.* 172:1409–17
69. El-Banoby, F. E., Rudolph, K. 1979. Induction of water-soaking in plant leaves by extracellular polysaccharides from phytopathogenic pseudomonads and xanthomonads. *Physiol. Plant Pathol.* 15:341–49
70. Fett, W. F., Osman, S. F., Dunn, M. F. 1989. Characterization of exopolysaccharides produced by plant-associated fluorescent pseudomonads. *Appl. Environ. Microbiol.* 55:579–83
71. Fett, W. F., Osman, S. F., Fishman, M. L., Siebles T. S. III, 1986. Alginate production by plant-pathogenic pseudomonads. *Appl. Environ. Microbiol.* 52:466–73
72. Fialho, A. M., Zielinski, N. A., Fett, W. F., Chakrabarty, A. M., Berry, A. 1990. Distribution of alginate gene sequences in the *Pseudomonas* rRNA homology group I-Azomonas-Azotobacter lineage of superfamily B procaryotes. *Appl. Environ. Microbiol.* 56:436–43
73. Finan, T. M., Kunkel, B., De Vos, G. F., Signer, E. R. 1986. Second symbiotic megaplasmid in *Rhizobium meliloti* carrying exopolysaccharide and thiamine synthesis genes. *J. Bacteriol.* 167:66–72
74. Geider, K. P., Bellemann, P., Bernhard, F., Chang, J. C., Geier, G., et al. 1991. Exopolysaccharides in the interaction of the fire-blight pathogen *Erwina amylovora* with its host cells. See Ref. 87a, pp. 99–93
75. Gil-Serrano, A., del Junco, A. S., Tejero-Mateo, P., Megias, M., Caviedes, M. A. 1990. Structure of the extracellular polysaccharide secreted by *Rhizobium legunimosarum* var. *phaseoli* CIAT 899. *Carbohydr. Res.* 204:103–7
76. Glazebrook, J., Walker, G. C. 1989. A novel exopolysaccharide can function in place of the Calcofluor-binding exopolysaccharide in nodulation of alfalfa by Rhizobium meliloti. *Cell* 56:661–72
77. Goodman, R. N., Huang, J. S., Huang, P. Y. 1974. Host specific phytotoxic polysaccharide from apple tissue infected by *Erwina amylovora. Science* 183:1081–82
78. Gorin, J. A. P., Spencer, T. F. 1961. Structural relationship of extracellular polysaccharides from phytopathogenic *Xanthomonas* sp. Part 1: Structure of extracellular polysaccharide from *X. stewartii. Can. J. Chem.* 39:2282–89
79. Gottesman, S., Stout, V. 1991. Regulation of capsular polysaccharide synthesis in *Escherichia coli* K12. *Mol. Microbiol.* 5:1599–1606
80. Gray, J. X., Djordjevic, M. A., Rolfe, B. G. 1990. Two genes that regulate exopolysaccharide production in *Rhizobium* sp. strain NGR234: DNA sequences and resultant phenotypes. *J. Bacteriol.* 172:193–203
81. Gray, J. X., Rolfe, B. G. 1990. Exopolysaccharide production in *Rhizobium* and its role in invasion. *Mol. Microbiol.* 4:1425–31
82. Gray, J. X., Zhan, H., Levery, S. B., Battisti, L., Rolfe, B. G., Leigh, J. A. 1991. Heterologous exopolysaccharide production in *Rhizobium* sp. strain NGR234 and consequences for nodule development. *J. Bacteriol.* 173:3066–77
83. Gross, M., Geier, G., Geider, K., Rudolph, K. 1989. Levan and levansucrase from the fireblight pathogen *Erwina amylovora*. In *Proc. 7th Int. Conf. on Plant Pathogenic Bacteria,* pp 81–84. Budapest: Academiai Kiado
84. Gross, M., Rudolph, K. 1987. Studies on the extracellular polysaccharide produced in vitro by *Pseudomonas phaseolicola. J. Phytopathology* 119:289–97
85. Halverson, L. J., Stacey, G. 1986. Signal exchange in plant-microbe interactions. *Microbiol. Rev.* 50:193–225
86. Harding, N. E., Cleary, J. M., Cabañas, D. K., Rosen, R. G., Kang, K. S. 1987. Genetic and physical analyses of a cluster of genes essential for xanthan gum biosynthesis in *Xanthomonas campestris. J. Bacteriol.* 169:2854–61
87. Henikoff, S., Wallace, J. C., Brown, J. P. 1990. Finding protein similarities with nucleotide sequence databases. *Methods Enzymol.* 183:111–32

87a. Hennecke, H., Verma, D. P. S., eds. 1991. *Advances in Molecular Genetics of Plant-Microbe Interactions,* Vol. 1. Dordrecht: Kluwer Academic

88. Her, G. R., Glazebrook, J., Walker, G. C., Reinhold, V. N. 1990. Structural studies of a novel exopolysaccharide produced by a mutant of *Rhizobium meliloti* strain Rm1021. *Carbohydr. Res.* 198:305–12
89. Hisamatsu, M., Abe, J., Amamura, A., Harada, T. 1980. Structural elucidation

on succinoglycan and related polysaccharides from *Agrobacterium* and *Rhizobium* by fragmentation with two special β-D-glycanases and methylation analysis. *Agric. Biol. Chem.* 44:1049–55
90. Hisamatsu, M., Yamada, T., Higashiura, T., Ikeda, M. 1987. The production of acidic, *O*-acetylated cyclosophorans [cyclic(1–2)β-D-glucans] by *Agrobacterium* and *Rhizobium* species. *Carbohydr. Res.* 163:115–22
91. Hollingsworth, R. I., Dazzo, F. B., Hallenga, K., Musselman, B. 1988. The complete structure of the trifoliin A lectin-binding capsular polysaccharide of *Rhizobium trifolii* 843. *Carbohydr. Res.* 172:97–112
92. Hötte, B., Ruth-Arnold, I., Pühler, A., Simon, R. 1990. Cloning and analysis of a 35.3-kilobase DNA region involved in exopolysaccharide production in *Xanthomonas campestris* pv. *campestris*. *J. Bacteriol.* 172:2804–7
93. Hotter, G. S., Scott, D. B. 1991. Exopolysaccharide mutants of *Rhizobium loti* are fully effective on a determinate nodulating host but are ineffective on an indeterminate nodulating host. *J. Bacteriol.* 173:851–59
94. Houng, H. S. H., Cook, T. M. 1986. Cloning of the galactose utilization genes of *Vibrio cholerae*. First colloquium in biological sciences. *Ann. N. Y. Acad. Sci.* 435:601–3
95. Huang, J., Sukordhaman, M., Schell, M. A. 1989. Excretion of the *egl* gene product of *Pseudomonas solanacearum* utilizes an unusual signal sequence. *J. Bacteriol.* 171:3767–74
96. Husain, A., Kelman, A. 1958. Relation of slime production to mechanism of wilting and pathogenicity of *Pseudomonas solanacearum*. *Phytopathology* 48:155–65
97. Hynes, M. F., Simon, R., Müller, P., Niehaus, K., Labes, M., Pühler, A. 1986. The two megaplasmids of *Rhizobium meliloti* are involved in the effective nodulation of alfalfa. *Mol. Gen. Genet.* 202:356–62
98. Ielpi, L., Couso, R., Dankert, M. 1981. Lipid-linked intermediates in the biosynthesis of xanthan gum. *FEBS Lett.* 130:253–56
99. Ielpi, L., Couso, R. O., Dankert, M. A. 1981. Xanthan gum biosynthesis: pyruvic acid acetal residues are transferred from phosphoenolpyruvate to the pentasaccharide-P-P-lipid. *Biochem. Biophys. Res. Commun.* 102:1400–8
100. Jann, K., Jann, B. 1991. Genetic approaches to oligosaccharide metabolism. *Biochem. Soc. Trans London* 19: 623–28
101. Jannson, P. E., Kenne, L., Lindberg, B. 1975. Structure of the extracellular polysaccharide from *Xanthomonas campestris*. *Carbohydr. Res.* 45:275–82
102. Jansson, P., Kenne, L., Lindberg, B., Ljunngren, H., Lonngren, J., et al. 1977. Demonstration of an octasaccharide repeating unit in the extracellular polysaccharide of *Rhizobium meliloti* by sequential degradation. *J. Am. Chem. Soc.* 99:3812–15
103. Kamoun, S., Cooley, M. B., Rogowsky, P. M., Kado, C. I. 1989. Two chromosomal loci involved in production of exopolysaccharide in *Agrobacterium tumefaciens*. *J. Bacteriol.* 171:1755–59
104. Kao, C. C., Barlow, E., Sequeira, L. 1992. Extracellular polysaccharide consistently is required for wildtype virulence of *Pseudomonas solanacearum*. *J. Bacteriol.* 174:1068–71
105. Kao, C. C., Sequeira, L. 1992. A gene cluster required for the coordinated biosynthesis of lipopolysaccharide and extracellular polysaccharide also affects virulence of *Pseudomonas solanacearum*. *J. Bacteriol.* 173:7841–47
106. Kapp, D., Niehaus, K., Quandt, J., Müller, P., Pühler, A. 1990. Cooperative action of *Rhizobium meliloti* nodulation and infection mutants during the process of forming mixed infected nodules. *Plant Cell* 2:139–51
107. Keenleyside, W. J., Jayaratne, P., MacLachlan, P. R., Whitfield, C. 1992. The *rcsA* gene of *Escherichia coli* O9:K30:H12 is involved in the expression of the serotype-specific group I K (capsular) antigen. *J. Bacteriol.* 174:8–16
108. Keller, M. P., Müller, P., Simon, R., Pühler, A. 1988. *Rhizobium meliloti* genes for exopolysaccharide synthesis and nodule infection located on megaplasmid 2 are actively transcribed during symbiosis. *Mol. Plant-Microbe Interact.* 1:267–74
109. Kelman, A., Hruschka, J. 1973. The role of motility and aerotaxis in the selective increase of avirulent bacteria in still broth cultures of *Pseudomonas solanacearum*. *J. Gen. Microbiol.* 76:177–88
110. Kim, C. H., Tully, R. E., Keister, D. L. 1989. Exopolysaccharide-deficient mutants of *Rhizobium fredii* HH303 which are symbiotically effective. *Appl. Environ. Microbiol.* 55:1852–54
111. Klein, S., Hirsch, A. M., Smith, C. A., Signer, E. R. 1988. Interaction of *nod*

and *exo Rhizobium meliloti* in alfalfa nodulation. *Mol. Plant-Microbe Interact.* 1:94–100

112. Klein, S., Walker, G. C., Signer, E. R. 1988. All *nod* genes of *Rhizobium meliloti* are involved in alfalfa nodulation by *exo* mutants. *J. Bacteriol.* 170:1003–6
113. Ko, Y. H., Gayda, R. 1990. Nodule formation in soybeans by exopolysaccharide mutants of *Rhizobium fredii* USDA 191. *J. Gen. Microbiol.* 136:105–13
114. Kröncke, K.-D., Golecki, J. R., Jann, K. 1990. Further electron microscopic studies on the expression of *Escherichia coli* group II capsules. *J. Bacteriol.* 172:3469–72
115. Latchford, J. W., Borthakur, D., Johnston, A. W. B. 1991. The products of *Rhizobium* genes, *psi* and *pss,* which affect exopolysaccharide production, are associated with the bacterial cell surface. *Mol. Microbiol.* 5:2107–14
116. Leigh, J. A., Lee, C. C. 1988. Characterization of polysaccharides of *Rhizobium meliloti exo* mutants that form ineffective nodules. *J. Bacteriol.* 170:3327–32
117. Leigh, J. A., Reed, J. W., Hanks, J. F., Hirsch, A. M., Walker, G. C. 1987. Rhizobium meliloti mutants that fail to succinylate their Calcofluor-binding exopolysaccharide are defective in nodule invasion. *Cell* 51:579–87
118. Leigh, J. A., Signer, E. R., Walker, G. C. 1985. Exopolysaccharide-deficient mutants of *Rhizobium meliloti* that form ineffective nodules. *Proc. Natl. Acad. Sci. USA* 82:6231–35
119. Levery, S. B., Zhan, H., Lee, C. C., Leigh, J. A., Hakomori, S. 1991. Structural analysis of a second acidic exopolysaccharide of *Rhizobium meliloti* that can function in alfalfa root nodule invasion. *Carbohydr. Res.* 210:339–47
120. Long, S. R. 1989. *Rhizobium*-legume nodulation: life together in the underground. *Cell* 56:203–14
121. Long, S., Reed, J. W., Himawan, J., Walker, G. C. 1988. Genetic analysis of a cluster of genes required for synthesis of the Calcofluor-binding exopolysaccharide of *Rhizobium meliloti. J. Bacteriol.* 170:4239–48
122. Matthysse, A. G. 1983. Role of bacterial cellulose fibrils in *Agrobacterium tumefaciens* infections. *J. Bacteriol.* 154:906–15
123. McNeil, M., Darvill, J., Darvill, A. G., Albersheim, P., van Veen, R., et al. 1986. The discernible, structural features of the acidic polysaccharides secreted by different *Rhizobium* species are the same. *Carbohydr. Res.* 146:307–26
124. Mendoza, A., Chatterjee, A. K. 1990. Molecular cloning of an *Erwina amylovora rcsB* gene required in polysaccharide synthesis. *Phytopathology* 80:982 (Abstr.)
125. Morales, V. M., Stemmer, W. P. C., Sequeira, L. 1985. Genetics of avirulence in *Pseudomonas solanacearum*. In *Plant Cell/Cell Interactions,* ed. I. Sussex, A. Ellingboe, M. Crouch, R. Malmberg, pp. 89–96. Cold Spring Harbor, NY: Cold Spring Harbor Lab.
126. Müller, P., Hynes, M., Kapp, D., Niehaus, K., Pühler, A. 1988. Two classes of *Rhizobium meliloti* infection mutants differ in exopolysaccharide production and in coinoculation properties with nodulation mutants. *Mol. Gen. Genet.* 211:17–26
127. Mulligan, J. T., Long, S. R. 1989. A family of activator genes regulates expression of *Rhizobium meliloti* nodulation genes. *Genetics* 122:7–18
128. Noel, K. D., Vandenbosch, K. A., Kulpaca, B. 1986. Mutations in *Rhizobium phaseoli* that lead to arrested development of infection threads. *J. Bacteriol.* 168:1392–1401
129. Norris, J. H., Macol, L. A., Hirsch, A. M. 1988. Nodulin gene expression in effective alfalfa nodules and nodules arrested at three different stages of development. *Plant Physiol.* 88:321–28
130. O'Connell, K. P., Araujo, R. S., Handelsman, J. 1990. Exopolysaccharide-deficient mutants of *Rhizobium* sp. strain CIAT899 induce chlorosis in common bean (*Phaseolus vulgaris*). *Molecular Plant-Microbe Interactions* 3:424–28
131. O'Neill, M. A., Darvill, A. G., Albersheim, P. 1991. The degree of esterification and points of substitution by *O*-acetyl and *O*-(3-hydroxybutanoyl) groups in the acidic extracellular polysaccharides secreted by *Rhizobium leguminosarum* biovars *viciae, trifolii* and *phaseoli* are not related to host range. *J. Biol. Chem.* 266:9549–55
132. O'Neill, M. A., Robison, P. D., Chou, K. J., Darvill, A. G., Albersheim, P. 1991. Evidence that the acidic polysaccharide secreted by *Agrobacterium radiobacter* (ATCC 53271) has a seventeen glycosyl-residue repeating unit. *Carbohydr. Res.* In press
133. Orgambide, G., Montrozier, H., Servin, P., Roussel, J., Trigalet-Demery, D., Trigalet, A. 1991. High heterogeneity of the exopolysaccharides of *Pseudomonas solanacearum* strain GMI 1000 and the

complete structure of the major polysaccharide. *J. Biol. Chem.* 266:8312–21

134. Osbourne, A. E., Clarke, B. R., Stevens, B. J. H., Daniels, M. J. 1990. Use of oligonucleotide probes to identify members of two-component regulatory systems in *Xanthomonas campestris* pv. *campestris*. *Mol. Gen. Genet.* 222:145–51
135. Philip-Hollingsworth, S., Hollingsworth, R. I., Dazzo, F. B. 1989. Host-range related structural features of the acidic extracellular polysaccharides of *Rhizobium trifolii* and *Rhizobium leguminosarum*. *J. Biol. Chem.* 264:1461–66
136. Philip-Hollingsworth, S., Hollingsworth, R. I., Dazzo, F. B., Djordjevic, M. A., Rolfe, B. G. 1989. The effect of interspecies transfer of *Rhizobium* host-specific nodulation genes on acidic polysaccharide structure and in situ binding by host lectin. *J. Biol. Chem.* 264:5710–14
137. Poetter, K., Coplin, D. L. 1991. Structural and functional analysis of the *rcsA* gene from *Erwina stewartii*. *Mol. Gen. Genet.* 229:155–60
138. Priefer, U. B. 1989. Genes involved in lipopolysaccharide production and symbiosis are clustered on the chromosome of *Rhizobium leguminosarum* biovar *viciae* VF39. *J. Bacteriol.* 171:6161–68
139. Pühler, A., Arnold, W., Buendia-Claveria, A., Kapp, D., Keller, M., et al. 1991. The role of the *Rhizobium meliloti* exopolysaccharides EPS I and EPS II in the infection process of alfalfa nodules. See Ref. 87a, pp. 189–94
140. Putnoky, P., Petrovics, G., Kereszt, A., Grosskopf, E., thi Kam Ha, D., et al. 1990. *Rhizobium meliloti* lipopolysaccharide and exopolysaccharide can have the same function in the plant-bacterium interaction. *J. Bacteriol.* 172:5450–58
141. Puvanesarajah, V., Schell, F. M., Stacey, G., Douglas, C. J., Nester, E. W. 1985. A role for 2-linked-β-D-glucan in the virulence of *Agrobacterium tumefaciens*. *J. Bacteriol.* 164:102–6
142. Ramírez, M. E., Fucikovsky, L., Garcia-Jimenez, F., Quintero, R., Galindo, E. 1988. Xanthan gum production by altered pathogenicity variants of *Xanthomonas campestris*. *Appl. Microbiol. Biotechnol.* 29:5–10
143. Reed, J. W., Walker, G. C. 1991. The *exoD* gene of *Rhizobium meliloti* encodes a novel function needed for alfalfa nodule invasion. *J. Bacteriol.* 173:664–77
144. Reed, J. W., Walker, G. C. 1991. Acidic conditions permit effective nodulation of alfalfa by invasion-deficient *Rhizobium meliloti exoD* mutants. *Genes Dev.* 5:2274–87
145. Reed, J. W., Capage, M., Walker, G. C. 1991. *Rhizobium meliloti exoG* and *exoJ* mutations affect the ExoX-ExoY system for modulation of exopolysaccharide production. *J. Bacteriol.* 173:3776–88
146. Reed, J. W., Glazebrook, J., Walker, G. C. 1991. The *exoR* gene of *Rhizobium meliloti* affects RNA levels of other *exo* genes but lacks homology to known transcriptional regulators. *J. Bacteriol.* 173:3789–94
147. Reuber, T. L., Long, S., Walker, G. C. 1991. Regulation of *Rhizobium meliloti exo* genes in free-living cells and in planta examined by using Tn*phoA* fusions. *J. Bacteriol.* 173:426–34
148. Reuber, T. L., Reed, J., Glazebrook, J., Glucksman, M. A., Ahmann, D., et al. 1991. *Rhizobium meliloti* exopolysaccharides: genetic analyses and symbiotic importance. *Biochem. Soc. Trans.* 19:636–41
149. Roberts, I. S., Coleman, M. J. 1991. The virulence of *Erwina amylovora:* molecular genetic perspectives. *J. Gen. Microbiol.* 137:1453–57
150. Robertsen, B. K., Aman, P., Darvill, A. G., McNeil, M., Albersheim, P. 1981. Host-symbiont interactions. V. The structure of acidic extracellular polysaccharides secreted by *Rhizobium leguminosarum* and *Rhizobium trifolii*. *Plant Physiol.* 67:389–400
151. Romeiro, R., Karr, A., Goodman, R. 1981. Isolation of a factor from apple that agglutinates *Erwina amylovora*. *Plant Physiol.* 68:772–77
152. Ronson, C. W., Nixon, B. T., Ausubel, F. M. 1987. Conserved domains in bacterial regulatory proteins that respond to environmental stimuli. *Cell* 49:579–81
153. Rudolph, K. W. E., Gross, M., Neugebauer, M., Hokawat, S., Zachowski, A., et al. 1989. Extracellular polysaccharides as determinants of leaf spot diseases caused by pseudomonads and xanthomonads. In *Phytotoxins and Plant Pathogenesis, NATO ASI Series,* ed. A. Graniti, R. D. Durbin, A. Ballio. 27:177–218. Berlin/Heidelberg: Springer Verlag
154. Schouten, H. J. 1989. A possible role in pathogenesis for the swelling of extracellular slime of *Erwina amylovora* at increasing water potential. *Neth. J. Plant Pathol.* 95:169–74

155. Sijam, K., Goodman, R. N., Karr, A. L. 1985. The effect of salts on the viscosity and wilt-inducing capacity of the capsular polysaccharide of *Erwina amylovora*. *Physiol. Plant Pathol.* 22: 231–39
156. Smit, G., Kijne, J. W., Lugtenberg, B. J. J. 1987. Involvement of both cellulose fibrils and a Ca^{2+}-dependent adhesin in the attachement of *Rhizobium leguminosarum* to pea root hair tips. *J. Bacteriol.* 169:4294–4301
157. Smith, A. R. W., Rastall, R. A., Rees, N. H., Hignett, R. C. 1990. Structure of the extracellular polysaccharide of *Erwina amylovora:* a preliminary report. *Acta Hortic.* 273:211–19
158. Stacey, G., So, J.-S., Roth, L. E., Lakshmi S. K. B., Carlson, R. W. 1991. A lipopolysaccharide mutant of *Bradyrhizobium japonicum* that uncouples plant from bacterial differentiation. *Mol. Plant-Microbe Interact.* 4:332–40
159. Staskawicz, B. J., Dahlbeck, D., Miller, J., Damm, D. 1983. Molecular analysis of virulence genes in *Pseudomonas solanacearum*. In *Molecular Genetics of the Bacteria-Plant Interaction,* ed. A. Pühler, pp. 345–52. Heidelberg: Springer
160. Steinberger, E. M., Beer, S. V. 1988. Creation and complementation of pathogenicity mutants of *Erwina amylovora*. *Mol. Plant-Microbe Interact.* 1:135–44
161. Stevenson, G., Lee, S. J., Romana, L. K., Reeves, P. R. 1991. The *cps* gene cluster of *Salmonella* strain LT2 includes a second mannose pathway: sequence of two genes and relationship to genes in the *rfb* gene cluster. *Mol. Gen. Genet.* 227:173–80
162. Stock, J. B., Stock, A. M., Mottonen, J. M. 1990. Signal transduction in bacteria. *Nature* 344:395–400
163. Stout, V., Gottesman, S. 1990. RcsB and RcsC, a two component regulator of capsule synthesis in *Escherichia coli*. *J. Bacteriol.* 172:659–69
164. Stout, V., Torres-Cabassa, A., Maurizi, M. R., Gutnick, D., Gottesman, S. 1991. RcsA, an unstable positive regulator of capsular polysaccharide synthesis. *J. Bacteriol.* 173:1738–47
165. Sutherland, I. W. 1977. Bacterial exopolysaccharides—their nature and production. In *Surface Carbohydrates of the Prokaryotic Cell,* ed. I. W. Sutherland, pp. 27–96. New York: Academic
166. Sutherland, I. W. 1985. Biosynthesis and composition of Gram-negative bacterial extracellular and wall polysaccharides. *Annu. Rev. Microbiol.* 39: 243–70
167. Sutherland, I. W. 1988. Bacterial surface polysaccharides: structure and function. *Int. Rev. Cytol.* 113:187–231
168. Sutton, J. C., Williams, P. H. 1970. Relation of xylem plugging to black rot lesion development in cabbage. *Can. J. Bot.* 48:391–401
169. Tait, M. I., Sutherland, I. W. 1989. Synthesis and properties of a mutant type of xanthan. *J. Appl. Bacteriol.* 66:457–60
170. Tang, J.-L., Gough, C. L., Daniels, M. J. 1990. Cloning of genes involved in negative regulation of production of extracellular enzymes and polysaccharide of *Xanthomonas campestris* pv. *campestris*. *Mol. Gen. Genet.* 222:157–60
171. Tang, J.-L., Liu, Y.-N., Barber, C. E., Dow, J. M., Wootton, J. C., Daniels, M. J. 1991. Genetic and molecular analysis of a cluster of *rpf* genes involved in positive regulation of synthesis of extracellular enzymes and polysaccharide in *Xanthomonas campestris* pv. *campestris*. *Mol. Gen. Genet.* 226:409–17
172. Tolmasky, M. E., Staneloni, R. J., Leloir, L. F. 1982. Lipid-bound saccharides in *Rhizobium meliloti*. *J. Biol. Chem.* 257:6751–57
173. Torres-Cabassa, A. S., Gottesman, S. 1987. Capsule synthesis in *Escherichia coli* K-12 is regulated by proteolysis. *J. Bacteriol.* 169:981–89
174. Torres-Cabassa, A., Gottesman, S., Frederick, R. D., Dolph, P. J., Coplin, D. L. 1987. Control of extracellular polysaccharide biosynthesis in *Erwina stewartii* and *Escherichia coli* K-12: a common regulatory function. *J. Bacteriol.* 169:4525–31
175. Trisler, P., Gottesman, S. 1984. *lon* transcriptional regulation of genes necessary for capsular polysaccharide synthesis in *Escherichia coli* K-12. *J. Bacteriol.* 160:184–91
176. Troy, F. A. 1979. The chemistry and biosynthesis of selected bacterial capsular polymers. *Annu. Rev. Microbiol.* 33:519–60
176a. Urzainqui, A., Ahmann, D., Walker, G. C., 1992. Exogenous suppression of the symbiotic deficiencies of *Rhizobium meliloti exo* mutants *J. Bacteriol.* In press
177. Uttaro, A. D., Cangelosi, G. A., Geremia, R. A., Nester, E. W., Ugalde, R. A. 1990. Biochemical characterization of avirulent *exoC* mutants of *Agrobacterium tumefaciens*. *J. Bacteriol.* 172: 1640–46

178. Van Alfen, N. K. 1989. Reassessment of plant wilt toxins. *Annu. Rev. Phytopathol.* 27:551–50
179. Van Alfen, N. K., McMillan, B. D., Dryden, P. 1987. The multi-component extracellular polysaccharide of *Clavibacter michiganense* subsp. *insidiosum. Phytopathology* 77:496–501
180. Van Alfen, N. K., McMillan, B. D., Wang, Y. 1987. Properties of the extracellular polysaccharides of *Clavibacter michiganense* subsp. *insidiosum* that may affect pathogenesis. *Phytopathology* 77:501–5
181. van den Bulk, R. W., Löffler, H. J. M., Dons, J. J. M. 1990. Inhibition of callus development from protoplasts of *Lycopersicon peruvianum* by extracellular polysaccharides of *Clavibacter michiganense* subsp. *michiganense. Plant Science* 71:105–12
182. Vanderslice, R. W., Doherty, D. H., Capage, M. A., Betlach, M. R., Hassler, R. A., et al. 1989. Genetic engineering of polysaccharide in *Xanthomonas campestris.* In *Recent Developments in Industrial Polysaccharides: Biomedical and Biotechnological Advances,* ed. V. Crescenzi, I. C. M. Dea, S. S. Stivola, pp. 145–56. New York: Gordon & Breach Science
183. Vimr, E. R., Aaronson, W., Silver, R. P. 1989. Genetic analysis of chromosomal mutations in the polysialic acid gene cluster of *Escherichia coli* K1. *J. Bacteriol.* 171:1106–17
184. Whitfield, C. 1988. Bacterial extracellular polysaccharides. *Can. J. Microbiol.* 34:415–20
185. Whitfield, C., Sutherland, I. W., Cripps, R. E. 1981. Surface polysaccharides in mutants of *Xanthomonas campestris. J. Gen. Microbiol.* 124:385–92
186. Williams, M. N. W., Hollingsworth, R. I., Klein, S., Signer, E. R. 1990. The symbiotic defect of *Rhizobium meliloti* exopolysaccharide mutants is suppressed by *lpsZ*$^+$, a gene involved in lipopolysaccharide biosynthesis. *J. Bacteriol.* 172:2622–32
187. Willis, D. K., Rich, J. J., Hrabak, E. M. 1991. *hrp* genes of phytopathogenic bacteria. *Mol. Plant-Microbe Interact.* 4:132–38
188. Xu, P., Iwata, M., Leong, S., Sequeira, L. 1990. Highly virulent strains of *Pseudomonas solanacearum* that are defective in extracellular polysaccharide production. *J. Bacteriol.* 172:3946–51
189. Yang, C., Signer, E. R., Hirsch, A. M. 1992. Nodules initiated by *R. meliloti* exopolysaccharide (*exo*) mutants lack a discrete, persistent meristem. *Plant Physiol.* 98:143–51
190. Young, D. H., Sequeira, L. 1986. Binding of *Pseudomonas solanacearum* fimbriae to tobacco leaf cell walls and its inhibition by bacterial extracellular polysaccharides. *Physiol. Mol. Plant Pathol.* 28:393–402
191. Yu, N., Hisamatsu, M., Amemura, A., Harada, T. 1983. Structural studies on an extracellular polysaccharide (APS-I) of *Rhizobium meliloti* 201. *Agric. Biol. Chem.* 47:491–98
192. Zevenhuizen, L. P. T. M. 1990. Recent developments in *Rhizobium* polysaccharides. In *Novel Biodegradable Microbial Polymers,* ed. E. A. Dawes, pp. 387–402. Dordrecht: Kluwer Academic
193. Zevenhuizen, L. P. T. M., van Neerven, A. R. W. 1983. (1-2)-β-D-glucan and acidic oligosaccharides produced by *Rhizobium meliloti. Carbohydr. Res.* 118:127–34
194. Zhan, H., Gray, J. X., Levery, S. B., Rolfe, B. G., Leigh, J. A. 1990. Functional and evolutionary relatedness of genes for exopolysaccharide synthesis in *Rhizobium meliloti* and *Rhizobium* sp. strain NGR234. *J. Bacteriol.* 172:5245–53
195. Zhan, H., Lee, C. C., Leigh, J. A. 1991. Low phopsphate regulation of the second exopolysaccharide (EPSb) in *Rhizobium meliloti* strain SU47. *J. Bacteriol.* 173:7391–94
196. Zhan, H., Leigh, J. A. 1990. Two genes that regulate exopolysaccharide production in *Rhizobium meliloti. J. Bacteriol.* 172:5254–59
197. Zhan, H., Levery, S. B., Lee, C. C., Leigh, J. A. 1989. A second exopolysaccharide of *Rhizobium meliloti* strain SU47 that can function in root nodule invasion. *Proc. Natl. Acad. Sci. USA* 86:3055–59
198. Belleman, P., Geider, K. 1992. Localization of transposon insertions in pathogenicity mutants of *Erwinia amylovora* and their biochemical characterization. *J. Gen. Microbiol.* 138:931–40

Annu. Rev. Microbiol. 1992. 46:347–75

DOUBLE-STRANDED AND SINGLE-STRANDED RNA VIRUSES OF *SACCHAROMYCES CEREVISIAE*[1]

Reed B. Wickner

Section on Genetics of Simple Eukaryotes, Laboratory of Biochemical Pharmacology, National Institute of Diabetes, Digestive and Kidney Diseases, National Institutes of Health, Bethesda, Maryland 20892

KEY WORDS: ribosomal frameshifting, replication in vitro, packaging site, *KEX* genes, N-acetyltransferase, killer toxin

CONTENTS

Abstract

Yeast RNA viruses include L-A (and its toxin-encoding satellites M_1, M_2,. . .) and L-BC dsRNA viruses and the single-stranded replicons 20S RNA and 23S RNA. L-A has a single-segment 4.6-kb linear genome encoding a major coat protein (*gag*) and its RNA-dependent RNA polymerase (*pol*), the latter expressed as a *gag-pol* fusion protein formed by a −1 ribosomal frameshift. In vitro replication, transcription, and binding systems for L-A have been used to define *cis* sites necessary for packaging and replication of viral RNA. Cellular functions that promote viral replication include the *MAK3*-encoded N-acetyltransferase whose modification of the *gag* N terminus is necessary for L-A virus assembly. The toxins encoded by the M satellite RNAs are processed by enzymes (*KEX1* and *KEX2*, for killer expression) whose study led to discovery of mammalian hormone-processing enzymes.

20S RNA is an apparently naked circular RNA replicon (with a dsRNA form called W) encoding a RNA polymerase-like molecule. Its copy number is induced 10,000-fold in 1% potassium acetate, and it is subject to the same *SKI* antiviral system that represses L-A, L-BC, and M dsRNA copy number.

INTRODUCTION

For yeasts and other fungi, cytoplasmic genetic elements are the biological equivalent of viruses for higher eukaryotes. Because of the high frequency of mating among yeasts and of hyphal fusion among fungi in nature, spread by cell fusion is very efficient. As a result, these elements need not go outside of the cell. Thus, the *Saccharomyces cerevisiae* dsRNA viruses, L-A and L-BC, are present in most strains, and strains lacking the yeast retroviruses Ty1 or Ty2 have not been found in spite of a careful search. No virus with a natural extracellular route of infection has been described in a yeast or fungus, although many viruses are widely distributed in their respective hosts.

Although a natural extracellular infectious cycle for yeast viruses is unknown, El-Sherbeini & Bostian (28) have shown that viral particles can be introduced into cells along with plasmids in the process of transformation and can become established in the recipient cells. Recently, Russell et al (97) showed that RNA can be directly introduced into yeast cells and is expressed there, a method that shows great promise for development of a RNA transfection procedure.

Yeast virology (Table 1) began with the discovery of the cytoplasmically inherited killer phenomenon (75, 108) and its association with dsRNA (4, 124). Some strains of *Saccharomyces cerevisiae* secrete a protein toxin that is lethal to other strains but not to the toxin-secreting strain, which is said to be immune or resistant. The toxin and immunity protein are encoded by M_1, a dsRNA satellite of the major yeast dsRNA virus, L-A (9–11). Several other satellites of L-A, each encoding a toxin-immunity system, have been described and are called M_2, M_3, M_{28}, etc (89, 138). L-A itself is a 4.6-kb linear single-segment dsRNA virus. L-A represents a family of closely related viruses distinguishable on the basis of sequence, function, or both. Another single-segment dsRNA virus whose genome is the same size as L-A, but unrelated in sequence at the level of hybridization, is L-BC (109). L-BC is

Table 1 Families of yeast viruses

dsRNA	Size (kb)	Encodes	References
L-A	4.6	*gag:* Major coat protein, 76 kDa *gag-pol:* Minor viral protein, 176 kDa; homology with RNA-dependent RNA polymerases; ssRNA-binding activity	42, 58, 62
M_1, M_2, M_{28}, . . .	1.0–1.8	Killer toxin, immunity protein precursors	9, 104
L-BC	4.6	Major coat protein	29, 109, 111
ssRNA (circular?)			
20S RNA (=W dsRNA)	2.6	90-kDa possible RNA-dependent RNA polymerase	48, 79, 95, 126
23S RNA (=T dsRNA)	2.8	Possible RNA-dependent RNA polymerase	126; personal communication[a]
Retroviruses			
Ty1	5.9	LTRs: δ elements TYA: *gag* TYB: protease, integrase, reverse transcriptase, RNase H	5, 23
Ty2	5.9	LTRs: δ elements TYA: *gag* TYB: protease, integrase, reverse transcriptase, RNase H	5, 23
Ty3	5.4	LTRs: σ elements TYA3: *gag* TYB3: protease, reverse transcriptase, endonuclease	5, 53
Ty4	6.3	LTRs: τ elements	5, 23

[a] R. Esteban, L. Esteban & N. Rodriguez-Cousino, personal communication.

like L-A in many respects but is a distinct system, and, having been studied little, is not discussed further in this review.

20S RNA was first described as a species of stable RNA whose synthesis was induced by the culture conditions used to induce sporulation, namely, absence of a nitrogen source and the presence of acetate as the carbon source (65, 66). Although investigators first believed this RNA was a form of rRNA or mRNA involved in yeast meiosis or sporulation, they found that the ability to induce 20S RNA was inherited in a non-Mendelian fashion (48). 20S RNA is now known to be an independent replicon, probably a circular ssRNA (78). W dsRNA, a minor species present in many strains (126), is apparently a replicative intermediate of 20S RNA (79, 95). The relationship between the recently discovered 23S RNA (R. Esteban, N. Rodriguez-Cousino & L. Esteban, personal communication) and another minor dsRNA, called T (126), is similar to that between 20S RNA and W (see below).

The L-A virus and its satellites have been the most intensively studied of the many fungal dsRNA viruses and therefore occupy most of our attention (also reviewed in 18, 70, 130, 131). With the in vitro systems for replication, transcription, and packaging of L-A and its satellites, of vectors expressing their proteins in vivo, and of molecular analysis of the extensive genetics of this system developed in the past, this area has begun to make contributions of interest beyond the boundaries of this system itself.

The studies of the processing, secretion, and mechanism of action of the killer toxins have led to the discovery of mammalian cellular genes, the *KEX2*-like proteases responsible for processing prohormones (also reviewed in 3, 77). Several other non-Mendelian elements have been described in yeast including [PSI] (22), [URE3] (1, 68), [eta] (73), and [D] (33). While their characterization is more preliminary, they promise to yield interesting results in the future.

VIRAL REPLICATION CYCLE OF L-A

Mature L-A virions contain a single molecule of L-A dsRNA in a 160S 40-nm isometric particle composed of about 120 molecules per particle of the 76-kilodalton (kDa) major coat protein (called *gag*) and of 1 or 2 molecules per particle of a 170-kDa *gag-pol* fusion protein (15, 17, 26, 32, 42, 56, 62). The virions have a conservative transcriptase activity that produces full-length viral (+) strands (38, 82, 100, 137). The (+) strands are all extruded from the viral particle (32) where they serve as mRNA to make the viral proteins and the species that is packaged to form new viral particles (Figure 1). The new viral particles, containing only a (+) strand, have a replicase activity that synthesizes a (−) strand on the (+) strand template to produce a single L-A dsRNA molecule inside the viral particle, thus completing the cycle (38, 40).

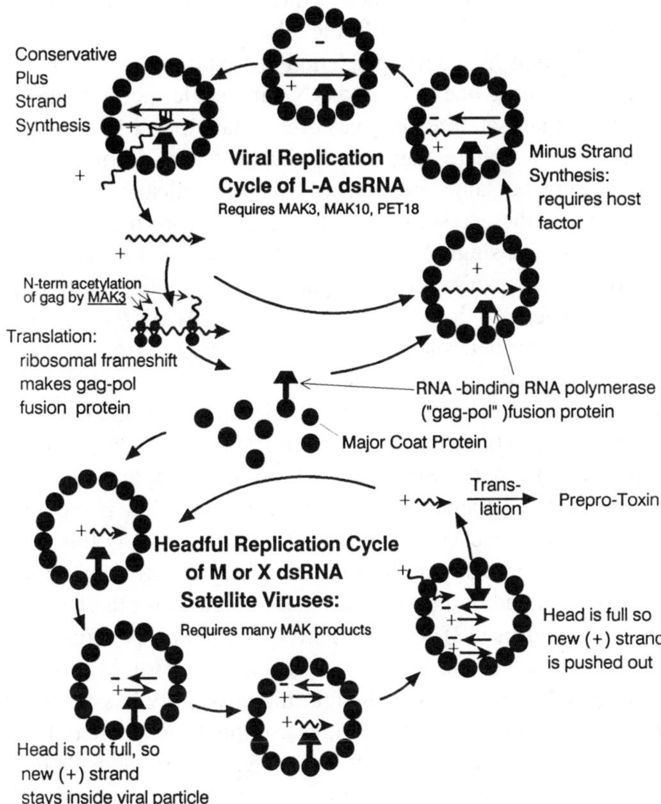

Figure 1 The viral replication cycles of L-A and its satellites.

Because the transcription step is conservative, the overall process is as well, like that in reovirus (reviewed in 102), but unlike that in the *Penicillium stoloniferum* PsV-S virus (16), *Aspergillus foetidus* AfV-S (90), and bacteriophage $\phi6$ (122), which are all semiconservative. The difference between conservative and semiconservative may simply reflect the fate of the pairing between the parental (−) strand and the (+) strand transcript just behind the growing point. If they are allowed to stay together, then the parental (+) strand is displaced. If they are forced to separate, then the parental strands are together when the transcript is finished. Because the pairing of nascent (+) strand and parental (−) strand at the growing point is probably the same, the difference between conservative and semiconservative systems is really a difference of the strand-segregation mechanism.

The satellite RNAs of L-A, namely the toxin-encoding M dsRNAs, and deletion mutants of M and L-A itself, have a replication cycle like that of

L-A, except for a single difference: while L-A particles have only one (+) ssRNA molecule or dsRNA molecule, those of M_1 may have one or two molecules (32), and X, a 530-bp deletion mutant derived from L-A, is found in particles containing one to eight dsRNA molecules per particle (34). Since particles with small satellite genomes often retain their (+) strand transcripts within the particle where they can be replicated to form another dsRNA molecule within the same particle, a "headful replication" model was proposed to explain these results (32). The particles are presumably primarily adapted to contain a single L-A dsRNA molecule, so that satellites less than half the size of L-A can, after the packaging of a single (+) strand, replicate within the particle until the head is full (Figure 1). When the packaging signal is incorporated into heterologous in vivo transcripts (see below), these are found to be packaged one per particle in L-A–encoded particles not along with a L-A genome (37). Thus, a single particle initially packages a single (+) strand per particle. The name "headful replication" was chosen to distinguish this mechanism from the Streisinger "headful packaging" mechanism that occurs in certain DNA phages. In those cases, the DNA is replicated in the cytoplasm and then packed into the head until the head is full. The headful replication model probably applies to other fungal dsRNA viruses, and data that can be interpreted in this way have been obtained for the *Ustilago maydis* killer-associated virus (12). Is this a purely mechanical phenomenon? So far, no evidence indicates that a specific mechanism extrudes new (+) strand transcripts from the particles, although this possibility has not been eliminated. Probably, when the head is full, transcripts have nowhere to go but out of the particle and are driven out by the energy of phosphodiester-bond formation. If the head is not full, the transcript may remain inside simply because there is nothing forcing it out.

It is not clear why all dsRNA viruses package their (+) ssRNA rather than the dsRNA and why replication of RNA viruses is, in general, asynchronous [(+) and (−) strands made in different reactions], unlike DNA replication, which is roughly synchronous in most systems.

That both (+) and (−) strands are synthesized inside viral particles seems to be a unique (and so far a general) feature of dsRNA viruses. Reoviruses and rotaviruses make (−) strands inside viral core particles, as do *Penicillium* viruses (16, 47, 102). Plus ssRNA viruses make both strands outside of particles, as do (−) strand viruses. Thus, the difference between a (+) ssRNA and a dsRNA virus is not simply which species leaves the cell to start a new infection. One could imagine a virus with the same replication cycle as L-A in which the (+) strand–containing form was infectious and was converted (inside the particle) into the dsRNA-containing form on entry into a new cell, where transcription ensued. Such a virus has not been described.

IN VITRO SYSTEMS: PACKAGING, REPLICATION, AND TRANSCRIPTION

The L-A and satellite virus replication cycles were largely worked out based on in vivo experiments and experiments using isolated viral particles where both enzymes and templates were supplied by the particles. These might be called *in viro* (in the virus) experiments. But to carry out detailed studies of the interaction of enzymes and templates, one must have a template-dependent system. Treatment of L-A dsRNA-containing particles under low-ionic-strength conditions opened the particles, releasing their dsRNA, making them accessible to added (+) ssRNA templates (41). These opened empty particles were reisolated free of the dsRNA and used for enzymological studies. In the replication reaction, viral (+) ssRNA was bound specifically (30, 41, 42), and on addition of rNTPs and an extract of uninfected cells (host factor), it was converted to the dsRNA form (41). The transcription reaction, the synthesis of viral (+) ssRNA from an added dsRNA template, was also carried out by these opened empty particles (43).

Binding of Viral (+) ssRNA: the Packaging Signal

The sites recognized in the binding reaction were determined using a 530-bp deletion mutant of L-A called X (34). X dsRNA lacks almost all of the internal coding sequences of L-A but has 25 bp from the 5' end of L-A's (+) strand a 17-bp inverted repeat of the 5' end sequence and the 3' 488 bases of L-A's (+) strand (34). Because X is packaged, replicated, and transcribed in vivo, it must contain all the essential sites necessary for these processes. The binding sites in X and M_1 are similar: a stem-loop structure with an A residue protruding from the 5' side of the stem (Figure 2) (31, 37). In vitro binding studies showed that the stem structure, but not the stem sequence, was important for the binding as was the protruding A residue. The loop sequence was also important (37). In each case, the binding site is located about 400 bases from the 3' end of the (+) strand.

The in vivo significance of this binding reaction was found when either binding site was inserted in a yeast expression vector so that the RNA produced from the vector would include this signal. This RNA, produced in a strain carrying the L-A virus, was packaged inside L-A virions (37). Thus, this is a portable packaging signal for the L-A virus and its satellites. Moreover, the density of the viral particles carrying the vector transcripts showed that they were not packaged along with L-A (+) strands in the majority of the particles, but rather were packaged one molecule per particle, as predicted by the headful replication model.

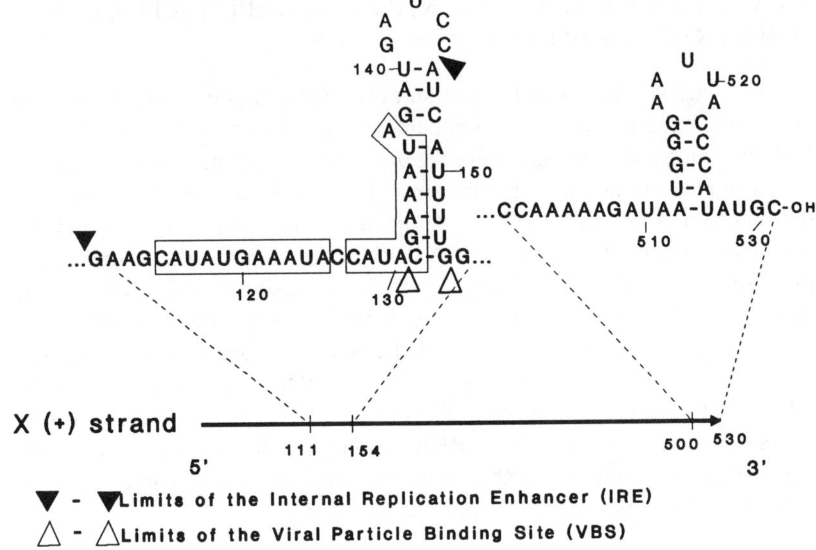

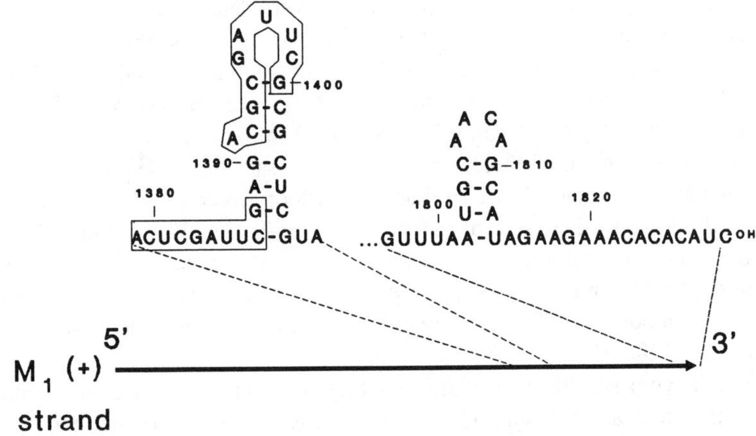

Figure 2 Viral packaging site (viral binding site) and sites necessary for replication on L-A and M_1 (+) ssRNAs. The region between the open triangles is necessary and sufficient for binding of L-A (+) ssRNA [or X (+) ssRNA] to opened empty particles but the binding is enhanced by inclusion of 10 more nucleotides upstream (30, 37). The 3' end site of M_1 can fully substitute for that of L-A, although they have little similarity in sequence or secondary structure (31). The boxed sequences are direct repeats present in the packaging sites of L-A and M_1. Their significance as repeats has not been examined. The sequence of this part of M_1 is from Georgopoulos et al (49).

The Replication Reaction

The opened empty particles convert added viral (+) ssRNA to the dsRNA form (41). This reaction requires Mg^{2+}, NTPs, a crude fraction from uninfected cells (called host factor), and a low concentration of polyethylene glycol (41). The template specificity is determined by two sites, one at the 3' end where synthesis begins, and the other at an internal site (the internal replication enhancer) that largely overlaps with and may be identical to the viral-binding site discussed above (31). The 3'-end site consists of the last four nucleotides at the 3' end and an adjacent stem-loop structure. While the sequence of the stem is not important, the stem structure and the sequence of the loop are critical (31).

The nature of the interaction of the polymerase with the internal site and the 3'-end site has been investigated (44) using as a model enzymes that interact with two DNA sites such as the type I restriction endonucleases (the recognition site and the cut site) and DNA-dependent RNA polymerases (upstream transcription activation sequences and promoter sites). Because synthesis must begin at the 3' end and the internal site overlaps with or is identical to the binding site, the polymerase probably first binds to the internal site. The internal site does not alter the polymerase affinity for the 3' end by an allosteric mechanism, nor does the polymerase track down the RNA chain from the internal site to the 3' end. Rather, the binding of the polymerase to the internal site probably brings it spatially close to the 3' end and thus facilitates its finding the specific structure on the 3' end where it will start synthesis (44).

The host factor (41) has not yet been characterized. It is known, however, that it is not determined by any of the host genes, *MAK3, MAK10,* or *PET18,* that are needed for the propagation of L-A in vivo (T. Fujimura & R. B. Wickner, unpublished data).

The Transcription Reaction

Adding viral dsRNA to the opened empty particles, with the host factor(s) and high concentrations of polyethylene glycol, results in the conservative synthesis of viral (+) ssRNA (43). This reaction, like the replication reaction, is specific for viral templates, but the recognized *cis*-acting signals have not been determined. The enzyme converts much of the (+) strands synthesized to the dsRNA form so that the complete cycle from dsRNA to dsRNA goes on in vitro (43).

L-A–ENCODED INFORMATION AND ITS EXPRESSION

Sequencing of full-length cDNA clones of L-A dsRNA (62) showed that L-A has two open reading frames (ORFs): the 5' ORF (*gag*) encodes the 76-kDa major coat protein and the 3' ORF (*pol*) overlaps *gag* by 130 nucleotides and encodes a protein with sequence patterns (67) diagnostic of the RNA-dependent RNA polymerases of (+) strand RNA viruses and dsRNA viruses (Figure 3) (62). Localized mutagenesis of the region surrounding the most conserved of the motifs typical of the RNA-dependent RNA polymerases has shown that there are wider domains essential for the propagation of the M_1 satellite virus that include the conserved motifs (93). These domains have a largely conserved predicted secondary structure, as one might expect (93). The *pol* ORF is in the −1 frame with respect to *gag* (62) and is only expressed as a *gag-pol* fusion protein, present in the virions at about one to two copies per particle (42).

gag-pol *Fusion Protein Formed by −1 Ribosomal Frameshifting*

Most retroviruses form their *gag-pol* fusion proteins by −1 ribosomal frameshifting using a mechanism called simultaneous slippage (63; reviewed in 55). When the ribosomes encounter a sequence of the form . . .X XXY YYZ QRS. . ., where the *gag* (unshifted) reading frame is shown, then the peptidyl tRNA in the ribosomal P site bound to XXY and the aminoacyl tRNA in the ribosomal A site bound to YYZ can both slip back one nucleotide on the mRNA and still have their nonwobble bases properly paired with the mRNA (Figure 4). Then the next tRNA bound will read the triplet ZQR instead of

Two L-A ORFs Encode Chimeric RNA Polymerase - RNA Binding Protein with Major Coat Protein Domain

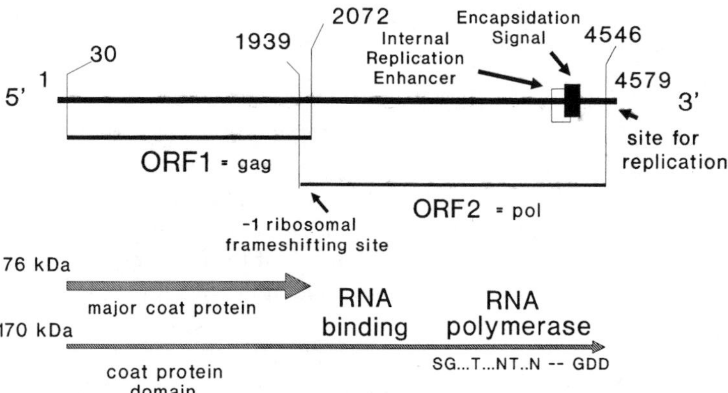

Figure 3 L-A–encoded proteins and their expression. The 170-kDa *gag-pol* fusion protein is formed by ribosomal frameshifting fusing ORF1 and ORF2. The internal replication enhancer and the encapsidation signal are inside the *pol* ORF.

QRS, and translation will continue in the −1 frame to form the fusion protein. The heptamer sequence X XXY YYZ is called the slippery site, and the presence of a strong stem-loop or pseudoknot structure just 3' to this site is generally necessary to produce efficient frameshifting (13, 63). The overlap region of L-A's *gag* and *pol* ORFs contains a sequence of this form, . . .G GGU UUA GGA. . ., which is followed by a potential pseudoknot (62). This region is now known to be precisely that necessary and sufficient to insure frameshifting by L-A (26). Efficient frameshifting in this system requires, as expected, that the Xs all be the same, but X can be G, A, C, or U, and viral slippery sites occur with each of these possibilities. Y, however, can only be A or U in L-A (26) and only A or U are found in nature (55). In addition to the importance of the ease with which the shifting tRNAs can re-pair after shifting back one base, the ease with which the tRNAs can unpair may be important, and A-site tRNA-mRNA pairing may be stronger than P-site pairing (26, 26a). This possibility would account for the inability of slippery sites in which Y = G or C to shift efficiently, even though repairing ought to be quite good.

Although ribosomal frameshifting has been demonstrated or plausibly postulated in many retroviruses, coronaviruses, the L-A dsRNA virus, several plant (+) ssRNA viruses, and bacteriophages T7 and λ (reviewed in 55), the importance of the efficiency of frameshifting has not been assessed in any of these systems. However, an assessment was made recently in the L-A system, and it was found that the efficiency of frameshifting was critical (26a). Using a system in which M_1 propagation was supported by an L-A cDNA expression plasmid (132), the ratio of *gag* to *gag-pol* was altered by changes in the

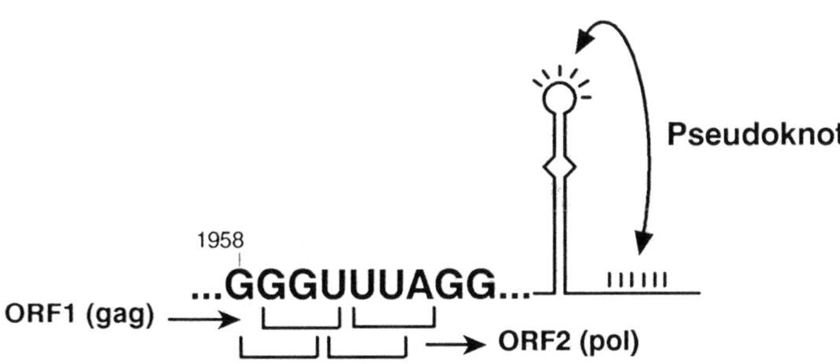

Figure 4 The signals in L-A responsible for −1 ribosomal frameshifting to make the *gag-pol* fusion protein. The figure shows the slippery site (GGGUUUA) and the nearby pseudoknot in the region of overlap of the *gag* and *pol* ORFs. A pseudoknot is a stem-loop structure whose loop can base-pair with a region down-stream of the stem. The advance of the ribosome is slowed by the pseudoknot, facilitating the ribosome's movement in the reverse direction. The ability of the tRNAs to correctly re-pair their nonwobble bases in the −1 frame is essential for efficient frameshifting (see text).

slippery site, by chromosomal mutations affecting the efficiency of frameshifting in the normal plasmid, or by making synthesis of *gag-pol* dependent on a +1 shift (26a). In all cases, M_1 propagation was supported if the efficiency of shifting was within twofold of the normal rate. In light of the parallels of this system with retroviruses, this suggests that drugs affecting ribosomal frameshifting might be useful as antiretroviral agents.

Why do L-A, many retroviruses, and some (+) ssRNA viruses use ribosomal frameshifting (55, 63) [or in some cases translational read-through of termination codons (87, 88, 121)] to make their *gag-pol* or other fusion proteins, and not splicing or RNA editing? All retroviruses, dsRNA viruses, and (+) ssRNA viruses use their (+) strands (*a*) for translation, (*b*) as the species packaged to form new virions, and (*c*) as a template for replication. If, for purposes of translation, they splice some of the (+) strands, they will be generating mutants of the virus, unless they remove, in the spliced-out region, some signal necessary for packaging or replication of the genome. Perhaps for this reason, splicing [and RNA editing (85), to which this same logic applies] is unknown among dsRNA and (+) ssRNA viruses. Although retroviruses splice their (+) ssRNA to make *env* mRNA and, in the case of HIV, the mRNAs for *tat, rev,* and other regulatory proteins, all these splice events remove the packaging site just downstream of the splice donor site. Ribosomal frameshifting and translational readthrough do not change the (+) strands and so do not generate mutations.

L-A and M_1 (+) ssRNA Are Uncapped

The L-A and M_1 transcripts are uncapped (81) and thus ought to be poorly translated. But L-A is well translated and the major coat protein can constitute 5–10% of cell protein (25). Both L-A and M_1 (+) strands have very short 5' untranslated regions that may make the caps unnecessary by altering translation initiation (50, 52, 54, 111). In the RNA transient expression studies of Russell et al (96), capped messages were translated more efficiently than those that were uncapped. However, the 5' ends of L-A and M_1 did not promote translation of luciferase mRNA in yeast better than nonviral 5' ends. The poliovirus 5' end is similarly uncapped and the virus turns off host translation by inducing the cleavage of a key cap-binding protein and by having a special 5' structure that facilitates the internal entry of ribosomes, thus bypassing the cap requirement.

Assembly Model for L-A Virus

The *gag-pol* fusion protein has a ssRNA-binding activity that the *gag* protein does not have (42). This activity was demonstrated on Northwestern blots using protein that had been denatured with SDS and urea. Perhaps for this reason, the binding is not specific for viral (+) strands. The existence of a *gag-pol* fusion protein and the ssRNA binding activity of the *pol* domain of this protein suggested a model for how the L-A virus is assembled (37, 42) (Figure 5). First the *pol* domain recognizes and binds the packaging signal on the L-A (+) strands, and then the *gag* domain of the fusion protein primes the assembly of *gag* units to form the coat. This model explains, in a very natural way, how particles containing one viral (+) strand and the polymerase can be formed, as well as how a coat can be formed.

This model can be used to explain why the efficiency of ribosomal frameshifting is so critical for the propagation of M_1 (26a). An increased proportion of *gag-pol* fusion proteins might start the assembly of more particles than there is *gag* protein to supply, so that only incomplete particles are assembled. Because there are about 120 *gag* molecules per particle and the efficiency of frameshifting is normally about 1.8%, each particle should contain two *gag-pol* fusion molecules. This suggests that it may be a dimer of *gag-pol* that binds to the viral (+) strands initiating packaging. Reducing the proportion of *gag-pol* fusion proteins might result in a much greater reduction in the rate of dimer formation and thus inefficient assembly. Alternatively, a relative excess of *gag* may result in particles completing assembly and closing before the *gag-pol* fusion protein can find a viral (+) strand.

Recently, mutations in chromosomal genes affecting ribosomal frameshifting were isolated (26a). These mutations can adversely affect M_1 virus propagation without apparently affecting cell growth, again arguing that drugs that affect frameshifting might be well tolerated. The vector used to find these chromosomal genes is also being used in a plate assay to screen for such drugs (26).

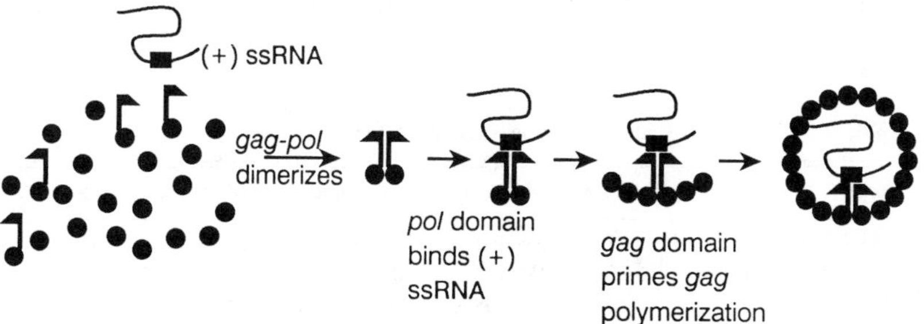

Figure 5 Assembly model for L-A (26a, 37, 42).

CHROMOSOMAL GENES AFFECTING L-A AND M dsRNA VIRUS PROPAGATION

Genetic studies of the L-A and M_1 viral system showed that the expression of viral products and propagation of viral genomes requires an array of chromosomal genes (Table 2). *MAK* (maintenance of killer) genes are defined by mutations that result in loss of viral RNAs. M_1 needs 30 such genes (reviewed in 131), of which only *MAK3, MAK10,* and *PET18* are required by L-A. The *SKI* (for superkiller, the phenotype of the mutants) gene products repress the replication of several viral genomes, and the *KEX1* and *KEX2* products are needed for proteolytic processing of the killer preprotoxin (see Killer Toxin Processing, Secretion, and Action).

MAK3 *Is a N-Acetyltransferase Whose Acetylation of the* gag *N Terminus Is Necessary for Viral Assembly*

The sequence of the *MAK3* gene product shows considerable homology with the *rimI* gene product of *Escherichia coli,* a N-acetyltransferase that modifies the N terminus of the ribosomal protein S18, as well as with other N-acetyltransferases (J. C. Tercero, L. Riles & R. B. Wickner, in preparation). The *gag* protein, produced from a L-A cDNA expression plasmid, was N-terminally blocked when produced in a wild-type strain, but unblocked when produced in a *mak3-1* host (J. C. Tercero & R. B. Wickner, in preparation). The *gag* and *gag-pol* proteins form particles and encapsidate the L-A (+) ssRNA that is produced by the expression plasmid in a wild-type host but do not detectably form particles or encapsidate (+) ssRNA in the mutant. This suggests that the *MAK3* protein blocks the N terminus by acetylation and that blocking is necessary for proper particle assembly (J. C. Tercero & R. B. Wickner, in preparation). This situation parallels that in Rous sarcoma virus and many plant RNA viruses whose coat proteins are N-terminally acetylated. Apparently, this is the first time that the physiological role of this modification has been tested. The myristylation of the N

Table 2 Chromosomal genes affecting yeast RNA replicons[a]

Name	Number	M_1 replic	L-A replic	L-BC replic	20S RNA
MAK3, 10, PET18	3	Needed	Needed	NE	NE
Other *MAK* genes	>20	Needed	NE	NE	NE
SKI	6	Represses	Represses	Represses	Represses
Porin, *NUC1*	2	NE	Represses	?	?
CLO1	1	NE	NE	Needed	?

[a] The effects of mutations in the indicated genes on the different RNA replicons are indicated. NE, no effect; ?, not tested. The effects of these genes on 23S RNA (T dsRNA) have not been tested.

terminus of many mammalian retroviruses, however, is known to be important for proper membrane association of retroviral *gag* proteins (91), and myristylation of poliovirus is required for proper assembly (19, 86).

MAK10 *and* PET18 *Affect Viral Stability*

The *MAK10* and *PET18* genes are both necessary for L-A replication, and therefore also for M. The *pet18* mutations consist of large deletions of several open reading frames, and two distinct regions are needed for L-A propagation (116). Although the *pet18* mutants are deletions of both of these regions (116), the propagation of L-A is only defective above 30°C (72). The effects on cell growth and mitochondrial DNA replication result from genes different from those needed for L-A (116).

The *MAK10* gene encodes a 733–amino acid protein whose production is repressed by the presence of glucose (Y.-J. Lee & R. B. Wickner, unpublished). Mutants whose propagation of L-A is temperature sensitive have been used to study the role of *MAK10* and *PET18* in L-A virus replication. Viral particles isolated from the mutants maintained at the permissive temperature were structurally unstable, suggesting that these proteins have a role in viral assembly or are actually part of the virion (39, 40).

MAK Genes Needed by M, but not by L-A

Among the over 20 chromosomal genes in this class, *MAK1* is DNA topoisomerase I (114), *MAK8* is ribosomal protein L3 (133), and *MAK11* is an essential membrane-associated protein (61). *MAK16* encodes a nuclear protein, and *mak16*^{ts} mutants arrest at a point in G_1 at which they can mate (like *cdc28*^{ts} mutants) (129). The polyamines, putrescine and spermidine (or spermine), are also necessary for M propagation (20, 119). Several genes involved in translational control of the *GCN4* gene also affect M_1 propagation (54). While these studies have yielded information about the host role of some of the *MAK* gene products, they have not explained how these genes are involved in M_1 replication.

X dsRNA is a 530-bp deletion mutant of L-A lacking nearly all of the L-A coding sequences. Although X is derived entirely from L-A, it requires for its replication many chromosomal genes that M_1 needs but that L-A itself does not (34). The satellite RNA M_1 and the defective interfering RNA X follow essentially the same replication cycle as L-A (see above), so it is puzzling that they need the products of many chromosomal genes that L-A does not need. We know of two differences between M_1 or X and L-A: (*a*) both of the former must borrow viral proteins from L-A and (*b*) M_1 and X are smaller than L-A and so do not fill the particles. If the L-A (+) strand is packaged by the protein synthesized from it, acting as mRNA (*cis* packaging), then the M_1 and X (+) strands might be more exposed to the ravages of the yeast antiviral *SKI* system (see below). Indeed, *ski*$^-$ mutations suppress M_1's need for many *MAK* products. But nothing of what is known of the *MAK* proteins suggests that they could have such a protective effect. The packaging of smaller genomes might produce relatively unstable viral particles, and these might require the products of the *MAK* genes for stability. Further work will be needed to resolve this problem.

The SKI *Antiviral System*

Six genes (*SKI2,3,4,6,7,8;* superkiller) act together to repress the propagation of several completely independent replicons in *S. cerevisiae* (78, 94, 115). The *ski*$^-$ mutants show 4- to 10-fold elevated copy numbers of L-A, M, L-BC, and 20S RNA. However, only the effect on the M dsRNA replicon appears to be critical to the cell (94). The *ski*$^-$ mutants, including deletion-insertion mutations of the *SKI8* and *SKI3* genes (92, 110), if they do not contain M or one of the deletion derivatives of M (called S dsRNAs), are apparently completely healthy and show normal growth, mating, and meiosis. However, in the presence of a M replicon, *ski*$^-$ mutants are at least cold-sensitive for growth (8°C) (94). A non-Mendelian element called [D] (for disease) makes *ski*$^-$ killer strains high-temperature sensitive as well and slow growing even at 30°C (33). [D] is not located on ψ, L-A, M_1, or L-BC, but appears to depend on L-A for its propagation. The cytopathology caused by the derepressed replication of M in *ski*$^-$ mutants does not result from the load of dsRNA or viral particles, because elimination of only M from such strains, by removing the repression of L-A copy number, results in a massive increase in dsRNA and viral particles (2), and yet the cells become completely healthy (94). The derepressed M replication may overutilize one or more of the products of the several chromosomal genes needed for M and not for L-A.

The *ski*$^-$ mutations also dramatically affect the requirement of M_1 for the chromosomal *MAK* gene products. Except for *mak3, mak10, mak16,* and *pet18*, the *ski*$^-$ mutations suppress all *mak*$^-$ mutations tested (118). This suggests that the *SKI* gene products act more directly on M_1 propagation than do these *MAK* products.

The *SKI2, SKI3,* and *SKI8* genes have been cloned and sequenced (92, 110; W. R. Widner & R. B. Wickner, unpublished data; Y. Matsumoto, G. Sarkar, S. S. Sommer & R. B. Wickner, unpublished data). SKI3 is a 163-kDa nuclear protein, but its function is completely unknown (92). It has an amino acid repeat pattern (103) also found in CDC23, CDC16, and CYC8 of *S. cerevisiae* and nuc2$^+$ of *Schizosaccharomyces pombe*. This repeat may be concerned with intramolecular interactions (57) or may mediate interaction of proteins with the cytoskeleton and nuclear scaffold (6). The conservation (albeit weak) of a common sequence pattern argues in favor of the interaction of these proteins with a common structure, thus favoring the latter hypothesis.

The *MAK11* gene product (61) has a different repeat pattern first described in β-transducin (35). Although its function is unknown in any of the several proteins in which it has been found, Williams et al (136) have found this pattern in the *TUP1* gene and showed that the TUP1 protein and the CYC8 protein are present together in a nuclear complex. Thus, proteins with one repeat may associate with proteins having the other (136). There is as yet no information on the association of *MAK11* with *SKI3*.

How can the action of the *SKI* proteins be explained? As discussed earlier, L-A mRNA is uncapped. The (presumably) capped L-A mRNA made from a full-length cDNA clone of L-A driven by the PGK1 promoter bypasses many of the chromosomal mutations that would otherwise result in the loss of M_1 (*mak* mutations) and makes the cell a superkiller. Both of these effects also result from *ski* mutations, suggesting that the *SKI* system is a general system whose function is to search out and destroy uncapped RNAs (132). This hypothesis would explain the action of *SKI* on the L-A, L-BC, and 20S RNA systems, all of which replicate in the cytoplasm and so probably have uncapped mRNAs.

Mitochondrial Proteins That Repress L-A Copy Number

Although the mitochondrial genome is not necessary for replication or expression of L-A or M dsRNA, growth on ethanol elevates L-A copy number (83). Because *MAK10* is necessary for L-A propagation and *MAK10* is repressed by glucose (Y.-J. Lee & R. B. Wickner, unpublished data), decreased availability of *MAK10* may be the reason for glucose repression of L-A copy number.

Recently, other connections of L-A with mitochondrial functions have appeared. A disruption of the gene for the major mitochondrial outer membrane porin causes large amounts of L-A viral particles to accumulate (25). Furthermore, mutation of *NUC1,* the major mitochondrial nuclease (123), likewise results in a dramatic increase in L-A copy number (74). The *nuc1* mutations do not produce the superkiller phenotype or detectable increase in M dsRNA copy number, so this effect appears to be distinct from that of the *SKI* genes. Both *mak3* and *mak10* mutations slow growth on glycerol or ethanol (25), but their loss of L-A dsRNA does not cause the slowing (Y.-J. Lee & R. B. Wickner, unpublished data).

L-A GENETICS

Several natural variants of L-A have been recognized based on their genetic interactions with M_1, M_2, *mak,* and *ski* mutations. Due to its complexity, this section is not recommended for those with a weak stomach.

L-A-E cannot maintain M_1 or M_2 in a wild-type strain, but can do so in a *ski*$^-$ mutant (51, 94, 109, 117, 128). Supplying only *gag* protein from an

L-A-H cDNA expression clone is sufficient to enable L-A-E to support M_1 in a wild-type strain, indicating that L-A-E's *pol* domain is normal and its *gag-pol* fusion protein adequate in amount, but that *gag* is either undersupplied or qualitatively abnormal (132). This result also suggests that the *gag* protein protects the viral genome from the *SKI* system.

L-A-H, L-A-HN, and L-A-HNB all can maintain M_1 or M_2 in a wild-type host (the HOK function or H) (51, 109, 134). They differ from each other, however, in their interactions with chromosomal genes. L-A-H can maintain M_2 in an *mkt*$^-$ host at any temperature, but L-A-HN and L-A-HNB can do so only below 30°C (128). Also, L-A-E lowers the copy number of L-A-H, but not of L-A-HN and L-A-HNB (51).

L-A-HNB can make M_1 propagation completely independent of several of the MAK proteins [the B (bypass) function] (120). In experiments using the *mak11-1* mutation, mutant cells carrying L-A-HN lost M_1 rapidly, while those carrying L-A-HNB maintained M_1 stably and at the same copy number as *MAK*$^+$ strains. Expressing only *gag* from the cDNA clone of L-A-HNB was sufficient to allow *mak11-1* mutant cells carrying L-A-HN to act as if they had L-A-HNB and to stably propagate M_1 (132). This suggests that the *MAK11* product interacts with the major coat protein or that it helps the major coat protein protect the viral genome from the *SKI* system. A better understanding of these interactions awaits elucidation of the molecular basis of the antiviral action of the *SKI* system.

DEFECTIVE INTERFERENCE

Deletion mutants of M_1 and of L-A have been described (34, 36, 107) that, like M_1, depend on L-A for their propagation. These mutants repress the copy number of L-A itself (as does M_1) (2), and also tend to interfere with the propagation of M_1. Three such mutants have been cloned and sequenced (30, 69, 112). Using a clone of S14, one such defective interfering deletion mutant of M_1, the region responsible for interference with M_1 propagation, was limited to a 132-nucleotide region that includes the packaging signal for M_1 discussed above. A smaller fragment that also included the packaging signal did not interfere, indicating that more than this site is involved (59).

KILLER TOXIN PROCESSING, SECRETION, AND ACTION

The killer toxins are small proteins initially synthesized as preprotoxins, which are then processed by proteolysis and secreted (9, 104, reviewed in 18). They bind to cell-wall receptors, and the K_1 toxin makes proton pores in the membrane of sensitive cells, ultimately killing them (76). The study of this aspect of the system has proven of great interest as a model for prohormone processing and secretion and as a means of studying the genetics and biochemistry of yeast cell wall constituents.

KEX *Genes and Mammalian Prohormone Processing*

The cleavages undergone by the K_1 preprotoxin (9, 140) closely resemble those seen with preproinsulin and other mammalian hormones in which cleavage C-terminal to a pair of basic residues is followed by removal of those basic residues (Figure 6). Two genes are necessary for the secretion of active killer toxin, called *KEX1* and *KEX2* (for killer expression) (71, 127). *KEX2* is also necessary for secretion of α-pheromone, a hormone-like peptide secreted by cells of the α mating type that signals their presence to cells of the *a* mating type (71). The key discovery that *KEX2* (46, 64) and *KEX1* (21, 27, 125) are proteases and that their specificity precisely explained these defects led to a successful search in several labs for mammalian homologues. Using this approach, investigators have found three proteases that are potential pre-prohormone processing enzymes. The human furin gene had already been cloned and sequenced (96), and its homology with *KEX2* made it a candidate for this role (45). The PC2 and PC3 genes were cloned using PCR and primers based on domains conserved in *KEX2* and furin (101, 105, 106). While furin

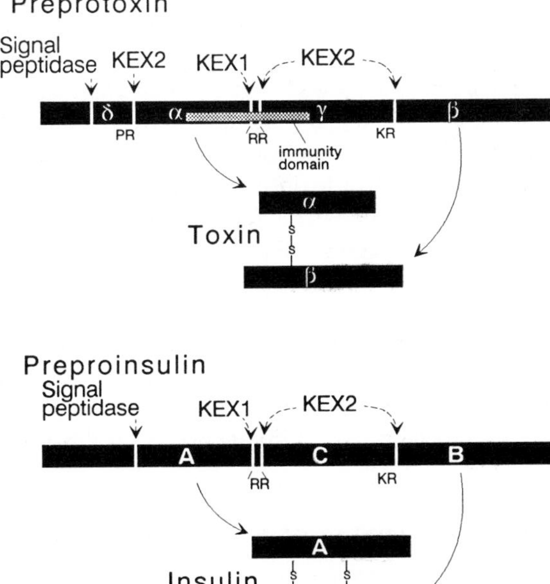

Figure 6 Processing of K_1 killer preprotoxin (9, 140) is analogous and homologous to processing of mammalian prohormones (see text).

expression is widely distributed, expression of PC2 and PC3 is largely restricted to neuroendocrine tissues. The ability of yeast KEX2 protein to function in mammalian cells in prohormone processing also supports the parallel between these systems (113).

Toxin Immunity and Mechanism of Action

Each killer toxin has an essential cell-wall binding site, mutations in which lead to resistance to toxin action. For the K_1 toxin, this site is the β(1-6) glucan, a polymer found as an important constituent of the cell walls of many fungi (60). A study of the multiple genes controlling the synthesis of β(1-6) glucan has been developed starting with the isolation of K_1 toxin-resistant mutants (8, 80). Similarly, the K_{28} toxin binds to mannan, a mannose-rich glycoprotein, and selection of K_{28}-resistant mutants yields mutants in the mannan biosynthesis pathway (98). Other killer toxins likely target β(1-3) glucan or other cell wall components.

Killer strains have an immunity that makes their spheroplasts resistant to toxin and does not affect their cell walls' affinity for toxin. Although mutations affecting β(1-6) glucan biosynthesis affect resistance to the killer toxin, spheroplasts made from such mutants are still as sensitive as wild-type cells, suggesting the toxin has a second receptor, perhaps in the membrane (139). That this immunity may result from the action of the protoxin is suggested by the normal immunity of *kex1* and *kex2* mutants (127) and the presence of the unprocessed protoxin in such mutants (7). Moreover, localized mutagenesis studies of the preprotoxin expressed from a cDNA expression plasmid have shown that determinants of immunity are spread over the α and γ domains (7) (see Figure 6). The protoxin may bind to the toxin membrane receptor in such a way as to prevent the active toxin from binding.

While immunity is a function of the α and γ domains, the binding of toxin to (1-6)-β-D-glucan is a function of the α and β domains (subunits), and the ion channel formation (toxic action) (76) is confined to the α subunit (139).

Many other yeast killer toxins have been described (reviewed in 131). Of these, the K_2 and K_{28} toxin genes, located on M_2 and M_{28} dsRNAs, have been characterized. M_2 (24) has no amino acid sequence homology with M_1 (9), in spite of significant parallels between the toxins, such as processing by the KEX1 and KEX2 proteases. M_2 (128) and M_{28} (99) dsRNAs are inherited much as is M_1.

20S RNA AND 23S RNA: SINGLE-STRANDED RNA REPLICONS

20S RNA was first described as a stable RNA appearing in sporulating yeast (65, 66). It was later found that some strains could not synthesize 20S RNA, and that 20S RNA amplification was not necessarily connected with sporulation, only that the same culture conditions (potassium acetate medium) induced both processes (48). Crosses of strains that could amplify 20S RNA with those that could not produced meiotic segregants, all of which could amplify. This non-Mendelian trait controlling 20S RNA does not reside on the mitochondrial genome and is not associated with L-A or M dsRNA (48). It is also distinct from other known non-Mendelian elements of yeast (48).

Later, two minor dsRNAs, called T and W, estimated to be 2.7 and 2.25 kb, respectively, were described (126). The copy number of each was amplified 10-fold by growth of cells at 37°C compared to that in stationary phase cells grown at 30°C. T and W did not hybridize with L-A, L-BC, or M dsRNAs or with any cellular DNAs. Evidence was also presented that neither T nor W were related to other yeast non-Mendelian elements (126).

Recently, most of 20S RNA was cloned and shown to be amplified 10,000-fold on acetate medium and to be completely absent from strains that do not amplify it (78). It is not homologous to chromosomal, 2μ, or mitochondrial DNA, nor to L-A, L-BC, or M dsRNAs (78). In addition, 20S RNA does not associate in cell extracts with enough of any protein to form a capsid (135).

Three lines of evidence indicate that 20S RNA is a circular molecule: (*a*) attempts to label the ends of highly purified 20S RNA produced only labeling of contaminating 18S rRNA; (*b*) on two-dimensional gel electrophoresis, 20S RNA showed aberrant migration similar to that shown by circular, branched, and lariat molecules; (*c*) electron microscopy of 20S RNA that had been coated with phage T4 gene 32 protein, crosslinked with glutaraldehyde, and spread, showed about 50% circular molecules (Figure 7) (78).

20S RNA (79) and W dsRNA (95) were independently sequenced in different labs, and when the sequences were compared, they proved to be essentially identical. Since the copy number of 20S RNA vastly exceeds that of W dsRNA, the latter is probably a replicative form of the former. The sequence contains one long ORF, encoding a 90-kDa protein with some resemblance to the RNA-dependent RNA polymerases of (+) single-stranded RNA viruses (79, 95).

Rodriguez et al (95) found that the end-labeled (+) strand of W comigrates on a strand-separation gel with 20S RNA and have thus suggested that 20S RNA is linear, not circular as indicated above. The fact that the sequence of 20S RNA still has a gap (79) also raises doubts about the circularity of the molecule. There must at least be a special sequence or structure at one point in 20S RNA that is resistant to reverse transcriptase. Further work will be needed to determine the structure of these molecules.

Circular RNAs such as viroids, satellite RNAs, and the hepatitis delta virus are thought to replicate by a rolling-circle mechanism, in which self-cleavage

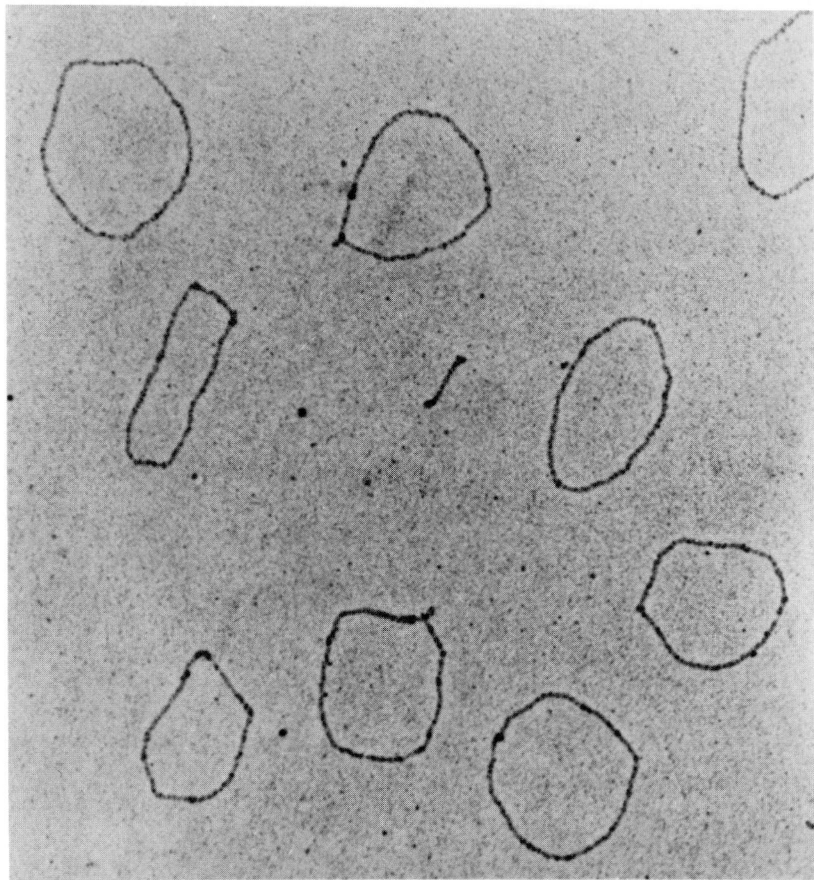

Figure 7 20S RNA appears circular with electron microscopy (78).

and self-ligation reactions convert oligomers into monomeric circles (14, 84). Apparently, dimer-length 20S RNA molecules of both (+) and (−) strand polarity have been observed (79). These dimer-length RNAs comprise only a very small proportion of (+) strands, but close to half of (−) strand molecules. Their presence again suggests the existence of a circular form.

The sequence of T dsRNA, like W dsRNA (= 20S RNA), has a single long open reading frame (R. Esteban, N. Rodriguez-Cousino & L. Esteban, personal communication). While the protein sequence has some similarity to RNA-dependent RNA polymerases, it shows marked similarity to W. A high-copy-number single-stranded form of T dsRNA, called 23S RNA, has also been found (R. Esteban, N. Rodriguez-Cousino & L. Esteban, personal communication). All strains carrying T also carry W, suggesting the possibility that T requires W (126). Also, all killer strains carry 20S RNA (W), suggesting that M or L-A may require 20S RNA (48). However, these possibilities have not yet been critically tested, and there is no known physiological effect of carrying either 20S or 23S RNA.

The control of synthesis of 20S RNA by growth conditions is a point of some interest. The intracellular signal transmitting the absence of a nitrogen source and the presence of acetate as a carbon source appears to be a drop in cAMP concentration. While the 20S RNA ORF has a potential A-kinase phosphorylation site, as yet no evidence indicates that the protein is phosphorylated, or even that cAMP controls 20S replication.

CONCLUSIONS

To the extent that the fundamental mechanisms by which the yeast viruses propagate and by which this propagation is controlled come to be understood, these systems may be useful to those studying viruses of mammalian cells. Of course, the killer systems are already used in brewing to exclude contaminants, and efforts are underway in many labs to use them as virus vectors for producing proteins and peptides.

The L-A and 20S RNA systems that constitute the bulk of the above discussion are typical mature viruses that have reached a balance between virus and host. The adjective "delicate" is often applied to such balances (as to other balances found in nature). Perhaps this term is not as appropriate as "inevitable." Most balances in nature are not delicate, or they would not have survived as long as they have. Viruses that kill their hosts the most ruthlessly are in the greatest danger of running out of hosts and ceasing to exist. Only those that temper their demands will survive. From the host viewpoint, one could likewise say that only those hosts that learn to control viral replication will survive. Clearly, one of the central features of these systems is the *SKI* system that moderates the replication of the L-A, L-BC, and 20S RNA systems. Because the *SKI* genes are essential only for their antiviral action, and *ski*$^-$ mutants appear to have no other phenotype, we suggest that the *SKI* genes constitute a dedicated antiviral system. An understanding of its mechanism of action would clarify many aspects of these systems and perhaps provide a starting point for the search for a homologous system in mammals, much as the *KEX* genes have facilitated the hunt for mammalian prohormone processing proteases.

ACKNOWLEDGMENTS

The author is greatly indebted to his colleagues, past and present, for their efforts in the study of these systems. Particular thanks go to Stephen Ball,

Jonathan Dinman, Rosa Esteban, Tsutomu Fujimura, Tateo Icho, Yang-Ja Lee, Michael Leibowitz, Yutaka Matsumoto, Sang-Ki Rhee, Juan Carlos Ribas, Susan Porter Ridley, Steve Sommer, Juan Carlos Tercero, Akio Toh-e, Hiroshi Uemura, Rosaura Valle, Michelene Wesolowski, and William R. Widner

Literature Cited

1. Aigle, M., Lacroute, F. 1975. Genetical aspects of [URE3], a non-mitochondrial, cytoplasmically inherited mutation in yeast. *Mol. Gen. Genet.* 136:327–35
2. Ball, S. G., Tirtiaux, C., Wickner, R. B. 1984. Genetic control of L-A and L-(BC) dsRNA copy number in killer systems of *Saccharomyces cerevisiae*. *Genetics* 107:199–217
3. Barr, P. J. 1991. Mammalian subtilisins: the long-sought dibasic processing endoproteases. *Cell* 66:1–3
4. Bevan, E. A., Herring, A. J., Mitchell, D. J. 1973. Preliminary characterization of two species of dsRNA in yeast and their relationship to the "killer" character. *Nature* 245:81–86
5. Boeke, J. D. 1989. Transposable elements in *Saccharomyces cerevisiae*. In *Mobile DNA*, ed. D. Berg, M. Howe, pp. 335–74. Washington, DC: Am. Soc. Microbiol.
6. Boguski, M. S., Sikorski, R. S., Hieter, P., Goebl, M. 1990. Expanding family. *Nature* 346:114
7. Boone, C., Bussey, H., Greene, D., Thomas, D. Y., Vernet, T. 1986. Yeast killer toxin: site-directed mutations implicate the precursor protein as the immunity component. *Cell* 46:105–13
8. Boone, C., Sommer, S. S., Hensel, A., Bussey, H. 1990. Yeast *KRE* genes provide evidence for a pathway of cell wall β-glucan assembly. *J. Cell Biol.* 110: 1833–43
9. Bostian, K. A., Elliott, Q., Bussey, H., Burn, V., Smith, A., Tipper, D. J. 1984. Sequence of the preprotoxin dsRNA gene of type 1 killer yeast: multiple processing events produce a two-component toxin. *Cell* 36:741–51
10. Bostian, K. A., Hopper, J. E., Rogers, D. T., Tipper, D. J. 1980. Translational analysis of the killer-associated s virus-like particle dsRNA genome of S. cerevisiae: M dsRNA encodes toxin. *Cell* 19:403–14
11. Bostian, K. A., Sturgeon, J. A., Tipper, D. J. 1980. Encapsidation of yeast killer dsRNAs: dependence of M on L. *J. Bacteriol.* 143:463–70
12. Bozarth, R. F., Koltin, Y., Weissman, M. B., Parker, R. L., Dalton, R. E., Steinlauf, R. 1981. The molecular weight and packaging of dsRNAs in the mycovirus from *Ustilago maydis* killer strains. *Virology* 113:492–502
13. Brierley, I., Digard, P., Inglis, S. C. 1989. Characterization of an efficient coronavirus ribosomal frameshifting signal: requirement for an RNA pseudoknot. *Cell* 57:537–47
14. Bruening, G., Buzayan, J. M., Hampel, A., Gerlach, W. L. 1988. Replication of small satellite RNAs and viroids: possible participation of nonenzymic reactions. See Ref. 26b, pp. 127–45
15. Bruenn, J. A. 1980 Virus-like particles of yeast. *Annu. Rev. Microbiol.* 34:49–68
16. Buck, K. W. 1979. Replication of double-stranded RNA mycoviruses. In *Viruses and Plasmids in Fungi*, ed. P. A. Lemke, pp. 93–160. New York: Marcel Dekker
17. Buck, K. W., Lhoas, P., Street, B. K. 1973. Virus particles in yeast. *Biochem. Soc. Trans.* 1:1141–42
18. Bussey, H. 1988. Proteases and the processing of precursors to secreted proteins in yeast. *Yeast* 4:17–26
19. Chow, M., Newman, J. F., Filman, D., Hogle, J. M., Rowlands, D. J., Brown, F. 1987. Myristylation of picornavirus capsid protein VP4 and its structural significance. *Nature* 327:482–86
20. Cohn, M. S., Tabor, C. W., Tabor, H., Wickner, R. B. 1978. Spermidine or spermine requirement for killer double-stranded RNA plasmid replication in yeast. *J. Biol. Chem.* 253:5225–27
21. Cooper, A., Bussey, H. 1989. Characterization of the yeast *KEX1* gene product: a carboxypeptidase involved in processing secreted precursor proteins. *Mol. Cell. Biol.* 9:2706–14
22. Cox, B. S., Tuite, M. F., McLaughlin, C. S. 1988. The Psi factor of yeast: a problem in inheritance. *Yeast* 4:159–79
23. Curcio, M. J., Garfinkel, D. J. 1991. Regulation of retrotransposition in *Saccharomyces cerevisiae*. *Mol. Microbiol.* 5:1823–29

24. Dignard, D., Whiteway, M., Germain, D., Tessier, D., Thomas, D. Y. 1991. Expression in yeast of a cDNA copy of the K2 killer toxin gene. *Mol. Gen. Genet.* 227:127–36
25. Dihanich, M., Van Tuinen, E., Lambris, J. D., Marshallsay, B. 1989. Accumulation of viruslike particles in a yeast mutant lacking a mitochondrial pore protein. *Mol. Cell. Biol.* 9:1100–8
26. Dinman, J. D., Icho, T., Wickner, R. B. 1991. A −1 ribosomal frameshift in double-stranded RNA virus of yeast forms a gag-pol fusion protein. *Proc. Natl. Acad. Sci. USA* 88:174–78
26a. Dinman, J. D., Wickner, R. B. 1992. Ribosomal frameshifting efficiency and *gag/gag-pol* ratio are critical for yeast M_1 dsRNA virus propogation. *J. Virol.* In press
26b. Domingo, E., Holland, J. J., Ahlquist, P., eds. 1988. *RNA Genetics*. Boca Raton, FL: CRC Press
27. Dmochowska, A., Dignard, D., Henning, D., Thomas, D. Y., Bussey, H. 1987. Yeast KEX1 gene encodes a putative protease with a carboxypeptidase B-like function involved in killer toxin and alpha factor precursor processing. *Cell* 50:573–84
28. El-Sherbeini, M., Bostian, K. A. 1987. Viruses in fungi: infection of yeast with the K1 and K2 killer virus. *Proc. Natl. Acad. Sci. USA* 84:4293–97
29. El-Sherbeini, M., Tipper, D. J., Mitchell, D. J., Bostian, K. A. 1984. Virus-like particle capsid proteins encoded by different L double-stranded RNAs of *Saccharormyces cerevisiae:* their roles in maintenance of M double-stranded killer plasmids. *Mol. Cell. Biol.* 4:2818–27
30. Esteban, R., Fujimura, T., Wickner, R. B. 1988. Site-specific binding of viral plus single-stranded RNA to replicase-containing open virus-like particles of yeast. *Proc. Natl. Acad. Sci. USA* 85:4411–15
31. Esteban, R., Fujimura, T., Wickner, R. B. 1989. Internal and terminal *cis*-acting sites are necessary for in vitro replication of the L-A double-stranded RNA virus of yeast. *EMBO J.* 8:947–54
32. Esteban, R., Wickner, R. B. 1986. Three different M1 RNA-containing viruslike particle types in *Saccharomyces cerevisiae:* in vitro M1 dsRNA synthesis. *Mol. Cell. Biol.* 6:1552–61
33. Esteban, R., Wickner, R. B. 1987. A new non-Mendelian genetic element of yeast that increases cytopathology produced by M1 dsRNA in ski strains. *Genetics* 117:399–408
34. Esteban, R., Wickner, R. B. 1988. A deletion mutant of L-A dsRNA replicates like M1 dsRNA. *J. Virol.* 62:1278–85
35. Fong, H. K. W., Hurley, J. B., Hopkins, R. S., Miake-Lye, R., Johnson, M. S., et al. 1986. Repetive segmental structure of the transducin β subunit: homology with the *CDC4* gene and identification of related mRNAs. *Proc. Natl. Acad. Sci. USA* 83:2162–66
36. Fried, H. M., Fink, G. R. 1978. Electron microscopic heteroduplex analysis of "killer" double-stranded RNA species from yeast. *Proc. Natl. Acad. Sci. USA* 75:4224–28
37. Fujimura, T., Esteban, R., Esteban, L. M., Wickner, R. B. 1990. Portable encapsidation signal of the L-A double-stranded RNA virus of S. cerevisiae. *Cell* 62:819–28
38. Fujimura, T., Esteban, R., Wickner, R. B. 1986. In vitro L-A dsRNA synthesis in virus-like particles from *Saccharomyces cerevisiae. Proc. Natl. Acad. Sci. USA* 83:4433–37
39. Fujimura, T., Wickner, R. B. 1986. Thermolabile L-A virus-like particles from pet18 mutants of *Saccharomyces cerevisiae. Mol. Cell. Biol.* 6:404–10
40. Fujimura, T., Wickner, R. B. 1987. L-A double-stranded RNA viruslike particle replication cycle in *Saccharomyces cerevisiae:* particle maturation in vitro and effects of mak10 and pet18 mutations. *Mol. Cell. Biol.* 7:420–26
41. Fujimura, T., Wickner, R. B. 1988. Replicase of L-A virus-like particles of *Saccharomyces cerevisiae*. In vitro conversion of exogenous L-A and M1 single-stranded RNAs to double-stranded form. *J. Biol. Chem.* 263:454–60
42. Fujimura, T., Wickner, R. 1988. Gene overlap results in a viral protein having an RNA binding domain and a major coat protein domain. *Cell* 55:663–71
43. Fujimura, T., Wickner, R. B. 1989. Reconstitution of template-dependent in vitro transcriptase activity of a yeast double-stranded RNA virus. *J. Biol. Chem.* 264:10872–77
44. Fujimura, T., Wickner, R. B. 1992. Interaction of two *cis* sites with the RNA replicase of the yeast L-A virus. *J. Biol. Chem.* 267:2708–13
45. Fuller, R. S., Brake, A. J., Thorner, J. 1989. Intracellular targeting and structural conservation of a prohormone-processing endoprotease. *Science* 246: 482–86
46. Fuller, R. S., Stearne, R. E., Thorner, J. 1988. Enzymes required for yeast pro-

hormone processing. *Annu. Rev. Physiol.* 50:345–62

47. Gallegos, C. O., Patton, J. T. 1989. Characterization of rotavirus replication intermediates: a model for the assembly of single-shelled particles. *Virology* 172:616–27
48. Garvik, B., Haber, J. E. 1978. New cytoplasmic genetic element that controls 20S RNA synthesis during sporulation in yeast. *J. Bacteriol.* 134:261–69
49. Georgopoulos, D. E., Hannig, E. M., Leibowitz, M. J. 1986. Sequence of the M1-2 region of killer virus double-stranded RNA. In *Extrachromosomal Elements in Lower Eukaryotes,* ed. R. B. Wickner, A. Hinnebusch, A. M. Lambowitz, I. C. Gunsalus, A. Hollaender, pp. 203–13. New York: Plenum
50. Hannig, E. M., Leibowitz, M. J. 1985. Structure and expression of the M2 genomic segment of a type 2 killer virus of yeast. *Nucleic Acids Res.* 13:4379–4400
51. Hannig, E. M., Leibowitz, M. J., Wickner, R. B. 1985. On the mechanism of exclusion of M2 double-stranded RNA by L-A-E dsRNA in *Saccharomyces cerevisiae. Yeast* 1:57–65
52. Hannig, E. M., Thiele, D. J., Leibowitz, M. J. 1984. *Saccharomyces cerevisiae* killer virus transcripts contain template-coded polyadenylate tracts. *Mol. Cell. Biol.* 4:101–9
53. Hansen, L. J., Chalker, D. L., Sandemeyer, S. B. 1988. Ty3, a yeast retrotransposon associated with tRNA genes, has homology to animal retroviruses. *Mol. Cell. Biol.* 8:5245–56
54. Harashima, S., Hinnebusch, A. 1986. Multiple GCD genes required for repression of GCN4, a transcriptional activator of amino acid biosynthetic genes in *Saccharomyces cerevisiae. Mol. Cell. Biol.* 6:3990–98
55. Hatfield, D. L., Levin, J. G., Rein, A., Oroszlan, S. 1992. Translational suppression in retroviral gene expression. *Adv. Virus Res.* 41:193–239
56. Herring, A. J., Bevan, A. E. 1974. Virus-like particles associated with the double-stranded RNA species found in killer and sensitive strains of the yeast *Saccharomyces cerevisiae. J. Gen. Virol.* 22:387–94
57. Hirano, T., Kinoshita, N., Morikawa, K., Yanagida, M. 1990. Snap helix with knob and hole: essential repeats in S. pombe nuclear protein nuc2$^+$. *Cell* 60:319–28
58. Hopper, J. E., Bostian, K. A., Rowe, L. B., Tipper, D. J. 1977. Translation of the L-species dsRNA found in killer and sensitive strains of the yeast *Saccharomyces cerevisiae. J. Biol. Chem.* 252: 9010–17
59. Huan, B., Shen, Y., Bruenn, J. A. 1991. In vivo mapping of a sequence required for interference with the yeast killer virus. *Proc. Natl. Acad. Sci. USA* 88:1271–75
60. Hutchins, K., Bussey, H. 1983. Cell wall receptor for yeast killer toxin: involvement of (1→6)-β-D-glucan. *J. Bacteriol.* 171:2842–49
61. Icho, T., Wickner, R. B. 1988. The MAK11 protein is essential for cell growth and replication of M double-stranded RNA and is apparently a membrane-associated protein. *J. Biol. Chem.* 263:1467–75
62. Icho, T., Wickner, R. B. 1989. The double-stranded RNA genome of yeast virus L-A encodes its own putative RNA polymerase by fusing two open reading frames. *J. Biol. Chem.* 264:6716–23
63. Jacks, T., Madhani, H. D., Masiarz, F. R., Varmus, H. E. 1988. Signals for ribosomal frameshifting in the Rous sarcoma virus gag-pol region. *Cell* 55:447–58
64. Julius, D., Brake, A., Blair, L., Kunisawa, R., Thorner, J. 1984. Isolation of the putative structural gene for the lysine-arginine-cleaving endopeptidase required for the processing of yeast prepro- alpha-factor. *Cell* 36:309–18
65. Kadowaki, K., Halvorson, H. O. 1971. Appearance of a new species of ribonucleic acid during sporulation in *Saccharomyces cerevisiae. J. Bacteriol.* 105:826–30
66. Kadowaki, K., Halvorson, H. O. 1971. Isolation and properties of a new species of ribonucleic acid synthesized in sporulating cells of *Saccharomyces cerevisiae. J. Bacteriol.* 105:831–36
67. Kamer, G., Argos, P. 1984. Primary structural comparison of RNA-dependent polymerases from plant, animal and bacterial viruses. *Nucleic Acids Res.* 12:7269–82
68. Lacroute, F. 1971. Non-mendelian mutation allowing ureidosuccinic acid uptake in yeast. *J. Bacteriol.* 106:519–22
69. Lee, M., Pietras, D. F., Nemeroff, M. E., Corstanje, B. J., Field, L. J., Bruenn, J. A. 1986. Conserved regions in defective interfering viral double-stranded RNAs from a yeast virus. *J. Virol.* 58:402–7
70. Leibowitz, M. J., Koltin, Y., Rubio, V., eds. 1992. *Viruses of Simple Eukaryotes: Molecular Genetics and Applications to Biotechnology and Medi-*

cine. Newark, DE: Univ. Delaware Press

71. Leibowitz, M. J., Wickner, R. B. 1976. A chromosomal gene required for killer plasmid expression, mating and spore maturation in *Saccharomyces cerevisiae*. *Proc. Natl. Acad. Sci. USA* 73:2061–65

72. Leibowitz, M. J., Wickner, R. B. 1978. pet18: a chromosomal gene required for cell growth and for the maintenance of mitochondrial DNA and the killer plasmid of yeast. *Mol. Gen. Genet.* 165: 115–21

73. Liebman, S. W., All-Robyn, J. A. 1984. A non-Mendelian factor, η, causes lethality of yeast omnipotent suppressor strains. *Curr. Genet.* 8:567–73

74. Liu, Y., Dieckmann, C. L. 1989. Overproduction of yeast virus-like particle coat protein genome in strains deficient in a mitochondrial nuclease. *Mol. Cell. Biol.* 9:3323–31

75. Makower, M., Bevan, E. A. 1963. The inheritance of a killer character in yeast (*Saccharomyces cerevisiae*). *Proc. Int. Congr. Genet., 11th* 1:202

76. Martinac, B., Zhu, H., Kubalski, A., Zhou, X., Culbertson, M., et al. 1990. Yeast K1 killer toxin forms ion channels in sensitive yeast spheroplasts and in artificial liposomes. *Proc. Natl. Acad. Sci. USA* 87:6228–32

77. Marx, J. 1991. How peptide hormones get ready for work. *Science* 252:779–80

78. Matsumoto, Y., Fishel, R., Wickner, R. B. 1990. Circular single-stranded RNA replicon in *Saccharomyces cerevisiae*. *Proc. Natl. Acad. Sci. USA* 87:7628–32

79. Matsumoto, Y., Wickner, R. B. 1991. Yeast 20S RNA replicon. Replication intermediates and encoded putative RNA polymerase. *J. Biol. Chem.* 266: 12779–83

80. Meaden, P., Hill, K., Wagner, J., Slipetz, D., Sommer, S. S., Bussey, H. 1990. The yeast *KRE5* gene encodes a probable endoplasmic reticulum protein required for (1→6)-β-D-glucan synthesis and normal cell growth. *Mol. Cell. Biol.* 10:3013–19

81. Nemeroff, M. E., Bruenn, J. A. 1987. Initiation by the yeast viral transcriptase *in vitro*. *J. Biol. Chem.* 262:6785–87

82. Newman, A. M., Elliott, S. G., McLaughlin, C. S., Sutherland, P. A., Warner, R. C. 1981. Replication of dsRNA of the virus-like particles in *Saccharomyces cerevisiae*. *J. Virol.* 38:263–71

83. Oliver, S. G., McCready, S. J., Holm, C., Sutherland, P. A., McLaughlin, C. S., Cox, B. S. 1977. Biochemical and physiological studies of the yeast virus-like particle. *J. Bacteriol.* 130:1303–9

84. Owens, R. A., Hammond, R. W. 1988. Structure and function relationships in plant viroid RNAs. See Ref. 26b, pp. 107–125

85. Paterson, R. G., Lamb, R. A. 1990. RNA editing by G-nucleotide insertion in mumps virus P-gene mRNA transcripts. *J. Virol.* 64:4137–45

86. Paul, A. V., Schultz, A., Pincus, S. E., Oroszlan, S., Wimmer, E. 1987. Capsid protein VP4 of poliovirus is N-myristoylated. *Proc. Natl. Acad. Sci. USA* 84:7827–31

87. Pelham, H. R. B. 1978. Leaky UAG termination codon in tobacco mosaic virus RNA. *Nature* 272:469–71

88. Pelham, H. R. B. 1979. Translation of tobacco rattle virus RNAs in vitro: four proteins from three RNAs. *Virology* 97:256–65

89. Pfeiffer, P., Radler, F. 1984. Comparison of the killer toxin of several yeasts and the purification of a toxin of type K2. *Arch. Microbiol.* 137:357–61

90. Ratti, G., Buck, K. W. 1978 Semiconservative transcription in particles of a double-stranded RNA mycovirus. *Nucleic Acids Res.* 5:3843–54

91. Rein, A., McClure, M. R., Rice, N. R., Luftig, R. B., Schultz, A. M. 1986. Myristylation site in $Pr65^{gag}$ is essential for virus particle formation by Moloney murine leukemia virus. *Proc. Natl. Acad. Sci. USA* 83:7246–50

92. Rhee, S.-K., Icho, T., Wickner, R. B. 1989. Structure and nuclear localization signal of the SKI3 antiviral protein of *Saccharomyces cerevisiae*. *Yeast* 5:149–58

93. Ribas, J. C., Wickner, R. B. 1992. RNA-dependent RNA polymerase consensus sequence of the L-A dsRNA virus: definition of essential domains. *Proc. Natl. Acad. Sci. USA*. In press

94. Ridley, S. P., Sommer, S. S., Wickner, R. B. 1984. Superkiller mutations in *Saccharomyces cerevisiae* supress exclusion of M2 double-stranded RNA by L-A-HN and confer cold-sensitivity in the presence of M and L-A-HN. *Mol. Cell. Biol.* 4:761–70

95. Rodriguez-Cousino, N., Esteban, L. M., Esteban, R. 1991. Molecular cloning and characterization of W double-stranded RNA, a linear molecule present in *Saccharomyces cerevisiae:* identification of its single-stranded RNA form as 20S RNA. *J. Biol. Chem.* 266:12772–78

96. Roebroek, A. J., Schalken, J. A., Leunissen, J. A., Onnekink, C., Bloemers, H. P., Van de Ven, W. J.

1986. Evolutionary conserved close linkage of the c-fes/fps proto-oncogene and genetic sequences encoding a receptor-like protein. *EMBO J.* 5:2197–2202

97. Russell, P. J., Hambidge, S. J., Kirkegaard, K. 1991. Direct introduction and transient expression of capped and non-capped RNA in *Saccharomyces cerevisiae. Nucleic Acids Res.* 19:4949–53
98. Schmitt, M. J., Radler, F. 1987. Mannoprotein of the yeast cell wall as primary receptor for the killer toxin of *Saccharomyces cerevisiae* strain 28. *J. Gen. Microbiol.* 133:3347–54
99. Schmitt, M. J., Tipper, D. J. 1990. K_{28}, a unique double-stranded RNA killer virus of *Saccharomyces cerevisiae. Mol. Cell. Biol.* 10:4807–15
100. Sclafani, R. A., Fangman, W. L. 1984. Conservative replication of dsRNA in *Saccharomyces cerevisiae* by displacement of progeny single strands. *Mol. Cell. Biol.* 4:1618–26
101. Seidah, N. G., Gaspar, L., Mion, P., Marcinkiewicz, M., Mbikay, M., Chretien, M. 1990. cDNA sequence of two distinct pituitary proteins homologous to Kex2 and Furin gene products: tissue-specific mRNAs encoding candidates for pro-hormone processing proteinases. *DNA Cell Biol.* 9:415–24
102. Shatkin, A. J., Kozak, M. 1983. Biochemical aspects of reovirus transcription and translation. In *The Reoviridae,* ed. W. K. Joklik, pp. 79–106. New York: Plenum
103. Sikorski, R. S., Boguski, M. S., Goebl, M., Hieter, P. 1990. A repeating amino acid motif in CDC23 defines a family of proteins and a new relationship among genes required for mitosis and RNA synthesis. *Cell* 60:307–17
104. Skipper, N., Thomas, D. Y., Lau, P. C. K. 1984. Cloning and sequencing of the preprotoxin-coding region of the yeast M1 double-stranded RNA. *EMBO J.* 3:107–11
105. Smeekens, S. P., Avruch, A. S., LaMendola, J., Chan, S. J., Steiner, D. F. 1991. Identification of a cDNA encoding a second putative prohormone convertase related to PC2 in AtT20 cells and islets of Langerhans. *Proc. Natl. Acad. Sci. USA* 88:340–44
106. Smeekens, S. P., Steiner, D. F. 1990. Identification of a human insulinoma cDNA encoding a novel mammalian protein structurally related to the yeast dibasic processing protease, KEX2. *J. Biol. Chem.* 265:2997–3000
107. Somers, J. M. 1973. Isolation of suppressive mutants from killer and neutral strains of *Saccharomyces cerevisiae. Genetics* 74:571–79
108. Somers, J., Bevan, E. A. 1969. The inheritance of the killer character in yeast. *Genet. Res.* 13:71–83
109. Sommer, S. S., Wickner, R. B. 1982. Yeast L dsRNA consists of at least three distinct RNAs; evidence that the non-Mendelian genes [HOK], [NEX] and [EXL] are on one of these dsRNAs. *Cell* 31:429–41
110. Sommer, S. S., Wickner, R. B. 1987. Gene disruption indicates that the only essential function of the *SKI8* chromosomal gene is to protect *Saccharomyces cerevisiae* from viral cytopathology. *Virology* 157:252–56
111. Thiele, D. J., Hannig, E. M., Leibowitz, M. J. 1984. Multiple L double-stranded RNA species of *Saccharomyces cerevisiae:* evidence for separate encapsidation. *Mol. Cell. Biol.* 4:92–100
112. Thiele, D. J., Hannig, E. M., Leibowitz, M. J. 1984. Genome structure and expression of a defective interfering mutant of the killer virus of yeast. *Virology* 137:20–31
113. Thomas, G., Thorne, B. A., Thomas, L., Allen, R. G., Hruby, D. E., et al. 1988. Yeast *KEX2* endopeptidase correctly cleaves a neuroendocrine prohormone in mammalian cells. *Science* 241:226–30
114. Thrash, C., Voelkel, K., DiNardo, S., Sternglanz, R. 1984. Identification of *Saccharomyces cerevisiae* mutants deficient in DNA topoisomerase I. *J. Biol. Chem.* 259:1375–79
115. Toh-e, A., Guerry, P., Wickner, R. B. 1978. Chromosomal superkiller mutants of *Saccharomyces cerevisiae. J. Bacteriol.* 136:1002–7
116. Toh-e, A., Sahashi, Y. 1985. The PET18 locus of *Saccharomyces cerevisiae:* a complex locus containing multiple genes. *Yeast* 1:159–72
117. Toh-e, A., Wickner, R. B. 1979. A mutant killer plasmid whose replication depends on a chromosomal "superkiller" mutation. *Genetics* 91:673–82
118. Toh-e, A., Wickner, R. B. 1980. "Superkiller" mutations suppress chromosomal mutations affecting double-stranded RNA killer plasmid replication in *Saccharomyces cerevisiae. Proc. Natl. Acad. Sci. USA* 77:527–30
119. Tyagi, A. K., Wickner, R. B., Tabor, C. W., Tabor, H. 1984. Specificity of polyamine requirements for the replication and maintenance of different double-stranded RNA plasmids in *Sacchar-*

omyces cerevisiae. *Proc. Natl. Acad. Sci. USA* 81:1149–53

120. Uemura, H., Wickner, R. B. 1988. Suppression of chromosomal mutations affecting M1 virus replication in *Saccharomyces cerevisiae* by a variant of a viral RNA segment (L-A) that encodes coat protein. *Mol. Cell. Biol.* 8:938–44
121. Valle, R. P. C., Morch, M.-D. 1988. Stop making sense or regulation at the level of termination in eukaryotic protein synthesis. *FEBS Lett.* 235:1–15
122. Van Etten, J. L., Burbank, D. E., Cuppels, D. A., Lane, L. C., Vidaver, A. K. 1980. Semiconservative synthesis of single-stranded RNA by bacteriophage phi 6 RNA polymerase. *J. Virol.* 33:769–73
123. Vincent, R. D., Hofmann, T. J., Zassenhaus, H. P. 1988. Sequence and expression of NUC1, the gene encoding the mitochondrial nuclease in *Saccharomyces cerevisiae*. *Nucleic Acids Res.* 16:3297–3312
124. Vodkin, M., Katterman, F., Fink, G. R. 1974. Yeast killer mutants with altered double-stranded RNA. *J. Bacteriol.* 117:681–86
125. Wagner, J.-C., Wolf, D. H. 1987. Hormone (pheromone) processing enzymes in yeast. The carboxy-terminal processing enzyme of the mating pheromone alpha-factor, carboxypeptidase yscalpha is absent in alpha-factor maturation-defective *kex1* mutant cells. *FEBS Letts.* 221:423–26
126. Wesolowski, M., Wickner, R. B. 1984. Two new double-stranded RNA molecules showing non-Mendelian inheritance and heat-inducibility in *Saccharomyces cerevisiae*. *Mol. Cell. Biol.* 4:181–87
127. Wickner, R. B. 1974. Chromosomal and nonchromosomal mutations affecting the "killer character" of *Saccharomyces cerevisiae*. *Genetics* 76:423–32
128. Wickner, R. B. 1980. Plasmids controlling exclusion of the K_2 killer double-stranded RNA plasmid of yeast. *Cell* 21:217–26
129. Wickner, R. B. 1988. Host function of *MAK16:* G_1 arrest by a *mak16* mutant of *Saccharomyces cerevisiae*. *Proc. Natl. Acad. Sci. USA* 85:6007–11
130. Wickner, R. B. 1989. Yeast virology. *FASEB J.* 3:2257–65
131. Wickner, R. B. 1991. Yeast RNA virology: the killer systems. In *The Molecular and Cellular Biology of the Yeast Saccharomyces,* ed. J. Broach, E. Jones, J. Pringle, pp. 263–95. Cold Spring Harbor: Cold Spring Harbor Lab.
132. Wickner, R. B., Icho, T., Fujimura, T., Widner, W. R. 1991. Expression of yeast L-A double-stranded RNA virus proteins produces derepressed replication: a ski-phenocopy. *J. Virol.* 65:155–61
133. Wickner, R. B., Ridley, S. P., Fried, H. M., Ball, S. G. 1982. Ribosomal protein L3 is involved in replication or maintenance of the killer double-stranded RNA genome of *Saccharomyces cerevisiae*. *Proc. Natl. Acad. Sci. USA* 79:4706–8
134. Wickner, R. B., Toh-e, A. 1982. [HOK], a new yeast non-mendelian trait, enables a replication-defective killer plasmid to be maintained. *Genetics* 100:159–74
135. Widner, W. R., Matsumoto, Y., Wickner, R. B. 1991. Is 20S RNA naked? *Mol. Cell. Biol.* 11:2905–8
136. Williams, F. E., Varanasi, U., Trumbly, R. J. 1991. The CYC8 and TUP1 proteins involved in glucose repression in *Saccharomyces cerevisiae* are associated in a protein complex. *Mol. Cell. Biol.* 11:3307–16
137. Williams, T. L., Leibowitz, M. J. 1987. Conservative mechanism of the *in vitro* transcription of killer virus of yeast. *Virology* 158:231–34
138. Young, T. W. 1987. Killer yeasts. *The Yeasts,* ed. A. H. Rose, J. S. Harrison, pp. 131–64. New York: Academic
139. Zhu, H., Bussey, H. 1991. Mutational analysis of the functional domains of yeast K1 killer toxin. *Mol. Cell. Biol.* 11:175–81
140. Zhu, H., Bussey, H., Thomas, D. Y., Gagnon, J., Bell, A. 1987. Determination of the carboxyl termini of the α and β subunits of yeast K1 killer toxin: requirement of a carboxypeptidase B-like activity for maturation. *J. Biol. Chem.* 262:10728–32

Annu. Rev. Microbiol. 1992. 46:377–98

AUTOREGULATORY FACTORS AND COMMUNICATION IN ACTINOMYCETES

Sueharu Horinouchi and Teruhiko Beppu

Department of Agricultural Chemistry, The University of Tokyo, Bunkyo-ku, Tokyo 113, Japan

KEY WORDS: A-factor, microbial hormone, signal transduction, antibiotic production, cell differentiation

---

CONTENTS

Abstract

The ability to produce a wide variety of secondary metabolites and a mycelial form of growth that develops into spores are two biological aspects characteristic of the gram-positive bacterial genus *Streptomyces*. Secondary

0066-4227/92/1001-0377$02.00

metabolism and cell differentiation are controlled by diffusible low-molecular-weight chemical substances called autoregulators. A-factor, the representative of the autoregulators, triggers streptomycin production and aerial-mycelium formation in *Streptomyces griseus*. A-factor exerts its regulatory function with the aid of a receptor protein that itself acts as a repressor-type regulator. The A-factor signal via the A-factor-receptor protein is transferred to downstream genes, such as streptomycin-production genes and sporulation genes, through multiple regulatory genes in a complex regulatory cascade. Thus, A-factor can be termed a "microbial hormone." This review deals with the A-factor-regulatory cascade as a model system for other autoregulators. The biosynthesis of A-factor, the structures and characteristics of other autoregulators, and the importance of these autoregulators in the ecosystem are also included.

INTRODUCTION

Actinomycetes have characteristic biological aspects such as a mycelial form of growth that culminates in sporulation and the ability to form a wide variety of secondary metabolites including most of the antibiotics. Complex morphological development in these genera is phenotypically related to secondary metabolism. Recent development of recombinant DNA technology in *Streptomyces* spp. has enabled investigators to clone individual genes controlling simultaneously both processes, thus providing direct genetic evidence for their close relationship. Both differentiation and secondary metabolism are probably responses to environmental conditions that must be performed so that a signal is transferred in succession to many genes in the complex regulatory relay.

One of the important steps in such a regulatory cascade for differentiation and secondary metabolism in actinomycetes includes autoregulatory factors, which are low-molecular-weight chemical substances effective at extremely low concentrations and are essentially required as intrinsic factors for triggering secondary metabolite formation and/or morphogenesis. These properties of autoregulators are akin to those of hormones in eukaryotic organisms. A-factor, the focus of this review, is representative of such autoregulators. Hormonal control by A-factor as well as by various other autoregulators is one of the most characteristic features of actinomycetes, our discussion of which will reveal a new aspect of the regulatory system in prokaryotes.

CHARACTERISTICS OF AUTOREGULATORS

A-Factor

A-factor [2-(6'-methylheptanoyl)-3*R*-hydroxymethyl-4-butanolide] was originally discovered by Khokhlov and coworkers (30) in the culture broth of

Streptomyces griseus as a factor that induced streptomycin (Sm) production in a mutant strain of *S. griseus*. They also found that A-factor was essentially required for aerial mycelium formation (29). Genetic studies in our laboratory showed that some of Sm-nonproducing *S. griseus* strains obtained by conventional mutagenesis actually required a certain diffusible factor for recovery of both Sm production and sporulation (15). On the basis of the observations of Khokhlov's group, we chemically synthesized A-factor and found that the *R*-form of A-factor at a concentration as low as 10^{-9} M completely restored the wild-type phenotype (15). We also found that a decreased level of resistance to Sm in these A-factor-deficient mutants was restored to that of the wild-type strain through the stimulatory production of Sm-6-phosphotransferase (16). In addition to these phenotypes, A-factor induced the production of a diffusible yellow pigment that contains an aminosugar moiety (21). Figure 1 illustrates the pleiotropic effect of A-factor, together with a gene, *afsA*, responsible for A-factor synthesis and the A-factor-binding protein, which is described below.

Consistent with the pleiotropic effect of A-factor, the profiles of total proteins of *S. griseus,* as determined by sodium dodecyl sulfate-polyacrylamide gel electrophoresis, are dramatically different in the presence and absence of A-factor. More than 10 proteins widely ranging in molecular size are produced only in the wild-type strain, while a few proteins are present specifically in an A-factor-deficient mutant strain. The addition of A-factor to the mutant during growth made its protein profile become almost the same as that of the wild-type strain. Although the exogenous supply of an excess of A-factor slightly stimulates biomass formation in the cultures of both an A-factor-deficient and the wild-type strains, the yields of Sm per gram of mycelium do not change. However, Sm production begins earlier by one day when A-factor is added at the time of inoculation. These observations reflect the role of A-factor as a trigger of Sm production and sporulation but not as an enhancer for the yield of Sm. In relation to this, A-factor is produced just prior to Sm production and it rapidly disappears before Sm production reaches a maximum, which may suggest the presence of an A-factor-inactivating enzyme because A-factor itself is rather stable chemically.

The extremely low effective concentration of A-factor led us to establish a sensitive bioassay system for A-factor (15). Briefly, a test strain grown on an agar plug (5 mm in diameter by 3 mm in height) at 28°C for 2 days, or a paper disc containing the culture broth of a test strain, is transferred to a soft agar layer seeded with the A-factor-deficient *S. griseus* and incubated at 28°C for 2 days. Then nutrient soft agar, which contains spores of the indicator strain *Bacillus subtilis* is overlaid and the plate is further incubated overnight at 37°C. A-factor diffused from the agar plugs or paper discs into the soft agar causes the mutant *S. griseus* strain to produce streptomycin, which in turn is detected by growth inhibition of the indicator strain. Approximately 0.5 ng of

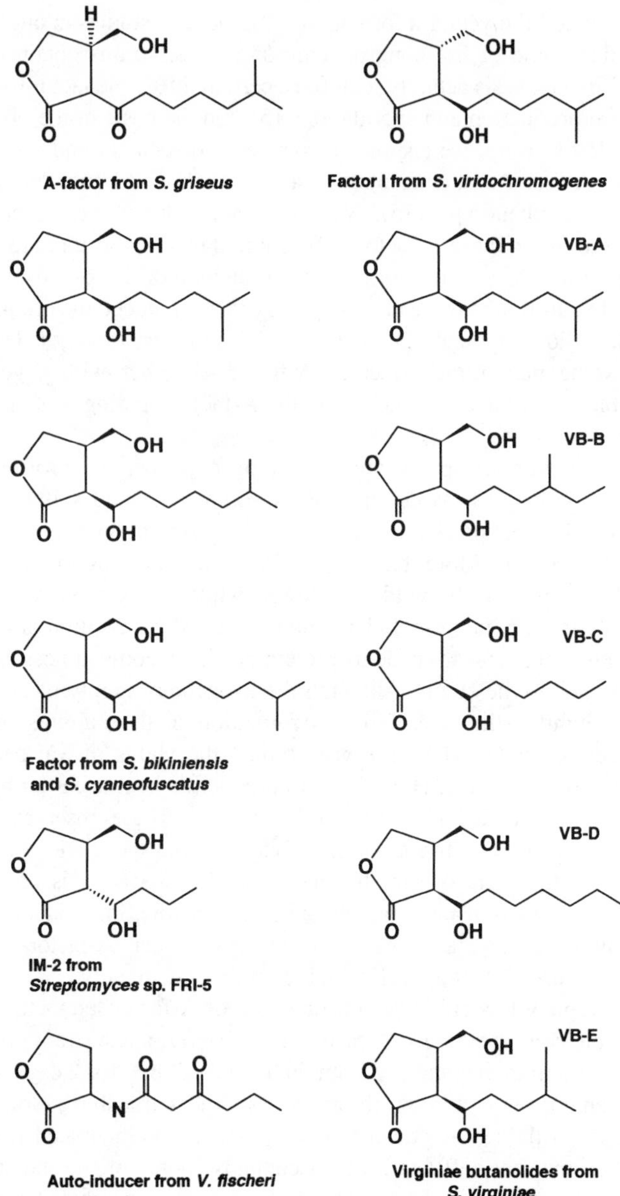

Figure 1 Chemical structures of autoregulatory factors with a butyrolactone ring.

A-factor in a paper disc is sufficient to form a detectable inhibitory zone. In relation to the ease in detection of A-factor production, the extremely low effective concentration of A-factor can make detecting the A-factor-deficient mutants by conventional plating techniques difficult. This sensitive assay system has shown that A-factor production is widely distributed among strains of actinomycetes (15). The autoregulatory role of A-factor, however, has been observed almost exclusively in the Sm-producing *S. griseus*. The response to A-factor has so far been found only in anthracycline production in *S. griseus* and in nosiheptide production in *Streptomyces actuosus* (47). On the other hand, when the cultured media of various *Streptomyces* spp. were analyzed as to the A-factor activity for the *S. griseus* mutant, an active spot with a R_f value different from that of A-factor was detected using thin-layer chromatography (2). We assume that A-factor and its analogs act as a chemical signal in a wider variety of species. If a mutant deficient in A-factor of a certain strain could be obtained, then we might see the effect of A-factor on antibiotic production and cell differentiation by just examining the phenotypes of the mutants.

A-Factor Homologues

VIRGINIAE BUTANOLIDES Yanagimoto et al (57) reported the presence of a regulatory factor for virginiamycin (staphylomycin) production in *Streptomyces virginiae*. Further studies by Yamada et al (56) revealed that the factor actually consists of a mixture of compounds, named virginiae butanolides (VBs) (Figure 2). Although they are very similar to A-factor, no effect of VBs on morphogenesis of the producing strain has so far been observed. The specific activities of VB-A and VB-C are almost the same, and they are active at concentrations of a few ng/ml. VB-B is less active than VBs A and C. The study of structure-function relationships with chemically synthesized analogs having acyl moieties of different chain lengths at the C-2 position revealed that a 1'-hydroxyheptyl and a 1'-hydroxyoctyl at this position were most effective with a minimum effective concentration of 0.8 ng/ml (45). In addition, γ-nonalactone and γ-undecalactone have a similar activity, but much less activity than the above factors. The structure-activity relationship in VBs seems to be slightly looser than in the case of A-factor; A-factor analogs with a side chain that was one carbon atom shorter or longer exhibited only 10% activity (28). A-factor is not active for virginiamycin production in this strain, and vice versa: VBs show almost no effect on Sm production in *S. griseus*. For example, 1 ng of A-factor caused detectable streptomycin production in an A-factor-deficient mutant of *S. griseus,* while more than 3 μg of VB-C caused streptomycin production alone when tested using the streptomycin-consynthesis method. These results indicate that the A-factor family

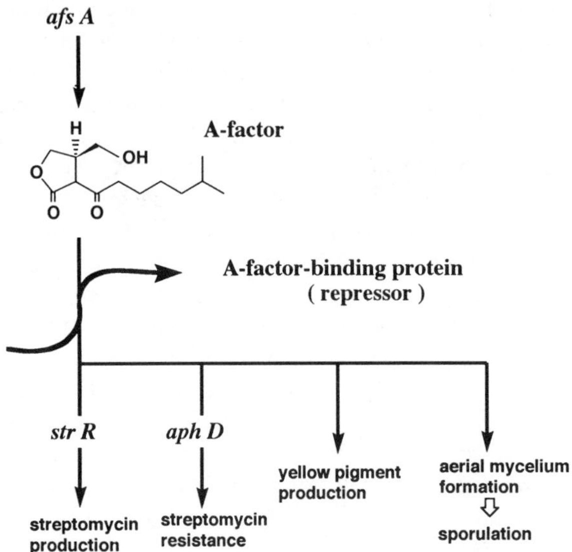

Figure 2 Regulatory cascade including A-factor and the A-factor-binding protein. A-factor is probably synthesized from a β-keto acid and a glycerol derivative with the AfsA protein encoded by *afsA*. Instability of Sm production and sporulation is ascribed to the extrachromosomal nature of *afsA*. The positive A-factor signal is transferred, via some additional regulatory proteins, to *strR*, a regulatory gene in the Sm biosynthetic cluster, to *aphD* encoding Sm-6-phosphotransferase and residing just downstream of the *strR* gene, and to the genes responsible for yellow-pigment production and sporulation.

possessing a β-keto group at the 6-position and the VB family with an OH group at this position are strictly discriminated from each other, but specificity within each of the families depending on the difference in the acyl side chains is rather loose.

I-FACTOR Gräfe's group has been working with a mutant strain derived from *S. griseus* ZIMET 43682 whose anthracycline production and sporulation are restored by A-factor (12, 14). A wide screening test with this mutant strain, based on the assumption that regulatory compounds are not necessarily specific for a given strain, led to the identification of I-factor from *Streptomyces viridochromogenes* (Figure 2). The factor possessing an OH group at the 6-position showed an inducing activity similar to, but about 10 times less than, A-factor's. A similar compound with a 6-OH group obtained by reduction of racemic A-factor, however, had almost no activity on the Sm-producing *S. griseus* strain with which Khokhlov's group worked. Gräfe et al (13) also isolated similar regulators as a mixture of homologues from *Streptomyces bikiniensis* and *Streptomyces cyaneofuscatus* (Figure 2) (13). The minimum amount of each compound required to induce anthracycline production and sporulation was about 10 times more than that of A-factor. In regard to the fact that these A-factor analogs have an OH group at the 6-position, it still remains unclear whether these compounds actually function as a chemical signal for secondary metabolism and morphogenesis in the original strains. Genetic studies including the isolation of mutants deficient in these substances are obviously necessary.

IM-2 Sato et al (51) isolated an A-factor analog, named IM-2, which triggered blue pigment production in *Streptomyces* sp. FRI-5 at a concentration of 0.6 ng/ml. The structure was determined to be 2,3-*trans*-2-(1'-hydroxybutyl)-3-(hydroxymethyl)butanolide.

AN AUTOREGULATOR OF BIOLUMINESCENCE A compound similar to A-factor was reported to be a regulator in a phylogenically different prokaryote: an autoinducer of bioluminescence in the marine luminous bacterium *Vibrio fischeri* (10). Synthesis of the luciferase in *V. fischeri* is induced only when the concentration of the autoinducer produced by the same organism reaches a certain critical value, about one to two molecules per cell. The structure of the autoinducer is similar to that of A-factor (Figure 1) (9). This autoregulator is inactive on other species of luminous bacteria, probably because the autoinducers of other species are different from that of *V. fischeri*. In spite of the similarity in structure, A-factor has no activity in this bioluminescence system (A. Eberhard, personal communication).

B-Factor

B-factor was isolated from commercial yeast extract on the basis of the observation that the addition of yeast extract to the culture of a mutant strain of *Nocardia* sp. restored rifamycin B production. B-factor purified from yeast extract is a butyl ester of 3'-AMP (Figure 3) and is active at a very low concentration of 10 ng/ml (26). Because rifamycin production in the mutant strain was partially restored by the exogenous supply of 3-amino-5-hydroxybenzoic acid, an intermediate in the biosynthetic pathway, B-factor seems to control a step upstream from this intermediate. The study on the structure-function relationship in B-factor with chemically synthesized homologues showed that all the homologues with acyl moieties with different chain lengths (C-2 to C-12) were active (27). The highest activity was obtained with the *n*-octyl ester, which caused half maximum induction at 3.1×10^{-10} M. In addition to the side-chain lengths, the adenosine moiety could be replaced by guanosine; the butyl ester of 3'-GMP also had the almost same activity as B-factor. An intrinsic substance with potent B-factor activity was also isolated from the parental *Nocardia* strain. The UV spectrum of the purified fraction

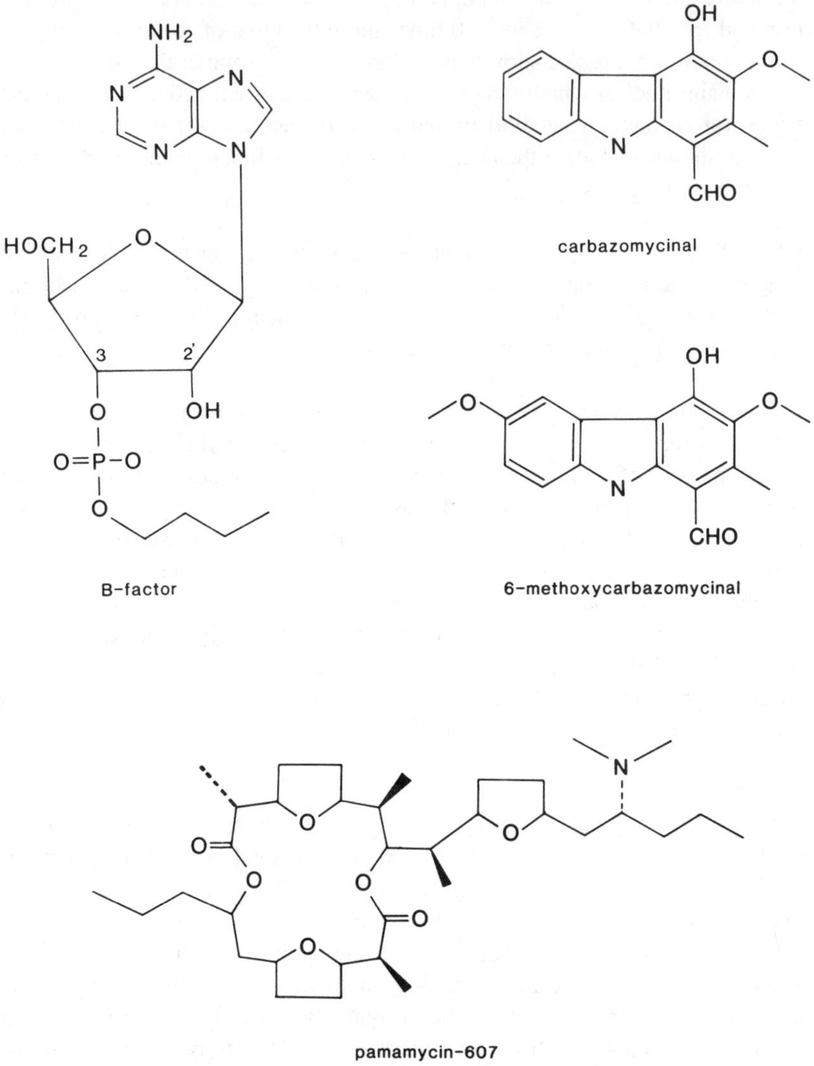

B-factor

carbazomycinal

6-methoxycarbazomycinal

pamamycin-607

Figure 3 Chemical structures of autoregulatory factors whose structures are determined.

showed its absorption maxima at 260 nm. These data suggest that a nucleotide derivative, possibly a B-factor homologue, is present in the *Nocardia* strain. Our recent findings (1) that a slight amount of B-factor is synthesized in the cell extracts of the *Nocardia* strain from 2',3'-cyclic AMP and butanol, and that several microbial RNases produce B-factor from RNA in the presence of butanol, suggest that B-factor homologues are widely distributed in both prokaryotes and eukaryotes. The structural similarity of B-factor to 3',5'-cAMP and guanosine oligophosphates, both of which are important in the regulation of multiple cellular functions in a wide variety of organisms, leads us to assume that B-factor or its homologues exert some physiologically significant function in general.

Factor C

Szabo's group (54) reported factor C from *S. griseus,* which restored sporulation of a mutant 52-1 strain in the submerged culture. Factor C produced at a late stage of growth is a protein of molecular mass of 34,500 (5) that is localized mainly in the membrane fraction. The factor C activity was distributed in a wide variety of *Streptomyces* spp. Immunoblot experiments showed that the non-spore-forming mutant, *S. griseus* 52-1, still produced a small amount of factor C. On addition of factor C to the mutant strain, it disappeared rapidly from the medium, but the sporulated cells thus obtained contained a large amount of factor C. Factor C therefore seems to induce further production of itself in a larger amount than when it is added. Szabo et al (54) proposed the hypothesis that factor C affects a cell membrane receptor and by changing its function evokes a chain of events leading to differentiation. For further elucidation of the mechanism of regulation displayed by factor C, genetic studies, including the cloning of the factor C gene, are obviously necessary.

In relation to the proteinous regulator, we observed a potential low-molecular-weight autoregulator that appears to bind to a protein (42). We isolated more than 50 carbapenem-nonproducers from *Streptomyces fulvoviridis* N1501, which were divided into 6 cosynthesis groups. Two mutants, FN27 and FN104, which were not directly involved in this biosynthetic pathway, showed a bald phenotype. Both of their defects in the abilities to sporulate and to produce carbapenem were restored by the addition of the culture broths of the parental strain and all the mutants belonging to the other five groups (41). Purification of the active compound required pronase or urea treatments of a concentrated culture broth to separate the factor from the protein. This protein may modulate the regulatory effect of the low-molecular-weight substance, although no genetic or biochemical evidence to support this idea has been obtained.

Pamamycin and Others

McCann & Pogell (37) reported on three differentiation factors of *Streptomyces alboniger:* an aerial mycelium stimulator, named pamamycin, which also showed antibacterial and antifungal activities, and two inhibitors of aerial mycelium differentiation. Structural elucidation of pamamycin by Marumo's group (33, 34) revealed that it is a mixture consisting of at least 8 homologues with a 16-membered macrodiolide ring ranging in molecular weight from 593 to 691. Pamamycin-607 is one component of the mixture with a molecular weight of 607, and its absolute stereochemistry was recently determined (Figure 3) (43). Pamamycin-607 showed a concentration-dependent effect on aerial mycelium formation, whereas it inhibited the growth of substrate mycelium at a higher concentration. Although aerial mycelium formation in this organism was also induced by exogenous supply of Ca^{2+} (44), involvement of the cation in the action of pamamycin has not yet definitively been concluded.

Kondo et al (32) isolated aerial mycelium inhibitors in a *Streptoverticillium* sp. during their screening for a mycelium-inducing substance. These compounds, carbazomycinal and 6-methoxycarbazomycinal (Figure 3), inhibit aerial mycelium formation at concentrations of 0.5 to 1.0 μg/ml. No antibacterial or antifungal activity of these compounds has been detected. Because of lack of genetic or biochemical studies on these substances, how they are involved in the morphogenesis of the producing organism is still unknown.

Similarly, no further information on an additional possible factor, named a sporulation pigment, which was reported as a characteristic pigment associated with sporulation of a *Streptomyces venezuelae* strain (52), has been obtained, nor has any information appeared regarding a possible autoregulator and its antagonist for aerial mycelium formation in *Streptomyces cattleya* (36). The dialyzable stimulator was detected only in a late stage of growth. Interestingly, in liquid fermentation, the stimulator suppressed thienamycin production when added at inoculation, although no inhibition was observed when the stimulator was added 24 h after the inoculation.

BIOSYNTHESIS OF BUTYROLACTONE-TYPE REGULATORS

Recent biosynthetic studies of virginiae butanolides by Yamada and coworkers (50) have established the synthetic pathway of γ-butyrolactone-type regulators. A *Streptomyces antibioticus* strain that produces several milligrams of VB-A per liter enabled them to elucidate the biosynthetic pathway using feeding experiments with ^{13}C-labeled precursors. Figure 4 shows the pathway these authors have proposed (50). The VB-A molecule is assembled from two acetate, one isovalerate, and one glycerol molecule. The β-ketoacyl CoA is synthesized from isovaleryl CoA as a starter and from two malonyl CoAs derived from two acetate molecules. This type of elongation of carbons is the same as in polyketide biosynthesis. Then, the β-keto acid couples with a three-carbon compound derived from glycerol. At the last step, the keto group at the 6-position is reduced in the presence of NADPH.

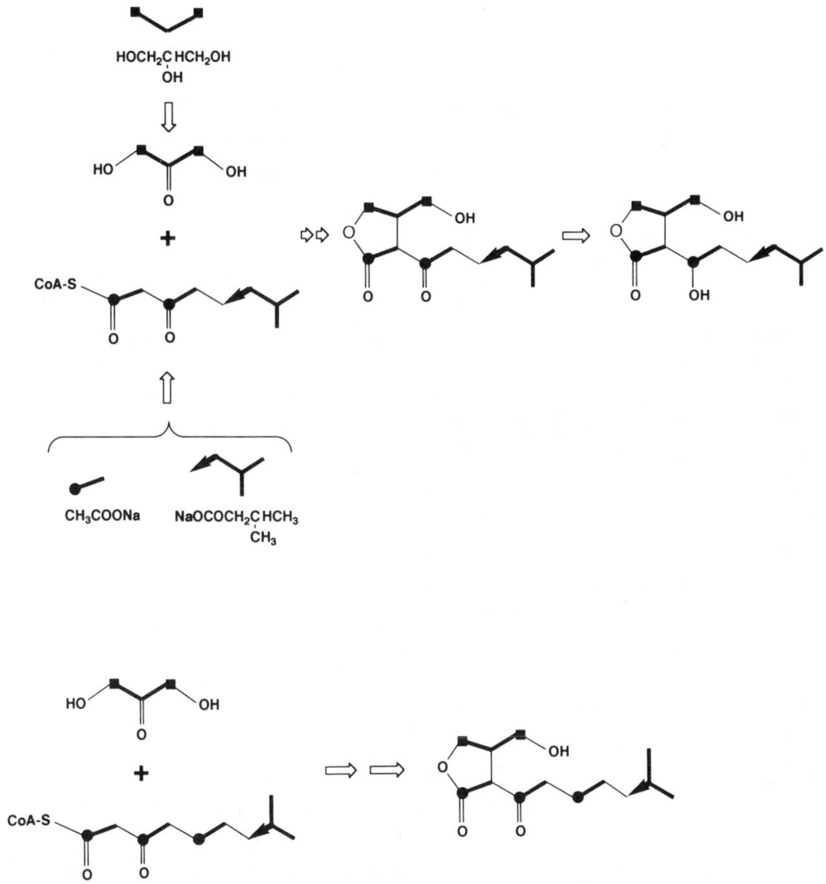

Figure 4 Biosynthetic pathway of virginiae butanolide A (*top*) and A-factor (*bottom*). According to this pathway proposed by Yamada's group (50), the β-ketoacyl CoA in VB-A is synthesized from two malonyl CoAs derived from two acetate molecules and one isovaleryl CoA. The C-3 unit is a glycerol derivative. By analogy with the pathway for VB-A, the β-ketoacyl CoA in A-factor is synthesized from three malonyl CoAs and one isobutyryl CoA.

According to this scheme, the lengths and branching of acyl side chains at the 2-position of γ-butyrolactone-type regulators are determined by the number of malonyl CoAs and the variety of starter molecules. When isobutyryl CoA is the starter instead of isovaleryl CoA, and three malonyl CoAs are used, the β-ketoacyl CoA is formed as the key precursor for the synthesis of A-factor. In relation to the proposed biosynthetic pathway of A-factor, we speculate that the *afsA* gene cloned from *S. griseus* as the gene complementing the A-factor deficiency of mutant strains (21) codes for an enzyme catalyzing the condensation between the above β-keto acid and the glycerol derivative. The *afsA* gene exhibits the following properties: (*a*) A trimmed 1.2-kb fragment encoding a single open reading frame of 301 amino acids confers A-factor production to A-factor-deficient mutant strains of *S. griseus*. (*b*) The 1.2-kb fragment shows a remarkable gene dosage effect on A-factor production when it is carried on a high-copy-number plasmid, and (*c*) the 1.2-kb fragment can confer A-factor production with a marked gene dosage effect to all the *Streptomyces* strains so far tested (22). We suppose that almost all *Streptomyces* spp. contain the glycerol derivative as a common metabolite and a pathway for polyketide biosynthesis that endows the β-keto acid as a common metabolite as well. We therefore speculate that the AfsA protein is one of the key enzymes responsible for the condensation, although further studies are obviously necessary to clarify the function of AfsA.

MOLECULAR MECHANISMS INVOLVED IN REGULATION BY AUTOREGULATORS

Genetics of A-Factor Biosynthesis

A-factor is the only autoregulator whose biosynthesis is genetically studied. A-factor-deficient mutants of *S. griseus* are easily obtained by so-called plasmid-curing treatments, such as acridine orange treatments, incubation at high temperatures (35–37°C), and protoplast regeneration. Consistent with this observation, genetic mapping by protoplast fusion techniques with two mutants containing double auxotrophic markers in addition to the A-factor deficiency suggested that the A-factor mutation had no fixed position on the chromosomal linkage map (17). Furthermore, 91–93% of the recombinants obtained from these crosses were A-factor positive. These observations suggested the extrachromosomal nature of the determinant of A-factor biosynthesis (*afsA*). Additional evidence for this instability of the A-factor determinant was obtained from Southern blot hybridization experiments using as a probe the *afsA* gene probably encoding an A-factor biosynthetic enzyme, as already mentioned before. A considerably long DNA stretch of more than 10 kilobase pairs covering the *afsA* gene was completely deleted in the A-factor-deficient mutants. These data imply that the *afsA* gene is on an unstable extrachromosomal element such as a plasmid or a transposon in *S. griseus*. However, our attempts to detect plasmid DNA physically by conventional and pulse-field agarose gel electrophoresis have failed so far. The mechanism of deletion of *afsA* at a very high frequency is therefore still unknown. On the other hand, the *afsA* gene in *Streptomyces coelicolor* is mapped at a fixed position on the chromosome, and the mutation frequency of *afsA* is as low as those for auxotrophic markers (17). In *S. coelicolor,* a mutation in *afsA* does not cause any detectable alteration in antibiotic production or morphogenesis.

Nucleotide sequence analysis revealed that the *afsA* gene coded for a protein of 301 amino acid residues and 32.6 kilodaltons (kDa) (23) whose codon usage pattern was in agreement with the general pattern of *Streptomyces* genes with an extremely high G+C content (3). It is interesting that the transcriptional start point, as determined by S1 nuclease mapping, was the adenine residue, the first position of the ATG translational initiation codon. This presents a striking contrast to the conventional interaction between ribosomes and Shine-Dalgarno sequences in translational initiation in other prokaryotes (11). A similar feature of the transcription initiation has been observed with several antibiotic-resistance determinants that are supposedly enhanced at the onset of antibiotic biosynthesis because of a requirement for self-protection (4, 19, 24). As described before, A-factor production by *S. griseus* is enhanced exactly at the onset of streptomycin production. The unusual feature of transcription and translation of the *afsA* gene may be associated with its differential expression during the growth phases.

The Binding Proteins for A-Factor and Virginiae Butanolides

A-FACTOR-BINDING PROTEIN The similarity in the functional features of A-factor to eukaryotic hormones prompted us to examine the presence of a specific binding protein in *S. griseus*. We synthesized the optically active form of tritium-labeled A-factor, [1'-^{3}H]-3*R*-hydroxymethyl-2-(6-methylheptanoyl)-4-butanolide, according to the method of Mori & Yamane (40) and used it as the probe for detecting the presumptive binding protein. As expected, a binding protein specific to A-factor was found in the cytoplasmic fraction, but not in the membrane fraction (38). The binding protein had an apparent molecular weight of approximately 26,000, as determined by gel filtration chromatography. Scatchard analysis showed that the binding protein bound A-factor in the molar ratio of 1:1 with a dissociation constant, K_d, of 0.7 nM. The number of the binding protein was roughly estimated to be 40 per genome. The extremely small K_d value is consistent with the extremely low effective concentration of A-factor, 10^{-9} M in vivo. The absence of any binding protein in the membrane fraction is consistent with a model in which A-factor diffuses freely into the cell and requires no surface receptor.

Our recent success in isolating mutants deficient in the A-factor-binding protein resulted from an incidental finding that a *S. griseus* mutant strain 2247 requiring no A-factor for its Sm production or sporulation had a defective binding protein (39). This observation implied that the A-factor-binding protein in the absence of A-factor repressed the expression of both the phenotypes in the wild-type strain. Screening among mutagenized *S. griseus* colonies for strains producing Sm and sporulating in the absence of A-factor yielded three mutants that were also deficient in the A-factor-binding protein. The reversal of the defect in the binding protein of these mutants led to simultaneous loss of Sm production and sporulation. All these data suggested that the A-factor-binding protein played a role in repressing both Sm production and sporulation and that the binding of A-factor to the protein released its repression. The mutants deficient in the A-factor-binding protein began to produce Sm and sporulate at an earlier stage of growth than the wild-type strain. The amount of Sm produced by these mutants was approximately 10 times as large as that produced by the parental strain. The conidial chain length of the binding protein–deficient mutants was distinctly longer than that in the parental strain. Functional involvement of the binding protein in A-factor regulation convinces us that the protein is a real functional receptor and that A-factor can be termed a prokaryotic hormone. Cytoplasmic localization of the binding protein resembles the regulatory system of steroid hormones in eukaryotes. However, a vivid contrast is that the complex between steroid hormones and their receptor binds to specific promoter sequences and acts as an activator for transcription, whereas in the A-factor system, the binding protein itself may bind to some gene(s) and act as a repressor.

As already mentioned, A-factor and VBs are strictly discriminated by *S. griseus* and *S. virginiae,* respectively. Consistent with these observations in vivo, the A-factor-binding protein from *S. griseus* did not bind VB-C significantly, and conversely, only very weak competition with unlabeled A-factor was observed in binding of radioactive VBs. On the other hand, the VB-binding protein, as described below, showed rather loose specificity as to the difference in the acyl chains in the VB family. The structure-activity relationship of the binding proteins so far observed is consistent with that observed in vivo.

VIRGINIAE BUTANOLIDE-BINDING PROTEIN A similar binding protein for virginiae butanolides was detected in *S. virginiae* by using [^{3}H]VB-C (31). The VB-binding protein present at a copy number of 30–40 per genome had a dissociation constant of 1.1 nM. Okamoto et al (48) purified the binding protein to homogeneity and characterized its properties. The binding protein showed a molecular mass of 36,000 on sodium dodecyl sulfate-polyacrylamide gel electrophoresis and bound to VB in a 1:1 binding stoichiometry. These workers also cloned the gene (*vbrA*) coding for the binding protein by the DNA-probing method with oligonucleotides designed for amino acid sequences of the protein. The nucleotide sequence of the *vbrA* gene predicted a 319–amino acid protein (M_r, 34,676) whose COOH-terminal half showed considerable homology with the NusG protein, a putative antitermination protein of *Escherichia coli*. Just downstream from the *vbrA* gene resided a gene homologous with the *E. coli rplK* gene coding for a ribosomal protein. These data strongly suggest that the *vbrA* gene is the counterpart of the *E. coli nusG* gene. The function of the NusG protein in *E. coli* is not adequately understood, except that it appears to participate in the *N*-mediated transcriptional antitermination in phage λ (8). The success in cloning the gene encoding the binding protein and subsequent overexpression of the gene product will facilitate the elucidation of binding between the protein and the autoregulators, and further clarify the regulatory mechanism.

A REGULATORY CASCADE FOR STREPTOMYCIN PRODUCTION

Identification of an A-Factor-Dependent Promoter

Sm production is one of the most remarkable of those phenotypes in *S. griseus* that are controlled by A-factor. In the absence of A-factor, neither Sm nor biosynthetic intermediates were detected in the culture broth when analyzed using high-performance liquid chromatography (15). Induced expression of the Sm-biosynthetic genes by A-factor is clearly demonstrated by measuring an increased level of self-resistance to Sm due to the Sm-6-phosphotransferase, which is involved in the last stages of the Sm biosynthetic pathway. Extensive genetic analyses hitherto performed have provided a reliable base for elucidating the regulatory mechanism working at a step downstream from the binding protein.

A detailed work from Piepersberg's laboratory (7) on the transcriptional organization of part of the Sm biosynthetic gene cluster showed that at least three mRNAs covering a regulatory gene (*strR*), the Sm-6-phosphotransferase gene (*aphD*), and the aminocyclitol amidinotransferase gene (*strB*) were detectable only in the presence of A-factor. We therefore conducted experiments to determine which promoter(s) is directly responsive to A-factor. Promoter-probing experiments together with S1 nuclease mapping revealed that the transcription of only the *strR* promoter was enhanced by A-factor (55). In addition, the *aphD* gene that encodes the major Sm resistance determinant was found to be transcribed mainly by read-through from the A-factor-dependent *strR* promoter. This accounts for the prompt induction of Sm resistance upon the addition of A-factor. Actually, Sm-6-phosphotransferase is produced 3–4 h after the addition of A-factor, irrespective of the timing of the addition. Such direct dependence of the transcription of *aphD* on the *strR* promoter would help achieve a rapid increase in self-resistance just prior to induction of Sm biosynthesis by A-factor.

The above results indicate that the stimulatory effect of A-factor is amplified through increased production of StrR, which in turn stimulates the transcription of other Sm-production genes. This assumption is supported by the observation that the presence of the *strR* gene at a high copy number causes an A-factor-deficient mutant to produce Sm. Distler et al (7) and Pissowotzki et al (49) have shown that the StrR product stimulates the transcription of the *strB* coding sequence, probably by an antitermination mechanism. They assume that most of the mRNA transcribed constitutively from the *strB* promoter is prematurely terminated before reaching the end of the gene and that the StrR product suppresses the premature termination, leading to generation of the complete transcript. Whether the proposed antitermination effect of the StrR product is also involved in regulation of other genes in the Sm biosynthetic cluster is not clear.

An Activator Protein for Streptomycin Production

Further subcloning experiments indicated that the region 400 to 300 bp upstream from the transcriptional start points of *strR* is essential for its A-factor dependence. Removal of the essential region resulted in loss of the response to A-factor. These features are reminiscent of those of upstream activating sequences (UAS) of prokaryotes (53) and of enhancers in eukaryotes (25, 35). These findings imply that some sequence in this region is recognized by a regulatory protein, either an activator or a repressor, which stimulates the *strR* promoter. The following observations led to the conclusion that the putative regulator was an activator: (*a*) The presence of the *strR* promoter including the UAS-like region at a high copy number did not lead to a corresponding increase in promoter activity but rather caused a significant reduction when compared to the case of the *aphD* and *strB* promoters, probably because of titration of an activator assumed to be present in a limiting amount in the cell, and (*b*) the wild-type *S. griseus* strain containing the *strR* promoter on a high-copy-number plasmid produced little or no Sm, probably because of poor transcription of the chromosomal *strR* promoter resulting from the titration of an activator. The putative activator protein is therefore distinct from the A-factor-binding protein that acts as a repressor-type regulator for Sm production. Consistent with this observation, the activator appears to control only Sm production but not sporulation, because the transformants containing the *strR* promoter at a high copy number, like the wild-type strain, still form abundant spores.

We next tried to detect the activator able to bind to the UAS-like sequence using a gel retardation assay with partially purified cell extracts. As expected, a protein that specifically bound to the UAS sequence was detected only in the wild-type *S. griseus* strain, but not in an A-factor-negative mutant (D. V̄ujaklija, S. Horinouchi & T. Beppu, unpublished data). The exogenous supply of A-factor to the culture of the mutant made the protein appear. These data clearly indicate that the transcription of the *strR* promoter is controlled by the activator protein whose production depends on A-factor. Similar gel retardation assays showed the presence of several kinds of proteins capable of binding to different DNA fragments derived from the upstream region of the transcriptional start point of *strR*. These proteins were produced by both the wild-type and the A-factor-negative mutant strains.

A Model for Regulation of Streptomycin Biosynthesis

Figure 5 illustrates a regulatory cascade model including the putative activator and the A-factor-binding protein. On binding A-factor, the binding protein or the receptor is released from the promoter of a still unknown gene *X*, leading to the production of a putative activator protein X, which then causes elevated transcription of *strR*. The StrR protein in turn causes general stimulation of the Sm biosynthetic pathway. Although this simplest model consists of a single activator protein for *strR*, an additional regulatory protein(s) may be present that transfers the A-factor signal from protein X to the activator. The fact that the activator protein does not affect sporulation, as described below, may support this possibility. Activation of the whole Sm biosynthetic gene cluster by A-factor may ultimately be achieved through the positive regulatory effect of the StrR product. Such a regulatory cascade may result in highly pleiotropic control as well as marked amplification of the initial positive signal of A-factor.

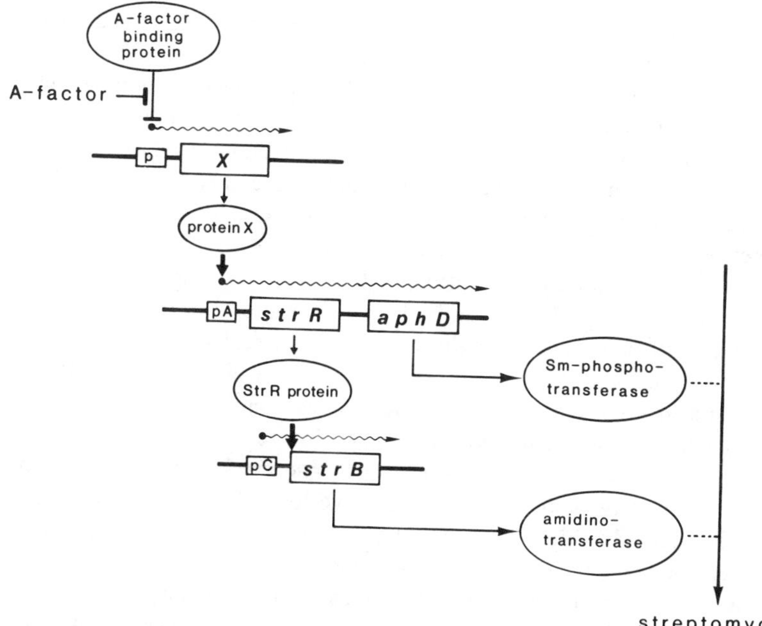

Figure 5 Regulatory cascade for streptomycin biosynthesis in *S. griseus*. A-factor produced just before Sm production binds to the A-factor-binding protein, as a result of which the binding protein present on the promoter (p) of a still unknown gene *X* is released, leading to transcription of *X*. The protein X then binds to the promoter (pA) of *strR*, and activates the synthesis of mRNA covering both *strR* and *aphD*. The StrR protein thus produced activates the expression of *strB*, probably by an antitermination mechanism, in such a way that StrR allows the mRNA starting at the *strB* promoter (pC) to read through a putative termination point.

A similar regulatory relay may function for sporulation. The transformants of the wild-type strain containing the *strR* promoter on a high-copy-number vector, which produced less or no Sm, still formed abundant spores just like the wild-type strain. We therefore speculate that a counterpart of the above activator is available, and it is responsible for activating regulatory genes necessary for sporulation. Independence of Sm production and sporulation in these transformants suggests that the activator protein for sporulation is present and is distinct from the activator for *strR*. To find the putative regulatory protein for sporulation, we recently cloned four different DNA fragments that caused sporulation, but not Sm production, in an A-factor-deficient *S. griseus* strain. Each of these fragments supposedly codes for a certain protein responsible for a certain step in sporulation, one of which may correspond to the putative regulator.

BIOLOGICAL IMPLICATIONS AND FUTURE PERSPECTIVES

The features of A-factor and virginiae butanolides convince us to define these γ-butyrolactones as hormones for secondary metabolism and/or cell differentiation in *Streptomyces*. The presence of γ-butyrolactone homologues in various actinomycetes indicates the generality of these compounds as a hormone in actinomycetes. On the other hand, several secondary metabolites with autoregulatory activities, such as pamamycin and carbazomycinal, cannot yet be termed real intrinsic regulators because of a lack of genetic evidence. Although actinomycetes belong to the prokaryotes, their filamentous morphology with differentiated mycelia and spores is obviously similar to that of fungi. We can find several examples in filamentous fungi showing that secondary metabolites act as a chemical signal for the fungi's mating and sexual differentiation, such as sirenin, which acts as an attractant for fusion in *Allomyces* spp. (46), and trisporic acids, which trigger sexual morphogenesis in *Blakeslea* spp. (6). These observations suggest that some of the diverse secondary metabolites, especially γ-butyrolactones, are the consequence of convergent evolution of actinomycetes as filamentous microorganisms, in which the diffusible factors have evolved to act as chemical signals between the cells at a distance in the hyphae.

Exerting the regulatory functions for each of the γ-butyrolactone autoregulators apparently requires binding proteins with a ligand specificity different from strain to strain. The binding proteins for A-factor in *S. griseus* and virginiae butanolides in *S. virginiae* can obviously distinguish the functional groups, either a hydroxyl or a keto group, at the 6-position. In addition, the virginiae butanolide–binding protein shows a slightly loose specificity to the lengths and branching of the acyl side chain at the 2-position. Even among *S.*

griseus strains, the ligand specificity appears to differ from strain to strain, because several A-factor homologues with different acyl chain lengths or a keto group at the 6-position are still active in the strain with which Gräfe's group has been working. The specific pair of an autoregulator and its binding protein in a given strain should have simultaneously evolved and should function as a potent system for discriminating different chemical signals from other species to prevent miscommunication among various actinomycetes species in the ecosystem.

A still unanswered question is how the biosyntheses of autoregulators are triggered and controlled in the life cycle of actinomycetes. In *S. coelicolor,* the expression of *afsA* probably encoding the key enzyme for A-factor biosynthesis is controlled by *afsR* (20), whose gene product is phosphorylated by a specific membrane-bound kinase named AfsK (18). The AfsR and AfsK proteins as a regulator and a sensor, respectively, supposedly compose a two-component regulatory system. The AfsK sensor may recognize a certain environmental signal and transfer it, via protein phosphorylation, to the AfsR protein serving as a positive regulator for *afsA*. Although no counterpart for *afsR* in *S. griseus* has yet been detected, a regulatory system similar to this AfsR-AfsK system may function as a switch for the *afsA* expression in *S. griseus*. In addition to the control of biosynthesis of autoregulators, a signal transfer to the downstream genes required for morphogenesis and secondary metabolism remains to be explored in detail. The γ-butyrolactone-type regulators conceivably transfer their positive signal in a manner similar to the relay of the A-factor-positive signal to the Sm biosynthetic genes, as shown in Figure 5. Because we have observed a change in the profile of phosphorylated proteins in the presence and absence of A-factor, a signal relay through phosphorylation of intermediate regulatory proteins may play an important role in the signal transfer downstream from the step involving autoregulators and their binding proteins.

Literature Cited

1. Azuma, M., Nishi, K., Horinouchi, S., Beppu, T. 1990. Ribonucleases catalyze the synthesis of B-factor (3'-butylphosphoryl AMP), an inducer of rifamycin production in a *Nocardia* sp. *J. Antibiot.* 43:321–23
2. Beppu, T. 1992. Secondary metabolites as chemical signals for cellular differentiation. *Gene*. In press
3. Bibb, M. J., Findlay, P. R., Johnson, M. W. 1984. The relationship between base composition and codon usage in bacterial genes and its use for the simple and reliable identification of protein-coding sequences. *Gene* 30:157–66
4. Bibb, M. J., Janssen, G. R., Ward, J. M. 1986. Cloning and analysis of the promoter region of the erythromycin resistance gene (*ermE*) of *Streptomyces erythraeus*. *Gene* 41:357–68
5. Biro, S., Bekesi, I., Vitalis, S., Szabo, G. 1980. A substance affecting differentiation in *Streptomyces griseus*. Purification and properties. *Eur. J. Biochem.* 103:359–63
6. Bu'Lock, J. D. 1976. Hormones in fungi. In *The Filamentous Fungi,* ed. J. E. Smith, D. R. Berry, 2:345–68. London: Edward Arnold
7. Distler, J., Ebert, A., Mansouri, K.,

Pissowotzki, K., Stockmann, M., et al. 1987. Gene cluster for streptomycin biosynthesis in *Streptomyces griseus:* nucleotide sequence of three genes and analysis of transcriptional activity. *Nucleic Acids Res.* 15:8041–56

8. Downifig, W. L., Sullivan, S. L., Gottesman, M. E., Dennis, P. P. 1990. Sequence and transcriptional pattern of the essential *Escherichia coli secE-nusG* operon. *J. Bacteriol.* 172:1621–27
9. Eberhard, A., Burlingame, A. L., Eberhard, C., Kenyon, G. L., Nealson, K. H., et al. 1981. Structural identification of autoinducer of *Photobacterium fischeri* bioluminescence. *Biochemistry* 20:2444–49
10. Eberhard, A., Widrig, C. A., McBath, P., Schineller, J. B. 1986. Analogs of the autoinducer of bioluminescence in *Vibrio fischeri. Arch. Microbiol.* 146: 35–40
11. Gold, L., Pribnow, D., Schneider, T., Shinedling, S., Singer, B. S., et al. 1981. Translational initiation in prokaryotes. *Annu. Rev. Microbiol.* 35: 365–403
12. Gräfe, U., Eritt, I., Hänel, F., Friedrich, W., Roth, M., et al. 1986. Factors governing polyketide and glycopeptide production by streptomycetes. In *Regulation of Secondary Metabolite Formation,* ed. H. Kleinkauf, H. von Döhren, H. Dornauer, G. Nesemann, pp. 225–48. Weinheim, FRG: VCH Verlagsgesellschaft mbH
13. Gräfe, U., Reinhardt, G., Schade, W., Eritt, I., Fleck, W. F., et al. 1983. Interspecific inducers of cytodifferentiation and anthracycline biosynthesis from *Streptomyces bikiniensis* and *S. cyaneofuscatus. Biotechnol. Lett.* 5: 591–96
14. Gräfe, U., Schade, W., Eritt, I., Fleck, W. F. 1982. A new inducer of anthracycline biosynthesis from *Streptomyces viridochromogenes. J. Antibiot.* 35:1722–23
15. Hara, O., Beppu, T. 1982. Mutants blocked in streptomycin production in *Streptomyces griseus*—the role of A-factor. *J. Antibiot.* 35:349–58
16. Hara, O., Beppu, T. 1982. Induction of streptomycin inactivating enzyme by A-factor in *Streptomyces griseus. J. Antibiot.* 35:1208–15
17. Hara, O., Horinouchi, S., Uozumi, T., Beppu, T. 1983. Genetic analysis of A-factor synthesis in *Streptomyces coelicolor* A3(2) and *Streptomyces griseus. J. Gen. Microbiol.* 129:2939–44
18. Hong, S.-K., Kito, M., Beppu, T., Horinouchi, S. 1991. Phosphorylation of the AfsR product, a global regulatory protein for secondary-metabolite formation in *Streptomyces coelicolor* A3(2). *J. Bacteriol.* 173:2311–18
19. Horinouchi, S., Furuya, K., Nishiyama, M., Suzuki, H., Beppu, T. 1987. Nucleotide sequence of the streptothricin acetyltransferase gene from *Streptomyces lavendulae* and its expression in heterologous hosts. *J. Bacteriol.* 169: 1929–37
20. Horinouchi, S., Kito, M., Nishiyama, M., Furuya, K., Hong, S.-K., et al. 1990. Primary structure of AfsR, a global regulatory protein for secondary metabolite formation in *Streptomyces coelicolor* A3(2). *Gene* 95:49–56
21. Horinouchi, S., Kumada, Y., Beppu, T. 1984. Unstable genetic determinant of A-factor biosynthesis in streptomycin-producing organisms: cloning and characterization. *J. Bacteriol.* 158:481–87
22. Horinouchi, S., Nishiyama, M., Suzuki, H., Kumada, Y., Beppu, T. 1985. The cloned *Streptomyces bikiniensis* A-factor determinant. *J. Antibiot.* 38:636–41
23. Horinouchi, S., Suzuki, H., Nishiyama, M., Beppu, T. 1989. Nucleotide sequence and transcriptional analysis of the *Streptomyces griseus* gene (*afsA*) responsible for A-factor biosynthesis. *J. Bacteriol.* 171:1206–10
24. Janssen, G. R., Ward, J. M., Bibb, M. J. 1989. Unusual transcriptional and translational features of the aminoglycoside phosphotransferase gene (*aph*) from *Streptomyces fradiae. Genes Dev.* 3: 415–29
25. Johnson, P. F., McKnight, S. L. 1989. Eukaryotic transcriptional regulatory proteins. *Annu. Rev. Biochem.* 58:799–839
26. Kawaguchi, T., Asahi, T., Satoh, T., Uozumi, T., Beppu, T. 1984. B-factor, an essential regulatory substance inducing the production of rifamycin in a *Nocardia* sp. *J. Antibiot.* 37:1587–95
27. Kawaguchi, T., Azuma, M., Horinouchi, S., Beppu, T. 1988. Effect of B-factor and its analogues on rifamycin biosynthesis in *Nocardia* sp. *J. Antibiot.* 41:360–65
28. Khokhlov, A. S. 1982. Low molecular weight microbial bioregulators of secondary metabolism. In *Overproduction of Microbial Product,* ed. V. Krumphanzl, B. Sikyta, Z. Vanek, pp. 97–109. London: Academic
29. Khokhlov, A. S., Anisova, L. N., Tovarova, I. I., Kleiner, E. Y., Kovalenko, I. V., et al. 1973. Effect of A-factor on the growth of asporogenous

mutants of *Streptomyces griseus,* not producing this factor. *Z. Allg. Mikrobiol.* 13:647–55

30. Khokhlov, A. S., Tovarova, I. I., Borisova, L. N., Pliner, S. A., Schevchenko, L. A., et al. 1967. A-factor responsible for the biosynthesis of streptomycin by a mutant strain of *Actinomyces streptomycini. Dokl. Akad. Nauk SSSR* 177:232–35
31. Kim, H. S., Nihira, T., Tada, H., Yanagimoto, M., Yamada, Y. 1989. Identification of binding protein of virginiae butanolide C, an autoregulator in virginiamycin production from *Streptomyces virginiae. J. Antibiot.* 42:769–78
32. Kondo, S., Katayama, M., Marumo, S. 1986. Carbazomycinal and 6-methoxycarbazomycinal as aerial mycelium formation–inhibitory substances of *Streptoverticillium* species. *J. Antibiot.* 39:727–30
33. Kondo, S., Yasui, K., Katayama, M., Marumo, S., Kondo, T., et al. 1987. Structure of pamamycin-607, an aerial mycelium-inducing substance of *Streptomyces alboniger. Tetrahedron Lett.* 27:5861–64
34. Kondo, S., Yasui, K., Natsume, M., Katayama, M., Marumo, S. 1988. Isolation, physico-chemical properties and biological activity of pamamycin-607, an aerial mycelium-inducing substance from *Streptomyces alboniger. J. Antibiot.* 41:1196–1204
35. Marriott, S. J., Brady, J. N. 1989. Enhancer function in viral and cellular gene regulation. *Biochim. Biophys. Acta* 989:97–110
36. McCann-McCormick, P. 1981. Endogenous factors involved in aerial mycelium formation in actinomycetes. In *Microbiology—1981,* ed. D. Schlessinger, pp. 348–51. Washington, DC: Am. Soc. Microbiol.
37. McCann, P. A., Pogell, B. M. 1979. Pamamycin: a new antibiotic and stimulator of aerial mycelia formation. *J. Antibiot.* 32:673–78
38. Miyake, K., Horinouchi, S., Yoshida, M., Chiba, N., Mori, K., et al. 1989. Detection and properties of A-factor-binding protein from *Streptomyces griseus. J. Bacteriol.* 171:7298–7302
39. Miyake, K., Kuzuyama, T., Horinouchi, S., Beppu, T. 1990. The A-factor-binding protein of *Streptomyces griseus* negatively controls streptomycin production and sporulation. *J. Bacteriol.* 172:3003–8
40. Mori, K., Yamane, K. 1982. Synthesis of optically active forms of A-factor, the inducer of streptomycin biosynthesis in inactive mutants of *Streptomyces griseus. Tetrahedron* 38:2919–21
41. Nakata, K., Horinouchi, S., Beppu, T. 1989. Cloning and characterization of the carbapenem biosynthetic genes from *Streptomyces fulvoviridis. FEMS Microbiol. Lett.* 57:51–56
42. Nakata, K., Horinouchi, S., Beppu, T. 1989. The carbapenem biosynthetic gene cluster from *Streptomyces fulvoviridis* and a potential regulatory substance controlling carbapenem production and sporulation. In *Trends in Actinomycetology in Japan,* ed. Y. Koyama, pp. 79–82. Tokyo: Soc. Actinomycetes
43. Natsume, M., Kondo, S., Marumo, S. 1989. The absolute stereochemistry of pamamycin-607, an aerial mycelium-inducing substance of *Streptomyces alboniger. J. Chem. Soc. Chem. Commun.* 24:1911–13
44. Natsume, M., Yasui, K., Marumo, S. 1989. Calcium ion regulates aerial mycelium formation in actinomycetes. *J. Antibiot.* 42:440–47
45. Nihira, T., Shimizu, Y., Kim, H. S., Yamada, Y. 1988. Structure-activity relationships of virginiae butanolide C, an inducer of virginiamycin production in *Streptomyces virginiae. J. Antibiot.* 41:1828–37
46. Nutting, W. H., Rapaport, H., Machlis, L. 1968. The structure of sirenin. *J. Am. Chem. Soc.* 90:6434–38
47. Ohkishi, H., Miyasaka, K., Horisaka, T., Chou, H., Watanabe, Y. 1988. Strain improvement of a nosiheptide producer, *Streptomyces actuosus* 40037. In *Proc. 3rd Franco-Japanese Symp. The Impact of Genetics on Industrial Microorganisms,* ed. A. Fujiwara, pp. 47–49. Tokyo: Sansei
48. Okamoto, S., Nihira, T., Kataoka, H., Suzuki, A., Yamada, Y. 1992. Purification and molecular cloning of a butyrolactone autoregulator receptor from *Streptomyces virginiae. J. Biol. Chem.* 267:1093–98
49. Pissowotzki, K., Mansouri-Taleghani, K., Piepersberg, W. 1990. Streptomycin biosynthesis. *UCLA Symp. Mol. Cell. Biol. Suppl.* 14A:93
50. Sakuda, S., Higashi, A., Tanaka, S., Nihira, T., Yamada, Y. 1992. Biosynthesis of virginiae butanolide A, a butyrolactone autoregulator from *Streptomyces. J. Am. Chem. Soc.* 114:663–68
51. Sato, K., Nihira, T., Sakuda, S., Yanagimoto, M., Yamada, Y. 1989. Isolation and structure of a new butyrolactone autoregulator from *Streptomyces* sp. FRI-5. *J. Ferment. Bioeng.* 68:170–73

52. Scribner, H. E., Tang, T., Bradley, S. G. 1973. Production of a sporulation pigment by *Streptomyces venezuelae*. *Appl. Microbiol.* 25:873–79
53. Struhl, K. 1989. Molecular mechanisms of transcriptional regulation in yeast. *Annu. Rev. Biochem.* 58:1051–77
54. Szabo, G., Szeszak, F., Vitalis, S., Toth, F. 1988. New data on the formation and mode of action of factor C. In *Biology of Actinomycetes '88,* ed. Y. Okami, T. Beppu, H. Ogawara, pp. 324–29. Tokyo: Jpn. Sci. Soc. Press
55. Vujaklija, D., Ueda, K., Hong, S.-K., Beppu, T., Horinouchi, S. 1991. Identification of an A-factor-dependent promoter in the streptomycin biosynthetic gene cluster of *Streptomyces griseus*. *Mol. Gen. Genet.* 229:119–28
56. Yamada, Y., Sugamura, K., Kondo, K., Yanagimoto, M., Okada, H. 1987. The structure of inducing factors for virginiamycin production in *Streptomyces virginiae*. *J. Antibiot.* 40:496–504
57. Yanagimoto, M., Yamada, Y., Terui, G. 1979. Extraction and purification of inducing material produced in staphylomycin fermentation. *Hakko Kogaku Kaishi* 57:6–14

Annu. Rev. Microbiol. 1992. 46:399–428

GENETICS OF COMPETITION FOR NODULATION OF LEGUMES

Eric W. Triplett

Department of Agronomy and the Center for the Study of Nitrogen Fixation, University of Wisconsin, Madison, Wisconsin 53706

Michael J. Sadowsky

Departments of Soil Science and Microbiology, University of Minnesota, St. Paul, Minnesota 55108

KEY WORDS: nodulation, legumes, molecular genetics, competitiveness, *Rhizobium*

CONTENTS

0066-4227/92/1001-0399$02.00

Abstract

An economically important problem in microbial ecology concerns the efficacy of rhizobial inoculants for the formation of nitrogen-fixing root nodules on legume crop plants such as soybean, alfalfa, and clover. Some strains of rhizobia can increase symbiotic nitrogen fixation under controlled conditions. However, attempts to improve nitrogen fixation under agricultural conditions with such strains often fail, usually as a result of the presence of indigenous rhizobia limiting nodulation by the inoculum strains. This problem is referred to as the *Rhizobium* competition problem, and molecular genetics is being used to address the problem from two perspectives. First, the host specificity of rhizobia is being characterized with the long term goal of developing strains that can nodulate a very strain-specific host-legume genotype. Second, the genetic basis of competitiveness in several strains is being examined. Genetic determinants of nodulation competitiveness have been isolated and mechanisms for their stable integration into the genome of superior nitrogen-fixing strains have been developed. Several phenotypes have been identified as playing an important role in nodulation competitiveness including antibiosis, motility, speed of nodulation, cell-surface characteristics, and nodulation efficiency. Several solutions to this problem are likely to result from these strategies and will be useful for certain legumes in specific locations.

STATEMENT OF THE COMPETITION PROBLEM

The interactions of microbes in soil are extremely complex. Soils and their microbial inhabitants vary to such a great extent that generalizations made about microbial ecology cannot usually be made without significant qualifications. Nevertheless, the ability to predict and/or improve the survival and competitiveness of beneficial microbes in soil is crucial to solving economically important problems in microbial ecology. These problems concern degradation of organic wastes, biological control of plant diseases, or the improvement of symbiotic nitrogen fixation. This review discusses recent attempts to improve our knowledge of genetic determinants involved in *Rhizobium* spp. ecology with special reference to the use of genetics to improve the ability of *Rhizobium* inoculant strains to occupy that niche which is most important to agriculture, the nitrogen-fixing root nodule of leguminous plants. Because of limited space allotted to this review, we intend to cite only those papers that serve as examples for certain topics rather than attempt to cite all papers in the literature on the *Rhizobium* competition problem. Where we refer to *Rhizobium* or rhizobia in general terms, we are describing work on both fast-growing *Rhizobium* and slow-growing *Bradyrhizobium* species.

Since the 1920s, researchers have known that soils often contain indigenous strains of *Rhizobium* that often limit legume yield (8, 52, 93, 125). Once this problem was identified, strains were identified that provided higher legume productivity under controlled conditions. These strains were then used as inoculants in an attempt to improve legume yields in soils containing inferior strains of rhizobia. Often these inoculation attempts failed to improve legume productivity because the indigenous strains occupied the root nodules rather than the inoculum strains. When inoculation with *Rhizobium* does improve yield, the indigenous population is very small (143, 145, 200, 223). Investigators have known for many years that the size of indigenous populations of *Rhizobium* is influenced strongly by the cropping history of the soil. As early as 1926, Wilson (224) showed that *Rhizobium* populations were larger in soils with a recent planting of the homologous legume host. Many reports have confirmed this result, including a very thorough study by Woomer et al (226), who found that mean annual rainfall, legume cover and biomass, and soil nutrients accounted for 90% of the variation in the population size of *Rhizobium* spp. in soils.

Thus, indigenous rhizobia appear to be well adapted to their niche and can occupy legume nodules to the exclusion of inoculum strains. This occurs even when levels of inoculum far exceed the levels of indigenous rhizobia in the soil (217, 218). This situation is known as the *Rhizobium* competition problem. Several factors enable indigenous strains to occupy nodules to the exclusion of the inoculum strains, and Dowling & Broughton (48) discussed these factors in their review. We only discuss those factors that have been described or further substantiated since that review was published.

ECONOMIC IMPORTANCE OF THE PROBLEM

In recent years, several strategies have been developed to construct strains of *Rhizobium* and *Bradyrhizobium* that provide increased symbiotic nitrogen fixation under controlled conditions. However, these studies were done by culturing plants in sterile media and inoculating with a single strain. These experiments commonly show a 5–15% increase in biomass or grain yield when plants are inoculated with the superior strain compared with the wild-type strain. The results of such experiments, however, are only valid when near-isogenic strains are being compared. Some examples of the construction of superior strains are described below.

Chemical mutagenesis has been used to induce mutations in *Bradyrhizobium japonicum* that cause increased nitrogenase activity and/or increased plant growth (139, 155, 156, 186) compared with the wild-type strains. In soils with a low *B. japonicum* population, one of these chemical mutants did significantly increase soybean yield under field conditions (223). These chemical mutants also have very little value in determining which genes in *B. japonicum* are beneficial or detrimental to soybean productivity. Mutagenesis with chemical mutagens can cause multiple, undefined mutations in a cell. As a result, one cannot compare single point mutants to the wild-type strain with this technique.

Uptake hydrogenase activity in root nodule bacteroids improves yield of soybean using near-isogenic strains of *B. japonicum* that differ only in their ability to oxidize the dihydrogen evolved from the nitrogenase reaction (59, 60, 89, 103). The evidence for uptake hydrogenase–mediated yield enhancement is less clear in other legumes, because the literature contains no examples in which near-isogenic strains have been used to examine this question. Thus, experiments that lack the appropriate near-isogenic strain comparisons must be viewed as inconclusive.

Recently, Chen et al (24) constructed an acid-tolerant strain of *Rhizobium leguminosarum* bv. *trifolii* that expressed increased nitrogen fixation compared with the acid-sensitive parent strain. They developed this strain by curing the acid-sensitive strain of its Sym plasmid and replacing it with pBRIAN, a self-mobilizable Sym plasmid from *R. leguminosarum* bv. *trifolii*.

Cannon, Ronson, and colleagues (22, 166) have proposed that increased expression of *nifA* and *dctA* can increase symbiotic nitrogen fixation in alfalfa and soybean. This hypothesis is intriguing, but the necessary definitive experiments to demonstrate this possibility have not been published.

Spaink et al (190) have constructed a hybrid *nodD* gene that does not require a flavonoid for *nod* gene expression. A *Rhizobium* strain with this hybrid *nodD* gene expresses increased nitrogenase activity in the nodule. These and other examples show that genetic alterations can increase the ability of rhizobia to fix nitrogen. Some of these genetic alterations have been drastic, such as replacement of Sym plasmids or chemical mutagenesis, while others are more discrete, such as the addition of another copy of *nifA* to the genome.

The ability to genetically enhance symbiotic nitrogen fixation further encourages rhizobiologists to find solutions to the competition problem. Until such solutions are found, attempts to improve the microsymbiont's ability to fix nitrogen will be of limited agronomic benefit.

APPROACHES TO ADDRESS THE COMPETITION PROBLEM

Mass Inoculation of Bradyrhizobia

One method to overcome the competition problem is to massively inoculate legumes with superior inoculum strains many times over a period of several years. The hope is that such an inoculation would eventually permit the inoculum strain to occupy the soil in sufficient numbers to limit nodule occupancy by indigenous strains. Weaver & Frederick (218) have estimated that to obtain 50% occupancy of soybean nodules by the inoculum strain, an inoculant rate 1000 times the soil population would be required. In 1976, Dunigan et al (53) began a seven-year study to determine whether *B. japonicum* strain 110 could survive in Louisiana soils following inoculation for a three-year period and a subsequent four-year period. This method not only permitted the permanent establishment of strain 110 in the soil, but the strain became more competitive in succeeding years. Similar results were found with repeated inoculations of red clover with *R. leguminosarum* bv. *trifolii* (140).

Isolation of Genetic Determinants for Competitiveness

Several groups have isolated genetic determinants thought to confer increased nodulation competitiveness. These include genetic determinants for bacteriocin production and resistance (204, 205), nodulation efficiency (180), and extracellular polysaccharide production (14). All of these are described in detail below. Determinants of nodulation competitiveness have been identified by transposon Tn*5* mutagenesis. Moreover, the transposon Tn*phoA* (199) shows promise in the identification of extracellular proteins involved in nodulation competitiveness.

Chemical Control of Nodulation

Many compounds have been identified that inhibit flavonoid induction of *nod* genes in *Rhizobium* and *Bradyrhizobium* spp. (31, 43, 65, 118). Formation of soybean nodules can be prevented by the use of such inhibitors (31, 118). According to this strategy, legume seeds would be inoculated with a strain possessing a flavonoid-independent *nodD* gene, such as that constructed by Spaink et al (190), along with an inhibitor to prevent *nod* gene induction in the indigenous strains. A flavonoid-independent *nodD*-containing strain would constitutively express *nod* genes and not be affected by inhibitors of *nod* gene expression. The commercial utility of such inhibitors is limited by the observation that strains of *Bradyrhizobium* differ in the types of inhibitor necessary to prevent *nod* gene induction (31). To completely prevent nodulation of soybean in the field would require a minimum of four inhibitors.

TYPES OF NODULATION DETERMINANTS

Because the competitiveness of indigenous strains of *Rhizobium* and *Bradyrhizobium* is influenced by their interaction with host legumes and by the process of nodulation itself, it is important to understand the molecular determinants involved in the relationship. Three types of nodulation loci have been identified in several rhizobia and bradyrhizobia: common *nod* genes, host-specific *nod* genes (*hsn*), and the genotype- or cultivar-specific *nod* genes (GSN or CSN). These genetic determinants are discussed below.

Common Nodulation Genes

Elucidation of the molecular genetics of nodulation has been achieved mostly with the fast-growing rhizobia, *Rhizobium meliloti, R. leguminosarum* bvs. *viceae* and *trifolii,* and *Rhizobium* sp. strain NGR234 [see reviews by Denarie et al (37), Long (134, 135), Martinez et al (141), and Stanley & Cervantes (195)], mostly because nodulation and nitrogen fixation genes reside on the large indigenous symbiotic plasmids in these species (135). Due to the large number of *nod* genes thus far identified (*nodA* through *nodZ*), currently designated nodulation loci are referred to as *nol* genes (173). In *R. meliloti,* one of the better-studied species, the essential nodulation genes, *nodD*$_1$*ABC,* are closely linked (Figure 1) and located in a gene cluster on the symbiotic plasmid (36, 37, 113, 135). Hybridization and transcription fusion analyses indicate that the *nodABC* genes are transcribed in the same direction and that *nodD*$_1$ is located upstream of *nodA* and is divergently transcribed (55, 102, 169). Interstrain and interspecies complementation studies have demonstrated that *nodABC* genes are functionally similar among rhizobial species and, as such, have been termed common nodulation genes (113). In *R. leguminosarum* bvs. *viceae* and *trifolii* and in *R. meliloti,* mutations in the common *nod* genes result in a root-hair curling–minus (Hac$^-$) phenotype (43–46, 55, 169, 201, 210). These genes are involved in the very early stages of symbiosis. The inducible expression of common nodulation genes requires *nodD,* which encodes a plant substance–inducible, positive regulatory protein (42, 65, 135, 137, 147, 150, 158, 163). In fast-growing *Rhizobium* species, *nodD* genes (specifically *nodD*$_1$) are constitutively expressed (150, 165, 170) and, in some instances, have been shown to be autoregulatory (147, 170).

Although earlier studies suggested that *nodD* genes from different rhizobial

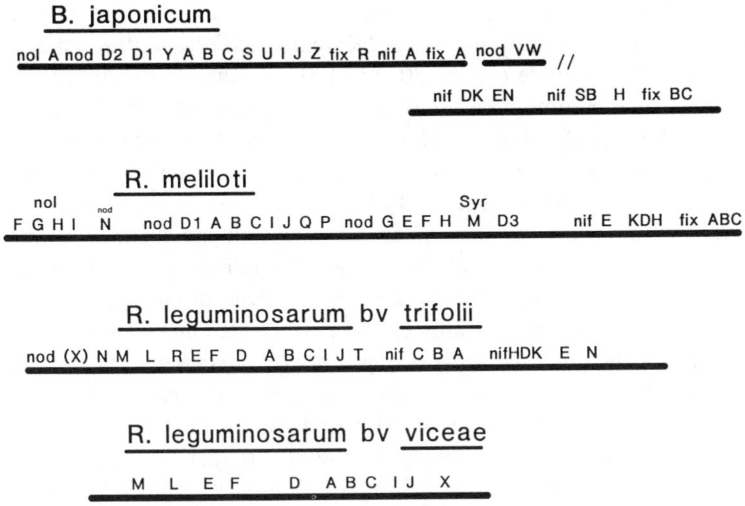

Figure 1 Physical relationship of nodulation (*nod* and *nol*), N_2-fixation (*nif*), and fixation-related genes in *Bradyrhizobium japonicum, Rhizobium meliloti,* and *R. leguminosarum* bvs. *trifolii* and *viceae*. Adapted from Long (135), Lewis-Henderson & Djordjevic (129, 130), Martinez et al (141), and Sadowsky et al (175).

(or bradyrhizobial) species were functionally similar, recent studies (9, 101, 192) have shown that the *nodD* proteins from different rhizobia do not behave identically. NodD has been shown to have a regulatory function in the nodule bacteria.

Most information about the regulation of *nod* gene expression comes from *lacZ*-transcriptional- and translational-fusion studies done with *R. meliloti* and *R. leguminosarum* bvs. *trifolii* and *viceae* (43, 135, 158, 229). Similar studies done with other fast growers (176) and with *B. japonicum* (117) indicate that the induction of many *nod* genes requires the presence of plant-derived flavonoid signal molecules. The induction, which can be seen at the level of transcription, depends on the *nodD* gene product (6, 65–67, 81, 101, 104, 114, 135, 147, 150, 158, 165, 168, 170, 187, 192). Current models suggest that the *nodD* gene product directly interacts with plant-derived signal molecules (66, 99, 135, 171). Moreover, the *nodD* protein binds to putative promoter areas, termed the *nod* box sequence (171), of inducible *nod* genes (66, 99, 135). Biochemical and genetic studies suggest that the *nodD* protein is a DNA-binding, transcriptional activator (135, 171).

The identities of many of the *nod* gene–inducing substances have been determined (43, 117, 158, 176, 229) and, in the case of the fast-growing rhizobia, belong to classes of compounds called flavanols, flavones, and flavanones. In *B. japonicum,* the major inducing substances are isoflavones (9, 82, 117). *Rhizobium leguminosarum* bvs. *viceae* and *trifolii* usually have single copies of the *nodD* gene, while *Rhizobium meliloti* and *Bradyrhizobium japonicum* have multiple copies (80, 81, 100). Spaink et al (189, 191, 192) have shown that in some cases *nodD* can act as a host-specificity determinant.

Although many of the same nodulation genes have been found in *Bradyrhizobium japonicum,* two major organizational differences separate this bacterium from the fast growers (122, 151, 172, 193). First, in *B. japonicum,* the nodulation genes are not located on plasmids but are distributed on the chromosome (142), and second, the nitrogen-fixation structural *nifDK* and *H* genes are physically separated by a relatively large distance from the *nod* genes (Figure 1). Recently, Müller et al (149a) reported that *nif* and *nod* genes are located on the same 100 kb *Spe*I fragment in *B. japonicum* strains 110*spc*4 and 61A24. In addition, *nodD*$_1$ expression in *B. japonicum* strain I-110 is induced by isoflavonoids (9, 216). Mutations in common nodulation genes affect competitiveness in a relatively straightforward manner; a strain that can no longer nodulate or has a delay in nodulation on its legume host is often rendered noncompetitive.

Host-Specific Nodulation Genes

In *R. meliloti,* another nodulation gene cluster is located near the common *nod* region (102). This gene cluster is involved in host-specific nodulation (*hsn*) and has been designated *hsnABCD* [also termed *nodFEGH,* respectively (36, 102)]. Host-specific nodulation genes are required for nodulation of specific host plants in different legume genera. The *hsn* loci have been so named because mutations in these regions cannot be functionally complemented by analogous regions from other *Rhizobium* species. Rhizobial host-range determinants have been defined by two means (42, 43): first, by the examination of nodulation mutants that have altered host-range properties. These are mutants that cannot be complemented by DNA regions from other rhizobia. Second, *hsn* loci have been defined as those genes that, when transferred to a suitable strain, render the recipient capable of nodulating the donor's host plant. Analogous *hsn* genes (termed *nodFEG*) have also been reported in *R. leguminosarum* bvs. *trifolii* (46, 164, 182) and *viceae* (50, 187, 189, 220) and in the wide-host-range *Rhizobium* strain MPIK3030 (6, 7, 12, 112, 128, 157, 195).

Nieuwkoop et al (151) reported the identification of a *hsn* gene linked to the common nodulation region in *Bradyrhizobium japonicum* strain USDA 110. While this sequence has been shown to be involved in the host-specific nodulation of siratro (*Macroptilium atropurpureum*), mutations in this region do not affect the nodulation of soybean. In addition, Hahn & Hennecke (86) identified another *hsn* locus in *B. japonicum* strain USDA 110 that was not linked to the common nodulation genes. This locus contains two nodulation genes, *nodV* and *nodW* (80) and is apparently essential for the nodulation of siratro, mungbean, and cowpea, but not soybean. As with common nodulation genes, mutations or lesions in *hsn* loci affect competitiveness by inhibiting or delaying nodulation. Because these loci affect nodulation of particular legume genera, competitiveness is only reduced with respect to specific legumes.

Genotype- or Cultivar-Specific Nodulation Determinants

Genotype-specific nodulation (GSN) genes are those bacterial sequences that allow nodulation of specific plant genotypes within a given legume species (70, 129, 130, 173). If the plant genotype is a cultivated variety, the genes are referred to as cultivar-specific nodulation (CSN) determinants (94, 129, 130). The GSN-like genes were initially identified in *R. leguminosarum* bv. *viceae* strain TOM. Strain TOM nodulates the pea genotype *Pisum sativum* cv. Afghanistan (17, 131), but European *R. leguminosarum* bv. *viceae* strains fail to nodulate this host (17). Genes enabling strain TOM to nodulate Afghanistan pea have been localized on the symbiotic plasmid, pRL5JI (83, 98). Davis and colleagues (34) subsequently identified a single gene in this region, *nodX*, which is necessary for the nodulation of cv. Afghanistan peas. The *nodX* gene resides downstream of the *nodIJ* genes in strain TOM (Figure 1). Data from *lacZ*-fusion studies indicated that *nodX* is moderately induced by pea root exudate and by the flavanone hesperetin (34). Interestingly, there are no homologues to *nodX* in conventional *R. leguminosarum* bv. *viceae* strains, although several *R. leguminosarum* bv. *viceae* strains reportedly contain the *nodX* gene (138). While mutations in *nodX* result in a *nod*$^-$ phenotype on Afghanistan pea, they do not affect nodulation of European pea genotypes. Moreover, transfer of *nodX* to conventional strains allows the recipients to nodulate Afghanistan peas.

Several other GSN (or CSN) genes have also been reported. In *Sinorhizobium fredii* (formally *Rhizobium fredii*) strain USDA 257, a single, chromosomally located CSN gene, *nolC*, has been shown to control nodulation of a commercial soybean cultivar (94, 119). In this case, Tn*5* insertions in the gene regions allow *S. fredii* to nodulate commercial soybean cultivars. Phenotypically this region is similar to that reported by Djordjevic et al (46) and Innes et al (104). These authors identified a gene, *hsnA*, on a 14-kb fragment of the *R. leguminosarum* bv. *trifolii* Sym plasmid that encodes host-specific nodulation functions. Specifically, Tn*5* mutations in the *hsnA* region of *R. leguminosarum* bv. *trifolii* strain ANU843 displayed extended host range. Phenotypically, these mutants could not nodulate *Trifolium repens*, but could nodulate *Trifolium subterraneum*. More importantly, unlike the wild-type parent, the *hsnA* mutants could nodulate *P. sativum*.

Lewis-Henderson & Djordjevic (129) reported that in *R. leguminosarum* bv. *trifolii, nodM* is a CSN gene for *T. subterraneum* (clover) cv. Woogenellup. Recently, the *nolA* gene was found in *B. japonicum* strain USDA 110. This GSN gene allows serocluster 123 isolates to form nodules on serogroup 123–restricting plant genotypes (173). The 710-bp open reading frame is located approximately 3.6 kb transcriptionally downstream of *nodD* and is presumably transcribed from its own promoter. Translational/transcriptional *lacZ* fusion experiments indicated that *nolA* was moderately induced by *nod* gene transcriptional activators, such as soybean seed extract and the isoflavone genistein (173). More recently, a second putative GSN gene was identified in *B. japonicum* strain USDA 438 (175a). This GSN nodulation locus, termed MAP1.4, extends the host range of a serogroup 123 isolate. The DNA region was isolated by random Tn*5* mutagenesis and a mutation in the locus only affects nodulation of soybean genotype PI 377578. The GSN and CSN genes can affect nodulation competitiveness of an organism by eliminating or inhibiting nodulation ability (which obviously reduces its competitiveness) or by blocking nodulation of other nodulation-competent strains (23, 49).

INTERACTION OF HOST AND MICROBE NODULATION DETERMINANTS

Gene-for-Gene Interactions

Many plant pathogen races elicit a hypersensitive response on a particular host plant depending on the presence of single dominant alleles in the plant host and corresponding dominant alleles in the pathogen. This relationship is referred to as a gene-for-gene interaction (68). In the host plant, these alleles are termed resistance genes, while in the pathogen they are referred to as avirulence genes (57, 58, 68). If either bacterial or plant allele is recessive, the plant does not recognize the pathogen and plant defense mechanisms are not activated. The end result of this interaction is that the plant becomes diseased (74). On the other hand, if the pathogen is recognized, then the plant activates a series of disease-defense mechanisms that culminate in what is termed a hypersensitive reaction (108, 197). The hypersensitive response itself can be viewed as the end product of the interaction of the alleles of both partners. In a like manner, legume nodule initiation and formation is the end result of the interaction of specific plant alleles (nodulation-restriction genes) and specific bacterial genes (host-range determinants) (42, 74, 109, 175), and in many instances, the corresponding host-specificity determinants act in a gene-for-gene manner (47, 70, 74, 94, 129, 130, 173, 175). Thus, regardless of the system (biotroph or necrotroph), the overall interaction between pathogen or symbiont and its host can be viewed as the genetic interrelationship of the host-range determinants of both partners.

Hierarchy and Types of Nodulation Determinants

The simultaneous interaction of plant and bacterial genes affects competition for nodulation among bacterial strains. Several recent studies have shown that the combined action of genes present in legume host and bacterium control, to a significant degree, the outcome of the genetic interactions of the symbiotic partners (27, 143). Many of the same mechanisms operating to control specificity in plant pathogen–host interactions also appear to be operating in legume-*Rhizobium* symbioses (73, 175; for review, see 42). Several studies have indicated that gene-for-gene recognition in plant-pathogen interactions is superimposed on an organism's ability to parasitize (58, 72, 194). The same overall concept is apparently true for *Rhizobium*-legume interactions; the specificity of host-specific nodulation is superimposed on the ability to nodulate itself (42, 43, 175). That is, rhizobial host-range determinants appear to function in concert with those genes controlling nodulation itself.

Taken together, results from several studies suggest a hierarchy to nodulation determinants. While some host-range determinants affect nodulation at the legume genus level, such as the *hsn* genes of *R. meliloti* (102), others affect nodulation of a single genotype within a given species and involve CSN or GSN determinants (34, 94, 129, 130, 173). Although the avirulence gene-resistance gene model (68) elegantly describes the genetic interaction between the dominant alleles of the flax plant and its fungal pathogen, flax rust, results from several studies indicate that host and microbe genetic specificity determinants can interact in more that one way. To date, microbial host species–specific genes have been identified in *Agrobacterium, Bradyrhizobium, Rhizobium, Pseudomonas,* and *Xanthomonas* spp. (74, 173, 175, 194), and while some of the bacterial-specificity determinants act negatively to control plant-microbe interactions (43, 46, 94, 127, 129, 130), others act positively (34, 101, 109, 173, 196, 227). Insertions in negatively acting elements extend host range and insertions in positively acting genes limit host range. For example, Djordjevic and coworkers (46) reported that a Tn*5* mutation in a *R. leguminosarum* bv. *trifolii* gene region caused an extended host-range phenotype. Also, Heron et al (94) reported on the isolation of *S. fredii* Tn*5* mutants with extended host-range. Similarly, Horvath and coworkers (102) indicated that *hsnD* (*nodH*) mutants of *R. meliloti* have an extended host range that now includes two *Vicia* species. The *R. meliloti nodH* and *nodQ* genes also appear to act as dominant suppressors of nodulation of clover and vetch (35, 61).

In addition to these types of negative factors, in several clear examples, host range is controlled by positively acting host-range determinants. In these cases, Tn*5* insertions in specific gene regions narrow host range. For example, Tn*5* insertions in *hsn* or *nod-1* loci of *B. japonicum* affect nodulation of siratro or cowpea, but not soybean (86, 151). Positively acting host-range determinants are also clearly exemplified in experiments in which host range is extended by the transfer of new genes into an organism (101).

While some plant host–specificity determinants act dominantly (26, 29, 30, 69, 149, 213) to control nodulation or pathogenesis, others are recessive (97, 129). However, in some instances, both resistance and avirulence determinants can act codominantly (162), resulting perhaps from instances in which the avirulence and resistance genes are inherited as oligogenic traits (54, 74, 105). Moreover, in numerous examples, single resistance genes recognize multiple avirulence genes (see 74).

Host Genes Affecting Nodulation and Competition

In many symbiotic partnerships, the host plant exerts a major influence on initiation of symbiosis (116). Because the competitiveness of indigenous and introduced microorganisms is tied to an organism's ability to nodulate a given legume, host genes influencing nodulation affect competition in a primary manner. In symbioses in which nodulation is restricted or inhibited, a very early step of the symbiotic sequence must be defective. The host genome, the rhizobial genome, and the environment all influence the early stages of infection (215). Mutants with lesions in common nodulation genes cannot curl root hairs or nodulate (61, 62, 126) and many nonnodulating plant mutants do not exhibit root hair curling or cortical cell divisions (51, 144).

Host-Controlled Nodulation Restriction

To date, there are several clear examples of host genes directly affecting nodulation ability and hence competition (25, 97, 129, 130, 173, 214). One of the early reported cases of host-controlled nodulation restriction was found in the *Pisum sativum–R. leguminosarum* bv. *viceae* strain TOM symbiosis. While strain TOM nodulates the pea genotype *P. sativum* cv. Afghanistan (17, 70, 111, 132, 138, 228), most European and North American strains of *R. leguminosarum* bv. *viceae* fail to nodulate this host (17). On the other hand, several Middle Eastern strains of *R. leguminosarum* bv. *viceae* nodulate commercial pea cultivars in addition to cv. Afghanistan (111, 131, 138, 225). A single recessive host gene, *sym-2*, found in cv. Afghanistan, was subsequently shown to condition nodulation restriction (97, 132). A CSN gene, *nodX*, which specifically interacts with the *sym-2* locus, has been isolated from strain TOM (34).

Results from many studies have indicated that soybean genotypes are differentially nodulated by *B. japonicum* strains (27, 28, 38, 110, 214). Soybean genotypes restricting nodulation by specific strains or serogroups of *Bradyrhizobium* have been reported. The genes restricting nodulation can be either dominant or recessive and affect the nodulation of all bradyrhizobia or specific strains (213). In 1954, a single recessive gene, termed rj_1, was reported that produces a nonnodulating condition with all *B. japonicum* strains (222). The single dominant gene Rj_2 (19) conditions restricted nodulation with all tested strains of the 122 and c1 serogroups (20). The single dominant gene Rj_3 conditions restricted nodulation with only USDA 33 and not with other strains that are serologically related (212). A fourth single dominant gene, Rj_4, conditions restricted nodulation with strain USDA 61 (40, 213). The reaction of Rj_4 with strains serologically similar to USDA 61 has not been reported, although other strains are reportedly restricted for nodulation (39, 172a).

Several plant-introduction genotypes (PI) and cultivars have also been identified that restrict the nodulation and reduce the competitiveness of *B. japonicum* strain USDA 123 (110). While the nodulation of isolates belonging to serogroup 123 was restricted by the PI genotypes, those isolates belonging to serogroups 127 or 129 were only partially restricted or not restricted at all. The PIs were well nodulated and effectively fixed N_2 with inoculant-quality strains, and, in relation to the commonly grown soybean cultivar Williams, dramatically reduced the competitiveness of strain USDA 123. A single GSN gene, *nolA,* was recently identified and allows serocluster 123 isolates to form nodules on serogroup 123–restricting plant genotypes (173).

Although results from initial studies suggested that the PI genotypes restrict the nodulation of *B. japonicum* strains in a serogroup-specific manner, more recent studies indicated that the relationship between nodulation restriction and serogroup may be more complex than originally thought (175, 177). Additional soybean genotypes have also been reported that also restrict the nodulation of several other isolates in *B. japonicum* serocluster 123 (27). Subsequent studies indicated that three of the genotypes, PIs 377578, 371607, and 417566, restricted nodulation by different serocluster 123 isolates (27, 28, 110, 177). Although preliminary studies indicate that nodulation restriction in one of the PI genotypes (PI 417566) is conditioned by a single dominant gene (26), the genetics of the host determinants conditioning restricted nodulation in the other PI genotypes have not been determined.

Host-controlled nodulation restriction can also be found in the *R. leguminosarum* bv. *trifolii*–clover symbiosis (70, 75, 77). Several *R. leguminosarum* bv. *trifolii* strains have been described (such as TA1 and UNZ29) that are specifically restricted for nodulation by *Trifolium subterraneum* (subclover) cv. Woogenellup. Strain TA1 can form effective nodules on other *Trifolium subterraneum* cultivars (75), and cv. Woogenellup is nodulated by other *R. leguminosarum* bv. *trifolii* strains (79). A recent report by Lewis-Henderson & Djordjevic (129) indicated that a single recessive gene in cv. Woogenellup, *rwt1,* is responsible for conditioning nodulation restriction with strain TA1. Two negatively acting *R. leguminosarum* bv. *trifolii* strain TA1 genes, *nodM* and *csn-1*, have been shown to specifically interact (in a gene-for-gene manner) with cv. Woogenellup (129).

Breeding for Increased Competitiveness

The breeding of plant varieties for increased nodulation competitiveness, with particular rhizobia or bradyrhizobia, has essentially relied on the use of plant germplasms that specifically exclude nodulation by specific bacterial strains (26, 76, 84, 85, 90). This indirect approach has been used with soybean (26, 84, 85, 124) and white clover (90). In all cases, the germplasm used for breeding purposes was identified by selection of lines showing restricted nodulation with particular rhizobial strains (25, 90, 120). In soybean, studies have been done to enhance nodule occupancy by strain USDA 110 (84, 85) and to inhibit nodulation by competitive strains in serocluster 123 (25, 27, 28). While Greder and coworkers (84) did not examine competitiveness per se, they did report that breeding of germplasm with high recovery rates for strain USDA 110 may be difficult because of erratic heritabilities and other problems. Cregan and coworkers (26), on the other hand, reported that nodulation competitiveness in one PI genotype (417566) was conditioned by a single dominant allele. However, the PI used only conditioned reduced competitiveness of a few serocluster 123 strains. Since serocluster 123 contains genetically diverse strains (177) that are restricted for nodulation by different soybean genotypes, these authors suggested that the genetic factors present in three to four sources will be needed to restrict nodulation by all strains in serocluster 123.

In the clover–*R. leguminosarum* bv. *trifolii* system, Hardarson & Jones (90) reported that the preference of white clover for *R. leguminosarum* bv. *trifolii* strains is a heritable, additive trait without a dominance or maternal effect. The authors did not determine the genetics of the inheritance or determine the number of genes involved. Taken together, results from all these studies suggest that much work remains to be done concerning breeding plant lines for preference for particular rhizobial strains.

MEASURING COMPETITIVENESS

Mathematical Evaluation of Competitiveness

Quantification of competitiveness has been a problem when investigators need to obtain a quantitative comparison of the competitiveness of two or more strains. Mathematical models of competitiveness are also potentially useful for the prediction of the behavior of inoculants in soil. Amarger & Lobreau (3) published an excellent model, which describes the relationship between the proportion of nodules occupied by the inoculum strain and the number of cells in the inoculum and the number of indigenous rhizobia.

Beattie et al (13) published a refinement of this model allowing for the effects of dual nodule occupancy.

Gene-Probe Techniques for Strain Identification in Soils and Nodules

Traditional methods for strain identification in soil and nodules include evaluations of intrinsic or acquired antibiotic resistance, serology, colony morphology, and symbiotic effectiveness (2, 154, 188, 203, 205). Recombinant DNA technology complements these techniques. Gene probes have been used to identify strains with specific host specificities in *B. japonicum* (174) and the presence of specific insertion-sequence elements in *R. meliloti* (219), as well as for strain identification in *R. leguminosarum* bv. *trifolii* (95).

PHENOTYPES INVOLVED IN NODULATION COMPETITIVENESS

Several characteristics of root-nodule bacteria play a significant role in nodulation competitiveness. These include antibiosis, cell-surface characteristics, motility, speed of nodulation and the autoregulatory response, and symbiotic effectiveness. Each of these are described below, but none have been shown to be important under field conditions using near-isogenic strains.

Antibiosis

Some strains of *Rhizobium* produce antibiotics or possess phages (91, 92, 96, 106, 183–185). Some of these specifically inhibit other *Rhizobium* strains. The most highly characterized anitbiotic is trifolitoxin, a small peptide produced by *R. leguminosarum* bv. *trifolii* strain T24 (184, 206, 207). This peptide is bacteriostatic to most strains of *S. fredii* and *R. leguminosarum* bvs. *trifolii, phaseoli,* and *viceae* at low concentrations (184, 206). At levels higher than that typically produced by T24, the growth of *R. meliloti* strains is restrained (209). Strains in other bacterial genera, including *Bradyrhizobium, Agrobactertium, Pseudomonas, Xanthomonas, Escherichia,* and *Bacillus,* are not inhibited. Strain T24 limits nodulation by trifolitoxin-sensitive strains but not by trifolitoxin-resistant strains (184, 206). Transposon Tn*5* mutants of T24 lacking trifolitoxin production do not inhibit nodulation by trifolitoxin-sensitive strains (206).

Using T24 as an inoculant strain may seem to be an easy solution to the problem of nodulation competitiveness on clover roots. However, T24 nodules do not fix nitrogen. Thus, the trifolitoxin production and resistance genes (*tfx*) had to be isolated and stably integrated into the genome of *R. leguminosarum* bv. *trifolii* TA1, an effective strain (204, 205, 208). The resulting exconjugant showed increased nodulation competitiveness compared with the wild-type strain (205). However, all of the competition assays published to date with this system have been done in sterile media. It is not known whether the *tfx* genes will be useful in inoculants under agricultural conditions. Recent evidence does suggest that trifolitoxin is produced in soil (A. H. Bosworth, B. T. Breil & E. W. Triplett, unpublished results).

Cell-Surface Characteristics

Cell-surface composition plays a role in nodulation competitiveness. In *S. fredii,* nonmucoid mutants of strain USDA208 are more competitive for nodulation of "Peking" soybean roots than is the wild-type strain (231). In contrast, Tn*5* mutants of *R. leguminosarum* bv. *phaseoli, Rhizobium tropici,* and *B. japonicum* deficient in exopolysaccharide (EPS) production are less able to compete for nodule occupancy than are the corresponding wild-type strains (5, 14). Bhagwat et al (14) isolated and mapped the wild-type region of DNA homologous to the Tn*5* insertion site of an EPS-deficient mutant of *B. japonicum* strain USDA 110. A 3.6-kb region was identified that complements the mucoid and competitiveness phenotypes in the mutant (14).

Motility

Several reports suggest that motility is required for nodulation competitiveness (4, 18, 133, 148). The first such report was by Ames & Bergman (4), who examined nonmotile mutants of *R. meliloti* that were either flagellated or nonflagellated. Both types of nonmotile mutants were less competitive for nodulation than the wild-type strain but were identical to the wild-type in growth rate and nodule formation. Liu et al (133) examined one nonmotile Tn*7* mutant of *B. japonicum.* Although this mutant was less competitive for nodulation than the wild-type strain, it also had a slower growth rate. This observation suggests that the mutation may have had effects other than just motility. Motility mutants unimpaired in growth rate are necessary to ensure that pleiotropic effects do not confound the data.

Caetano-Anollés et al (18) found that nonmotile spontaneous mutants of *R. meliloti* had lower nodulation competitiveness and reduced efficiency of nodule initiation compared with the wild-type parent strain. The mutants were not impaired in growth rates or speed of nodulation. Transposon Tn*5*-induced mutants of *R. leguminosarum* bv. *trifolii* TA1 that were defective in motility were also less competitive than the wild-type strain but were not impaired in growth rate (148).

Speed of Nodulation

Early rhizobial infection of legume roots induces an autoregulatory response in the plant that prevents infection by subsequent inoculations (15, 16, 115, 159, 198). This has also been demonstrated in split-root systems in which two sides of a root are spatially separated and inoculated at different time intervals

(115, 181). Nodulation is prevented on that side of the split root that is inoculated 24 hours after the other. Suppression of nodulation increases as the time interval between inoculation of the two sides increases (116). The suppression of late nodulation occurs at the nodule-meristem stage of development prior to nodule growth (21). The split-root system appears to be a useful screening method for determining the competitiveness of a group of strains (181).

Several investigators have shown a correlation between speed of nodulation and competitiveness (88, 116, 121, 146, 153, 198, 202, 211a). However, in each of these studies, unrelated strains were used rather than genetically defined isogenic ones. As a result, no definitive conclusion about the role of speed of nodulation in nodulation competitiveness can be ascertained. Moreover, there are some conflicting reports about the correlation between speed of nodulation and nodulation competitiveness (71, 230). However, these studies also used genetically unrelated strains.

Symbiotic Effectiveness: Effect of Mutations in nod *and Structural* nif *Genes on Nodulation Competitiveness*

Mutations in *nod* gene regions can result in a lack of stimulation of the autoregulatory response, slower infection rates, decreased nodulation competitiveness, and impaired nodule development (86, 181). In contrast, mutations affecting nitrogen fixation do affect nodulation competitiveness. Hahn & Studer (87) produced a mutant of *B. japonicum* 110*spc*4 with a Tn*5* insertion in the *nifD* gene necessary for the synthesis of one of the subunits of nitrogenase. Although this mutant was Fix$^-$, phenotypically it still competed for nodule occupancy in dual inoculations with the wild-type strain.

nifA-*Dependent Nodulation Efficiency genes*

Sanjuan & Olivares (178, 180) have identified a 5-kb region on a cryptic plasmid of *R. meliloti* GR4, pRmeGR4b, that influences the kinetics of nodulation. That is, mutations in this region, referred to as *nfe,* delay nodule formation by one or two days. Insertions of Tn*3*-HoHo1 in the *nfe* gene region also reduced nodulation competitiveness of strain GR4. Using *lacZ* fusions, the authors found that expression of *nfe* genes required a functional *nifA* gene. Southern hybridization with total DNA of four other strains of *R. meliloti* showed no hybridization homology with a *nfe* probe (180). When *nfe* was transferred to these four strains by conjugation, the nodulation competitiveness of two of these strains improved (180).

nifA-*Dependent Genes of Unknown Function*

The discovery of *nfe* led Sanjuan & Olivares (179) to determine whether *nifA* affects nodulation competitiveness in *R. meliloti* strains lacking *nfe*. They found that a NifA$^-$ mutant of *R. meliloti* had reduced nodulation competitiveness with normal *nodC* gene expression. This decrease in nodulation competitiveness was not attributed to a loss of symbiotic effectiveness. Addition of multiple copies of *nifA* from *Klebsiella pneumoniae* also resulted in increased nodulation competitiveness of *R. meliloti* strains. Because the *R. meliloti* strain tested did not possess a sequence homologous to *nfe* detectable by southern hybridization, the *nifA* gene is probably regulating other genes involved in nodulation competitiveness. Since *nifA* expression is repressed in the presence of oxygen in *R. meliloti* (33, 41), it is surprising that *nifA* can so profoundly affect a phenotype such as nodulation competitiveness.

Thus, in this review, we have discussed two important roles of multiple copies of *nifA* in *Rhizobium* spp. This gene serves as the positive regulator of other *nif* genes and, when over-expressed, is thought by Cannon and colleagues (22, 166) to result in increased symbiotic nitrogen fixation followed by increased yield. Also, when present in multiple copies, *nifA* can improve the nodulation competitiveness of a strain. Thus, strains with an additional copy of *nifA* stably integrated into the genome may be useful in improving an inoculant's ability to occupy nodules to the exclusion of indigenous rhizobia and in increasing nitrogen fixation by the legume-*Rhizobium* symbiosis. Critical experiments to test this hypothesis, however, remain to be done.

ISSUES CONCERNING THE FIELD TESTING OF RECOMBINANT RHIZOBIA

Stability of Introduced Genes

To construct strains useful for field studies, one must integrate genes of interest into a symbiotically silent region of the genome. This is especially true in light of the fact that plasmid-encoded genes are frequently lost during replication in the absence of selection pressure (123, 136, 205). Furthermore, recombinant plasmids can transfer into other bacteria by conjugation in soil (110a, 161). If the recombinant plasmids contain genes that improve nodulation competitiveness, improvement of the competitiveness of the less desirable indigenous strains may occur. Various methods for stable integration of foreign genes into *Rhizobium* and *Bradyrhizobium* spp. have been developed. Most of these are based on transposon-mediated or marker-exchange insertion.

Several vectors are available for transposon-mediated gene insertion. These are based on Tn*5* (70), Tn*7* (10, 11), or IS1 (63, 107). All of these vector systems have been used in *Rhizobium* spp. or closely related bacteria. Other vectors are available for insertion of genes by marker exchange into symbiotically silent regions of the genome, such as inositol metabolism (221) or adenylate cyclase (152) in *R. meliloti* or in nonessential, but otherwise unidentified, regions in *B. japonicum* (1).

Introduction of Antibiotic-Resistance Genes

Each of the above vectors requires antibiotic-resistance markers to monitor the insertion event. However, field testing of strains containing antibiotic-resistance markers is controversial, particularly in Europe. Dale & Ow (32) have described a mechanism to avoid field inoculation of antibiotic-resistance markers in transgenic plants. In this system, the antibiotic-resistance gene is flanked by recombination sites that are excised by Cre recombinase in plants. A similar recombination system would be very useful for the planned introduction of recombinant bacteria.

Ross et al (167) have reported a unique system for the stable maintenance of plasmids in the absence of selection pressure. This system is based on the use of thymidine/thymine auxotrophs as recipient strains for the introduced genes. The foreign genes are introduced into a plasmid containing a gene essential for thymidine synthesis. This plasmid lacks antibiotic resistance markers and is stably maintained in rhizobia thymidine/thymine auxotrophs.

Although technology apparently can enable us to avoid the use of antibiotic-resistance markers, such markers are very useful in monitoring the efficacy of the introduced strains, particularly during early phases of field testing.

Genetic Load of Introduced Genes

In any experiment designed to test the efficacy of recombinant strains, the genetic load of the added genes on the cell must be determined. Such a load may prevent optimum growth, infection, nodulation, and/or nitrogen fixation. To determine whether a genetic load exists, the symbiotic properties of the recombinant strain along with wild-type and a second recombinant strain with either an insertion or deletion in the region of interest must be compared. That is, test strains that are as genetically identical as possible save for the phenotype of interest should be tested. For example, trifolitoxin production and resistance genes were tested for their effect on nodulation competitiveness (205).

Although competitiveness genes can be over-expressed with a strong promoter such as the *trp* promoter from *Salmonella typhimurium* (56), such over-expression may cause a severe load on the metabolism of the cell, affecting its symbiotic properties.

Choice of Recipient Strain for the Cloned Genes

Another important issue in the planned introduction of recombinant strains of *Rhizobium* spp. is the choice of host strain in which to insert the genes of interest. This choice may be more important to the success of a planned introduction than the genes themselves. Thus, a successful inoculant must not only provide superior symbiotic nitrogen fixation and be very competitive for nodulation in the presence of a large population of indigenous rhizobia but must also be able to survive as a saprophyte in the soil.

FUTURE PROSPECTS

Much progress has been made in our understanding of the mechanisms of nodulation competitiveness and host-specificity in rhizobia since the advent of molecular biology. However, much remains to be done. This improved state of our knowledge should lead to practical solutions to the *Rhizobium* competition problem in the near future. Several solutions will likely be necessary given the number of legume-rhizobia symbioses of interest and the wide variety of environments in which they are cultured. For example, altered EPS production may enhance the competitiveness of strains of *B. japonicum* (14) and *R. leguminosarum* bv. *phaseoli* (5) but may be useless for the alfalfa-*R. meliloti* symbiosis because EPS is required for infection (64). Similarly, antibiosis may be a useful strategy for *R. leguminosarum* but is not likely to be useful for *B. japonicum* as strains of this species have high intrinsic resistance to many antibiotics.

The use of a legume-*Rhizobium* pair with an infectivity sufficiently restrictive to prevent nodulation by the indigenous strains may be the best solution. Pairs that can prevent nodulation by all indigenous strains are not yet available. Increased research in this area will lead to the development of the appropriate combinations necessary to restrict nodulation by indigenous strains.

Recent progress in the molecular biology of nodulation competitiveness has answered several important questions in this area that are also relevant to microbial ecology in general. All of the appropriate genetic tools are available to address other problems and develop long-term solutions to the competition problem that will provide *Rhizobium* inoculants capable of nodulating legumes to the exclusion of the indigenous strains under agricultural conditions.

Literature Cited

1. Acuna, G., Alvarez-Morales, A., Hahn, M., Hennecke, H. 1987. A vector for the site-directed, genomic integration of foreign DNA into soybean root-nodule bacteria. *Plant Mol. Biol.* 9:41–50
2. Amarger, N. 1981. Competition for nodule formation between effective and ineffective strains of *Rhizobium meliloti*. *Soil Biol. Biochem.* 13:475–80
3. Amarger, N., Lobreau, J. P. 1982. Quantitative study of nodulation competitiveness in *Rhizobium* strains. *Appl. Environ. Microbiol.* 44:583–88
4. Ames, P., Bergman, K. 1981. Competitive advantage provided by bacterial motility in the formation of nodules by *Rhizobium meliloti*. *J. Bacteriol.* 148: 728–29
5. Araujo, R. S., Handelsman, J. 1990. Characteristics of exopolysaccharide-deficient mutants of *Rhizobium* spp. with altered nodulation competitiveness. See Ref. 84a, p. 247
6. Bachem, C. W., Banfalvi, Z., Kondorosi, E., Schell, J., Kondorosi, A. 1986. Identification of host range determinants

in the *Rhizobium* species MPIK3030. *Mol. Gen. Genet.* 203:42–48

7. Bachem, C., Kondorosi, E., Banfalvi, Z., Horvath, B., Kondorosi, A., Schell, J. 1985. Identification and cloning of nodulation genes from the wide host range *Rhizobium* strain MPIK3030. *Mol. Gen. Genet.* 199:271–78
8. Baldwin, I. L., Fred, E. B. 1929. Strain variation in the root-nodule bacteria of clover, *Rhizobium trifolii*. *J. Bacteriol.* 17:17–18
9. Banfalvi, Z., Nieuwkoop, A., Schell, M., Besl, L., Stacey, G. 1988. Regulation of *nod* gene expression in *Bradyrhizobium japonicum*. *Mol. Gen. Genet.* 214:420–24
10. Bao, Y., Lies, D. P., Fu, H., Roberts, G. P. 1991. An improved Tn7-based system for the single-copy insertion of cloned genes into chromosomes of Gram-negative bacteria. *Gene* 109:167–68
11. Barry, G. F. 1988. A broad-host-range shuttle system for gene insertion into the chromosomes of Gram-negative bacteria. *Gene* 71:75–84
12. Bassam, B. J., Rolfe, B. G., Djordjevic, M. A. 1986. *Macroptilium atropurpureum* (siratro) host specificity genes are linked to a *nodD*-like gene in the broad host range *Rhizobium* strain NGR234. *Mol. Gen. Genet.* 203:49–57
13. Beattie, G. A., Clayton, M. K., Handelsman, J. 1989. Quantitative comparison of the laboratory and field competitiveness of *Rhizobium leguminosarum* bv. *phaseoli*. *Appl. Environ. Microbiol.* 55:2755–61
14. Bhagwat, A. A., Tully, R. E., Keister, D. L. 1991. Isolation and characterization of a competition-defective *Bradyrhizobium japonicum* mutant. *Appl. Environ. Microbiol.* 57:3496–3501
15. Bhuvaneswari, T. V., Bhagwat, A. A., Bauer, W. D. 1981. Transient susceptibility of root cells in four common legumes to nodulation by rhizobia. *Plant Physiol.* 68:1144–49
16. Bhuvaneswari, T. V., Turgeon, B. G., Bauer, W. D. 1980. Early events in the infection of soybean (*Glycine max* L. Merr.) by *Rhizobium japonicum*. I. Localization of infectible root cells. *Plant Physiol.* 66:1027–31

16a. Bothe, H., de Bruijn, F. J., Newton, W. E., eds. 1988. *Nitrogen Fixation: Hundred Years After*. Stuttgart: Gustav Fischer

17. Brewin N. J., Beringer, J. E., Johnston, A. W. B. 1980. Plasmid mediated transfer of host range specificity between two strains of *Rhizobium leguminosarum*. *J. Gen. Microbiol.* 128:1817–27
18. Caetano-Anollés, G., Wall, L. G., De Micheli, A. T., Macchi, E. M., Bauer, W. D., Favelukes, G. 1988. Role of motility and chemotaxis in efficiency of nodulation by *Rhizobium meliloti*. *Plant Physiol.* 86:1228–35
19. Caldwell, B. E. 1966. Inheritance of a strain-specific ineffective nodulation in soybeans. *Crop Sci.* 6:427–28
20. Caldwell, B. E., Hinson, K., Johnson, H. W. 1966. A strain-specific ineffective nodulation reaction in the soybean *Glycine max* L. Merrill. *Crop Sci.* 6:495–96
21. Calvert, H. E., Pence, M. K., Pierce, M., Malik, N. S. A., Bauer, W. D. 1984. Anatomical analysis of the development and distribution of *Rhizobium* infections in soybean roots. *Can. J. Bot.* 62:2375–84
22. Cannon, F. C., Beynon, J., Hankinson, T., Kwiatkowski, R., Legocki, R. P., et al. 1988. Increasing biological nitrogen fixation by genetic manipulation. See Ref. 16a, pp. 735–40
23. Chatterjee, A., Balatti, P. A., Gibbons, W., Pueppke, S. G. 1990. Interaction of *Rhizobium fredii* USDA257 and nodulation mutants derived from it with agronomically improved soybean cultivar McCall. *Planta* 180:303–11
24. Chen, H., Richardson, A. E., Gartner, E., Djordjevic, M. A., Roughley, R. J., Rolfe, B. G. 1991. Construction of an acid-tolerant *Rhizobium leguminosarum* bv. *trifolii* strain with enhanced capacity for nitrogen fixation. *Appl. Environ. Microbiol.* 57:2005–11
25. Cregan, P. B., Keyser, H. H. 1986. Host restriction of nodulation by *Bradyrhizobium japonicum* strain USDA 123 in soybean. *Crop Sci.* 26:911–16
26. Cregan, P. B., Keyser, H. H., Sadowsky, M. J. 1988. A single dominant gene in soybean genotype PI 417566 controls competition for nodulation of *Bradyrhizobium japonicum* isolate MN1–1c. *Abstr. The American Society of Agronomy Annu. Meeting, Anaheim, Calif.*, p. 78
27. Cregan, P. B., Keyser, H. H., Sadowsky, M. J. 1989. Soybean genotype restricting nodulation of a previously unrestricted serocluster 123 bradyrhizobia. *Crop Sci.* 29:307–12
28. Cregan, P. B., Keyser, H. H., Sadowsky, M. J. 1989. A soybean genotype that restricts nodulation of a previously unrestricted isolate of *Bradyrhizobium japonicum* serocluster 123. *Crop Sci.* 29:307–12
29. Cregan, P. B., Keyser, H. H., Sadowsky, M. J. 1990. Gene-for-gene interac-

tion in the legume-*Rhizobium* symbiosis. In *Belstville Symposia in Agricultural Research XIV, The Rhizosphere and Plant Growth,* ed. D. L. Keister, P. B. Cregan, pp. 163–71. Dordrecht: Kluwar

30. Crute, I. R. 1986. Investigations of gene-for-gene relationships: the need for genetic analyses of both host and parasite. *Plant Pathol.* 35:15–17
31. Cunningham, S. D., Kollmeyer, W. D., Stacey, G. 1991. Chemical control of interstrain competition for soybean nodulation by *Bradyrhizobium japonicum. Appl. Environ. Microbiol.* 57: 1886–92
32. Dale, E. C., Ow, D. W. 1991. Gene transfer with subsequent removal of the selection gene from the host genome. *Proc. Natl. Acad. Sci. USA* 88:10558–62
33. David, M., Daveran, M., Balut, J., Dedieu, A., Domergue, O, et al. 1988. Cascade regulation of nif gene expression in Rhizobium meliloti. *Cell* 54:671–83
34. Davis, E. O., Evans, I. J., Johnston, A. W. B. 1988. Identification of *nodX,* a gene that allows *Rhizobium leguminosarum* biovar *iciae* strain TOM to nodulate Afghanistan peas. *Mol. Gen. Genet.* 212:531–35
35. Debelle, F., Maillet, F., Vasse, J., Rosenberg, C., De Billy, F., et al. 1988. Interference between *Rhizobium meliloti* and *Rhizobium trifolii* nodulation genes: genetic basis of *Rhizobium meliloti* dominance. *J. Bacteriol.* 170:5718–27
36. Debelle, F., Rosenberg, C., Vasse, J., Maillet, F., Martinez, E., et al. 1986. Assignment of symbiotic developmental phenotypes to common and specific nodulation (*nod*) genetic loci of *Rhizobium meliloti. J. Bacteriol.* 168:1075–86
37. Denarié, J., Boistard, P., Casse-Delbart, F., Atherly, A. G., Berry, J., Russell, P. 1981. Indigenous plasmids of *Rhizobium.* In *Biology of the Rhizobiaceae,* ed. K. L. Giles, A. G. Atherly, pp. 191–219. Orlando: Academic
38. Devine, T. 1984. Genetics and breeding of nitrogen fixation. In *Biological Nitrogen Fixation,* ed. M. Alexander, pp. 127–54. New York: Plenum
39. Devine, T. E., Kuykendall, L. D., O'Neil, J. J. 1990. The Rj4 allele in soybean represses nodulation by chlorosis-inducing bradyrhizobia classified as DNA homology group II by antibiotic resistance profiles. *Theor. Appl. Genet.* 80:33–37
40. Devine, T. E., O'Neil, J. J. 1986. Registration of BARC-2 (Rj4) and BARC-3(rj4) soybean germplasm. *Crop Sci.* 26:1263–64
41. Ditta, G., Virts, E., Palomares, A., Kim, C.-H. 1987. The *nifA* gene of *Rhizobium meliloti* is oxygen regulated. *J. Bacteriol.* 169:3217–23
42. Djordjevic M. A., Gabriel, D. W., Rolfe, B. G. 1987. *Rhizobium*—the refined parasite of legumes. *Annu. Rev. Phytopathol.* 25:145–68
43. Djordjevic, M. A., Redmond, J. W., Batley, M., Rolfe, B. G. 1987. Clovers secrete specific phenolic compounds which either stimulate or repress *nod* gene expression in *Rhizobium trifolii. EMBO J.* 6:1173–79
44. Djordjevic, M. A., Sargent, C. L., Innes, R. W., Kuempel, P. L., Rolfe, B. G. 1985. Host-range genes also affect competitiveness in *Rhizobium trifolii.* See Ref. 58a, p. 117
45. Djordjevic, M. A., Schofield, P. R., Ridge, R. W., Morrison, N. A., Bassam, B. J., et al. 1985. *Rhizobium* nodulation genes involved in root hair curling (Hac) are functionally conserved. *Plant Mol. Biol.* 4:147–60
46. Djordjevic, M. A., Schofield, P. R., Rolfe, B. G. 1985. Tn*5* mutagenesis of *Rhizobium trifolii* host-specific nodulation genes results in mutants with altered host-range ability. *Mol. Gen. Genet.* 200:463–71
47. Djordjevic, S. P., Ridge, R. W., Chen, H., Redmond, J. W., Bately, M., Rolfe, B. G. 1988. Induction of pathogenic-like responses in the legume *Macroptilium atropurpureum* by transposon-induced mutant of the fast-growing, broad-host-range *Rhizobium* strain NGR234. *J. Bacteriol.* 170:1848–57
48. Dowling, D. N., Broughton, W. J. 1986. Competition for nodulation of legumes. *Annu. Rev. Microbiol.* 40: 131–57
49. Dowling, D. N., Stanley, J., Broughton, W. J. 1989. Competitive nodulation blocking of Afghanistan pea is determined by *nodDABC* and *nodFE* alleles in *Rhizobium leguminosarum. Mol. Gen. Genet.* 216:170–74
50. Downie, J. A., Hombrecher, G., Ma, Q. S., Knight, C. D., Wells, B., Johnston, A. W. B. 1983. Cloned nodulation genes of *Rhizobium leguminosarum* determine host-range specificity. *Mol. Gen. Genet.* 190:359–65
51. Dudley, M. E., Long, S. R. 1989. A non-nodulating alfalfa mutant displays neither root hair curling nor early cell division in response to *Rhizobium meliloti. Plant Cell* 1:65–72
52. Dunham, D. H., Baldwin, I. L. 1931.

Double infection of leguminous plants with good and poor strains of rhizobia. *Soil Sci.* 32:235–49

53. Dunigan, E. P., Bollich, P. K., Hutchinson, R. L., Hicks, P. M., Zaunbrecher, F. C., et al. 1984. Introduction and survival of an inoculant strain of *Rhizobium japonicum* in soil. *Agron. J.* 76:463–66
54. Ebba, T., Person, C. 1975. Genetic control of virulence in *Ustilago hordei*. IV. Duplicate genes for virulence and genetic and environmental modifications of a gene-for-gene relationship. *Can. J. Genet. Cytol.* 17:631–36
55. Egelhoff T. T., Fisher, R. F., Mulligan, J. T., Long, S. R. 1985. Nucleotide sequence of *Rhizobium meliloti* 1021 nodulation genes: *nodD* is read divergently from *nodABC*. *DNA* 4:241–48
56. Egelhoff, T. T., Long, S. R. 1985. *Rhizobium meliloti* nodulation genes: identification of *nodDABC* gene products, purification of *nodA* protein, and expression of *nodA* in *Rhizobium meliloti*. *J. Bacteriol.* 164:591–99
57. Ellingboe, A. H. 1981. Changing concepts in host-pathogen genetics. *Annu. Rev. Phytopath.* 19:125–45
58. Ellingboe, A. H. 1976. Genetics of host-parasite interactions. In *Encyclopedia of Plant Pathology,* Vol. 4, *Physiological Plant Pathology,* ed. R. Heitfuss, P. H. Williams, pp. 761–78. New York: Springer-Verlag

58a. Evans, H. J., Bottomley, P. J., Newton, W. E., eds. 1985. *Nitrogen Fixation Research Progress*. Boston: Martinus Nijhoff

59. Evans, H. J., Hanus, F. J., Haugland, R. A., Cantrell, M. A., Xu, L.-S., et al. 1985. Hydrogen recycling in nodules affects nitrogen fixation and growth of soybeans. *Proc. World Soybean Conf.* 3:935–42
60. Evans, H. J., Harker, A. R., Papen, H., Russell, S. A., Hanus, F. J., Zuber, M. 1987. Physiology, biochemistry, and genetics of the uptake hydrogenase in rhizobia. *Annu. Rev. Microbiol.* 41:335–61
61. Faucher, C., Camut, S., Denarie, J., Truchet, G. 1989. The *nodH* and *nodQ* host range genes of *Rhizobium meliloti* behave as avirulence genes in *R. leguminosarum* bv *viceae* and determine changes in the production of plant-specific extracellular signals. *Mol. Plant-Microb. Interact.* 2:291–300
62. Faucher, F., Maillet, F., Vasse, J., Rosenberg, C., van Brussel, A. A. N., et al. 1988. *Rhizobium meliloti* host range *nodH* gene determines production of an alfalfa-specific extracellular signal. *J. Bacteriol.* 170:5489–99
63. Fellay, R., Kirsch, H. M., Prentki, P., Frey, J. 1989. Omegon-Km: a transposable element designed for in vivo insertional mutagenesis and cloning of genes in Gram-negative bacteria. *Gene* 76:215–26
64. Finan, T. M., Hirsch, A. M., Leigh, J. A., Johansen, E., Kuldau, G. A., et al. 1985. Symbiotic mutants of *Rhizobium meliloti* that uncouple plant from bacterial differentiation. *Cell* 40:869–77
65. Firmin, J. L., Wilson, K. E., Rossen, L., Johnston, A. W. B. 1986. Flavonoid activation of nodulation genes in *Rhizobium* reversed by other compounds present in plants. *Nature* 324:90–92
66. Fisher R. F., Egelhoff, T. T., Mulligan, J. T., Long, S. R. 1988. Specific binding of proteins from *Rhizobium meliloti* cell-free extracts containing *nodD* to DNA sequences upstream of inducible nodulation genes. *Genes Dev.* 2:282–93
67. Fisher, R. F., Swanson, J. A., Mulligan, J. T., Long, S. R. 1987. Extended region of nodulation genes in *Rhizobium meliloti* 1021. II. Nucleotide sequence, transcription sites, and protein products. *Genetics* 117:1432–35
68. Flor, H. H. 1946. Genetics of pathogenicity in *Melamspora lini*. *J. Agric. Res.* 73:335–57
69. Flor, H. 1971. Current status of the gene-for-gene concept. *Annu. Rev. Phytopath.* 9:275–96
70. Fobert, P. R., Roy, N., Nash, J. H., Iyer, V. N. 1991. Procedure for obtaining efficient root nodulation of a pea cultivar by a desired *Rhizobium* strain and preempting nodulation by other strains. *Appl. Environ. Microbiol.* 57:1590–94
71. Franco, A. A., Vincent, J. M. 1976. Competition amongst rhizobial strains for the colonization and nodulation of two tropical legumes. *Plant Soil* 45:27–48
72. Gabriel, D. W. 1986. Specificity and gene function in plant-pathogen interactions. *Am. Soc. Microbiol. News.* 52:19–25
73. Gabriel, D. W., Burges, A., Lazo, G. R. 1986. Gene-for-gene recognition of five cloned avirulence genes from *Xanthomonas campestris* pv. *malvacearum* by specific resistance genes in cotton. *Proc. Natl. Acad. Sci. USA* 83:6415–19
74. Gabriel, D. W., Rolfe, B. G. 1990. Working models of specific recognition in plant-microbe interactions. *Annu. Rev. Phytopath.* 28:365–91

75. Gibson, A. 1968. Nodulation failure in *Trifolium subterraneum* L. cv Woogenellup (Syn Marrar). *Aust. J. Agric. Sci.* 19:907–18
76. Gibson, A. H. 1988. Host genetics in symbiotic nitrogen fixation by legumes, In *Microbiology in Action,* ed. W. G. Murrell, I. R. Kennedy, pp. 177–92. Letchworth, UK: Research Studies Press; New York: Wiley
77. Gibson, A. H., Brockwell, J. 1968. Symbiotic characteristics of subspecies of *Trifolium subterraneum* L. *Aust. J. Agric. Res.* 19:891–905
78. Gibson, A. H., Curnow, B. C., Bergersen, F. J., Brockwell, J., Robinson, A. C. 1975. Studies of field populations of *Rhizobium:* effectiveness of strains of *Rhizobium trifolii* associated with *Trifolium subterraneum* L. pastures of south eastern Australia. *Soil Biol. Biochem.* 7:95–102
79. Gibson, A. H., Date, R. A., Ireland, J. A., Brockwell, J. 1976. A comparison of competitiveness and persistence amongst five strains of *Rhizobium trifolii. Soil Biol. Biochem.* 8:395–401
80. Gottfert, M., Grob, P., Hennecke, H. 1990. Proposed regulatory pathway encoded by the *nodV* and the *nodW* genes, determinants of host specificity in *Bradyrhizobium japonicum. Proc. Natl. Acad Sci. USA* 87:2680–84
81. Gottfert, M., Horvath, B., Kondorosi, E., Putnoky, P., Rodriguez-Quniones, F., Kondorosi, A. 1986. At least two *nodD* genes are necessary for efficient nodulation of alfalfa by *Rhizobium meliloti. J. Mol. Biol.* 191:411–20
82. Gottfert, M., Webber, J., Hennecke, H. 1988. Induction of a *nodA-lacZ* fusion in *Bradyrhizobium japonicum* by an isoflavone. *J. Plant Physiol.* 132:394–97
83. Gotz, R., Evans, I. J., Downie, J. A., Johnston, A. W. B. 1985. Identification of host-range DNA which allows *Rhizobium leguminosarum* strain TOM to nodulate cv. Afghanistan peas. *Mol. Gen. Genet.* 201:296–300
84. Greder, R. R., Orff, J. H., Lambert, J. W. 1986. Heritabilities and associations of nodule mass and recovery of *Bradyrhizobium japonicum* serogroup USDA 110 in soybean. *Crop Sci.* 26:33–37
84a. Gresshoff, P. M., Roth, L. E., Stacey, G., Newton, W. E., eds. 1990. *Nitrogen Fixation: Achievements and Objectives.* New York: Chapman and Hall
85. Gupta, V. P., Garg, I. K., Rana, N. D., Singh, J. M. 1982. Variation and heritability for leaf and root characteriztics in soybeans, across locations. *Soybean Genet. Newslett.* 9:71–74
86. Hahn, M., Hennecke, H. 1988. Cloning and mapping of a novel nodulation region from *Bradyrhizobium japonicum* by genetic complementatiion of a deletion mutant. *Appl. Environ. Microbiol.* 54: 55–61
87. Hahn, M., Studer, D. 1986. Competitiveness of a *nif-Bradyrhizobium japonicum* mutant against the wild-type strain. *FEMS Microbiol. Lett.* 33:143–48
88. Handelsman, J., Ugalde, R. A., Brill, W. J. 1984. *Rhizobium meliloti* competitiveness and the alfalfa agglutinin. *J. Bacteriol.* 157:703–7
89. Hanus, F. J., Albrecht, S. L., Zablotowicz, R. M., Emerich, D. W., Russell, S. A. 1981. Yield and N content of soybean seed as influenced by *Rhizobium japonicum* inoculants possessing the hydrogenase characteristic. *Agron. J.* 73:368–72
90. Hardarson, G., Jones, D. G. 1979. The inheritance of preference for strains of *Rhizobium trifolii* by white clover *Trifolium repens. Ann. Appl. Biol.* 92:329–33
91. Hashem, F. M., Angle, J. S. 1988. Rhizobiophage effects on *Bradyrhizobium japonicum*, nodulation and soybean growth. *Soil Biol. Biochem.* 20:69–73
92. Hashem, F. M., Angle, J. S., Ristiano, P. A. 1986. Isolation and characterization of rhizobiophages specific for *Bradyrhizobium japonicum* USDA 117. *Can. J. Bot.* 32:326–29
93. Helz, G. E., Baldwin, I. L., Fred, E. B. 1927. Strain variations and host specificity of root-nodule bacteria of the pea group. *J. Agric. Res.* 35:1039–55
94. Heron D. S., Ersek, T., Krishan, H. B., Pueppke, S. G. 1989. Nodulation mutants of *Rhizobium fredii* USDA 257. *Mol. Plant-Microbe Interact.* 2:4–10
95. Hodgson, A. L. M., Roberts, W. P. 1983. DNA colony hybridization to identify *Rhizobium* strains. *J. Gen. Microbiol.* 129:207–12
96. Hodgson, A. L. M., Roberts, W. P., Waid, J. S. 1985. Regulated nodulation of *Trifolium subterraneum* inoculated with bacteriocin-producing strains of *Rhizobium trifolii. Soil Biol. Biochem.* 17:475–78
97. Holl, F. B. 1975. Host plant control of the inheritance of dinitrogen fixation in the *Pisum-Rhizobium* symbiosis. *Euphytica* 24:767–70
98. Hombrecher, G., Gotz, R., Dibb, N. J., Downie, J. A., Johnston, A. W. B., Brewin, N. J. 1984. Cloning and mutagenesis of nodulation genes from *Rhizobium leguminosarum* TOM, a

strain with extended host range. *Mol. Gen. Genet.* 194:293–98

99. Hong, G.-F., Burn, J. E., Johnston, A. W. B. 1987. Evidence that DNA involved in the expression of nodulation (*nod*) genes in *Rhizobium* binds to the regulatory gene *nodD*. *Nucleic Acids Res.* 15:9677–90
100. Honma, M. A., Ausubel, F. M. 1987. *Rhizobium meliloti* has three functional copies of the *nodD* symbiotic regulatory element. *Proc. Natl. Acad. Sci. USA* 84:8558–62
101. Horvath, B., Bachem, C. W., Schell, J., Kondorosi, A. 1987. Host-specific regulation of nodulation genes in *Rhizobium* is mediated by a plant-signal, interacting with the *nodD* gene products. *EMBO J.* 6:841–48
102. Horvath, B., Kondorosi, E., John, M., Schmidt, J., Torok, I., et al. 1986. Organization, structure, and symbiotic function of Rhizobium meliloti nodulation genes determining host specificity for alfalfa. *Cell* 46:335–43
103. Hungria, M., Neves, M. C. P., Dobereiner, J. 1989. Relative efficiency, ureide transport and harvest index in soybeans inoculated with isogenic HUP mutants of *Bradyrhizobium japonicum*. *Biol. Fertil. Soils* 7:325–29
104. Innes, R. W., Kuempel, P. L., Plazinski, J., Canter-Cremers, H., Rolfe, B. G., Djordjevic, M. A. 1985. Plant factors induce expression of nodulation and host-range genes in *R. trifolii*. *Mol. Gen. Genet.* 201:426–32
105. Jones, D. A. 1988. Genetic properties of inhibitor genes in flax rust that alter avirulence to virulence on flax. *Phytopathology* 78:342–44
106. Joseph, M. V., Desai, J. D., Desai, A. J. 1983. Production of antimicrobial and bacteriocin-like substances by *Rhizobium trifolii*. *Appl. Environ. Microbiol.* 45:532–35
107. Joseph-Liauzun, E., Fellay, R., Chandler, M. 1989. Transposable elements for efficient manipulation of a wide range of Gram-negative bacteria: promoter probes and vectors for foreign genes. *Gene* 85:83–89
108. Keen, N. T. 1982. Specific recognition in gene-for-gene host parasite systems. *Adv. Plant Pathol.* 1:35–82
109. Keen, N. T., Staskawicz, B. 1988. Host range determinants in plant pathogens and symbionts. *Annu. Rev. Microbiol.* 42:421–40
110. Keyser, H. H., Cregan, P. B. 1987. Nodulation and competition for nodulation of selected soybean genotypes among *Bradyrhizobium japonicum* serogroup 123 isolates. *Appl. Environ. Microbiol.* 53:2631–35

110a. Kinkle, B.K., Schmidt, E. L. 1991. Transfer of the pea symbiotic plasmid pJB5JI in nonsterile soil. *Appl. Environ. Microbiol.* 57:3264–69

111. Kneen, B. E., LaRue, T. A. 1984. Peas (*Pisum sativum* L.) with strain specificity for *Rhizobium leguminosarum*. *Heredity* 52:383–89
112. Kondorosi, A., Horvath, B., Gottfert, M., Putnoky, P., Rostas, K., et al. 1985. Identification and organization of *Rhizobium meliloti* genes relevant to the initiation and development of nodules. See Ref. 58a, pp. 73–78
113. Kondorosi, E., Banfalvi, Z., Kondorosi, A. 1984. Physical and genetic analysis of a symbiotic region of *Rhizobium meliloti:* identification of nodulation genes. *Mol. Gen. Genet.* 193:443–52
114. Kondorosi, E., Kondorosi, A. 1986. Nodule induction on plant roots by *Rhizobium*. *Trends Biol. Sci.* 11:296–99
115. Kosslak, R. M., Bohlool, B. B. 1984. Suppression of nodule development of one side of a split-root system of soybeans caused by prior inoculation of the other side. *Plant Physiol.* 75:125–30
116. Kosslak, R. M., Bohlool, B. B., Dowdle, S. F., Sadowsky, M. J. 1983. Competition of *Rhizobium japonicum* strains in early stages of soybean nodulation. *Appl. Environ. Microbiol.* 46:870–73
117. Kosslak, R. M., Bookland, R., Barkei, J., Paaren, H. E., Appelbaum, E. R. 1987. Induction of *Bradyrhizobium japonicum* common *nod* genes by isoflavones isolated from *Glycine max*. *Proc. Natl. Acad. Sci. USA* 84:7428–32
118. Kosslak, R. M., Joshi, R. S., Bowen, B. A., Paaren, H. E., Appelbaum, E. R. 1990. Strain-specific inhibition of *nod* gene induction in *Bradyrhizobium japonicum* by flavonoid compounds. *Appl. Environ. Microbiol.* 56:1333–41
119. Krishnan, H. B., Pueppke, S. G. 1991. *nodC,* a *Rhizobium fredii* gene involved in cultivar-specific nodulation of soybean, shares homology with a heat-shock gene. *Mol. Microbiol.* 5:737–45
120. Kvien, C. S., Ham, G. E., Lambert, J. W. 1981. Recovery of introduced *Rhizobium japonicum* strains by soybean genotypes. *Agron. J.* 73:900–5
121. Labandera, C. A., Vincent, J. M. 1975. Competition between an introduced strain and native Uruguayan strains of *Rhizobium trifolii*. *Plant and Soil* 42:327–47
122. Lamb, J. W., Hennecke, H. 1986. In *Bradyrhizobium japonicum* the common nodulation genes, *nodABC,* are linked to

nifA and *fixA*. *Mol. Gen. Genet.* 202: 512–17

123. Lambert, G. R., Harker, A. R., Cantrell, M. A., Hanus, F. J., Russell, S. A., et al. 1987. Symbiotic expression of cosmid-borne *Bradyrhizobium japonicum* hydrogenase genes. *Appl. Environ. Microbiol.* 53:422–28
124. Lawson, R. M. 1980. *Genetic variability in soybeans for nodule number and weight, and recovery of* Rhizobium japonicum *strain 110*. PhD dissertation. Univ. Minn. St. Paul, Minn. Diss. Abstr. 80–19540
125. Leonard, L. T. 1930. A failure of Austrian winter peas apparently due to nodule bacteria. *J. Am. Soc. Agron.* 22:277–80
126. Lerouge, P., Roche, P., Faucher, C., Maillet, F., Truchet, G., et al. 1990. Symbiotic host specificity of *Rhizobium meliloti* is determined by a sulphated and acylated glucoisamine oligosaccharide signal. *Nature* 344:781–84
127. Leroux, B., Yanofsky, M. F., Winans, S. C., Ward, J. E., Ziegler, S. F., Nester, E. 1987. Characterization of the *virA* locus of *Agrobacterium tumefaciens:* a transcriptional regulator and host-range determinant. *EMBO J.* 6: 849–56
128. Lewin, A., Rossenberg, C., Meyer, H., Wong, A. C. H., Nelson, L., et al. 1987. Multiple host-specificity loci of the broad host-range *Rhizobium* sp. NGR234 selected using the widely compatible legume *Vigna unguiculata*. *Plant Mol. Biol.* 8:447–59
129. Lewis-Henderson, W. R., Djordjevic, M. A. 1991. A cultivar-specific interaction between *Rhizobium leguminosarum* bv *trifolii* and subterranean clover is controlled by *nodM,* other bacterial cultivar specificity genes, and a single recessive allele. *J. Bacteriol.* 173:2791–99
130. Lewis-Henderson, W. R., Djordjevic, M. A. 1991. *nodT,* a positively acting cultivar specificity determinant controlling nodulation of *Trifolium subterraneum* by *Rhizobium leguminosarum* bv *trifolii*. *Plant Mol. Biol.* 16:515–26
131. Lie, T. A. 1978. Symbiotic specialization in pea plants: the requirement of specific *Rhizobium* strains for peas from Afghanistan. *Ann. Appl. Biol.* 88:462–65
132. Lie, T. A. 1984. Host genes in *Pisum sativum* L. conferring resistance to European *Rhizobium leguminosarum* strains. *Plant Soil* 82:415–25
133. Liu, R., Tran, V. M., Schmidt, E. L. 1989. Nodulating competitiveness of a nonmotile Tn*7* mutant of *Bradyrhizobium japonicum* in nonsterile soil. *Appl. Environ. Microbiol.* 55:1895–1900
134. Long, S. R. 1984. The genetics of *Rhizobium* nodulation. In *Plant Microbe Interactions,* ed. T. Kosuge, E. W. Nester, pp. 265–306. New York: Macmillan
135. Long, S. R. 1989. Rhizobium-legume nodulation: life together in the underground. *Cell* 56:203–14
136. Long, S. R., Buikema, W. J., Ausubel, F. M. 1982. Cloning of *Rhizobium meliloti* nodulation genes by direct complementation of Nod^- mutants. *Nature* 298:485–88
137. Long, S. R., Fisher, R. F., Ogawa, J., Swanson, J., Ehrhardt, D. W., et al. 1991. *Rhizobium meliloti* nodulation gene regulation and molecular signals. In *Advances in Molecular Genetics of Plant-Microbe Interactions,* ed. H. Hennecke, D. P. S. Verma, 1:127–33. Dordrecht: Kluwer Academic
138. Ma, S.-W., Iyer, V. N. 1990. New field isolates of *Rhizobium leguminosarum* biovar viciae that nodulate the promative pea cultivar Afghanistan in addition to modern cultivars. *Appl. Environ. Microbiol.* 56:2206–12
139. Maier, R., Brill, W. J. 1978. Mutant strains of *Rhizobium japonicum* with increased ability to fix nitrogen for soybean. *Science* 201:448–50
140. Martensson, A. M. 1990. Competitiveness of inoculant strains of *Rhizobium leguminosarum* bv. *trifolii* in red clover using repeated inoculation and increased inoculum levels. *Can. J. Microbiol.* 36:136–39
141. Martinez, E., Romero, D., Palacios, R. 1990. The *Rhizobium* genome. *Crit. Rev. Plant Sci.* 9:59–93
142. Masterson, R. V., Russell, P. R., Atherly, A. G. 1982. Nitrogen fixation (*nif*) genes and large plasmids of *Rhizobium japonicum*. *J. Bacteriol.* 152:928–31
143. Materon, L. A., Hagedorn, C. 1983. Competitiveness and symbiotic effectiveness of five strains of *Rhizobium trifolii* on red clover. *Soil Sci. Soc. Am. J.* 47:491–95
144. Mathews, A., Kosslak, R. M., Sengupta-Gopalan, C., Appelbaum, E. R., Carroll, B. J., Gresshoff, P. M. 1989. Biological characterization of root exudates and extracts from nonnodulating and supernodulating soybean mutants. *Mol. Plant-Microbe Interact.* 2:283–90
145. May, S. N., Bohlool, B. B. 1982. Competition among *Rhizobium leguminosarum* strains for nodulation of lentils (*Lens*

esculenta). *Appl. Environ. Microbiol.* 45:960–65

146. McDermott, T. R., Graham, P. H. 1990. Competitive ability and efficiency in nodule formation of strains of *Bradyrhizobium japonicum*. *Appl. Environ. Microbiol.* 56:3035–39
147. McIver, J. M., Djordjevic, M. A., Weinman, J. J., Bender, G. L., Rolfe, B. G. 1989. Extension of host-range in *Rhizobium leguminosarum* bv *trifolii* caused by point mutations in *nodD* that result in alterations in regulatory function and recognition of inducer molecules. *Mol. Plant-Microbe Interact.* 2:97–106
148. Mellor, H. Y., Glenn, A. R., Arwas, R., Dilworth, M. J. 1987. Symbiotic and competitive properties of motility mutants of *Rhizobium trifolii* TA1. *Arch. Microbiol.* 148:34–39
149. Mukherjee, D., Lambert, J., Cooper, R., Kennedy, B. 1966. Inheritance of resistance to bacterial blight of soybeans. *Crop Sci.* 6:324–26

149a. Müller, P., Schultes, A., Werner, D. 1990. *Abstr. of the 5th Int. Symp. on the Molecular Genetics of Plant-Microbe Interactions, Interlaken, Switzerland,* ed. M. Gottfert, H. Hennecke, H. Paul. Zurich: Eidgenössiche Technische Hochschule

150. Mulligan, J. T., Long, S. R. 1985. Induction of *Rhizobium meliloti nodC* expression by plant exudate requires *nodD*. *Proc. Natl. Acad. Sci. USA* 82:6609–13
151. Nieuwkoop, A. J., Banfalvi, Z., Deshmane, N., Gerhold, D., Schell, M., et al. 1987. A locus encoding host range is linked to the common nodulation genes of *Bradyrhizobium japonicum*. *J. Bacteriol.* 169:2631–38
152. O'Gara, F., Boesten, B., Fanning, S. 1988. The development and exploitation of "marker genes" suitable for risk evaluation studies on the release of genetically engineered microorganisms in soil. In *Risk Assessment for Deliberate Release,* ed. W. Klingmuller, pp. 50–64. Berlin: Springer Verlag
153. Oliveira, L. A., Graham, P. H. 1989. Speed of nodulation and competitive ability among strains of *Rhizobium leguminosarum* bv. *phaseoli*. *Arch. Microbiol.* 153:311–15
154. Oliveira, L. A., Graham, P. H. 1990. Evaluation of strain competitiveness in *Rhizobium leguminosarum* bv. *phaseoli* using a *nod*$^{+}$ and *fix*$^{-}$ natural mutant. *Arch. Microbiol.* 153:305–10
155. Ozawa, T., Yamguchi, M. 1986. Enhancement of symbiotic nitrogen fixation of *Bradyrhizobium japonicum* by acridine orange. *Agric. Biol. Chem.* 50:1061–62
156. Paau, A. S. 1989. Improvement of *Rhizobium* inoculants. *Appl. Environ. Microbiol.* 55:862–65
157. Perret, X., Broughton, W. J., Brenner, S. 1991. Canonical ordered cosmid library of the symbiotic plasmid of *Rhizobium* species NGR234. *Proc. Natl. Acad. Sci. USA* 88:1923–27
158. Peters, N. K., Frost, J. W., Long, S. R. 1986. A plant flavone, luteolin, induces expression of *Rhizobium meliloti* nodulation genes. *Science* 233:977–80
159. Pierce, M., Bauer, W. D. 1983. A rapid regulatory response governing nodulation in soybean. *Plant Physiol.* 73:286–90
160. Deleted in proof
161. Richaume, A., Angle, J. S., Sadowsky, M. J. 1989. Influence of soil variables on in situ plasmid transfer from *Escherichia coli* to *Rhizobium fredii*. *Appl. Environ. Microbiol.* 55:1730–34
162. Roelfs, A. P. 1988. Resistance of leaf and stem in wheat. In *Breeding Strategies for Resistance to the Rusts of Wheat,* ed. N. W. Simmonds, S. Rajaram, pp. 10–22. Mexico, District Federal: Centro Internacional de Mejoramiento del Maiz y Trigo
163. Rolfe, B. G. 1988. Flavones and isoflavones as inducing substances of legume nodulation. *Biofactors.* 1:3–10
164. Rolfe, B. G., Innes, R. W., Schofield, P. R., Watson, J. W., Sargent, C. L., et al. 1985. Plant-secreted factors influence the expression of *R. trifolii* nodulation and host-range genes. See Ref. 58a, pp. 79–85
165. Rolfe, B. G., Redmond, J. W., Batley, M., Chen, H., Djordjevic, S. P., et al. 1986. Intercellular communication and recognition in the *Rhizobium*-legume symbiosis. In *Recognition in Microbe-Plant Symbiotic and Pathogenic Interactions,* ed. B. Lugtenberg, pp. 39–51. Berlin/Heidelberg: Springer-Verlag
166. Ronson, C. W., Bosworth, A., Genova, M., Gudbrandsen, S., Hankinson, T., et al. 1990. Field release of genetically-engineered *Rhizobium meliloti* and *Bradyrhizobium japonicum* strains. See Ref. 84a, pp. 397–403
167. Ross, P., O'Gara, F., Condon, S. 1990. Thymidylate synthase gene from *Lactococcus lactis* as a genetic marker: an alternative to antibiotic resistance genes. *Appl. Environ. Microbiol.* 56:2164–69
168. Rossen, L., Davis, E., Johnston, A. W. B. 1987. Plant induced expression of *Rhizobium* genes involved in host

specificity and early stages of nodulation. *Trends Biol. Sci.* 12:191–99

169. Rossen, L., Johnston, A. W. B., Downie, J. A. 1984. DNA sequence of the *Rhizobium leguminosarum* nodulation genes *nodAB* and *C* required for root hair curling. *Nucleic Acid Res.* 12:9497–9508

170. Rossen, L., Shearman, C. A., Johnston, A. W. B., Downie, J. A. 1985. The *nodD* gene of *Rhizobium leguminosarum* is autoregulatory and in the presence of plant exudate induces the *nodABC* genes. *EMBO J.* 4:3369–73

171. Rostas, K., Kondorosi, E., Horvath, B., Simoncsits, A., Kondorosi, A. 1986. Conservation of extended promoter regions of nodulation genes in *Rhizobium*. *Proc. Natl. Acad. Sci. USA* 83:1757–61

172. Sadowsky, M. J., Bohlool, B. B. 1983. Possible involvement of a megaplasmid in nodulation of soybeans by fast-growing rhizobia from China. *Appl. Environ. Microbiol.* 46:906–11

172a. Sadowsky, M. J., Cregan, P. B. 1992. The soybean *Rj4* allele restricts nodulation by *Bradyrhizobium japonicum* serogroup 123 strains. *Appl. Environ. Microbiol.* 58:720–23

173. Sadowsky, M. J., Cregan, P. B., Gottfert, M., Sharma, A., Gerhold, D., et al. 1991. The *Bradyrhizobium japonicum nolA* gene and its involvement in the genotype-specific nodulation of soybeans. *Proc. Natl. Acad. Sci. USA* 88:637–41

174. Sadowsky, M. J., Cregan, P. B., Keyser, H. H. 1989. DNA hybridization probe for use in determining restricted nodulation among *Bradyrhizobium japonicum* serocluster 123 field isolates. *Appl. Environ. Microbiol.* 56:1768–74

175. Sadowsky, M. J., Cregan, P. B., Rodriguez-Quinones, F., Keyser, H. H. 1990. Microbial influence on gene-for-gene interactions in legume-*Rhizobium* symbioses. *Plant Soil* 129:53–60

175a. Sadowsky, M. J., Liu, R. L., Bhagwat, A. A., Cregan, P. B. 1990. Identification and isolation of *aur*-like, genotype-specific nodulation (GSN) determinant from *Bradyrhizobium japonicum*. In *Abstr. of the 5th Int. Symp. on the Molecular Genetics of Plant-Microbe Interactions, Interlaken, Switzerland,* ed. M. Gottfert, H. Hennecke, H. Paul. Zurich: Eidgenössiche Technische Hochschule

176. Sadowsky, M. J., Olson, E. R., Foster, V. E., Kosslak, R. M., Verma, D. P. S. 1988. Two host-inducible genes of *Rhizobium fredii* and the characterization of the inducing compound. *J. Bacteriol.* 170:171–78

177. Sadowsky, M. J., Tully, R. E., Cregan, P. B., Keyser, H. H. 1987. Genetic diversity in *Bradyrhizobium japonicum* serogroup 123 and its relation to genotype-specific nodulation of soybean. *Appl. Environ. Microbiol.* 53:2624–30

178. Sanjuan, J., Olivares, J. 1989. Implication of *nifA* in regulation of genes located on a *Rhizobium meliloti* cryptic plasmid that affect nodulation efficiency. *J. Bacteriol.* 171:4154–61

179. Sanjuan, J., Olivares, J. 1991. Multicopy plasmids carrying the *Klebsiella pneumoniae nifA* gene enhance *Rhizobium meliloti* nodulation competitiveness on alfalfa. *Mol. Plant-Microb. Interact.* 4:365–69

180. Sanjuan, J., Olivares, J. 1991. NifA-NtrA regulatory system activates transcription of *nfe,* a gene locus involved in nodulation competitiveness of *Rhizobium meliloti*. *Arch. Microbiol.* 155:543–48

181. Sargent, L., Huang, S. Z., Rolfe, B. G., Djordjevic, M. A. 1987. Split-root assays using *Trifolium subterraneum* show that *Rhizobium* infection induces a systemic response that can inhibit nodulation of another invasive *Rhizobium* strain. *Appl. Environ. Microbiol.* 53:1611–19

182. Schofield, P. R., Watson, J. M. 1986. DNA sequence of *Rhizobium trifolii* nodulation genes reveals a reiterated and potentially regulated sequence preceeding *nodABC* and *nodFE*. *Nucleic Acids Res.* 14:2891–2903

183. Schwinghamer, E. A. 1971. Antagonism between strains of *Rhizobium trifolii* in culture. *Soil Biol. Biochem.* 3:355–63

184. Schwinghamer, E. A., Belkengren, R. P. 1968. Inhibition of rhizobia by a strain of *Rhizobium trifolii:* some properties of the antibiotic and of the strain. *Arch. Mikrobiol.* 64:130–45

185. Schwinghamer, E. A., Pankhurst, C. E., Whitfield, P. R. 1973. A phage-like bacteriocin of *Rhizobium trifolii*. *Can. J. Microbiol.* 19:359–68

186. Scott, D. B., Hennecke, H., Lim, S. T. 1979. The biosynthesis of nitrogenase MoFe protein polypeptides in free-living cultures of *Rhizobium japonicum*. *Biochim. Biophys. Acta* 565:365–78

187. Shearman, C. A., Rossen, L., Johnston, A. W. B., Downie, J. A. 1986. The *Rhizobium leguminosarum* nodulation gene *nodF* encodes a polypeptide similar to acyl carrier protein and is regulated by

nodD plus a factor in pea root exudate. *EMBO J.* 5:647–52

188. Somasegaran, P., Hoben, H. J. 1985. *Methods in Legume*-Rhizobium *Technology*. Paia, HI: NifTAL. 367 pp.
189. Spaink, H. P., Okker, R. J. H., Wijffelman, C. A., Pees, E., Lugtenberg, B. J. J. 1987. Promoters in the nodulation region of the *Rhizobium leguminosarum* Sym plasmid pRL1JI. *Plant Mol. Biol.* 9:27–39
190. Spaink, H. P., Okker, R. J. H., Wijffelman, C. A., Tak, T., Roo, L. G., et al. 1989. Symbiotic properties of rhizobia containing a flavonoid-independent hybrid *nodD* product. *J. Bacteriol.* 171:4045–53
191. Spaink, H. P., Wijffelman, C. A., Okker, R. J. H., Lugtenberg, B. J. J. 1989. Localization of functional regions of the *Rhizobium nodD* product using hybrid *nodD* genes. *Plant Mol. Biol.* 12:59–73
192. Spaink, H. P., Wijffelman, C. A., Pees, E., Okker, R. J. H., Lugtenberg, B. J. J. 1987. *Rhizobium* nodulation gene *nodD* as a determinant of host specificity. *Nature* 328:337–40
193. Stacey, G. 1988. Genetics of symbiotic nitrogen fixation. In The *Impact of Chemistry on Biotechnology—Muiltidisciplinary Discussions*, ed. M. Phillips, S. P. Shoemaker, R. D. Middlekauff, R. M. Ottenbrite, pp. 262–78. Washington, DC: Am. Chem. Soc.
194. Stall, R. E. 1986. Plasmid-specified host-specificity in *Xanthomonas campestris* pv. *vesicatoria*. In *Proc. 6th Int. Conf. on Plant Pathogenic Bacteria, 2–7 June 1985*. Boston: Martin Nijhoff
195. Stanley, J., Cervantes, E. 1991. Biology and genetics of the broad host range *Rhizobium* sp. NGR234. *J. Appl. Bacteriol.* 70:9–19
196. Staskawicz, B., Dahlbeck, D., Keen, N. 1984. Cloned avirulence gene of *Pseudomonas syringae* pv. *glycinea* determines race-specific incompatibility on *Glycine max* (L.) Merr. *Proc. Natl. Acad. Sci. USA* 81:6024–28
197. Staskawicz, B., Dahlbeck, D., Keen, N., Napoli, C. 1987. Molecular characterization of cloned avirulence genes from race 0 and race 1 of *Pseudomonas syringae* pv. *glycinea*. *J. Bacteriol.* 169:5789–94
198. Stephens, P. M., Cooper, J. E. 1988. Variation in speed of infection of "no root hair zone" of white clover and nodulating competitiveness among strains of *Rhizobium trifolii*. *Soil Biol. Biochem.* 20:465–70
199. Taylor, R. K., Manoil, C., Mekalanos, J. J. 1989. Broad-host-range vectors for delivery of Tn*phoA*: use in genetic analysis of secreted virulence determinants of *Vibrio cholerae*. *J. Bacteriol.* 171:1870–78
200. Thies, J. E., Singleton, P. W., Bohlool, B. B. 1991. Modeling symbiotic performance of introduced rhizobia in the field based on indices of indigenous population size and nitrogen status of the soil. *Appl. Environ. Microbiol.* 57:29–37
201. Torok, I., Kondorosi, E., Stepkowski, T., Posfai, J., Kondorosi, A. 1984. Nucleotide sequence of *Rhizobium meliloti* nodulation genes. *Nucleic Acid Res.* 12:9509–24
202. Trinick, M. J., Hadobas, P. A. 1989. Competition by *Bradyrhizobium* strains for nodulation of the nonlegume *Parasponia andersonii*. *Appl. Environ. Microbiol.* 55:1242–48
203. Trinick, M. J., Hadobas, P. A. 1989. Effectiveness and competition for nodulation of *Vigna unguiculata* and *Macroptilium atropurpureum* with *Bradyrhizobium* from *Parasponia*. *Can. J. Microbiol.* 35:1156–63
204. Triplett, E. W. 1988. Isolation of genes involved in nodulation competitiveness from *Rhizobium leguminosarum* bv. *trifolii* T24. *Proc. Natl. Acad. Sci. USA* 85:3810–14
205. Triplett, E. W. 1990. Construction of a symbiotically effective strain of *Rhizobium leguminosarum* bv. *trifolii* with increased nodulation competitiveness. *Appl. Environ. Microbiol.* 56:98–103
206. Triplett, E. W., Barta, T. M. 1987. Trifolitoxin production and nodulation are necessary for the expression of superior nodulation competitiveness by *Rhizobium leguminosarum* bv. *trifolii* strain T24 on clover. *Plant Physiol.* 85:335–42
207. Triplett, E. W., Lethbridge, B., Midland, S. L., Tate, M. E., Sims, J. J. 1988. Cloning of genes from *Rhizobium leguminosarum* bv. *trifolii* T24 responsible for the production of trifolitoxin, an anti-rhizobial peptide involved in nodulation competitiveness. See Ref. 16a, p. 788
208. Triplett, E. W., Schink, M. J., Noeldner, K. L. 1989. Mapping and subcloning of the trifolitoxin production and resistance genes from *Rhizobium leguminosarum* bv. *trifolii* T24. *Mol. Plant-Microb. Interact.* 2:202–8
209. Triplett, E. W., Vogelzang, R. D. 1989. A rapid bioassay for the activity of the

anti-rhizobial peptide, trifolitoxin. *J. Microbiol. Methods* 10:177–82

210. Truchet, G., Debelle, G. F., Vasse, J., Terzaghi, B., Garnerone, A., et al. 1985. Identification of a *Rhizobium meliloti* pSym 2011 region controlling the host specificity of root curling and nodulation. *J. Bacteriol.* 164:1200–10

211. van Brussel, A., Recourt, K., Pees, E., Spaink, H. P., Tak, T., et al. 1990. A biovar specific signal of *Rhizobium leguminosarum* bv. *viceae* induces increased nodulation gene-inducing activity in root exudate of *Vicia sativa* subsp. *nigra. J. Bacteriol.* 172:5394–5401

211a. van Rensburg, J., Strijdom, B. W. 1982. Root surface association in relation to nodulation of *Medicago sativa. Appl. Environ. Microbiol.* 44:93–97

212. Vest, G. 1970. Rj3—a gene conditioning ineffective nodulation in soybean. *Crop Sci.* 10:34–35

213. Vest, G., Caldwell, B. E. 1972. Rj4—a gene conditioning ineffective nodulation in soybean. *Crop Sci.* 12:692–93

214. Vest G., Weber, D. F., Sloger, C. 1973. Nodulation and nitrogen fixation. In *Soybeans: Improvement, Production and Uses, Agronomy No. 16,* ed. B. Caldwell, pp. 353–90. Madison, WI: Am. Soc. Agronomy

215. Vincent, J. M. 1980. Factors controlling the legume-*Rhizobium* symbiosis. In *Nitrogen Fixation,* ed. W. E. Newton, W. H. Orme-Johnson, 2:103–29. Baltimore: University Park Press

216. Wang, S.-P., Stacey, G. 1991. Studies of the *Bradyrhizobium japonicum nodD1* promoter: a repeated structure for the *nod* box. *J. Bacteriol.* 173:3356–65

217. Weaver, R. W., Frederick, L. R. 1974. Effect of inoculum rate on competitive nodulation of *Glycine max* L. Merrill. I. Greenhouse studies. *Agron. J.* 66:229–32

218. Weaver, R. W., Frederick, L. R. 1974. Effect of inoculum rate on competitive nodulation of *Glycine max* L. Merrill. II. Field studies. *Agron. J.* 66:233–36

219. Wheatcroft, R., Watson, R. J. 1988. A positive strain identification method for *Rhizobium meliloti. Appl. Environ. Microbiol.* 54:574–76

220. Wijffelman, C. A., Pees, E., van Brussel, A. A. Priem, M., Okker, R., Lugtenberg, B. J. 1985. Analysis of the nodulation region of the *Rhizobium leguminosarum* Sym plasmid pRL1JI. See Ref. 58a, p. 127

221. Williams, M. K., Cannon, F., McLean, P., Beynon, J. 1988. Vector for the integration of genes into a defined site in the *Rhizobium meliloti* genome. In *Molecular Genetics of Plant-Microbe Interactions—1988,* ed. R. Palacios, D. P. S. Verma, pp. 198–99. St. Paul, MN: APS Press

222. Williams, L. F., Lynch, D. L. 1954. Inheritance of non-nodulating character in the soybean. *Agron. J.* 46:28–29

223. Williams, L. F., Phillips, D. A. 1983. Increased soybean productivity with a *Rhizobium japonicum* mutant. *Crop Sci.* 23:246–50

224. Wilson, J. K. 1926. Legume bacteria population of the soil. *J. Am. Soc. Agron.* 18:911–19

225. Winarno, R., Lie, T. A. 1979. Competition between *Rhizobium* strains in nodule formation: interaction between nodulating and non-nodulating strains. *Plant Soil* 51:135–42

226. Woomer, P., Singleton, P. W., Bohlool, B. B. 1988. Ecological indicators of native rhizobia in tropical soils. *Appl. Environ. Microbiol.* 54: 1112–16

227. Yanofsky, M. F., Nester, E. W. 1986. Molecular characterization of a host-range determining locus from *Agrobacterium tumefaciens. J. Bacteriol.* 168: 244–50

228. Young, J. P. W., Johnston, A. W. B., Brewin, N. J. 1982. A search for peas (*Pisum sativum* L.) showing strain specificity for symbiotic *Rhizobium leguminosarum. Heredity* 50:197–201

229. Zaat, S. A. J., Wijffelman, C. A., Spaink, H. P., van Brussel, A. A. N., Okker, R. J. H., Lugtenberg, B. J. J. 1987. Induction of the *nodA* promoter of *Rhizobium leguminosarum* Sym plasmid pRL1JI by plant flavanones and flavones. *J. Bacteriol.* 169:198–204

230. Zdor, R. E., Pueppke, S. G. 1988. Early infection and competition for nodulation of soybean by *Bradyrhizobium japonicum* 123 and 138. *Appl. Environ. Microbiol.* 54:1996–2002

231. Zdor, R. E., Pueppke, S. G. 1991. Nodulation competitiveness of Tn*5*-induced mutants of *Rhiozbium fredii* 208 that are altered in motility and extracellular polysaccharide production. *Can. J. Microbiol.* 37:52–58

Annu. Rev. Microbiol. 1992. 46:429–59

POSITIVE REGULATION IN THE GRAM-POSITIVE BACTERIUM: *Bacillus subtilis*

A. Klier, T. Msadek, and G. Rapoport

Unité de Biochimie Microbienne, Institut Pasteur, Centre National de la Recherche Scientifique, URA 1300, 25, rue du Docteur Roux, 75724 Paris Cedex 15, France

KEY WORDS: regulatory polypeptides, DNA binding, antitermination, two-component system, secondary σ factors

---

CONTENTS

0066-4227/92/1001-0429$02.00

Abstract

Temporally and environmentally regulated gene expression in prokaryotes occurs primarily at the level of transcription initiation. Two main modes of regulation have been described, including either the binding of a repressor that blocks transcription or the interaction of a positive regulator with the transcription complex, leading to transcription initiation. Several classes can be distinguished among positive regulators according to their mechanisms of action.

This review describes the different types of positive regulators identified in *Bacillus subtilis,* a gram-positive bacterium. These include accessory regulatory polypeptides, classical positive regulators that bind to target sites located just upstream from the promoter, ambiactive regulators that can act both positively and negatively, antiterminators, two-component signal transduction systems, and positive regulators associated with specific secondary σ factors.

INTRODUCTION

A dominant theme in regulation of gene expression is how a gene is expressed in a selective manner in response to an external or internal signal. Clearly, the elements that ultimately respond to such stimuli are genes encoding regulatory proteins, which *in fine* can act by binding to specific sites on DNA. These sites can be located either close to the regulated promoters or at a distance. Interaction between the regulatory protein and DNA leads to the activation or enhancement of transcription initiation or elongation by RNA polymerase. This is called positive control.

A common pattern in responding to environmental changes through positive control is the sensing of a signal and its transduction to the transcription machinery in order to stimulate the transcription of the target genes. The general outline of the participants involved in positive control is the following. The signal, which may correspond to the concentration of a small intra- or extracellular molecule, is perceived by the sensor. Next, the signal is directly or indirectly transmitted to a regulatory protein. This signal transduction leads to a change in the conformation of the regulatory protein, either by covalent or noncovalent modifications. The altered activator protein binds to a specific DNA site or in some cases to a specific RNA site. The nucleic acid–protein interaction catalyzes transcription initiation or elongation by RNA polymerase.

Among positive regulators, several groups can be distinguished according to their mechanism of action. This review deals with the description of the different classes of positive regulators identified in *Bacillus subtilis,* which is the best-studied organism among bacilli. *B. subtilis* is a soil bacterium that has developed many regulatory mechanisms in order to adapt to environmental changes. Growth in the environment is very often limited by the availability of basic nutrients, and both specific and pleiotropic regulatory pathways allow the bacterium to either use alternate substrates or, as a last resort, to sporulate. Several examples of positive control have been described in *B. subtilis,* making this organism one of the best examples for the study of the different classes of this type of regulation. This review discusses six types of positive regulation: regulation by accessory polypeptides, regulators that bind to DNA near the promoter region, ambiactive regulators that can act both positively and negatively, antiterminators, two-component signal transduction systems, and regulators associated with specific secondary σ factors.

POSITIVE REGULATION BY ACCESSORY POLYPEPTIDES

Bacillus subtilis can secrete a wide variety of extracellular enzymes, including proteases, α-amylase, levansucrase, and β-glucanases (96). Expression of genes encoding extracellular enzymes is controlled by many transcriptional regulatory factors (60). The cloning of several of these regulatory genes from *B. subtilis, B. natto, B. sterothermophilus, B. licheniformis,* and *B. amyloliquefaciens* enhances the production of extracellular enzymes when these genes are present on a multicopy plasmid. These genes include *sacV, senN, senS, degR, degQ, degT,* and *tenA*. The size of their protein products varies from small polypeptides (46–65 amino acids for SacV, Sen, DegR, and DegQ) to larger polypeptides (372 and 205 amino acids for the *degT* and *tenA* products, respectively). Although these regulators show no sequence homology with each other, they all stimulate the production of extracellular enzymes at the transcriptional level. Most of these regulators share the following characteristics: although they act as positive regulators when overproduced using multicopy plasmids, deletion of the corresponding genes from the *B. subtilis* chromosome does not lead to any detectable phenotype. The target sites of these regulatory polypeptides are often located upstream from the promoters of genes encoding degradative enzymes.

SacV

The *sacV* gene, when cloned on a high-copy-number plasmid, leads to levansucrase overproduction, probably because of an increased rate of transcription of the *sacB* gene (71). The cloned DNA fragment contains several putative open reading frames (ORFs). One of them (ORF1) contains a TTG start codon preceded by a ribosome-binding site, and the encoded 64–amino acid polypeptide seems to be the most likely candidate responsible for the increased synthesis of levansucrase (71). Although no definitive evidence indicates that this polypeptide is involved, the homology found between SacV and ORF2 of the sporulation inhibition sequence (Sin), which encodes a DNA binding protein (38), is interesting.

SenN and SenS

While trying to improve the production of subtilisin by *B. subtilis,* Wong et al identified a DNA fragment from *B. natto* able to stimulate the expression of neutral protease (*nprE*), alkaline protease or subtilisin (*aprE*), α-amylase, and alkaline phosphatase at a moderate but significant level of two- to fourfold (149). This new gene, *senN* (for subtilisin secretion enhancer), codes for a small highly basic polypeptide of 60 amino acids in length, which shares no similarity with the products encoded by *degQ, sacV,* or *degR*. SenN stimulates gene expression at the transcriptional level. A homologous locus has been identified in *B. subtilis* (142). The *senS* gene encodes a homologous but slightly larger polypeptide than SenN and maps genetically at a different position than other known regulatory genes. The *senS* gene and its corresponding protein have several interesting features. A transcription terminator sequence is present between the promoter region and the *senS* open reading frame, suggesting that *senS* expression may be regulated by an antitermination mechanism. Additionally, a most interesting feature of this region is the presence of sequences that are very similar to the *nusA* consensus Box-A and potential Box-B sequences (76). This suggests that a *B. subtilis* putative NusA-like protein could be involved in regulation by antitermination and that *senS* expression could itself be highly regulated. Deletion of the terminator-like structure was shown to be lethal for *B. subtilis*.

The deduced amino acid sequence revealed that SenS contains 65 amino acids with numerous Lys residues (142). Both SenN and SenS have partial homology with the N-terminal regions of minor σ factors and with region 2.1 of σ^{43}, the major *B. subtilis* σ factor (44, 142). Furthermore, a helix-turn-helix (HTH) motif similar to that found in σ factors and in the Sin protein (see below) was present in the N-terminal half of SenS. These properties suggest that SenS may act by binding to DNA and interacting with the RNA polymerase core enzyme.

DegR

First identified in *Bacillus natto,* the *degR* gene (formerly *prtR*) enhances the production of the *B. subtilis* extracellular proteases and levansucrase, but not α-amylase, RNase, nor alkaline phosphatase (85, 133, 154). Genetic mapping localized this gene near *metB,* far from other genes causing similar effects. The *degR* gene encodes a 60–amino acid polypeptide that activates the target genes at the transcriptional level. The enhancement did not result from stabilization of the mRNA. However, deletion of this gene from the *B. subtilis* chromosome did not lead to any obvious phenotype. Helmann et al also showed that the *degR* gene is expressed from a σ^D-type promoter (46).

DegQ

The *B. subtilis degQ* gene (formerly *sacQ*) was initially characterized using a single mutation, *degQ36,* which leads to overproduction of levansucrase and proteases (65). This gene encodes a 46–amino acid polypeptide, which is not related by sequence to other regulators acting on exocellular enzyme production (6, 153). Homologous *degQ* genes have been characterized in *Bacillus amyloliquefaciens* (135) and *B. licheniformis* (6, 90, 145). As shown for *degR,* the *degQ* gene is also dispensable because its deletion did not lead to any recognizable phenotype. However, cloning of the *degQ* gene onto a multicopy plasmid led to a high level of expression of the polypeptide, which then stimulated the rate of synthesis of a class of secreted enzymes including levansucrase, proteases, β-glucanases, and xylanase. A similar phenotype is obtained with the *B. subtilis* strain harboring the *degQ36* mutation. This mutation is located in the promoter site of the *degQ* gene, leading to a −10 sequence that is closer to the consensus sequence of a typical vegetative *B. subtilis* promoter (153). Using *lacZ* fusions with the promoter regions of the target genes, several groups of investigators showed that the *degQ* gene product acts at the transcriptional level, upstream from the promoters of the target genes (10, 48, 50, 57, 113).

Expression of *degQ* increased significantly when cells were grown under conditions of limiting carbon or phosphate sources or during amino acid starvation (81). Moreover, this expression also increased during growth in the presence of decoyinine, a specific inhibitor of GMP synthetase. This result suggests that the expression of *degQ* is submitted to a complex regulatory system that responds to a nutritional-deprivation signal.

DegT

A regulatory gene originating from *B. sterothermophilus* was identified by selection of recombinant plasmids leading to enhanced production of extracellular alkaline protease (*aprE*) in *B. subtilis* (130). *B. subtilis* strains carrying the *degT* gene on a multicopy plasmid displayed a pleiotropic phenotype: (*a*) enhancement of production of extracellular enzymes such as levansucrase, (*b*) repression of autolysin activity, (*c*) decrease of transformation frequency, (*d*) altered control of sporulation, (*e*) loss of flagella, and (*f*) abnormal cell division. This pleiotropic phenotype is similar to those of *B. subtilis* strains overexpressing *degQ* or *degR*. However, the *degT* gene product corresponds to a polypeptide of 372 amino acids, proposed to be composed of three domains, which could explain some properties of the molecule. Region I corresponding to the N-terminal part of the molecule (amino acid positions 51–160) is highly hydrophobic, and it is assumed that this hydrophobic core might be located in the cytoplasmic membrane and function as a membrane sensor protein for environmental stimuli. The central domain of the protein contains a putative HTH sequence, suggesting that DegT could interact with DNA. Moreover, an amphipathic α-helix structure rich in aspartic and glutamic acid residues is located in the C-terminal part of the molecule. This structure has been identified in several transcriptional activators.

These observations and the pleiotropic phenotype induced by *degT* suggest that DegT functions as a membrane sensor for environmental stimuli, such as nutrient-source starvation, and interacts with DNA to activate or repress transcription of several target genes.

TenA and TenI

The *tenA* and *tenI* genes of *B. subtilis* were first identified by using a shotgun cloning approach and were selected for their abilities to stimulate the production of alkaline protease up to 55-fold (91). The two genes are organized as part of an operon. When the genes are cloned on a multicopy plasmid, the first one (*tenA*) stimulates alkaline-protease production whereas the second one (*tenI*) exerts an opposite effect. TenA acts at the transcriptional level; however, neither *tenA* nor *tenI* are essential for cell growth or for production of extracellular enzymes. A slight delay in sporulation was observed in a *B. subtilis* strain deleted for *tenA* and *tenI*. A terminator-like structure resides upstream from the *tenA* ORF, suggesting that expression of this operon might be controlled by an antitermination mechanism. This structure is similar to that reported for the *senS* gene. Deletion of this structure in the case of *tenA* led to an increase of the expression of *tenA*. The overall organization of the operon therefore suggests the involvement of a complex regulatory network. However, as for the other regulators of this class, the role of TenA and TenI is not well defined.

The manner in which the extracellular enzyme genes are controlled by these dispensable regulatory polypeptides is not well understood at present. In all cases, these polypeptides affect the transcriptional efficacy of several target genes, but only when they are overexpressed. Deletion of the corresponding regulatory genes does not lead to any obvious phenotype. One possibility is that some cross complementation occurs between these regulators. Interestingly, expression of the genes encoding these regulators is often highly regulated and in some cases dependent on nutritional signals. In the case of *degT*, investigators assumed that this membrane-associated polypeptide itself could act as an environmental sensor. However, no direct evidence supports this hypothesis and no *B. subtilis degT* equivalent has yet been identified.

Several hypotheses can be made concerning the mode of action of these regulators. They could bind to DNA at target sites adjacent to structural genes of secreted proteins. Alternatively, they could either bind to RNA polymerase, thus modifying its specificity, or to other regulators or genes involved in the transcription mechanism. However, these different regulators probably do not act upon the same target genes in the same manner.

POSITIVE REGULATION INVOLVING DNA-BINDING SITES LOCATED CLOSE TO THE PROMOTER

In several prokaryotic systems, proteins have been shown to bind just upstream from the RNA polymerase recognition site and to stimulate transcription. One of the best-studied examples is the λ CI repressor, which activates transcription by binding to DNA and interacting with RNA polymerase (97, 98). This specific interaction is absolutely required to activate the transcription of the target gene. In *B. subtilis*, several systems are positively controlled by a similar mechanism.

GltC

In *B. subtilis*, glutamate synthase is a key enzyme for nitrogen metabolism. Along with glutamine synthetase, it directs the assimilation of ammonium ion into glutamine and glutamate. Synthesis of both of these enzymes is regulated at the transcriptional level in response to the availability of nitrogen sources. In the case of glutamate synthase, the genes that encode the large and small subunits of the enzyme (*gltA* and *gltB*), are expressed at a high level when cells are grown in the presence of ammonium, but at a low level in cells growing in the presence of glutamate (12). Analysis of Tn*917* insertions leading to a Glt$^-$ phenotype indicates that some of the insertions are located far upstream from *gltA*, in a locus designated *gltC*. These *gltC* mutants were completely auxotrophic for glutamate (13).

Complementation tests showed that a 1.1-kb DNA fragment that overlaps the *gltC* locus codes for a diffusible factor that restores glutamate synthesis expression in the *gltC* mutant. This region encodes a 35-kilodalton (kDa) polypeptide that is required *in trans* for *gltA* and *gltB* gene expression. The *gltC* gene is itself transcribed from a promoter that is divergent from, but overlaps with, the *gltA* promoter (14). In addition to increasing expression from the *gltA* promoter, the *gltC* product negatively regulates its own transcription in a manner independent of the nitrogen source.

The sequence of the GltC polypeptide suggests that it is a member of a family of regulatory proteins, most of which are involved in regulation of amino acid biosynthesis. These LysR-family members are especially similar in their amino-terminal region, which is thought to represent a DNA-binding domain. Interestingly, a 9-bp sequence that occurs several times in the *gltA-gltC* regulatory region is similar to sequences found in the target DNA of several other LysR-family proteins (14).

The mechanism of activation of *gltA* expression by GltC is not well understood. Under conditions of limiting glutamate, a conformational change may be induced in GltC, such that it stimulates the association of RNA polymerase with the *gltA* promoter, increasing the rate of transcription of *gltA*. It was assumed that this conformational change does not alter either its binding properties or its ability to repress *gltC* transcription. However, the identity of the metabolite that acts to inform GltC of the availability of glutamate within the cell remains to be determined.

CitR and AlsS

The LysR family has been extended to gram-positive bacteria not only by the addition of GltC, but also by the characterization of CitR and AlsC from *B. subtilis*. The *citA* gene encodes citrate synthase, which catalyzes the first step in the TCA cycle. The gene has been cloned and its DNA sequence determined (56). Upstream from the *citA* gene, a regulatory gene (*citR*) was identified. It encodes a product homologous to the LysR family that apparently positively controls *citA* expression at the transcriptional level.

Acetolactate synthase is synthesized only after the onset of the stationary phase. The structural gene, *alsS,* has been cloned and sequenced, as has an overlapping and divergently transcribed gene, *alsC* (100). The *alsD* gene, encoding acetolactate decarboxylase, is located downstream from *alsS*. The product of the *alsC* gene is homologous to members of the LysR family of bacterial regulators. Insertional inactivation of *alsC* leads to a complete loss of *alsS* transcription.

MerR

B. subtilis sp. RC607 was originally isolated by virtue of its resistance to mercury and cadmium (68). Cloning of the *mer* operon on a multicopy plasmid in *B. subtilis* was found to confer inducible mercury resistance. The nucleotide sequence of the entire *mer* operon was determined (143, 144). Five contiguous ORFs were predicted from the sequence, one of which clearly corresponds to mercuric ion reductase (*merA*). A sixth, distally located open reading frame that corresponds to organomercurial-lyase (*merB*) was also found. The first ORF of the operon is believed to encode a regulatory gene product. This *merR* gene is cotranscribed with the structural genes. The *B.*

subtilis merR gene is similar in sequence to other characterized *merR* genes including those of *Staphylococcus aureus* and of the gram-negative-derived transposons Tn*501* and Tn*21* (47). The MerR proteins appear to be evolutionarily related and share several conserved sequence features. Most notably, there are three invariant cysteine residues, implicated as ligands for metal binding, and a conserved amino-terminal region with homology to HTH DNA-binding motifs. Moreover, all MerR responsive operator sequences have a conserved palindromic core structure of GTAC N4 GTAC (43). The MerR protein acts as a dimer of identical subunits (132 amino acids each) and binds to the *mer* operator region with high affinity (47). The MerR protein bound less tightly to its operator region in the presence of mercuric ion; this reduced affinity was largely accounted for by an increased rate of dissociation of the MerR protein from the operator. Despite the reduced DNA-binding affinity, genetic and biochemical evidence support a model in which the MerR protein–mercuric ion complex is a positive regulator of operon transcription. Unlike the majority of described transcription activator proteins, the MerR proteins bind between the −35 and −10 consensus elements of the target promoter DNA. This promoter is distinguished by exceptionally long spacer regions between these two consensus elements (19 or 20 bp). A role for the MerR–mercuric ion complex may be to compensate for this unfavorable spacing by altering local DNA structure. The presence of the mercuric ion may favor an allosteric conformational change of MerR into an active form, allowing transcription of the operon (47).

AMBIACTIVE OR BIFUNCTIONAL REGULATORS

Regulatory proteins can act positively or negatively depending upon the gene targets. Some of these dual function regulatory proteins have been characterized in *Escherichia coli* such as AraC or λ CI (64, 98). In *B. subtilis,* few cases have been described.

Sin

The *sin* gene was identified by its ability to affect late-growth processes in *B. subtilis* when it is overexpressed (38). The *sin* gene was cloned and sequenced. It encodes a polypeptide of 111 amino acid residues. Overexpression of Sin by placing the gene on a multicopy plasmid represses sporulation. Under the same conditions, more than 90% repression of the *aprE* gene was observed. The inactivation of the *sin* gene in the chromosome causes a loss of competence and motility. In addition, the disruption of the *sin* gene results in a slight increase (three- to fourfold) of exoenzyme production (levansucrase, alkaline protease, neutral protease, and α-amylase) and in a deficiency of autolysin production leading to cell filamentation (38, 112) (T. Msadek & F.

Kunst, unpublished results). The *sin* gene is identical to the *flaD* locus of *B. subtilis* and *B. licheniformis* (111, 112). Expression of the *sin* gene occurs at a relatively low level throughout growth and remains largely unaffected by *spo0A* and *spo0H* mutations (37). Sin is required for the expression of the late competence genes and motility genes (27, 115) as well as for autolysin production (111, 112), suggesting that Sin is both a positive and a negative regulator. The *sin* gene resides downstream from ORF1, which is under developmental control and requires functional *spo0A* and *spo0H* gene products. Three transcripts were correlated with Sin synthesis and corresponding promoter-like sequences were identified. The P3 promoter is located between ORF1 and *sin* and is responsible for the most abundant mRNA. ORF1 was not required for multicopy inhibition of sporulation and protease production.

The overexpression of Sin did not affect the expression of early sporulation genes (*spo0B, spo0F,* and *spo0H*) or that of some other late growth–regulated genes (*citB* and *isp*). However, *spoIIA, spoIIG,* and *spoIIF* gene expression is dramatically reduced when the *sin* gene is present on a multicopy plasmid. Moreover, expression of the *spoIIE* gene is slightly higher when the *sin* gene is disrupted (115).

Sin has been purified to homogeneity and binds to two sites upstream from the *aprE* gene in vitro (39). The stronger Sin-binding site (SBS-1) is located between positions −265 and −220 from the start point of transcription. This region is required for Sin repression because a strain deleted for this region is not affected by overexpression of the *sin* gene. A second weaker binding site was detected near the *aprE* promoter site. The SBS-1 was further analyzed, and it was suggested that it contains two adjacent binding sites that contain two different sequences of partial dyad symmetry. DNA looping, caused by Sin binding, may be involved in repression of *aprE* as has been documented for other systems. Recently, Sin was shown to be able to bind to the *spoIIA* promoter. There is also some evidence that expression of the *sin* gene is negatively autoregulated. Interestingly, Sin can control target genes transcribed by RNA polymerase with different σ factors. Whether DNA-binding occurs in a similar fashion for each of the target genes remains to be seen.

An analysis of the predicted amino acid sequence of Sin revealed a complex structure, which may account for the binding properties of this dual regulatory protein. A potential leucine-zipper motif flanked by two HTH motifs has been proposed (39). The leucine-zipper motif could be responsible for the dimerization of the molecule, which is consistent with the observation that Sin exists as a tetramer in solution. The two HTH motifs would be involved in recognizing the two different regions of dyad symmetry. However, the role of these motifs in switching the Sin regulatory protein from a negative to a positive regulator at the end of the vegetative growth phase is not understood.

AbrB

The AbrB protein of *B. subtilis* regulates a wide variety of genes whose expression is associated with the end of vegetative growth and the onset of sporulation (95, 128, 129). The *abrB* locus was first identified as a suppressor that overcomes sporulation's *spo0A* dependence (140, 156). It is involved in the negative control of the *aprE, spo0H, spo0E, spoVG,* and *tycA* genes (30, 35, 69, 102, 129). It has also been implicated in the positive control of the *hpr* locus (93, 129).

The *abrB* gene encodes a 96–amino acid polypeptide that contains a typical HTH domain common to DNA-binding proteins. Two promoter sites were localized upstream from the *abrB* gene. The downstream promoter site is negatively controlled by the *spo0A* gene product, and two 0A boxes were identified between the downstream promoter and the *abrB*-coding sequence (127).

The binding of the AbrB protein to the subtilisin *aprE* promoter is representative of the results obtained using footprinting and gel retardation assays with other target DNAs (129). In all cases, large protected areas of 50–120 nucleotides overlapping the promoter region were detected. AbrB protects an area of the *aprE* promoter between −59 and +25. Two binding sites within the leader and promoter regions of *tycA* were also identified. For the *spoVG* gene, the binding sites are located in the AT-rich region upstream from the promoter (−67 to −51). Two mutations within this region render transcription from the *spoVG* promoter insensitive to repression by *abrB* in vivo (35, 102). The binding of AbrB to the *spo0E* promoter also occurs in a region that includes the promoter. AbrB binds to its own promoter, and the *abrB* gene is negatively autoregulated. This binding affects both promoters of the *abrB* gene (128).

Although no conserved recognition consensus site for AbrB binding has been identified, some common features of these sites can be pointed out. All AbrB protected sites are small AT-rich sequences located about one helical turn apart, indicating that the polypeptide binds to one face of the DNA helix. Conserved among these sequences is the C(G)AAAA sequence found in the −70 to −40 region of all AbrB target sites. Gel retardation experiments suggest that AbrB binding is cooperative and involves site-specific DNA bending as judged from the abnormally slow migration during gel electrophoresis (35). As observed with Sin, AbrB repression is not RNA polymerase specific. This protein represses *spoVG,* which is transcribed by the $E\sigma^H$ form of RNA polymerase holoenzyme, and also inhibits expression of *aprE, spo0H, spo0E,* and *tycA*. The *aprE* and *spo0H* genes are both transcribed by $E\sigma^A$; this is probably true for *spo0E* and *tycA* as well.

AbrB acts also as a positive regulator of the *hpr* locus (129). The AbrB protein binds to a DNA fragment containing the *hpr* promoter. However, the sequence of the binding site has not been determined. Strauch et al (129) suggested that the mechanism of AbrB binding to the *hpr* DNA is in some way different from its binding to negatively regulated genes, which could reflect its role as a positive regulator. The relative distance between the promoter and the binding site of the regulatory protein may be critical in determining its role as a positive or negative effector.

GerE

The *gerE* gene encodes a regulatory protein that governs the synthesis of the spore-coat proteins (18, 19). Its predicted product, an 8.5-kDa polypeptide, contains a region of significant homology to the HTH structure characteristic of the DNA binding domain of many prokaryotic regulatory proteins. The absence of the *gerE* gene product causes overexpression of the *cotA* gene and prevents expression of two other coat genes, *cotB* and *cotC*. This suggests that GerE is both a negative and positive regulator of mother cell–specific gene expression. The GerE protein has been shown to bind to the promoter region of the *cotB* and *cotC* genes and stimulates their transcription by the σ^K-RNA polymerase holoenzyme in vitro (59a).

REGULATION BY ANTITERMINATION

Attenuation control by *trans*-acting binding factors is a common regulatory mechanism. These factors regulate the formation of RNA structures that serve as transcriptional terminators. The expression level of genes located downstream from these RNA structures depends on the formation of the transcription terminator. The first polypeptide identified as an antiterminator factor is the N protein from the bacteriophage λ. N works by enabling RNA polymerase to transcribe regions of DNA that would otherwise cause the mRNA to terminate. In the presence of N, the N and Cro mRNAs are extended, allowing transcription of the flanking genes (101). In *B. subtilis*, several examples of regulation by antitermination are well documented.

SacY and SacT

Sucrose induces the synthesis of at least three proteins in *B. subtilis:* an intracellular sucrase (*sacA* gene), an extracellular levansucrase (*sacB* gene) and an enzyme II^{Scr} (*sacP* gene) (58). The *sacP* and *sacA* genes are organized in an operon and are not linked to *sacB* (33). The *sacB* gene and the *sacPA* operon are controlled by specific regulators: SacY and SacT, respectively. Both of these regulatory genes have been cloned and sequenced (11, 23, 24, 158).

The coding region of *sacB* is preceded by a 400-bp regulatory region called *sacR,* which contains the promoter and the target sites for specific and pleiotropic regulators (10, 57, 120). A transcriptional terminator is involved in *sacB* induction by sucrose and resides just downstream from the promoter. Deletion of the termination structure or the introduction of single-base changes that alter the dyad symmetry of the palindromic structure led to constitutive synthesis of levansucrase. The *sacB* promoter is constitutive and the transcripts extend past the terminator only in the presence of sucrose (113).

Downstream from the *sacPA* promoter is located a palindromic sequence that showed significant homology with the transcriptional terminator of the *sacB* gene (23). Fifty out 53 bases are identical. Presumably, expression of the *sacPA* operon is also regulated by an antitermination mechanism. Insertion of a 4-bp sequence in the loop of the putative terminator structure led to constitutive expression of the *sacPA* gene (7a).

The *sacY* and *sacT* genes appear to encode antiterminator proteins required for the expression of *sacB* and *sacPA*, respectively. Because SacT is strongly similar in amino acid sequence to SacY and because the two palindromic sequences present downstream from the *sacB* and *sacPA* promoters are highly related, it was proposed that there is some cross-talk between both systems (23, 121). The first genetic evidence for this cross-talk came from the isolation of *sacY* mutations in which sucrase and levansucrase are both produced constitutively. Studies also demonstrated that SacT can act on *sacB* expression and SacY on the expression of the *sacPA* operon, leading to low levels of induction by sucrose (121).

Another similar antitermination mechanism, involving the BglG protein, regulates the expression of the *E. coli bgl* operon. BglG, SacY, and SacT are strongly homologous. Furthermore, within the 5' regulatory regions of *sacB, sacPA,* and *bgl,* a highly conserved 32-bp sequence can be detected, located upstream from and partially overlapping the palindromic structure (23, 54). In the *bgl* system, research demonstrated that BglG is a RNA-binding protein (54). It was proposed that the RNA target can form a hairpin structure able to prevent formation of the terminator hairpin. A similar model can be proposed for the other two systems. The antiterminator could bind its mRNA 32-bp sequence target in such a way that the secondary structure determining transcription termination is not formed.

The *sacY* gene is the second of an operon (formerly named the *sacS* locus) whose expression is induced by sucrose and increased in a *degU*(Hy) mutant (119; A. Klier, unpublished results). The first gene, *sacX,* codes for a polypeptide that is 56% identical to the *sacP* product (157, 158). The *sacX* gene product is probably an inefficient enzyme II^{Scr}, acting as a sucrose sensor, which seems to be involved in negative control of the SacY antiterminator through PTS-dependent phosphorylation (17). The *sacX* gene is not required for the control of the SacT antiterminator. SacT is also expected to be controlled by phosphorylation since it requires both enzyme I and Hpr of the phosphotransferase system (PTS) in order to act as an antiterminator (7a). However, the sucrose sensor for the induction of the *sacPA* operon expression remains to be identified.

HutP

The genes responsible for histidine catabolism are organized as an operon (*hut* operon) (9, 89). The *hut* regulatory region, the *hutH* gene encoding histidinase, and part of the *hutU* gene of *B. subtilis* have been cloned and sequenced (89). The promoter of the *hut* operon shares sequence characteristics with promoters recognized by RNA polymerase σ^{43} holoenzyme. The *hutP* mutation lies within the first open reading frame, and the HutP protein may positively regulate *hut* expression. Because the noncoding region between the *hutP* and *hutH* genes contains a sequence that could form a stem and loop structure, transcriptional antitermination has been proposed to be involved in induction of the *hut* operon by histidine. Moreover, a *hutC* mutation leading to constitutive expression of the *hut* operon consists of a single change in the palindromic sequence that could destabilize the terminator structure. Because the *hutP* gene product controls expression of the *hut* operon, *in trans,* the HutP protein probably acts as an antiterminator. Moreover, the expression of this operon is highly repressed in the presence of carbon sources and amino acids and derepressed during stationary growth. However, the targets of this regulation by nutrient sources have not yet been defined (9, 150).

Attenuation control by a *trans*-acting RNA-binding factor appears to be quite common in gram-positive organisms. In the *sac* antiterminators, regulation probably occurs through phosphorylation mediated by the PTS. Regulation by antitermination has also been proposed to be involved in expression of the *trp, ilv-leu, pur,* and *pyr* operons of *B. subtilis*. However, in the biosynthetic operons, the final product of the operon seems to be necessary to activate a negative regulatory factor, which could then bind to the leader mRNA, thus promoting the formation of a terminator structure.

SIGNAL TRANSDUCTION BY TWO-COMPONENT SYSTEMS

The involvement of signal transduction components in cell differentiation and growth control is true for both bacterial and eukaryotic cells. This control is critical for prokaryotic cells, for which growth must be continually adjusted to environmental changes. Many bacterial signal transduction systems contain members of two families of homologous proteins, one of which controls the activity of the other. Typically, the first component, or modulator, receives an extracellular signal, directly or indirectly, possibly through its N-terminal receiver domain (5, 59, 104, 123). Studies of the nitrogen regulation, chemotaxis, osmoregulation, and phosphorous regulation systems in *E. coli* have demonstrated that the modulators belong to a family of histidine-kinases (123). The signal is then transduced via the C-terminal domain of the modulator to the second component, the response regulator or effector, which is generally a transcriptional regulator. The phosphorylation of the response regulator at a conserved aspartate residue is an essential feature of the signal transduction mechanism (105, 122, 123).

B. subtilis normally lives in the soil where nutrients are limited and conditions tend to fluctuate widely. Growth in this environment is generally limited by the availability of essential nutrients. Molecular genetic analysis has allowed the identification of a network of interacting signal-transduction proteins. To date, five known histidine-kinase proteins in *B. subtilis* (CheA, ComP, SpoIIJ, DegS, and PhoR) are thought to interact with six cognate response regulators (CheY, ComA, Spo0A, Spo0F, DegU, and PhoP).

KinA-KinB/Spo0F-Spo0A

Endospore formation in response to nutrient deprivation by *B. subtilis* is a paradigm for developmental regulation in prokaryotes, and well over 100 genes affecting this process have been identified. Sporulation and competence represent two alternative developmental pathways available to *B. subtilis* at the onset of the stationary phase. The initiation of sporulation usually occurs at the time of transition from the exponential to the stationary phase, when cell growth is limited by the depletion of essential nutrients. Several mutants have been identified that fail to initiate the sporulation process: among these are *spo0A, spo0B, spo0E, spo0F,* and *spo0H* (52). The corresponding genes have been cloned and sequenced: *spo0A* and *spo0F* encode proteins containing the characteristic N-terminal domain of the phosphorylated response regulators (31, 139, 155). The product of the *spo0H* gene was found to be a σ factor, σ^H, providing promoter specificity to RNA polymerase (29). The *spo0E* gene product is a negative regulator of sporulation (94) and the *spo0B* gene is part of an operon that includes the *obg* gene, which encodes a putative G protein (15, 138).

The predicted *spo0F* product is similar to the CheY response regulator in that it consists only of the conserved N-terminal response-regulator domain (123). Overproduction of Spo0F from a multicopy plasmid effectively inhibits sporulation (114). Mutants lacking Spo0F fail to initiate sporulation and this phenotype is suppressed by secondary mutations altering the N-terminal domain of Spo0A. These mutations have all been localized to Spo0A sequences that would be expected to be adjacent to the response regulator acidic phosphoacceptor site (118, 136). These results implicated the Spo0A protein as the ultimate regulator of the initiation of sporulation. Spo0A was known to be a DNA-binding protein that regulated *abrB* gene expression (127).

The phosphorylation of Spo0A results from a series of reactions that have been termed a phosphorelay (15, 137). Stimulation of a kinase that phosphorylates Spo0F is the initial event of this phosphorylation cascade. Spo0F is postulated to act as a secondary messenger that can be phosphorylated by a variety of kinases in response to specific environmental signals. At least two kinases, KinA, the product of the *spoIIJ* locus, and KinB, are involved in the phosphorylation of Spo0F (7, 15, 92, 136). The key enzyme in the phosphorelay is Spo0B, which is a phosphotransferase linking the phosphorylation of Spo0F to that of Spo0A. Obg, which is probably a G protein, might control the activity of Spo0B in some unknown manner. Genetic evidence also suggests that the product of the *spo0E* gene acts as a negative regulator of flux through the phosphorelay (94). Controls of the flow of phosphate to Spo0A are certainly of great importance in coordinating the developmental process (137).

Spo0A is a transcriptional regulator that represses expression of the *abrB* gene (127). AbrB controls many genes whose functions are not required during conditions of nutrient excess. The end product of the phosphorelay, Spo0A~P, has a greater affinity for the Spo0A binding site on the *abrB* promoter than unphosphorylated Spo0A. This implies that phosphorylated Spo0A is a more effective repressor of *abrB* expression and is responsible for the earliest events in stationary phase and sporulation (137).

Spo0A~P, but not Spo0A, was shown to be an activator of transcription of the *spoIIA* operon, which encodes the specific sporulation σ factor σ^F (32, 137). The upstream region of the *spoIIA* promoter contains a Spo0A binding site located at position −65 with respect to the transcription start point (137, 151). The nonphosphorylated form of Spo0A can also bind to this site, but it is inefficient as an activator (137). This probably means that Spo0A~P is required to interact with the transcriptional machinery or to bind to the DNA in order for activation of transcription from the *spoIIA* promoter to occur. The relative position of the Spo0A binding site upstream or downstream from the promoter would be the key determinant for the final role of Spo0A~P.

ComP-ComA and ComQ

In *B. subtilis,* the development of competence, the ability to take up exogenous DNA, involves an environmentally induced and temporally regulated series of events whose control overlaps with the control of sporulation and the regulation of certain genes that are usually expressed only during the stationary phase. Several different competence (*com*) genes have been identified, and their regulation has been studied by using Tn*917 lac* operon fusions (2–4). The *com* genes were classified into two groups on the basis of their expression. The phenotypes of the mutants belonging to the first group suggest that the products of the corresponding genes (*comP, comA, comQ, srfA, comK*) act early in the competence pathway and that they have a regulatory role. The second group of genes (*comC, comDE, comF, comG*) are expressed only during the stationary phase in competence medium, but not in complex media, and are involved in the late phase of the competence pathway (27).

The *comP* and *comA* genes were cloned and sequenced and shown to encode a two-component signaling system (146, 148). The sensor kinase, ComP, promotes the establishment of competence. The stimulus detected by the membrane receptor domain of ComP remains to be determined. One possibility is that ComP senses the level of a crucial nutrient. The predicted protein sequence of the *comA* gene product was found to be similar to that of several members of the response regulator class of prokaryotic signal transducers. Particularly striking is the relationship to the DegU protein (see below). The ComA protein contains three highly conserved aspartic acid residues in the N-terminal portion of the protein, which form a putative phosphorylation site. Also present within the predicted ComA sequence is a possible HTH motif. The similarity to the histidine kinase and response regulator proteins, together with a putative DNA binding domain, suggest that the ComP protein may relay information about the nutritional or growth states via ComA to the transcriptional machinery of the cell.

Just upstream from the *comP* and *comA* genes lies the *comQ* gene (81, 147), whose product is thought to act together with the ComP-ComA regulators to positively control expression of the *srfA* operon (27, 147). The *srfA* operon (formerly *comL*) is required for both synthesis of surfactin, an extracellular peptide antibiotic, and expression of *comK* and late competence genes (86, 141). Both ComQ and ComK act as positive regulators but show no amino acid sequence similarities with other known regulators. Several lines of evidence suggest that ComA does not interact directly with the promoters of late competence genes. It appears that the only role of ComP, ComA, and ComQ in competence development is to positively regulate expression of *srfA,* which in turn leads to expression of *comK* and of late competence genes (42, 88). A ComA box, TTGCGGN$_2$TCCCGCAN was identified upstream from *degQ, gsiA,* and *srfA,* three genes controlled by ComA and expressed during the late vegetative phase (81, 83, 87). This observation suggests that ComA~P probably acts directly on these three gene targets. Two mutations designated *mecA* and *mecB* allow a complete bypass of null mutations in *comP, comA, comQ, srfA,* and *comK* for late-competence gene expression (27, 28, 103). These results imply that these regulatory proteins exert their effects prior to the action of the *mecA* and *mecB* gene products on the expression of late competence genes.

One approach used to identify competence-specific transcription factors was based on determining the effects of placing fragments carrying the *comG* or *comC* promoter regions on multicopy plasmids. A derepression of expression of chromosomal *lacZ* fusions with late competence genes was observed, as well as an inhibition of competence development, suggesting that a positive effector was being titrated (78). The active site in the *comC* promoter region was shown to reside between −97 and −79 from the start point of transcription. Deletion to −79 markedly reduced the level of competence. These results suggested that the binding site for a competence transcription factor (CTF) overlapped or was identical to the binding site required for *comC* expression. Interestingly, examination of the *comC* and *comG* promoters reveals sequence elements with axes of dyad symmetry located at −84 and −81, respectively, which may represent part of the CTF recognition site. In addition, gel shift experiments revealed the presence of a protein that specifically binds to the *comC* promoter fragments (78). This DNA-binding activity is lost in a *comP-comA* mutant and restored in *mec* mutants. Whether CTF is the product of a *mec* gene or of another gene under the control of *mec* remains to be determined (27, 78).

Several other regulators are involved in the expression of competence in *B. subtilis*. The expression of late *com* genes also requires the sporulation regulator Spo0A and, to a lesser extent, Spo0H. The development of competence is also affected by mutations in *degS* and *degU,* which regulate the expression of degradative enzymes. Null mutations in *sin* confer nonmotility in addition to competence deficiency. The *abrB* gene is a regulatory gene for certain sporulation genes, but also for competence (27). This list demonstrates the interrelationships of the various postexponential responses, and Dubnau (27) has proposed the existence of a signal transduction network that relays information to several response-specific effector pathways. This signal transduction network might consequently permit the combinational processing of environmental information input, in order to specify particular responses and to activate specific pathways.

DegS-DegU

B. subtilis produces numerous degradative enzymes, including levansucrase, α-amylase, and proteases. The expression level of secreted proteins is globally controlled by the products of at least three genes: *degQ, degS,* and *degU*. Regulation is at the level of transcription (48, 49, 113).

The *degS* and *degU* genes form an operon encoding a two-component system (51, 61, 131). Indeed, strong similarities were found between DegS and the histidine-kinase modulators and between DegU and the response-regulator proteins, suggesting that DegS may transfer a phosphate group to

DegU. Recent studies showed that such transfer occurs in vitro (20, 84, 132). The DegS protein is probably a cytoplasmic protein, because it does not contain hydrophobic domains required for insertion into the membrane. However, the nature of the signal received directly or indirectly by DegS is not known.

Different missense mutations have been identified in the *degS* and *degU* genes (51, 80). The *degS*(Hy) and *degU*(Hy) mutations lead to an increase in degradative enzyme synthesis. However, these mutations have a pleiotropic effect, since the corresponding mutants have the capacity to sporulate in the presence of glucose and are deficient in transformability, autolysin synthesis, and motility. A second class of missense mutations in the *degS* and *degU* genes leads to a deficiency of degradative enzyme synthesis without affecting the competence pathway. One of these mutations, *degU146,* replaces a conserved aspartate residue at position 56 with an asparagine, presumably inactivating the phosphorylation site of DegU (20, 80, 160). In addition, deletion of the *degS-degU* operon leads to a deficiency in both degradative synthesis as well as competence. Deletion of the *degS* gene alone, however, also leads to a deficiency in degradative enzyme synthesis but has no significant effect on the development of competence (80, 81).

The interpretation of these phenotypes led to the hypothesis that the DegU effector could exist under two active conformations: the phosphorylated form would be required for degradative enzyme synthesis while the unphosphorylated form would be necessary for the expression of competence genes. The *degS*(Hy) and *degU*(Hy) mutations would lead to increased amounts of the phosphorylated form of DegU. Mutations leading to a deficiency of degradative enzyme synthesis, such as *degS42* or *degU146,* are thought to promote accumulation of the unphosphorylated form of the DegU effector (80). In vitro phosphorylation experiments using modified DegS and DegU proteins support this hypothesis (20, 132, 160). Interestingly, both forms of DegU are required for two distinct functions and apparently act as positive regulators.

Several reports have shown that (Hy) mutations in the *degU* and *degQ* genes lead to an increase in the level of *sacB* and *aprE* mRNA (10, 48, 113). Thus, in every case examined, these pleiotropic mutations appear to increase the transcription initiation of their target genes. The sites at which these pleiotropic mutations stimulate the transcription of the *sacB* and *aprE* promoters were localized by deletion analysis. The stimulation effect of the *degU*(Hy) and *degQ*(Hy) mutations requires regions located between positions −164 and −141 upstream from the *aprE* transcription start site (48–50). Comparison of the corresponding regions in the *sacB* and *aprE* promoters showed that 27 bases out of 48 are identical in both upstream promoter regions (48, 49). The *sacX-sacY* operon is also under the control of DegS-

DegU (119; A. Klier, unpublished results). A region necessary for stimulation by DegU was located using deletion analysis within 75 nucleotides upstream from the −35 consensus promoter region. This region displays some weak homology with the target sites found upstream from the *sacB* and *aprE* genes (16).

However, no biochemical evidence yet supports specific binding of DegU~P to these target sites. Possibly, some as yet unindentified intermediates are involved in the regulation of degradative enzyme synthesis. As the similarity between the putative targets is slight, comparison of these regions with other known targets for DegU and DegQ may allow a consensus sequence to be established. In two of these target genes, *sacB* and *aprE,* the same region is involved in stimulation by either the *degU32*(Hy) mutation or the *degQ36*(Hy) mutation (48–50). It remained to be determined whether the products of the two regulatory genes act independently or upon each other. Expression of the *degS* and *degU* genes was not affected by a deletion of the *degQ* gene (80, 82). However, the *degQ*(Hy) phenotype cannot be expressed in a *degS42*(−) strain deficient for degradative enzyme synthesis (6). In addition, the *degQ36*(Hy) phenotype is lost in a strain deleted for the *degS* and *degU* genes (81). Therefore DegS and DegU seem to be required for DegQ to act upon the target genes.

The expression of *degQ* is subject to growth-phase regulation and was reduced in a strain deleted for the *degS-degU* genes or carrying the *degU32*(Hy) mutation. Interestingly, the phosphorylated form of DegU may also act as a repressor, because DegU(Hy) mutations, which lead to the accumulation of DegU~P, repress expression of *degQ* and *degR* and synthesis of flagella (50, 81). Expression of *degQ* is controlled by both the DegS-DegU and ComP-ComA two-component systems. Several regulatory targets were identified upstream from the *degQ* gene. Two of them were shown to reside between positions −393 and −186 and between positions −78 and −40 for regulation by DegS-DegU and ComP-ComA, respectively (81).

A *degS-degU* deletion mutant can be bypassed by *mec* mutations for both the competence pathway and for degradative enzyme synthesis (27, 103; F. Kunst & T. Msadek, unpublished results). This result strongly suggests that the *mecA* gene product could act as an intermediate between nonphosphorylated DegU and late competence genes. A similar assumption can be made to help one understand the role of the phosphorylated form of DegU on the expression of genes encoding degradative enzymes, suggesting that the *mecA* gene product plays a central role in this regulatory network. Alternatively, the *mec* mutations may lead to cross-talk between the *mec* gene products and genes encoding degradative enzymes.

In several prokaryotic systems, proteins have been shown to bind upstream from the RNA polymerase recognition site to repress or stimulate transcription. The best-studied examples are the λ CI repressor, CAP, MalT, and

NtrC, which activate transcription by binding to DNA and interacting with RNA polymerase (1, 40, 99). DNA looping may allow such interactions to take place. Whether DNA looping is in fact a universal mechanism for regulation at a distance remains to be determined.

PhoR-PhoP

B. subtilis produces phosphate-repressible alkaline phosphatase and phosphodiesterase during the vegetative growth phase. The synthesis of these enzymes is controlled by *phoS, phoT,* and two closely linked genes *phoP* and *phoR* (77, 152). Because both negative and constitutive mutations are located in the *phoR* locus, the *phoR* gene product is considered to act as a positive and negative regulator. The *phoP* and *phoR* genes were cloned and sequenced (109, 110). Both gene products belong to the family of two-component systems. Seki et al (110) suggest that PhoR of *B. subtilis* is the sensor histidine-kinase for the PhoP response regulator. The phosphate-regulated gene systems therefore seem to be organized in a similar fashion between gram-positive and gram-negative bacteria. However, several alkaline phosphatases are encoded by a multigene family and are synthesized in phosphate-starved vegetative cells and in sporulating cells (55). PhoP and PhoR are only required for the production of the vegetative alkaline phosphatase but were not required for *phoA* transcription during sporulation (55).

POSITIVE REGULATION INVOLVING SECONDARY σ FACTORS

In addition to the primary σ factor, both gram-negative and gram-positive bacteria employ alternative σ factors that confer different promoter specificities on RNA polymerase core enzyme. Several bacteriophages that infect each of these groups also encode specific σ factors. Some of the alternative σ factors allow transcription of a particular set of genes whose products contribute to a common physiological response (44). For example, σ^F of enteric bacteria and σ^D of *B. subtilis,* both of which confer the same promoter specificity on core polymerase, allow transcription of genes whose products are required for motility and chemotaxis (8, 45). Some σ factors induce the transcription of genes whose products are required at a precise time, for example at a particular time after bacteriophage infection or during the sporulation phase of *B. subtilis*. It was originally proposed that developmental gene expression during sporulation is controlled by a cascade of σ factors (66). This model was based on the discovery of a cascade of σ factors that govern the process of infection of *B. subtilis* by phage SPO1. It is now well established that the sporulation program is largely governed by the regulated appearance of new σ factors, which confer the capacity to use a specific class of promoters on RNA polymerase core enzyme. Five developmental σ factors have now been characterized and identified as the products of known *spo* genes (79, 125, 126). Interestingly, auxiliary control mechanisms that cause interdependence between the successive transcription pathways were also identified. These mechanisms allow the coupling of the activation of developmental σ factors to the course of sporulation. Moreover, the expression of the different σ factors is coordinated between both compartments of sporulating cells. Because several excellent reviews have already been devoted to *B. subtilis* sporulation, the cascade of σ factors is not described further here (114, 117, 124, 126).

σ^{54} *of* B. subtilis *and LevR*

Most of the identified σ factors share several regions of amino acid similarity (44). Two well-conserved subregions (4.2 and 2.4) have been proposed to be involved in the recognition of the −35 and −10 DNA sequences of the promoter region. However, a new class of σ factors, σ^{54}, shows little amino acid sequence similarity to other σ factors (63, 134). The lack of the conserved 4.2 and 2.4 subregions in the σ^{54} proteins is not surprising because RNA polymerase-σ^{54} holoenzyme allows efficient transcription only from −24/−12 promoters. This new type of bacterial promoter has been identified in the last few years in gram-negative bacteria. The general features of this −24/−12 promoter are the conserved TGGYRYRN4TTGGA consensus sequence, its recognition by a specific σ^{54} factor, and the requirement for an additional transcriptional regulatory protein that hydrolyzes ATP in order to activate the expression of the associated genes. In addition, most of these genes possess a promoter upstream activator sequence (UAS), which is the target site for the activating protein. This kind of promoter has been identified in connection with genes for nitrogen assimilation, pilin formation, motility, and pathogenicity in different gram-negative bacteria. To date, only one example has been described in gram-positive bacteria, controlling expression of the *B. subtilis* levanase operon.

Levanase is an exocellular enzyme that hydrolyzes sucrose and two fructose polymers, levan and inulin (62). The levanase gene (*sacC*) is the distal gene of an operon whose expression is inducible by fructose, the final product of levan hydrolysis, and is subject to catabolite repression (70, 72). The first four genes of the operon, *levD, levE, levF,* and *levG,* encode fructose-specific PTS enzymes (74). The fructose-inducible promoter of the levanase operon was mapped by primer extension analysis and two sequences were identified, TGGCAC and TTGCA at positions −12/−24, which are identical to those found in σ^{54}-dependent promoters (25, 72). The *levR* gene, located just upstream from the levanase operon, encodes a large polypeptide of M_r 106,063, which acts as a positive regulator of the expression of the operon. Two domains, A and B containing 200 and 161 residues, respectively, were identified in LevR. Domain A of LevR shares similarities with the well-conserved central domain of the NifA-NtrC family of bacterial activators (25). This domain is required for the formation of open complexes between σ^{54} RNA polymerase holoenzyme and the promoter. The reaction is ATP-dependent and requires the presence of a specific activator (41, 63). A putative ATP binding site is present in LevR (25). Mutations leading to constitutive expression of the levanase operon were identified in the *levD* and *levE* genes, whose products are involved in the fructose-PTS and which act as negative regulators of the expression of the levanase operon (74). The involvement of a specific component of the PTS in induction has also been shown for the *E. coli bgl* operon (67, 108). Enzyme II^{Bgl}, which is involved in β-glucoside transport, may exert its negative-regulator effect by phosphorylating the positive regulator BglG and thereby abolishing its activity (67, 108). A similar mechanism was proposed for the levanase operon (73, 74). The induction of the operon may be controlled by a PTS-mediated phosphorylation of LevR, because the B domain of LevR shows significant homology with SacT, SacY, and BglG, suggesting that LevR activity may be controlled in the same manner as that of the other three regulators (25). In the presence of fructose, the *levD, E, F,* and *G* gene products along with the general proteins of the PTS (enzyme I and Hpr) are involved in the phosphotransfer reactions leading to fructose phosphorylation and transport. In the absence of the inducer, the phosphate group would be transferred to LevR, probably via the *levD* and *levE* gene products. Inactivation of these genes or of the *ptsI* gene leads to a constitutive phenotype, presumably by preventing the inactivation of LevR through phosphorylation (25, 74).

Recently, a new *B. subtilis* gene, called *sigL,* was identified, which encodes a polypeptide containing 463 residues, the sequence of which is similar to the sequences of σ^{54} factors from gram-negative bacteria (26). The characteristic domain organization of σ^{54} factors is well conserved in SigL. Two different structural motifs may be involved in DNA sequence recognition by σ^{54}. A leucine-zipper motif is involved in contacting the -12 region, and a HTH motif in contacting the -24 region (107). These two motifs, as well as an acidic region presumably involved in the melting step allowing the initiation of transcription, are present in SigL. The expression of the levanase operon was abolished in a strain containing a *sigL* null mutation. This mutation did not affect sporulation, competence, nor motility in *B. subtilis*. However, the disruption of the *sigL* gene leads to the inability of the corresponding strains to grow in minimal medium containing arginine, ornithine, isoleucine, or valine as the sole nitrogen sources (26).

For most of the σ^{54}-dependent promoters, upstream activating sequences (UASs) are located more than 100 bp upstream from the transcription start site. Binding of the specific activator protein to these sites was proposed to allow the correct binding of the RNA polymerase and stimulate the isomerization of closed complexes to open complexes (63, 134). An activating DNA sequence required for activation of the levanase operon promoter by LevR was identified. This DNA sequence is 50 bp long and is centered 125 bp upstream from the transcription start site in a region containing a 16-bp palindromic structure. Interestingly, this region of dyad symmetry continues to act as a regulatory element when placed up to at least 0.6 kb upstream from the promoter (75). However, the efficacy of activation is lowered. The integrity of the palindromic structure is necessary for promoter activation, and this structure can titrate a diffusible factor, which is probably the LevR polypeptide (75). However, the ability of LevR to bind to this palindromic structure remains to be demonstrated.

The interaction between LevR and the RNA polymerase–σ^{54} complex I may be facilitated by a DNA loop. In gram-negative bacteria, the integration host factor (IHF) facilitates productive contacts between NifA-type activators and σ^{54} holoenzyme by stimulating DNA bending (22, 53, 106). Interestingly, a DNA sequence that shares some similarities with the IHF binding site (34) was found between the promoter and the UAS of the levanase operon (75). A similar mechanism could occur in *B. subtilis*.

CONCLUDING REMARKS

A significant majority of regulatory mechanisms in *Bacillus subtilis* involve positive regulation, in contrast to gram-negative bacteria. These regulators appear to interact and cross-talk to a large extent, forming a signal transduction network. Postexponential growth responses such as competence, degradative enzyme production, sporulation, and motility are thus controlled by many common regulators. Regulation by antitermination is also frequently found in *B. subtilis*.

At least one type of positive regulation appears to be specific to gram-positive bacteria: small accessory regulatory polypeptides such as DegQ, DegR, or Sen, which can act as positive transcriptional activators when they are overexpressed, but whose absence does not lead to any detectable phenotype.

ACKNOWLEDGMENTS

We thank M. Arnaud, M. K. Dahl, M. Débarbouillé, D. Dubnau, F. Kunst, I. Martin-Verstraete, I. Smith, and A. L. Sonenshein for helpful discussions and for providing us with unpublished information. We are grateful to Christine Dugast for expert secretarial assistance. The work from our laboratory reported in this review was supported by funds from Institut Pasteur, Centre National de la Recherche Scientifique, Université Paris 7, and Fondation pour la Recherche Médicale.

Literature Cited

1. Adhya, S., Garges, S. 1990. Positive control. *J. Biol. Chem.* 265:10797–800
2. Albano, M., Breitling, R., Dubnau, D. A. 1989. Nucleotide sequence and genetic organization of the *Bacillus subtilis comG* operon. *J. Bacteriol.* 171:5386–5404
3. Albano, M., Dubnau, D. A. 1989. Cloning and characterization of a cluster of linked *Bacillus subtilis* late competence mutations. *J. Bacteriol.* 171: 5376–85
4. Albano, M., Hahn, J., Dubnau, D. 1987. Expression of competence genes in *Bacillus subtilis*. *J. Bacteriol.* 169: 3110–17
5. Albright, L. M., Huala, E., Ausubel, F. M. 1989. Prokaryotic signal transduction mediated by sensor and regulator pairs. *Annu. Rev. Genet.* 23:311–36
6. Amory, A., Kunst, F., Aubert, E., Klier, A., Rapoport, G. 1987. Characterization of the *sacQ* genes from *Bacillus licheniformis* and *Bacillus subtilis*. *J. Bacteriol.* 169:324–33
7. Antoniewski, C., Savelli, B., Stragier, P. 1990. The *spoIIJ* gene, which regulates early development steps in *Bacillus subtilis,* belongs to a class of environmentally responsive genes. *J. Bacteriol.* 172:86–93

7a. Arnaud, M., Vary, P., Zagorec, M., Klier, A., Débarbouillé, M., et al. 1992. Regulation of the *sacPA* operon of *Bacillus subtilis:* identification of PTS components involved in SacT activity. *J. Bacteriol.* 174:3161–70

8. Arnosti, D. N., Chamberlin, M. J. 1989. Secondary σ factor controls transcription of flagellar and chemotaxis genes in *E. coli*. *Proc. Natl. Acad. Sci. USA* 86:830–34
9. Atkinson, M. R., Wray, L. V. Jr., Fisher, S. H. 1990. Regulation of histidine and proline degradation enzymes by amino acid availability in *Bacillus subtilis*. *J. Bacteriol.* 172:4758–65
10. Aymerich, S., Gonzy-Treboul, G., Steinmetz, M. 1986. 5'-Noncoding region *sacR* is the target of all identified regulation affecting the levansucrase gene in *Bacillus subtilis*. *J. Bacteriol.* 166:993–98
11. Aymerich, S., Steinmetz, M. 1987. Cloning and preliminary characterization of the *sacS* locus from *Bacillus subtilis* which controls the regulation of the exoenzyme levansucrase. *Mol. Gen. Genet.* 208:114–20
12. Bohannon, D. E., Rosenkrantz, M. S., Sonenshein, A. L. 1985. Regulation of *Bacillus subtilis* glutamate synthase genes by the nitrogen source. *J. Bacteriol.* 163:957–64
13. Bohannon, D. E., Sonenshein, A. L. 1989. Positive regulation of glutamate biosynthesis in *Bacillus subtilis*. *J. Bacteriol.* 171:4718–27
14. Bohannon, D. E., Sonenshein, A. L. 1990. GltC, the positive regulator of glutamate synthase gene expression. See Ref. 159, pp. 141–45
15. Burbulys, D., Trach, K. A., Hoch, J. A. 1991. Initiation of sporulation in B. subtilis is controlled by a multicomponent phosphorelay. *Cell* 64:545–52
16. Crutz, A.-M., Steinmetz, M. 1991. *Transcriptional control of the* sacX-sacY *operon involved in induction by sucrose of* Bacillus subtilis *levansucrase*. Presented at the Int. Conf. on Bacilli, 6th, Stanford, Calif.
17. Crutz, A.-M., Steinmetz, M., Aymerich, S., Richter, R., Le Coq, D. 1990. Induction of levansucrase in *Bacillus subtilis:* an antitermination mechanism negatively controlled by the phosphotransferase system. *J. Bacteriol.* 172: 1043–50
18. Cutting, S., Mandelstam, J. 1986. The nucleotide sequence and the transcription during sporulation of the *gerE* gene of *Bacillus subtilis*. *J. Gen. Microbiol.* 132:3013–24
19. Cutting, S., Panzer, S., Losick, R. 1989. Regulatory studies on the promoter for a gene governing synthesis and assembly of the spore coat in *Bacillus subtilis*. *J. Mol. Biol.* 207:393–404
20. Dahl, M. K., Msadek, T., Kunst, F., Rapoport, G. 1991. Mutational analysis of the *Bacillus subtilis* DegU regulator and its phosphorylation by the DegS protein kinase. *J. Bacteriol.* 173:2539–47
21. Deleted in proof
22. de Lorenzo, V., Herrero, M., Metzke, M., Timmis, K. N. 1991. An upstream XylR- and IHF-induced nucleoprotein complex regulates the σ^{54}-dependent Pu promoter of TOL plasmid. *EMBO J.* 10:1159–67
23. Débarbouillé, M., Arnaud, M., Fouet, A., Klier, A., Rapoport, G. 1990. The *sacT* gene regulating the *sacPA* operon in *Bacillus subtilis* shares strong homology with transcriptional antiterminators. *J. Bacteriol.* 172:3966–73
24. Débarbouillé, M., Kunst, F., Klier, A., Rapoport, G. 1987. Cloning of the *sacS* gene encoding a positive regulator of the

sucrose regulon in *Bacillus subtilis*. *FEMS Microbiol. Lett.* 41:137–40
25. Débarbouillé, M., Martin-Verstraete, I., Klier, A., Rapoport, G. 1991. The levanase regulator LevR of *Bacillus subtilis* has domains homologous to both σ^{54}- and PTS-dependent regulators. *Proc. Natl. Acad. Sci. USA* 88:2212–16
26. Débarbouillé, M., Martin-Verstraete, I., Kunst, F., Rapoport, G. 1991. The *Bacillus subtilis sigL* gene encodes an equivalent of σ^{54} from Gram-negative bacteria. *Proc. Natl. Acad. Sci. USA* 88:9092–96
27. Dubnau, D. 1991. Genetic competence in *Bacillus subtilis*. *Microbiol. Rev.* 55:395–424
28. Dubnau, D., Roggiani, M. 1990. Growth medium-independent genetic competence mutants of *Bacillus subtilis*. *J. Bacteriol.* 172:4048–55
29. Dubnau, E., Weir, J., Nair, G., Carter, L. III, Moran, C. Jr., et al. 1988. *Bacillus* sporulation gene *spo0H* codes for σ^{30} (σ^{H}). *J. Bacteriol.* 170:1054–62
30. Dubnau, E. J., Cabane, K., Smith, I. 1987. Regulation of *spo0H*, an early sporulation gene in *Bacilli*. *J. Bacteriol.* 169:1182–91
31. Ferrari, F. A., Trach, K., Le Coq, D., Spence, J., Ferrari, E., et al. 1985. Characterization of the *spo0A* locus and its deduced product. *Proc. Natl. Acad. Sci. USA* 82:2647–51
32. Fort, P., Piggot, P. J. 1984. Nucleotide sequence of sporulation locus *spoIIA* in *Bacillus subtilis*. *J. Gen. Microbiol.* 130:2147–53
33. Fouet, A., Arnaud, M., Klier, A., Rapoport, G. 1987. *Bacillus subtilis* sucrose-specific enzyme II of the phosphotransferase system: expression in *Escherichia coli* and homology to enzymes II from enteric bacteria. *Proc. Natl. Acad. Sci. USA* 84:8773–77
34. Friedman, D. I. 1988. Integration host factor: a protein for all reasons. *Cell* 55:545–54
35. Fürbass, R., Gocht, M., Zuber, P., Marahiel, M. A. 1991. Interaction of AbrB, a transcriptional regulator from *Bacillus subtilis* with the promoters of the transition state-activated genes *tycA* and *spoVG*. *Mol. Gen. Genet.* 225:347–54
36. Ganesan, A. T., Hoch, J. A., eds. 1988. *Genetics and Biotechnology of Bacilli*, Vol. 2. San Diego: Academic. 430 pp.
37. Gaur, N. K., Cabane, K., Smith, I. 1988. Structure and expression of the *Bacillus subtilis sin* operon. *J. Bacteriol.* 170:1046–53
38. Gaur, N. K., Dubnau, E., Smith, I. 1986. Characterization of a cloned *Bacillus subtilis* gene that inhibits sporulation in multiple copies. *J. Bacteriol.* 168:860–69
39. Gaur, N. K., Oppenheim, J., Smith, I. 1991. The *Bacillus subtilis sin* gene, a regulator of alternate developmental processes, codes for a DNA-binding protein. *J. Bacteriol.* 173:678–86
40. Gralla, J. D. 1989. Bacterial gene regulation from distant DNA sites. *Cell* 57:193–95
41. Gussin, G. N., Ronson, C. W., Ausubel, F. M. 1986. Regulation of nitrogen fixation genes. *Annu. Rev. Genet.* 20:567–91
42. Hahn, J., Dubnau, D. 1991. Growth stage signal transduction and the requirements for *srfA* induction in development of competence. *J. Bacteriol.* 173:7275–82
43. Helmann, J. D., Ballard, B. T., Walsh, C. T. 1990. A mercury-regulated transcriptional activator from *Bacillus* SP RC607. See Ref. 159, pp. 23–32
44. Helmann, J. D., Chamberlin, M. J. 1988. Structure and function of bacterial sigma factors. *Annu. Rev. Biochem.* 57:839–72
45. Helmann, J. D., Marquez, L. M., Chamberlin, M. J. 1988. Cloning, sequencing and disruption of the *B. subtilis* σ^{28} gene. *J. Bacteriol.* 170:1568–74
46. Helmann, J. D., Marquez, L. M., Singer, V. L., Chamberlin, M. J. 1988. Cloning and characterization of the *Bacillus subtilis* sigma-28 gene. See Ref. 36, pp. 189–93
47. Helmann, J. D., Wang, Y., Mähler, I., Walsh, C. T. 1989. Homologous metalloregulatory proteins from both Gram-positive and Gram-negative bacteria control transcription of mercury resistance operons. *J. Bacteriol.* 171:222–29
48. Henner, D. J., Ferrari, E., Perego, M., Hoch, J. A. 1988. Location of the targets of the *hpr-97*, *sacU32*(Hy), and *sacQ36*(Hy) mutations in upstream regions of the subtilisin promoter. *J. Bacteriol.* 170:296–300
49. Henner, D. J., Ferrari, E., Perego, M., Hoch, J. A. 1988. Upstream activating sequences in *Bacillus subtilis*. See Ref. 36, pp. 3–9
50. Henner, D. J., Yang, M., Band, L., Shimotsu, H., Ruppen, M., et al. 1987. Genes of *Bacillus subtilis* that regulate the expression of degradative enzymes. In *Genetics of Industrial Microorganisms. Proc. Fifth Int. Symp. on the Genetics of Industrial Microorganisms*, ed. M. Alacevic, D. Hranueli, Z. Toman, pp. 81–90. Zagreb: Pliva

51. Henner, D. J., Yang, M., Ferrari, E. 1988. Localization of *Bacillus subtilis sacU*(Hy) mutations to two linked genes with similarities to the conserved procaryotic family of two-component signaling systems. *J. Bacteriol.* 170: 5102–9
52. Hoch, J. A. 1976. Genetics of bacterial sporulation. *Adv. Genet.* 18:69–99
53. Hoover, T. R., Santero, E., Porter, S., Kustu, S. 1990. The integration host factor stimulates interaction of RNA polymerase with NIFA, the transcriptional activator for nitrogen fixation operons. *Cell* 63:11–22
54. Houman, F., Diaz-Torres, M. R., Wright, A. 1990. Transcriptional antitermination in the bgl operon of E. coli is modulated by a specific RNA binding protein. *Cell* 62:1153–63
55. Hulett, F. M., Bookstein, C., Edwards, C., Jensen, K., Kapp, N., et al. 1990. Structural similarities and regulation of *Bacillus subtilis* alkaline phosphatases. See Ref 159, pp. 163–69
56. Jin, S., Sonenshein, A. L. 1991. Bacillus subtilis *genes that code for enzymes involved in acetyl-CoA metabolism.* Presented at the Int. Conf. on Bacilli, 6th, Stanford, Calif.
57. Klier, A., Fouet, A., Débarbouillé, M., Kunst, F., Rapoport, G. 1987. Distinct control sites located upstream from the levansucrase gene of *Bacillus subtilis. Mol. Microbiol.* 1:233–41
58. Klier, A. F., Rapoport, G. 1988. Genetics and regulation of carbohydrate catabolism in *Bacillus. Annu. Rev. Microbiol.* 42:65–95
59. Kofoid, E. C., Parkinson, J. S. 1988. Transmitter and receiver modules in bacterial signaling proteins. *Proc. Natl. Acad. Sci. USA* 85:4981–85
59a. Kroos, L. 1991. Gene regulation in the mother-cell compartment of sporulating *Bacillus subtilis. Semin. Dev. Biol.* 2: 63–71
60. Kunst, F., Amory, A., Débarbouillé, M., Martin, I., Klier, A., et al. 1988. Polypeptides activating the synthesis of secreted enzymes. See Ref. 36, pp. 27–31
61. Kunst, F., Débarbouillé, M., Msadek, T., Young, M., Mauël, C., et al. 1988. Deduced polypeptides encoded by the *Bacillus subtilis sacU* locus share homology with two-component sensor-regulator systems. *J. Bacteriol.* 170: 5093–5101
62. Kunst, F., Lepesant, J.-A., Dedonder, R. 1977. Presence of a third sucrose hydrolyzing enzyme in *Bacillus subtilis:* constitutive levanase synthesis by mutants of *Bacillus subtilis* 168. *Biochimie* 59:287–92
63. Kustu, S., Santero, E., Keener, J., Popham, D., Weiss, D. 1989. Expression of σ^{54}(*ntrA*)-dependent genes is probably united by a common mechanism. *Microbiol. Rev.* 53:367–76
64. Lee, N., Gielow, W., Wallace, R. 1981. Mechanism of *araC* autoregulation and the domains of two overlapping promoters, P_C and P_{BAD}, in the L-arabinose regulatory region of *E. coli. Proc. Natl. Acad. Sci. USA* 78:752–56
65. Lepesant, J.-A., Kunst, F., Lepesant-Kejzlarova, J., Dedonder, R. 1972. Chromosomal location of mutations affecting sucrose metabolism in *Bacillus subtilis* Marburg. *Mol. Gen. Genet.* 118:135–60
66. Losick, R., Pero, J. 1981. Cascades of sigma factors. *Cell* 25:582–84
67. Mahadevan, S., Wright, A. 1987. A bacterial gene involved in transcription antitermination: regulation at a rho-independent terminator in the bgl operon of E. coli. *Cell* 50:485–94
68. Mähler, I., Levinson, H. S., Wang, Y., Halvorson, H. O. 1986. Cadmium and mercury-resistant *Bacillus* strains from a salt marsh and from Boston Harbor. *Appl. Environ. Microbiol.* 52:1293–98
69. Marahiel, M. A., Zuber, P., Czekay, G., Losick, R. 1987. Identification of the promoter for a peptide antibiotic biosynthesis gene from *Bacillus brevis* and its regulation in *Bacillus subtilis. J. Bacteriol.* 169:2215–22
70. Martin, I., Débarbouillé, M., Ferrari, E., Klier, A., Rapoport, G. 1987. Characterization of the levanase gene of *Bacillus subtilis* which shows homology to yeast invertase. *Mol. Gen. Genet.* 208:177–84
71. Martin, I., Débarbouillé, M., Klier, A., Rapoport, G. 1987. Identification of a new locus, *sacV,* involved in the regulation of levansucrase synthesis in *Bacillus subtilis. FEMS Microbiol. Lett.* 44:39–43
72. Martin, I., Débarbouillé, M., Klier, A., Rapoport, G. 1989. Induction and metabolite regulation of levanase synthesis in *Bacillus subtilis. J. Bacteriol.* 171:1885–92
73. Martin, I., Débarbouillé, M., Klier, A., Rapoport, G. 1990. The levanase operon of *Bacillus subtilis* includes regulatory genes involved in a fructose-specific PTS. See Ref. 159, pp. 69–79
74. Martin-Verstraete, I., Débarbouillé, M., Klier, A., Rapoport, G. 1990. Levanase operon of *Bacillus subtilis* includes a fructose-specific phosphotransferase

system regulating the expression of the operon. *J. Mol. Biol.* 214:657–71

75. Martin-Verstraete, I., Débarbouillé, M., Klier, A., Rapoport, G. 1992. Mutagenesis of the *Bacillus subtilis* "−12, −24" promoter of the levanase operon and evidence for the existence of an upstream activating sequence. *J. Mol. Biol.* 226: In press
76. McCready, P., Doi, R. H. 1990. Possible regulation of *senS* by a *nusA*-related mechanism. See Ref. 159, pp. 393–95
77. Miki, T., Minami, Z., Ikeda, Y. 1965. The genetics of alkaline phosphatase formation in *Bacillus subtilis*. *Genetics* 52:1093–1100
78. Mohan, S., Dubnau, D. 1990. Transcriptional regulation of *comC:* evidence for a competence-specific transcription factor in *Bacillus subtilis*. *J. Bacteriol.* 172:4064–71
79. Moran, C. P. Jr. 1989. Sigma factors and the regulation of transcription. See Ref. 116, pp. 167–84
80. Msadek, T., Kunst, F., Henner, D., Klier, A., Rapoport, G., et al. 1990. Signal transduction pathway controlling synthesis of a class of degradative enzymes in *Bacillus subtilis:* expression of the regulatory genes and analysis of mutations in *degS* and *degU*. *J. Bacteriol.* 172:824–34
81. Msadek, T., Kunst, F., Klier, A., Rapoport, G. 1991. DegS-DegU and ComP-ComA modulator-effector pairs control expression of the *Bacillus subtilis* pleiotropic regulatory gene *degQ*. *J. Bacteriol.* 173:2366–77
82. Msadek, T., Kunst, F., Klier, A., Rapoport, G., Dedonder, R. 1990. The Deg signal transduction pathway: mutations and regulation of expression of *degS, degU* and *degQ*. See Ref. 159, pp. 245–55
83. Mueller, J. P., Sonenshein, A. L. 1991. *Control of the* Bacillus subtilis gsiA *operon by the* comP-comA *signal transduction system and the role of GsiA in the developmental pathway*. Presented at the Int. Conf. on Bacilli, 6th, Stanford, Calif.
84. Mukai, K., Kawata, M., Tanaka, T. 1990. Isolation and phosphorylation of the *Bacillus subtilis degS* and *degU* gene products. *J. Biol. Chem.* 265:20000–6
85. Nagami, Y., Tanaka, T. 1986. Molecular cloning and nucleotide sequence of a DNA fragment from *Bacillus natto* that enhances production of extracellular proteases and levansucrase in *Bacillus subtilis*. *J. Bacteriol.* 166:20–28
86. Nakano, M. M., Magnuson, R., Myers, A., Curry, J., Grossman, A. D., Zuber, P. 1991. *srfA* is an operon required for surfactin production, competence development, and efficient sporulation in *Bacillus subtilis*. *J. Bacteriol.* 173: 1770–78
87. Nakano, M. M., Xia, L. A., Zuber, P. 1991. Transcription initiation region of the *srfA* operon, which is controlled by the *comP-comA* signal transduction system in *Bacillus subtilis*. *J. Bacteriol.* 173:5487–93
88. Nakano, M. M., Zuber, P. 1991. The primary role of ComA in establishment of the competent state in *Bacillus subtilis* is to activate expression of *srfA*. *J. Bacteriol.* 173:7269–74
89. Oda, M., Sugishita, A., Furukawa, K. 1988. Cloning and nucleotide sequences of histidinase and regulatory genes in the *Bacillus subtilis hut* operon and positive regulation of the operon. *J. Bacteriol.* 170:3199–3205
90. Okada, J., Shimogaki, H., Murata, K., Kimura, A. 1984. Cloning of the gene responsible for the extracellular proteolytic activities of *Bacillus licheniformis*. *Appl. Microbiol. Biotechnol.* 20: 406–12
91. Pang, A. S. H., Nathoo, S., Wong, S.-L. 1991. Cloning and characterization of a pair of novel genes that regulate production of extracellular enzymes in *Bacillus subtilis*. *J. Bacteriol.* 173:46–54
92. Perego, M., Cole, S. P., Burbulys, D., Trach, K., Hoch, J. A. 1989. Characterization of the gene for a protein kinase which phosphorylates the sporulation-regulatory proteins Spo0A and Spo0F of *Bacillus subtilis*. *J. Bacteriol.* 171:6187–96
93. Perego, M., Hoch, J. A. 1988. Sequence analysis and regulation of the *hpr* locus, a regulatory gene for protease production and sporulation in *Bacillus subtilis*. *J. Bacteriol.* 170:2560–67
94. Perego, M., Hoch, J. A. 1991. Negative regulation of *Bacillus subtilis* sporulation by the *spo0E* gene product. *J. Bacteriol.* 173:2514–20
95. Perego, M., Spiegelman, G. B., Hoch, J. A. 1988. Structure of the gene for the transition state regulator, *abrB:* regulator synthesis is controlled by the *spo0A* sporulation gene in *Bacillus subtilis*. *Mol. Microbiol.* 2:689–99
96. Priest, F. G. 1977. Extracellular enzyme synthesis in the genus *Bacillus*. *Bacteriol. Rev.* 41:711–53
97. Ptashne, M. 1986. Gene regulation by proteins acting nearby and at a distance. *Nature (London)* 322:697–701

98. Ptashne, M., Jeffrey, A., Johnson, A. D., Maurer, R., Meyer, B. J., et al. 1980. How the λ repressor and Cro work. *Cell* 19:1–11
99. Raibaud, O. 1989. Nucleoprotein structures at positively regulated bacterial promoters: homology with replication origins and some hypotheses on the quaternary structure of the activator proteins in these complexes. *Mol. Microbiol.* 3:455–58
100. Renna, M., Najimudin, N., Zahler, S. A. 1991. *Regulation of the expression of* alsS, *the structural gene for acetolactate synthase in* B. subtilis. Presented at the Int. Conf. on Bacilli, 6th, Stanford, Calif.
101. Roberts, J. W. 1988. Phage lambda and the regulation of transcription termination. *Cell* 52:5–6
102. Robertson, J. B., Gocht, M., Marahiel, M. A., Zuber, P. 1989. AbrB, a regulator of gene expression in *Bacillus,* interacts with the transcription initiation regions of a sporulation gene and an antibiotic biosynthesis gene. *Proc. Natl. Acad. Sci. USA* 86:8457–61
103. Roggiani, M., Hahn, J., Dubnau, D. 1990. Suppression of early competence mutations in *Bacillus subtilis* by *mec* mutations. *J. Bacteriol.* 172:4056–63
104. Ronson, C. W., Nixon, B. T., Ausubel, F. M. 1987. Conserved domains in bacterial regulatory proteins that respond to environmental stimuli. *Cell* 49:579–81
105. Sanders, D. A., Gillece-Castro, B. L., Stock, A. M., Burlingame, A. L., Koshland, D. E. Jr. 1989. Identification of the site of phosphorylation of the chemotaxis response regulator protein, CheY. *J. Biol. Chem.* 264:21770–78
106. Santero, E., Hoover, T., Keener, J., Kustu, S. 1989. In vitro activity of the nitrogen fixation regulatory protein NIFA. *Proc. Natl. Acad. Sci. USA* 86: 7346–50
107. Sasse-Dwight, S., Gralla, J. D. 1990. Role of eucaryotic-type functional domains found in the procaryotic enhancer receptor factor σ^{54}. *Cell* 62:945–54
108. Schnetz, K., Rak, B. 1988. Regulation of the *bgl* operon of *Escherichia coli* by transcriptional antitermination. *EMBO J.* 7:3271–77
109. Seki, T., Yoshikawa, H., Takahashi, H., Saito, H. 1987. Cloning and nucleotide sequence of *phoP,* the regulatory gene for alkaline phosphatase and phosphodiesterase in *Bacillus subtilis*. *J. Bacteriol.* 169:2913–16
110. Seki, T., Yoshikawa, H., Takahashi, H., Saito, H. 1988. Nucleotide sequence of the *Bacillus subtilis phoR* gene. *J. Bacteriol.* 170:5935–38
111. Sekiguchi, J., Ezaki, B., Kodama, K., Akamatsu, T. 1988. Molecular cloning of a gene affecting the autolysin level and flagellation in *Bacillus subtilis*. *J. Gen. Microbiol.* 134:1611–21
112. Sekiguchi, J., Ohsu, H., Kuroda, A., Moriyama, H., Akamatsu, T. 1990. Nucleotide sequences of the *Bacillus subtilis flaD* and a *Bacillus licheniformis* homologue affecting the autolysin level and flagellation. *J. Gen. Microbiol.* 136:1223–30
113. Shimotsu, H., Henner, D. J. 1986. Modulation of *Bacillus subtilis* levansucrase gene expression by sucrose and regulation of the steady-state mRNA level by *sacU* and *sacQ* genes. *J. Bacteriol.* 168:380–88
114. Smith, I. 1989. Initiation of sporulation. See Ref. 116, pp. 185–210
115. Smith, I., Mandic-Mulec, I., Gaur, N. 1991. The role of negative control in sporulation. *Res. Microbiol.* 142:831–39
116. Smith, I., Slepecky, R. A., Setlow, P., eds. 1989. *Regulation of Procaryotic Development. Structural and Functional Analysis of Bacterial Sporulation and Germination*. Washington, DC: Am. Soc. Microbiol. 304 pp.
117. Sonenshein, A. L. 1989. Metabolic regulation of sporulation and other stationary-phase phenomena. See Ref. 116, pp 109–30
118. Spiegelman, G., Van Hoy, B., Perego, M., Day, J., Trach, K., et al. 1990. Structural alterations in the *Bacillus subtilis* Spo0A regulatory protein which suppress mutations at several *spo0* loci. *J. Bacteriol.* 172:5011–19
119. Steinmetz, M., Aymerich, S. 1990. The *Bacillus subtilis sac-deg* constellation: how and why? See Ref 159, pp. 303–11
120. Steinmetz, M., Le Coq, D., Aymerich, S., Gonzy-Treboul, G., Gay, P. 1985. The DNA sequence of the gene for the secreted *Bacillus subtilis* enzyme levansucrase and its genetic control sites. *Mol. Gen. Genet.* 200:220–28
121. Steinmetz, M., Le Coq, D., Aymerich, S. 1989. Induction of saccharolytic enzymes by sucrose in *Bacillus subtilis:* evidence for two partially interchangeable regulatory pathways. *J. Bacteriol.* 171:1519–23
122. Stock, A. M., Mottonen, J. M., Stock, J. B., Schutt, C. E. 1989. Three-dimensional structure of CheY, the response regulator of bacterial chemotaxis. *Nature (London)* 337:745–49
123. Stock, J. B., Ninfa, A. J., Stock, A. M.

1989. Protein phosphorylation and regulation of adaptive responses in bacteria. *Microbiol. Rev.* 53:450–90

124. Stragier, P. 1989. Temporal and spatial control of gene expression during sporulation: from facts to speculation. See Ref. 116, pp. 243–54
125. Stragier, P. 1991. Dances with sigmas. *EMBO J.* 10:3559–66
126. Stragier, P., Losick, R. 1990. Cascades of sigma factors revisited. *Mol. Microbiol.* 4:1801–6
127. Strauch, M., Webb, V., Spiegelman, G., Hoch, J. A. 1990. The Spo0A protein of *Bacillus subtilis* is a repressor of the *abrB* gene. *Proc. Natl. Acad. Sci. USA* 87:1801–5
128. Strauch, M. A., Perego, M., Burbulys, D., Hoch, J. A. 1989. The transition state transcription regulator AbrB of *Bacillus subtilis* is autoregulated during vegetative growth. *Mol. Microbiol.* 3:1203–9
129. Strauch, M. A., Spiegelman, G. B., Perego, M., Johnson, W. C., Burbulys, D., et al. 1989. The transition state transcription regulator *abrB* of *Bacillus subtilis* is a DNA binding protein. *EMBO J.* 8:1615–21
130. Takagi, M., Takada, H., Imanaka, T. 1990. Nucleotide sequence and cloning in *Bacillus subtilis* of the *Bacillus stearothermophilus* pleiotropic regulatory gene *degT*. *J. Bacteriol.* 172:411–18
131. Tanaka, T., Kawata, M. 1988. Cloning and characterization of *Bacillus subtilis iep,* which has positive and negative effects on production of extracellular proteases. *J. Bacteriol.* 170:3593–3600
132. Tanaka, T., Kawata, M., Mukai, K. 1991. Altered phosphorylation of *Bacillus subtilis* DegU caused by single amino acid changes in DegS. *J. Bacteriol.* 173:5507–15
133. Tanaka, T., Kawata, M., Nagami, Y., Uchiyama, H. 1987. *prtR* enhances the mRNA level of the *Bacillus subtilis* extracellular proteases. *J. Bacteriol.* 169: 3044–50
134. Thony, B., Hennecke, H. 1989. The −24/−12 promoter comes of age. *FEMS Microbiol. Rev.* 63:341–58
135. Tomioka, N., Honjo, M., Funahashi, K., Manabe, K., Akaoka, A., et al. 1985. Cloning, sequencing, and some properties of a novel *Bacillus amyloliquefaciens* gene involved in the increase of extracellular protease activities. *J. Biotechnol.* 3:85–96
136. Trach, K., Burbulys, D., Spiegelman, G., Perego, M., Van Hoy, B., et al. 1990. Phosphorylation of the Spo0A protein: a cumulative environsensory activation mechanism. See Ref. 159, pp. 357–65
137. Trach, K., Burbulys, D., Strauch, M., Wu, J.-J., Dhillon, N., et al. 1991. Control of the initiation of sporulation in *Bacillus subtilis* by a phosphorelay. *Res. Microbiol.* 142:815–23
138. Trach, K., Hoch, J. A. 1989. The *Bacillus subtilis spo0B* stage 0 sporulation operon encodes an essential GTP-binding protein. *J. Bacteriol.* 171:1362–71
139. Trach, K. A., Chapman, J. W., Piggot, P. J., Hoch, J. A. 1985. Deduced product of the stage 0 sporulation gene *spo0F* shares homology with Spo0A, OmpR, and SfrA proteins. *Proc. Natl. Acad. Sci. USA* 82:7260–64
140. Trowsdale, J., Chen, S. M. H., Hoch, J. A. 1979. Genetic analysis of a class of polymyxin-resistant partial revertants of stage 0 sporulation mutants of *Bacillus subtilis:* map of the chromosome region near the origin of replication. *Mol. Gen. Genet.* 173:61–70
141. van Sinderen, D., Withoff, S., Boels, H., Venema, G. 1990. Isolation and characterization of *comL,* a transcription unit involved in competence development of *Bacillus subtilis*. *Mol. Gen. Genet.* 224:396–404
142. Wang, L.-F., Doi, R. H. 1990. Complex character of *senS,* a novel gene regulating expression of extracellular-protein genes of *Bacillus subtilis*. *J. Bacteriol.* 172:1939–47
143. Wang, Y., Mähler, I., Levinson, H. S., Halvorson, H. O. 1987. Cloning and expression in *Escherichia coli* of chromosomal mercury resistance genes from a *Bacillus* sp. *J. Bacteriol.* 169:4848–51
144. Wang, Y., Moore, M., Levinson, H. S., Silver, S., Walsh, C., et al. 1989. Nucleotide sequence of a chromosomal mercury resistance determinant from a *Bacillus* sp. with broad-spectrum mercury resistance. *J. Bacteriol.* 171:83–92
145. Watanabe, K., Sato, N., Asano, K., Hatanaka, Y., Okada, J., et al. 1987. Nucleotide sequence of the gene increasing the extracellular proteolytic activities of *Bacillus licheniformis;* comparison with similar phenotypic genes from other *Bacillus* sp. *Agric. Biol. Chem.* 51:2807–9
146. Weinrauch, Y., Guillen, N., Dubnau, D. A. 1989. Sequence and transcription mapping of *Bacillus subtilis* competence genes *comB* and *comA,* one of which is related to a family of bacterial regulatory determinants. *J. Bacteriol.* 171:5362–75
147. Weinrauch, Y., Msadek, T., Kunst, F.,

Dubnau, D. 1991. Sequence and properties of *comQ,* a new competence regulatory gene of *Bacillus subtilis. J. Bacteriol.* 173:5685–93

148. Weinrauch, Y., Penchev, R., Dubnau, E., Smith, I., Dubnau, D. 1990. A *Bacillus subtilis* regulatory gene product for genetic competence and sporulation resembles sensor protein members of the bacterial two-component signal-transduction systems. *Genes Dev.* 4:860–72
149. Wong, S.-L., Wang, L.-F., Doi, R. H. 1988. Cloning and nucleotide sequence of *senN,* a novel *"Bacillus natto"* (*B. subtilis*) gene that regulates expression of extracellular protein genes. *J. Gen. Microbiol.* 134:3269–76
150. Wray, L. V., Rice, P., Atkinson, M., Fisher, S. 1991. *Regulation of the histidine utilization operon in* Bacillus subtilis. Presented at the Int. Conf. on Bacilli, 6th, Stanford, Calif.
151. Wu, J.-J., Piggot, P. J., Tatti, K. M., Moran, C. P. Jr. 1991. Transcription of the *Bacillus subtilis spoIIA* locus. *Gene* 101:113–16
152. Yamane, K., Maruo, B. 1987. Alkaline phosphatase possessing alkaline phosphodiesterase activity and other phosphodiesterases in *Bacillus subtilis. J. Bacteriol.* 134:108–14
153. Yang, M., Ferrari, E., Chen, E., Henner, D. J. 1986. Identification of the pleiotropic *sacQ* gene of *Bacillus subtilis. J. Bacteriol.* 166:113–19
154. Yang, M., Shimotsu, H., Ferrari, E., Henner, D. J. 1987. Characterization and mapping of the *Bacillus subtilis prtR* gene. *J. Bacteriol.* 169:434–37
155. Yoshikawa, H., Kazami, J., Yamashita, S., Chibazakura, T., Sone, H., et al. 1986. Revised assignment for the *Bacillus subtilis spo0F* gene and its homology with *spo0A* and with two *Escherichia coli* genes. *Nucleic Acids Res.* 14:1063–72
156. Zuber, P., Losick, R. 1987. Role of AbrB in Spo0A- and Spo0B-dependent utilization of a sporulation promoter in *Bacillus subtilis. J. Bacteriol.* 169: 2223–30
157. Zukowski, M., Miller, L., Cogswell, P., Chen, K. 1988. Inducible expression system based on sucrose metabolism genes of *Bacillus subtilis.* See Ref. 36, pp. 17–22
158. Zukowski, M., Miller, L., Cogswell, P., Chen, K., Aymerich, S., et al. 1990. Nucleotide sequence of the *sacS* locus of *Bacillus subtilis* reveals the presence of two regulatory genes. *Gene* 90:153–55
159. Zukowski, M. M., Ganesan, A. T., Hoch, J. A., eds. 1990. *Genetics and Biotechnology of Bacilli,* Vol. 3. San Diego: Academic. 420 pp.
160. Dahl, M. K., Msadek, T., Kunst, F., Rapoport, G. 1992. The phosphorylation state of the DegU response regulator acts as a molecular switch allowing either degradative enzyme synthesis or expression of genetic competence in *Bacillus subtilis. J. Biol. Chem.* 267: In press

Annu. Rev. Microbiol. 1992. 46:461–95

PENICILLIN AND CEPHALOSPORIN BIOSYNTHETIC GENES: Structure, Organization, Regulation, and Evolution

Y. Aharonowitz and G. Cohen

The George S. Wise Faculty of Life Sciences, Department of Molecular Microbiology and Biotechnology, Tel Aviv University, Ramat Aviv, 69978, Israel

J. F. Martin

Faculty of Biology, Section of Microbiology, Department of Ecology, Genetics, and Microbiology, University of Leon, 24071 Leon, Spain

KEY WORDS: β-lactam antibiotics, isopenicillin N synthase (IPNS), δ-(L-α aminoadipyl)-L-cysteinyl-D-valine (ACV) synthetase, cephamycins, β-lactam antibiotics, horizontal gene transfer

---

CONTENTS

0066-4227/92/1001-0461$02.00

Abstract

Penicillins and cephalosporins are produced by a wide variety of microorganisms, including some filamentous fungi, many gram-positive streptomycetes, and a few gram-negative unicellular bacteria. All produce these β-lactam antibiotics by essentially the same biosynthetic pathway. Recently, most of the penicillin and cephalosporin biosynthetic genes have been cloned, sequenced, and expressed. The biosynthetic genes code for enzymes that possess multifunctional peptide synthetase, cyclase, epimerase, expandase, hydroxylase, lysine aminotransferase, and acetyltransferase activities and are organized in chromosomal gene clusters and coordinately expressed. DNA hybridization screens of streptomycetes demonstrate that β-lactam biosynthetic genes may be more widespread in nature than is indicated by conventional antibiotic screens. They offer the possibility of expanding the search for organisms with potential to make new β-lactam antibiotics. Attempts to improve current yields of β-lactams in production strains by introducing into them additional copies of biosynthetic genes have been partially successful. Comparative sequence analysis of bacterial and fungal β-lactam biosynthetic genes show they share very high sequence identity. A model that explains the similarity of biosynthetic genes from an evolutionary standpoint assumes horizontal gene-transfer between the two groups of organisms. Indirect evidence suggests the transfer occurred from the bacteria to the fungi.

INTRODUCTION

Molecular genetics of antibiotic production is currently one of the most challenging and rapidly expanding areas of research in bioactive microbial products. This is particularly true for the penicillin and cephalosporin class of β-lactam antibiotics with which this review is concerned. To a large extent the chemical nature of these clinically important compounds, their mode of action, and the biochemical pathways by which they are assembled are now well understood. Several articles (4, 45, 107) document these studies. Within the past few years, intensive efforts have been made to identify and characterize the genetic systems determining penicillin and cephalosporin biosynthesis. The impetus for these investigations has been mainly twofold. First, a comprehensive knowledge of the genetic basis of β-lactam production may open the way for the rational design of new and superior penicillins and cephalosporins and permit improved production. Second, we may acquire new insights into some fundamental biological questions concerning the role and origin of these antibiotics. Unquestionably, the single key factor responsible for these developments has been the application of powerful molecular genetic tools for the cloning of antibiotic-specific genes and for genetic manipulation of antibiotic-producing organisms. In this review, we record

Table 1 Microorganisms that produce β-lactam antibiotics[a]

Class of β-lactam	Structure	Fungi	Bacteria	
			Gram-positive	Gram-negative
Penam	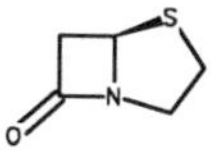	*Aspergillus* *Penicillium* *Epidermophyton* *Trichphyton* *Polypaecilum* *Malbranchea* *Sartorya* *Pleurophomopsis*		
Cephem	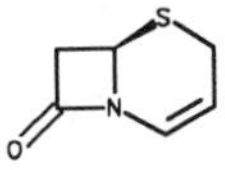	*Cephalosporium* *Anixiopsis* *Arachnomyces* *Spiroidium* *Scopulariopsis* *Diheterospora* *Paecilomyces*	*Streptomyces* *Nocardia*	*Flavobacterium* *Xanthomonas* *Lysobacter*
Clavam	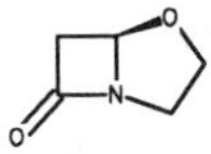		*Streptomyces*	
Carbapenem	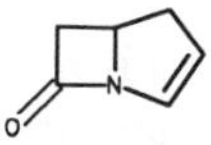		*Streptomyces*	*Seratia* *Erwinia*
Monobactam	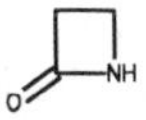		*Nocardia*	*Pseudomonas* *Gluconobacter* *Chromobacter* *Agrobacter* *Acetobacter*

[a] Only certain species within the microorganisms listed make the indicated β-lactam antibiotic (35).

progress in the molecular genetics of penicillin and cephalosporin biosynthesis. In particular, we focus on the structure, organization, regulation, and evolution of the biosynthetic genes. Other reviews emphasizing different aspects of this field have recently appeared (24, 44, 74, 91).

BIOSYNTHESIS OF β-LACTAM ANTIBIOTICS

Penicillins and cephalosporins are members of the β-lactam class of antibiotics. Originally discovered in some filamentous fungi, they were subsequently found to occur in a much larger group of microorganisms (Table 1). Chief among these are mycelial forming gram-positive bacteria belonging to the genus *Actinomyces* and some unicellular gram-negative bacteria. Fungi make either penicillins or cephalosporins, whereas the bacteria synthesize a multitude of cephalosporins and cephamycins but do not produce penicillins as end products (35). All these antibiotics contain a β-lactam ring fused to a second sulfur-containing ring (Figure 1). The presence or absence of the sulfur atom is what distinguishes the penicillins and cephalosporins from several other classes of β-lactam antibiotics. In the following, we confine our discussion to the sulfur-containing β-lactams.

Pioneering studies by Abraham and colleagues (36) on the biosynthesis of β-lactams in cell free systems derived from fungi led to the crucial observation that a cysteine-containing tripeptide, δ-(L-α-aminoadipyl)-L-cysteinyl-D-valine (ACV), originally detected in *Penicillium chrysogenum* by Arnstein & Morris (2), was the earliest direct precursor of these antibiotics. Similar findings were subsequently reported for bacterial producers of β-lactams (47). The elucidation of the main details of the biosynthetic pathway made clear that prokaryotes and eukaryotes synthesize penicillins and cephalosporins in essentially the same way (Figure 1). This view of β-lactam biosynthesis suggests that the microbial antibiotic-specific genes and the enzymes they encode must be related. As we shall see, comparative analysis of β-lactam biosynthetic genes supports this notion and has provided new ideas on the evolution of the biosynthetic pathway.

Inspection of Figure 1 reveals that many genes are involved in the overall biosynthesis of penicillins, cephalosporins, and cephamycins. Genes associated with β-lactam production are referred to collectively as β-lactam synthetase genes. They may conveniently be divided into two groups. Two early biosynthetic genes participate in the synthesis of all the penicillins and cephalosporins (Figure 1). We follow the practice of Ingolia & Queener (44) and Skatrud (91) in denoting these common genes by the abbreviation *pcb* for penicillin and cephalosporin biosynthesis. Other biosynthetic genes that generate diverse β-lactams are denoted by the abbreviations *pen, cef,* and *cmc* to indicate their roles in synthesis of penicillins, cephalosporins, and cephamycins respectively.

GENES COMMON TO ALL PENICILLIN AND CEPHALOSPORIN PRODUCERS

pcbAB: *ACV Synthetase*

Penicillin and cephalosporin biosynthesis commences with the condensation of three amino acids, L-α-aminoadipic acid, L-cysteine, and L-valine, to form the linear tripeptide ACV. Banko et al (12) first demonstrated that crude cell-free extracts of *Cephalosporium acremonium* (syn. *Acremonium chrysogenum*) contain an activity that carries out the stepwise synthesis of ACV.

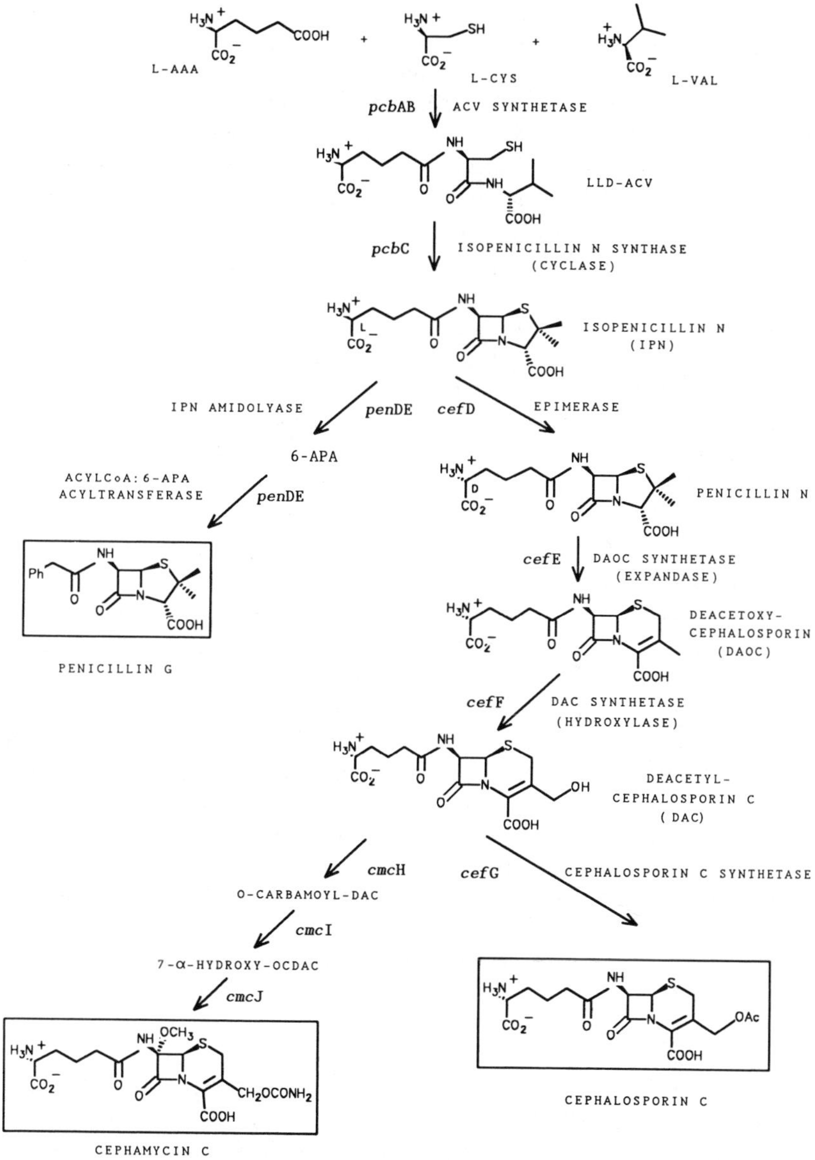

Figure 1 Biosynthesis of penicillin and cephalosporin classes of β-lactam antibiotics in bacteria and fungi. Gene designations are shown in bold face.

They suggested that the activity results from a single multienzyme, ACV synthetase (ACVS), that functions in a similar manner to the peptide synthetases that mediate the nonribosomal synthesis of peptide antibiotics. ACVS has been purified from *Aspergillus nidulans* (102), *C. acremonium* (6), and *Streptomyces clavuligerus* (50, 110) and was shown to catalyze the ATP-dependent activation of each of the L-amino acids, to bind the activated amino acids as thiolesters, to epimerize L-valine, and to polymerize the amino acids to make ACV (102). A multienzyme thiotemplate mechanism has been proposed for the ACVS reaction (53).

Historically, the gene coding for ACVS was designated *pcbAB* because two genetic loci were supposed responsible for synthesis of the α-aminoadipyl-cysteine (AC) dipeptide and the ACV tripeptide (73). However, genetic evidence indicates that a single gene encodes all the activities needed to form ACV (28). Diez et al (29) cloned the *Penicillium chrysogenum* by complementation of mutants blocked in penicillin synthesis. They assumed that the β-lactam genes are clustered and that cosmid and phage clones previously shown to contain the *pcbC* and *penDE* genes (Figure 2) might also contain the *pcbAB* gene. Several such clones could complement two ACVS defective mutants, *npe5* and *npe10,* and restored penicillin production to the *npe5* mutant. Cell-free extracts of the complemented mutants contained a protein of molecular weight of at least 250,000 that was absent in the mutants but present in the parental strain. Transcriptional mapping studies revealed a long transcript of about 11.5 kb. The approximate initiation and terminator sites were identified and the entire DNA region covering them sequenced. An open reading frame (ORF) was found containing 11,376 bp that encodes a protein of 3792 amino acids with a deduced molecular weight of 425,971. Smith et al (97, 98) independently reported similar findings. No introns were found in the *P. chrysogenum* gene (or other fungal *pcbAB* genes), although some small in-frame introns lacking termination signals may be present. A conspicuous feature of the ACVS protein sequence is the presence of three repeated regions, or domains, denoted A, C, and V and containing more than 500 amino acids. They show extensive sequence similarity with the *Bacillus brevis* peptide synthetases, gramicidin S synthetase I and tyrocidin synthetase I, and two other enzymes that carry out ATP-pyrophosphate exchange reactions, luciferase and 4-coumerate-CoA ligase (53). All of these enzymes activate carboxyl groups present in amino acids, hydroxy acids, or simple acids. The A, C, and V regions probably specify, therefore, the amino acid–activating domains of ACVS. The *P. chrysogenum pcbAB* gene is separated from the two remaining genes needed for penicillin production, *pcbC* and *penDE,* by about 1100 bp and is transcribed in opposite orientation to them (Figure 2).

The *C. acremonium* and *A. nidulans* ACVS genes were isolated in a similar

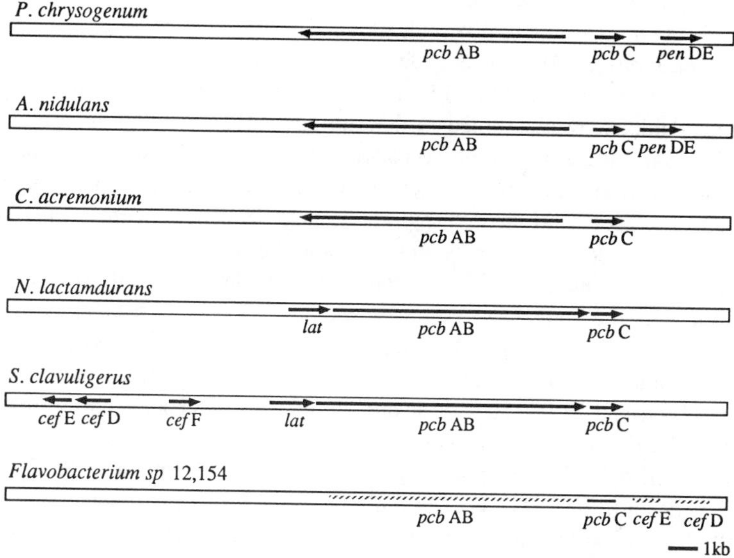

Figure 2 Physical map of the β-lactam biosynthetic gene cluster in bacteria and fungi. The linear maps have been arbitrarily aligned at the start site of the *pcbC* gene. The genes and their products are: *penDE*, acyl-CoA:isopenicillin N acyltransferase; *pcbC*, IPNS; *pcbAB*, ACVS; *lat*, lysine ε-aminotransferase; *cefF*, DACS; *cefD*, IPN epimerase; *cefE*, DAOCS; see also Figure 1. Arrows indicate sizes of structural genes; arrowheads show directions of transcription where known. Only approximate positions and sizes of the *Flavobacterium pcbAB*, *cefE*, and *cefD* genes have been reported (96). The *S. clavuligerus cmcI* gene (OCDAC) has been located in the region between the *pcbC* and *cefD* genes (17). The *S. clavuligerus cefD* and *cefF* genes are expressed from a single transcript that extends beyond these genes into an ORF of unknown function (58).

fashion to that described above. In the case of *C. acremonium*, a 24-kb DNA fragment was identified in a cosmid that hybridized with the *pcbAB* and *pcbC* genes of *P. chrysogenum* (40). The two genes are closely linked. Complementation studies with the *nep5* ACVS-defective *P. chrysogenum* mutant showed that a 15.6-kb subfragment contained a functional *pcbAB* gene. Northern analysis of *C. acremonium* RNA employing probes internal to the *pcbAB* and *pcbC* genes identified two transcripts of 11.4 kb and 1.15 kb respectively. An ORF of 11,136 bp encoding a protein of 3712 amino acids was located upstream of the *pcbC* gene in agreement with the size of the transcript. The deduced ACVS molecular weight is 414,791. In isolating the *A. nidulans pcbAB* gene (*acvA*), MacCabe et al (67) predicted from genetic and complementation analyses of mutants defective in penicillin production that it resides just upstream of the *pcbC* gene (*ipnA*). DNA sequencing of that region showed a large ORF that was identified as belonging to ACVS by positively matching the amino acid sequence deduced from that ORF with that of a 15–amino acid internal peptide isolated from the purified enzyme (68). The *A. nidulans pcbAB* gene encodes for a protein of 3770 amino acids with a predicted molecular weight of 422,486. Because no amino-terminal sequence was available, it was assumed in determining the size of the gene that the translational initiation methionine codon is the first such codon in frame and downstream from the major transcription start point. Three highly repetitive domains are present in the deduced amino acid sequences of the *C. acremonium* and *A. nidulans* enzymes and resemble the three repetitive domains in the *P. chrysogenum* ACVS. Similar functions are ascribed to them. In addition, all the fungal ACVS enzymes possess a consensus thioesterase domain that is like the motif found in fatty acid synthetases. The *P. chrysogenum* and *C. acremonium pcbAB* genes and the enzymes they encode share 62.9% and 54.9% nucleotide and amino acid sequence identity, respectively. The *C. acremonium pcbAB* gene is separated by 1233 bp and transcribed in the opposite direction from the *pcbC* gene (40) and has been mapped to chromosome VI (91). The *A. nidulans pcbAB* and *pcbC* genes are also tightly linked and are divergently transcribed from an intergenic region of 872 bp (Figure 2). Disruption of the *A. nidulans* and *C. acremonium pcbAB* genes resulted in loss of ACVS function without impairing other penicillin biosynthetic genes (43, 96).

ACVS genes from two prokaryotic gram-positive cephamycin-producing species were recently cloned and sequenced. Coque et al (26) used heterologous hybridization probes derived from the *pcbAB* gene of *P. chrysogenum* and the *pcbC* gene of *Streptomyces griseus* to identify the corresponding *Nocardia lactamdurans* genes in a cosmid containing a 34-kb DNA fragment. The *N. lactamdurans pcbAB* gene possesses an ORF of 10,947 bp that codes for a protein of 3649 amino acids with a deduced molecular weight of 404,134. The bacterial enzyme possesses three repeated domains and a thioesterase motif similar to that of the fungal ACVS enzymes. Doran et al (32) reported that an unknown ORF in *Streptomyces clavuligerus* located just upstream of the *pcbC* gene is similar to the nucleotide sequence coding for the carboxy terminal part of the fungal ACVS genes. Moreover, a 712-bp segment of that ORF showed 72% sequence identity at the nucleotide level with the 3' end of the *N. lactamdurans* ACVS gene, establishing that it encodes the *pcbAB* gene (26). Furthermore, the nucleotide sequence coding for the amino terminal part of the *S. clavuligerus* ACVS was identified in a region immediately downstream of the *lat* gene (Figure 2). These results localize the *pcbAB* gene in a DNA segment of about 12 kb between *lat* and *pcbC* (100). The *N. lactamdurans* and *S. clavuligerus pcbAB* and *pcbC* genes are closely linked, and in contrast to the divergent expression of these genes in the fungal gene clusters, they are transcribed in the same orientation (Figure 2). Some partial information is also available for ACVS genes from two prokaryotic gram-negative cephalosporin-producing species. The *pcbAB* gene of *Flavobacterium* sp. 12,154 was cloned and mapped in a cluster of β-lactam biosynthetic genes (96), and a Japanese patent application by Takeda Industries, No. 2-291274, 1990, presents sequence data for the *pcbAB* gene of *Lysobacter lactamgenus*.

To complete this account of the ACVS genes, we note that sequence similarity among the *pcbAB* genes is high. This is evident from a comparison of the nucleotide sequences encoding the three repeated domains in the prokaryotic and eukaryotic genes. Parallel, interfungal domains share on average about 71% nucleotide sequence identity, whereas parallel fungal-bacterial domains share on average about 48% sequence identity. Although the numbers differ somewhat for the individual A, C, and V domain comparisons, they imply that all the *pcbAB* genes are closely related. Significantly, sequence identities among the three domains within fungal genes, or within bacterial genes, or between them, are not appreciably different, averaging about 37%. In contrast, very little similarity is found between regions separating domains. The relevance of these results in connection with the origin of the relatedness among β-lactam genes is discussed below.

pcbC*: Isopenicillin N Synthase*

The second step in the biosynthetic route to penicillins and cephalosporins is the conversion of ACV to isopenicillin N (IPN). A single enzyme, isopenicillin N synthase (IPNS), catalyzes oxidation (desaturation) of the linear tripeptide to form the bicyclic isopenicillin N, which possesses weak but significant antibiotic properties. The enzyme, frequently referred to as cyclase, is of particular interest both because of its unusual mechanism of action, which results in the stepwise formation of the β-lactam and thiazolidine fused rings, and its broad substrate specificity that has found an ingenious use in creating novel penicillins from ACV analogs (4, 107). The IPNS reaction requires ferrous iron, molecular oxygen, and an electron donor usually provided as ascorbate (79). A critical role for the iron is suggested by a variety of physical studies that indicate that it binds directly to ACV through its cysteinyl thiol (21, 75). In fact, only the free thiol form of ACV will serve as a substrate; the *bis*-disulfide dimer is inactive (79). Reduction of the latter is conveniently carried out in vitro by dithiothreitol (DTT). However, *S. clavuligerus*, a potent producer of β-lactams, possesses a disulfide reductase that recognizes *bis*-ACV as a substrate. Its properties resemble those of the thioredoxin family of oxidoreductases (23).

In 1985, Queener and his colleagues at Eli Lilly in collaboration with Baldwin & Abraham at Oxford University described the first isolation of a β-lactam synthetase gene, that of the IPNS gene of the eukaryotic filamentous fungus *C. acremonium* (83). Since then, IPNS structural genes have been cloned from eight more β-lactam-producing organisms. These include two additional fungal species, *P. chrysogenum* (13, 19) and *A. nidulans* (81, 105), several gram-positive species of *Streptomyces* (37, 65, 89, 105) and *Nocardia* (26), and a single species of the gram-negative, unicellular, *Flavobacterium* (88). IPNS genes are given the gene designation *pcbC*. All of the *pcbC* genes were isolated in essentially the same way, using synthetic or heterologous DNA probes to screen by hybridization genomic libraries of the organisms. Thus, in the cloning of the *pcbC* gene of *C. acremonium* (83), two pools of 17-mer oligodeoxynucleotides were synthesized, both containing 32 different oligomers, such that they comprised all possible nucleotide sequences encoded by two short peptides located at the amino-terminal portion of the IPNS protein. The DNA probes were then used to screen a cosmid genomic library. In this way, a clone was identified that contained the nucleotide sequence predicted for the peptides and was found to possess an ORF coding for a protein of 338 amino acids with a deduced molecular weight of 38,416. When that ORF was expressed in *Escherichia coli,* IPNS activity was readily detected. The *C. acremonium pcbC* gene was mapped to chromosome VI (92) and subsequently found to be closely linked to the *pcbAB* gene (40). Essentially the same approach was used to isolate the *pcbC* genes of two *Streptomyces* species, *S. clavuligerus* (65) and *S. lipmanii* (105), except that the synthetic DNA probes employed were based on a single peptide and were much longer, 48- and 60-mer, respectively, than those used to isolate the *C. acremonium* gene.

Once the IPNS gene of *C. acremonium* was available, the next step was to use it as a heterologous probe to isolate the corresponding genes from other fungal β-lactam-producing species (19, 81, 105). Likewise, DNA probes based on the *Streptomyces pcbC* genes led to the isolation of several other bacterial IPNS genes (26, 37, 89). Thus, the *pcbC* gene from *S. jumonjinensis,* a gram-positive bacterium, could be employed to clone the related gene from a cephalosporin-producing species of *Flavobacterium* (88). Growing awareness that fungal and bacterial β-lactam biosynthetic genes are organized in clusters led to the subsequent exploitation of *pcbC* genes as tools to clone other β-lactam biosynthetic genes.

Cross-hybridization experiments of the type described above have also been used to examine the distribution of β-lactam biosynthetic genes in nature. In one study (77), the *pcbC* gene of *C. acremonium* was used as a probe to detect related genes in chromosomal DNA in a variety of archaebacteria including halophiles, methanogens, and thermophiles; gram-positive eubacteria including bacilli and streptomycetes; some higher eukaryotes, metazoa, and plants; and several lower eukaryotes. In a separate study (H. Koltai, Y. Aharonowitz & G. Cohen, unpublished results), genomic DNA from a representative collection of gram-negative bacteria including *E. coli, Myxococcus xanthus, Acinetobacter calcoaceticus,* and *Agrobacterium, Gluconobacter, Seratia, Xanthomonas,* and *Flavobacterium* species were screened using hybridization employing a *pcbC* probe derived from *Flavobacterium*. This group of microorganisms was chosen to include some producers of penicillins and cephalosporins, several species that produce unrelated monobactam and carbapenem β-lactam antibiotics, and a few species that produce no known β-lactam antibiotics. The results of both of these studies showed that clear-cut positive responses to the probes appeared only in those species known to synthesize penicillin and cephalosporin antibiotics. Therefore, *pcbC* genes are apparently confined to a rather limited although diverse group of microorganisms.

A similar hybridization analysis was carried out among *Streptomyces* species to determine how prevalent β-lactam genes are within this group (24, 89). Malpartida et al (71) first described such an approach to demonstrate homology between *Streptomyces* genes coding for synthesis of different polyketide antibiotics. As was mentioned earlier, the *Streptomyces* are especially abundant in penicillin and cephalosporin β-lactam-producing species. Whereas all the β-lactam-producing strains that were tested responded to several different *Streptomyces pcbC* probes, unexpectedly some of the non-β-lactam producers also responded to the probe. Furthermore, this response depended on the conditions employed and on the particular *Streptomyces* probe used. Thus, two non-β-lactam-producing species *S. lividans* and *S. coelicolor* gave weak but distinct hybridization signals in stringent conditions with the *S. clavuligerus pcbC* probe but not with the *S. lipmanii* and *S. jumonjinensis* probes. It is noteworthy, therefore, that a recent study in which the *P. chrysogenum pcbC* gene was employed as a hybridization probe confirmed the existence in *S. coelicolor* and *S. lividans* of IPNS-related sequences (37). Also, when three related strains of *S. griseus* were examined, all were found to respond to *pcbC* probes despite the fact that only one of them was recognized as a β-lactam producer (89).

These observations imply that *pcbC* genes may occur in many streptomycetes that have, until now, been considered nonproducers of penicillins and cephalosporins. They also raise the possibility that some *Streptomyces* species, for example, might be unable to make β-lactam antibiotics because they contain a silent or defective IPNS gene. A positive hybridization response does not signify a functional gene. The concept of a silent antibiotic synthetase gene was first reported in a study of the gene coding for phenoxazinone synthetase, a key enzyme in the biosynthesis of actinomycin (51), and was subsequently reported for other *Streptomyces* species that make polyketide and carbapenam antibiotics. In some cases, these silent genes could be activated to recover their function (51, 71). One explanation for the inability of certain streptomycetes, such as the two *S. griseus* strains mentioned above, to make β-lactams is that they lack a regulatory element necessary for IPNS expression. In fact, a recent study of the *S. griseus pcbC* gene suggests that its expression may depend on a positively acting regulator (37). When that gene was introduced into an IPNS-defective mutant of *S. clavuligerus,* cephalosporin synthesis was restored; no expression was detected, however, when the same *S. griseus* gene with its own promoter was put into *S. lividans*. Therefore, DNA hybridization screens of culture collections designed to detect IPNS and other β-lactam biosynthetic genes could prove a valuable and sensitive approach for identifying new potential antibiotic-producing organisms. To effectively carry out such a program, however, will require the use of a variety of probes from different producer organisms.

The microbial IPNS genes form a closely related family of genes, as seen in Table 2, which lists the percentage nucleotide sequence identities found from pairwise comparison of *pcbC* genes from fungi, gram-positive, and gram-negative microorganisms. Clearly, the three groups of *pcbC* genes are, rather surprisingly, almost equally similar to one another. Percentage identities of greater than 60% between functionally related prokaryotic and eukaryotic genes is not common and suggests some underlying important feature. This point is strikingly demonstrated from a further comparison in which the deduced primary amino acid sequences coded by the nine *pcbC* genes have been aligned to display maximum similarity (Figure 3). In addition to the very high incidence of identical amino acids present at the same site in all nine sequences, approximately 40%, amino acid changes at numerous additional sites are conservative. The overall effect is one of extensive sequence similarity. Results of computer analysis of the amino acid sequence data in terms of predicted secondary structure and hydrophobicity properties of the IPNS proteins further emphasizes the structural similarity of these proteins (87, 105).

A closer examination of Figure 3 reveals that most of the sequence identity is scattered throughout the protein rather than confined to a few well-defined regions. Unfortunately, this situation precludes the possibility of identifying functionally important domains, such as those involved in substrate binding and catalytic activity. Moreover, no recognizable sequence motifs provide a clue to their role in enzyme function, nor is there any discernible sequence relatedness of the IPNS proteins to any other known characterized proteins with the exception of the β-lactam synthetases themselves. S. Kovacevic & J. R. Miller (personal communication) have searched for conserved amino acid sequence motifs in the β-lactam synthetases. They found several examples of relatedness. Thus, the Cys106-containing region of the cyclases is similar to two regions of expandase-hydroxylase (see below). When the analysis is

Table 2 Percent sequence identity between *pcbC* genes

Species	*A. nidulans*	*C. acremonium*	*P. chrysogenum*	*Flavo-bacterium*	*S. clavuligerus*	*S. jumonjinensis*	*S. griseus*	*S. lipmanii*	*N. lactamdurans*	
A. nidulans		77.3	83.1	59.0	60.6	62.3	59.0	60.9	62.3	
C. acremonium	72.3		78.3	58.4	59.7	63.6	57.1	58.7	60.0	
P. chrysogenum	76.6	75.7		57.7	58.7	60.3	57.1	57.1	59.4	Amino acid
Flavobacterium	62.9	66.7	65.2		62.3	62.9	61.3	58.1	61.0	identities
S. clavuligerus	63.7	66.6	65.6	70.7		82.9	74.8	73.6	76.5	
S. jumonjinensis	64.1	69.7	70.0	70.2	85.5		77.7	72.6	78.7	
S. griseus	63.3	66.3	64.7	70.5	79.9	82.9		76.1	75.2	
S. lipmanii	64.5	67.8	65.8	68.9	80.0	81.0	83.6		73.9	
N. lactamdurans	63.7	67.2	64.5	69.7	79.4	82.0	79.8	79.6		

Nucleotide identities

```
--MGSVS-KANVPKIDVSPLFGDDQAAKMRVAQQIDAASRDTGFFYAVNHGINVQRLSQKTKEFHMSITP     67
--MASTP-KANVPKIDVSPLFGDNMEEKMKVARAIDAASRDTGFFYAVNHGVDVKRLSNKTREFHFSITD     67
MGSVPVP-VANVPRIDVSPLFGDDKEKKLEVARAIDAASRDTGFFYAVNHGVDLPWLSRETNKFHMSITD     69
----MNR-HADVPVIDISGLSGNDMDVKKDIAARIDRACRGSGFFYAANHGVDLAALQKFTTDWHMAMSA     65
---MKMP-SAEVPTIDVSPLFGDDAQEKVRVGQEINKACRGSGFFYAANHGVDVQRLQDVVNEFHRTMSP     66
-MPILMP-SAEVPTIDISPLSGDDAKAKQRVAQEINKAARGSGFFYASNHGVDVQLLQDVVNEFHRNMSD     68
-MPVLMP-SAHVPTIDISPLFGTDAAAKKRVAEEIHGACRGSGFFYATNHGVDVQQLQDVVNEFHGAMTD     68
-MPVLMP-SADVPTIDISPLFGTDPDAKAHVARQINEACRGSGFFYASHHGIDVRRLQDVVNEFHRTMTD     68
-MPIPML-PAHVPTIDISPLSGGDADDKKRVAQEINKACRESGFFYASHHGIDVQLLKDVVNEFHRTMTD     68

EEKWDLAIRAYNKEHQDQVRAGYYLSIPGKKAVESFCYLNPNFTPDHPRIQAKTPTHEVNVWPDETKHPG    137
EEKWDLAIRAYNKEHQDQIRAGYYLSIPEKKAVESFCYLNPNFKPDHPLIQSKTPTHEVNVWPDEKKHPG    137
EEKWQLAIRAYNKEHESQIRAGYYLPIPGKKAVESFCYLNPSFSPDHPRIKEPTPMHEVNVWPDEAKHPG    139
EEKWELAIRAYNPANP-RNRNGYYMAVEGKKANESFCYLNPSFDADHATIKAGLPSHEVNIWPDEARHPG    134
QEKYDLAIHAYNKNNS-HVRNGYYMAIEGKKAVESFCYLNPSFSEDHPEIKAGTPMHEVNSWPDEEKHPS    135
QEKHDLAINAYNKDNP-HVRNGYYKAIKGKKAVESFCYLNPSFSDDHPMIKSETPMHEVNLWPDEEKHPR    137
QEKHDLAIHAYNPDNP-HVRNGYYKAVPGRKAVESFCYLNPDFGEDHPMIAAGTPMHEVNLWPDEERHPR    137
QEKHDLAIHAYNENNS-HVRNGYYMARPGRKTVESWCYLNPSFGEDHPMIKAGTPMHEVNVWPDEERHPD    137
EEKYDLAINAYNKNNP-RTRNGYYMAVKGKKAVESWCYLNPSFSEDHPQIRSGTPMHEGNIWPDEKRHQR    137

FQDFAEQYYWDVFGLSSAL-LKGYALALGKEENFFARHFKPDDTLASVVL-IRYPYLDPYPEAAIKTAAD    205
FREFAEQYYWDVFGLSSAL-LRGYALALGKEEDFFSRHFKKEDALSSVVL-IRYPYLNPIPPAAIKTAED    205
FRAFAEKYYWDVFGLSSAV-LRGYALALGRDEDFFTRHSRRDTTLSSVVL-IRYPYLDPYPEPAIKTADD    207
MRRFYEAYFSDVFDVAAVI-LRGFAIALGREESFFERHFSMDDTLSAVSL-IRYPFLENYPP--LKLGPD    200
FRPFCEEYYWTMHRLSKVL-MRGFALALGKDERFFEPELKEADTLSSVSL-IRYPYLEDYPP--VKTGPD    201
FRPFCEDYYRQLLRLSTVI-MRGYALALGRREDFFDEALAEADTLSSVSL-IRYPYLEEYPP--VKTGAD    203
FRPFCEGYYRQMLKLSTVL-MRGLALALGRPEHFFDAALAEQDSLSSVSL-IRYPYLEEYPP--VKTGPD    203
FRSFGEQYYREVFRLSKVLLLRGFALALGKPEEFFENEVTEEDTLSCRSLMIRYPYLDPYPEAAIKTGPD    207
FRPFCEQYYRDVFSLSKVL-MRGFALALGKPEDFFDASLSLADTLSAVTL-IHYPYLEDYPP--VKTGPD    203

GTKLSFEWHEDVSLITVLYQSNVQNLQVETAAGYQDIEADDTGYLINCGSYMAHLTNNYYKAPIHRVKWV    275
GTKLSFEWHEDVSLITVLYQSDVANLDVEMPQGYLDIEADDNAYLVNCGSYMAHITNNYYPAPIHRVKWV    275
GTKLSFEWHEDVSLITVLYQSDVQNLQVKTPDGWQDIQADDTGFLINCGSYMAHITDDYYPAPIQRVKWV    277
GEKLSFEHHQDVSLITVLYQTAIPNLQVETAEGYLDIPVSDEHFLVNCGTYMAHITNGYYPAPVHRVKYI    270
GEKLSFEDHFDVSMITVLYQTQVQNLQVETVDGWRDLPTSDTDFLVNAGTYLGHLTNDYFPSPLHRVKFV    271
GTKLSFEDHLDVSMITVLYQTEVQNLQVETVDGWQDIPRSDEDFLVNCGTYMGHITHDYFPAPNHRVKFI    273
GQLLSFEDHLDVSMITVLFQTQVQNLQVETVDGWRDIPTSENDFLVNCGTYMAHVTNDYFPAPNHRVKFV    273
GTRLSFEDHLDVSMITVLFQTEVQNLQVETVDGWQSLPTSGENFLINCGTYLGYLTNDYFPAPNHRVKYV    277
GTKLSFEDHLDVSMITVLFQTEVQNLQVETADGWQDLPTSGENFLVNCGTYMGYLTNDYFPAPNHRVKFI    273

NAERQSLPFFVNLGYDSVIDPFDP-REPNGKSDR------EPLSYGDYLQNGLVSLINKNGQT  A.nid  332
NEERQSLPFFVNLGFNDTVQPWDP-SKEDGKTDQ------RPISYGDYLQNGLVSLINKNGQT  P.chr  332
NEERQSLPFFVNLGWEDTIQPWDP-ATAKDGAKDAAKDK-PAISYGEYLQGGLRGLMKKNGQT  C.acr  339
NAERLSIPFFANLSHASAIDPFAP-PPYAPPGGN------PTVSYGDYLQHGLLDLIRANGQT  Flav   327
NAERLSLPFFFHAGQHTLIEPFFP-DGAPEGKQGN-----EAVRYGDYLNHGLHSLIVKNGQT  N.lac  329
NAERLSLPFFLNAGHNSVIEPFVP-EGAAGTVKN------PTTSYGEYLQHGLRALIVKNGQT  S.jum  330
NAERLSLPFFLNGGHEAVIEPFVP-EGASEEVRN------EALSYGDYLQHGLRALIVKNGQT  S.cla  330
NAERLSLPFFLHAGQNSVMKPFHP-EDTGDRKLN------PAVTYGEYLQEGFHALIAKNVQT  S.lip  334
NAERLSLPFFLHAGHTTVMEPFSP-EDTRGKELN------PPVRYGDYLQQASNALIAKNGQT  S.gri  330
```

Figure 3 Deduced amino acid sequences of nine bacterial and fungal IPNS genes. The sequences were aligned to achieve maximal similarity. Sites at which the same amino acids occur in each sequence are shaded. Arrows indicate two conserved cysteines.

extended to include the deduced amino acid sequences from cyclase, epimerase, expandase, hydroxylase, expandase-hydroxylase, and acetyltransferase genes, quite unexpectedly a conserved motif S------Q----N is found in 9 out of 10 of the sequences. The block of amino acids surrounding this motif is similar in all the enzymes. Kovacevic & Miller propose the intriguing possibility that all the β-lactam genes may have arisen from a single ancestral gene.

Nevertheless, two highly conserved domains bracketing two conserved cysteine residues (see Figure 3) have been the main focus of attention primarily because of their potential role in enzyme activity. Several observations suggest that free cysteine thiols may be required for activity. Thus, the *C. acremonium* enzyme is inactivated when oxidized and its activity is reversibly restored if it is treated with a disulfide reductant such as DTT (9). Also, chemical blocking of cysteine thiols results in loss of activity (46, 79). The *C. acremonium* enzyme, like the other fungal IPNS proteins, has just two cysteines. To explore the role of each of the cysteines, site-directed mutagenesis was used to exchange the cysteine codons in the *pcbC* gene to serine codons (84). Replacement of the cysteine at position 106 resulted in an approximately 20-fold drop in IPNS specific activity, whereas the effect of a similar replacement at position 255 was much less pronounced, about two-fold. When both cysteines were changed to serines, the specific activity was only marginally different from that of the single change at Cys106. Cysteine thiols are therefore important for activity but not essential. In fact, Coque et al (26) have shown that one of the two conserved cysteines in the IPNS proteins, Cys249 (equivalent to Cys255 in *C. acremonium*) is absent in the *N. lactamdurans* enzyme and is replaced by an alanine residue (Figure 3). Curiously, the one or two additional nonconserved cysteine residues present in the bacterial IPNS proteins are confined to just two sites (Figure 3), at positions 38 and 144 (*C. acremonium* numbering), and when missing, they are replaced by nonbulky glycine, alanine, and serine residues. Although several IPNS-defective mutants have been isolated from fungal and bacterial β-lactam producers, only in one case has the mutation been characterized. The N-2 strain of *C. acremonium* contains a single amino acid alteration at position 285, which results in proline being replaced by leucine (82). Figure 3 shows that the mutation resides in a highly conserved region of IPNS.

Baldwin et al (8) have used photoaffinity labeling to explore the IPNS regions involved in substrate binding and catalytic activity. Two cysteine-containing peptides, Asp40–Arg78 and Thr237–Gly256 (Figure 3) are labeled by a diazirinyl-containing analog of ACV. According to these authors' model, Cys255 and the two peptides are involved in binding of ACV at the α-aminoadipyl moiety; Cys106 may interact with the thiol group of the cysteine residue of ACV or perhaps bind to the iron cofactor. The model is consistent with the site-directed mutagenesis studies (84) and spectroscopic studies (75).

Most of the fungal and bacterial IPNS genes have been expressed in *E. coli,* and sometimes the investigator obtains very high yields of enzyme (7, 64). The fungal *pcbC* genes do not possess introns. The recombinant proteins appear to be indistinguishable in biochemical properties from those of the parent β-lactam producers, except that the *C. acremonium* protein lacks the amino terminal methionine and glycine residues predicted from its open reading frame (11). Currently, several laboratories are using recombinant

IPNS for studies on mechanism of action and crystal-structure determination. Recombinant IPNS may also prove valuable for creating, in vitro, new and unusual β-lactam antibiotics.

GENES SPECIFIC FOR THE SYNTHESIS OF PENICILLIN

penDE: *Acyl-CoA:Isopenicillin N Acyltransferase*

Some fungi, notably *P. chrysogenum* and *A. nidulans,* can replace the α-aminoadipyl side chain of isopenicillin N with nonpolar side chains such as phenylacetyl and phenoxyacetyl groups (to produce penicillin G and penicillin V respectively) when the corresponding acids are added to the culture medium. Queener & Neuss (80) proposed a two-step enzymatic process for conversion of isopenicillin N to penicillin G. In the first step, the L-α-aminoadipyl side chain is cleaved to yield 6-aminopenicillanic acid (6-APA), which is then converted, in the second step, to penicillin G through addition of a phenylacetyl group from its CoA derivative. These activities, isopenicillin N amidohydrolase and acyl-CoA:6-APA acyltransferase (AAT) are encoded by a single gene designated *penDE*.

To clarify the molecular basis for these final events in the biosynthesis of penicillins, Alvarez et al (1) purified the *P. chrysogenum* AAT and showed that it was a protein with a molecular weight of about 29,000. Purified AAT recognizes 6-APA as a substrate and, to a much lesser extent, it accepts IPN, implying an IPN acetyltransferase activity. To establish unequivocally whether conversion of IPN to penicillin G requires one or two distinct enzymes, the gene encoding AAT was isolated using a synthetic oligo probe based on the amino terminal portion of the protein (15). Clones were isolated that complemented two *P. chrysogenum* mutants, *npe6* and *npe8,* that have little or no AAT activity and, when introduced into a cephalosporin-producing species of *C. acremonium,* enabled it to make benzylpenicillin (39). The *penDE* ORF was found to reside downstream of the *pcbC* gene and encoded a protein with a deduced molecular weight of 39,943. Unexpectedly, the amino acid terminus of that ORF did not match that coded by the synthetic probe used to isolate the gene. The latter was located within the gene. The *A. nidulans penDE* (*acyA*) gene was also cloned and found to encode a protein with a deduced molecular weight of 39,240 (76). Comparison of the two fungal *penDE* genes shows that they share 73% sequence identity at the nucleotide level and 76% identity at the amino acid level. Tobin et al (99) have reported similar results for the *P. chrysogenum* and *A. nidulans* genes. Three introns are present in the amino-terminal half of both fungal genes and occur in the same positions (15, 76, 99). This is the first example of a fungal β-lactam gene containing introns.

Characterization of the two fungal AAT enzymes by Whiteman et al (106) demonstrated that they consist of two subunits of approximate molecular weights 11,000 and 29,000. These authors showed that active AAT exists as a heterodimer with a total molecular weight of about 40,000, in agreement with the deduced value for the *penDE* gene. The amino-terminal sequence of the smaller subunit corresponded to that predicted for the amino-terminal sequence of the *penDE* ORF. When that ORF was expressed in *E. coli,* a single polypeptide with a molecular weight of about 40,000 was obtained that could convert both 6-APA and IPN to penicillin G (99). Therefore, conversion of IPN to penicillin G is probably catalyzed by a single enzyme consisting of two subunits that are derived from the full-length preprotein by a posttranslation processing event (76, 99; F. J. Fernandez & J. F. Martin, unpublished results).

GENES SPECIFIC FOR THE SYNTHESIS OF CEPHALOSPORINS

cefD: *Epimerase*

With the synthesis of isopenicillin N, the biosynthetic route to penicillins and cephalosporins diverges into two branches (Figure 1). In *P. chrysogenum,* the α-aminoadipyl side chain of isopenicillin N is exchanged for one of several hydrophobic side chains derived from the corresponding acyl-CoA derivatives. In cephalosporin-producing organisms, the L-α-aminoadipyl side chain is first isomerized to the D-enantiomer to produce penicillin N; penicillin N and not isopenicillin N is a substrate for ring-expansion enzymes. Epimerase activity was first shown in cell-free extracts of *C. acremonium* by Konomi et al (56), who detected trace amounts of cephalosporin (i.e. a penicillinase-resistant antibiotic) in reaction mixtures containing ACV. Cephalosporin was found only if penicillin N was produced in the reaction. This observation was confirmed by Baldwin et al (10), who showed conversion of isopenicillin N into the penicillinase-resistant deacetoxycephalosporin C. They proposed that *C. acremonium* possessed an epimerase activity that was highly labile in cell-free preparations. In contrast, the corresponding enzyme from *S. clavuligerus* was found to be much more stable and became, therefore, the target for further investigations. Jensen et al (48) purified the *S. clavuligerus* enzyme and found it was stabilized by pyridoxal phosphate and had a molecular weight of about 60,000. An epimerase of similar molecular weight was also reported for *N. lactamdurans* (62). However, Usui & Yu (101) purified the *S. clavuligerus* isopenicillin N epimerase and determined its molecular weight to be approximately 47,000, a value subsequently confirmed by Kovacevic et al (58). The bacterial epimerases are reported to be racemases, catalyzing a reversible isomerization between isopenicillin N and penicillin N (48, 62, 101). To isolate the *S. clavuligerus* epimerase gene, a 63-base oligodeoxynucleotide probe was synthesized based on the amino terminal sequence of the purified enzyme (58). This probe was used to identify the *cefD* gene in cosmid clones that had previously been shown to contain several other biosynthetic genes (Figure 2). The *S. clavuligerus* epimerase gene is located immediately upstream of the expandase gene (*cefE*, see below). A single ORF coding for a protein of 398 amino acid residues, corresponding to a molecular weight of 43,497, is separated from the expandase gene by just 81 bp. The *cefD* and *cefE* genes are expressed from the same transcript (58). Skatrud (91) recently cloned the *cefD* gene of *S. lipmanii* and showed that the two *Streptomyces* epimerase genes share 78% sequence identity at the nucleotide level and 71% at the amino acid level.

cefE, cefF, *and* cefEF: *Expandase, Hydroxylase, and Expandase-Hydroxylase*

Penicillin N is the direct precursor of all the cephalosporins and cephamycins. In the biosynthesis of cephalosporins (Figure 1), the five-membered thiazolidine ring of penicillin N is expanded to a six-membered dihydrothiazine ring through an oxidative reaction catalyzed by deacetoxycephalosporin C synthetase (DAOCS), also referred to as expandase. Deacetoxycephalosporin C (DAOC) is then hydroxylated to form deacetylcephalosporin C (DAC) by deacetylcephalosporin C synthase (DACS) or DAOC hydroxylase. Enzymatic expansion of the penicillin five-member ring was first observed in cell-free extracts of *C. acremonium* (54, 109) and shown to require ferrous iron, molecular oxygen, DTT, and 2-oxoglutarate as cofactors (61). With the exception of the latter, the same components are required in the oxidative cyclization of ACV to isopenicillin N. Attempts to purify expandase and hydroxylase proteins showed that the prokaryotic and eukaryotic enzymes have important structural differences. In the case of *S. clavuligerus,* the expandase and hydroxylase activities were clearly separable and were associated with proteins of molecular weights of 29,500 and 26,200, respectively (49). In *C. acremonium,* the two activities remained physically linked during purification and behaved as a monomeric protein with a molecular weight of about 41,000 (5, 33, 86). The possibility could not be ruled out that in the fungal system two proteins of very similar size copurified and that both could catalyze their reactions independently. Alternatively, a single protein could possess both expandase and hydroxylase activities. To address this issue, the gene or genes coding for expandase-hydroxylase activities were cloned.

The approach used to isolate the genes encoding these activities involved the simultaneous use of two separate probes to screen a cosmid genomic library of *C. acremonium* (85). Because an amino-terminal sequence of the protein could not be obtained, probably due to blockage, different proteolytic methods were tried to obtain peptide-cleavage fragments. One internal peptide was convenient in that it allowed construction of a mixed probe with a few oligomers. Another peptide sequence was chosen to make a single long probe of 60 bases. The codon usage for this probe was assumed to be that found in the *C. acremonium* IPNS gene (83). Two cosmids were isolated from the library that hybridized to both probes. An ORF coding for a polypeptide of molecular weight 36,462 matched the sequence predicted for the peptide fragments. When that ORF was expressed in *E. coli,* the cell extracts contained both expandase and hydroxylase activities. Biochemical analysis of the recombinant protein showed that it was indistinguishable from the native expandase-hydroxylase enzyme, thereby providing unambiguous proof of the bifunctional nature of the polypeptide from *C. acremonium*. The gene coding for these two activities in *C. acremonium* is denoted *cefEF* and is located on chromosome II (92). The nucleotide sequences of the *C. acremonium cefEF* and *pcbC* structural genes possess 11% sequence identity (44). A decapeptide GK----S-CY in the expandase-hydroxylase, including the cysteine residue at position 100, has 50% identity with an IPNS peptide that contains the cysteine residue 106 and that may be involved in substrate binding and catalytic activity (84). Also, a second sequence in the middle of the expandase-hydroxylase enzyme is similar to the Cys106 cyclase region, R--YF/Y--V/IP--K/R--E (S. Kovacevic & J. R. Miller, personal communication).

Because *S. clavuligerus* is predicted to have separate genes that code for expandase and hydroxylase, sequence analysis of these genes and comparison with the fungal *cefEF* gene might provide some insights into the protein domains responsible for the ring expansion and hydroxylation activities. The *S. clavuligerus* expandase gene, *cefE,* was isolated using a strategy rather different from that used to clone the fungal gene (59). It was assumed that the bacterial β-lactam biosynthetic genes were clustered. Accordingly, a 66-base guessmer probe was synthesized to match the first 22 amino acids at the amino-terminal portion of the enzyme and used to hybridize to several cosmid clones that contained DNA regions covering the *pcbC* gene. One of the clones that hybridized strongly with the synthetic probe contained an ORF coding for a protein of 311 amino acids, corresponding to a molecular weight of 34,519, and was some 21 amino acids shorter than the fungal protein. The fungal *cefEF* and bacterial *cefE* genes and proteins possess 65% and 57% sequence identity, respectively, and are, therefore, closely related. The *S. clavuligerus cefE* gene was expressed in *E. coli,* and the recombinant protein had a molecular weight of 34,600 (34) and could catalyze only the expandase reaction. This enzyme is indistinguishable from the native expandase of *S. clavuligerus* as judged by physical and biochemical properties.

A short digression at this point will enable the reader to appreciate the potential importance of these studies for the synthesis of cephalosporins.

Medically important cephalosporins are produced by derivatizing the 7-amino group of 7-aminodeacetoxy-cephalosporanic acid (7-ADCA) (16). Currently, the method involves chemical-ring expansion of penicillin G and enzymatic hydrolysis of the side chain. Because it was not possible to separate the *C. acremonium* hydroxylase and expandase activities, the desired enzymatic process could not be developed due to the formation of hydroxylated products. Cantwell et al (18) proposed introducing into *P. chrysogenum* a genetically modified bacterial expandase gene (*cefE*) that would recognize penicillin G or V. If expressed, it would enable *P. chrysogenum* to produce the direct precursor for making 7-ADCA. A first step toward achieving this goal was recently reported (18). The *cefE* ORF of *S. clavuligerus* was fused to the promoter of the *P. chrysogenum pcbC* gene and inserted into a vector containing an *amdS* gene as a dominant selectable marker. Cultures of *P. chrysogenum* were transformed with the vector, and a significant number of transformants possessing expandase activity were obtained. The low enzyme activity found in these transformants led to further successful efforts to construct hybrid genes that produce levels of expandase comparable to that found in *C. acremonium* strains (91). Whether the *Streptomyces cefE* gene product can be genetically engineered to accept penicillin G or V as a substrate remains to be seen.

The *S. clavuligerus* hydroxylase gene (*cefF*) was isolated in a fashion analogous to the *cefE* gene by taking advantage of the fact that the biosynthetic genes are clustered (57). A cosmid previously shown (58) to contain the epimerase and expandase genes was subcloned and a DNA fragment was identified that responded in hybridization to a 60-base synthetic guessmer probe. The latter was synthesized according to the amino-terminal sequence of the purified hydroxylase. An ORF coding for a protein of 318 amino acids with a molecular weight of 34,584 was found and expressed in *E. coli*. The *cefF* gene was shown to be closely linked to the epimerase and expandase genes but is transcribed in the opposite orientation (Figure 2) (57).

Comparison of the nucleotide sequences of the *S. clavuligerus cefE* and *cefF* structural genes show that they share 71% sequence identity; the deduced amino acid sequences possess 59% sequence identity (57). This revealing result provides a clue to the possible origin of the *cefE* and *cefF* genes in *Streptomyces*. Not only are the sequences of the genes alike, but careful studies with highly purified recombinant proteins indicate that the *S. clavuligerus* hydroxylase possesses a low but significant ring-expansion activity (3). Similarly, the *S. clavuligerus* expandase exhibits weak but reproducible hydroxylation activity. Thus, both expandase and hydroxylase are bifunctional enzymes and are analogous to the fungal expandase-hydroxylase. These findings fit a picture of a gene duplication occurring in the actinomycetes' bifunctional gene followed by the loss of most of the expandase function encoded in one copy and loss of most of the hydroxylase function encoded in the other copy.

Other β-Lactam Biosynthetic Genes

The β-lactam biosynthetic route in *C. acremonium* terminates in cephalosporin C. *Streptomyces* and *Flavobacterium* spp. produce a more diverse group of cephalosporin and cephamycin β-lactam antibiotics. Cephalosporin C synthetase catalyzes the conversion of DAC to cephalosporin C by transferring an acetyl group from acetyl-CoA to the hydroxyl moiety on the sulfur-containing ring. Although the gene responsible for this activity has not been isolated, it has been given the designation *cefG* because it is presumably required only for cephalosporin synthesis. Other genes are required for the synthesis of cephamycins. In *S. clavuligerus,* three biochemical steps are recognized in the conversion of DAC to cephamycin C; carbamylation of DAC to *O*-carbamyl-DAC (OCDAC), hydroxylation of OCDAC to 7-α-hydroxy OCDAC, and O-methylation of the latter by cephamycin synthetase (methyltransferase). On the assumption that each of these activities is encoded by a single gene, the genes are designated as *cmcH, cmcI,* and *cmcJ,* respectively. The genes have not been characterized. The *cmcI* hydroxylase and *cefF* hydroxylase share the same cofactors in the reactions they catalyze (44); both enzymes recognize the same substrate, DAOC, but cause hydroxylation at different positions. Both hydroxylase genes are therefore likely to be closely related. The fact that the β-lactam biosynthetic genes are clustered should facilitate the cloning of these modifying genes. Indeed, Burnham et al (17) and Xiao et al (108) reported the cloning of a DNA fragment from *S. clavuligerus* that contains several biosynthetic genes including the OCDAC hydroxylase gene, and Chen et al (20) have reported that a 22-kb fragment from *Streptomyces cattleya,* a producer of cephamycin C, encodes all the information needed to make that antibiotic. Other modifying β-lactam synthetase genes must be postulated to account for the production by some streptomycetes of related cephamycins A and B (27) and by *Flavobacterium* sp. 12,154 a complex variety of 7-formamido derivatives of cephalosporins (90).

We turn now to a different class of β-lactam biosynthetic genes that may play a role in connecting primary metabolism with antibiotic synthesis. Cephalosporin- and cephamycin C–producing strains of actinomycetes possess an enzyme, L-lysine ε-aminotransferase, that mediates formation of α-aminoadipic acid, a precursor of ACV, by removal of the ε-amino group from lysine (52, 69). Fungal β-lactam-producing strains, on the other hand, make α-aminoadipic acid as an intermediate in the synthesis of lysine and, therefore, do not seem to require this enzyme for ACV synthesis and β-lactam production. Madduri et al (70) described the cloning of a gene responsible for lysine aminotransferase activity in *S. clavuligerus* and showed that it maps in the cluster of β-lactam biosynthetic genes (Figure 2). The gene, designated *lat,* is found exclusively in *Streptomyces* and *Nocardia* β-lactam-producing species (25, 70). Coque et al (25) sequenced the DNA region located upstream of the *pcbAB-pcbC* gene cluster in *N. lactamdurans* and found an unidentified ORF of 1353 bp, coding for a protein of 451 amino acids, and separated by 64 bp from the 5' end of the *pcbAB* gene (25). The deduced molecular weight based on the open reading frame is 48,811. When that ORF was expressed in *S. lividans,* high levels of lysine ϵ-aminotransferase activity accumulated. No activity could be detected in the untransformed cells. The enzyme was purified from the *S. lividans* transformants and shown to have a molecular weight of 52,800. Similar results were obtained for the *S. clavuligerus lat* gene (100). DNA probes from the *N. lactamdurans lat* gene hybridized strongly with DNA from four *Streptomyces* species that produce cephamycins but not with four other species that do not produce these β-lactams. These results and those of Madduri et al (70) employing a *S. clavuligerus lat* probe suggest that in *Streptomyces* spp. the *lat* gene is associated uniquely with β-lactam biosynthesis and is not a component of general lysine metabolism. The amino acid sequence encoded in the *lat* ORF was compared with sequences in a protein data bank and found to be similar to that of ornithine-5-aminotransferase and N-acetyl ornithine-5-aminotransferase. A consensus pyridoxal phosphate-binding site was noted around the *N. lactamdurans* and *S. clavuligerus* lysine residues at positions 300 and 304, respectively (25, 100). Both *lat* genes are closely linked to and transcribed in the same direction as the *pcbAB* and *pcbC* genes (Figure 2) (see also 100; P. Liras, M. A. Lumbreras, T. J. R. Coque & J. F. Martin, unpublished data). Expression of all three genes (and possibly other late genes in the pathway) may therefore be under coordinate control. In support of this idea is the fact that in *N. lactamdurans* no recognizable transcription termination signals were found in the *lat, pcbAB,* and *pcbC* intergenic regions (25).

How did the *lat* gene become linked to the β-lactam biosynthetic genes? Clearly, the coordinated synthesis of α-aminoadipate with its conversion to ACV by ACVS would be beneficial in terms of efficient production of β-lactams and might confer an ecological advantage to that producer organism. Gene clustering is one mechanism for achieving this end. In this respect, the β-lactam-producing fungi would seem to be at a disadvantage, because their source of α-aminoadipic acid derives from a primary metabolic process. If the β-lactam biosynthetic genes were transferred in totality from bacteria to fungi, an issue discussed shortly, then the β-lactam-producing fungi might harbor *lat* genes. Also, if a gene rearrangement were to have subsequently occurred, then the putative *lat* gene in fungi would not necessarily be linked to other β-lactam biosynthetic genes (25). Preliminary studies (C. Esmahan, E.

Alverez, E. Montenegro, S. Gil & J. F. Martin, unpublished data) suggest that *P. chrysogenum* and *C. acremonium* possess a lysine ε-aminotransferase activity. Whether the gene coding for this activity is related to the *lat* gene of β-lactam–producing actinomycetes, and whether it plays any role in β-lactam synthesis in fungi, remains to be clarified.

ORGANIZATION AND REGULATION OF PENICILLIN AND CEPHALOSPORIN BIOSYNTHETIC GENES

Clustering of Biosynthetic Genes

Researchers now commonly assume that antibiotic biosynthetic genes tend to exist in gene clusters. Malpartida & Hopwood (72) first demonstrated gene clustering in *Streptomyces* for actinorhodin biosynthetic genes. More recent work has shown that the β-lactam biosynthetic genes present in several *Streptomyces* and *Nocardia* species that produce cephamycin antibiotics are arranged in a cluster on the chromosome (17, 20, 26, 57–59, 70, 96, 100). Several studies (28, 29, 67, 68, 96, 98) establish that the *pcbAB, pcbC,* and *penDE* genes in *P. chrysogenum* and *A. nidulans* are also tightly clustered. These three genes entirely determine penicillin synthesis. A somewhat different situation is found in *C. acremonium,* in which the *pcbAB* and *pcbC* genes are linked on chromosome VI (91), whereas *cefEF* has been mapped to chromosome II (92). Figure 2 summarizes information on the organization and direction of transcription of β-lactam biosynthetic genes in *P. chrysogenum, A. nidulans, C. acremonium, S. clavuligerus, N. lactamdurans,* and *Flavobacterium* sp. 12,154.

Regulation of β-Lactam Genes and Application to Strain Improvement

In the preceding sections, we have surveyed the salient features of the structure and organization of β-lactam genes. Little, however, is known about the molecular mechanisms that regulate the expression of these genes despite the fact that many studies show the biosynthetic pathway is subject to numerous metabolic controls. Recent efforts in this direction have focused on characterizing the DNA regions controlling gene expression and analyzing transcription events in terms of critical cell growth parameters that affect antibiotic formation.

Several groups have reported transcription analysis of the upstream regions of the *pcbC* and *pcbAB* genes. In the fungi, the divergently transcribed genes are separated by about 1 kb (Figure 2). Smith et al (94) used S1 mapping and primer extension to identify transcription initiation sites in the 5' region of the *C. acremonium pcbC* gene. Major and minor pairs of mRNA start sites were found on either side of a pyrimidine-rich block in the promoter region at positions −64 and −72, relative to the first base of the ATG initiation codon. A consensus TATA box was observed 68 bp upstream of the first major transcription start site. A similar motif was found at position −147 in the 5' region of the *A. nidulans pcbC* gene. Its position matches that found in yeast promoters. Also, the sequence flanking the translation initiation codon matches the consensus fungal sequence. Barredo et al (13) mapped the start site of *P. chrysogenum pcbC* mRNA by primer extension and showed that a single transcript was made that originated close to the structural gene, starting at position −11. Similar studies by Kolar et al (55) with a penicillin-production strain of *P. chrysogenum* revealed two major transcription initiation sites, at −131 and −132 as well as a more distal site at −397. Primer extension studies by McCabe et al (68) of the *A. nidulans pcbAB* gene demonstrated a major mRNA start point at −230 bp (68). These authors found no recognizable core promoter sequences, a situation frequently encountered in fungal genes. Because the *pcbC* and *pcbAB* genes may be regulated in a coordinate fashion, a search was made for potential regulatory elements, such as receptor sites for *trans*-acting proteins, within the intergenic region separating the *pcbAB* and *pcbC* genes. A 53-bp region of dyad symmetry is located equidistant from the two genes, but no other extensive sequence identities were detected (68).

Transcriptional regulation appears to be an important feature of β-lactam gene expression in some but not all producer strains. Analysis of *pcbC* mRNA during a *C. acremonium* seven-day fermentation showed a large accumulation of a 1.5-kb transcript between the second and fourth days; this correlated with the appearance of products of the pathway after isopenicillin N (94). The fact that mRNA levels decreased after the fifth day when antibiotics peaked was attributed to stability of the enzyme. Regulation of *pcbC* expression in *C. acremonium* occurs, therefore, primarily at the transcriptional level. Similar studies of the *S. clavuligerus pcbC* gene show its expression to be under transcriptional control. When cultures of *S. clavuligerus* were grown in rich or defined media, the amounts of *pcbC* mRNA correlated well with the normalized IPNS enzyme activity and antibiotic production; in defined media, peak values of both occurred much earlier than in rich media (S. Mendelowitz, G. Cohen & Y. Aharonowitz, unpublished results). A rather different picture was found in *P. chrysogenum*. Levels of *pcbC* mRNA and IPNS stayed about the same throughout the fermentation, both in a wild-type strain and in a highly mutated overproducer strain (60). The latter exhibited 32- to 64-fold more mRNA than the wild-type strain. Penalva et al (38, 78) showed that in *A. nidulans* the *pcbC* gene was transcribed only after arrest of cell growth and only then was penicillin detected in the fermentation broth. Also, analysis of mRNA levels of penicillin biosynthetic genes in *A. nidulans* showed that in conditions in which penicillin synthesis was repressed, no transcripts were detected, suggesting common regulation of these genes at the transcriptional level (67).

The interpretation of regulatory mechanisms in mutant, high-β-lactam-producing strains is complicated by possible chromosomal aberrations in the cluster of biosynthetic genes. For example, one *P. chrysogenum* overproducer strain had 8–10 copies of the *pcbC* gene (95) and another contained the *pcbC* and *penDE* genes in a DNA segment of at least 35 kb amplified 14-fold (14). The significance of such findings is relevant to attempts to genetically manipulate high producer strains, either through introducing additional copies of β-lactam genes to overcome pathway blocks or by altering regulatory elements. Indeed, attempts to increase cephalosporin C yields in *C. acremonium* and penicillin in *P. chrysogenum* by inserting multiple copies of the *pcbC* gene were unsuccessful (93). A similar result was obtained in terms of penicillin production in a wild-type strain of *A. nidulans* (78). Nevertheless, the chromosomal integration of a single extra copy of the *cefDE* expandase-hydroxylase gene in a high-producing strain of *C. acremonium* significantly improved production of cephalosporin C (93). In this case, a twofold increase in expandase-hydroxylase activity was apparently sufficient to overcome a rate-limiting step in processing penicillin N. Furthermore, amplification of the *pcbC-penDE* gene cluster of *P. chrysogenum* Wis 54-1255 led to as much as a 40% improvement in production yields (103). Increased antibiotic yields were also reported in *A. nidulans* transformants containing multiple copies of *pcbAB* and *pcbC* genes (66).

THE ORIGIN AND EVOLUTION OF THE BIOSYNTHETIC GENES

In several places in this review, we allude to the relevance of studies on β-lactam genes for understanding the evolution of the biosynthetic pathway. In this respect, three aspects of β-lactam genes are especially noteworthy. The first is the remarkable sequence similarity among the genes. The second is the unusual and narrow distribution of microorganisms that make β-lactam antibiotics, and the third is the physical clustering of β-lactam genes. The interpretation of these findings from an evolutionary standpoint has been a fertile area for new ideas in the biology of β-lactams.

Sequence similarity among the bacterial and fungal β-lactam IPNS genes is exceptionally high when viewed with respect to other related prokaryotic and eukaryotic genes (Figure 3) (31, 41). Two plausible explanations have been advanced to account for these findings (105). The similarity might simply result from a slow but constant rate of evolutionary change of the gene, perhaps reflecting strict functional constraints on the IPNS protein and thus precluding many amino acid changes. A more elaborate version of this idea is to suppose varying rates of evolutionary change of the *pcbC* gene. Alternatively, the high similarity may have resulted from a horizontal gene-transfer event, plausibly from the bacterial lineage to the fungal lineage after the prokaryotic-eukaryotic split. This latter explanation was preferred for two reasons; first, the rate of molecular evolution of the fungal IPNS gene is not appreciably different from that of other fungal genes (22, 77, 105), and second, no evidence suggests that the presence of that gene, and by inference β-lactam antibiotics, confers any selective advantage for survival of the organism. According to this interpretation of the similarity among microbial IPNS genes, Weigel et al (105) estimated that a gene-transfer event took place about 370 million years ago, i.e. well after the proposed time of the prokaryotic-eukaryotic split some 1.8 billion years ago (30).

In their analysis of the evolutionary history of the *pcbC* gene, Weigel et al (105) used sequence data from fungal and gram-positive *Streptomyces* species. Subsequently, the *pcbC* gene of a β-lactam-producing species of *Flavobacterium* was cloned and sequenced (88). *Flavobacterium* spp. belong to a third major group of β-lactam producers, the gram-negative bacteria. This information has been used to reexamine the gene-transfer hypothesis. Two main issues need to be clarified in reconstructing the evolution of the *pcbC* gene, the rate of evolutionary change of the gene and the nature of the horizontal gene-transfer event or events. Figure 4 shows the unrooted phylogenetic tree of nine *pcbC* genes. It clearly distinguishes the three taxonomic classes of microorganisms. If a constant-rate hypothesis is assumed, then the root and relative lengths of the branches of the gene tree can be determined. When this is done (63, 77), the ratio between the distance separating the gram-positive and gram-negative *pcbC* genes and the distance separating the fungal and bacterial *pcbC* genes is much higher in comparison to the ratio expected from a phylogenetic tree for eubacteria and fungi as determined from sequence analysis of 5S RNA (22). Therefore, the bacterial and fungal IPNS genes are much more closely related to each other than would have been expected on the basis of their rates of evolution. This surprising conclusion is the more so because the 5S RNA genes are considered a slowly evolving group of genes (42). The constant-rate hypothesis is therefore incompatible with the data. The possibility that the similarity among the IPNS genes may result from varying rates of evolution among the lineages has also been examined (63). Although no conclusive evidence was found for this view, it cannot be entirely ruled out.

The above discussion indicates that an evolutionary scenario involving gene transfer between prokaryotes and eukaryotes can best explain the similarity of the *pcbC* genes. Other β-lactam genes in the biosynthetic pathway would probably also have been transmitted in this way and, therefore, would be expected to display high sequence similarity. This prediction has recently

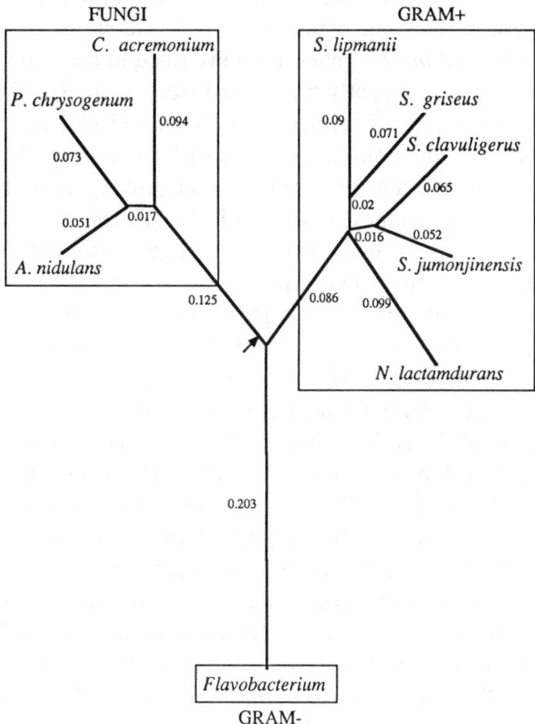

Figure 4 Unrooted phylogenetic tree derived from analysis of the nucleotide sequences of nine IPNS genes. The tree was constructed according to the neighbor-joining method (63). Branch lengths are proportional to the number of nonsynonymous substitutions per site. The arrow indicates the position of the root that is obtained by dividing into two equal parts the longest pathway in the tree.

received strong support from two results: (*a*) the *cefEF* gene of *C. acremonium* and the *cefE* gene of *S. clavuligerus* share almost the same percentage sequence identity as do the *pcbC* genes from these organisms, and (*b*) high sequence similarity is found between the three repeated domains of the fungal and bacterial *pcbAB* genes.

What other evidence is there for horizontal gene transfer? The peculiar and narrow distribution in nature of β-lactam-producing species strongly suggests horizontal inheritance of the biosynthetic genes rather than vertical transmission from common ancestral genes (24, 77). Also, we have seen that the biosynthetic genes are invariably clustered. Whereas widespread physical association of the β-lactam biosynthetic genes does not of course, by itself, constitute proof for horizontal gene transfer, it is consistent with this hypothesis and provides a rationale for explaining the narrow distribution of β-lactam genes in microorganisms. Nevertheless, two features of the clustering of β-lactam genes need some explanation. In *S. clavuligerus* and *N. lactamdurans,* the *pcbC* and *pcbAB* genes are transcribed in the same direction, yet in the fungi they are divergently transcribed (see Figure 2). Also, the *pcbAB* gene of the gram-positive *S. clavuligerus* is located between the *pcbC* and *cefE* genes, while in the gram-negative *Flavobacterium* sp. 12,154, it maps outside these genes. If a transfer event from the prokaryotes to the eukaryotes is the source of the relatedness of their β-lactam genes, then gene rearrangements must have occurred while maintaining tight clustering. Possibly, such alterations reflect in some unknown way the particular needs and circumstances of these microorganisms for β-lactam antibiotics.

The question of the time and direction of the assumed gene-transfer event still needs to be addressed. Weigel et al (105) postulated a single transfer event to have taken place from *Streptomyces* to fungi and estimated this happened well after the split leading to the gram-positive and gram-negative bacteria some 1–1.5 million years ago (42). However, in this case, the gram-positive *Streptomyces* IPNS genes should more closely resemble the fungal genes than does the gram-negative *Flavobacterium* gene. Thus, the *Streptomyces* and fungal IPNS genes would have evolved over a relatively short time span whereas the *Streptomyces* and *Flavobacterium* genes would have evolved over a much longer time corresponding to the split between the two groups of eubacteria. The fact that the topology of the gene tree and species tree is the same rules out this possibility. Otherwise, one must postulate the existence of multiple gene-transfer events at about the same time, between gram-positive, gram-negative, and fungal species. Alternatively, if we suppose the putative transfer event occurred well before the gram-positive–gram-negative split, the sequence data would permit reconstruction of an IPNS gene tree that would be identical to the species tree in topology and free of the distortion at its distal end. These qualitative arguments suggest that the transfer event could not have taken place at a time very much earlier or very much later than the gram-positive–gram-negative split.

What can we say about the direction of the hypothetical gene-transfer event? The sequence data provide no evidence regarding the direction. A priori, we have no reason to assume one direction over the other, although several independent observations suggest the putative transfer occurred from bacteria to fungi. First, the bacteria possess a more elaborate capacity to make β-lactams than do the fungi, suggesting that the transfer event resulted in the latter acquiring only part of the full complement of biosynthetic genes. Second, a comparison of the G+C base composition of bacterial and fungal IPNS genes suggests that codon usage in fungal IPNS genes is reminiscent of a higher G+C ancestor (77). Such an ancestor could be the *Streptomyces* that have a high G+C content, one greater than 70%. Another structural feature of the fungal IPNS genes that implies a bacterial origin is that they all lack introns. Fungal genes do occasionally contain introns and in fact the gene for acetyltransferase, uniquely associated with the fungal penicillin producers, contains three introns (99, 104). Third, an argument for a bacterial origin of the transfer event depends on the generalization that because the *Streptomyces* are gram-positive organisms and are susceptible to β-lactam antibiotics, they must have evolved specific resistance mechanisms to avoid self-toxification. Fungi are resistant to β-lactams. Therefore, β-lactam biosynthetic genes would probably have arisen first in the *Streptomyces* (or some other prokaryote) together with their resistance genes, rather than the other way round.

If we accept the notion of gene transfer as a key feature in the evolution of β-lactam biosynthetic genes in microorganisms, what can we say about how this took place? The only statement that can be made with some certainty is that such an event or events would require the bacterial and fungal participants to be in close proximity. Indeed, we know that all three groups of β-lactam producers inhabit soil environments. Also, a variety of genetic-transfer mechanisms, such as conjugative plasmids, DNA transformation, and bacteriophage transduction, in principle could be mediators of gene transfer. In summary, we can say that in its simplest form the horizontal gene-transfer hypothesis supposes that a single gene-transfer event occurred from the actinomycetes to the fungi close to the time of the gram-positive–gram-negative split. If, however, IPNS and other β-lactam genes should turn out to be more widespread among, for example, the prokaryotic organisms than is presently indicated, then this account of the evolution of β-lactam genes may need significant revision.

CONCLUDING REMARKS

Molecular characterization of the known β-lactam genes present in the biosynthetic cluster in fungi and bacteria is nearing completion. Genes encoding ACV synthetase, cyclase, epimerase, expandase, and hydroxylase have all been cloned, sequenced, and mapped. Perhaps the most striking result to emerge is the extremely high sequence similarity found among the prokaryotic and eukaryotic biosynthetic β-lactam genes and their occurence in gene clusters. These findings are compatible with a horizontal gene transfer of the biosynthetic genes from the prokaryotes to the eukaryotes. Despite these impressive advances in our understanding of the structural and organizational features of β-lactam biosynthetic genes and of how they evolved, we have only a very superficial insight into the control mechanisms that operate on these genes. Undoubtedly, one of the key tasks in future studies will be to identify the *cis*- and *trans*-acting DNA elements that govern expression of the biosynthetic genes. A start in this direction has already been made. Another will be to identify and map the remaining β-lactam biosynthetic genes. In conjunction with such studies, we anticipate a major effort to modify β-lactam biosynthetic genes so as to alter enzyme characteristics such as substrate specificity and catalytic activity. Structure-function analysis of the antibiotic synthetases is crucial to this endeavor. Now that adequate amounts of highly purified enzymes are available from heterologous expression systems, these enzymes are the focus of rigorous physical study. From a fundamental point of view, the mechanism of action of, for example, ACVS and IPNS is of great intrinsic interest. From a practical point of view, it may be possible to rationally design and create producer strains that make new and improved antibiotics. Amplification of specific biosynthetic genes promises to be a powerful tool to overcome metabolic blocks in producer strains. Reconstruction of the β-lactam gene cluster may lead to more efficient gene expression and increased antibiotic yields. Along rather different lines, we expect that a search for new sources of β-lactam genes employing sensitive nucleic acid hybridization probes and polymerase chain reaction methodologies will result in the discovery of β-lactam genes with potentially new and useful properties.

ACKNOWLEDGMENTS

We are grateful to James R. Miller and Steven Kovacevic for providing us with material prior to publication and to Paul Skatrud for sending us in advance a copy of his review. We thank Dan Graur and Ilan Yudkewitz for help with the evolutionary analysis. Some of the research carried out at Tel Aviv University was supported in part by a joint grant from the National Council for Research and Development, Israel, and the GBF, FRG, and by the German Israeli Foundation, GIF.

Literature Cited

1. Alvarez, E., Cantoral, J. M., Barredo, J. L., Diez, B. Martin, J. F. 1987. Purification to homogeneity and characterization of acyl coenzyme A:6-aminopenicillanic acid acyltransferase of *Penicillium chrysogenum*. *Antimicrob. Agents. Chemother*. 31:1675–82
2. Arnstein, H. R. V., Morris, D. 1960. The structure of a peptide containing α-aminoadipic acid, cysteine, and valine, present in the mycelium of *Penicillium chrysogenum*. *Biochem. J.* 76:357–61
3. Baker, B. J., Dotzlaf, J. E., Yeh, W.-K. 1991. Deacetoxycephalosporin C hydroxylase of *Streptomyces clavuligerus:* purification, characterization, bifunctionality, and evolutionary implication. *J. Biol. Chem.* 266:5087–93
4. Baldwin, J. E., Abraham, E. P. 1988. The biosynthesis of penicillins and cephalosporins. *Nat. Prod. Rep.* 5:129–45
5. Baldwin, J. E., Adlington, R. M., Coates, B., Crabbe, M. J. C., Crouch, N. P., et al. 1987. Purification and initial characterization of an enzyme with deacetoxycephalosporin C synthetase and hydroxylase activities. *Biochem. J.* 245:831–41
6. Baldwin, J. E., Bird, J. W., Field, R. A., O'Callaghan, N. M., Schofield, C. J. 1990. Isolation and partial characterisation of ACV synthetase from *Cephalosporium acremonium* and *Streptomyces clavuligerus*. *J. Antibiot.* 43:1055–57

7. Baldwin, J. E., Blackburn, J. M., Schofield, C. J., Sutherland, J. D. 1990. High level expression in *Escherichia coli* of a fungal gene under the control of strong promoters. *FEMS Microbiol. Lett.* 68:45–52
8. Baldwin, J. E., Coates, J. B., Moloney, M. G., Pratt, A. J., Willis, A. C. 1990. Photoaffinity labelling of isopenicillin N synthetase. *Biochem. J.* 266:561–67
9. Baldwin, J. E., Gagnon, J., Ting, H.-H. 1985. N-terminal amino acid sequence and some properties of isopenicillin-N synthetase from *Cephalosporium acremonium*. *FEBS Lett.* 188:253–58
10. Baldwin, J. E., Keeping, J. W., Singh, P. D., Vallejo, C. A. 1981. Cell-free conversion of isopenicillin N into deacetoxycephalosporin C by *Cephalosporium acremonium* mutant M-0198. *Biochem. J.* 194:649–51
11. Baldwin, J. E., Killin, S. J., Pratt, A. J., Sutherland, J. D., Turner, N. J., et al. 1987. Purification and characterization of cloned isopenicillin N synthetase. *J. Antibiot.* 40:652–59
12. Banko, G., Demain, A. L., Wolfe, S. 1987. ACV synthetase: a multifunctional enzyme with broad specificity for the synthesis of penicillin and cephalosporin precursors. *J. Am. Chem. Soc.* 109:2858–60
13. Barredo, J. L., Cantoral, J. M., Alvarez, E., Diez, B., Martin, J. F. 1989. Cloning, sequence analysis and transcriptional study of the isopenicillin N synthase of *Penicillium chrysogenum* AS-P-78. *Mol. Gen. Genet.* 216:91–98
14. Barredo, J. L., Diez, B., Alvarez, E., Martin, J. F. 1989. Large amplification of a 35-kb DNA fragment carrying two penicillin biosynthetic genes in high penicillin producing strains of *Penicillium chrysogenum*. *Curr. Genet.* 16: 453–59
15. Barredo, J. L., van Soligen, P., Diez, B., Alvarez, E., Cantoral, J. M., et al. 1989. Cloning and characterization of acyl-CoA:6-APA acyltransferase gene of *Penicillium chrysogenum*. *Gene* 83:291–300
16. Bunnell, C. A., Luke, W. D., Perry, F. M. 1986. In *Beta-Lactam Antibiotics for Clinical Use,* ed. S. F. Queener, J. E. Webber, S. W. Queener, pp. 225–84. New York: Dekker
17. Burnham, M. K. R., Hodgson, J. E., Normansell, I. D. 1987. Isolation and expression of genes involved in the biosynthesis of beta-lactams. *Eur. Patent Application, No. 0233715*
18. Cantwell, C. A., Beckmann, R. J., Dotzlaf, J. E., Fisher, D. L., Skatrud, P. L., et al. 1990. Cloning and expression of a hybrid *Streptomyces clavuligerus cefE* gene in *Penicillium chrysogenum*. *Curr. Genet.* 17:213–21
19. Carr, L. G., Skatrud, P. L., Scheetz, M. E., Queener, S. W., Ingolia, T. D. 1986. Cloning and expression of the isopenicillin N synthetase gene from *Penicillium chrysogenum*. *Gene* 48: 257–66
20. Chen, C. W., Lin, H.-F., Kuo, C. L., Tsai, H.-L., Tsai, J. F.-Y. 1988. Cloning and expression of a DNA sequence conferring cephamycin C production. *Bio/Technology* 6:1222–24
21. Chen, J. V., Orville, A. M., Harpel, M. R., Frolik, C. A., Surerus, K. K., et al. 1989. Spectroscopic studies of isopenicillin N synthase. *J. Biol. Chem.* 264:21677–81
22. Chen, M.-W., Anne, J., Volckaert, G., Huysmans, E., Vandenberghe, A., et al. 1984. The nucleotide sequences of the 5SrRNAs of seven molds and a yeast and their use in studying ascomycete phylogeny. *Nucleic Acids Res.* 12:4881–92
23. Cohen, G., Newton, G. L., Fahey, R., Aharonowitz, Y. 1991. Thioredoxin like disulfide reductase system in Streptomyces. *Int. Symp. Biol. Actinomycetes, 8th, Madison, Wis.* (Abstr.)
24. Cohen, G., Shiffman, D., Mevarech, M., Aharonowitz, Y. 1990. Microbial isopenicillin N synthase genes: structure, function, diversity and evolution. *Trends Biotechnol.* 8:105–11
25. Coque, J. J. R., Liras, P., Laiz, L., Martin, J. F. 1991. A gene encoding lysine 6-aminotransferase which forms α-aminoadipic acid, a precursor of beta-lactam antibiotics, is located in the cluster of cephamycin biosynthetic genes in *Nocardia lactamdurans*. *J. Bacteriol.* 173:6258–64
26. Coque, J. J. R., Martin, J. F., Calzada, J. G., Liras, P. 1991. The cephamycin biosynthetic genes *pcbAB,* encoding a large multidomain peptide synthetase, and *pcbC* of *Nocardia lactamdurans* are clustered together in an organization different from the same genes in *Acremonium chrysogenum* and *Penicillium chrysogenum*. *Mol. Microbiol.* 5:1125–33
27. Demain, A. L. 1983. Biosynthesis of beta-lactam antibiotics. See Ref. 27a, 1:189–228

27a. Demain, A. L., Solomon, N. A., eds. 1983. *Antibiotics Containing the Beta-Lactam Structure*. Berlin/Heidelberg: Springer-Verlag

28. Diez, B., Barredo, J. L., Alvarez, E.,

Cantoral, J. M., van Solingen, P., et al. 1989. Two genes involved in penicillin biosynthesis are linked in a 5.1 kb *Sal*I fragment in the genome of *Penicillium chrysogenum*. *Mol. Gen. Genet.* 218: 572–76

29. Diez, B. S., Gutierrez, S., Barrredo, J. L., van Soligen, P., van der Voort, L. H. M., et al. 1990. The cluster of penicillin biosynthetic genes. Identification and characterization of the *pcbAB* gene encoding the α-aminoadipyl-cysteinyl-valine synthetase and linkage to the *pcbC* and *penDE* genes. *J. Biol. Chem.* 265:16358–65
30. Doolittle, R. F., Anderson, K. L., Feng, D.-F. 1989. Estimating the prokaryote-eukaryote divergence time from protein sequences. In *The Hierarchy of Life*, ed. B. Fernholm, K. Bremer, H. Jornvall, pp. 73–85. Amsterdam: Elsevier
31. Doolittle, R. F., Feng, D. F., Johnson, M. S., McClure, M. A. 1986. Relationship of human protein sequences to those of other organisms. *Cold Spring Harbor Symp. Quant. Biol.* 51:447–55
32. Doran, J. L., Leskiw, B. K., Petrich, A. K., Westlake, D. W. S., Jensen, S. E. 1991. Production of *Streptomyces clavuligerus* isopenicillin N synthase in *Escherichia coli* using two-cistron expression system. *J. Ind. Microbiol.* 5:197–206
33. Dotzlaf, J. E., Yeh, W.-K. 1987. Copurification and characterization of deacetoxycephalosporin C synthetase/hydroxylase from *Cephalosporium acremonium*. *J. Bacteriol.* 169:1611–18
34. Dotzlaf, J. E., Yeh, W.-K. 1989. Purification and properties of deacetoxycephalosporin C synthase from recombinant *Escherichia coli* and its comparison with the native enzyme purified from *Streptomyces clavuligerus*. *J. Biol. Chem.* 264:10219–27
35. Elander, R. P. 1983. Strain improvement and preservation of beta-lactam-producing microorganisms. See Ref. 27a, 1:97–146
36. Fawcett, P. A., Usher, J. J., Huddleston, J. A., Bleaney, R. C., Nisbet, J. J., et al. 1976. Synthesis of δ-(L-α-aminoadipyl)cysteinylvaline and its role in penicillin biosynthesis. *Biochem. J.* 157:651–60
37. García-Dominguez, M., Liras, P., Martin, J. F. 1991. Cloning and characterization of the isopenicillin N synthase gene of *Streptomyces griseus* NRRL 3851 and studies of expression and complementation of the cephamycin pathway in *Streptomyces clavuligerus*. *Antimicrob. Agents Chemother.* 35:44–52
38. Gomez-Pardo, E., Penalva, M. A. 1990. The upstream region of the IPNS gene determines expression during secondary metabolism in *Aspergillus nidulans*. *Gene* 89:109–15
39. Gutierrez, S., Diez, B., Alvarez, E., Barredo, J. L., Martin, J. F. 1991. Expression of the *penDE* gene of *Penicillium chrysogenum* encoding isopenicillin-N acyltransferase in *Cephalosporium acremonium* production of benzylpenicillin by the transformants. *Mol. Gen. Genet.* 225:56–64
40. Gutierrez, S., Diez, B., Montenegro, E., Martin, J. F. 1991. Characterization of the *Cephalosporium acremonium pcbAB* gene encoding α-aminoadipyl-cysteinyl-valine synthetase, a large multidomain peptide synthetase: linkage to the *pcbC* gene as a cluster of early cephalosporin-biosynthetic genes and evidence of multiple functional domains. *J. Bacteriol.* 173:2354–65
41. Hensel, R., Zwickl, P., Fabry, S., Lang, J., Palm, P. 1989. Sequence comparison of glyceraldehyde-3-phosphate dehydrogenase from three urkingdoms: evolutionary implications. *Can. J. Microbiol.* 35:81–85

41a. Hershberger, C. L., Queener, S. W., Hegeman, G., eds. 1989. *Genetics and Molecular Biology of Industrial Microorganisms*. Washington, DC: Am. Soc. Microbiol.

42. Hori, H., Osawa, S. 1987. Origin and evolution of organisms as deduced from 5S ribosomal RNA sequences. *Mol. Biol. Evol.* 4:445–72
43. Hoskins, J. A., O'Callaghan, N., Queener, S. W., Cantwell, C. A., Wood, J. S. 1990. Gene disruption of the *pcbAB* gene encoding ACV synthetase in *Cephalosporium acremonium*. *Curr. Genet.* 18:523–30
44. Ingolia, T. D., Queener, S. W. 1989. Beta-lactam biosynthetic genes. *Med. Res. Rev.* 9:245–64
45. Jensen, S. E. 1986. Biosynthesis of cephalosporins. *CRC Crit. Rev. Biotechnol.* 3:277–310
46. Jensen, S. E., Leskiw, B. K., Vining, L. C., Aharonowitz, Y., Westlake, D. W., et al. 1986. Purification of isopenicillin N synthetase from *Streptomyces clavuligerus*. *Can. J. Microbiol.* 32: 953–58
47. Jensen, S. E., Westlake, D. W., Wolfe, S. 1982. Cyclization of δ-(L-α-aminoadipyl)-L-cysteinyl-D-valine to penicillins by cell-free extracts of *Streptomyces clavuligerus*. *J. Antibiot.* 35: 483–90
48. Jensen, S. E., Westlake, D. W., Wolfe,

S. 1983. Partial purification and characterization of isopenicillin N epimerase activity from *Streptomyces clavuligerus*. *Can. J. Microbiol.* 29:1526–31

49. Jensen, S. E., Westlake, D. W. S., Wolfe, S. 1985. Deacetoxycephalosporin C synthetase and deacetoxycephalosporin C hydroxylase are two separate enzymes in *Streptomyces clavuligerus*. *J. Antibiot.* 38:263–65
50. Jensen, S. E., Wong, A., Rollins, M. J., Westlake, D. W. S. 1990. Purification and partial characterization of δ-(L-α-aminoadipyl)-L-cysteinyl-D-valine synthetase from *Streptomyces clavuligerus*. *J. Bacteriol.* 172:7269–71
51. Jones, G. H., Hopwood, D. A. 1984. Activation of phenoxazinone synthase expression in *Streptomyces lividans* by cloned DNA sequence from *Streptomyces antibioticus*. *J. Biol. Chem.* 259: 14158–64
52. Kern, B. A., Hendlin, D., Inamine, E. 1980. L-lysine ε-aminotransferase involved in cephamycin C synthesis in *Streptomyces lactamdurans*. *Antimicrob. Agents Chemother.* 17:676–85
53. Kleinkauf, H., Von Dohren, H. 1990. Nonribosomal biosynthesis of peptide antibiotics. *Eur. J. Biochem.* 192:1–15
54. Kohsaka, M., Demain, A. L. 1976. Conversion of penicillin N to cephalosporin(s) by cell-free extracts of *Cephalosporium acremonium*. *Biochem. Biophys. Res. Commun.* 70:465–73
55. Kolar, M., Holzmann, K., Weber, G., Leitner, E., Schwab, H. 1991. Molecular characterization and functional analysis in *Aspergillus nidulans* of the 5'-region of the *Penicillium chrysogenum* isopenicillin N synthetase gene. *J. Biotechnol.* 17:67–80
56. Konomi, T., Herchen, S., Baldwin, J. E., Yoshida, M., Hunt, N. A., et al. 1979. Cell-free conversion of δ-(L-α-aminoadipyl)-L-cysteinyl-D-valine into antibiotic with the properties of isopenicillin N in *Cephalosporium acremonium*. *Biochem. J.* 184:427–30
57. Kovacevic, S., Miller, J. R. 1991. Cloning and sequencing of the β-lactam hydroxylase gene (*cefF*) from *Streptomyces clavuligerus:* gene duplication may have led to separate hydroxylase and expandase activities in the actinomycetes. *J. Bacteriol.* 173:398–400
58. Kovacevic, S., Tobin, M. B., Miller, J. R. 1990. The β-lactam biosynthesis genes for isopenicillin N epimerase and deacetoxycephalosporin C synthetase are expressed from a single transcript in *Streptomyces clavuligerus*. *J. Bacteriol.* 172:3952–58
59. Kovacevic, S., Weigel, B. J., Tobin, M. B., Ingolia, T. D., Miller, J. R. 1989. Cloning, characterization, and expression in *Escherichia coli* of the *Streptomyces clavuligerus* gene encoding deacetoxycephalosporin C synthetase. *J. Bacteriol.* 171:754–60
60. Kuck, U., Walz, M., Mohr, G., Mracek, M. 1989. The 5'-sequence of the isopenicillin N-synthetase gene (*pcbC*) from *Cephalosporium acremonium* directs the expression of the prokaryotic hygromycin B phosphotransferase gene (*hph*) in *Aspergillus niger*. *Appl. Microbiol. Biotechnol.* 31:358–65
61. Kupka, J., Shen, Y. Q., Wolfe, S., Demain, A. L. 1983. Partial purification and properties of the α-ketoglutarate-linked ring-expansion enzyme of β-lactam biosynthesis of *Cephalosporium acremonium*. *FEMS Microbiol. Lett.* 16:1–6
62. Laiz, L., Liras, P., Castro, J. M., Martin, J. F. 1990. Purification and characterization of the isopenicillin N epimerase from *Nocardia lactamdurans*. *J. Gen. Microbiol.* 136:663–71
63. Landan, G., Cohen, G., Aharonowitz, Y., Shuali, Y., Graur, D., et al. 1990. Evolution of isopenicillin N synthase genes may have involved horizontal gene transfer. *Mol. Biol. Evol.* 7:399–406
64. Landman, O., Shiffman, D., Av-Gay, Y., Aharonowitz, Y., Cohen, G. 1991. High level expression in *Escherichia coli* of isopenicillin N synthase genes from *Flavobacterium* and *Streptomyces* and recovery of active enzyme from inclusion bodies. *FEMS Microbiol. Lett.* 84:239–44
65. Leskiw, B. K., Aharonowitz, Y., Mevarech, M., Wolfe, S., Vining, L. C., et al. 1988. Cloning and nucleotide sequence determination of the isopenicillin N synthetase gene from *Streptomyces clavuligerus*. *Gene* 62:187–96
66. MacCabe, A. P., Riach, M. B. R., Kinghorn, J. R. 1991. Identification and expression of the ACV synthetase gene. *J. Biotechnol.* 17:91–97
67. MacCabe, A. P., Riach, M. B. R., Unkles, S. E., Kinghorn, J. R. 1990. The *Aspergillus nidulans npeA* locus consists of three contiguous genes required for penicillin biosynthesis. *EMBO J.* 9:279–87
68. MacCabe, A. P., van Liempt, H., Palissa, H., Unkles, S. E., Riach, M. B. R., et al. 1991. δ-(L-α-Aminoadipyl)-L-cysteinyl-D-valine synthetase from *Asper-*

gillus nidulans: molecular characterization of the *acvA* gene encoding the first enzyme of the penicillin biosynthetic pathway. *J. Biol. Chem.* 266:12646–54

69. Madduri, K., Stuttard, C., Vining, L. C. 1989. Lysine catabolism in *Streptomyces* sp. is primarily through cadaverine: β-lactam producers also make α-aminoadipate. *J. Bacteriol.* 171:299–302
70. Madduri, K., Stuttard, C., Vining, L. C. 1991. Cloning and location of a gene governing lysine ϵ-aminotransferase, an enzyme initiating beta-lactam biosynthesis in *Streptomyces* sp. *J. Bacteriol.* 173:985–88
71. Malpartida, F., Hallam, S. E., Kieser, H. M., Motamedi, H., Hutchinson, C. R., et al. 1987. Homology between *Streptomyces* genes coding for synthesis of different polyketides used to clone antibiotic biosynthetic genes. *Nature* 325:818–21
72. Malpartida, F., Hopwood, D. A. 1984. Molecular cloning of the whole biosynthetic pathway of a *Streptomyces* antibiotic and its expression in a heterologous host. *Nature* 309:462–64
73. Martin, J. F., Ingolia, T. D. Queener, S. W. 1991. Molecular genetics of penicillin and cephalosporin antibiotic biosynthesis. In *Molecular Industrial Mycology: Systems and Applications for Filamentous Fungi,* ed. S. A. Leong, R. M. Berka, pp. 149–96. New York: Dekker
74. Miller, J. R., Ingolia, T. D. 1989. Cloning and characterization of beta-lactam biosynthetic genes. *Mol. Microbiol.* 3:689–95
75. Ming, L.-J., Que, L. Jr., Kriauciunas, A., Frolik, C. A., Chen, V. A. 1990. Coordination chemistry of the metal binding site of isopenicillin N synthase. *Inorg. Chem.* 29:1111–12
76. Montenegro, E., Barredo, J. L., Gutierez, S., Diez, B., Alvarez, E., et al. 1990. Cloning, characterization of the acyl-CoA:6-amino penicillanic acid acyltransferase gene of *Aspergillus nidulans* and linkage to the isopenicillin N synthase gene. *Mol. Gen. Genet.* 221: 322–30
77. Penalva, M. A., Moya, A., Dopazo, J., Ramon, D. 1990. Sequences of isopenicillin N synthetase gene suggest horizontal gene transfer from prokaryotes to eukaryotes. *Proc. R. Soc. London Ser. B* 241:164–69
78. Penalva, M. A., Vian, A., Patino, C., Perez-Aranda, A. Ramon, D. 1989. Molecular biology of penicillin production in *Aspergillus nidulans.* See Ref. 41a, pp. 256–61
79. Perry, D., Abraham, E. P., Baldwin, J. E. 1988. Factors affecting the isopenicillin N synthetase reaction. *Biochem. J.* 255:345–51
80. Queener, S. W., Neuss, N. 1982. The biosynthesis of beta-lactam antibiotics. In *The Chemistry and Biology of Beta-Lactam Antibiotics,* ed. E. B. Morin, M. Morgan, 3:1–81. London: Academic
81. Ramon, D., Carramolino, L., Patino, C., Sanchez, F., Penalva, M. A. 1987. Cloning and characterization of the isopenicillin N synthetase gene mediating the formation of the β-lactam ring in *Aspergillus nidulans. Gene* 57:171–81
82. Ramsden, M., McQuade, B. A., Saunders, K., Turner, M. K., Harford, S. 1989. Characterization of a loss-of-function in the isopenicillin N synthetase gene of *Acremonium chrysogenum. Gene* 85:267–73
83. Samson, S. M., Belagaje, R., Blankenship, D. T., Chapman, J. L., Perry, D., et al. 1985. Isolation, sequence determination and expression in *Escherichia coli* of the isopenicillin N synthetase gene from *Cephalosporium acremonium. Nature* 318:191–94
84. Samson, S. M., Chapman, J. L., Belagaje, R., Queener, S. W., Ingolia, T. D. 1987. Analysis of the role of cysteine residues in isopenicillin N synthetase activity by site-directed mutagenesis. *Proc. Natl. Acad. Sci. USA* 84:5705–9
85. Samson, S. M., Dotzlaf, J. E., Slisz, M. L., Becker, G. W., van Frank, R. M., et al. 1987. Cloning and expression of the fungal expandase/hydroxylase gene involved in cephalosporin biosynthesis. *Bio/Technology* 5:1207–16
86. Scheidegger, A., Kuenzi, M. T., Nuesch, J. 1984. Partial purification and catalytic properties of a bifunctional enzyme in the biosynthetic pathway of beta-lactams in *Cephalosporium acremonium. J. Antibiot.* 37:522–31
87. Shiffman, D. 1989. *Cloning, sequencing, expression and comparative analysis of microbial isopenicillin N synthase genes.* PhD thesis. Tel Aviv University, Tel Aviv Israel. 92 pp.
88. Shiffman, D., Cohen, G., Aharonowitz, Y., Palissa, H., VonDohren, H., et al. 1990. Nucleotide sequence of the isopenicillin N synthase gene (*pcbC*) of the gram negative *Flavobacterium* sp. SC 12,154. *Nucleic Acids Res.* 18:660
89. Shiffman, D., Mevarech, M., Jensen, S. E., Cohen, G., Aharonowitz, Y. 1988. Cloning and comparative sequence analysis of the gene coding for isopenicillin N synthase in *Streptomyces. Mol. Gen. Genet.* 214:562–69
90. Singh, P. D., Young, M. G., Johnson,

J. H., Cimarusti, C. M., Sykes, R. 1984. Bacterial production of 7-formamidocephalosporins, isolation and structure determination. *J. Antibiot.* 37:773–80

91. Skatrud, P. L. 1992. Molecular biology of the beta-lactam producing fungi. In *More Gene Manipulations in Fungi,* ed. T. W. Bennett, L. L. Lashre. New York: Academic. In press
92. Skatrud, P. L., Queener, S. W. 1989. An electrophoretic molecular karyotype for an industrial strain of *Cephalosporium acremonium. Gene* 78:331–38
93. Skatrud, P. L., Tietz, A. J., Ingolia, T. D., Cantwell, C. A., Fisher, D. L., et al. 1989. Use of recombinant DNA to improve production of cephalosporin C by *Cephalosporium acremonium. Bio/Technology* 7:477–86
94. Smith, A. W., Ramsden, M., Dobson, M. J., Harford, S., Peberdy, J. F. 1990. Regulation of isopenicillin N synthetase (IPNS) gene expression in *Acremonium chrysogenum. Bio/Technology* 8:237–40
95. Smith, D. J., Bull, J. H., Edwards, J., Turner, G. 1989. Amplification of the isopenicillin N synthetase gene in a strain of *Penicillium chrysogenum* producing high levels of penicillin. *Mol. Gen. Genet.* 216:492–97
96. Smith, D. J., Burnham, M. K. R., Bull, J. H., Hodgson, J. E., Ward, J. M., et al. 1990. β-Lactam antibiotic biosynthetic genes have been conserved in clusters in prokaryotes and eukaryotes. *EMBO J.* 9:741–47
97. Smith, D. J., Burnham, M. K. R., Edwards, J., Earl, A. J., Turner, G. 1990. Cloning and heterologous expression of the penicillin biosynthetic gene cluster from *Penicillium chrysogenum. Bio/Technology* 8:39–41
98. Smith, D. J., Earl, A. J., Turner, G. 1990. The multifunctional peptide synthetase performing the first step in penicillin biosynthesis in *Penicillium chrysogenum* is a 421,073 dalton protein similar to *Bacillus brevis* peptide antibiotic synthetases. *EMBO J.* 9:2743–50
99. Tobin, M. B., Fleming, M. D., Skatrud, P. L., Miller, J. R. 1990. Molecular characterization of the acyl-coenzyme A:isopenicillin N acyltransferase gene (penDE) from *Penicillium chrysogenum* and *Aspergillus nidulans* and activity of recombinant enzyme in *Escherichia coli. J. Bacteriol.* 172:5908–14
100. Tobin, M. B., Kovacevic, S., Madduri, K., Hoskins, J. A., Skatrud, P. L., et al. 1991. Localization of lysine ϵ-aminotransferase (*lat*) and δ-(L-α-aminoadipyl)-L-cysteinyl-D-valine (ACV) synthetase (*pcbAB*) genes from *Streptomyces clavuligerus* and production of lysine ϵ-aminotransferase activity in *Escherichia coli. J. Bacteriol.* 173:6223–29
101. Usui, S., Yu, C.-A. 1989. Purification and properties of isopenicillin N epimerase from *Streptomyces clavuligerus. Biochim. Biophys. Acta* 999:78–85
102. Van Liempt, H., VonDohren, H., Kleinkauf, H. 1989. δ-(L-α-aminoadipyl)-L-cysteinyl-D-valine synthetase from *Aspergillus nidulans. J. Biol. Chem.* 264:3680–84
103. Veenstra, A. E., van Solingen, P., Bovenberg, R. A. L., van der Voort, L. H. M. 1991. Strain improvement of *Penicillium chrysogenum* by recombinant DNA techniques. *J. Biotechnol.* 17:81–90
104. Veenstra, A. E., van Solingen, P., Huininga-Muurling, H., Koekman, B. P., Groenen, M. A. M., et al. 1989. Cloning of penicillin biosynthetic genes. See Ref. 41a, pp. 262–69
105. Weigel, B. J., Burgett, S. G., Chen, V. J., Skatrud, P. L., Frolik, C., et al. 1988. Cloning and expression in *Escherichia coli* of isopenicillin N synthetase genes from *Streptomyces lipmanii* and *Aspergillus nidulans. J. Bacteriol.* 170:3817–26
106. Whiteman, P. A., Abraham, E. P., Baldwin, J. E., Fleming, M. D., Schofield, C. J., et al. 1990. Acyl coenzyme A:6-aminopenicillanic acid acyltransferase from *Penicillium chrysogenum* and *Aspergillus nidulans. FEBS Lett.* 262:342–44
107. Wolfe, S., Demain, A. L., Jensen, S. E., Westlake, D. W. S. 1984. Enzymatic approach to synthesis of unnatural beta-lactams. *Science* 226:1386–92
108. Xiao, X., Hintermann, G., Piret, J., Hausler, A., Demain, A. L. 1991. Purification of cephalosporin 7-α-hydroxylase from *Streptomyces clavuligerus* and cloning of its gene. *Int. Symp. Biol. Actinomycetes, 8th, Madison, Wis.* p. 60. (Abstr.)
109. Yoshida, M., Konomi, T., Kohsaka, M., Baldwin, J. E., Herchen, S., et al. 1978. Cell-free ring expansion of penicillin N to deacetoxycephalosporin C by *Cephalosporium acremonium* CW-19 and its mutant. *Proc. Natl. Acad. Sci. USA* 75:6253–57
110. Zhang, J. Y., Demain, A. L. 1990. Purification of ACV synthetase from *Streptomyces clavuligerus. Biotechnol. Lett.* 12:649–54

Annu. Rev. Microbiol. 1992. 46:497–531

SIGNALING AND HOST RANGE VARIATION IN NODULATION

Jean Dénarié, Frédéric Debellé, and Charles Rosenberg

Laboratoire de Biologie Moléculaire des Relations Plantes-Microorganismes, CNRS-INRA, BP27, 31326 Castanet-Tolosan Cedex, France

KEY WORDS: nitrogen fixation, symbiosis, legumes, nodulation factors, host specificity, *Rhizobium, Bradyrhizobium*

CONTENTS

0066-4227/92/1001-0497$02.00

Abstract

Rhizobium, Bradyrhizobium, and *Azorhizobium* strains, collectively referred to as rhizobia, elicit on their leguminous hosts, in a specific manner, the formation of nodules in which they fix nitrogen. Rhizobial *nod* genes, which determine host specificity, infection, and nodulation, are involved in the exchange of low molecular weight signal molecules between the plant and the bacteria as follows. Transcription of the *nod* operons is under the control of NodD regulatory proteins, which are specifically activated by plant flavonoid signals. The common and species-specific structural *nod* genes are involved in turn in the synthesis of specific lipo-oligosaccharides that signal back to the plant to elicit root-hair deformations, cortical-cell divisions, and nodule-meristem formation.

INTRODUCTION

Legumes and their symbiotic bacteria (collectively referred to as rhizobia) make the largest contribution to global biological nitrogen fixation (62). Rhizobia elicit on their hosts, in a specific manner, the formation of specialized organs, the nodules, in which they reduce molecular nitrogen into ammonia (11). The *Rhizobium*-legume symbiosis, because of its agricultural importance, has ensured continuing research support world-wide (62) and is presently the best understood of all plant-microbe interactions (73, 154).

The symbiosis is specific: every strain has a definable host range (164). Considerable progress has been made during the past few years in understanding the mechanisms underlying this specificity thanks to interdisciplinary approaches bringing together bacterial molecular genetics, plant physiology and cytology, and biochemistry. These studies have revealed that in the course of infection and nodulation both partners exchange low-molecular-weight signal molecules. The plant controls the expression of bacterial nodulation genes via phenolic compounds (93, 108) whereas the rhizobia signal back to the plant by secreting specific lipooligosaccharides (27, 118).

Diversity of Legumes and Rhizobia

The ability to establish a nitrogen-fixing symbiosis with rhizobia is restricted to legumes with one exception, the genus *Parasponia* of the Ulmaceae family (164). The Leguminosae family comprises three subfamilies, Caesalpinioideae, Mimosoideae, and Papilionoideae, each of which contains genera able to form root nodules (2, 124). Leguminous plants (approximately 15,000 species) exhibit very diverse morphology, habitat, and ecology ranging from arctic annuals to tropical trees. As the great majority of legumes are nodulated by rhizobia, the symbiosis with rhizobia is apparently not an adaptation to a specialized ecological niche but rather depends on some genetic particularity of legumes that is sufficiently complex that it has rarely evolved elsewhere in the plant kingdom (164).

Rhizobia were first classified, according to their host range, into cross-inoculation groups. The use of modern methods of bacterial systematics such as numerical taxonomy and nucleic acid hybridization and sequencing led to the definition of three genera, *Rhizobium* (fast growing), *Bradyrhizobium* (slow growing), and *Azorhizobium* (N_2-dependent growth in free-living conditions) (39, 84, 164) (Table 1). These genera are quite distinct and are much closer to nonsymbiotic relatives than they are to each other. For example, *Rhizobium* is closely related to the plant-associated *Agrobacterium*, while *Rhodopseudomonas palustris*, a soil phototroph, and *Xanthobacter* spp. are the closest relatives of *Bradyrhizobium* and *Azorhizobium*, respectively (39, 72, 84). Thus, rhizobia do not form a monophyletic group. The reason why such diverse bacteria are grouped into a single family, the Rhizobiaceae, is their common ability to establish a nitrogen-fixing symbiosis with legumes. For the sake of brevity, in the following text, *Rhizobium leguminosarum* bv. *phaseoli*, bv. *trifolii*, and bv. *viciae* (see Table 1) are referred to as *R. phaseoli*, *R. trifolii*, and *R. leguminosarum*, respectively.

Infection and Nodulation Patterns

Rhizobia multiply in the rhizosphere and on the root surface of legumes. The mechanisms of root infection vary and include simple entry through intercellular spaces in the epidermis or in the middle lamella (crack entry) as in *Arachis* (peanut) and *Stylosanthes*, or entry through root hairs as in alfalfa and vetch (see 11, 30). In the latter case, rhizobia elicit root-hair curling, the

Table 1 Rhizobia-plant associations

Rhizobia	Host plants
Rhizobium meliloti	Alfalfa *(Medicago)*
Rhizobium leguminosarum	
biovar *viciae*	Pea *(Pisum)*, vetch *(Vicia)*
biovar *trifolii*	Clover *(Trifolium)*
biovar *phaseoli*	Bean *(Phaseolus)*
Rhizobium loti	*Lotus*
Rhizobium fredii	Soybean *(Glycine)*
Rhizobium sp. NGR234	Tropical legumes, *Parasponia* (nonlegume)
Rhizobium tropici	Bean *(Phaseolus)*, *Leucaena*
Bradyrhizobium japonicum	Soybean *(Glycine)*
Bradyrhizobium "cowpea"	Tropical legumes
Azorhizobium caulinodans	*Sesbania* (stem-nodulating)

formation of an infection focus within the curl, and the development of a tubular structure, the infection thread, which grows through the root-hair cell and into the root cortex where it ramifies. Associated with the infection is the induction of cell division in the cortex and the formation of nodule primordia. The development of primordia gives rise to nodules that are genuine organs, not mere tumors or deformed roots. Bacteria multiply in infection threads and are then released into plant cells where they differentiate into nitrogen-fixing bacteroids. Nodules are always formed on the roots; however, some aquatic legumes belonging to the genera *Aeschynomene* and *Sesbania* also exhibit stem nodulation. Nitrogen-fixing nodules of a given plant species have a characteristic ontogeny, morphology, and anatomy. They also have a defined development that is either indeterminate (with a persistent meristem) as in alfalfa, pea, or vetch, or determinate (with a transient meristem) in soybeans or *Phaseolus* (see recent reviews 11, 16).

The route of infection is a characteristic of the host because the same bacteria can penetrate different host species by either crack entry or root-hair infection threads, and a given legume is infected by the same type of mechanism whatever the strain (see 27). Similarly, the structural and developmental characteristics of an efficient nodule are specified by the plant and not by the rhizobial strain, indicating that the host possesses the genetic information for symbiotic infection and nodulation and that the role of the bacteria is to turn on this plant symbiotic program (see 27).

A Variety of Host Ranges

The bacterial symbiont forms nodules only on a restricted number of hosts, and each host is only nodulated by a restricted number of microsymbionts. However, the degree of specificity varies greatly among rhizobia (for a review, see 164). Some isolates can have a broad host range. For example, some tropical *Bradyrhizobium* strains nodulate legumes in different tribes and subfamilies (Papilionoideae and Mimosoideae) (164), and *Rhizobium* sp. strain NGR234 nodulates at least 35 different legume genera, belonging to 13 tribes, as well as the nonlegume *Parasponia* (S. G. Pueppke & W. J. Broughton, unpublished data). In contrast, some isolates have a narrow host range. For example, *Rhizobium meliloti* strains elicit nitrogen-fixing nodules only on species of the genera *Medicago, Melilotus,* and *Trigonella,* and *R. trifolii* only on species of *Trifolium* (clover). Some strains even discriminate between genotypes within a legume species. For example, most isolates of *R. leguminosarum* nodulate European pea varieties but not certain peas from Afghanistan that require special strains with an extended host range (106). Rhizobial strains have been reported to form effective nodules on one plant species (or genus) and ineffective ones on another, showing that specificity is not limited to nodulation but may also affect the late stages of nodule development and the establishment of a nitrogen-fixing symbiosis. In this chapter, however, we only consider the specificity of the early steps of symbiosis, rhizospheric life, infection, and nodulation.

Rhizobial Genes Controlling Infection, Nodulation, and Host Range

Molecular genetics, combined with the study of the symbiotic behavior of bacterial mutant strains, has led to the identification of two major classes of rhizobial genes involved in infection and nodulation (for reviews see 8, 30, 108, 109, 112, 153, 154). One class includes genes in which mutations result in both: (*a*) alterations in the synthesis, transport, and assembly of surface components such as exopolysaccharides, lipopolysaccharides, and β-1,2-glucans and (*b*) alterations in the infection process, such as the inability to elicit the formation of infection threads, resulting in the formation of nonfixing empty nodules (Nod$^+$Fix$^-$ phenotype) (119, 127). These results show that these polysaccharides are involved in the infection process, and a possible role of exopolysaccharides and lipopolysaccharides in the determination of host specificity has been suggested (60, 61, 119, 127). However, no clear genetic evidence, such as gain-of-function experiments, has yet shown that rhizobial surface components are major determinants of host range. The role of surface components in infection (60, 61, 100a, 119) and root adhesion (90–92) has been reviewed recently and is not discussed here.

The second class consists of the nodulation (*nod* and *nol*) genes (27, 93, 108). Because of the many different nodulation genes described in the various rhizobial species, all the letters of the alphabet have been used for the nomenclature of *nod* genes. Therefore, researchers have designated the recently identified nodulation genes as *nol* genes. The *nod* and *nol* genes are involved in infection and nodulation, and their inactivation can result in various *in planta* phenotypes such as the absence of nodules (Nod$^-$) or a delayed but effective (nitrogen-fixing) nodulation (Nod^d, Fix$^+$) on homologous hosts, or changes in host range. Some *nod* genes are common to all rhizobia whereas others are specific. The *nod* genes are regulated in a similar way in all rhizobia (27, 46, 47, 50, 93, 108, 111, 144). Their regulation involves the following three elements. (*a*) Transcription is activated by *trans*-acting regulatory NodD proteins; (*b*) NodD proteins bind to promoter regions that contain extended conserved sequences, the *nod* boxes; (*c*) activation of *nod* gene transcription by the NodD proteins requires the presence of plant signals, flavonoids, or other phenolic compounds exuded by legume roots. The presence of highly-conserved *nod* boxes upstream of all known common and host-specific *nod* operons of *R. leguminosarum* and *R. meliloti* and the general presence of *nod* boxes among rhizobia has led to a definition of *nod* genes as those genes that are part of an operon preceded by a *nod* box (51, 135). This definition, however, may not be valid for all rhizobia because in *Bradyrhizobium japonicum* and *Rhizobium* sp. NGR234, some nodulation genes are indeed preceded by a *nod* box, whereas other genes specifying host range do not seem to be preceded by these conserved sequences (8, 12). In those *Rhizobium* species studied to date, the *nod* genes reside on large symbiotic plasmids (pSym) that also carry the *fix* and *nif* nitrogen-fixation genes. In *Bradyrhizobium* and *Azorhizobium,* the symbiotic genes are likely to be chromosomal.

Both regulatory and structural *nod* genes are major determinants of host range, as determined with gain-of-function experiments. For example, mutations in the regulatory *nodD* gene of *R. trifolii* (115) and transfer of the *nodD1* gene of *Rhizobium* sp. NGR234 into other *Rhizobium* species (79) result in an extension of the host range. Similarly, mutations in the structural *nodE* gene of *R. trifolii* (33) and *nodQ* gene of *R. meliloti* (18, 43, 142) extend the host range. Moreover, the transfer of species-specific structural *nod* genes between rhizobia is associated with the transfer of specificity for particular hosts (24, 31, 43, 105, 156). In this chapter, we review the present knowledge of the mechanisms by which the regulatory and structural *nod* genes determine the host specificity of infection and nodulation.

REGULATORY *nod* GENES AS HOST-RANGE DETERMINANTS

Several lines of genetic evidence have established that *nodD* genes are determinants of host specificity. (*a*) Some mutations in *nodD* cannot be complemented by *nodD* DNA from other *Rhizobium* species (79, 152). (*b*) Some point mutations in the *nodD* gene of *R. trifolii* result in an extended host-range (115). (*c*) A *nodD* hybrid gene, constructed in vitro from *R. meliloti* and *R. trifolii nodD* genes, causes constitutive expression of *nod* genes and extends the host range to include tropical legumes (148). (*d*) The transfer of the *nodD1* gene from *Rhizobium* sp. NGR234 into *R. meliloti* results in the transfer of the ability to nodulate Siratro (79). We now examine how these regulatory genes are involved in the control of host specificity.

Basic Elements of the nodD *Regulatory Circuit*

The regulatory *nodD* genes are found in all of the strains of *Rhizobium, Bradyrhizobium,* and *Azorhizobium* and are required for nodulation (108) because mutants that do not contain a functional *nodD* gene cannot nodulate their hosts (78, 94). Some species like *R. leguminosarum* and *R. trifolii* have only one *nodD* gene (94), while *R. meliloti* (57, 78), *R. phaseoli* (21), *R. fredii* (3), and *B. japonicum* (58) carry two or three copies of *nodD*. The different *nodD* genes are conserved at the nucleotide sequence level and share homology with the *lysR* family of prokaryotic regulatory genes (71). In common with the LysR-family proteins, the N-terminal end of NodD contains a helix-turn-helix motif characteristic of DNA-binding proteins (71, 168). The NodD proteins are transcriptional activators that bind to the *nod* box, a consensus sequence present in the promoter of the *nod* genes inducible by plant exudates (94). This *nod* box was originally defined in *R. meliloti* as a 47-bp consensus sequence required for *nod* gene induction (135). In the genetically distant *Azorhizobium caulinodans* and *B. japonicum,* less conserved *nod* boxes have been identified and new shorter consensus sequences have been proposed to account for these divergent *nod* boxes (54, 163). These short consensus sequences are also found in promoter sequences of genes regulated by proteins of the LysR family. The long *nod* boxes contain repetitions of these short consensus sequences and could be bound by multimeric forms of NodD. The activation of *nod* gene expression by NodD proteins requires the presence of compounds present in plant exudates (108). The inducing molecules have been purified from seed or root exudates. They are flavonoids or related compounds derived from phenylpropanoid metabolism, which is also known to provide molecules involved in plant defense (132). The *nod*-inducing compounds can be active at concentrations as low as 10^{-9} M (113).

NodD Proteins Are Specifically Activated by Flavonoids

The *R. meliloti nodD1* gene does not restore the ability to nodulate Siratro in a *nodD1* mutant of the broad host range strain *Rhizobium* sp. MPIK3030 (= NGR234) (79). Similarly, the *R. meliloti nodD1* gene does not restore the ability to nodulate red clover in a *nodD* mutant of *R. trifolii* (152). This lack of complementation results from the inability of *R. meliloti nodD1* to respond to the compounds exuded by Siratro (79) and red clover (152). In addition, when different *nodD* genes are introduced into the same *Rhizobium* strain, the structural *nod* genes are induced by different sets of flavonoids (152). Such a phenomenon is likely responsible for the fact that *R. meliloti* or *R. trifolii* strains carrying the *nodD1* gene of MPIK3030 exhibit an extended host range including Siratro (79). These experiments indicate that different NodD proteins have different inducer specificities. Generally, a NodD protein from a given *Rhizobium* species is most active in the presence of exudates from the plants nodulated by that *Rhizobium* species (152, 166).

Using synthetic or natural compounds of known structure investigators have determined the structural features of flavonoids necessary for *nod* gene induction by different NodD proteins (Table 2). NodD proteins are generally active in the presence of a family of related flavonoids (Table 2). NodDs from narrow-host-range rhizobia (*R. meliloti, R. leguminosarum, R. trifolii*) respond to few flavonoids, while NodDs from the broad-host-range *Rhizobium*

Table 2 Effects of flavonoids and chalcones, in conjunction with various NodD proteins, on *nod* gene expression

COMPOUNDS	SUBSTITUTION POSITION						NodD PROTEIN[a]					
	3	5	7	3'	4'	5'	Rt	Rl	Rm D1	Rm D2	NGR	Bj
Flavones												
luteolin		OH	OH	OH	OH		++	++	++	-	++	-
apigenin		OH	OH	OH			++	++	+	-	++	+
---			OH	OH	OH		n	++	++	-	n	n
chrysoeriol		OH	OH	OCH_3	OH		n	n	++	-	n	n
---				OH	OH		n	n	-	-	n	n
---			OH		OH		++	n	+	n	++	+
chrysin		OH	OH				+	-	-	-	++	-
Flavonols												
myricetin	OH	OH	OH	OH	OH	OH	-	-	-	-	+	n
quercetin	OH	OH	OH	OH	OH		-	-	-	-	++	-
kaempferol	OH	OH	OH		OH		-	-	-	-	++	+
Flavanones												
eriodictyol		OH	OH	OH	OH		+	++	+	-	n	-
naringenin		OH	OH		OH		++	++	-	-	++	-
hesperetin		OH	OH	OH	OCH_3		-	++	-	-	++	n
Isoflavones												
genistein		OH	OH		OH		-	-	-	-	++	++
daidzein			OH		OH		-	-	-	-	++	++
Chalcone												
4,4'dihydroxy-2'-methoxychalcone			OH		OH		n	n	++	++	n	n

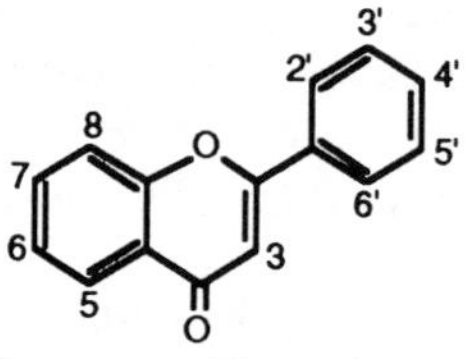

flavones and flavonols

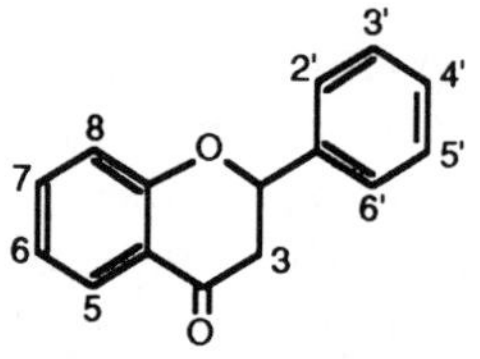

flavanones

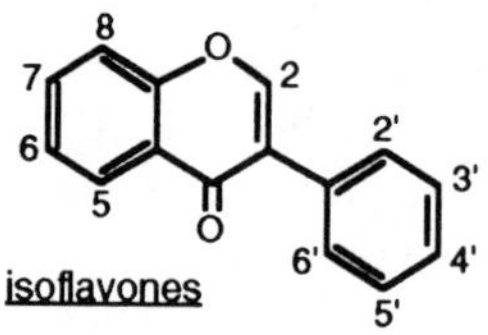

isoflavones

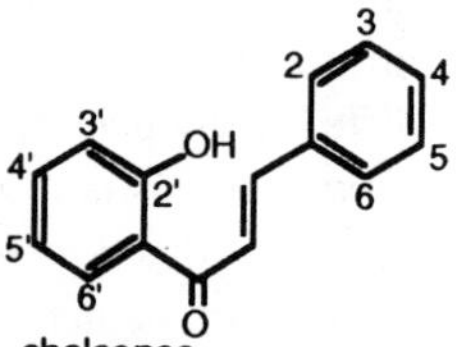

chalcones

[a] The *nod* inducing activities of various flavonoids and chalcone were recorded for rhizobial strains harboring the *nodD* genes of *R. trifolii* (Rt), *R. leguminosarum* (Rl), *R. meliloti* (Rm), NGR234 (NGR), and *B. japonicum* (Bj). The different activities are related to the maximum inducing activity recorded for these strains. ++ = 50–100%; + = 10–50%; − = 0–10% approximately of the maximum induction. n = not determined. (After 7, 10, 32, 45, 66, 98, 99, 103, 111, 113, 122, 165, 167.)

NGR234 have a larger spectrum of inducing molecules including even monocyclic aromatic compounds (103) (Table 2). The activity of a given flavonoid compound varies widely according to the rhizobial species considered. For example, the isoflavone daidzein is a strong activator of the *B. japonicum nod* genes but does not activate the *R. meliloti, R. leguminosarum,* or *R. trifolii nod* genes.

Though the specific binding of flavonoids to NodD has not been biochemically demonstrated, genetic data suggest a direct interaction of these compounds with NodD. Point mutations in *nodD* have been obtained that result in *nod* gene activation with a broader range of inducers (13, 14, 115). Furthermore, hybrid *nodD* genes have been constructed from *nodD* genes of different *Rhizobium* species and show modified flavonoid sensitivity (79, 151). The hybrid NodD proteins exhibit the flavonoid specificity of the NodD product constituting their C-terminal end (79, 151). These experiments indicate that the C-terminal part of NodD, which is less conserved than the N-terminal end, is involved in determining flavonoid specificity (168). In addition, this C-terminal region shows homology to ligand-binding domains of animal steroid receptors that are known to bind flavonoids (65, 168). However, some of the mutations that modify the flavonoid specificity map in the N-terminal region (115, 151). Thus, the flavonoid specificity also depends on the overall tertiary structure of NodD.

Naringenin, which is an inducer of *R. leguminosarum nod* genes, has been shown to accumulate in the cytoplasmic membrane (125) where the *R. leguminosarum* NodD protein has been localized (138). This observation has led to the suggestion that the site of interaction between NodD and the flavonoids is the bacterial inner membrane (138). Experiments with *R. leguminosarum* and *R. meliloti* NodD have shown that binding of NodD to the *nod* box does not require the presence of flavonoids (46, 76). However, the binding of *A. caulinodans* NodD to the *nod* box is enhanced by the flavonoid naringenin (54). Therefore, the mechanism of activation of NodD proteins by flavonoids is still unknown.

Plant Exudates Contain Inducers and Antiinducers

A given legume releases several flavonoids with or without inducing properties for its specific *Rhizobium*. The nature and amounts of the compounds exuded depend on the plant and its stage of development. For alfalfa, bean, or soybean, the spectrum of flavonoids present in seed exudates is different from that present in root exudates (59, 68, 69, 81, 82, 121). Flavonoids are released as aglycones or glycosidic conjugates of lower activity but higher solubility (69, 114). These glycoconjugates can be converted to active forms by bacterial glycosidases. The availability of *nod*-inducing flavonoids can sometimes limit nodulation: Nodulation of a cultivar of alfalfa was enhanced when the inducer luteolin was supplied to the plant (88).

The different flavonoids present in exudates can interfere with each other's ability to induce *nod* gene expression. In several cases, flavonoids without inducing properties have been shown to inhibit *nod* gene activation by effective inducers (32, 45, 66, 99, 122). The antiinducers generally have structures similar to those of the inducers, and the inhibition can be overcome by increasing the concentration of inducers (122). Thus, one can consider the antiinducers to be competitive inhibitors. Some of these inhibitors are strain specific, acting only on a few strains belonging to the same cross-inoculation group (99). Synergistic interactions between inducers have also been observed (68, 69). This effect could be explained if one considers that NodD acts as a multimer: A multimeric complex interacting with different flavonoids could have a higher activity.

The spatio-temporal distribution of flavonoids in the rhizosphere and at the root surface is likely to determine the levels of induction of the *Rhizobium nod* genes. This effect can be directly observed by placing a legume seedling in a Petri dish on a lawn of bacteria carrying a *nod-lacZ* fusion (32, 53, 122, 126). Zones of *nod* gene induction or inhibition can be distinguished. Nodules generally appear in the zone of maximum induction corresponding to the zone of emerging root hairs.

Flavonoids present in root exudates might influence rhizobial infection by mechanisms other than *nod* gene regulation. Rhizobia respond by positive chemotaxis to plant-root exudates and move towards localized sites on legume roots (34, 49, 63). Both *Bradyrhizobium* and *Rhizobium* spp. are attracted by phenolic compounds that are potential nutrients (120, 123), but also, at very low concentrations, by compounds that may not be of nutritional value, such as flavonoids (1, 4, 15, 87). In *R. meliloti,* the chemotactic effect of flavonoids and their *nod* gene–inducing ability are closely correlated, and this chemo-attraction requires functional *nodD* and *nodABC* genes (15). In *R. leguminosarum* and *R. phaseoli,* this correlation is less pronounced as is the dependence of chemotaxis upon *nod* genes (1, 4). Chemotaxis toward flavonoids may facilitate the contact between the bacterium and the root surface of its specific host.

Surprisingly, growth of *R. meliloti* in a minimal medium is strongly stimulated by micromolar concentrations of the *nod* gene inducer luteolin and other 5,7-dihydroxyl substituted flavones naturally released in alfalfa exudates (67). This intriguing growth stimulation does not depend on the presence of the *nod* genes and is also observed for flavonoids, such as quercetin, that are not inducers of the *R. meliloti nod* genes (67). Quercetin and luteolin also increase the growth rate of *R. trifolii* and *Pseudomonas putida,* but not of *Agrobacterium tumefaciens,* which is taxonomically close to *R. meliloti*. The mechanism by which micromolar concentrations of specific flavonoids could reduce the bacterial doubling time so drastically (by a factor of three to five times) is not known. Like chemotaxis, this growth-stimulating phenomenon may play a critical role in structuring the microbial community around the developing root (67).

Additional Elements of the nodD *Regulatory Circuit*

R. meliloti carries three functional *nodD* genes whose relative importance in the nodulation process depends on the host plant considered (57, 64, 78). NodD proteins have different sensitivities to the root exudates of various hosts that contain different sets of flavonoids (64, 66, 77). Studies using purified compounds showed that *nodD1* but not *nodD2* responds to luteolin (77, 111, 117) whereas *nodD2* is active in the presence of 4,4' dihydroxy 2' methoxy chalcone (70). In addition, the three *nodD* genes interfere with each other in the activation of the common and host-specific *nod* genes (77, 111, 117). The significance, in terms of control of specificity, of these *nodD* reiterations is not clear, because there is no correlation between the number of *nodD* genes and the broadness of the host range: The narrow-host-range *R. meliloti* species has three copies, whereas the broad-host-range strain NGR234 possesses two copies. Note that the role of NodD proteins is not limited to regulating *nod* gene expression in response to flavonoid signal molecules. For example, in *R. meliloti, nodD3* is also involved in the regulation of *nod* gene expression in the presence of an excess of combined nitrogen (40) (see Figure 1).

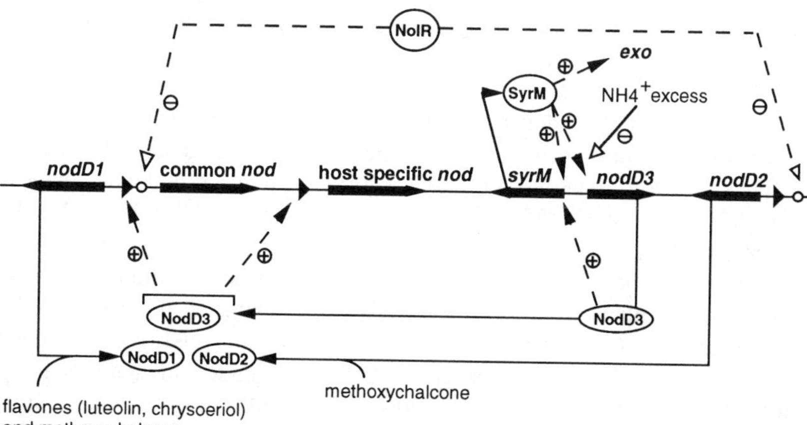

Figure 1 A model for the regulation of *R. meliloti* nodulation genes. The black triangles represent the *nod* boxes; the open circles indicate the repressor-binding sites (after 40, 64, 70, 77, 96, 97, 117). *exo:* genes involved in acidic exopolysaccharide synthesis.

The various rhizobial *nodD* genes also differ in their regulation (168). The *R. leguminosarum* single *nodD* gene negatively regulates its own transcription (134), whereas the *R. meliloti nodD1* gene is constitutively expressed (116). In *B. japonicum, R. phaseoli,* and *R. meliloti,* the expression of one of the *nodD* genes is specifically activated in the presence of plant inducers (7, 22, 40, 111). This represents another mechanism by which plant signals can influence the level of *nod* gene expression in a specific manner.

In *R. meliloti,* the *syrM* gene, another member of the *lysR* family, activates the expression of the *nodD3* gene, and *nodD3* activates in turn the expression of *syrM* (96, 136) (Figure 1). When both *nodD3* and *syrM* are carried on a multicopy plasmid, they can induce *nod* gene expression at a high level even without plant inducers (64, 77, 96, 111, 117). Interestingly, *syrM* also regulates the expression of *exo* genes involved in exopolysaccharide (EPS) synthesis (9, 136), indicating that *syrM* could coordinately regulate the metabolisms of EPS and of the Nod factors, both of which are involved in the infection process. Because *syrM* transcription is controlled by *nodD3* and *nodD2,* which are activated by flavonoids (40, 96, 111), the presence of specific plant inducers could influence both EPS and Nod factor synthesis.

A repressor of *nod* gene expression, NolR, has been identified in several strains of *R. meliloti* (97). This repressor binds to the *nodD1* and *nodD2* promoters and is thought to regulate the expression of the inducible *nod* genes by inhibiting the transcription of *nodD1* and *nodD2* (94) (see Figure 1). A mutation in *nolR* results in a slight delay in nodulation, which suggests that NolR is required for optimizing *nod* gene expression during infection and nodulation (97). It is not known whether the effector that interacts with the repressor to modulate its activity during symbiosis is a plant signal.

Other nod *Regulatory Genes*

Recent results suggest that in *B. japonicum* other *nod* regulatory genes are determinants of host specificity, in addition to the *nodD* genes. The *nodVW* mutants are Nod^+ on soybean but have lost the ability to nodulate cowpea, mungbean, and Siratro (55). The amino acid sequences of NodV and NodW suggest that they are members of the family of two-component regulatory systems (55). The current hypothesis is that NodV (the membranous sensor) responds to a plant signal and that NodW (the regulator) regulates one or several genes involved in nodulation (55). Another gene, *nolA,* is essential for the nodulation of soybean genotypes that are restricted in nodulation by particular serogroups of *Bradyrhizobium* (137). NolA shows sequence similarity to the N terminus of MerR transcriptional regulatory proteins (137). Two hypotheses can explain how *nod* regulatory genes can be determinants of host specificity: (*a*) Inducers may vary with hosts, as demonstrated in the case of *nodD* genes, and (*b*) genes may regulate different specific *nod* genes.

Consistent with the finding of these new putative regulatory *nod* genes, several *B. japonicum* and *Rhizobium* sp. NGR234 nodulation genes are not preceded by a *nod* box (8, 104, 105).

The interaction between *nod* regulatory proteins and plant signals contributes to the specificity of the symbiotic relationship. However, the presence of a *nodD* gene capable of interacting with a signal from a given host plant does not guarantee nodulation of that host. The transfer of the appropriate *nodD* gene is not generally sufficient to transfer the host range from one *Rhizobium* species to another. The transfer of host-specific structural *nod* genes is usually required, showing that these genes play a major role in the control of host specificity.

STRUCTURAL *nod* GENES AS HOST-RANGE DETERMINANTS

Common nod *Genes*

The structural *nod* genes are classified into two categories, the common and the specific. The common *nodABC* genes have been found in all *Rhizobium, Bradyrhizobium,* and *Azorhizobium* isolates studied so far (52, 112, 153). These genes are functionally interchangeable between *Rhizobium* and *Bradyrhizobium* species without altering the host range (8, 112). Transposon-induced mutations in *nodABC* genes suppress the ability to elicit any detectable plant response such as root-hair curling (Hac$^-$ phenotype), infection-thread formation (Inf$^-$), cortical cell divisions, and nodule formation (Nod$^-$), regardless of the host (109, 112). That three conserved bacterial genes are essential for specifying quite different types of infection and nodulation on a wide variety of hosts is intriguing and suggests that the plant possesses the genetic program for symbiotic infection and nodulation and that the role of the rhizobial *nodABC* genes is to trigger this program (27, 160).

The *nodIJ* genes are present in *R. leguminosarum, R. trifolii,* and *B. japonicum* (42, 56, 157), and partial sequence data suggest their presence in *R. meliloti* and *A. caulinodans* (52, 83). They reside downstream of *nodC* and seem to be part of the same operon as *nodC* (Figure 2). Mutations in *nodIJ* result in a nodulation delay with *R. leguminosarum,* but have no detectable effect with *B. japonicum* (42, 56). The presence of the *nodABCIJ* genes in the genetically distant *Rhizobium, Bradyrhizobium,* and *Azorhizobium* isolates suggests a common origin for these genes.

Specific nod *Genes*

Some *nod* genes are not functionally or structurally conserved among rhizobia (95). These genes have been termed host-specific *nod* genes when they satisfy at least one of the following criteria: (*a*) Mutations in these genes cannot be

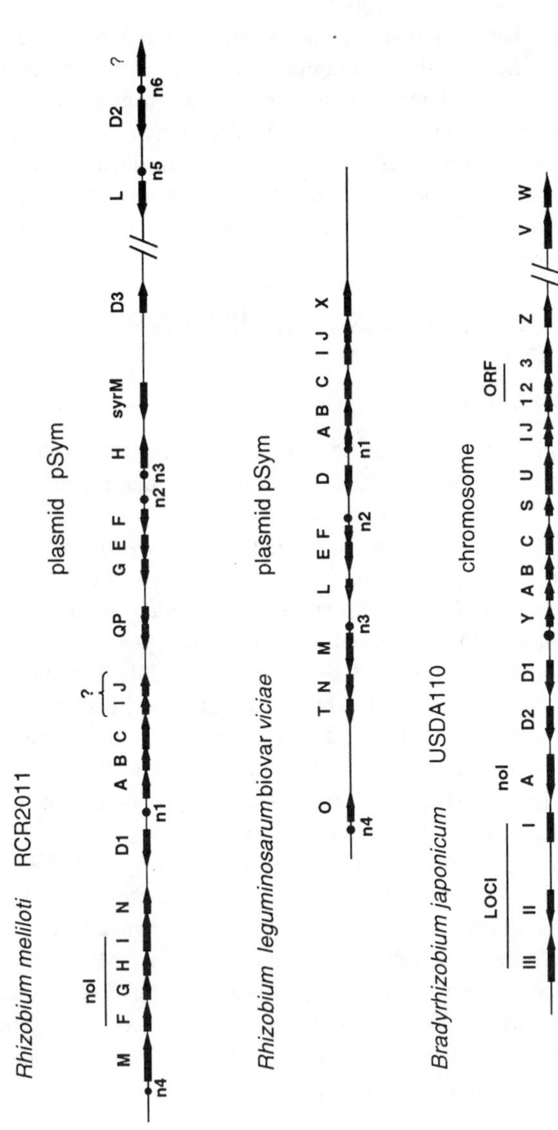

Figure 2 Organization of nodulation genes in *R. meliloti* and *R. leguminosarum* (after 27), and *B. japonicum* (after 8). The capital letters represent *nod* genes, unless otherwise stated. The arrows indicate the direction of gene transcription. Black circles represent the *nod* boxes.

fully complemented by DNA from other rhizobial species or biovars (25, 33, 80, 95, 104). (*b*) Mutations result in a narrowing of the host range. For example, in *B. japonicum, nodZ* and *nodVW* mutants have lost the ability to nodulate Siratro but not soybean (8, 55), and *Rhizobium* sp. NGR234 *nodSU* mutants are Nod$^-$ on *Leucaena leucocephala* but Nod$^+$ on Siratro (104). (*c*) Mutations result in the ability to infect or nodulate new hosts. In *R. trifolii,* in contrast to wild-type strains, *nodFE*$^-$ mutants poorly nodulate white and red clovers but have acquired the ability to infect and nodulate peas (33). In *R. meliloti, nodH*$^-$ mutants have lost the ability to infect and nodulate the homologous host alfalfa but have acquired the ability to infect and nodulate heterologous hosts such as vetch species, while *nodQ* mutants can infect both alfalfa and vetch (25, 43, 80). (*d*) These genes are necessary in interspecific crosses to extend the host range of the recipient bacterium to include plants nodulated by the donor strain (24, 31, 43, 105, 150, 156).

Whereas mutations in the *nodABC* genes totally block nodulation, mutations in individual host-specific *nod* genes do not abolish the ability to nodulate all the homologous hosts. Rather, they cause a nodulation delay or a decrease in the number of nodules formed. For example in *R. leguminosarum* and *R. meliloti,* individual mutations in the host-range *nodF, E, L, M,* and *N* genes do not block nodulation (5, 25, 35, 37, 80, 158). However, these genes are collectively essential because the deletion of all of the host-specific genes results in a completely Nod$^-$ phenotype (37, 38).

The host-range variation mediated by host-specific *nod* genes seems to rely on two types of genetic mechanisms: allelic variability due to different allelic forms of a gene and variability due to combinations of different nonallelic genes. A clear case of allele-dependent variation occurs with the *nodE* gene, common to *R. leguminosarum, R. trifolii,* and *R. meliloti* (Figure 2). The *nodE* alleles present in *R. leguminosarum* and *R. trifolii* are highly homologous, but the *nodE* gene is the major determinant of the host range difference between these two biovars, and a region of the *nodE* protein has been localized that determines its host-specific properties (150).

In contrast, a particular combination of genes present in one species and not in others can provide a given host specificity, as illustrated with the *nodPQ* and *nodH* genes, which determine alfalfa specificity and are present in *R. meliloti* but not in *R. leguminosarum, R. trifolii,* and *B. japonicum* (Figure 2) (25, 80, 131). The *nodSU* genes, present in *B. japonicum* and in the broad-host-range tropical *Rhizobium* sp. NGR234, have not been detected in *R. leguminosarum* and *R. meliloti* (56, 104). The *nodT* gene, present in *R. leguminosarum* and *R. trifolii,* has not been identified in *R. meliloti* (157). Similarly, host-specific *nod* genes have been described that can be present only in some biovars or races within a biovar. The *nodO* gene was found in *R. leguminosarum* biovar *viciae* but not in the closely related biovar *trifolii* (41).

Of an even narrower specificity is the *nodX* gene, which is present in a particular race of *R. leguminosarum* biovar *viciae* and confers the ability to nodulate Afghanistan lines of pea containing a single nodulation-restriction gene (20). Similarly, the *nolA* gene of *B. japonicum* is essential for nodulation of soybean genotypes restrictive to a particular *Bradyrhizobium* serocluster (137). A single dominant plant gene appears to be involved in this nodulation restriction (8, 137). Thus, the rhizobial *nodX* and *nolA* genes might represent the bacterial counterpart of gene-to-gene interaction systems analogous to those found in many pathogen-plant relationships (8, 89).

Sequence comparisons have been fruitful, and significant homologies have been found between most of host-specific *nod* gene products and known proteins (see Table 3), suggesting enzymatic functions for particular host-

Table 3 Products of the nodulation genes of rhizobia

Genes	Species biovar[a]	Cellular localization[b]	Homologies	References
nodA	Common	Cytoplasmic		86, 133, 139, 159
nodB	Common	Cytoplasmic		133, 159
nodC	Common	Outer mb	Chitin synthases	83, 85, 86, 133, 159
nodD	Common	Cytoplasmic mb	DNA-binding proteins	117
nodE	Rl, Rt, Rm	Cytoplasmic mb	β-ketoacyl synthases	150
nodF	Rl, Rt, Rm	Cytoplasmic	Acyl carrier proteins	50, 144
nodG	Rm		Alcohol dehydrogenases	26, 48
			3-Oxoacyl-(acyl-carrier protein) reductase	145
nodH	Rm		Sulfotransferases	26, 80, 128
nodI	Rl, Rt, Bj	Cytoplasmic mb	ATP-binding proteins	42, 157
nodJ	Rl, Rt, Bj	Cytoplasmic mb	Hydrophobic domains	42, 157
nodL	Rl, Rt, Rm	Cytoplasmic mb	Acetyltransferases	35, 155
nodM	Rl, Rt, Rm		D-glucosamine synthase	5, 37, 155
nodN	Rl, Rt, Rm			5, 155
nodO	Rl	Secreted	Hemolysin, Ca^{2+}-binding	23, 41
nodP	Rm		ATP-sulfurylase	18, 128, 142, 143
nodQ	Rm		ATP-sulfurylase and APS kinase	18, 128, 142, 143; unpubl.[c]
nodSU	Bj, NGR			56, 104
nodT	Rl, Rt	Outer mb	Transit sequences	157
nodV	Bj	Cytoplasmic mb	Sensor, two component regulatory family	55
nodW	Bj	Cytoplasmic	Regulator, two component regulatory family	55
nodX	Rl (Tom)			20
nolA	Bj		DNA-binding proteins	137

[a] Rhizobia species and biovars: Rm, *R. meliloti;* Rl, *R. leguminosarum* bv. *viciae;* Rt, *R. leguminosarum* bv. *trifolii;* Bj, *B. japonicum;* NGR, *Rhizobium* sp. NGR234.

[b] mb = membrane.

[c] J. Schwedock & S. R. Long, unpublished data.

range proteins and raising the question of whether these Nod enzymes are involved in the modification of surface macromolecular bacterial components or in the production of extracellular low-molecular-weight signals.

Extracellular Specific nod *Factors*

The use of various plant bioassays has revealed that the *nod* genes are involved in the production of extracellular Nod factors. First, the sterile supernatant of *R. leguminosarum* cultures grown in conditions of *nod* gene induction elicits a thick and short root (Tsr) phenotype on *Vicia sativa* subsp. *nigra* (162). Mutations in the *nodABC* genes result in a loss of Tsr activity of the filtrates. The sterile supernatants of *nod*-induced *R. leguminosarum* cultures also have a spectacular effect on vetch root-hair initiation (Hai phenotype) and hair deformations (Had) (167). Filtration through molecular sieves indicates that the Tsr/Had factors have a low molecular mass. Investigations have also shown that the *nodAB* genes of *R. meliloti* specify the synthesis of a Nod factor that stimulates mitosis of cultured plant protoplasts. This factor is heat stable, partially hydrophobic, and has a low molecular weight (140).

The sterile supernatant of *R. meliloti* cultures, induced by the flavone luteolin, elicits hair deformation on alfalfa (44). Inactivation of the *nodABC* genes suppresses this Had activity. Use of vetch (Tsr and Hai) and alfalfa (Had) bioassays revealed that the biological activity of sterile supernatants is specific and exhibits the same host specificity as observed for nodulation by the bacteria (44). These results suggest a role of the host range *nod* genes in specifying extracellular factor production. *R. meliloti nodH* mutants have an altered host-range: They have lost the ability to infect and nodulate the homologous host alfalfa and have acquired the ability to infect and nodulate vetch, a heterologous host (18, 25, 80). The sterile filtrate of a *nodH* mutant is active on vetch and not on alfalfa, implying that the *nodH* gene is modifying the specificity of an extracellular Nod factor (44). Similar results are obtained when the *R. meliloti nodABC* and *nodH* genes are introduced into *Escherichia coli:* The *E. coli* transconjugants produce an extracellular alfalfa-specific factor only in the presence of an active *nodH* gene (6).

When the *R. meliloti* host-range *nodFEG, nodH,* and *nodPQ* genes are conjugatively transferred into *R. leguminosarum,* the transconjugants gain the ability to nodulate alfalfa and exhibit a strongly reduced nodulation of vetch (43). Mutations in either *nodH* or *nodQ* restore the normal *R. leguminosarum* specificity, indicating an epistasy of these *R. meliloti* genes over the host-specific *nod* genes of *R. leguminosarum*. The specificity of the filtrates of these different strains, as revealed by the Had and Tsr biological assays, strictly parallels the symbiotic specificity (43). These results indicate that the *nodQ* gene is also involved in the modification of the Nod factor specificity. A *R. meliloti* derivative lacking the whole *nod* region and possessing a plasmid carrying only the common *nodABCIJ* genes and the regulatory *nodD1* gene produces a Nod factor that is not active on alfalfa but is active on vetch (43, 44) and clover (6). To interpret all these data, the following model was proposed (43, 44, 102). The common *nodABC* genes determine the production of a Nod factor precursor(s), which can be detected with the vetch bioassays. Host-range genes, such as the *nodH* and *nodQ* genes in *R. meliloti*, specify the modification of this precursor(s) into host-specific signal(s).

Nod-Factor Chemical Characterization and Biological Activity

R. MELILOTI When the supernatant of a luteolin-induced *R. meliloti* culture is extracted with organic solvents of various polarity, most of the Had activity is found in the butanol fraction, suggesting an amphiphilic nature for the Nod factors. High-performance liquid chromatography (HPLC) fractionation of the butanol extract, followed by Had bioassays, revealed that the concentration of the active molecule(s) was extremely low and insufficient for complete structural analysis. Thus an increase in the production of the Nod factors was necessary (101).

The first method used to overproduce the factors involved increasing the gene dosage of *nod* genes by constructing strains merodiploid for the *nod* region. Indeed, the introduction of the recombinant plasmid pGMI149, carrying the common *nodABCIJ* genes, the *nodFEG*, *nodH*, and *nodPQ* host-range genes, and the *nodD1*, *syrM*, and *nodD3* regulatory genes (see Figure 2), into *R. meliloti* strains resulted in a >100-fold increase in the Had activity of the culture supernatant on alfalfa. This amplification made possible the observation of far-UV (220 nm) absorbing peaks comigrating with the Had activity, when the butanol-soluble fraction was chromatographed by reverse phase HPLC (RP-HPLC) (101). Tn*5* insertions into either the *nodA* or *nodC* genes resulted in the simultaneous disappearance of these UV-absorbing peaks and of the Had activity, indicating that these compounds were Nod factors. For structural analysis of the Nod factors, large-scale cultures were performed. The Nod factors were purified by a series of RP-HPLC, gel permeation, and ion-exchange chromatography. The approximate yield of Nod factors varied from 0.4 to 2.5 mg/liter of induced culture (101, 129). The structure of the major Nod factors was established using mass spectrometry, NMR spectroscopy, chemical modification, and radioactive labeling. The major compounds are α and β anomers of the same molecule NodRm-1 (101, 129), which was recently renamed NodRm-IV(S) (for nomenclature, see Figure 3 and also 128). NodRm-IV(S) is a N-acyl-tri-N-acetyl-β-1,4-D-glucosamine tetrasaccharide bearing a sulfate group on C6 of the reducing sugar (M_r = 1102). The acyl moiety, carried by the nonreducing terminal sugar, is a C16 chain with two double bonds at positions 2 and 9. A related compound, NodRm-IV(Ac,S) (M_r = 1144), bearing an O-acetate group on the C6 of the

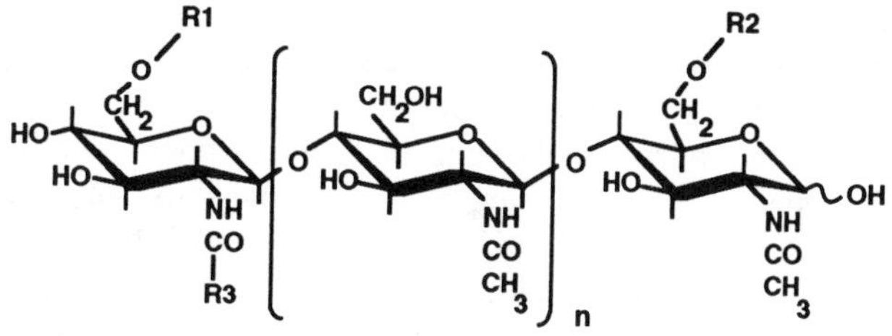

R. meliloti

R1= CH3CO or H
R2= SO3H
R3= C16: 2 (2,9)
n= 2 or 3

R. leguminosarum

R1= CH3CO
R2= H
R3= C18:4 (2,4,6,11) or C18:1 (11)
n= 2 or 3

Figure 3 Structure of the major Nod factors of *R. meliloti* and *R. leguminosarum*. The following nomenclature is used (128). Chitin oligomer backbone: $n = 2$, four glucosamine residues (= IV); $n = 3$, five glucosamine residues (= V). Substitutions on the chitin oligomer backbone: Ac, O-acetate; S, sulfate. The N-acyl chain is described following the usual lipid nomenclature. The four major factors of *R. meliloti* are called NodRm-IV(Ac,S), NodRm-IV(S), NodRm-V(Ac,S), and NodRm-V(S) with a C16:2 N-acyl chain for all factors (128). The four major factors from *R. leguminosarum* are called NodRlv-IV(Ac,C18:4), NodRlv-IV(Ac,C18:1), NodRlv-V(Ac,C18:4), and NodRlv-V(Ac,C18:1) (149).

nonreducing terminal sugar was also found (129). Chitin also consists of linear chains of N-acetyl-D-glucosamine residues joined by β-1,4 linkages.

The second method used to increase the Nod-factor production consisted of amplifying only the regulatory gene couple *syrM-nodD3* (128). The major compound was NodRm-IV(Ac,S), which represented more than 70% of the total amount of Nod factors. New compounds, NodRm-V(Ac,S) ($M_r = 1347$) and NodRm-V(S) ($M_r = 1305$), were detected that contained five glucosamine residues instead of four (Figure 3). The N-acyl chain was the same (C16:2) in all major compounds. Other minor compounds were detected that differed slightly from NodRm-IV(S) in the length of the aliphatic chain (C18) and in the number and location of the double bonds. Although the Nod factors obtained by the two overproduction strategies are similar, quantitative and qualitative differences are observed. These differences probably result from a change in the stoichiometry of Nod enzymes when only a subset of the structural *nod* genes present on pGMI149 is amplified (see 128) (Figure 2). The plasmid pGMI149 used in the first method contains *nodABCIJ, nodFEG, nodPQ,* and *nodH* but not *nodL* nor the *nodMNnolFGHI* operon nor the n6 locus (see Figure 2). Thus, the Nod factors from overproducing strains carrying only an amplification of the regulatory *syrM-nodD3* genes will probably more closely reflect those produced by the wild-type strain.

Purified NodRm-IV(Ac,S) and NodRm-IV(S) factors exhibit biological activity on alfalfa, a *R. meliloti* host. They elicit root-hair deformations at concentrations down to 10^{-12} M (101, 128), cortical cell divisions, and the formation of genuine nodules at concentrations down to 10^{-9} M (161). Thus, lipooligosaccharides of rhizobial origin elicit plant responses at different steps of the symbiotic process, including plant organogenesis, and are active at concentrations clearly lower than those at which the plant hormones operate.

Chemical and genetic modifications of Nod factors have been performed in order to study structure-function relationships. Removal of the O-acetate group slightly decreases nodule-formation ability (128, 161). Each of the following modifications, reduction of the anomeric carbon of the reducing sugar, removal of the sulfate group, or hydrogenation of the double bonds of the N-acyl chain, strongly decreases the ability to induce nodules (161). These results show that the anomeric carbon, the sulfate group, and at least one of the double bonds of the fatty acid chain are essential for organogenetic activity on alfalfa. Biological effects of Nod factors were recently reviewed (27, 118).

R. LEGUMINOSARUM Methods based on radioactive labeling of Nod factors followed by thin-layer chromatography (TLC) have been used to detect Nod factors from the supernatant of various strains (130, 149; N. Price, personal communication). Radioactive acetate was used as a precursor to label Nod factors of *R. leguminosarum*. After butanol extraction of the supernatant of naringenin-induced cultures, and RP-TLC, three major spots could be detected that were absent in *nodC* mutants (147, 149). HPLC separates the compounds corresponding to these spots into four peaks. Large-scale cultures were used for structural determination. Mass and NMR spectrometry analysis showed that the four major compounds have a general structure similar to the Nod factors found in *R. meliloti* (see Figure 3). They contain a β-1-4-linked poly-N-acetyl-glucosamine moiety; the terminal nonreducing residue is O-acetylated on C6 and is N-acylated (149). As in *R. meliloti,* compounds were found in which the oligosaccharide backbone contains either four or five sugar residues. But the *R. leguminosarum* factors are not sulfated, and the N-acyl chains (C18:1 and C18:4) are different from those found in *R. meliloti* (C16:2). Two compounds, NodRlv-IV(Ac,C18:1) (M_r = 1094) and NodRlv-V(Ac,C18:1) (M_r = 1297), are N-acylated by vaccenic acid, a C18 fatty acid with a *cis* double bond in position 11, which is the most abundant fatty acid present in cells of *R. leguminosarum*. More interestingly, two other compounds, NodRlv-IV(Ac,C18:4) (M_r = 1088) and NodRlv-V(Ac,C18:4) (M_r = 1291), are N-acylated by a highly unsaturated fatty acid that contains three *trans* double bonds in conjugation with the carbonyl group and a fourth double bond at position 11 (149).

Purified NodRlv-IV(Ac,C18:4) and NodRlv-V(Ac,C18:4) factors elicit various responses on vetch, a *R. leguminosarum* host. They induce root-hair deformations (10^{-12} M), thickening and shortening of roots (10^{-10} M), and nodule-meristem formation (5×10^{-8} M) (147). The nodule meristems do not give rise to well-developed nodules containing vascular bundles as in alfalfa. NodRlv factors elicit (at 5×10^{-8} M) a change in the pattern of the flavonoids released in the root exudates. This response is also seen with *R. leguminosarum* cells and has been called INI, for increased *nod* gene–inducing activity (147, 161a). Its physiological significance is not clear. In addition, NodRlv factors elicit the transcription of the early nodulin genes ENOD5 and ENOD12 in root hairs, at concentrations down to 10^{-13} M (118).

OTHER RHIZOBIAL SPECIES *Rhizobium* sp. strain NGR234 has an unusually broad host range (105). The production of extracellular Nod factors by this strain was detected by a root-hair deformation test with the tropical legume Siratro (12). Mass and NMR spectrometry revealed that NodNGR compounds have the same general structure as the NodRm and NodRlv factors: They are N-acylated chitin oligomers. The major compounds are pentameric. Interestingly, these molecules also have structural features clearly different from the other factors. The terminal nonreducing glucosamine residue is substituted by one or two carbamoyl groups located at various positions. The reducing glucosamine residue carries a methyl-fucose sugar substituted by a sulfate or an acetyl group (N. Price, W. J. Broughton & J. C. Promé, personal communication). Purified NodNGR factors exhibit a broad host range of activity and elicit root-hair deformations on Siratro, cowpea, alfalfa, and vetch, at very low concentrations.

^{14}C-acetate labeling of cultures of various wild-type *B. japonicum* strains, followed by TLC fractionation of butanol extracts, has revealed the presence of different Nod factors (146). However, a spot appeared to be common to the various strains, and further studies showed that this spot corresponds to a compound with the same general structure as the Nod factors of other rhizobial species: N-acylated and substituted chitin oligomers. The compound is a pentamer of β-1-4-linked N-acetyl-D-glucosamine, N-acylated on the terminal nonreducing residue, and it carries a sugar substituent on the reducing end (H. Spaink & G. Stacey, personal communication). These chemical results support the previously proposed model (43, 44) that the common *nodABC* genes determine the synthesis of a common precursor(s) and that the enzymes encoded by the host-range *nod* genes specify various decorations of this precursor to make it plant specific.

The Common nod *Genes and the Molecule Backbone*

What could be the biochemical function of the common and specific *nod* genes in the biosynthesis and transport of the Nod factors? To address this question, three approaches have been used: (*a*) determination of the structure of the molecules produced by *Rhizobium nod* mutant strains, (*b*) prediction of the function of *nod* gene products by the search for homology with already known enzymes, and (*c*) physiological or biochemical study of the *nod* gene products.

In *R. leguminosarum* and *R. meliloti*, a Tn*5* insertion in the *nodABC* operon suppresses the production of the lipooligosaccharidic Nod factors (101, 147). Indeed, mutations in *nodABC* are the only ones that completely suppress both any detectable Nod-factor production and symbiotic activity; the proteins encoded by these genes play a crucial role and cannot be functionally replaced by other rhizobial proteins.

A *R. leguminosarum* strain cured of the pSym plasmid and containing only the *nodDABC* genes produces two major lipooligosaccharides (M_r = 1052 and 1255), which have the general structure of the wild-type Nod factors but are not O-acetylated and are N-acylated by vaccenic acid (147). The chitin oligomer backbone could be generated either by the degradation of preexisting chitin polymers or by the synthesis from N-acetyl-D-glucosamine monomers. Sequence comparisons have revealed that NodC shares homology with yeast chitin synthases, suggesting that NodC is involved in the synthesis of the chitin oligomer backbone (M. Atkinson & S. Long, personal communication; F. Debellé, C. Rosenberg & J. Dénarié, unpublished results). In contrast, no significant homology could be found with chitinases. Biochemical and immunological studies have shown that NodA and NodB are cytosolic whereas NodC is a membrane protein (85, 86, 139) like chitin synthases. NodAB proteins could be involved in the synthesis of the amino sugar backbone together with NodC, where the N-acylation of the molecule is mediated by housekeeping enzymes, or alternatively, NodAB could specify this N-acylation.

Supporting the hypothesis of a polymerization is the recent finding that the *nodM* gene, found in *R. leguminosarum* and *R. meliloti*, shares homology with an *E. coli* glucosamine synthase (5, 37, 155). Physiological experiments have shown directly that *nodM* codes for a glucosamine-phosphate synthase involved in glucosamine synthesis (37, 93). The presence of a housekeeping glucosamine-phosphate synthase isoenzyme (37) could explain why *nodM*$^-$ mutants exhibit only a moderate delay in infection and nodulation (5, 93, 155). Though the *nodIJ* genes are commonly present in rhizobia and are usually in the same operon as *nodC*, no influence of these genes on the abundance and structure of *nod* metabolites in various cell compartments or in the growth medium could be detected (146).

Synthesis of Specific N-Acyl Chains

The operon containing the *nodFE* genes is involved in the control of host specificity in *R. leguminosarum*, *R. trifolii*, and *R. meliloti* (25, 33, 36, 150).

The *nodE* gene is the main factor determining the difference in host-specific characteristics of the *R. leguminosarum* biovars *viciae* and *trifolii* (150). NodF shows homology to acyl carrier proteins (159). Purified NodF protein from *R. leguminosarum* carries a 4'-phosphopantetheine moiety, which suggests that this protein might be involved in the biosynthesis of a fatty acid (50). The NodE protein shares homology with a group of β-ketoacylsynthases such as the *E. coli* condensing enzyme of fatty acid biosynthesis, FabB, and those presumed to be involved in the synthesis of β-ketide antibiotics in *Streptomyces* species (150). Indeed, a *nodE*$^-$ mutant of *R. leguminosarum* secretes Nod factors that are N-acylated by vaccenic acid (C18:1), whereas the wild-type strain also secretes Nod factors with a highly unsaturated fatty acid (C18:4) (149). Thus, the *nodE* gene is likely to specify the synthesis of the highly unsaturated fatty acid moiety of lipooligosaccharide factors. The fact that a *R. leguminosarum* strain carrying only the *nodDABC* genes (and not the *nodFE* genes) produces Nod factors that are N-acylated by vaccenic acid suggests that in the absence of the C18:4 fatty acid the N-acyl transferase would acylate the oligosaccharidic backbone with the fatty acids prevalent in the bacterial cell (149). In *R. meliloti,* the *nodG* gene, which is not found in *R. leguminosarum,* resides downstream from *nodFE* (26, 48, 80). NodG shares homology with a 3-oxoacyl (acyl-carrier protein) reductase and several alcohol dehydrogenases and might be involved in the synthesis of the acyl moiety.

Purified highly unsaturated Nod factors (C18:4), produced by *R. leguminosarum nodFE*$^+$ strains, elicit on vetch, a homologous host, Hai and Tsr phenotypes at concentrations down to 10^{-11} and 10^{-10} M, respectively. At higher concentrations (5×10^{-8} M), they induce the secretion in root exudates of new flavonoids (the INI phenotype) and the formation of nodule meristems in the root cortex (149). In contrast, the Nod factors produced by the *nodE*$^-$ derivatives, which are N-acylated by the monounsaturated vaccenic acid (C18:1), induce Hai and Tsr responses but do not induce an INI reaction nor the formation of nodule meristems (149). These results indicate that in the vetch system, the various symbiotic steps do not exhibit the same specific requirement; the root-hair deformation is less specific than the nodule-meristem initiation. The presence of conjugated double bonds on the N-acyl chain seems to be important for nodule formation and nodule meristem formation in alfalfa and vetch, respectively (149, 161). These results indicate that the *nodE* gene determines *R. leguminosarum* host specificity by mediating the synthesis of the unsaturated fatty acid moiety of the lipooligosaccharidic Nod factors (149).

Sulfation of the R. meliloti *Nod Factors*

In *R. meliloti,* the *nodH* and *nodQ* genes determine the alfalfa specificity of Nod factors (43, 44, 128). By which mechanism do they operate? *R. meliloti* strains overproducing Nod factors and that have a mutation in *nodH* do not produce detectable amounts of sulfated NodRm factors (128). Instead, they secrete lipooligosaccharides that differ from the NodRm factors only by the absence of the sulfate group. Surprisingly, mutants with a Tn*5* insertion in *nodP* or *nodQ* produce a mixture of sulfated and nonsulfated factors (128). This leaky phenotype seems to result from the presence of a second copy of the *nodPQ* genes present on the second megaplasmid (142, 143). Double mutants with a Tn*5* insertion in both *nodQ* copies excrete only nonsulfated factors (128). Thus *nodH* and *nodPQ* genes are involved in the sulfation of the lipooligosaccharide factors.

In gram-negative bacteria, the first steps of sulfate metabolism involve the synthesis of activated forms of sulfate derived from ATP, adenosine 5'-phosphosulfate (APS), and 3'-phosphoadenosine 5'-phosphosulfate (PAPS) (100). The NodP product and a region of NodQ have strong homologies with CysD and CysN of *E. coli,* which are subunits of an ATP sulfurylase (143). Another region of *nodQ* is homologous to CysC, an APS kinase (J. Schwedock & S. R. Long, unpublished data). NodP and NodQ products have ATP sulfurylase and APS kinase activities both in vitro and in vivo, and their function is probably to generate PAPS (143). The *nodH* gene product has a significant homology with steroid sulfotransferases from mammals and probably specifies the transfer of the sulfate group from an activated donor onto lipooligosaccharide NodRm precursors (128).

Purified sulfated NodRm factors are Had^+Nod^+ on alfalfa, the homologous host, and are Hai^-Tsr^- on vetch, a nonhomologous host. In contrast, the nonsulfated NodRm factors are Hai^+Tsr^+ on vetch and Had^-Nod^- on alfalfa (128). These results indicate that in alfalfa quite distinct symbiotic steps such as root-hair deformation and nodule induction seem to exhibit similar specific requirements for sulfated molecules. The sulfation of NodRm factors determines their specificity in bioassays, and in *R. meliloti* strains, the production of sulfated or nonsulfated Nod factors is correlated with the host specificity of infection and nodulation (128). All these correlations strongly indicate that the *nodH* and *nodPQ* genes specify the host range by determining the sulfation of lipooligosaccharidic Nod factors. Two tropical *Rhizobium* species were recently found to also secrete sulfated Nod factors: *R. tropici,* which nodulates *Phaseolus* and *Leucaena* (R. Poupot, E. Martinez & J. C. Promé, personal communication), and the broad-host-range *Rhizobium* sp. NGR234 (N. Price, W. J. Broughton & J. C. Promé, personal communication).

O-Acetylation and Other Substitutions of Nod Factors

Mutations within the *nodL* gene of *R. leguminosarum* strongly reduce nodulation of peas, *Lens,* and *Lathyrus* but have little effect on nodulation of *Vicia* species (156). Thus, the *nodL* gene is considered a host-specific gene. The sequence of the *nodL* gene product of *R. leguminosarum* bv. *viciae* shows homology to the acetyl transferases encoded by the *lacA* and *cysE* genes of *E. coli* (35). *R. leguminosarum nodL* mutants secrete Nod factors lacking the O-acetate group, indicating that the *nodL* gene determines the O-acetylation on C6 of the terminal nonreducing glucosamine (149). Recently, a *nodL* gene was also identified in *R. meliloti,* and a mutant with a deletion that removes the *nodL* region produces NodRm factors lacking the O-acetate group (M. Ardourel & C. Rosenberg, personal communication). Thus, in both *R. leguminosarum* and *R. meliloti,* the NodL protein specifies the same biochemical function: O-acetylation of C6 of the terminal nonreducing glucosamine. *R. leguminosarum* NodRlv factors deprived of the O-acetyl group retain the ability to induce Had and Tsr responses on vetch, but they have lost the ability to elicit the INI reaction and to induce nodule-meristem formation (149). With *R. meliloti* NodRm factors, the absence of the O-acetate group results in a slight decrease of both Had and Nod activities on alfalfa (128, 161).

The structural differences between Nod factors from the two temperate rhizobia studied so far, *R. leguminosarum* and *R. meliloti,* rely on the presence (or absence) of an O-sulfate group and on the number and positions of the double bonds of the N-acyl chain. Recent results suggest that tropical and subtropical rhizobia seem to use different types of decoration for their Nod factors. The reducing glucosamine residues of Nod factors are ornamented with a different and uncommon sugar in *Rhizobium* sp. NGR234 (N. Price, W. J. Broughton & J. C. Promé, personal communication), *B. japonicum* (H. Spaink & G. Stacey, personal communication) and *A. caulinodans* (P. Mergaert, M. Holsters & J. C. Promé, personal communication). Moreover, the nonreducing terminal glucosamine residue of *Rhizobium* sp. NGR234 carries two new substituents in addition to the N-acyl group (N. Price, W. J. Broughton & J. C. Promé, personal communication). Thus, the wide variety of host ranges found among rhizobia could be determined by a variety of combinations of diverse substituents on the common chitin oligomer backbone.

A Secreted Nod Protein

One *nod* gene product seems to not be involved in the synthesis of Nod factors but rather seems to facilitate directly or indirectly the interaction between a *Rhizobium* Nod factor and a plant receptor. The *nodO* gene is specific for *R. leguminosarum* bv. *viciae* and has not been detected in the related biovars *phaseoli* and *trifolii* (37). A mutation in *nodO* causes a slight delay in nodulation (157). A *R. leguminosarum* strain in which a deletion has removed all of the nodulation region except the *nodDABCIJ* genes is completely Nod$^-$ (38). This type of strain produces lipooligosaccharidic Nod-factor precursors, NodRlv-IV(C18:1) and NodRlv-V(C18:1), which do not induce nodule-meristem formation (149). The introduction into this strain of clones carrying either *nodFE, nodFEL,* or *nodO* genes partially restores nodulation (37, 38). In contrast to *nodFEL* genes, the *nodO* gene does not seem to modify quantitatively or qualitatively the Nod-factor production (146, 147), but codes for a secreted Ca^{2+}-binding protein homologous to *E. coli* hemolysin and related proteins (23, 41). The function of NodO could be to facilitate Nod factor–plant receptor interactions at the root surface. Whether similar Nod secreted proteins are produced by other rhizobia is not known.

Signal Exchange and Plant Determinants of Specificity

Genetic studies of both symbiotic partners have identified some examples of gene-for-gene interactions. Molecular analysis of the bacterial genes responsible for these highly specific relationships suggests the possible exchange of specific signal molecules between plant and bacteria. Several soybean genotypes have been identified that restrict nodulation by *B. japonicum* strain USDA123 but are nodulated normally by other non-123 serogroup *B. japonicum* strains. A single dominant gene appears to be involved in soybean (19). In *B. japonicum,* the nodulation gene *nolA* is essential for nodulation of these resistant genotypes and shares homology with regulatory proteins (137). These results suggest either that these soybean lines secrete a signal recognized directly or indirectly by the *nolA* gene product, or that *nolA* regulates a specific *nod* gene involved in the production of a factor recognized by these soybean lines (see Other *nod* Regulatory Genes).

Most *R. leguminosarum* strains that nodulate European peas cannot nodulate Afghanistan peas. A single plant gene determines this nodulation restriction (74, 106). The *R. leguminosarum* strain TOM can overcome this restriction and efficiently nodulate Afghanistan lines because of the presence of a single gene, *nodX* (20, 75). This gene is located immediately downstream of *nodABCIJ* (see Figure 2) and might be involved in a modification of the Nod factors, which could then be recognized by resistant Afghanistan peas. Recently, alfalfa genotypes were selected that suppress the Nod^- phenotype of *R. meliloti nodH* mutants by a dominant gene (17). Such genotypes might have a modified interaction with the nonsulfated Nod factors secreted by the *nodH* mutants.

The development of plant genetic engineering methods has opened the way to new experimental approaches. The introduction of a pea lectin gene into white clover seedlings results in the transfer of the ability to be infected and nodulated by a pea-specific *Rhizobium* (28). This finding indicates that root lectin is a plant determinant of the *Rhizobium*-legume specific recognition. Root lectin is particularly abundant at the tip of root hairs in the root zone susceptible to *Rhizobium* infection and is predominantly present at the external surface of the plasma membrane (29, 110). The Nod signals have structural properties that are characteristic of effective lectin ligands such as hexosamine oligosaccharides (107). A plant-root lectin could thus be part of the Nod signal–receptor complex (110). Whatever the nature and the location of the Nod factor receptor(s), the possibility of preparing signal derivatives for radioactive and/or photoaffinity labeling should allow the isolation of Nod factor–binding plant molecules.

CONCLUSIONS AND PERSPECTIVES

The early steps of the rhizobia-legumes symbioses involve a series of complex and sequential exchanges of low-molecular-weight signals between the two partners. Plant flavonoids secreted in root exudates induce the expression of the bacterial *nod* genes that specify infection and nodulation. Flavonoids activate, in a specific manner, the regulatory NodD proteins that control the transcription of common and specific *nod* genes. The *nod* genes determine the synthesis of lipooligosaccharidic signal molecules (N-acylated chitin oligomers), the Nod factors, which in turn activate the transcription of plant symbiotic genes. The common *nodABC* genes code for the synthesis of the molecule backbone, which is similar in all rhizobia, whereas the host-range *nod* genes mediate the decoration of the molecules with various substituents (N-acyl chain, sulfate, acetate and carbamoyl groups, sugars) to make them plant specific. The purified Nod factors elicit root-hair deformations, cortical cell divisions, and the formation of nodule meristems in a specific manner, indicating that rhizobia determine root-hair curling, nodulation, and host range by the production of these signal molecules. Thus, the recent findings provide a general model to describe signal exchange during the early steps of the symbiotic interactions.

However, we are far from having a clear picture of the molecular mechanisms underlying this signal exchange. Not much is known about the synthesis and secretion of flavonoids in legume roots. On the bacterial side, recent progress has revealed the complexity of the regulatory circuits that control *nod* gene expression in *R. meliloti* and *B. japonicum*. Future research will complete the identification of the regulatory elements and analyze their role in the control of *nod* gene expression both in vitro and *in planta*. The molecular mechanisms by which the flavonoid plant signals activate specifically the regulatory NodD proteins are still unknown and will be analyzed using in vitro systems. The genetic (numerous regulatory genes) and molecular (multimeric regulatory proteins) complexity of the circuits is intriguing and probably reflects the complexity of the control of *nod* gene expression during the various steps of the symbiotic process.

Future work on the biochemical function of structural *nod* genes will dissect the biosynthetic pathways of the lipooligosaccharidic Nod factors and identify the various mechanisms that have evolved among rhizobia to generate a great variety of decorations of the core molecule, giving rise to the wide variety of rhizobial host-range combinations. On the plant side, the next step is to identify and characterize the receptors, and possibly other molecules, responsible for the specific plant response to the bacterial Nod factors. The availability of signal molecules that can be modified at will, both genetically and chemically, should allow the study of Nod signal perception, transduction, and activation of the expression of plant genes involved in the plant responses to Nod factors. For deciphering the molecular basis of host specificity and signal exchange in plant-bacteria interactions in general as well as for the induction of a plant organogenesis, rhizobia-legume associations will continue to provide very useful systems.

ACKNOWLEDGMENTS

We sincerely thank our colleagues J. C. Promé, G. Truchet, and P. Roche for stimulating discussions and W. J. Broughton, S. Long, N. J. P. Price, H. Spaink, and G. Stacey for providing unpublished information. We are also extremely grateful to the numerous colleagues who supplied us with reprints and preprints. We thank Julie V. Cullimore and Anne Stanford for critical reading of the manuscript. This work was supported in part by a grant from the Conseil Régional Midi-Pyrénées.

Literature Cited

1. Aguilar, J. M. M., Ashby, A. M., Richards, A. J. M., Loake, G. J., Watson, M. D., Shaw, C. H. 1988. Chemotaxis of *Rhizobium leguminosarum* biovar *phaseoli* towards flavonoid inducers of the symbiotic nodulation genes. *J. Gen Microbiol.* 134:2741–46
2. Allen, O. N., Allen, E. K. 1981. *The Leguminosae. A Source Book of Characteristics, Uses, and Nodulation.* Madison: Univ. Wisconsin Press
3. Appelbaum, E. R., Thompson, D. V., Idler, K., Chartrain, N. 1988. *Bradyrhizobium japonicum* USDA 191 has two *nodD* genes that differ in primary structure and function. *J. Bacteriol.* 170:12–20
4. Armitage, J. P., Gallagher, A., Johnston, A. W. B. 1988. Comparison of the chemotactic behaviour of *Rhizobium leguminosarum* with and without the nodulation plasmid. *Mol. Microbiol.* 2:743–48
5. Baev, N., Endre, G., Petrovics, G., Banfalvi, Z., Kondorosi, A. 1991. Six nodulation genes of *nod* box locus-4 in *Rhizobium meliloti* are involved in nodulation signal production—*nodM* codes for D-glucosamine synthetase. *Mol. Gen. Genet.* 228:113–24
6. Banfalvi, Z., Kondorosi, A. 1989. Production of root hair deformation factors by *Rhizobium meliloti* nodulation genes in *Escherichia coli:* HsnD (*nodH*) is involved in the plant host-specific modification of the NodABC-factor. *Plant Mol. Biol.* 13:1–12
7. Banfalvi, Z., Nieuwkoop, A., Schell, M., Besl, L., Stacey, G. 1988. Regulation of *nod* gene expression in *Bradyrhizobium japonicum. Mol. Gen. Genet.* 214:420–24
8. Barbour, W. M., Wang, S. P., Stacey, G. 1992. Molecular genetics of *Bradyrhizobium* symbiosis. See Ref. 154, pp. 645–81
9. Barnett, M. J., Long, S. R. 1990. DNA sequence and translational product of a new nodulation-regulatory locus: *syrM* has sequence similarity to NodD proteins. *J. Bacteriol.* 172:3695–3700
10. Bassam, B. J., Djordjevic, M. A., Redmond, J. W., Batley, M., Rolfe, B. G. 1988. Identification of a *nodD*-

dependent locus in the *Rhizobium* strain NGR234 activated by phenolic factors secreted by soybeans and other legumes. *Mol. Plant-Microb. Interact.* 1:161–68
11. Brewin, N. J. 1991. Development of the legume root nodule. *Annu. Rev. Cell Biol.* 7:191–226
12. Broughton, W. J., Krause, A., Lewin, A., Perret, X., Price, N. P. J., et al. 1991. Signal exchange mediates host-specific nodulation of tropical legumes by the broad host-range *Rhizobium* species NGR234. See Ref. 73, pp. 162–67
13. Burn, J. E., Hamilton, W. D., Wootton, J. C., Johnston, A. W. B. 1989. Single and multiple mutations affecting properties of the regulatory gene *nodD* of *Rhizobium*. *Mol. Microbiol.* 3:1567–77
14. Burn, J. E., Rossen, L., Johnston, A. W. B. 1987. Four classes of mutations in the *nodD* gene of *Rhizobium leguminosarum* biovar. *viciae* that affect its ability to autoregulate and/or activate other *nod* genes in the presence of flavonoid inducers. *Genes Dev.* 1:456–64
15. Caetano-Anollés, G., Christ-Estes, D. K., Bauer, W. D. 1988. Chemotaxis of *Rhizobium meliloti* to the plant flavone luteolin requires functional nodulation genes. *J. Bacteriol.* 170:3164–69
16. Caetano-Anollés, G., Gresshoff, P. M. 1991. Plant genetic control of nodulation. *Annu. Rev. Microbiol.* 45:345–82
17. Caetano-Anollés, G., Gresshoff, P. M. 1992. Plant genetic suppression of the non-nodulation phenotype of *Rhizobium meliloti* host range *nodH* mutants: gene-for-gene interaction in the alfalfa-*Rhizobium* symbiosis? *Theor. Appl. Genet.* In press
18. Cervantès, E., Sharma, S. B., Maillet, F., Vasse, J., Truchet, G., Rosenberg, C. 1989. The product of host-specific *nodQ* gene of *Rhizobium meliloti* shares homology with translation, elongation and initiation factors. *Mol. Microbiol.* 3:745–55
19. Cregan, P. B., Keyser, H. H., Sadowsky, M. J. 1989. Host plant effects on nodulation and competitiveness of the *Bradyrhizobium japonicum* serotype strains constituting serocluster 123. *Appl. Environ. Microbiol.* 55:2532–36
20. Davis, E. O., Evans, I. J., Johnston, A. W. B. 1988. Identification of *nodX*, a gene that allows *Rhizobium leguminosarum* biovar *viciae* strain TOM to nodulate Afghanistan peas. *Mol. Gen. Genet.* 212:531–35
21. Davis, E. O., Johnston, A. W. B. 1990. Analysis of 3 *nodD* genes in *Rhizobium leguminosarum* biovar *phaseoli—nodD1* is preceded by *nolE*, a gene whose product is secreted from the cytoplasm. *Mol. Microbiol.* 4:921–32
22. Davis, E. O., Johnston, A. W. B. 1990. Regulatory functions of the 3 *nodD* genes of *Rhizobium leguminosarum* biovar *phaseoli*. *Mol. Microbiol.* 4:933–41
23. De Maagd, R. A., Wijfjes, A. H. M., Spaink, H. P., Ruiz-Sainz, J. E., Wijffelman, C. A., et al. 1990. *nodO*, a new *nod* gene of the *Rhizobium leguminosarum* biovar viciae Sym plasmid encodes a secreted protein. *J. Bacteriol.* 171:6764–70
24. Debellé, F., Maillet, F., Vasse, J., Rosenberg, C., de Billy, F., et al. 1988. Interference between *Rhizobium meliloti* and *Rhizobium trifolii* nodulation genes: genetic basis of the *R. meliloti* dominance. *J. Bacteriol.* 170:5718–27
25. Debellé, F., Rosenberg, C., Vasse, J., Maillet, F., Martinez, E., et al. 1986. Assignment of symbiotic developmental phenotypes to common and specific nodulation (*nod*) genetic loci of *Rhizobium meliloti*. *J. Bacteriol.* 168:1075–86
26. Debellé, F., Sharma, S. B. 1986. Nucleotide sequence of *Rhizobium meliloti* RCR2011 genes involved in host specificity of nodulation. *Nucleic Acids Res.* 14:7453–72
27. Dénarié, J., Roche, P. 1992. *Rhizobium* nodulation signals. See Ref. 162a, pp. 295–324
28. Diaz, C. L., Melchers, L. S., Hooykaas, P. J. J., Lugtenberg, B. J. J., Kijne, J. W. 1989. Root lectin as a determinant of host-specificity in the *Rhizobium*-legume symbiosis. *Nature* 338: 579–81
29. Diaz, C. L., Van Spronsen, P. C., Bakhuizen, R., Logman, G. J. J., Lugtenberg, B. J. J., Kijne, J. W. 1986. Correlation between infection by *Rhizobium leguminosarum* and lectin on the surface of *Pisum sativum* L. roots. *Planta* 168:530–39
30. Djordjevic, M. A., Gabriel, D. W., Rolfe, B. G. 1987. *Rhizobium*, the refined parasite of legumes. *Annu. Rev. Phytopathol.* 25:145–68
31. Djordjevic, M. A., Innes, R. W., Wijffelman, C. A., Schofield, P. R., Rolfe, B. G. 1986. Nodulation of specific legumes is controlled by several distinct loci in *Rhizobium trifolii*. *Plant Mol. Biol.* 6:389–401
32. Djordjevic, M. A., Redmond, J. W., Batley, M., Rolfe, B. G. 1987. Clovers secrete specific phenolic compounds which either stimulate or repress *nod*

gene expression in *Rhizobium trifolii*. *EMBO J.* 6:1173–79
33. Djordjevic, M. A., Schofield, P. R., Rolfe, B. G. 1985. Tn*5* mutagenesis of *Rhizobium trifolii* host specific nodulation genes result in mutants with altered host range ability. *Mol. Gen. Genet.* 200:463–71
34. Dowling, D. N., Broughton, W. J. 1986. Competition for nodulation of legumes. *Annu. Rev. Microbiol.* 40: 131–57
35. Downie, J. A. 1989. The *nodL* gene from *Rhizobium leguminosarum* is homologous to the acetyl transferase encoded by *lacA* and *cysE*. *Mol. Microbiol.* 3:1649–51
36. Downie, J. A., Hombrecher, G., Ma, Q. S., Knight, C. D., Wells, B., Johnston, A. W. B. 1983. Cloned nodulation genes of *Rhizobium leguminosarum* determine host-range specificity. *Mol. Gen. Genet.* 190:359–65
37. Downie, J. A., Marie, C., Scheu, A. K., Firmin, J. L., Wilson, K. E., et al. 1991. Genetic and biochemical studies on the nodulation genes of *Rhizobium leguminosarum* bv. *viciae*. See Ref. 73, pp. 134–41
38. Downie, J. A., Surin, B. P. 1990. Either of two *nod* gene loci can complement the nodulation defect of a *nod* deletion mutant of *Rhizobium leguminosarum* bv. *viciae*. *Mol. Gen. Genet.* 222:81–86
39. Dreyfus, B., Garcia, J. L., Gillis, M. 1988. Characterization of *Azorhizobium caulinodans* gen. nov., sp. nov., a stem-nodulating nitrogen-fixing bacterium isolated from *Sesbania rostrata*. *Int. J. Syst. Bacteriol.* 38:89–98
40. Dusha, I., Bakos, A., Kondorosi, A., de Bruijn, F. J., Schell, J. 1989. The *Rhizobium* meliloti early nodulation genes (*nodABC*) are nitrogen-regulated. Isolation of a mutant strain with efficient nodulation capacity on alfalfa in the presence of ammonium. *Mol. Gen. Genet.* 219:89–96
41. Economou, A., Hamilton, W. D. O., Johnston, A. W. B., Downie, J. A. 1990. The *Rhizobium* nodulation gene *nodO* encodes a Ca^{2+}-binding protein that is exported without N-terminal cleavage and is homologous to haemolysin and related proteins. *EMBO J.* 9:349–54
42. Evans, I. J., Downie, J. A. 1986. The NodI product of *Rhizobium leguminosarum* is closely related to ATP-binding bacterial transport proteins: nucleotide sequence of the *nodI* and *nodJ* genes. *Gene* 43:95–101
43. Faucher, C., Camut, S., Dénarié, J., Truchet, G. 1989. The *nodH* and *nodQ* host range genes of *Rhizobium meliloti* behave as avirulence genes in *R. leguminosarum* bv. *viciae* and determine changes in the production of plant-specific extracellular signals. *Mol. Plant-Microbe Interact.* 2:291–300
44. Faucher, C., Maillet, F., Vasse, J., Rosenberg, C., van Brussel, A. A. N., et al. 1988. *Rhizobium meliloti* host range *nodH* gene determines production of an alfalfa-specific extracellular signal. *J. Bacteriol.* 170:5489–99
45. Firmin, J. L., Wilson, K. E., Rossen, L., Johnston, A. W. B. 1986. Flavonoid activation of nodulation genes in *Rhizobium* reversed by other compounds present in plants. *Nature* 324:90–92
46. Fisher, R. F., Egelhoff, T. T., Mulligan, J. T., Long, S. R. 1988. Specific binding of proteins from *Rhizobium meliloti* cell-free extracts containing NodD to DNA sequences upstream of inducible nodulation genes. *Genes Dev.* 2:282–93
47. Fisher, R. F., Long, S. R. 1989. DNA footprint analysis of the transcriptional activator proteins NodD1 and NodD3 on inducible *nod* genes promoters. *J. Bacteriol.* 171:5492–5502
48. Fisher, R. F., Swanson, J. A., Mulligan, J. T., Long, S. R. 1987. Extended region of nodulation genes in *Rhizobium meliloti* 1021. II. Nucleotide sequence, transcription start sites and protein products. *Genetics* 117:191–201
49. Gaworzewska, E. T., Carlile, M. J. 1982. Positive chemotaxis of *Rhizobium leguminosarum* and other bacteria towards root exudates from legumes and other plants. *J. Gen Microbiol.* 128: 1179–88
50. Geiger, O., Spaink, H. P., Kennedy, E. P. 1991. Isolation of the *Rhizobium leguminosarum* NodF nodulation protein: NodF carries a 4'-phosphopantetheine prosthetic group. *J. Bacteriol.* 173:2872–78
51. Gerhold, D., Stacey, G., Kondorosi, A. 1989. Use of a promoter-specific probe to identify two loci from the *Rhizobium meliloti* nodulation regulon. *Plant Mol. Biol.* 12:181–88
52. Goethals, K., Gao, M., Tomekpe, K., Van Montagu, M., Holsters, M. 1989. Common *nodABC* genes in *nod* locus 1 of *Azorhizobium caulinodans:* nucleotide sequence and plant-inducible expression. *Mol. Gen. Genet.* 219:289–98
53. Goethals, K., Van Den Eede, G., Van Montagu, M., Holsters, M. 1990. Identification and characterization of a functional *nodD* gene in *Azorhizobium*

caulinodans ORS571. *J. Bacteriol.* 172:2658–66
54. Goethals, K., Van Montagu, M., Holsters, M. 1992. Conserved motifs in a divergent *nod* box of *Azorhizobium caulinodans* ORS571 reveal a common structure in promoters regulated by LysR-type proteins. *Proc. Natl. Acad. Sci. USA.* 89:1646–50
55. Göttfert, M., Grob, P., Hennecke, H. 1990. Proposed regulatory pathway encoded by the *nodV* and *nodW* genes, determinants of host specificity in *Bradyrhizobium japonicum*. *Proc. Natl. Acad. Sci. USA* 87:2680–84
56. Göttfert, M., Hitz, S., Hennecke, H. 1990. Identification of *nodS* and *nodU*, two inducible genes inserted between the *Bradyrhizobium japonicum nodYABC* and *nodIJ* genes. *Mol. Plant-Microbe Interact.* 3:308–16
57. Göttfert, M., Horvath, B., Kondorosi, E., Putnoky, P., Rodriguez-Quinones, F., Kondorosi, A. 1986. At least two *nodD* genes are necessary for efficient nodulation of alfalfa by *Rhizobium meliloti*. *J. Mol. Biol.* 191:411–20
58. Göttfert, M., Lamb, J. W., Gasser, R., Semenza, J., Hennecke, H. 1989. Mutational analysis of the *Bradyrhizobium japonicum* common *nod* genes and further *nod* box-linked genomic DNA regions. *Mol. Gen. Genet.* 215:407–15
59. Graham, T. L. 1991. Flavonoid and isoflavonoid distribution in developing soybean seedling tissues and in seed and root exudates. *Plant Physiol.* 95:594–603
60. Gray, J. X., De Maagd, R. A., Rolfe, B. G., Johnston, A. W. B., Lugtenberg, B. J. J. 1992. The role of the *Rhizobium* cell surface during symbiosis. See Ref. 162a, pp. 359–76
61. Gray, J. X., Rolfe, B. G. 1990. Exopolysaccharide production in *Rhizobium* and its role in invasion. *Mol. Microbiol.* 4:1425–31
62. Gresshoff, P. M., Roth, L. E., Stacey, G., Newton, W. E., eds. 1990. *Nitrogen Fixation: Achievements and Objectives*. New York: Chapman and Hall
63. Gulash, M., Ames, P., Larosiliere, R. C., Bergman, K. 1984. *Rhizobium* are attracted to localized sites on legume roots. *Appl. Environ. Microbiol.* 48: 149–52
64. Györgypal, Z., Iyer, N., Kondorosi, A. 1988. Three regulatory *nodD* alleles of diverged flavonoid-specificity are involved in host-dependent nodulation by *Rhizobium meliloti*. *Mol. Gen. Genet.* 212:85–92
65. Györgypal, Z., Kondorosi, A. 1991. Homology of the ligand-binding regions of *Rhizobium* symbiotic regulatory protein NodD and vertebrate nuclear receptors. *Mol. Gen. Genet.* 226:337–40
66. Györgypal, Z., Kondorosi, E., Kondorosi, A. 1991. Diverse signal sensitivity of NodD protein homologs from narrow and broad host range rhizobia. *Mol. Plant-Microbe Interact.* 4:356–64
67. Hartwig, U. A., Joseph, C. M., Phillips, D. A. 1991. Flavonoids released naturally from alfalfa seeds enhance growth rate of *Rhizobium meliloti*. *Plant Physiol.* 95:797–803
68. Hartwig, U. A., Maxwell, C. A., Joseph, C. M., Phillips, D. A. 1989. Interactions among flavonoid *nod* gene inducers released from alfalfa seeds and roots. *Plant Physiol.* 91:1138–42
69. Hartwig, U. A., Maxwell, C. A., Joseph, C. M., Phillips, D. A. 1990. Chrysoeriol and luteolin released from alfalfa seeds induce *nod* genes in *Rhizobium meliloti*. *Plant Physiol.* 92:116–22
70. Hartwig, U. A., Maxwell, C. A., Joseph, C. M., Phillips, D. A. 1990. Effects of alfalfa *nod* gene–inducing flavonoids on *nodABC* transcription in *Rhizobium meliloti* strains containing different *nodD* genes. *J. Bacteriol.* 172:2769–73
71. Henikoff, S., Haughn, G. W., Calvo, J. M., Wallace, J. C. 1988. A large family of bacterial activator proteins. *Proc. Natl. Acad. Sci. USA* 85:6602–6
72. Hennecke, H., Kaluza, K., Thöny, B., Fuhrmann, M., Ludwig, W., Stackebrandt, E. 1985. Concurrent evolution of nitrogenase genes and 16S rRNA in *Rhizobium* species and other nitrogen fixing bacteria. *Arch. Microbiol.* 142:342–48
73. Hennecke, H., Verma, D. P. S., eds. 1991. *Advances in Molecular Genetics of Plant-Microbe Interactions*. Dordrecht: Kluwer Academic
74. Holl, F. B. 1975. Host plant control of the inheritance of dinitrogen fixation in the *Pisum-Rhizobium* symbiosis. *Euphytica* 24:767–70
75. Hombrecher, G., Gotz, R., Dibb, N. J., Downie, J. A., Johnston, A. W. B., Brewin, N. J. 1984. Cloning and mutagenesis of nodulation genes from *Rhizobium leguminosarum* TOM, a strain with extended host range. *Mol. Gen. Genet.* 194:293–98
76. Hong, G. F., Burn, J. E., Johnston, A. W. B. 1987. Evidence that DNA involved in the expression of nodulation (*nod*) genes in *Rhizobium* binds to the product of the regulatory gene *nodD*. *Nucleic Acids Res.* 15:9677–90

77. Honma, M. A., Asomaning, M., Ausubel, F. M. 1990. *Rhizobium meliloti nodD* genes mediate host-specific activation of *nodABC*. *J. Bacteriol.* 172:901–11
78. Honma, M. A., Ausubel, F. M. 1987. *Rhizobium meliloti* has three functional copies of the *nodD* symbiotic regulatory gene. *Proc. Natl. Acad. Sci. USA* 84: 8558–62
79. Horvath, B., Bachem, C. W. B., Schell, J., Kondorosi, A. 1987. Host-specific regulation of nodulation genes in *Rhizobium* is mediated by a plant signal, interacting with the *nodD* gene product. *EMBO J.* 6:841–48
80. Horvath, B., Kondorosi, E., John, M., Schmidt, J., Török, I., et al. 1986. Organization, structure and symbiotic function of Rhizobium meliloti nodulation genes determining host specificity for alfalfa. *Cell* 46:335–43
81. Hungria, M., Joseph, C. M., Philips, D. A. 1991. Anthocyanidins and flavonols, major *nod* gene inducers from seeds of a black-seeded common bean (*Phaseolus vulgaris* L.). *Plant Physiol.* 97:751–58
82. Hungria, M., Joseph, C. M., Phillips, D. A. 1991. *Rhizobium nod* gene inducers exuded naturally from roots of common bean (*Phaseolus vulgaris* L.). *Plant Physiol.* 97:759–64
83. Jacobs, T. W., Egelhoff, T. T., Long, S. R. 1985. Physical and genetic map of a *Rhizobium meliloti* gene region and nucleotide sequence of *nodC*. *J. Bacteriol.* 162:469–76
84. Jarvis, B. D. W., Gillis, M., De Ley, J. 1986. Intra- and intergeneric similarities between the ribosomal ribonucleic acid cistrons of *Rhizobium* and *Bradyrhizobium* species and some related bacteria. *Int. J. Syst. Bacteriol.* 36:129–38
85. John, M., Schmidt, J., Wieneke, U., Krüssmann, H. D., Schell, J. 1988. Transmembrane orientation and receptor-like structure of the *Rhizobium meliloti* common nodulation protein NodC. *EMBO J.* 7:583–88
86. Johnson, D., Roth, E. L., Stacey, G. 1989. Immunogold localization of the NodC and NodA proteins of *Rhizobium meliloti*. *J. Bacteriol.* 171:4583–88
87. Kape, R., Parniske, M., Werner, D. 1991. Chemotaxis and *nod* gene activity of *Bradyrhizobium japonicum* in response to hydroxycinnamic acids and isoflavonoids. *Appl. Environ. Microbiol.* 57:316–19
88. Kapulnik, Y., Joseph, C. M., Phillips, D. A. 1987. Flavone limitations to root nodulation and symbiotic nitrogen fixation in alfalfa. *Plant Physiol.* 84:1193–96
89. Keen, N. T., Staskawicz, B. J. 1988. Host range determinants in plant pathogens and symbionts. *Annu. Rev. Phytopathol.* 42:421–40
90. Kijne, J. W. 1992. The *Rhizobium* infection process. See Ref. 154, pp. 348–97
91. Kijne, J. W., Bakhuizen, R., van Brussel, A. A. N., Canter-Cremers, H. C. G., Diaz, C. L., et al. 1992. The *Rhizobium* trap: root hair curling in the root nodule symbiosis. In *Perspectives in Plant Cell Recognition,* ed. J. R. Green, J. A. Callow, pp. 267–84. Cambridge: SEB Seminar Ser., Cambridge Univ. Press
92. Kijne, J. W., Lugtenberg, B. J. J., Smit, G. 1992. Attachment, lectin and initiation of infection in *(Brady) Rhizobium*-legume interactions. See Ref. 162a, pp. 281–94
93. Kondorosi, A. 1991. Overview on genetics of nodule induction: factors controlling nodule induction by *Rhizobium meliloti*. See Ref. 73, pp. 111–18
94. Kondorosi, A. 1992. Regulation of nodulation genes in rhizobia. See Ref. 162a, pp. 325–40
95. Kondorosi, E., Banfalvi, Z., Kondorosi, A. 1984. Physical and genetic analysis of a symbiotic region of *Rhizobium meliloti:* identification of nodulation genes. *Mol. Gen. Genet.* 193:445–52
96. Kondorosi, E., Buiré, M., Cren, M., Iyer, N., Hoffmann, B., Kondorosi, A. 1991. Involvement of the *syrM* and *nodD3* genes of *Rhizobium meliloti* in *nod* gene activation and in optimal nodulation of the plant host. *Mol. Microbiol.* 5:3035–48
97. Kondorosi, E., Gyuris, J., Schmidt, J., John, M., Duda, E., et al. 1989. Positive and negative control of *nod* gene expression in *Rhizobium meliloti* is required for optimal nodulation. *EMBO J.* 8:1331–40
98. Kosslak, R. M., Bookland, R., Barkei, J., Paaren, H. E., Appelbaum, E. R. 1987. Induction of *Bradyrhizobium japonicum* common *nod* genes by isoflavones isolated from Glycine max. *Proc. Natl. Acad. Sci. USA* 84:7428–32
99. Kosslak, R. M., Joshi, R. S., Bowen, B. A., Paaren, H. E., Appelbaum, E. R. 1990. Strain-specific inhibition of *nod* gene induction in *Bradyrhizobium* japonicum by flavonoid compounds. *Appl. Environ. Microbiol.* 56:1333–41
100. Kredich, N. M. 1987. Biosynthesis of cysteine. In Escherichia coli *and* Salmonella typhimurium *Cellular and*

Molecular Biology, ed. F. C. Neidhardt, pp. 419–28. Washington DC: Am. Soc. Microbiol.

100a. Leigh, J. A., Coplin, D. L. 1992. Exopolysaccharides in plant-bacterial interactions. *Annu. Rev. Microbiol.* 46: 307–46

101. Lerouge, P., Roche, P., Faucher, C., Maillet, F., Truchet, G., et al. 1990. Symbiotic host-specificity of *Rhizobium meliloti* is determined by a sulphated and acylated glucosamine oligosaccharide. *Nature* 344:781–84

102. Lerouge, P., Roche, P., Promé, J. C., Faucher, C., Vasse, J., et al. 1991. *Rhizobium meliloti* nodulation genes specify the production of an alfalfa-specific sulphated oligosaccharide signal. See Ref. 62, pp. 177–86

103. Le Strange, K. K., Bender, G. L., Djordjevic, M. A., Rolfe, B. G., Redmond, J. W. 1990. The *Rhizobium* strain NGR234 *nodD1* gene product responds to activation by the simple phenolic compounds vanillin and isovanillin present in wheat seedling extracts. *J. Bacteriol.* 3:214–20

104. Lewin, A., Cervantès, E., Wong, C. H., Broughton, W. J. 1990. *nodSU,* two new *nod* genes of the broad host-range *Rhizobium* strain NGR234 encode host-specific nodulation of the tropical tree *Leucaena leucocephala*. *Mol. Plant-Microbe Interact.* 3:317–26

105. Lewin, A., Rosenberg, C., Meyer, Z. A. H., Wong, C. H. Nelson, L., et al. 1987. Multiple host-specificity loci of the broad host-range *Rhizobium* sp. NGR234 selected using the widely compatible legume *Vigna unguiculata*. *Plant Mol. Biol.* 8:447–59

106. Lie, T. A. 1978. Symbiotic specialization in pea plants: the requirement of specific *Rhizobium* strains for peas from Afghanistan. *Annu. Appl. Biol.* 88:462–65

107. Lis, H., Sharon, N. 1986. Lectins as molecules and as tools. *Annu. Rev. Biochem.* 55:35–67

108. Long, S. R. 1989. Rhizobium-legume nodulation: life together in the underground. *Cell* 56:203–14

109. Long, S. R. 1989. *Rhizobium* genetics. *Annu. Rev. Genet.* 23:483–506

110. Lugtenberg, B. J. J., Diaz, C. L., Smit, G., De Pater, S., Kijne, J. W. 1991. Roles of the lectin in the *Rhizobium*-legume symbiosis. See Ref. 73, pp. 174–81

111. Maillet, F., Debellé, F., Dénarié, J. 1990. Role of the *nodD* and *SyrM* genes in the activation of the regulatory gene *nodD3,* and of the common and host-specific *nod* genes of *Rhizobium meliloti*. *Mol. Microbiol.* 4:1975–84

112. Martinez, E., Romero, D., Palacios, R. 1990. The *Rhizobium* genome. *Crit. Rev. Plant Sci.* 9:59–93

113. Maxwell, C. A., Hartwig, U. A., Joseph, C. M., Phillips, D. A. 1989. A chalcone and two related flavonoids released from alfalfa roots induce *nod* genes of *Rhizobium meliloti*. *Plant Physiol.* 91:842–47

114. Maxwell, C. A., Phillips, D. A. 1990. Concurrent synthesis and release of *nod*-gene-inducing flavonoids from alfalfa roots. *Plant Physiol.* 93:1552–58

115. McIver, J., Djordjevic, M. A., Weinman, J. J., Bender, G. L., Rolfe, B. G. 1989. Extension of host range of *Rhizobium leguminosarum* bv. *trifolii* caused by point mutations in *nodD* that result in alterations in regulatory function and recognition of inducer molecules. *Mol. Plant-Microbe Interact.* 2:97–106

116. Mulligan, J. T., Long, S. R. 1985. Induction of *Rhizobium meliloti nodC* expression by plant exudate requires *nodD*. *Proc. Natl. Acad. Sci. USA* 82:6609–13

117. Mulligan, J. T., Long, S. R. 1989. A family of activator genes regulates expression of *Rhizobium meliloti* nodulation genes. *Genetics* 122:7–18

118. Nap, J. P., Bisseling, T. 1990. Developmental biology of a plant-prokaryote symbiosis: the legume root nodule. *Science* 250:948–54

119. Noel, K. D. 1992. Rhizobial polysaccharides required in symbioses with legumes. See 162a, pp. 341–58

120. Parke, D., Rivelli, M., Ornston, L. N. 1985. Chemotaxis to aromatic and hydroaromatic acids: comparison of *Bradyrhizobium japonicum* and *Rhizobium trifolii*. *J. Bacteriol.* 163:417–22

121. Peters, N. K., Frost, J. W., Long, S. R. 1986. A plant flavone, luteolin, induces expression of *Rhizobium meliloti* nodulation genes. *Science* 233:977–80

122. Peters, N. K., Long, S. R. 1988. Alfalfa root exudates and compounds which promote or inhibit induction of *Rhizobium meliloti* nodulation genes. *Plant Physiol.* 88:396–400

123. Peters, N. K., Verma, D. P. S. 1990. Phenolic compounds as regulators of gene expression in plant-microbe interactions. *Mol. Plant-Microbe Interact.* 3:4–8

124. Polhill, R. M., Raven, P. H. 1981. *Advances in Legume Systematics*. Kew, England: Royal Botanic Gardens

125. Recourt, K., van Brussel, A. A. N., Driessen, A. J. M., Lugtenberg, B. J. J. 1989. Accumulation of a *nod* gene in-

ducer, the flavonoid naringenin, in the cytoplasmic membrane of *Rhizobium leguminosarum* biovar *viciae* is caused by the pH-dependent hydrophobicity of naringenin. *J. Bacteriol.* 171:4370–77

126. Redmond, J. W., Batley, M., Djordjevic, M. A., Innes, R. W., Kuempel, P. L., Rolfe, B. G. 1986. Flavones induce expression of nodulation genes in *Rhizobium. Nature* 323:632–35
127. Reuber, T. L., Reed, J. W., Glazebrook, J., Urzainqui, A., Walker, G. C. 1991. Analyses of the roles of *R. meliloti* exopolysaccharides in nodulation. See Ref. 73, pp. 182–88
128. Roche, P., Debellé, F., Maillet, F., Lerouge, P., Faucher, C., et al. 1991. Molecular basis of symbiotic host specificity in Rhizobium meliloti: nodH and nodPQ genes encode the sulfation of lipo-oligosaccharide signals. *Cell* 67: 1131–43
129. Roche, P., Lerouge, P., Ponthus, C., Promé, J. C. 1991. Structural determination of bacterial nodulation factors involved in the *Rhizobium meliloti*-alfalfa symbiosis. *J. Biol. Chem.* 266: 10933–40
130. Roche, P., Lerouge, P., Promé, J. C., Faucher, C., Vasse, J., et al. 1991. NodRm-1, a sulphated lipo-oligosaccharide signal of *Rhizobium meliloti* elicits hair deformation, cortical cell division and nodule organogenesis on alfalfa roots. See Ref. 73, pp. 119–26
131. Rodrigues-Quinones, F., Banfalvi, Z., Murphy, P., Kondorosi, A. 1987. Interspecies homology of nodulation genes in *Rhizobium. Plant Mol. Biol.* 8:61–76
132. Rolfe, B. G., Gresshoff, P. M. 1988. Genetic analysis of legume nodule initiation. *Annu. Rev. Plant Physiol. Plant Mol. Biol.* 39:297–319
133. Rossen, L., Johnston, A. W. B., Downie, J. A. 1984. DNA sequence of the *Rhizobium leguminosarum* nodulation genes *nodAB* and *C* required for root hair curling. *Nucleic Acids Res.* 12:9497–9508
134. Rossen, L., Shearman, C. A., Johnston, A. W. B., Downie, J. A. 1985. The *nodD* gene of *Rhizobium leguminosarum* is autoregulatory and in the presence of plant exudate induces the *nodA,B,C* genes. *EMBO J.* 4:3369–73
135. Rostas, K., Kondorosi, E., Horvath, B., Simoncsits, A., Kondorosi, A. 1986. Conservation and extended promoter regions of nodulation genes in *Rhizobium. Proc. Natl. Acad. Sci. USA* 83:1757–61
136. Rushing, B. G., Yelton, M. M., Long, S. R. 1991. Genetic and physical analysis of the *nodD3* region of *Rhizobium meliloti. Nucleic Acids Res.* 19:921–27
137. Sadowsky, M. J., Cregan, P. B., Gottfert, M., Sharma, A., Gorhold, D., et al. 1991. The *Bradyrhizobium japonicum nolA* gene and its involvement in the genotype-specific nodulation of soybeans. *Proc. Natl. Acad. Sci. USA* 88:637–41
138. Schlaman, H. R. M., Spaink, H. P., Okker, R. J. H., Lugtenberg, B. J. J. 1989. Subcellular localization of the *nodD* gene product in *Rhizobium leguminosarum. J. Bacteriol.* 171:4686–93
139. Schmidt, J., John, M., Wieneke, U., Krüssmann, H. D., Schell, J. 1986. Expression of the nodulation gene *nodA* in *Rhizobium meliloti* and localization of the gene product in the cytosol. *Proc. Natl. Acad. Sci. USA* 83:9581–85
140. Schmidt, J., Wingender, R., John, M., Wieneke, U., Schell, J. 1988. *Rhizobium meliloti nodA* and *nodB* genes are involved in generating compounds that stimulate mitosis of plant cells. *Proc. Natl. Acad. Sci. USA* 85:8578–82
141. Deleted in proof
142. Schwedock, J., Long, S. R. 1989. Nucleotide sequence and protein products of two new nodulation genes of *Rhizobium meliloti, nodP* and *nodQ. Mol. Plant-Microbe Interact.* 2:181–94
143. Schwedock, J., Long, S. R. 1990. ATP sulfurylase activity of the *nodP* and *nodQ* gene products of *Rhizobium meliloti. Nature* 348:644–47
144. Shearman, C. A., Rossen, L., Johnston, A. W. B., Downie, J. A. 1986. The *Rhizobium* gene *nodF* encodes a protein similar to acyl carrier protein and is regulated by *nodD* plus a factor in pea root exudate. *EMBO J.* 5:647–52
145. Sheldon, P. S., Kekwick, R. G. O., Sidebottom, C., Smith, C. G., Slabas, A. R. 1990. 3-oxoacyl-(acyl-carrier protein) reductase from avocado (*Persea americana*) fruit mesocarp. *Biochem. J.* 271:713–20
146. Spaink, H. P., Aarts, A., Stacey, G., Bloemberg, G. V., Lugtenberg, B. J. J., Kennedy, E. P. 1991. Detection and separation of *Rhizobium* and *Bradyrhizobium* Nod metabolites using thin layer chromatography. *Mol. Plant-Microbe Interact.* 5:72–80
147. Spaink, H. P., Geiger, O., Sheeley, D. M., van Brussel, A. A. N., York, W. S., et al. 1991. The biochemical function of the *Rhizobium leguminosarum* proteins involved in the production of host specific signal molecules. See Ref. 73, pp. 142–49
148. Spaink, H. P., Okker, R. J. H., Wijffelman, C. A., Tak, T., Goosen de Roo,

L., et al. 1989. Symbiotic properties of rhizobia containing a flavonoid-independent hybrid *nodD* product. *J. Bacteriol.* 171:4045–53

149. Spaink, H. P., Sheeley, D. M., van Brussel, A. A. N., Glushka, J., York, W. S., et al. 1991. A novel highly unsaturated fatty acid moiety of lipo-oligosaccharide signals determines host specificity of *Rhizobium*. *Nature* 354: 125–30

150. Spaink, H. P., Weinman, J., Djordjevic, M. A., Wijffelman, C. A., Okker, R. J. H., Lugtenberg, B. J. J. 1989. Genetic analysis and cellular localization of the *Rhizobium* host specificity-determining NodE protein. *EMBO J.* 8:2811–18

151. Spaink, H. P., Wijffelman, C. A., Okker, R. J. H., Lugtenberg, B. J. J. 1989. Localization of functional regions of the *Rhizobium nodD* product using hybrid *nodD* genes. *Plant Mol. Biol.* 12:59–73

152. Spaink, H. P., Wijffelman, C. A., Pees, E., Okker, R. J. H., Lugtenberg, B. J. J. 1987. *Rhizobium* nodulation gene *nodD* as a determinant of host specificity. *Nature* 328:337–40

153. Stacey, G. 1990. Workshop summary: compilation of the *nod, fix* and *nif* genes of Rhizobia and information concerning their function. See Ref. 62, pp. 239–44

154. Stacey, G., Evans, H. J., Burris, R. H., eds. 1992. *Biological Nitrogen Fixation*. New York: Chapman and Hall

155. Surin, B. P., Downie, J. A. 1988. Characterization of the *Rhizobium leguminosarum* genes *nodLMN* involved in efficient host-specific nodulation. *Mol. Microbiol.* 2:173–83

156. Surin, B. P., Downie, J. A. 1989. *Rhizobium leguminosarum* genes required for expression and transfer of host specific nodulation. *Plant Mol. Biol.* 12:19–29

157. Surin, B. P., Watson, J. M., Hamilton, W. D. O., Economou, A., Downie, J. A. 1990. Molecular characterization of the nodulation gene, *nodT,* from two biovars of *Rhizobium leguminosarum*. *Mol. Microbiol.* 4:245–52

158. Swanson, J., Tu, J. K., Ogawa, J. M., Sanga, R., Fisher, R. F., Long, S. R. 1987. Extended region of nodulation genes in *Rhizobium meliloti* 1021. Phenotypes of Tn*5* insertion mutants. *Genetics* 117:181–89

159. Török, I., Kondorosi, E., Strepkowski, T., Postfai, J., Kondorosi, A. 1984. Nucleotide sequence of *Rhizobium meliloti* nodulation genes. *Nucleic Acids Res.* 12:9509–24

160. Truchet, G., Barker, D. G., Camut, S., De Billy, F., Vasse, J., Huguet, T. 1989. Alfalfa nodulation in the absence of *Rhizobium*. *Mol. Gen. Genet.* 219: 65–68

161. Truchet, G., Roche, P., Lerouge, P., Vasse, J., Camut, S., et al. 1991. Sulphated lipo-oligosaccharide signals of the symbiotic procaryote *Rhizobium meliloti* elicit root nodule organogenesis on the host plant *Medicago sativa*. *Nature* 351:670–73

161a. van Brussel, A. A. N., Recourt, K., Pees, E., Spaink, H. P., Tak, T., et al. 1990. A biovar specific signal of *Rhizobium leguminosarum* bv. *viciae* induces increased nodulation gene–inducing activity in root exudate of *Vicia sativa* subsp. *nigra*. *J. Bacteriol.* 172:5394–5401

162. van Brussel, A. A. N., Zaat, S. A. J., Canter-Cremers, H. C. J., Wijffelman, C. A., Pees, E., et al. 1986. Role of plant root exudate and Sym plasmid-localized nodulation genes in the synthesis by *Rhizobium leguminosarum* of Tsr factor, which causes thick and short roots on common vetch. *J. Bacteriol.* 165:517–22

162a. Verma, D. P. S., ed. 1992. *Molecular Signals in Plant-Microbe Communication*. Boca Raton: CRC Press

163. Wang, S. P., Stacey, G. 1991. Studies of the *Bradyrhizobium japonicum nodD1* promoter: a repeated structure for the *nod* box. *J. Bacteriol.* 173:3356–65

164. Young, J. P. W., Johnston, A. W. B. 1989. The evolution of specificity in the legume-*Rhizobium* symbiosis. *Trends Ecol. Evol.* 4:331–49

165. Zaat, S. A. J., Schripsema, J., Wijffelman, C. A., Vanbrussel, A. A. N., Lugtenberg, B. J. J. 1989. Analysis of the major inducers of the *Rhizobium nodA* promoter from *Vicia sativa* root exudate and their activity with different *nodD* genes. *Plant Mol. Biol.* 13:175–88

166. Zaat, S. A. J., Wijffelman, C. A., Mulders, I. H. M., van Brussel, A. A. N., Lugtenberg, B. J. J. 1988. Root exudates of various host plants of *Rhizobium leguminosarum* contain different sets of inducers of *Rhizobium* nodulation genes. *Plant Physiol.* 86:1298–1303

167. Zaat, S. A. J., Wijffelman, C. A., Spaink, H. P., van Brussel, A. A. N., Okker, R. J. H., Lugtenberg, B. J. J. 1987. Induction of the *nodA* promoter of *Rhizobium leguminosarum* Sym plasmid pRl1JI by plant flavanones and flavones. *J. Bacteriol.* 169:198–204

168. Györgypal, Z., Kiss, G. B., Kondorosi, A. 1991. Transduction of plant signal molecules by the *Rhizobium* NodD proteins. *BioEssays* 13:575–81

Annu. Rev. Microbiol. 1992. 46:533–64

THE NATURAL HISTORY AND PATHOGENESIS OF HIV INFECTION

Haynes W. Sheppard and Michael S. Ascher

Viral and Rickettsial Disease Laboratory, Division of Laboratories, California Department of Health Services, Berkeley, California 94704

KEY WORDS: immunology of AIDS, predictors/correlates of progression, immune-system activation, alternative hypotheses

---

CONTENTS

0066-4227/92/1001-0533$02.00

Abstract

Infection with human immunodeficiency virus (HIV) results in progressive deterioration of the cell-mediated immune system characterized by T-helper-cell dysfunction and loss in the face of signs of generalized immune-system activation. The final stage of HIV disease, AIDS, has a myriad of opportunistic infections and malignancies as its hallmarks. The causal relationship between HIV and this complex disease pattern is clear but the mechanisms by which it occurs are not well understood. There are a number of new developments in our understanding of the natural history of HIV infection from a laboratory standpoint. Our review of this information raises further questions as to the validity of the conventional "cytopathic" model and all its direct descendants. In response to these conflicts, we have developed and present an alternative hypothesis in which AIDS pathogenesis, in all its manifestations, is seen as the outcome of one central process, excess immune activation generated by the interaction of virus with the CD4 receptor. The implications of this hypothesis on therapy of HIV infections are discussed.

INTRODUCTION AND BACKGROUND

The acquired immunodeficiency syndrome (AIDS) is the end result of infection with the human immunodeficiency virus (HIV-1) (21). The natural history of HIV disease is best described as a continuous process in which immune dysfunction and the loss of $CD4^+$ helper T-cells begins at the time of infection and progressively increases (41, 46, 203, 248) leading eventually to the wasting syndrome, opportunistic infections, or malignancies that constitute clinically defined AIDS (34, 113). The rate of this deterioriation is determined by complex interactions between a highly variable pathogen and variable host responses. The median incubation period from infection to AIDS is estimated to be 8–10 years (18, 54), but AIDS can occur in 2 to 4 months (102, 166), and some infected individuals have remained asymptomatic for more than 13 years (135).

The AIDS epidemic is occurring during a period of major breakthroughs in our understanding of the cellular immune system and of the specific T-cell compartment that is the target of HIV pathology. The T-cell receptor has been cloned and characterized (88); the presentation of peptide antigens by major histocompatability complex (MHC) has been discovered (242); the functional importance of other T-cell receptors (particularly CD4 and CD8) has been established (262); superantigens have provided an experimental model for immune activation and tolerance (1); and the powerful research tool of transgenic mice has provided insight into the molecular mechanisms of thymic selection and T-cell activation (114, 277). In spite of these advances, extensive characterization of HIV-1, and more than 10 years of intensive study, the underlying mechanism by which the virus destroys the immune system remains unclear.

When AIDS was first recognized, it appeared to have a relatively short natural history because the clinical observations were made during the late symptomatic phase. The concept that AIDS was a cytopathic infection arose directly from this clinical experience, the observation that CD4 was the virus receptor (162), and the ability of HIV to kill $CD4^+$ T-cells in vitro (163). The observations that the in vitro conditions were very unphysiologic and that, under more normal conditions, infected cells can grow and produce virus without cytopathic effects have been largely ignored (95, 127, 279). Furthermore, the occurrence of functional defects prior to cell loss, the extremely long disease process, and the demonstration that the viral burden is very low in vivo (238), have all raised doubts about the cytopathic model (6).

These paradoxical features of HIV infection have led one scientist to conclude that HIV cannot be the cause of AIDS (55) despite unassailable epidemiologic evidence that it is (205). In our opinion, the resolution of these paradoxical findings requires a new view of pathogenesis that places something other than direct cytopathic effects at the heart of HIV disease.

In 1988, we presented a hypothesis suggesting that HIV disease is the result of virus-mediated immune system activation (13), and subsequently we reviewed the experimental evidence for this mechanism (12). In the past few years, several other investigators have presented alternative hypotheses that focus on immunopathologic mechanisms (6, 91, 174), and the potential for a radical shift in perspective has been recognized (149). In this review, we present the natural history of HIV infection in the context of the debate over pathogenic mechanisms and evaluate the various alternatives. In the words of Lewis Thomas, "The record of the past half century has established, I think, two general principles about human disease. First, it is necessary to know a great deal about underlying mechanisms before one can really act effectively. . . .Second, for every disease there is a single key mechanism that dominates all others. If one can find it, and then think one's way around it, one can control the disorder" (252). In our view, the single key mechanism that unifies the disparate features of HIV disease can be found in the immunoregulatory consequences of physiologic activation signals that occur through the interaction of HIV with the CD4 receptor.

STAGING

Several schemes have been proposed to divide the continuum of HIV disease into identifiable stages (33, 110, 146, 220, 280). These rely heavily on lymphocyte subset abnormalities and have been used extensively in cross-sectional studies of natural history. While staging may be useful for epidemiologic classification of infected individuals, the ability to predict individual progression on the basis of stage is limited by the heterogeneity in disease progression rates, unknown infection times, and the broad distribution of lymphocyte subset values at which symptoms occur. For the purpose of this review, we divide the natural history of infection into only three phases, initial infection, the asymptomatic phase, and symptomatic disease [AIDS-related complex (ARC) and AIDS].

One of the most striking features of HIV disease is that most of the immune system damage, reflected in the selective loss of $CD4^+$ lymphocytes, occurs during the asymptomatic phase, the phase that constitutes most of the natural history. Although this review discusses the primary infection syndrome and the late symptomatic stages, the primary focus is the characterization of subtle changes during the asymptomatic period and their relationship to the underlying pathogenic mechanisms.

INITIAL INFECTION

Primary Clinical Syndrome

Primary infection with HIV is sometimes associated with an acute flu-like or mononucleosis-like syndrome (64, 69, 115, 253, 254). Detectable laboratory markers of this phase include transient appearance of p24 antigen (47) and acute elevations in β-2-microglobulin and neopterin (66). This has been interpreted as a virus dissemination phase that subsides as the HIV-specific immune response appears. In some cases, the symptoms can be sufficiently severe to require hospitalization, and despite resolution of the primary syndrome, such individuals tend to experience more rapid HIV disease progression (198). Studies with the polymerase chain reaction (PCR) and virus isolation have shown that such acute syndromes are associated with very high levels of infectious virus but that dramatic reductions in viremia occur at seroconversion (40, 47). These individuals illustrate the effectiveness of the immune response to HIV in controlling virus expression and help to explain why most people with primary HIV infections are asymptomatic and without detectable antigenemia (31).

A relatively acute loss of $CD4^+$ T-cells around the time of seroconversion has been reported (126). This, and the results described above, suggest that the first year of HIV infection is characterized by relatively effective immunologic control of viral pathology and development of an equilibrium between the HIV-specific immune response and the replication and dissemination of virus.

Initial Immune Response

Early studies of HIV-specific antibody suggested that virtually all infected individuals make a strong humoral response, associated with the loss of circulating viral antigen (3). However, the extreme sensitivity of the diagnostic antibody tests obscure the individual variation that provides insights into the natural history of infection. More quantitative studies have shown that HIV-specific immune responses are highly variable and may be an important determinant of disease progression rates (233). Cross-sectional studies showing that AIDS patients had lower levels of antibody against the major viral-core protein (p24) led to the conclusion that the loss of this antibody preceded the onset of symptomatic disease (20). However, longitudinal studies show that a broad range of inherent immune response capacity is reflected in the initial anti-p24 antibody titer (167), which is a powerful predictor of subsequent disease progression (233). This observation suggests that a deficient anti-HIV immune response is an underlying predisposition to disease progression, presumably through less efficient control of viral dissemination. This hypothesis is supported by studies showing higher viral burdens from onset of infection in those with rapidly progressive disease (15, 225) (see below).

The role of the initial cell-mediated immune response in the control of HIV has been less well characterized. A longitudinal study of one individual suggested that proliferative responses to HIV antigens might be present before seroconversion and even before detection of virus with PCR (42). Elevations in activated $CD8^+$ cells were also detected before seroconversion, but the antigen specificity of these cells was not determined (270). The elevated levels may represent an early anti-HIV cytotoxic response to the virus and/or nonspecific immune activation mediated by HIV (see below).

Susceptibility to Infection

As with most infectious diseases, the AIDS epidemic includes individuals who are exposed but not infected. Clearly, innoculum size is important as reflected by the relative efficiency of blood transfusion compared with other routes of infection (52). Having a sexual partner with AIDS was shown to be a risk factor for transmission (123), suggesting that infectivity might increase in later stages of immunodeficiency due to increased viral burden. However, one study found no association between stage of HIV disease and the amount of virus in semen (118). The relative efficiency of various modes of sexual transmission has been controversial, particularly in the case of oral intercourse, which until recently has been viewed as having minimal risk (136).

Despite these potential differences in exposure, prospective studies have clearly identified individuals with extremely high risk exposure who remain uninfected (194, 195).

Factors that might determine such innate resistance have not been well characterized. Peripheral blood mononuclear cell (PBMC) preparations from different seronegative donors vary widely in their susceptibility to HIV infection in vitro (264; H. W. Sheppard, unpublished observations). However, the basis for such resistance and its relevance to transmission are unclear. Studies of genetic loci of the immune system, including HLA, have shown no association between susceptibility and infection (71, 143). Weak HIV-specific T-cell responses have been observed in high-risk seronegative subjects (73), suggesting the possibility of protective cellular immunity in the absence of a humoral response. Alternatively, these observations could be unusually strong in vitro primary responses rather that the effects of viral exposure in vivo. The nature and importance of such innate or acquired resistance factors remains an important topic for future study.

In the unique case of perinatal transmission, early reports suggested that anti-gp120 or neutralizing antibody in the mother might protect neonates from HIV infection (50, 75, 218). However, conflicting results have been reported and the significance of antibody to the principal target of neutralization is uncertain (196).

Silent Infection

Early studies of primary HIV infection showed a conventional lag period from infection to seroconversion (3). Although viral core antigen can sometimes be detected shortly before seroconversion, HIV-1 specific antibody tests have become the gold standard for diagnosis of HIV infection in adults (29).

In the past two years, several studies have reported the detection of virus in antibody-negative individuals or in seroconvertors long before their first positive antibody test (56, 100, 199, 267). These reports have raised the concern that such individuals might be infectious during this period and represent a reservoir of potential transmission. Most of the evidence for such "silent" infection came from the detection of proviral DNA using the polymerase chain reaction (PCR) test. Additional studies, including repeat testing of the same population (99), failed to confirm these results (137, 195, 234, 271). Others have highlighted the known propensity for PCR testing to yield false positive results (121, 232, 266). In addition, a small proportion of asymptomatic seropositives are PCR negative, presumably due to extremely low viral burden (94). Therefore, with the clear exception of pediatric diagnosis in which serologic testing is complicated by maternal antibody, direct tests for virus (i.e. antigen, virus isolation, or PCR) have limited utility in the diagnosis of HIV infection.

ASYMPTOMATIC PHASE

By the time clinically defined AIDS appears, the immune system has collapsed and observations at that time are unlikely to provide insights into the proximate cause of the deterioration. For that reason, we emphasize descriptions of the period between early infection and clinical AIDS, when the gradual deterioration occurs. If any intervention is to be successful in combating AIDS, it is likely that it will have to be given early in the course of infection.

Correlates vs Predictors of Progression

The prospective studies of asymptomatic HIV-1 infected patients continue to enhance the ability to predict the outcome of HIV disease by combining variables (45, 59, 74, 81, 177, 201, 207, 224, 260, 276) or by the application of novel analytical methods (27, 65, 142, 214). In these studies, the distinction between cause and effect is frequently not made, and some of the laboratory markers predict the occurrence of AIDS because they reflect the consequences of progression. In our view, these predictions are better termed "correlates" or "outcomes" of progression. The case of CD4 counts is a clear example: the CD4 level at the onset of infection can hardly be said to predict AIDS. Indeed, the recently proposed AIDS case definition based on CD4 (see below) reflects the notion that the decline in $CD4^+$ cell number "is" AIDS. In our view, the term "predictors" of progression should be limited to intrinsic host or viral factors that are present before progression has occurred and truly predict progression. Several examples of such predictors, discussed above and below, include the magnitude of the initial immune response to HIV, the presence of an acute clinical syndrome, age at the time of infection, relative virulence of different strains of HIV, and host genetics.

Correlates and Clinical Trials

The protracted nature of the HIV disease process seriously impedes testing the efficacy of potential therapies. This problem has prompted attempts to use laboratory markers in place of clinical events as surrogate endpoints (147, 178). Correlates, as defined above, are more likely to change in response to effective therapy than true predictors, and CD4 has been appropriately used for this purpose. However, the relatively slow rate of CD4 decline during early asymptomatic disease still necessitates fairly long follow-up times. It has been suggested that the use of activation markers might substantially improve the ability to detect small but statistically significant effects of therapy (66, 155).

Early Functional Defects

Numerous defects in immune function are evident in the asymptomatic phase of HIV disease prior to substantial cell loss. Cell-mediated immune responses are progressively lost, first to microbial and viral recall antigens, then to the more potent stimulation by alloantigens, and finally to mitogens (41). These in vitro deficiencies are also reflected in the loss of delayed type hypersensitivity skin testing. Such anergy has been attributed to defective antigen-presenting capacity, suppressor factors, or the ability of HIV envelope glycoprotein gp120/160 to interfere with lymphocyte activation events (33, 39, 92, 193).

The selective infection and death of memory T-cells (226) or stem cells (63) would also explain early functional defects as well as the inability to regenerate immune capacity. Although this explanation is consistent with reports of early decline in $CD4^+$ $CD29^+$ memory cells (116), the mechanism for selective infection of this cell population is unclear. One hypothesis is that infected antigen-presenting cells might transmit virus to memory T-cells, leading to productive infection and cell death (197), but the evidence for this mechanism has been questioned (48).

Recent studies show that lymphocytes from HIV-infected individuals have evidence of chronic activation of inositol polyphosphate metabolism, which has been interpreted as an anergic state induced by aberrant signal transduction (188). An alternative interpretation is that chronically activated lymphocytes will appear defective in their responses when tested in vitro. This obviates the need to propose selective killing to explain anergy and is more consistent with the clinical and laboratory evidence for chronic activation in vivo (see below).

Patterns of CD4 Decline

The interest in CD4 as a surrogate marker has led to studies modeling the natural history of CD4 decline (246). Such analyses are particularly important since phase I/II clinical trials often attempt to assess efficacy on the basis of deviation from the expected CD4 decline (i.e. comparison with "historical controls") (213). Unfortunately, the inherent variability in CD4 measurements and the use of prevalent cohorts with largely unknown infection times has complicated this effort.

The simplest assumption in CD4 modeling is that there is a continuous but variable rate of decline that is directly related to the time between infection and AIDS. Longitudinal studies suggest that this may be the case for a large proportion of the natural history. However, there may be a period of accelerated decline associated with the late symptomatic phase of disease (132) in addition to the acute decline associated with seroconversion (126). Some have suggested that the period of relative stability ends as the result of decline in immune response to the virus, the emergence of viral escape mutants (187), and an acute increase in the viral burden (225). In our view, the accelerated cell loss and changes in viral properties associated with the symptomatic phase are consequences of immune-system collapse during the asymptomatic phase rather than the cause.

Humoral Immune Response

As described above, the humoral response to the p24 core antigen has early prognostic significance but shows relatively little change during the asymptomatic phase (233). Others (38, 106) have confirmed this work with particular evidence of a possible cellular defect in nonresponders (247) and genetic variation between continents (111).

For early prognosis, this represents an improvement over low prevalence p24 antigen detection because virtually all seropositives have some anti-p24 antibody and the titer is inversely related to the presence of circulating p24 antigen (167, 259). This observation has also led to the development of antigen assays with enhanced reactivity after acid dissociation of immune complexes (185). A possible genetic difference across continents has also been seen with antigen testing (26).

The significance of the immune response to components of the virus other than p24 has not been studied as extensively. In our experience, responses to other viral core components and polymerase gene products are also depressed. However, none of the depressed antibodies improved on the ability of anti-p24 to predict progression to AIDS (16). A review of this issue was recently published (2).

Neutralizing antibodies have also been studied extensively in the context of vaccine development, and the association between neutralization and resistance to viral challenge has been shown in experimental animals (23). It has also been suggested that neutralizing antibody titers are correlated with the stage of HIV disease (244). In our experience with several different assay systems, neutralizing antibody levels among infected individuals are highly variable but have no correlation with the rate of disease progression. Studies of enhancing antibody also show no prognostic value (176).

Cellular Immune Response

A vigorous cellular immune response to HIV infection has been clearly demonstrated. This response includes both $CD8^+$ cytotoxic T-cells (CTL) (265) and $CD16^+$ antibody-dependent cytotoxic cells (ADCC) (245, 257). Lymphocytes from infected individuals mediate the in vitro killing of HIV-infected cells as well as uninfected cells bearing cytophilic gp120 (108). Both responses are evident in the early stages of HIV infection and decline with disease progression, but HIV-specific CTL can still be identified even when other immune functions are severely impaired (265). HIV-specific T-cell proliferation, which is generally mediated by $CD4^+$ T-cells, can also be demonstrated but is generally lost early in the disease process along with responses to other recall antigens (255). $CD8^+$ T-cells from HIV infected adults can suppress viral growth in vitro through the elaboration of a soluble factor (148). While it is reasonable to assume that these mechanisms could play a role in the control of viral replication in vivo, their clinical significance remains unclear. As with humoral responses, cellular responses may fall victim to the HIV-mediated pathogenesis that occurs despite generally effective anti-viral responses.

Viral Burden

In general, the primary infection phase ends with seroconversion and the establishment of low-level chronic viremia (237). This has sometimes been interpreted as a classical latent phase when the disease process is arrested and the virus is present primarily as integrated proviral DNA. However, numerous studies have demonstrated that virus is expressed during this period and that gradual deterioration of immune system function and $CD4^+$ T-cell loss continue. Despite the generally low level of virus, quantitative differences in viral burden are associated with the rate of disease progression (158, 225). As with p24 antibody, the differences between individuals in early stages of infection are equal to or greater than the changes within an individual over time (15). Although much more work needs to be done in this area, these results again suggest that the increase in circulating virus may be a consequence rather than the cause of disease progression.

Cellular Reservoir

The possibility that the macrophage/monocyte population is the major reservoir of HIV infection has been widely proposed (22, 70, 219) with some disagreement as to its significance (164). More recent studies suggest that dendritic cells (128, 144) may be another important reservoir of HIV infection. These studies are reminiscent of earlier work comparing lymphocytes with monocytes and Langerhans cells showing very little virus replication associated with T lymphocytes both in vitro and in vivo (11, 212), further supporting the notion that the cytopathic effects of HIV-1 on $CD4^+$ lymphocytes may be an artifact of unphysiologic in vitro conditions and may not play a central role in pathogenesis.

Immune Activation

As we have previously reviewed, a host of clinical phenomena associated with immune activation are reported in the course of HIV disease (12). These include lymph node hyperplasia, the clinical manifestations of elevated cytokines (cachexia), hyperglobulinemia, autoimmune phenomena, and the development or exacerbation of inflammatory conditions. In addition, a growing number of laboratory markers provide evidence for high levels of generalized measurements of immune activation (98, 122, 184). The overall significance of immune activation is now well recognized (192, 233, 258) and the levels of such activation markers have emerged as both predictors and correlates of progression. Indeed, the strong correspondence between β-2/neopterin increase and CD4 decline has led to the suggestion that activation markers might be used in place of CD4 counts (66).

An adverse role for generalized immune activation must be reconciled with the independent and protective effects of the specific immune response to HIV (233, 239). In the cytopathic model, the immune stimulation is presumed to come from cytokines, induced by other infectious agents, which reactivate latent provirus. This would make the activation independent of the immune response to HIV. However, HIV-negative individuals with similar patterns of other infectious diseases do not have elevated activation markers. Curiously, of the viral agents tested for the ability to induce cytokine production, HIV and purified gp120 had the highest activity (43, 44). This observation leads to an autocatalytic model of pathogenesis in which HIV induces cytokine synthesis, which, in turn, induces the production of more HIV. However, cytokine production in macrophages generally requires signals from activated T-cells. Therefore, cytokine elevations in vivo may actually reflect the ability of HIV to amplify T-cell responses that in turn stimulate cytokine production in macrophages. The conflicting literature regarding the ability of HIV gp120 to deliver such signals and the overall effect on T-cell responses is reviewed below.

Virus and Host Variation

Interest is rapidly growing in the role of viral diversity in the natural history of HIV infection. The extensive heterogeneity of so-called HIV-1 "quasispecies" at the nucleotide level has been well documented among geographically diverse HIV-1 isolates (160, 180). In a few instances, the relationships between structure and function have been partially characterized. Examples include the primary neutralizing domain (V3) of the envelope glycoprotein (103), specific sequences of gp120 that effect binding to CD4 (10, 130), and primary-sequence epitopes, which induce CTL, T-cell proliferation (24), or ADCC (257).

Extensive heterogeneity can also occur within a single individual, and multiple variants can be present within an individual at any given time (77, 168). HIV-1 variants with greater in vitro growth potential and broader cell tropism have been recovered at later stages of HIV disease (133, 250, 251), leading to the hypothesis that the emergence of more virulent quasispecies leads to acceleration of immune system decline. The transmission of such "virulent" strains, or mixtures of quasispecies, leading to more rapid disease progression in the recipient, has also been reported (78). A study of two patients showed that diversity appeared to increase with disease progression, suggesting that increasing diversity may play a role in HIV pathogenesis (186). Acording to this model, successive viral variants arise in response to immune pressure and persist at a low level. When the diversity reaches a critical threshold, the immune system fails and an escape mutant overwhelms the host.

Host variations that affect the natural history of HIV infection are less well characterized. Several specific HLA genotypes have been associated with slower or with accelerated disease progression (113, 143), but no mechanism for these effects has been identified. It is tempting to speculate on a connection between these findings and the innate anti-HIV immune response capacity described above, but this has yet to be explored.

Because the degree of generalized immune activation is also a major determinant of prognosis, there may also be host factors that determine the sensitivity of an individual to both specific and nonspecific immune activation signals. Such differences could also explain the broad spectrum of disease seen in animal models (see below).

Cofactors

Despite extensive study, no specific infectious cofactors of HIV disease have been clearly identified. However, associations between various concomitant infections and disease progression have been suggested. A recent example is a study suggesting that early reactivation of Epstein-Barr virus (EBV) might play such a role (210). A novel mycoplasma has also been isolated from HIV-infected patients (140), and in vitro enhancement of HIV cytopathicity by mycoplasma has been demonstrated (19, 141, 173). This work has led to the suggestion that this or other mycoplasmas might be cofactors in HIV disease, perhaps through expression of a superantigen (see below). An alternative to the specific cofactor hypothesis is the notion that HIV-mediated immune system damage is accelerated by frequent immunologic challenge and that all infectious diseases are cofactors to some degree.

Age has been identified as a significant cofactor in adult hemophiliacs (209). This suggestion fits well with literature indicating that immune-system deterioration may well be a component of the normal aging process (76, 150). At the other extreme is the accelerated natural history in HIV-infected infants (see below). A more intensive study of the possible interaction between age-related immunologic changes and HIV infection might provide important insights into both processes.

LATE-STAGE DISEASE

Longer Survival Time

It has been suggested that the period of survival after the diagnosis of AIDS may have increased slightly over the course of the epidemic, possibly because of more effective treatment of opportunistic infections, antiviral therapy, or both. However, differences in demographics and accessibility of treatment makes comparisons difficult (65). In addition, many of the current AIDS cases have been infected longer than those who developed AIDS early in the epidemic. This may reflect a frailty selection in which those who are inherently more susceptible to the disease process are less represented in the surviving population (125). The most significant predictors of survival time are the nature of the presenting diagnosis and the age of the patient (65). Interestingly, the rate of progression prior to AIDS and survival time after the onset of AIDS do not appear to be related (89).

It has also been suggested that longer survival time and changes in the demographics of the epidemic will change the spectrum of AIDS-related clinical manifestations. The emergence of lymphoma as a more common AIDS-defining condition may be an example (37, 165). Also, some have suggested that because of the frequent occurrence of pelvic inflammatory disease and chronic vaginal candidiasis in HIV$^+$ women, these and other atypical conditions should be added to the AIDS case definition (25).

Case Definition

The Centers for Disease Control recently proposed that seropositivity for HIV-1 and < 200 CD4$^+$ lymphocytes/mm^3 be accepted as an AIDS-defining condition. This reflects concern over complexity in the present clinical definition and the perceived changes in the spectrum of late-stage clinical manifestations. This laboratory definition of immunodeficiency predicts the rapid onset of opportunistic infections, particularly *Pneumocystis carinii* pneumonia (PCP) (156, 200), and is the criterion for PCP prophylaxis (34). However, many individuals with < 200 CD4$^+$ lymphocytes do not have symptomatic disease. Therefore, implementation of this criterion would dramatically increase the point prevalence of AIDS cases by increasing the proportion of the HIV disease continuum defined as AIDS. The final cumulative incidence of AIDS would be relatively unchanged because the vast majority of individuals with profound CD4$^+$ cell loss eventually proceed to clinical AIDS as presently defined (235, 261). This modification will permit laboratory-based case surveillance and will be useful for epidemiologic modeling and medical care projections. However, the extensive overlap in the CD4 values of symptomatic and asymptomatic persons (190) means that many individuals who are excluded from the present clinical definition will still be excluded, while some asymptomatic individuals will be defined as AIDS cases. In addition, the median survival time after AIDS will be artificially increased by as much as 24 months.

NATURAL HISTORY OF INFECTION IN INFANTS

Although much less information is available, the natural history of HIV-1 infection in perinatally infected infants follows the same general pattern as seen in adults but at a much accelerated pace (7, 57, 272). Most infected infants develop AIDS within the first two years of life compared with the median of 10 years in adults. As in adults, both clinical and laboratory signs of immune activation are among the first manifestations of HIV infection. The changes in the major lymphocyte subsets follow the typical pattern seen in adults, which begins with normal CD4 levels and CD8 elevation, followed by decrease in the CD4 population, CD8 decline, and then finally general lymphopenia (172, 280). Higher viral burdens are associated with more rapid progression, and stronger HIV-specific immune responses are correlated with slower progression (139). However, high viral burdens, reflected in the ability to recover virus from plasma, are more frequent in children regardless of CD4 level (222), probably reflecting their less mature immune systems at the time of infection. As with adults, the survival time in infants after AIDS diagnosis is about 18 months.

ALTERNATIVE PATHOGENIC MECHANISMS

As described above, the viral cytopathic model has been modified to suggest an autocatalytic cycle in which cytokines reactivate latent provirus and virus induces the production of cytokines through molecular regulatory events (58, 86). This mechanism was based on the expectation that burden of proviral infection was high despite low levels of viral expression and was supported by in vitro experiments showing virus production and cell death after immunologic stimulation of latently infected lymphocytes (28, 134, 152, 274, 275). However, molecular detection methods such as the polymerase chain reaction (PCR) have failed to demonstrate a high level of proviral infection (227, 238). While viral regulatory genes probably influence viral burden (86, 268), the central paradox remains: the number of cells in vivo producing virus (1:10,000–1:100,000) or with integrated provirus (1:1,000–1:10,000) is not sufficient to account for the observed T-cell loss by direct killing (85). The fusion of many uninfected cells with a few infected cells (syncytia formation) might explain cell loss (30, 138), but there is no evidence for substantial syncytia formation in vivo.

A variety of alternatives to the direct cytopathic model have been proposed to explain the progressive immunodeficiency and $CD4^+$ cell loss leading to AIDS. In general, they fall into the two categories of autoimmunity and immune dysregulation.

Autoimmunity

Two major forms of autoimmunity have been suggested. One invokes the misdirection of HIV-specific immunity at viral components that have been adsorbed on the surface of uninfected cells (278). The second focuses on structural homology between viral and self antigens that results in the breakdown of self-tolerance (83, 84, 151). Such models require that dramatic antiself reactivity be generated by rather minor homologies between HIV and self molecules.

Another model has combined such crossreactive autoimmunity with an immune response to foreign alloantigens that mimics the presence of virus (91). This provides for a sustained autoimmunity even if virus is absent. However, virus is present during HIV disease (90), so the importance of such a complex autoimmune mechanism is unclear.

Immune Dysregulation

This class of pathogenic model focuses on the ability of HIV infection to disrupt immunologic mechanisms independent of any cytopathic effects. Passive blockade of immune function has been suggested based on the ability of gp120 to inhibit antigen-specific lymphocyte responses in vitro (119). Others have argued that the interaction of HIV or gp120 produces a negative signal that actively inhibits immune responses (92). This concept has been extended to suggest that such a negative or defective signal might induce programmed cell death through a mechanism known as apoptosis (6). Although such models would account for immune deficiency, they are inconsistent with the clinical and laboratory evidence for the generalized immune activation that accompanies progressive HIV disease.

In the model we have proposed, all three features of HIV disease (i.e. immune dysfunction, immune activation, and cell loss) are the result of activation signals, mediated by the gp120-CD4 interaction, which nonspecifically amplifies immune responses (14). A recent proposal, that HIV might encode a superantigen, draws a similar analogy to the activation-mediated immunopathology seen in the murine AIDS (MAIDS) model (see below). Finally, the cytopathic and immunoregulatory models have been combined in the proposal that certain mycoplasma infections may play a cofactor role by encoding a superantigen that would induce cytopathic virus through lymphocyte activation (79).

ANIMAL MODELS

Much has been written about the importance of experimental animal models for the study of HIV pathogenic mechanisms and potential treatment (181). Several species can be productively infected with HIV, but none have developed symptoms of immunodeficiency that parallel the human disease. However, several closely related animal retroviruses cause AIDS-like diseases in experimental animal models (60, 109, 131). Some simian immunodeficiency virus (SIV) variants cause immune activation, anergy, immunodeficiency, and death in appropriate primate species, no disease in other species, and rapid cytokine/cachectin-mediated death in very specific susceptible animals (49). These differences have raised some doubts about the direct application of findings in primates to human disease, but most investigators assume that simian and human AIDS have similar underlying pathogenic mechanisms. The wide spectrum of SIV disease probably reflects the diversity among different primate species and the degree to which a particular SIV variant is adapted to its host. Different progression rates among humans may result from more subtle genetic variation in both virus and host. In our view, the determining factor is the efficiency with which a given retrovirus delivers activation signals through its cellular receptor.

The relevance of MAIDS to HIV disease has been even more controversial, but a strong case can be made that MAIDS (109) and the rapidly progressive SIV model (51) reflect one extreme of the disease spectrum in which the virus acts as a mitogen causing lymphoproliferation directly. The other end of the spectrum consists of no signal and no disease (67). HIV may fall between these extremes, where any immune activation in response to an endemic pathogen is enhanced, causing disruption of immunoregulatory mechanisms and a slow progressive loss of function and eventually cells. Indeed, we suggest that treatment with antineoplastic or immunosuppressive drugs, which prevents progression of MAIDS, might be considered for the prevention and treatment of HIV disease (237).

TREATMENT

At the present time, the only approved therapies for HIV infection employ nucleoside analogs that block the HIV reverse transcriptase. However, many alternative approaches, targeting other phases of the HIV lifecycle, are under investigation or in limited trials through community-based research. In addition, therapy and prophylaxis for AIDS-related opportunistic infections have improved greatly. Although a review of this literature is beyond the scope of this review, it is important to point out that most natural-history studies are now done in the context of known and unknown chemotherapy. Therefore, the extensive experience with AZT warrants some discussion of its effects.

The trials of AZT show clear benefit, but several findings are paradoxical. First, although AZT appears to reduce circulating virus, either free or in immune complexes (61), the underlying cell-associated viral burden does not change detectably (53). Secondly, AZT causes a substantial reduction in activation markers, which is difficult to explain based on its antiviral activity alone (155). Finally, direct evidence suggests that AZT treatment reverses the chronic activation of the inositol phosphate pathway in the lymphocytes of HIV-infected individuals (188). This observation raises the possibility that the clinical benefit from AZT may result from antiactivation as well as antiviral effects.

The concept of using immunosuppression in treatment of HIV disease (87) has received very little attention despite the reported beneficial effects of steroids in *Pneumocystis* pneumonia (68, 157) and indomethacin in the inflammatory lesions of certain AIDS-related phenomena (211). An early trial of cyclosporin-A showed increased CD4 counts and loss of lymphadenopathy in early HIV disease (9). The effect was less pronounced in non-AIDS patients with less than 300 $CD4^+$ T-cells, and no improvement in patients with AIDS. A subsequent trial treated only AIDS patients and reported accelerated disease (202). This result and the prevailing notion that HIV disease is "immunosuppression" has discouraged further studies that could confirm the initial findings in asymptomatic patients. Also important is that both trials used the dose necessary for the maintenance of organ transplants. A much lower dose may eliminate excess activation and return immune-system activity to normal levels. In addition to such global treatment, therapy might be targeted symptomatically to the specific mediators that produce fever, cachexia, and diarrhea and thereby inhibit their action (159).

IMMUNOLOGY AND AIDS

An in-depth discussion of new developments in cellular immunology may seem out of place in a review of HIV disease. However, we feel that there are important new findings in immunology that have not been fully integrated into the natural history and pathogenesis of HIV infection and that should be more fully explored in developing models of HIV disease.

T-Cell Activation and Tolerance

After a quest of almost 20 years, the receptor on T lymphocytes that recognizes foreign antigens was identified in 1984 (88). The T-cell receptor (TCR) is a heterodimer that interacts with antigenic peptides presented in the context of a MHC molecule. However, the signal delivered by this interaction (signal 1) is insufficient, and an additional nonspecific costimulatory signal (signal 2) is required to activate T-cells (179).

While the delivery of both signals results in activation, each signal alone can induce a nonresponsive state (5, 107, 228, 243). This observation suggests that normal activation signals act cooperatively to exceed a critical intensity threshold (105, 243), while partial signals are below-threshold but still induce biochemical events that temporarily prevent the delivery of other above-threshold signals (i.e. anergy).

The key concept that integrates HIV pathogenesis and these immunologic phenomena is that the fate of a T-cell population is determined by the overall intensity and/or duration of the activation signals it receives. Unlike most other tissues, lymphoid cells undergo rapid proliferation and functional differentiation in response to receptor-mediated signals. Most of the progeny of such activated cells undergo programmed death, or apoptosis (4, 221, 241, 273). A few cells return to a resting state but retain the capacity to repeat the response cycle, i.e. serve as memory cells. This homeostatic mechanism prevents the net expansion or contraction of the immune system. Thus, programmed cell death is not a unique pathway initiated by aberrant signals but a normal component of immunologic regulation (230).

This process also provides an explanation for thymic selection in which self-reactive thymocytes are trapped in the presence of signal 1 and signal 2, cannot rest, and are driven to clonal deletion. Nonself reactive thymocytes receive signal 2, but signal 1 is below-threshold, reducing activation to a level that permits maturation (i.e. positive selection) (263). Peripheral tolerance seldom occurs because mature lymphocytes, unlike self-reactive thymocytes, can escape signal 1 through elimination of the antigen or migration away from the local site of infection. In addition, activation signals may have a less intense effect on mature T-cells so that escape from above-threshold signals is more easily achieved (206, 256). By analogy with this view of tolerance, we have proposed that the $CD4^+$ cell loss in HIV disease results from the presence of an extra virus-mediated activation signal that shifts clonal dynamics of antigen stimulation in the direction of excess programmed cell death (175, 231).

Role of CD4 in Activation and AIDS

In 1988 when this hypothesis was first proposed, CD4 was generally thought to be a cellular adhesion receptor with no established signal-transduction capacity (62). However, its participation in T-cell activation (8, 32) and immunologic tolerance (145) had been shown, and the possibility of a signal transduction role had been suggested (62). It is now generally accepted that CD4 and CD8 have critical coreceptor activity and transmit costimulatory signals (signal 2) that are critical in both T-cell activation (8, 32, 72, 169, 182, 256) and in the induction of tolerance (5, 80, 183, 249, 281).

The ability of HIV to transduce such signals through CD4 has been reported (117, 182, 188, 189, 223). Other studies have failed to find signal transduction (93, 124, 191), and the overall effect of HIV on lymphocyte activation has been claimed to be both positive (43, 120, 204, 215) and negative (36, 39, 82, 92, 119, 170, 217, 236). There may be several reasons for this discrepancy. Receptor ligation can produce opposite outcomes depending on the nature of the ligand, the timing and coordination of signal delivery, the degree of cross-linking, and the condition of the T-cell population. For example, costimulation through CD4 depends on its physical proximity to CD3, which is itself a function of prior activation (129, 216). In addition, many of the experiments testing HIV have been performed with a variety of native and recombinant gp120 preparations, which may vary dramatically in their structure and purity. A recent report of gp120-induced cytokine production showed dramatic differences in the activity of native gp120 or recombinant gp120 produced in different host-vector systems (44).

The in vitro effects of other CD4 ligands show a similar mixture of negative and positive effects on T-cell activation (5, 8, 32, 72, 169, 171, 179, 182, 240). In this context, negative in vitro effects have been interpreted as the blocking of a positive signal (236), but the overall consensus is that CD4-mediated signals participate positively in T-cell activation (262). In AIDS, the reverse is true, and the interaction of gp120 with CD4 is generally seen as negative because HIV disease is generally seen as immune suppression. Despite this controversy over the in vitro results, the association between signs of immune activation in vivo and progressive HIV disease argues that the primary effect of the gp120-CD4 interaction is stimulation and not suppression.

Superantigens

Superantigens have become relevant to the discussion of HIV pathogenesis because of their profound effects on the immune system and the fact that some are now known to be retroviral products (153). Superantigens transmit TCR signals (signal 1) by direct interaction with TCR β-chain variable regions (Vb) and, if present during fetal development, cause deletion of the T-cell clones that express those Vb molecules (112, 161). If a superantigen is not endogenous, the corresponding T-cell clones are retained, and later exposure to the superantigen induces polyclonal activation. In addition, some superantigens do not require CD4/CD8 costimulatory signals, suggesting that they can deliver a "super" signal 1 with above-threshold intensity (208, 229).

MAIDS, a lymphoproliferative/immunodeficiency disease of mice (17, 96), is caused by a virus with typical superantigen properties (97), providing a precedent for an immunodeficiency disease due strictly to excess activation signals. It has been suggested that HIV might also encode a Vb-binding superantigen (1, 101, 104, 269). However, a superantigen would be expected to cause very specific Vb deletions, rather than the general depletion of $CD4^+$ T-cells seen in HIV disease (154), or random holes in the Vb repertoire. A recent study reported selective Vb deletions in AIDS patients (101). However, the number of Vb families affected was unusually large, and there was an association between the level of expression in control subjects and the degree of Vb loss in the AIDS patients. In our view, the natural history of HIV disease is far more consistent with the effects of an aberrant costimulatory signal 2. Such a signal would require concomitant TCR signal (provided by foreign antigen presentation), but would have more generalized effects because all helper T-cells bear CD4, accounting for both the breadth of HIV-induced immunodeficiency and the length of the incubation period between HIV infection and AIDS.

Literature Cited

1. Acha-Orbea, H., Palmer, E. 1991. Mls—a retrovirus exploits the immune system. *Immunol. Today* 12:356–61
2. Allain, J. P., Laurian, Y., Einstein, M. H., Braun, B. P., Delaney, S. R., et al. 1991. Monitoring of specific antibodies to human immunodeficiency virus structural proteins: clinical significance. *Blood* 77:1118–23
3. Allain, J. P., Laurin, Y., Paul, D. A., Verroust, F., Leuther, M., et al. 1997. Long-term evaluation of HIV antigen and antibodies to p24 and gp41 in patients with hemophilia. *New Engl. J. Med.* 317:1114–21
4. Alles, A., Alley, K., Barrett, J. C., Buttyan, R., Columbano, A., et al. 1991. Apoptosis: a general comment. *FASEB J.* 5:2127–28
5. Alters, S. E., Shizuru, J. A., Ackerman, J., Grossman, D., Seydel, K. B., Fathman, C. G. 1991. Anti-CD4 mediates clonal anergy during transplantation tolerance induction. *J. Exp. Med.* 173:491–94
6. Ameisen, J. C., Capron, A. 1991. Cell dysfunction and depletion in AIDS: the programmed cell death hypothesis. *Immunol. Today* 12:102–5
7. Ammann, A. J. 1988. Immunopathogenesis of pediatric acquired immunodeficiency syndrome. *J. Perinatol.* 8:54–59
8. Anderson, P., Blue, M. L., Morimoto, C., Schlossman, F. 1987. Cross-linking of T3 (CD3) with T4 (CD4) enhances the proliferation of resting T lymphocytes. *J. Immunol.* 139:678
9. Andrieu, J.-M., Even, P., Venet, A., Tourani, J.-M., Stern, M., et al. 1988. Effects of cyclosporin on T-cell subsets in human immunodeficiency virus disease. *Clin. Immmunol. Immunopathol.* 46:181–98
10. Ardman, B., Kowalski, M., Bristol, J., Haseltine, W., Sodroski, J. 1990. Effects on CD4 binding of anti-peptide sera to the fourth and fifth conserved domains of HIV-1 gp120. *J. AIDS* 3: 206–14
11. Armstrong, J. A., Dawkins, R. L., Horne, R. 1985 Retroviral infection of accessory cells and the immunological paradox in AIDS. *Immunol. Today* 6: 121–22
12. Ascher, M. S., Sheppard, H. W. 1990. AIDS as immune system activation II: the panergic imnesia hypothesis. *J. AIDS* 3:177–91
13. Ascher, M. S., Sheppard, H. W. 1988. AIDS as immune system activation: a model for pathogenesis. *Clin. Exp. Immunol.* 73:165–67
14. Ascher, M. S., Sheppard, H. W. 1990. A unified hypothesis for three cardinal features of HIV immunology. *J. AIDS* 4:97–98
15. Ascher, M. S., Sheppard, H. W., Arnon, J. M., Lang, W. 1991. Viral burden in HIV disease. *J. AIDS* 4:824–30
16. Ascher, M. S., Sheppard, H. W.,

Bouvier, P. B. 1990. Significance of antibodies to multiple viral components in the prediction of HIV disease progression. *Int. Conf. AIDS, 6th, San Francisco, California,* Abstr. S. A. 287

17. Aziz, D. C., Hanna, Z., Jolicoeur, P. 1989. Severe immunodeficiency disease induced by a defective murine leukaemia virus. *Nature* 338:505–8
18. Bacchetti, P., Moss, A. R. 1989. Incubation period of AIDS in San Francisco. *Nature* 338:251–53
19. Balter, M. 1991. Montagnier pursues the mycoplasma-AIDS link. *Science* 251:71
20. Baltimore, D., Feinberg, M. B. 1989. HIV revealed—toward a natural history of the infection. *New Engl. J. Med.* 321:1673–75
21. Barre-Sinoussi, F., Chermann, J. C., Rey, F., Nugeyre, M. T., Chamaret, S., et al. 1983. Isolation of a T-lymphotropic retrovirus from a patient at risk for acquired immune deficiency syndrome (AIDS). *Science* 220:868–71
22. Bender, B. S., Davidson, B. L., Kline, R., Brown, C., Quinn, T. C. 1988. Role of the mononuclear phagocyte system in the immunopathogenesis of human immunodeficiency virus infection and the acquired immunodeficiency syndrome. *Rev. Infect. Dis.* 10:1142–54
23. Berman, P. W., Gregory, T. J., Riddle, L., Nakamura, G. R., Champe, M. A., et al. 1990. Protection of chimpanzees from infection by HIV-1 after vaccination with recombinant glycoprotein gp120 but not gp160. *Nature* 345:622–25
24. Berzofsky, J. A. 1991. Development of artificial vaccines against HIV using defined epitopes. *FASEB J.* 5:2412–18
25. Brettle, R. P., Leen, C. L. S. 1991. The natural history of HIV and AIDS in women. *AIDS* 5:1283–92
26. Brown, C., Kline, R., Atibu, L., Francis, H., Ryder, R., Quinn, T. C. 1991. Prevalence of HIV-1 p24 antigenemia in African and North American populations and correlation with clinical status. *AIDS* 5:89–92
27. Buckley, J. D., Pike, M. C., Mosley, J. W. 1991. Statistical approaches to evaluating prognostic indices in HIV infection. *AIDS* 5:1398
28. Bukrinsky, M. I., Stanwick, T. L., Dempsey, M. P., Stevenson, M. 1991. Quiescent T lymphocytes as an inducible virus reservoir in HIV-1 infection. *Science* 254:423–27
29. Burke, D. S., Brundage, J. F., Redfield, R. R., Damato, J. J., Schable, C. A., et al. 1988. Measurement of the false positive rate in a screening program for human immunodeficiency virus infections. *New Engl. J. Med.* 319:961–64
30. Burny, A., Bex, F., Brasseur, R., Khim, M. C. L., Delchambre, M., et al. 1988. Human immunodeficiency virus cell entry: new insights into the fusion mechanism. *J. AIDS* 1:579–82
31. Busch, M. P., El Amad, Z., Sheppard, H. W., Ascher, M. S., Lang, W. 1991. Primary HIV-1 infection. *New Engl. J. Med.* 325:733–35
32. Carrel, S., Moretta, A., Pantaleo, G., Tambussi, G., Isler, P., et al. 1988. Stimulation and proliferation of $CD4^+$ peripheral blood T lymphocytes induced by an anti-CD4 monoclonal antibody. *Eur. J. Immunol.* 18:333–39
33. Cefai, D., Debre, P., Kaczorek, M., Idziorek, T., Autran, B., Bismuth, G. 1990. Human immunodeficiency virus-1 glycoproteins gp120 and gp160 specifically inhibit the CD3/T cell-antigen receptor phosphoinositide transduction pathway. *J. Clin. Invest.* 86:2117–24
34. Centers for Disease Control. 1987. Revision of the CDC surveillance case definition for acquired immunodeficiency syndrome. *Mortal. Morb. Wkly. Rep.* 36(1S):1–15
35. Centers for Disease Control. 1989. Guidelines for prophylaxis against *Pneumocystis carinii* pneumonia for persons infected with human immunodeficiency virus. *Mortal. Morb. Wkly. Rep.* 38(S-5):1–9
36. Centers for Disease Control. 1990. Acquired immunodeficiency syndrome (AIDS)-interim proposal for a WHO staging system for HIV infection and disease. *WHO Wkly. Epidemiol. Rec.* 29:221–24
37. Centers for Disease Control. 1991. Opportunistic non-Hodgkins lymphomas among severely immunocompromised HIV-infected patients surviving for prolonged periods on antiretroviral therapy—United States. *Mortal. Morb. Wkly. Rep.* 40:591–600
38. Cheingsong-Popov, R., Panagiotidi, C., Bowcock, S., Aronstam, A., Wadsworth, J., Weber, J. 1991. Relation between humoral responses to HIV gag and env proteins at seroconversion and clinical outcome of HIV infection. *Br. Med. J.* 302:23–26
39. Chirmule, N., Kalyanaraman, V. S., Oyaizu, N., Slade, H. B., Pahwa, S. 1990. Inhibition of functional properties of tetanus antigen-specific T-cell clones by envelope glycoprotein gp120 of human immunodeficiency virus. *Blood* 75:152–59

40. Clark, S. J., Saag, M. S., Decker, W. D., Campbell-Hill, S., Roberson, J. L., et al. 1991. High titers of cytopathic virus in plasma of patients with symptomatic primary HIV-1 infection. *New Engl. J. Med.* 324:954–60
41. Clerici, M., Stocks, N. I., Zajac, R. A., Boswell, R. N., Lucey, D. R., et al. 1989. Detection of three distinct patterns of T helper cell dysfunction in asymptomatic, human immunodeficiency virus-seropositive patients. *J. Clin. Invest.* 84:1892–99
42. Clerici, M., Berzofsky, J. A., Shearer, G. M., Tacket, C. O. 1991. Exposure to human immunodeficiency virus type 1-specific T helper cell responses before detection of infection by polymerase chain reaction and serum antibodies. *J. Infect. Dis.* 164:178–82
43. Clouse, K. A., Robbins, P. B., Fernie, B., Ostrove, J. M., Fauci, A. S. 1989. Viral antigen stimulation of the production of human monokines capable of regulating HIV1 expression. *J. Immunol.* 143:470–75
44. Clouse, K. A., Cosentino, L. M., Weih, K. A., Pyle, S. W., Robbins, P. B., et al. 1991. The HIV-1 gp120 envelope protein has the intrinsic capacity to stimulate monokine secretion. *J. Immunol.* 147:2892–2901
45. Crowe, S. M., Carlin, J. B., Stewart, K. I., Lucas, C. R., Hoy, J. F. 1991. Predictive value of CD4 lymphocyte numbers for the development of opportunistic infections and malignancies in HIV-infected persons. *J. AIDS* 4:770–76
46. Cuthbert, R. J. G., Ludlam, C. A., Tucker, J., Steel, C. M., Beatson, D., et al. 1990. Five year prospective study of HIV infection in the Edinburgh haemophiliac cohort. *Br. Med. J.* 301: 956–61
47. Daar, E. S., Moudgil, T., Meyer, R. D., Ho, D. D. 1991. Transient high levels of viremia in patients with primary human immunodeficiency virus type 1 infection. *New Engl. J. Med.* 324:961–64
48. Dalgleish, A. G., Habeshaw, J., Manca, F. 1991. A unified hypothesis for three cardinal features of HIV immunology: authors' reply. *J. AIDS* 4:1165
49. Desrosiers, R. C. 1988. Simian immunodeficiency viruses. *Annu. Rev. Microbiol.* 42:607–25
50. Devash, Y., Calvelli, T. A., Wood, D. G., Reagan, K. J., Rubinstein, A. 1990. Vertical transmission of human immunodeficiency virus is correlated with the absence of high-affinity/avidity maternal antibodies to the gp120 principal neutralizing domain. *Proc. Natl. Acad. Sci. USA* 87:3445–49
51. Dewhurst, S., Embretson, J. E., Anderson, D. C., Mullins, J. I., Fultz, P. N. 1990. Sequence analysis and acute pathogenicity of molecularly cloned SIV. *Nature* 345:636–40
52. Donegan, E., Stuart, M., Niland, J. C., Sacks, H. S., Azen, S. P., et al. 1990. Infection with human immunodeficiency virus type 1 (HIV-1) among recipients of antibody-positive blood donations. *Ann. Intern. Med.* 113:733–39
53. Donovan, R. M., Dickover, R. E., Goldstein, E., Huth, R. G., Carlson, J. R. 1991. HIV-1 proviral copy number in blood mononuclear cells from AIDS patients on zidovudine therapy. *J. AIDS* 4:766–69
54. Downs, A. M., Ancelle-Park, R. A., Costagliola, D., Rigaut, J. P., Brunet, J. B. 1991. Transfusion-associated AIDS cases in Europe: estimation of the incubation period distribution and prediction of future cases. *J. AIDS* 4:805–13
55. Duesberg, P. H. 1989. Human immunodeficiency virus and acquired immunodeficiency syndrome: correlation but not causation? *Proc. Natl. Acad. Sci.* 86:755–62
56. Ensoli, F., Fiorelli, V., Mezzaroma, I., D'Offizi, G., Rainaldi, L., et al. 1991. Plasma viraemia in seronegative HIV-1-infected individuals. *AIDS* 5:1195–99
57. Espanol, T., Garcia, X., Caragol, I., Sauleda, S., Muntane, C. 1991. Immunological abnormalities in pediatric AIDS. *Immunol. Invest.* 20:215–21
58. Fernandez-Cruz, E., Desco, M., Montes, M. G., Longo, N., Gonzalez, B., Zabay, J. M. 1990. Immunological and serological markers predictive of progression to AIDS in a cohort of HIV-infected drug users. *AIDS* 4:987–94
59. Fiocre, B. 1987. Caprine retroviruses—experimental model for studying AIDS? *Bull. Acad. Vet. France*. 60:311–18
60. Fischl, M. A., Richman, D. D., Hansen, N., Collier, A. C., Carey, J. T., et al. 1990. The safety and efficacy of zidovudine (AZT) in the treatment of subjects with mildly symptomatic human immunodeficiency virus type 1 (HIV) infection. *Ann. Intern. Med.* 112:727–37
61. Fleischer, B., Schrezenmeier, H. 1988. Do CD4 or CD8 molecules provide a regulatory signal in T-cell activation? *Immunol. Today* 9:132–34
62. Folks, T. M., Kessler, S. W., Orenstein, J. M., Justement, J. S., Jaffe, E. S., Fauci, A. S. 1988. Infection and replication of HIV-1 in purified pro-

genitor cells of normal human bone marrow. *Science* 242:919–22

63. Fox, R., Eldred, L. J., Fuchs, E. J., Kaslow, R. A., Visscher, B. R., et al. 1987. Clinical manifestations of acute infection with human immunodeficiency virus in a cohort of gay men. *AIDS* 1:35–38
64. Franzetti, F., Cavalli, G., Foppa, C. U., Amprimo, M. C., Gaido, P., Lazzarin, A. 1988. Raised serum beta2 microglobulin levels in different stages of human immunodeficiency virus infection. *J. Clin. Lab. Immunol.* 27:133–37
65. Friedland, G. H., Saltzman, B., Vileno, J., Freeman, K., Schrager, L. K., Klein, R. S. 1991. Survival differences in patients with AIDS. *J. AIDS* 4:144–53
66. Fuchs, D., Kramer, A., Reibnegger, G., Werner, E. R., Dierich, M. P., et al. 1991. Neopterin and beta2-microglobulin as prognostic indices in human immunodeficiency virus type 1 infection. *Infection* 19:S98–S102
67. Fultz, P. N., Stricker, R. B., McClure, H. M., Anderson, D. C., Switzer, W. M., Horaist, C. 1990. Humoral response to SIV/SMM infection in macaque and mangabey monkeys. *J. AIDS* 3:319–29
68. Gagnon, S., Boota, A. M., Fischl, M. A., Baier, H., Kirksey, O. W., La Voie, L. 1990. Corticosteroids as adjunctive therapy for severe *Pneumocystis carinii* pneumonia in the acquired immunodeficiency syndrome. *New Engl. J. Med.* 323:1444–50
69. Gaines, H., von Sydow, M., Pehrson, P. O., Lundbergh, P. 1988. Clinical picture of primary HIV infection presenting as a glandular-fever-like illness. *Br. Med. J.* 297:1363–68
70. Gendelman, H. E., Orenstein, J. M., Martin, M. A., Ferrua, C., Mitra, R., et al. 1988. Efficient isolation and propagation of human immunodeficiency virus on recombinant colony-stimulating factor 1-treated monocytes. *J. Exp. Med.* 167:1428–41
71. Gilles, K., Louie, L., Newman, B., Crandall, J., King, M. C. 1988. Genetic susceptibility to AIDS: absence of an association with group-specific component (Gc). *New Engl. J. Med.* 317:630–31
72. Gilliland, L. K., Teh, H. S., Uckun, F. M., Norris, N. A., Teh, S. J., et al. 1991. CD4 and CD8 are positive regulators of T cell receptor signal transduction in early T cell differentiation. *J. Immunol.* 146:1759–65
73. Giorgi, J. V., Clerici, M., Berzofsky, J. A., Shearer, G. M. 1991. HIV-specific cellular immunity in high-risk HIV-1 seronegative homosexual men. *Int. Conf. AIDS, 7th, Florence, Italy.* Abstr. W. A. 1209
74. Goedert, J. J. 1990. Prognostic markers for AIDS. *Ann. Epidemiol.* 1:129–39
75. Goedert, J. J., Drummond, J. E., Minkoff, H. L. 1989. Mother-to-infant transmission of human immunodeficiency virus type 1: association with prematurity or low anti-gp120. *Lancet* 2:1351–54
76. Goidl, E. A. 1987. *Aging and the Immune Response-Cellular and Humoral Aspects.* New York: Marcel Dekker
77. Goodenow, M., Huet, T., Saurin, W., Kwok, S., Sninsky, J., Wain-Hobson, S. 1989. HIV-1 isolates are rapidly evolving quasispecies: evidence for viral mixtures and preferred nucleotide substitutions. *J. AIDS* 2:344–52
78. Goudsmit, J., Back, N. K. T., Nara, P. L. 1991. Genomic diversity and antigenic variation of HIV-1: links between pathogenesis, epidemiology and vaccine development. *FASEB J.* 5:2427–36
79. Gougeon, M. L., Olivier, R., Garcia, S., Guetard, D., Dragie, T., et al. 1991. Evidence for an engagement process towards apoptosis in lymphocytes of HIV-infected patients. *C. R. Acad. Sci.* 312:529–37
80. Grusby, M. J., Johnson, R. S., Papaioannou, V. E., Glimcher, L. H. 1991. Depletion of $CD4^+$ T cells in major histocompatibility complex class II-deficient mice. *Science* 253:1417–20
81. Gruters, R. A., Terpstra, F. G., De Goede, R. E. Y., Mulder, J. W., De Wolf, F., et al. 1991. Immunological and virological markers in individuals progressing from seroconversion to AIDS. *AIDS* 5:837–44
82. Gupta, S., Vayuvegula, B. 1987. Human immunodeficiency virus-associated changes in signal transduction. *J. Clin. Immunol.* 7:486–89
83. Habeshaw, J. A., Dalgleish, A. G. 1989. The relevance of HIV env/CD4 interactions to the pathogenesis of acquired immune deficiency syndrome. *J. AIDS* 2:457–68
84. Habeshaw, J. A., Dalgleish, A. G., Bountiff, L., Newell, A. L., Wilks, D., et al. 1990. AIDS pathogenesis: HIV envelope and its interaction with cell proteins. *Immunol. Today* 11:418–25
85. Harper, M. E., Marselle, L. M., Gallo, R. C., Wong-Staal, F. 1986. Detection of lymphocytes expressing human T-lymphotropic virus type III in lymph nodes and peripheral blood from infected individuals by in situ hybridiza-

tion. *Proc. Natl. Acad. Sci. USA* 83:772–76

86. Haseltine, W. A. 1988. Replication and pathogenesis of the AIDS virus. *J. AIDS* 1:217–40
87. Hausen, A., Dierich, M. P., Fuchs, D., Hengster, P., Reibnegger, G., et al. 1986. Immunosuppressants in patients with AIDS. *Nature* 320:114
88. Hedrick, S. M., Cohen, D., Nielsen, E., Davis, M. 1984. Isolation of cDNA clones encoding T cell specific membrane-associated proteins. *Nature* 308:149
89. Hessol, N. A., Byers, R. H., Lifson, A. R., O'Malley, P. M., Cannon, L., et al. 1990. Relationship between AIDS latency period and AIDS survival time in homosexual and bisexual men. *J. AIDS* 3:1078–85
90. Ho, D. D., Moudgil, T., Alam, M. 1989. Quantitation of human immunodeficiency virus type 1 in the blood of infected persons. *New Engl. J. Med.* 321:1621–25
91. Hoffmann, G. W., Kion, T. A., Grant, M. D. 1991. An idiotypic network model of AIDS immunopathogenesis. *Proc. Natl. Acad. Sci. USA* 88:3060–64
92. Hofmann, B., Nishanian, P., Baldwin, R. L., Insixiengmay, P., Nel, A., Fahey, J. L. 1990. HIV inhibits the early steps of lymphocyte activation, including initiation of inositol phospholipid metabolism. *J. Immunol.* 145:3699–3705
93. Horak, I. D., Popovic, M., Horak, E. M., Lucas, P. J., Gress, R. E., et al. 1990. No T-cell tyrosine protein kinase signalling or calcium mobilization after CD4 association with HIV-1 or HIV-1 gp120. *Nature* 348:557–60
94. Horsburgh, C. R., Ou, C. Y., Jason, J., Holmberg, S. D., Lifson, A. R., et al. 1990. Concordance of polymerase chain reaction with human immunodeficiency virus antibody detection. *J. Infect. Dis.* 162:542–45
95. Hoxie, J. A., Haggarty, B. S., Rackowski, J. L., Pillsbury, N., Levy, J. A. 1985. Persistent noncytopathic infection of normal human T lymphocytes with AIDS-associated retrovirus. *Science* 229:1400–2
96. Huang, M., Simard, C., Jolicoeur, P. 1989. Immunodeficiency and clonal growth of target cells induced by helper-free defective retrovirus. *Science* 246:1614–17
97. Hugin, A. W., Vacchio, M. S., Morse, H. C. 1991. A virus-encoded "superantigen" in a retrovirus-induced immunodeficiency syndrome of mice. *Science* 252:424–27
98. Iacobelli, S., Natoli, C., D'Egidio, M., Tamburrini, E., Antinori, A., Ortona, L. 1991. Lipoprotein 90K in human immunodeficiency virus-infected patients: a further serologic marker of progression. *J. Infect. Dis.* 164:819
99. Imagawa, D., Detels, R. 1991. HIV-1 in seronegative homosexual men. *New Engl. J. Med.* 325:1250–51
100. Imagawa, D. T., Lee, M. H., Wolinsky, S. M., Sano, K., Morales, F., et al. 1989. Human immunodeficiency virus type 1 infection in homosexual men who remain seronegative for prolonged periods. *New Engl. J. Med.* 320:1458–89
101. Imberti, L., Sottini, A., Bettinardi, A., Puoti, M., Primi, D. 1991. Selective depletion in HIV infection of T cells that bear specific T cell receptor V-beta sequences. *Science* 254:860–62
102. Isaksson, B., Albert, J., Chiodi, F., Furucrona, A., Krook, A., Putkonen, P. 1988. AIDS two months after primary human immunodeficiency virus infection. *J. Infect. Dis.* 158:866–68
103. Ivanoff, L. A., Looney, D. J., McDanal, C., Morris, J. F., Wong-Staal, F., et al. 1991. Alteration of HIV-1 infectivity and neutralization by a single amino acid replacement in the V3 loop domain. *AIDS Res. Hum. Retrovir.* 7:595–603
104. Janeway, C. 1991. Mls: makes a little sense. *Nature* 349:459–61
105. Janeway, C. A. 1991. The co-receptor function of CD4. *Sem. Immunol.* 3:153–60
106. Janvier, B., Baillou, A., Archinard, P., Mounier, M., Mandrand, B., et al. 1991. Immune response to a major epitope of p24 during infection with human immunodeficiency virus type 1 and implications for diagnosis and prognosis. *J. Clin. Microbiol.* 29:488–92
107. Jenkins, M. K., Ashwell, J. D., Schwartz, R. H. 1988. Allogenic non-T spleen cells restore the responsiveness of normal T cell clones stimulated with antigen and chemically modified antigen-presenting cells. *J. Immunol.* 140:3324–30
108. Jewett, A., Giorgi, J. V., Bonavida, B., 1990. Antibody-dependent cellular cytotoxicity against HIV-coated target cells by peripheral blood monocytes from HIV seropositive asymptomatic patients. *J. Immunol.* 145:4065–71
109. Jolicoeur, P. 1991. Murine acquired immunodeficiency syndrome (MAIDS): an animal model to study the AIDS pathogenesis. *FASEB J.* 5:2398–2405
110. Justice, A. C., Feinstein, A. R., Wells, C. K. 1989. A new prognostic staging system for the acquired immu-

nodeficiency syndrome. *New Engl. J. Med.* 320:1388–1416

111. Kaleebu, P., Cheingsong-Popov, R., Callow, D., Katabira, E., Mubiru, F., et al. 1991. Comparative humoral responses to HIV-1 p24gag and gp120env in subjects from East Africa and the UK. *AIDS* 5:1015–19
112. Kappler, J. W., Staerz, U., White, J., Marrack, P. C. 1988. Self-tolerance eliminates T cells specific for Mls-modified products of the major histocompatibility complex. *Nature* 332:35–40
113. Kaslow, R. A., Duquesnoy, R., Van Raden, M., Kingsley, L., Marrari, M., et al. 1990. A1, Cw7, B8, DR3 HLA antigen combination associated with rapid decline of T-helper lymphocytes in HIV-1 infection. *Lancet* 335:927–30
114. Kaye, J., Hsu, M. L., Sauron, M. E., Jameson, S. C., Gascoigne, N. R. J., Hedrick, S. M. 1989. Selective development of $CD4^+$ T cells in transgenic mice expressing a class II MHC-restricted antigen receptor. *Nature* 341:746–49
115. Kessler, H. A., Blaauw, B., Spear, J., Paul, D. A., Falk, L. A., Landay, A. 1987. Diagnosis of human immunodeficiency virus infection in seronegative homosexuals presenting with an acute viral syndrome. *J. Am. Med. Assoc.* 258:1196–99
116. Klimas, N. G., Caralis, P., LaPerriere, A., Antoni, M. H., Ironson, G., et al. 1991. Immunologic function in a cohort of human immunodeficiency virus type 1 seropositive and -negative healthy homosexual men. *J. Clin. Microbiol.* 29:1413–21
117. Kornfeld, H., Cruikshank, W. W., Pyle, S. W., Berman, J. S., Center, D. M. 1988. Lymphocyte activation by HIV-1 envelope glycoprotein. *Nature* 335:445–48
118. Krieger, J. N., Coombs, R. W., Collier, A. C., Ross, S. O., Chaloupka, K., et al. 1991. Recovery of human immunodeficiency virus type 1 from semen: minimal impact of stage of infection and current antiviral chemotherapy. *J. Infect. Dis.* 163:386–88
119. Krowka, J., Stites, D., Mills, J., Hollander, H., McHugh, T., et al. 1988. Effects of interleukin 2 and large envelope glycoprotein (gp120) of human immunodeficiency virus (HIV) on lymphocyte proliferative responses to cytomegalovirus. *Clin. Exp. Immunol.* 72:179–85
120. Kupcu, Z., Barrett, N., Dorner, F., Gallo, R. C., Eibl, M. M., Mannhalter, J. W. 1989. Effect of HIV on antigen-induced T cell proliferation. *Int. Conf. AIDS, Montreal, 5th, Canada,* Abstr. T. C. P. 38
121. Kwok, S., Higuchi, R. 1989. Avoiding false positives with PCR. *Nature* 339:237–38
122. Lafeuillade, A., Poizot-Martin, I., Quilichini, R., Gastaut, J. A., Kaplanski, S., et al. 1991. Increased interleukin-6 production is associated with disease progression in HIV infection. *AIDS* 5:1139–51
123. Laga, M., Taelman, H., Stuyft, P. V. D., Bonneux, L., Vercauteren, G., Piot, P. 1989. Advanced immunodeficiency as a risk factor for heterosexual transmission of HIV. *AIDS* 3:361–66
124. Lamarre, D., Capon, D. J., Karp, D. R., Gregory, T., Long, E. O., Sekaly, R. P. 1989. Class II MHC molecules and the HIV gp120 envelope protein interact with functionally distinct regions of the CD4 molecule. *EMBO J.* 8:3271–77
125. Lang, W. 1989. Frailty selection and HIV. *Lancet* 1:1397
126. Lang, W., Perkins, H., Anderson, R. E., Royce, R., Jewell, N., Winklestein, W. 1989. Patterns of T lymphocyte changes with human immunodeficiency virus infection: from seroconversion to the development of AIDS. *J. AIDS* 2:63–69
127. Langhoff, E., McElrath, J., Bos, H. J., Pruett, J., Granelli-Piperno, A., et al. 1989. Most $CD4^+$ T cells from human immunodeficiency virus-1 infected patients can undergo prolonged clonal expansion. *J. Clin. Invest.* 84:1637–43
128. Langhoff, E., Terwilliger, E. F., Bos, H. J., Kalland, K. H., Poznansky, M. C., et al. 1991. Replication of human immunodeficiency virus type 1 in primary dendritic cell cultures. *Proc. Natl. Acad. Sci. USA* 88:7998–8002
129. Ledbetter, J. A., June, C. H., Rabinovitch, S., Grossman, A., Tsu, T., Imboden, J. B. 1988. Signal transduction through CD4 receptors: stimulatory vs. inhibitory activity is regulated by CD4 proximity to the CD3/T cell receptor. *Eur. J. Immunol.* 18:525–32
130. Lekutis, C., Olshevsky, U., Furman, C., Thali, M., Sodroski, J. 1992. Contribution of disulfide bonds in the carboxyl terminus of the human immunodeficiency virus type I gp120 glycoprotein to CD4 binding. *J. AIDS* 5:78–81
131. Letvin, N. L., Daniel, D., Sehgal, P. K., Desrosiers, R. C., Hunt, R. D., et al. 1985. Induction of AIDS-like disease in Macaque monkeys with T-cell tropic retrovirus HTLV-III. *Science* 230:71–73
132. Levy, J. A. 1988. Mysteries of HIV:

challenges for therapy and prevention. *Nature* 333:519–22

133. Levy, J. A. 1990. Changing concepts in HIV infection: challenges for the 1990s. *AIDS* 4:1051–58
134. Lewis, D. E., Yoffe, B., Bosworth, C. G., Hollinger, F. B., Rich, R. R. 1988. Human immunodeficiency virus-induced pathology favored by cellular transmission and activation. *FASEB J.* 2:251–55
135. Lifson, A. R., Buchbinder, S. P., Sheppard, H. W., Mawle, A. C., Wilber, J. C., et al. 1991. Long-term human immunodeficiency virus infection in asymptomatic homosexual and bisexual men with normal CD4$^+$ lymphocyte counts: immunologic and virologic characteristics. *J. Infect. Dis.* 163:959–65
136. Lifson, A. R., O'Malley, P. M., Hessol, N. A., Buchbinder, S. P., Cannon, L., Rutherford, G. W. 1990. HIV seroconversion in two homosexual men after receptive oral intercourse with ejaculation: implications for counseling concerning safe sexual practices. *Am. J. Public Health* 80:1509–11
137. Lifson, A. R., Stanley, M., Pane, J., O'Mally, P. M., Wilber, J. C., et al. 1990. Detection of human immunodeficiency virus DNA using the polymerase chain reaction in a well-characterized group of homosexual and bisexual men. *J. Infect. Dis.* 161:436–39
138. Lifson, J. D., Feinberg, M. B., Reyes, G. R., Rabin, L., Banapour, B., et al. 1986. Induction of CD4-dependent cell fusion by the HTLV-III/LAV envelope glycoprotein. *Nature* 323:725–28
139. Ljunggren, K., Moschesee, V., Broliden, P. A., Giaquinto, C., Quinti, I., et al. 1990. Antibodies mediating cellular cytotoxicity and neutralization correlate with a better clinical stage in children born to human immunodeficiency virus–infected mothers. *J. Infect. Dis.* 161:198–202
140. Lo, S. C., Hayes, M. M., Wang, R. Y. H., Pierce, P. F., Kotani, H., Shih, J. W. K. 1991. Newly discovered mycoplasma isolated from patients infected with HIV. *Lancet* 338:1415–18
141. Lo, S. C., Tsai, S., Benish, J. R., Shih, J. W. K., Wear, D. J., Wong, D. M. 1991. Enhancement of HIV-1 cytocidal effects in CD4$^+$ lymphocytes by the AIDS-associated mycoplasma. *Science* 251:1074–76
142. Longini, I. M., Clark, W. S., Gardner, L. I., Brundage, J. F. 1991. The dynamics of CD4$^+$ t-lymphocyte decline in HIV-infected individuals: a Markov modeling approach. *J. AIDS* 4:1141–47
143. Louie, L. G., Newman, B., King, M. C. 1991. Influence of host genotype on progression to AIDS among HIV-infected men. *J. AIDS* 4:814–18
144. Macatonia, S. E., Lau, R., Patterson, S., Pinching, A. J., Knight, S. C. 1990. Dendritic cell infection, depletion and dysfunction in HIV-infected individuals. *Immunology* 71:38–45
145. MacDonald, H. R., Hengartner, H., Pedrazzini, T. 1987. Intrathymic deletion of self-reactive cells prevented by neonatal anti-CD4 antibody treatment. *Nature* 335:174–76
146. MacDonell, K. B., Chmiel, J. S., Goldsmith, J., Wallemark, C.-B., Steinberg, J., et al. 1988. Prognostic usefulness of the Walter Reed staging classification for HIV infection. *J. AIDS* 1:367–74
147. Machado, S. G., Gail, M. H., Ellenberg, S. S. 1990. On the use of laboratory markers as surrogates for clinical endpoints in the evaluation of treatment for HIV infection. *J. AIDS* 3:1065–73
148. Mackewicz, C. E., Ortega, H. W., Levy, J. A. 1991. CD8$^+$ cell anti-HIV activity correlates with the clinical state of the infected individual. *J. Clin. Invest.* 87:1462–66
149. Maddox, J. 1991. AIDS research turned upside down. *Nature* 353:297
150. Makinodan, T., Hahn, T. J., McDougall, S., Yamaguchi, D. T., Fang, M., Iida-Klein, A. 1991. Cellular immunosenescence: an overview. *Exp. Gerontol.* 26:281–88
151. Manca, F., Habeshaw, J. A., Dalgleish, A. G. 1990. HIV envelope glycoprotein, antigen specific T-cell responses, and soluble CD4. *Lancet* 335:811–15
152. Margolick, J. B., Volkman, D. J., Folks, T. M., Fauci, A. S. 1987. Amplification of HTLV-III/LAV infection by antigen-induced activation of T cells and direct suppression by virus of lymphocyte blastogenic responses. *J. Immunol.* 138:1719–23
153. Marrack, P., Kushnir, E., Kappler, J. 1991. A maternally inherited superantigen encoded by a mammary tumour virus. *Nature* 349:524–26
154. Marrack, P. 1991. Superantigens and immune system design. *Clin. Immunol. Spectr.* 3:8–10
155. Mastroianni, C. M., Paoletti, F., Vullo, V., Delia, S., Sorice, F. 1990. Serum beta2-M in asymptomatic HIV-infected patients treated with zidovudine. *AIDS* 4:1297
156. Masur, H., Ognibene, F. P., Yarchoan, R., Shelhamer, J. H., Baird, B. F., et al. 1989. CD4 counts as predictors of opportunistic pneumonias in human im-

munodeficiency virus (HIV) infection. *Ann. Intern. Med.* 111:223–30

157. Masur, H., Meier, P., McCutchan, J. A., Feinberg, J., Bernard, G., et al. 1990. Consensus statement on the use of corticosteroids as adjunctive therapy for pneumocystis pneumonia in the acquired immunodeficiency syndrome. *New Engl. J. Med.* 323:1500–4
158. Mathez, D., Paul, D., De Belilovsky, C., Sultan, Y., Deleuze, J., et al. 1990. Productive human immunodeficiency virus infection levels correlate with AIDS-related manifestations in the patient. *Proc. Natl. Acad. Sci. USA* 87:7438–42
159. Matsuyama, T., Kobayashi, N., Yamamoto, N. 1991. Cytokines and HIV infection: is AIDS a tumor necrosis factor disease? *AIDS* 5:1405–17
160. McCutchan, F. E., Sanders-Buell, E., Oster, C. W., Redfield, R. R., Hira, S. K., et al. 1991. Genetic comparison of human immunodeficiency virus (HIV-1) isolates by polymerase chain reaction. *J. AIDS* 4:1241–50
161. McDonald, H. R., Schneider, R., Lees, R. K., Howe, R. C., Acha-Orbea, H., et al. 1988. T-cell receptor V-beta use predicts reactivity and tolerance to Mls-1a encoded antigens. *Nature* 332:40–45
162. McDougal, J. S., Klatzmann, D. R., Maddon, P. J. 1991. CD4-gp120 interactions. *Curr. Opin. Immunol.* 3: 552–58
163. McDougal, J. S., Mawle, A., Cort, S. P., Nicholson, J. K. A., Cross, G. D., et al. 1985. Cellular tropism of the human retrovirus HTLV-III/LAV, I. Role of T-cell activation and expression of the T4 antigen. *J. Immunol* 135:3151–62
164. McElrath, M. J., Pruett, J. E., Cohn, Z. A. 1989. Mononuclear phagocytes of blood and bone marrow: comparative roles as viral reservoirs in human immunodeficiency virus type 1 infections. *Proc. Natl. Acad. Sci. USA* 86:675–79
165. McGrath, M. S., Shiramizu, B., Meeker, T. C., Kaplan, L. D., Herndier, B. 1991. AIDS-associated polyclonal lymphoma: identification of a new HIV-associated disease process. *J. AIDS* 4: 408–15
166. McLean, K. A., Holmes, D. A., Evans, B. A., McAlpine, L., Thorp, R., et al. 1990. Rapid clinical and laboratory progression of HIV infection. *AIDS* 4:369–71
167. McRae, B., Lange, J. A. M., Ascher, M. S., de Wolf, F., Sheppard, H. W., et al. 1991. Immune response to HIVp24 core protein during the early phases of human immunodeficiency virus infection. *AIDS Res. Hum. Retrovir.* 7:637–43
168. Meyerhans, A., Cheynier, R., Albert, J., Seth, M., Kwok, S., et al. 1989. Temporal fluctuations in HIV quasispecies in vivo are not reflected by sequential HIV isolations. *Cell* 58:901–10
169. Miceli, M. C., Parnes, J. R. 1991. The roles of CD4 and CD8 in T cell activation. *Sem. Immunol.* 3:133–41
170. Mittler, R. S., Hoffmann, M. K. 1989. Synergism between HIV gp120 and gp120-specific antibody in blocking human T cell activation. *Science* 245:1380–82
171. Mittler, R. S., Rankin, B. M., Kiener, P. A. 1991. Physical associations between CD45 and CD4 or CD8 occur as late activation events in antigen receptor-stimulated human T cells. *J. Immunol.* 147:3434–40
172. Monforte, A. D. A., Novati, R., Galli, M., Marchisio, P., Massironi, E., et al. 1990. T-cell subsets and serum immunoglobulin levels in infants born to HIV-seropositive mothers: a longitudinal evaluation. *AIDS* 4:1141–44
173. Montagnier, L., Berneman, D., Guetard, D., Blanchard, A., Chamaret, S., et al. 1990. Inhibition of HIV prototype strains infectivity by antibodies directed against a peptidic sequence of mycoplasma. *C. R. Acad. Sci. Paris* 311:425–30
174. Montagnier, L., Gougeon, M. L., Olivier, R., Garcia, S., Dauguet, C., et al. 1992. Factors and mechanisms of AIDS pathogenesis. In *Science Challenging AIDS*, ed. G. B. Rossi, E. Beth-Giraldo, L. Chieco-Bianchi, F. Dianzani, P. Verani, pp. 51–70. Basel: Karger
175. Montagnier, L., Guetard, D., Rame, V., Olivier, R., Adams, M. 1989. Virological and immunological factors of AIDS pathogenesis. *Quatr. Colloq. Cent Gardes* pp. 11–17
176. Montefiori, D. C., Lefkowitz, L. B., Keller, R. E., Holmberg, V., Sandstrom, E., et al. 1991. Absence of a clinical correlation for complement-mediated, infection-enhancing antibodies in plasma or sera from HIV-1-infected individuals. *AIDS* 5:513–17
177. Morfeldt-Manson, L., Bottiger, B., Nilsson, B., von Stedingk, L. V. 1991. Clinical signs and laboratory markers in predicting progression to AIDS in HIV-1 infected patients. *Scand. J. Infect. Dis.* 23:443–49
178. Moss, A. R. 1990. Laboratory markers as potential surrogates for clinical outcomes in AIDS trials. *J. AIDS* 3(2):S69–S71

179. Mueller, D. L., Jenkins, M. K., Schwartz, R. H. 1989. Clonal expansion versus functional clonal inactivation: a costimulatory signalling pathway determines the outcome of T cell receptor occupancy. *Annu. Rev. Immunol.* 7: 445–80
180. Myers, G., Berzofsky, J. A., Rabson, A. B., Smith, T. F., Wong-Staal, F., eds. 1990. *Human Retroviruses and AIDS 1990*. Los Alamos, NM: Theoretical Biology and Biophysics
181. Narayan, O., Zink, M. C., Huso, D., Sheffer, D., Crane, S., et al. 1988. Lentiviruses of animals are biological models of the human immunodeficiency viruses. *Microbiol. Pathogen.* 5:149–57
182. Neudorf, S. M. L., Jones, M. M., McCarthy, B. M., Harmony, J. A. K., Choi, E. M. 1990. The CD4 molecule transmits biochemical information important in the regulation of T lymphocyte activity. *Cell. Immunol.* 125:301–14
183. Newell, M. K., Haughn, L. J., Maroun, C. R., Julius, M. H. 1990. Death of mature T cells by separate ligation of CD4 and the T-cell receptor for antigen. *Nature* 347:286–89
184. Nishanian, P., Hofmann, B., Wang, Y., Jackson, A. L., Detels, R., Fahey, J. L. 1991. Serum soluble CD8 molecule is a marker of CD8 T-cell activation in HIV-1 disease. *AIDS* 5:805–12
185. Nishanian, P., Huskins, K. R., Stehn, S., Detels, R., Fahey, J. L. 1990. A simple method for improved assay demonstrates that HIV p24 antigen is present as immune complexes in most sera from HIV-infected individuals. *J. Infect. Dis.* 162:21–28
186. Nowak, M. A., May, R. M., Anderson, R. M. 1990. The evolutionary dynamics of HIV-1 quasispecies and the development of immunodeficiency disease. *AIDS* 4:1095–1103
187. Nowak, M. A., McLean, A. R. 1991. A mathematical model of vaccination against HIV to prevent the development of AIDS. *Proc. R. Soc. London Ser. B. Biol. Sci.* 246(1316):141–46
188. Nye, K. E., Knox, K. A., Pinching, A. J. 1991. Lymphocytes from HIV-infected individuals show aberrant inositol polyphosphate metabolism which reverses after zidovudine therapy. *AIDS* 5:413–17
189. Nye, K. E., Pinching, A. J. 1990. HIV infection of H9 lymphoblastoid cells chronically activates the inositol polyphosphate pathway. *AIDS* 4:41–45
190. Orholm, M., Nielsen, T. L., Nielsen, J. O., Lundgren, J. D. 1990. CD4 lymphocyte counts and serum p24 antigen of no diagnostic value in monitoring HIV-infected patients with pulmonary symptoms. *AIDS* 4:163–66
191. Orloff, G. M., Kennedy, M. S., Dawson, C., McDougal, S. 1991. HIV-1 binding to CD4 T cells does not induce a $Ca2^{+}$ influx or lead to activation of protein kinases. *AIDS Res. Hum. Retrovir.* 7:587–93
192. Osmond, D. H., Shiboski, S., Bacchetti, P., Winger, E. E., Moss, A. R. 1991. Immune activation markers and AIDS prognosis. *AIDS* 5:505–11
193. Oyaizu, N., Chirmule, N., Kalyanaraman, V. S., Hall, W. W., Good, R. A., Pahwa, S. 1990. Human immunodeficiency virus type 1 envelope glycoprotein gp120 produces immune defects in $CD4^{+}$ T lymphocytes by inhibiting interleukin 2 mRNA. *Proc. Natl. Acad. Sci. USA* 87:2379–83
194. Padian, N. S., Shiboski, S. C., Jewell, N. P. 1991. Female-to-male transmission of human immunodeficiency virus. *J. Am. Med. Assoc.* 266:1664–67
195. Pan, L. Z., Sheppard, H. W., Winkelstein, W., Levy, J. A. 1991. Lack of detection of human immunodeficiency virus in persistently seronegative homosexual men with high or medium risk for infection. *J. Infect. Dis.* 164:962–64
196. Parekh, B. S., Shaffer, N., Pau, C. P., Abrams, E., Thomas, P., et al. 1991. Lack of correlation between maternal antibodies to V3 loop peptides of gp120 and perinatal HIV-1 transmission. *AIDS* 5:1179–84
197. Pauza, C. D., Price, T. M. 1988. Human immunodeficiency virus infection of T cells and monocytes proceeds via receptor-mediated endocytosis. *J. Cell. Biol.* 107:959–68
198. Pedersen, C., Lindhardt, B. O., Jensen, B. L., Jensen, B. L., Lauritzen, E., et al. 1989. Clinical course of primary HIV infection: consequences for subsequent course of infection. *Br. Med. J.* 299:154–57
199. Pezzella, M., Rossi, P., Lombardi, V., Gemelli, V., Costantini, R. M., et al. 1989. HIV viral sequences in seronegative people at risk detected by in situ hybridisation and polymerase chain reaction. *Br. Med. J.* 298:713–16
200. Phair, J., Munoz, A., Detels, R., Kaslow, R., Rinaldo, C., Saah, A. 1990. The risk of *Pneumocystis carinii* pneumonia among men infected with human immunodeficiency virus type 1. *New Engl. J. Med.* 322:161–65
201. Phillips, A. N., Lee, C. A., Elfort, J.,

Webster, A., Janossy, G., Griffiths, P. D., Kernoff, P. B. A. 1991. p24 antigenaemia, CD4 lymphocyte counts and the development of AIDS. *AIDS* 5:1217–22

202. Phillips, A., Wainberg, M., Coates, R., Klein, M., Rachlis, A., et al. 1989. Cyclosporine-induced deterioration in patients with AIDS. *Can. Med. Assoc. J.* 140:1456–60
203. Phillips, A. N., Lee, C. A., Elford, J., Janossy, G., Timms, A., et al. 1991. Serial CD4 lymphocyte counts and development of AIDS. *Lancet* 337:389–92
204. Pinching, A. J. 1991. HIV/AIDS pathogenesis and treatment: new twists and turns. *Curr. Opin. Immunol.* 3:537–42
205. Pinching, A. J., Jeffries, D. J., Harris, J. R. W., Swirsky, D., Weber, J. N. 1990. HIV and AIDS. *Nature* 347: 324
206. Pircher, H., Rohrer, U. H., Moskophidis, D., Zinkernagel, R. M., Hengartner, H. 1991. Lower receptor avidity required for thymic clonal deletion than for effector T-cell function. *Nature* 351:482–85
207. Polis, M. A., Masur, H. 1990. Predicting the progression to AIDS. *Am. J. Med.* 89:701–5
208. Quaratino, S., Murison, G., Knyba, R. E., Verhoef, A., Londei, M. 1991. Human CD4− CD8− alpha-beta$^+$ T cells express a functional T cell receptor and can be activated by superantigens. *J. Immunol.* 147:3319–23
209. Ragni, M. V., Kingsley, L. A. 1990. Cumulative risk for AIDS and other HIV outcomes in a cohort of hemophiliacs in western Pennsylvania. *J. AIDS* 3:708–13
210. Rahman, M. A., Kingsley, L. A., Atchison, R. W., Belle, S., Breinig, M. C., et al. 1991. Reactivation of Epstein-Barr virus during early infection with human immunodeficiency virus. *J. Clin. Microbiol.* 29:1215–20
211. Ramirez, J., Alcami, J., Arnaiz-Villena, A., Regueiro, J. R., Rioperez, E., et al. 1986. Indomethacin in the relief of AIDS symptoms. *Lancet* 2:570–71
212. Rappersberger, K., Gartner, S., Schenk, P., Stingl, G., Groh, V., et al. 1988. Langerhans' cells are an actual site of HIV-1 replication. *Intervirology* 29: 185–94
213. Redfield, R. R., Birx, D. L., Ketter, N., Tramont, E., Polonis, V., et al. 1991. A phase I evaluation of the safety and immunogenicity of vaccination with recombinant gp160 in patients with early human immunodeficiency virus infection. *New Engl. J. Med.* 324:1677–84
214. Reibnegger, G., Spira, T. J., Fuchs, D., Werner-Felmayer, G., Dierich, M. P., Wachter, H. 1991. Individual probability for onset of full-blown disease in patients infected with human immunodeficiency virus type 1. *Clin. Chem.* 37:351–55
215. Rieckmann, P., Poli, G., Fox, C. H., Kehrl, J. H., Fauci, A. S. 1991. Recombinant gp120 specifically enhances tumor necrosis factor-alpha production and Ig secretion in B lymphocytes from HIV-infected individuals but not from seronegative donors. *J. Immunol.* 147:2922–27
216. Rivas, A., Takada, S., Koide, J., Sonderstrup-MacDevitt, G., Engleman, E. 1988. CD4 molecules are associated with the antigen receptor complex on activated but not resting T cells. *J. Immunol.* 140:2912–18
217. Rosenstein, Y., Burakoff, S. J., Herrmann, S. H. 1990. HIV-gp120 can block CD4-class II MHC-mediated adhesion. *J. Immunol.* 144:526–31
218. Rossi, P., Moschese, V., Broliden, P. A., Fundaro, C., Quinti, I., et al. 1989. Presence of maternal antibodies to human immunodeficiency virus 1 envelope glycoprotein gp120 epitopes correlates with the uninfected status of children born to seropositive mothers. *Proc. Natl. Acad. Sci. USA* 86:8055–58
219. Roy, S., Wainberg, M. A. 1988. Role of the mononuclear phagocyte system in the development of acquired immunodeficiency syndrome (AIDS). *J. Leukocyte Biol.* 43:91–97
220. Royce, R. A., Luckmann, R. S., Fusaro, R. E., Winkelstein, W. 1991. The natural history of HIV-1 infection: staging classifications of disease. *AIDS* 5:355–64
221. Russell, J. H., White, C. L., Loh, D. Y., Meleedy-Rey, P. 1991. Receptor-stimulated death pathway is opened by antigen in mature T cells. *Immunology* 88:2151–55
222. Saag, M. S., Crain, M. J., Decker, W. D., Campbell-Hill, S., Robinson, S., et al. 1991. High-level viremia in adults and children infected with human immunodeficiency virus: relation to disease stage and CD4$^+$ lymphocyte levels. *J. Infect. Dis.* 164:72–80
223. Sattentau, Q. J. 1988. The role of the CD4 antigen in HIV infection and immune pathogenesis. *AIDS* 2(1):S11–S16
224. Schinaia, N., Ghirardini, A., Chiarotti, F., Gringeri, A., Mannucci, P. M., et al. 1991. Progression to AIDS among Italian HIV-seropositive haemophiliacs. *AIDS* 5:385–91

225. Schnittman, S. M., Greenhouse, J. J., Psallidopoulos, M. C., Baseler, M., Salzman, N. P., et al. 1990. Increasing viral burden in CD4$^+$ T cells from patients with human immunodeficiency virus (HIV) infection reflects rapidly progressive immunosuppression and clinical disease. *Ann. Intern. Med.* 113:438–43
226. Schnittman, S. M., Lane, H. C., Greenhouse, J., Justement, J. S., Baseler, M., Fauci, A. S. 1990. 1990. Preferential infection of CD4$^+$ memory T cells by human immunodeficiency virus type 1: evidence for a role in the selective T-cell functional defects observed in infected individuals. *Proc. Natl. Acad. Sci. USA* 87:6058–62
227. Schnittman, S. M., Psallidopoulos, M. C., Lane, H. C., Thompson, L., Baseler, M., et al. 1989. The reservoir for HIV-1 in human peripheral blood is a T cell that maintains expression of CD4. *Science* 245:305–8
228. Schwartz, R. H. 1990. A cell culture model for T lymphocyte clonal anergy. *Science* 248:1349–56
229. Sekaly, R. P., Croteau, G., Bowman, M., Scholl, P., Burakoff, S., Geha, R. S. 1991 The CD4 molecule is not always required for the T cell response to bacterial enterotoxins. *J. Exp. Med.* 173:367–71
230. Sheppard, H. W., Ascher, M. S. 1991. AIDS and programmed cell death. *Immunol. Today* 12:423
231. Sheppard, H. W., Ascher, M. S. 1992. The relationship between AIDS and immunologic tolerance. *J. AIDS* 5:143–47
232. Sheppard, H. W., Ascher, M. S., Busch, M. P., Sohmer, P. R., Stanley, M., et al. 1991. A multicenter proficiency trial of gene amplification (PCR) for the detection of HIV-1. *J. AIDS* 4:277–83
233. Sheppard, H. W., Ascher, M. S., McRae, B., Anderson, R. E., Lang, W., Allain, J. P. 1991. The initial immune response to HIV and immune system activation determine the outcome of HIV disease. *J. AIDS* 4:704–12
234. Sheppard, H. W., Dondero, D., Arnon, J., Winkelstein, W. 1991. An evaluation of the polymerase chain reaction in HIV-1 seronegative men. *J. AIDS* 4:819–23
235. Sheppard, H. W., Winkelstein, W., Osmond, D., Moss, A. R. 1991. Effect of new AIDS case definition on numbers of cases among homosexual and bisexual men in San Francisco. *J. Am. Med. Assoc.* 266:2221
236. Silberman, S. L., Goldman, S. J., Mitchell, D. B., Tong, A. T., Rosenstein, Y., et al. 1991. The interaction of CD4 with HIV-1 gp120. *Sem. Immunol.* 3:187–92
237. Simard, C., Jolicoeur, P. 1991. The effect of anti-neoplastic drugs on murine acquired immunodeficiency syndrome. *Science* 251:305–8
238. Simmonds, P., Balfe, P., Peutherer, J. F., Ludlam, C. A., Bishop, J. O., Brown, A. J. L. 1990. Human immunodeficiency virus-infected individuals contain provirus in small numbers of peripheral mononuclear cells and at low copy numbers. *J. Virol.* 64:864–72
239. Simmonds, P., Beatson, D., Cuthbert, R. J. G., Watson, H., Reynolds, B., et al. 1991. Determinants of HIV disease progression: six-year longitudinal study in the Edinburgh haemophilia/HIV cohort. *Lancet* 338:1159–63
240. Sleckman, B. P., Rosenstein, Y., Igras, V. E., Greenstein, J. L., Burakoff, S. J. 1991. Glycolipid-anchored form of CD4 increases intercellular adhesion but is unable to enhance T cell activation. *J. Immunol.* 147:428–31
241. Smith, C. A., Williams, G. T., Kingston, R., Jenkinson, E. J., Owen, J. J. T. 1989. Antibodies to CD3/T-cell receptor complex induce death by apoptosis in immature T cells in thymic cultures. *Nature* 337:181–84
242. Solbach, W., Moll, H., Rollinghoff, M. 1991. Lymphocytes play the music but the macrophage calls the tune. *Immunol. Today* 12:4–6
243. St.-Pierre, Y., Watts, T. H. 1991. Characterization of the signaling function of MHC class II molecules during antigen presentation by B cells. *J. Immunol.* 147:2875–82
244. Szabo, B., Toth, F. D., Kiss, J., Ujhelyi, E., Fust, G., et al. 1990. Neutralizing antibodies and serum interferon levels in the different stages of HIV infection. *Acta Virol.* 34:164–70
245. Tanneau, F., McChesney, M., Lopez, O., Sansonetti, P., Montagnier, L., et al. 1990. Primary cytotoxicity against envelope glycoprotein of human immunodeficiency virus-1: evidence for antibody-dependent cellular cytotoxicity in vivo. *J. Infect. Dis.* 162:837–43
246. Taylor, J. M. G., Tan, S. J., Detels, R., Giorgi, J. V. 1991. Applications of a computer simulation model of the natural history of CD4 T-cell number in HIV-infected individuals. *AIDS* 5:159–67
247. Teeuwsen, V. J. P., Lange, J. M. A, Keet, R., Schattenkerk, J. K. M. E., Debouck, C., et al. 1991. Low numbers

of functionally active B lymphocytes in the peripheral blood of HIV-1-seropositive individuals with low p24-specific serum antibody titers. *AIDS* 5:971–79

248. Teeuwsen, V. J. P., Siebelink, K. H. J., de Wolf, F., Goudsmit, J., UytdeHaag, G. C. M., Osterhaus, A. D. M. E. 1990. Impairment of in vitro immune responses occurs within 3 months after HIV-1 seroconversion. *AIDS* 4:77–81
249. Teh, H. S., Garvin, A. M., Forbush, K. A., Carlow, D. A., Davis, C. B., et al. 1991. Participation of CD4 coreceptor molecules in T-cell repertoire selection. *Nature* 349:241–43
250. Tersmette, M., de Goede, R. E. Y., Lange, J. M. A., deWolfe, F., Eeftink-Schattenkerk, J. K. M., et al. 1989. Association between biological properties of human immunodeficiency virus variants and risk for AIDS and AIDS mortality. *Lancet* 1:983–85
251. Tersmette, M., Gruters, R. A., De Wolf, F., De Goede, R. E. Y., Lange, J. M. A., et al. 1989. Evidence for a role of virulent human immunodeficiency virus (HIV) variants in the pathogenesis of acquired immunodeficiency syndrome: studies on sequential HIV isolates. *J. Virol.* 63:2118–25
252. Thomas, L. 1979. *The Medusa and the Snail: More Notes of a Biology Watcher*. New York: Viking
253. Tindall, B., Barker, S., Donovan, B., Barnes, T., Roberts, J., et al. 1988. Characterization of the acute clinical illness associated with human immunodeficiency virus infection. *Arch. Int. Med.* 148:945–49
254. Tindall, B., Cooper, D. A. 1991. Primary HIV infection: host responses and intervention strategy. *AIDS* 5:1–14
255. Trauger, R., Lewis, D., Giermakowska, W., Wallace, M., Beecham, J., et al. 1991. Cell mediated immunity to HIV as it compares to Walter Reed stage 1–6 individuals: correlation with virus burden. *Int. Conf. AIDS, 7th, Florence, Italy*. Abstr. W. A. 1205
256. Turka, L. A., Linsley, P. S., Paine, R., Schieven, G. L., Thompson, C. B., Ledbetter, J. A. 1991. Signal transduction via CD4, CD8, and CD28 in mature and immature thymocytes. *J. Immunol.* 146:1428–36
257. Tyler, D. S., Stanley, S. D., Zolla-Pazner, S., Gorny, M. K., Shadduck, P. P., et al. 1990. Identification of sites within gp41 that serve as targets for antibody-dependent cellular cytotoxicity by using human monoclonal antibodies. *J. Immunol.* 145:3276–82
258. Ujhelyi, E., Fuchs, D., Krall, G., Zimonyi, I., Berkeessy, S., et al. 1990. Age dependency of the the progression of HIV disease in haemophiliacs; predictive value of T cell subset and neopterin measurements. *Immunol. Lett.* 26:67–74
259. Ujhelyi, E., Lange, J., Goudsmit, J., Salavecz, V., Buki, B., et al. 1990. Correlation of HIV core antigen, antibody and immune complex levels in sera of HIV-infected individuals. *AIDS* 4:928–29
260. van Griensven, G. J. P., de Vroome, E. M. M., de Wolf, F., Goudsmit, J., Roos, J., Coutinho, R. A. 1990. Risk factors for progression of human immunodeficiency virus (HIV) infection among seroconverted and seropositive homosexual men. *Am. J. Epidemiol.* 132:203–10
261. van Griensven, G. J. P., Boucher, E. C., Roos, M., Coutinho, R. A. 1991. Expansion of AIDS case definition. *Lancet* 338:1012–13
262. Veillette, A. 1991. Introduction: the functions of CD4 and D8. *Sem. Immunol.* 3:131–32
263. von Boehmer, H., Kisielow, P. 1990. Self-nonself discrimination by T cells. *Science* 248:1369–73
264. Wainberg, M. A., Blain, N., Fitz-Gibbon, L. 1987. Differential susceptibility of human lymphocyte cultures to infection by HIV. *Clin. Exp. Immunol.* 70:136–42
265. Walker, B. D., Plata, F. 1990. Cytotoxic T lymphocytes against HIV. *AIDS* 4:177–84
266. Winklestein, W., Sheppard, H. W. 1990. Median incubation time for human immunodeficiency virus (HIV). *Ann. Intern. Med.* 112:797–98
267. Wolinsky, S. M., Rinaldo, C. R., Kwok, S., Sninski, J. J., Gupta, P., et al. 1898. Human immunodeficiency virus type 1 (HIV-1) infection a median of 18 months before a diagnostic Western blot. *Ann. Intern. Med.* 111:961–72
268. Wong-Staal, F., Steffy, K. 1991. Genetic regulation of human immunodeficiency virus. *Microbiol. Rev.* 55:193–205
269. Yagi, J., Rath, S., Janeway, C. A. 1991. Control of T cell responses to staphylococcal enterotoxins by stimulator cell MHC class II polymorphism. *J. Immunol.* 147:1398–1405
270. Yagi, M. J., Joesten, M. E., Wallace, J., Roboz, J. P., Bekesi, J. G. 1991. Human immunodeficiency virus type 1 (HIV-1) genomic sequences and distinct changes in $CD8^+$ lymphocytes precede detectable levels of HIV-1 antibodies in

high-risk homosexuals. *J. Infect. Dis.* 164:183–88

271. Yerly, S., Chamot, E., Deglon, J. J., Hirschel, B., Perrin, L. H. 1991. Absence of chronic human immunodeficiency virus infection without seroconversion in intravenous drug users: a prospective and retrospective study. *J. Infect. Dis.* 164:965–68
272. Yogev, R., Connor, E. 1992. *Management of HIV Infection in Infants and Children*. St. Louis: Mosby/Year Book
273. Zacharchuk, C. M., Mercep, M., Chakraborti, P. K., Simons, S. S., Ashwell, J. D. 1990. Programmed T lymphocyte death. *J. Immunol.* 145:4037–45
274. Zack, J. A., Cann, A. J., Lugo, J. P., Chen, I. S. Y. 1988. HIV-1 production from infected peripheral blood T cells after HTLV-I induced mitogenic stimulation. *Science* 240:1026–29
275. Zagury, D., Bernard, J., Leonard, R., Cheynier, R., Feldman, M., et al. 1986. Long-term cultures of HTLV-III-infected T cells: a model of cytopathology of T-cell depletion in AIDS. *Science* 231:850–53
276. Zangerle, R., Fuchs, D., Reibnegger, G., Fritsch, P., Wachter, H. 1991. Markers for disease progression in intravenous drug users infected with HIV-1. *AIDS* 5:985–91
277. Zhou, P., Anderson, G. D., Savarirayan, S., Inoko, H., David, C. S. 1991. Thymic deletion of V-beta11$^+$, V-beta5$^+$ T cells in H-2E negative, HLA-DQbeta$^+$ single transgenic mice. *J. Immunol.* 146:854–59
278. Ziegler, J. L., Stites, D. P. 1986. Hypothesis: AIDS is an autoimmune disease directed at the immune system and triggered by a lymphotropic virus. *Clin. Immunol. Immunopathol.* 41:305–13
279. Zolla-Pazner, S. 1986. Immunologic abnormalities in infections with the human immunodeficiency virus. *Lab. Med.* 17:685–89
280. Zolla-Pazner, S., Des Jarlais, D. C., Friedman, S. R., Spira, T. J., Marmor, M., et al. 1987. Nonrandom development of immunologic abnormalities after infection with human immunodeficiency virus: implications for immunologic classification of the disease. *Proc. Natl. Acad. Sci. USA* 84:5404–8
281. Zuniga-Pflucker, J. C., Jones, L. A., Chin, T., Kruisbeek, A. M. 1991. CD4 and CD8 act as co-receptors during thymic selection of the T cell repertoire. *Sem. Immunol.* 3:167–75

Annu. Rev. Microbiol. 1992. 46:565–601

FUNCTIONAL AND EVOLUTIONARY RELATIONSHIPS AMONG DIVERSE OXYGENASES

S. Harayama and M. Kok

Department of Medical Biochemistry, University of Geneva, Geneva, Switzerland

E. L. Neidle

Department of Microbiology, University of Texas Medical School at Houston, Houston, Texas 77225

KEY WORDS: hydroxylase, biodegradation, electron transfer, *Pseudomonas,* aromatic

---

CONTENTS

0066-4227/92/1001-0565$02.00

Abstract

Oxygenases that incorporate one or two atoms of dioxygen into substrates are found in many metabolic pathways. In this article, representative oxygenases, principally those found in bacterial pathways for the degradation of hydrocarbons, are reviewed. Monooxygenases, discussed in this chapter, incorporate one hydroxyl group into substrates. In this reaction, two atoms of dioxygen are reduced to one hydroxyl group and one H_2O molecule by the concomitant oxidation of NAD(P)H. Dioxygenases catalyze the incorporation of two atoms of dioxygen into substrates. Two types of dioxygenases, aromatic-ring dioxygenases and aromatic-ring-cleavage dioxygenases, are discussed. The aromatic-ring dioxygenases incorporate two hydroxyl groups into aromatic substrates, and *cis*-diols are formed. This reaction also requires NAD(P)H as an electron donor. Aromatic-ring-cleavage dioxygenases incorporate two atoms of dioxygen into aromatic substrates, and the aromatic ring is cleaved. This reaction does not require an external reductant.

All the oxygenases possess a cofactor, a transition metal, flavin or pteridine, that interacts with dioxygen. The concerted reactions between dioxygen and carbon in organic compounds are spin forbidden. The cofactor is used to overcome this restriction.

For the oxygenases that require the NAD(P)H cofactor, the enzyme reaction is separated into two steps, the oxidation of NAD(P)H to generate two reducing equivalents, and the hydroxylation of substrates. Flavoprotein hydroxylases that catalyze the monohydroxylation of the aromatic ring carry out these two reactions on a single polypeptide chain. In other oxygenases, the NAD(P)H oxidation and a hydroxylation reaction are catalyzed by two separate polypeptides that are linked by a short electron-transport chain. Two reducing equivalents generated by the oxidation of NAD(P)H are transferred through the electron-transport chain to the cofactor on a hydroxylase component that they reduce. Dioxygen couples with the reduced cofactor and subsequently hydroxylates substrates. The electron-transport chains associated with oxygenases contain at least two redox centers. The first redox center is usually a flavin, while the second is an iron-sulfur cluster. The electron transport is initiated by a single two-electron transfer from NAD(P)H to a flavin, followed by two single-electron transfers from the flavin to an iron-sulfur cluster.

The primary sequences of many oxygenases have been determined, and according to their sequence similarities, the oxygenases can be grouped into several protein families. Among proteins of the same family, the sequences in regions involved in cofactor binding are strongly conserved. Local sequence similarities are also observed among oxygenases from different families, primarily in regions involved in cofactor binding.

INTRODUCTION

Dioxygen can be incorporated directly into organic compounds in reactions catalyzed by enzymes known as oxygenases or hydroxylases. These enzymes utilize transition metals or reduced flavin/pteridine to activate dioxygen, which in its ground state is unreactive. In these processes, highly reactive forms of oxygen, such as singlet oxygen and hydroxyl radicals, may be generated. The threat of reactive oxygen species has been proposed as one of the reasons why the use of dioxygen in general metabolism is limited, and why most biochemical oxidations are carried out by the transfer of electrons and/or reducing equivalents to FAD or $NAD(P)^+$, instead (17, 89).

Oxygenases play significant roles in microbial catabolic pathways. They initiate the degradation of aromatic compounds both by hydroxylating the aromatic ring in preparation for ring-cleavage and by catalyzing the ring-fission reaction (49). The initial reaction of alkane degradation is also catalyzed by an oxygenase. Microbial oxygenases in these pathways have been extensively studied in order to understand their reaction mechanisms, specificity, and regulation. These studies are also relevant to the application of microbial enzymes in environmental and industrial biotechnology.

Oxygenases in higher organisms, while less abundant, are by no means less important than those found in microorganisms. Mammalian oxygenases are involved in the hydroxylation of steroids, in the synthesis of neurotransmitters, and in the detoxication of poisonous compounds. Not all oxygenases, however, carry out essential, or even desirable, reactions. For example, plant lipoxygenases that oxidize polyunsaturated fatty acids in lipids have been studied because they produce both pleasant and unpleasant flavors in edible plant products (84).

The catabolic versatility of microbes plays an essential role in the carbon cycle and depends on the use of oxygenases in the initiation of degradative pathways. Recent studies have shed light on the structures and reaction mechanisms of some of these enzymes. Biochemical and biophysical studies have been combined with molecular genetic techniques. The deduced amino acid sequences of numerous oxygenases can now be compared. Also, patterns of evolutionary relationships are emerging, and concepts of gene families have developed, from these comparisons as well as from other studies. A discussion of every known oxygenase is beyond the scope of this review. Here, we focus on microbial oxygenases involved in the utilization of organic growth substrates, although selected mammalian enzymes related to their microbial counterparts are included.

GENERAL PROPERTIES OF OXYGENASES

Oxygenases that catalyze the incorporation of only one atom of dioxygen into substrates (S) are termed monooxygenases, also referred to as *mixed-function oxygenases*. The second atom of dioxygen is reduced to H_2O either by the substrates (S) themselves or by a cosubstrate reductant (H_2X):

$SH_2 + O_2 = SO + H_2O$ (catalyzed by an internal monooxygenase)
$S + O_2 + H_2X = SO + H_2O + X$ (catalyzed by an external monooxygenase).

Oxygenases that catalyze the incorporation of both atoms of dioxygen in their substrates are known as dioxygenases. In this article, we discuss two types of dioxygenases, aromatic-ring dioxygenases and aromatic-ring-cleavage dioxygenases. The aromatic-ring dioxygenases incorporate two hydroxyl groups on the aromatic ring at the expense of dioxygen and NAD(P)H, e.g.:

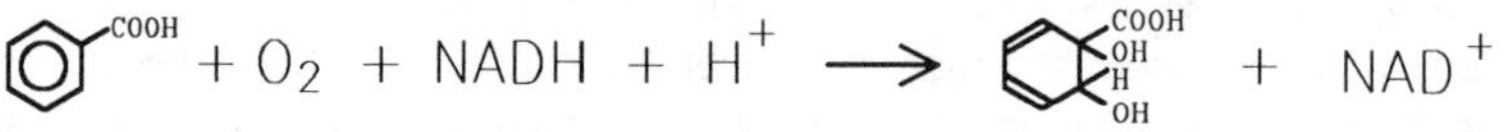

COOH $+ O_2 + NADH + H^+ \longrightarrow$ COOH OH H OH $+ NAD^+$

The aromatic-ring-cleavage dioxygenases open the aromatic ring by incorporating two atoms of dioxygen into substrates, e.g.:

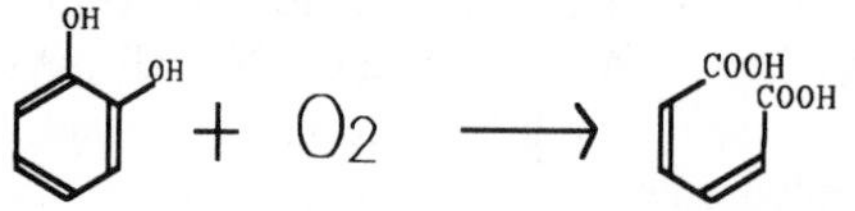

OH OH $+ O_2 \longrightarrow$ COOH COOH (Intradiol cleavage)

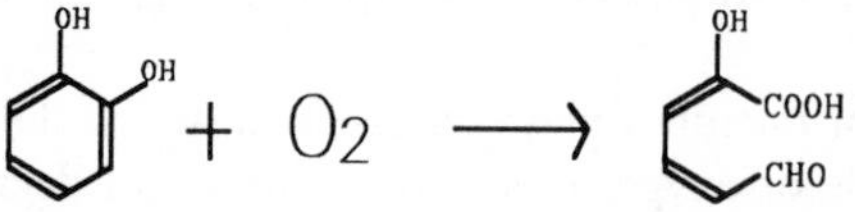

OH OH $+ O_2 \longrightarrow$ OH COOH CHO (extradiol cleavage)

Both monooxygenases and dioxygenases require cofactors capable of reacting with dioxygen because concerted reactions are spin-forbidden between paramagnetic dioxygen, with two unpaired electron spins, and carbon in organic compounds, which exists in a singlet state. Reactions of dioxygen with transition metals, which have incompletely filled *d* orbitals, are not spin-forbidden. Diamagnetic metal-oxygen complexes may be formed that can subsequently react with organic substrates (44).

Transition metals, such as iron, serve as catalysts in numerous oxygenases. Some oxygenases utilize nonheme and non-iron-sulfur Fe(II) or Fe(III), whereas others are heme enzymes. In some cases, other transition metals such as manganese, copper, and cobalt are used to bind dioxygen. Alternatively, or in addition to metal ions, flavins or pteridines may serve as cofactors in monooxygenases. These cofactors can exist in semiquinone forms in which the spin of one electron can be inverted (29). The semiquinones of flavins and pteridines can react with dioxygen, resulting in a covalently bound peroxide intermediate. In certain reactions initiated by Fe(II)-containing monooxygenases, α-ketoglutarate acts as a decarboxylating cosubstrate and ascorbic acid reduces the metal-containing catalyst (78).

AROMATIC-RING HYDROXYLASES

In the aerobic degradation of aromatic hydrocarbons, substrates are degraded via the cleavage of the aromatic ring by dioxygenases. Substrates of the aromatic-ring-cleavage dioxygenases possess two hydroxyl groups on the aromatic ring, either *ortho* or *para* to each other. One of the first steps in the degradation of the aromatic compounds, therefore, involves the introduction of one or two hydroxyl groups on the aromatic ring (49). Introduction of single hydroxyl groups on the aromatic ring, monohydroxylation, is generally catalyzed by monooxygenases, while simultaneous introduction of two hydroxyl groups, dihydroxylation, is catalyzed by dioxygenases. The majority of monooxygenases catalyzing monohydroxylation of the aromatic ring of substituted phenols are single-component enzymes, although multicomponent monooxygenases, such as phenol and toluene-4 monooxygenase have also been found.

p-*Hydroxybenzoate Hydroxylase, a Flavoprotein Monooxygenase*

The structure and reaction mechanisms of *p*-hydroxybenzoate hydroxylase have been extensively characterized (22, 117). Its reaction cycle is initiated by binding the substrate to the enzyme. The substrate-binding provokes a conformational change that increases the rate of the hydride transfer from

NADPH to the N5 position of the flavin isoalloxazine ring 10^5-fold. NADPH is oxidized and $FADH_2$ is formed. Dioxygen then binds to $FADH_2$ and is reduced by two single-electron transfers. The first electron transferred from the $FADH_2$ to dioxygen may produce a superoxide anion that may then be transformed to 4a-hydroperoxy-flavin by the second electron transfer. Unreactive dioxygen is thus converted to reactive 4a-hydroperoxy-flavin, which attacks the substrate aromatic ring.

In the next step, 4a-hydroperoxy-flavin is converted to a short-lived intermediate, and the substrate is hydroxylated. The chemical structure of this short-lived intermediate has not yet been identified (131, 132). If a substrate analog that does not possess an electron-donating substituent on the aromatic ring, such as benzoate or 6-hydroxynicotinate, is bound to the enzyme, the enzyme reaction is uncoupled: 4a-hydroperoxy-flavin is formed, but the substrate is not hydroxylated. The peroxide species on FAD then slowly decays to form hydrogen peroxide. This observation suggests that the hydroxylation reaction proceeds via an electrophilic attack of activated oxygen on the 3-position of the substrate. The phenolate form of the substrate is believed to be accessible to the electrophilic attack of activated oxygen (22). The products of the hydroxylation are 3,4-dihydroxybenzoate and flavin 4a-hydroxide. The latter compound is unstable, and spontaneously converts to FAD (116).

The 4-hydroxyl group of the substrate is hydrogen-bonded to the hydroxyl group of Tyr201 of the enzyme, which also hydrogen bonds with the hydroxyl group of Tyr385. Site-directed mutagenesis was used to examine the role of this hydrogen-bond network by substituting phenylalanine residues for each of these tyrosines. Both substitutions reduced the reduction rate of FAD by NADPH. In addition, substrate hydroxylation was uncoupled in the Tyr201-Phe mutant enzyme. Tyr201 therefore is assumed to stabilize the phenolate form of the substrate. The Tyr385-Phe mutant not only hydroxylates *p*-hydroxybenzoate, but also 3,4-dihydroxybenzoate. Therefore, Tyr385 appears to be a determinant of the substrate specificity (22).

Structural Relationships of Aromatic Flavoprotein Monooxygenases

The flavoprotein monooxygenases, exemplified by *p*-hydroxybenzoate hydroxylase, are classified into several subgroups according to their subunit sizes. 4-Hydroxybenzoate, 4-hydroxyphenylacetate 3, and salicylate hydroxylases are approximately 45 kilodaltons (kDa) in size (112, 116, 150), whereas other hydroxylases weigh between 60 and 80 kDa (100, 105, 151) (Table 1). Most flavoprotein monooxygenases incorporate a new hydroxyl group *ortho* to the existing hydroxyl group. Enzymes that catalyze *para* hydroxylation are also known. 3-Hydroxybenzoate-6-hydroxylase from

Table 1 Families of hydroxylase components

Enzyme	Organism or plasmid	Subunit size, cofactor	Reference
Aromatic-ring monooxygenases			
p-Hydroxybenzoate hydroxylase family		45 kDa, FAD	
p-Hydroxybenzoate 3-hydroxylase (EC 1.14.13.2)	*Pseudomonas fluorescens*		116
Salycilate hydroxylase (EC 1.14.13.1)	NAH7		149
Phenol 2-hydroxylase family		65 kDa	
2-4-Dichlorophenol 6-monooxygenase (EC 1.14.13.20)	pJP4		105
Phenol 2-hydroxylase (EC 1.14.13.7)	*Pseudomonas* EST1001		100
Unclassified flavoprotein monooxygenases			
o-Nitrophenol hydroxylase	*Pseudomonas putida* B2	65 kDa, FAD?	151
4-Hydroxyphenylacetate	*Pseudomonas putida*	44 kDa, FAD	112
3-Hydroxylase (EC 1.14.13.3)			
2,4-Dichlorophenoxyacetate monooxygenase	pJP4	32 kDa	127
3-Hydroxybenzoate-6-hydroxylase	*Pseudmonas cepacia*	44 kDa, FAD	140
3-Hydroxybenzoate-6-hydroxylase	*Micrococcus* sp.	70 kDa, FAD	111
Multicomponent phenol 2-hydroxylase family		Similarity to methane monooxygenase 34 kDa + 10 kDa + 58 kDa + 10 kDa, Fe(II)	
Phenol 2-hydroxylase (EC 1.14.13.7)	*Pseudomonas* CF600		98
Toluene 4-monooxygenase	*Pseudomonas mendocina*		148
Aromatic-ring dioxygenases			
Benzene 1,2-dioxygenase family		(50 kDa, 20 kDa)$_n$, Fe(II) + Rieske-type [2Fe-2S]	
Benzene 1,2-dioxygenase (EC 1.14.12.3)	*Pseudomonas putida*		56

Table 1 *Continued*

Enzyme	Organism or plasmid	Subunit size, cofactor	Reference
Toluene 2,3-dioxygenase	*Pseudomonas putida* F1		154
Benzoate 1,2-dioxygenase	*Acinetobacter calcoaceticus*		94
Toluate 1,2-dioxygenase	TOL		46
Naphthalene 1,2-dihydroxylase	NAH		66
Unclassified aromatic-ring dioxygenases			
Phthalate 4,5-dioxygenase (EC 1.14.12.7)	*Pseudomonas cepacia*	(50 kDa)$_4$, Fe(II) + Rieske-type [2Fe-2S]	8
4-Sulfonobenzoate 3,4-dioxygenase	*Comamonas testosteroni*	(50 kDa)$_2$, Fe(II) + Rieske-type [2Fe-2S]	71
4-Chlorophenylacetate 3,4-Dioxygenase	*Pseudomonas* CBS	(50 kDa)$_3$, Rieske-type [2Fe-2S]	80
Aromatic-ring-cleavage dioxygenases			
Catechol 2,3-dioxygenase I family		34 kDa, Fe(II)	
Catechol 2,3-dioxygenase I (EC 1.13.11.2)	TOL		88
2,3-Dihydroxybiphenyl dioxygenase	*Pseudomonas paucimobilis*		130
1,2-Dihydroxynaphthalene dioxygenase	NAH		45
Protocatechuate 4,5-dioxygenase family			
Protocatechuate 4,5-dioxygenase (EC 1.13.11.8)	*Pseudomonas paucimobilis*	(34 kDa, 15 kDa)$_n$, Fe(II)	91
Catechol 2,3-dioxygenase II (EC 1.13.11.2)	*Alcaligenes eutrophus*	(34 kDa)$_n$	60
Homoprotocatechuate dioxygenase	*Escherichia coli*	(30 kDa)$_n$	114
Catechol 1,2-dioxygenase family			
Catechol 1,2-dioxygenase I (EC 1.13.11.1)	*Pseudomonas arvilla*	(32 kDa, 30 kDa)$_n$, (30 kDa)$_n$, (32 kDa)$_n$	
Catechol 1,2-dioxygenase II (EC 1.13.11.1)	pP51	(28 kDa)$_n$	137

Protocatechuate 3,4-dioxygenase (EC 1.13.1.3)	*Acinetobacter calcoaceticus*	(23 kDa, 27 kDa)$_n$	52
Gentisate 1,2-dioxygenase (EC 1.13.11.4)	*Comamonas testosteroni*	(40 kDa)$_4$, Fe(II)	50
<u>Alkyl group hydroxylase</u>			
Alkane hydroxylase family		41 kDa, Membrane-bound, Fe(II)	
Alkane hydroxylase (EC 1.14.15.3)	ALK		65
Xylene monooxygenase	TOL		129
Methane monooxygenase	*Methylosinus trichosporium*	Component A = (54 kDa, 43 kDa, 23 kDa)$_2$, component B (= regulatory protein 16 kDa), binuclear iron cluster on component A	27
Vanillate demethylase	*Pseudomonas* sp.	(37 kDa)$_n$	13
Unclassified			
4-Methoxybenzoate monooxygenase (putidamonooxin) (EC 1.14.99.15)	*Pseudomonas putida*	(40 kDa)$_3$, Fe(II) + Rieske-type [2Fe 2S]	142
Toluene sulfonate methyl-monooxygenase	*Comamonas testosteroni*	(43 kDa)$_3$ or (43 kDa)$_4$	72
<u>Cytochrome P-450 family</u>			
Camphor 5-monooxygenase (EC 1.14.15.1)	CAM		109
Mitochondrial P-450	Eukaryotes		92
Microsomal P-450	Eukaryotes		92
Cytochrome P-450$_{bm-3}$	*Bacillus megaterium*	120 kDa, bifunctional (hydroxylase + electron-transport components)	68
<u>Aromatic amino acid hydroxylase family</u>		Fe(II), biopterin	
Phenylalanine 4-hydroxylase (EC 1.14.16.1)	Eukaryotes	52 kDa	35
Tyrosine 3-hydroxylase (EC 1.14.16.2)	Eukaryotes	59 kDa	35
Tryptophane hydroxylase (EC 1.14.16.4)	Eukaryotes	51 kDa	35

Pseudomonas cepacia is a monomer of 44 kDa whereas that from *Micrococcus* sp. has a molecular weight of 70,000 (111, 140).

Salicylate hydroxylase encoded on the NAH7 plasmid of *Pseudomonas putida,* which catalyzes the oxidative decarboxylation of 2-hydroxybenzoate, shares 25% amino acid sequence identity with *p*-hydroxybenzoate hydroxylase (149). The strongest sequence conservation is observed in and adjacent to the FAD-binding regions.

Phenol hydroxylase from the yeast *Trichosporon cutaneum* (118), 2,4-dichlorophenol hydroxylase encoded on the pJP4 plasmid of *Alcaligenes eutrophus* (105), and phenol hydroxylase from *Pseudomonas* species ES-T1001 (100) are 20–30 kDa heavier than *p*-hydroxybenzoate and salicylate hydroxylases. The dichlorophenol and phenol hydroxylases share 46% amino acid sequence identity, indicating that they are evolutionarily related.

Local sequence similarities between the group of dichlorophenol and phenol hydroxylases and that of *p*-hydroxybenzoate and salicylate hydroxylases are observed in two regions. In the first of these, at the amino-terminal regions of all four enzymes, a conserved amino acid sequence pattern forms an ADP-binding fingerprint associated with a $\beta\alpha\beta$ fold (144). In *p*-hydroxybenzoate hydroxylase, the conserved amino-terminal region binds the ADP portion of FAD (116). The global sequence similarity between *p*-hydroxybenzoate and salicylate hydroxylases indicates that the corresponding region in salicylate dehydrogenase also binds the ADP portion of FAD. In the dichlorophenol and phenol hydroxylases, however, the amino terminal $\beta\alpha\beta$ ADP-binding fold may be involved in NADPH rather than FAD binding because highly conserved asparatate or glutamate residues associated with NADH or FAD binding are absent in these domains.

The second region of homology among these hydroxylases, which corresponds to the amino acid residues, Met276 to Ser329, in *p*-hydroxybenzoate hydroxylase, may be involved in FAD binding because, in the three-dimensional structures of *p*-hydroxybenzoate hydroxylase, this region contains a FAD-binding β-strand (19). The sequence homology among the flavoprotein monooxygenases extends beyond this β-strand. Except for the two indicated regions, the amino acid sequences of phenol and dichlorophenol hydroxylases are different from those of salicylate and *p*-hydroxybenzoate hydroxylases. Possible evolutionary implications of the observed local sequence similarities in these enzymes are discussed below.

Multicomponent Phenol and Toluene-4 Monooxygenases

Two multicomponent aromatic-ring monooxygenases catalyze the first hydroxylation step in the degradation of phenol by *Pseudomonas* species CF600, and of toluene in *Pseudomonas mendocina* KR1 (93, 143). Phenol hydroxylase converts phenol into catechol, while toluene-4 monooxygenase catalyzes the insertion of one hydroxyl group into the *para* position on the aromatic ring. These reactions are similar to those catalyzed by flavoprotein monooxygenases, but unlike the single-component flavoproteins, these multicomponent enzymes are structurally related to methane monooxygenase, which contains a catalytic binuclear iron center (see below).

Genetic and biochemical analysis indicated that five different polypeptides are required for the phenol hydroxylase activity in vitro, whereas six polypeptides are required for the growth on phenolic substrates (108). One of these polypeptides has been purified and was characterized as an electron-transport component (see below). Genetic complementation and DNA sequencing analysis showed that, at least five, and probably six polypeptides are required for the activity of toluene-4 monooxygenase. Two of these, encoded by *tmoA* and *tmoE,* are presumably part of the terminal oxygenase (148), while the product of *tmoC* may be an electron-transport component (see below). The functions of two other polypeptides have not yet been identified. Interestingly, three components of toluene-4 monooxygenase were found to share significant sequence identity with three components of phenol hydroxylase (148).

AROMATIC-RING DIOXYGENASES

Bacterial aromatic-ring dioxygenases catalyze the dihydroxylation of the aromatic ring (30). In this reaction, both atoms of dioxygen are incorporated into substrates, and two hydroxyl groups are formed on the aromatic ring. The product is (substituted) *cis*-1,2-dihydroxycyclohexadiene, which is subsequently converted to (substituted) benzene glycol by a dehydrogenase (93). Components of bacterial aromatic-ring dioxygenases fall into two different functional classes: hydroxylase components and electron-transport components. In this section, the hydroxylase components are described; their electron-transport components are discussed separately in a later section.

All hydroxylase components of aromatic-ring dioxygenases are oligomers composed of either a single or two different tightly associated subunits in an α_n or $(\alpha\beta)_n$ configuration (Table 1). They possess two common cofactors, a [2Fe-2S] iron-sulfur center and one mononuclear nonheme iron, both of which are associated with the α-subunit in the $(\alpha\beta)_n$-type enzymes (21, 128, 147). The [2Fe-2S] iron-sulfur center has the spectral properties of the so-called Rieske iron-sulfur center that is found in respiratory and photosynthetic electron-transport proteins. In contrast to plant-type ferredoxins, in which four cysteine residues coordinate a [2Fe-2S] redox center, one iron of the Rieske-type [2Fe-2S] center coordinates with two cysteines while the other coordinates with two histidine residues. Upon reduction of the

Rieske-type [2Fe-2S] center, the iron coordinated with histidine is reduced (38).

In the $(\alpha\beta)_n$-type hydroxylases, the sizes of the two subunits, α and β, are roughly 50 and 20 kDa, respectively. A comparison of the amino acid sequences of the α- and β-subunits of benzoate, toluate, benzene, toluene, and naphthalene dioxygenases suggests that the α- and β-subunits are each derived from common ancestors (94). Alignment of the five α-subunits revealed five invariant histidines, two invariant cysteines, and two invariant tyrosines. The Rieske-type [2Fe-2S] cluster in the α-subunits may be coordinated with the two invariant cysteines and two of the invariant histidines. This [2Fe-2S] redox center probably accepts electrons from an electron-transport component. The binding of dioxygen is believed to be mediated by a mononuclear nonheme iron associated with the α-subunit (147). Some of the invariant tyrosines and histidines that are not involved in the [2Fe-2S] binding may coordinate the mononuclear iron.

The β-subunits share considerably less sequence similarity, although a global amino acid sequence pattern has been conserved. The lower degree of homology among the five β-subunits suggests that these subunits may not be directly involved in the common catalytic functions of the dioxygenases. Indeed, genetic studies indicate that the β-subunit may be important in the determination of substrate specificity (47). This observation suggests that the catalytic center of the hydroxylase component may be formed between the α- and β-subunits, with the α-subunit facing the oxidizable carbons of the substrate and the β-subunit recognizing the substrate structure (94).

The hydroxylase components of 4-sulphobenzoate-3,4-dioxygenase from *Comamonas testosteroni* (71), of phthalate dioxygenase from *Pseudomonas cepacia* (8), and of 4-chlorophenylacetate 3,4-dioxygenase from *Pseudomonas* sp. CBS (80) are all $(\alpha)_n$-type proteins with a subunit of approximately 50 kDa, similar to the α-subunits of the above-mentioned $(\alpha\beta)_n$ type enzymes. Unfortunately, no amino acid sequence information is available for these enzymes.

The hydroxylase component of 4-methoxybenzoate monooxygenase (putidamonooxin, see discussion below) also contains a Rieske-type [2Fe-2S] center and one nonheme iron (20).

AROMATIC-RING-CLEAVAGE DIOXYGENASES

Nearly all bacterial pathways for the degradation of aromatic compounds transform initial substrates into intermediates that carry two or more hydroxyl groups on the aromatic ring. These intermediates are substrates of aromatic-ring-cleavage dioxygenases. If two of the hydroxyl groups on a substrate are in the *ortho* position relative to each other, the ring fission by the aromatic-ring-cleavage dioxygenases occurs either between the two hydroxyl groups (intradiol cleavage) or at a bond proximal to one of the two hydroxyl groups (extradiol cleavage). For gentisate and homogentisate that carry two hydroxyl groups in the *para* position relative to each other, the ring cleavage occurs between the carboxyl or acetyl substituent and the proximal hydroxyl group.

Extradiol Cleavage Enzymes

The first aromatic-ring-cleavage dioxygenase for which the primary structure was determined is catechol 2,3-dioxygenase encoded by the *xylE* gene on the TOL catabolic plasmid (88). This enzyme consists of four identical subunits of 32 kDa, and contains one catalytically essential Fe(II) ion per subunit. The reaction product, 2-hydroxymuconic semialdehyde (or a substituted derivative), is yellow. Because of the intense color production, the structural gene for this enzyme, *xylE*, has been used as a reporter gene in molecular biology (41). The enzyme reaction proceeds by an ordered bi-uni mechanism: catechol first binds to the enzyme followed by the binding of dioxygen to form a ternary complex; the aromatic ring is then cleaved to produce 2-hydroxymuconic semialdehyde (99). The substrate range of catechol 2,3-dioxygenase is relatively broad: this enzyme oxidizes 3-methyl-, 3-ethyl-, 4-methyl-, and 4-chlorocatechol. 3-Chloro- and 4-ethylcatechol, in contrast, are not efficiently oxidized by this enzyme. 4-Ethylcatechol is a suicide inhibitor: this compound inactivates the enzyme in a mechanism-based fashion by oxidizing the Fe(II) cofactor to Fe(III). Mutant catechol 2,3-dioxygenases able to metabolize 4-ethylcatechol were obtained, and single amino acid substitutions were responsible for this phenotypic change (113; A. Wasserfallen, P. Cerdan, M. Rekik, K. N. Timmis & S. Harayama, in preparation). 3-Chlorocatechol is also a suicide inhibitor of catechol 2,3-dioxygenase (6). One mutant enzyme resistant to inactivation by 3-chlorocatechol was isolated in which the affinity to 3-chlorocatechol decreased but that to 3-methylcatechol increased.

The electronic structure of Fe(II) in the active site of catechol 2,3-dioxygenase was calculated from the data of magnetic circular dichroism in the near-infrared spectrum of this enzyme (77). Fe(II) apparently has five-coordinate square-pyramidal geometry. Catechol may bind as a bidentate ligand, occupying the axial and one equatorial ligation positions in the active site of the enzyme-substrate complex. The binding of catechol triggers a conformational change such that azide, an analog of dioxygen, can bind to an equatorial ligation site.

The comparison of the amino acid sequences of four catechol 2,3-dioxygenases from *Pseudomonas,* one 1,2-dihydroxynaphthalene dioxygenase, and three 2,3-dihydroxybiphenyl dioxygenases has revealed that they are members of the same superfamily (45). Catechol 2,3-dioxygenase from *A. eutrophus*, however, has a primary sequence quite different from those of the above-mentioned family (60). Therefore, catechol 2,3-dioxygenases appear to have at least two independent origins.

Protocatechuate 4,5-dioxygenase catalyzes the extradiol cleavage of protocatechuate. The enzyme consists of an equal number of two different subunits, α and β, 18 and 34 kDa, respectively, and its quaternary structure may be $(\alpha\beta)_2Fe^{2+}$ (4). The amino acid sequences of these subunits are not related to those of the main catechol 2,3-dioxygenase family, but the β-subunit of protocatechuate 4,5-dioxygenase (97) exhibits sequence similarity to that of catechol 2,3-dioxygenase from *A. eutrophus*. The Fe(II) environment of protocatechuate 4,5-dioxygenase from *C. testosteroni* was investigated using EPR spectroscopy (4). In a hypothetical reaction sequence, electron delocalization in the ternary complex, enzyme-Fe(II)-O-O, is assumed to polarize dioxygen, preparing the distal oxygen atom for nucleophilic attack on the aromatic ring of the substrates. The iron-peroxy-substrate intermediate, enzyme-Fe(II)-O-O-S, thus produced initiates a reaction sequence resulting in the ring fission of the substrate.

Homoprotocatechuate dioxygenase is a third kind of extradiol cleavage enzyme whose sequence indicates that it constitutes a discrete class of the extradiol cleavage enzymes (114).

Intradiol Cleavage Enzymes

In contrast to the extradiol cleavage enzymes, which contain Fe(II) as a cofactor, the intradiol cleavage enzymes, catechol 1,2-dioxygenase and protocatechuate 3,4-dioxygenase, contain a nonheme, non-iron-sulfur Fe(III) as a prosthetic group. Catechol 1,2-dioxygenases from many bacteria consist of nonidentical α- and β-subunits, $[\alpha\beta\text{-}Fe^{3+}]_n$, whereas those from other bacteria consist of single polypeptide species $[\alpha\alpha\text{-}Fe^{3+}]_n$. One *Pseudomonas* species produces two types of catechol 1,2-dioxygenase polypeptides, α and β, and three isozymes, $\alpha\alpha$, $\alpha\beta$ and $\beta\beta$ (87). Catechol 1,2-dioxygenase exhibits little or no activity towards chlorocatechols and is also called catechol dioxygenase I in order to distinguish it from chlorocatechol 1,2-dioxygenase (catechol dioxygenase II) found in degradative pathways for chlorinated aromatic compounds. The latter enzyme exhibits broader substrate specificity and cleaves both catechol and chlorocatechols. Type II catechol 1,2-dioxygenases have increased specificity for halogenated substrates. Only subtle differences in enzymatic activities have been found between type II catechol 1,2-dioxygenases (12, 137). The type I and II enzymes show a global sequence similarity (52).

Protocatechuate 3,4-dioxygenases thus far characterized contain equal numbers of two different subunits, α and β, and form different quaternary structures of $(\alpha\beta)_n$ (n = 3–12). The similarity in the primary sequences of catechol 1,2-dioxygenases I and II and of the α- and β-subunits of protocatechuate 3,4-dioxygenases indicates that these enzymes are derived from a common ancestor. The three-dimensional structure of the $(\alpha\beta)_2$-type protocatechuate 3,4-dioxygenase has been determined (102). The active site of the enzyme is located at the interface of the two structurally related subunits. Ferric ion is coordinated in the β-subunit to two tyrosine and two histidine residues in a near trigonal bypyramidal configuration. The fifth iron-coordination position is occupied by a solvent molecule. The α-subunit does not bind iron, and only two of the residues corresponding to the iron ligands on the β-subunit have been conserved. Protocatechuate-3,4-dioxygenase from *Brevibacterium fuscum* was used to examine changes in the Fe(III) coordination environment in the course of catalysis, leading to conclusions similar to those from the X-ray crystallographic study (134). Four amino acid ligands and most probably one hydroxyl ion are coordinated with Fe(III) in the free enzyme. Upon binding of the substrate to the active-site Fe(III), the hydroxyl ion and one histidine ligand may be displaced. Dioxygen subsequently attacks the activated aromatic ring of the substrate leading to aromatic-ring fission and the introduction of two oxygen atoms into the substrate.

Gentisate 1,2-Dioxygenase

Gentisate is a metabolite formed from anthranilate, β-naphthol, 3- and 4-hydroxybenzoate, salicylate, flavonones, and naphthalene disulfonate. Its aromatic-ring-cleavage enzyme, gentisate 1,2-dioxygenase, has been purified from *Pseudomonas acidovorans* and *C. testosteroni* and consists of a single polypeptide species of about 40 kDa with a quaternary structure of $(\alpha Fe^{2+})_4$. This enzyme contains a Fe(II) cofactor. Results of EPR studies suggest that H_2O coordinates with Fe(II) and that gentisate binds directly to the iron cofactor through two coordination bondings using one oxygen atom of the carbon-1 carboxylate and the oxygen atom of the carbon-2 hydroxyl group (50, 51).

ALKYL-GROUP HYDROXYLASES

Hydroxylation of an alkyl group is often the first step in the complete degradation of organic compounds. Such reactions are usually catalyzed by monooxygenases.

$$R\text{-}CH_3 + NAD(P)H + O_2 + H^+ = R\text{-}CH_2OH + NAD(P)^+ + H_2O$$

Alkyl-group monooxygenases are usually multicomponent and consist of a hydroxylase component and one or two electron-transport component(s). The hydroxylation occurs on the hydroxylase component while two reducing equivalents, [H], required for the hydroxylation reaction are generated by the electron-transport component. One can classify alkyl-group monooxygenases in several distinct categories according to their primary structures and/or biochemical properties. Cytochrome P-450s (discussed in the next section) comprise one of these categories. In this section, we describe hydroxylase components of alkyl-group hydroxylases other than cytochrome P-450.

Xylene Monooxygenase and Alkane Hydroxylase

The TOL plasmid of *Pseudomonas putida* encodes a well-characterized catabolic pathway for the degradation of toluene/xylenes. The first enzyme of the pathway is xylene monooxygenase, which oxidizes toluene and xylenes to (methyl)benzyl alcohols. This enzyme is composed of two different polypeptides that are encoded by *xylM* and *xylA* (48, 129). Nucleotide sequence analysis of the *xylMA* genes indicated that one of the gene products (XylM) shares significant amino acid sequence identity (27%) with the membrane-associated hydroxylase component of the alkane hydroxylase from *P. putida* (*oleovorans*) (65). The sequence of XylA indicates that it plays a role in electron transport (see below).

Alkane hydroxylase from *P. putida* catalyzes the hydroxylation of the terminal carbon of alkanes and the omegahydroxylation of fatty acids. This enzyme consists of three different polypeptides: the membrane-bound hydroxylase component encoded by *alkB* and two other polypeptides that together constitute an electron-transport chain (see below). The AlkB hydroxylase component has been purified, and Fe(II) and phospholipids were found to be required for its activity (115). Similarly, the XylM protein, the hydroxylase component of xylene monooxygenase, is membrane-bound and requires Fe(II) for its activity (J. P. Shaw & S. Harayama, unpublished).

Despite the observed structural similarities between the hydroxylase components of both enzymes, alkane hydroxylase and xylene monooxygenase do not have any substrates in common. Alkane hydroxylase seems to have a broader substrate specificity than xylene monooxygenase (145). All substrates are hydroxylated at the terminal carbon atom (62).

Methane Monooxygenase

Soluble methane monooxygenases from two obligate methylotrophs, *Methylococcus capsulatus* (Bath) and *Methylosinus trichosporium* OB3b, convert methane to methanol at the expense of NADH. These enzymes exhibit broad substrate specificity; they can hydroxylate a variety of alkanes, haloalkanes, alkenes, and aromatic and heterocyclic compounds (14, 34). They consist of three components, A, B, and C, which have been purified. Component A, the hydroxylase component of the enzyme, comprises three subunits: α (54 kDa), β (44 kDa), and γ (20 kDa) with the configuration of $(\alpha\beta\gamma)_2$. The EPR spectrum of the partially reduced component showed that two irons are connected via a μ-oxo bridge (bridge made by a carboxylic group: Fe-O-C-O-Fe). Each subunit contains two binuclear iron clusters (27).

NADH oxidation could be achieved by combining component A and component C (the electron-transport component, see below) in vitro. This preparation, however, did not catalyze the hydroxylation of substrates. For the coupling between NADH oxidation and substrate hydroxylation, component B (16 kDa) is required. Addition of component B prevents uncoupled NADH oxidation in the absence of substrates. Component B has no redox-active metals or cofactors, and therefore is considered to be an effector of component A (28).

The reaction cycle of the (A+B+C) holoenzyme is initiated by substrate binding to component A. The substrate binding changes the environment of the iron center in component A, and hence its redox potential. The redox potential of the [Fe(II) Fe(II)/Fe(III) Fe(II)] couple in component A is < -200 mV in the absence of the substrate, but becomes >150 mV in the presence of the substrate (70). The substrate binding therefore facilitates the reduction of the iron center by component C. A change in the redox potential of the catalytic iron after the substrate-binding is also observed in P-450_{cam}. After the reduction of the μ-oxo-bridged binuclear iron center by a single electron transfer, dioxygen binds to the reduced iron center. A second one-electron transfer to the iron-dioxygen complex then may yield an activated oxygen species: [Fe(II) Fe(II)-O-O-H]. This activated oxygen species is assumed to abstract a hydrogen on a methyl group of the substrate, forming a methyl-radical intermediate that will subsequently be oxygenated (70).

A μ-oxo bridged binuclear iron center is not only found in methane monooxygenase, but also in ribonucleotide reductase, hemerythrin, and purple acid phosphatase (11). The iron center of hemerythrin may be functionally different from those in other enzymes: hemerythrin reversibly binds dioxygen, whereas ribonucleotide reductase and methane monooxygenase bind and activate dioxygen irreversibly (76). Nevertheless, the studies on the binuclear iron centers in these enzymes may provide further information concerning the activity of methane monooxygenase.

The broad-substrate range of methane monooxygenase, like many cytochrome P-450s discussed below, is achieved at the expense of regiospecificity. Complex substrates are ususally converted into more than one oxidation product, some of which may be produced in excess over others. Such preferential hydroxylation at one position on the substrate molecule relative to another may result from either the architecture of the substrate-binding site, chemical constraints imposed by the reaction mechanism, or the combination of the two (34).

As mentioned earlier, some components of methane monooxygenase share sequence similarities with components of phenol hydroxylase, encoded by the *dmp* genes, and also with toluene-4 monooxygenase, encoded by the *tmo* genes (125). The sequences of the *tmoA* and *tmoE* and of the *dmpN* and *dmpL* gene products are similar to those of two subunits of methane monooxygenase encoded by the *mmoX* and *mmoY* genes. Because methane monooxygenase can hydroxylate the aromatic ring, it is not surprising to find that these three enzymes are derived from a common ancestor. In this respect, it would be interesting to determine whether the hydroxylase components of phenol hydroxylase and toluene-4 monooxygenase contain binuclear iron centers.

4-Toluene Sulfonate Methyl Monooxygenase and 4-Methoxybenzoate Demethylase

4-Toluene sulfonate methyl monooxygenase from *C. testosteroni* catalyzes the hydroxylation of the alkyl groups on alkylphenylcarboxylates and alkylphenylsulfonates (72). This enzyme consists of a 36-kDa electron-transport component and a 43-kDa hydroxylase component. The hydroxylase component contains a Rieske-type iron-sulfur center. The best substrates of the enzyme are *p*-toluate and 4-toluene sulfonate, but 4-methoxybenzoate is also hydroxylated and yields 4-hydroxybenzoate and formaldehyde (72). This reaction resembles the oxidative demethylation of aliphatic substrates by alkane hydroxylase (62) as well as enzymatic demethylation of aromatic methyl ethers described below.

4-Methoxybenzoate demethylase from *P. putida* is a monooxygenase and hydroxylates the methyl group of the substrate. The intermediate thus formed is unstable and spontaneously transformed to 4-hydroxybenzoate and formate. This enzyme consists of two subunits: the hydroxylase component called putidamonooxin and the electron-transport component called NADH-putidamonooxin reductase. The enzyme resembles 4-toluene sulfonate methyl monooxygenase in many respects: the subunit structures, the presence of FMN and ferredoxin in the electron-transport components (see below), the sizes of the electron-transport and hydroxylase components, and the presence of Rieske-type iron-sulfur centers and of one essential mononuclear nonheme iron in the hydroxylase components. The mononuclear Fe(II) of putidamonooxin binds dioxygen. In uncoupled reactions of the enzyme in the presence of "bad" substrates, hydrogen peroxide forms. Therefore, iron-peroxo-complex, $(FeO_2)^+$, is inferred to be an activated oxygen intermediate (142).

Vanillate demethylase, which is composed of two polypeptides, catalyzes a reaction in a manner similar to 4-toluene sulfonate methyl monooxygenase and 4-methoxybenzoate demethylase. Although the amino acid sequences of these components are known, they have not been biochemically characterized (13). 2,4-Dichlorophenoxyacetate monooxygenase catalyzes oxidative removal of acetate by a mechanism that may be similar to that of demethylases.

2,4-Dichlorophenoxyacetate monooxygenase, however, is composed of a single polypeptide species (127).

CYTOCHROME P-450 SYSTEMS

Cytochrome P-450 systems are multicomponent enzymes consisting of one terminal hydroxylase heme protein and one or two electron-transport component(s). They catalyze the hydroxylation of a variety of substrates (SH):

$$SH + O_2 + NAD(P)H + H^+ = SOH + H_2O + NAD(P)^+$$

The name cytochrome P-450 derives from the characteristic 450-nm absorption maximum of the enzyme bound to carbon monooxide. Many cytochrome P-450s have been identified from prokaryotes, yeast, fungi, plants, and insects, but most have been identified from mammals (92). Cytochrome P-450s are involved in the specific transformation of endogenous compounds, such as the biosynthesis of vitamin D and of sex-steroid and corticoid hormones (36, 101) or in the catabolism of a variety of natural and xenobiotic compounds. Enzymes that carry out the specific transformations have narrow substrate specificities, and their synthesis is tightly regulated. Enzymes that carry out the catabolic reactions are generally nonspecific, and these reactions protect mammals against the majority of chemicals they may encounter (36, 91, 106). Unfortunately, some chemicals are converted to toxic and/or carcinogenic compounds by several species of cytochrome P-450s (3).

Eukaryotic cytochrome P-450s located in the endoplasmic reticulum and mitochondria are membrane-bound, but bacterial P-450s are soluble (153). Membrane localization was once assumed to facilitate the binding of the typically water-insoluble substrates, but more recent research suggests that the substrate-binding site of membrane-bound enzymes faces the cytosol rather than the surface of the endoplasmic reticulum membrane (18). P-450_{cam}, the 414–amino acid bacterial enzyme that catalyzes the hydroxylation of camphor to 5-exo-hydroxycamphor has been crystallized and its three-dimensional structure determined (107, 109, 110). Amino acid sequence comparisons indicate that this enzyme diverged from other members of the cytochrome P-450 superfamily. Nevertheless, its three-dimensional folding pattern has been used to identify critical amino acid residues in other cytochrome P-450s, because all members of the cytochrome P-450 superfamily may have the same general folding pattern. The amino-terminal portions of the eukaryotic enzymes that mediate attachment to the microsomal or mitochondrial membrane (18, 153) are absent in cytochrome P-450_{cam}.

Substrate Specificity

The natural substrate of P-450_{cam} binds relatively loosely to the enzyme by seven hydrophobic contacts and one hydrogen bond. Other substrates can be accommodated in the binding site, but most of them cannot hydrogen bond to the enzyme. The regiospecificity for the hydroxylation of these substrates is low, and hydrogen peroxide is produced as a byproduct because of an increased mobility of the substrates in the binding pocket that uncouples the electron transfer and the substrate hydroxylation (110). Considerable sequence variation is found in the central portion of cytochrome P-450s, corresponding to residues 180 to 320 in P-450_{cam}. This heterogeneity may reflect substrate-specificity variation among the cytochrome P-450s (91). Indeed, five of the substrate-binding residues of P-450_{cam} have been identified in this region. Using the P-450_{cam} structure as a model, investigators have tentatively identified substrate-binding residues in other P-450s (153).

The substrate specificity of two closely related P-450 enzymes, P-450_{cou7a} and P-450_{15a}, was systematically investigated (58, 69). P-450_{cou7a} hydroxylates coumarin, while P-450_{15a} hydroxylates testosterone. Despite their different substrate specificities, the amino acid sequences of these two enzymes differ only by 11 residues. An experiment in which specific amino acid replacements were used to alter P-450_{cou7a} substrate specificity revealed that one amino acid substitution was sufficient to convert the cou7a into 15a specificity.

Reaction Mechanisms

The active site of P-450_{cam} is located on the top of the iron porphyrin IX moiety. One axial iron coordination position points downwards to Cys357 and the other points upwards into the oxygen-binding site (109). The residue Cys357 is invariant among all P-450s, and its replacement by a histidine or alanine residue eliminated the enzyme activity, suggesting a specific role for this residue in catalysis (136).

The catalytic cycle is initiated by substrate binding to the enzyme, which facilitates the first one-electron reduction of heme-Fe^{3+} to heme-Fe^{2+}. The catalytic iron in the substrate-free enzyme exhibits a low redox potential (about -300 mV). Upon the binding of a substrate, however, this redox potential increases to -173 mV, enabling the catalytic iron to be reduced by its electron-transport component (36). Subsequently, dioxygen binds to the free axial coordination position of Fe^{2+}, *trans* to the cysteinyl residue. In this $(FeOO)^{2+}$ complex, dioxygen occupies a position very close to the substrate, which moves away from the catalytic iron–dioxygen complex by approximately 1 Å (109). A second one-electron reduction results in the formation of an activated peroxide species, $(FeOOH)^{2+}$. Thr252 may be in contact with the oxygen atoms because of its polar character, and it may stabilize this intermediate form. Site-directed mutagenesis of P-450_{cam} showed that the activation of dioxygen and the substrate hydroxylation are uncoupled in the absence of Thr252 (55). Thr252 may be a proton donor to $(FeOOH)^{2+}$, thereby facilitating the cleavage of the O-O bond. Another possibility is that the substitution of Thr252 changes the structure and/or the solvent channel inside the substrate-binding pocket (110). Hydroxylation is believed to proceed via a biradical recombination mechanism that results in substrate activation (S*) (37):

$$(FeOOH)^{2+} + SH \rightarrow (FeO)^{3+}\ S^* + H_2O$$

Less is known about catalysis by mammalian cytochrome P-450s, although some catalytically important residues have been identified (32). The monooxygenation reaction catalyzed by microsomal cytochrome P-450s involves the transfer of electrons from NADPH-P-450 reductase or from cytochrome b_5 (see below). Electrostatic interactions may be involved in the association of the hydroxylase and electron-transport components. Site-directed mutagenesis showed that seven lysine and arginine residues are, in fact, important in the interaction between the hydroxylase and electron-transport components of P-450_d (123).

Evolutionary Relationships

Comparison of the primary structures of 154 members of a cytochrome P-450 superfamily allowed the composition of an evolutionary tree representing the divergent evolution of these enzymes. Sequences that diverged recently are grouped together into 27 families. Interestingly, within families, intron-exon structures have been fully conserved, suggesting that rearrangement of introns occurred early in the stages of the divergence of the P-450 sequences (92).

Genes for eukaryotic P-450s are often organized in tandem, perhaps the result of gene amplification (91, 101). In such forms, gene conversion, in which a part of a gene is copied to the equivalent part of another gene, may occur. In fact, a gene-conversion mechanism has been inferred from genetic exchanges between the structural gene for a steroid hydroxylase and its pseudogene homologue (53).

AROMATIC AMINO ACID HYDROXYLASES

The mammalian aromatic amino acid hydroxylases, phenylalanine hydroxylase, tryptophan hydroxylase, and tyrosine hydroxylase, share many characteristics. All are homotetrameric and utilize the cofactor, tetrahydrobiopterin

(16), as well as Fe(II) (24). This section describes the terminal hydroxylase components of these enzymes. Their electron-transport components, quinonoid dihydropteridine reductase, are described below. Some reaction steps of these enzymes are similar to those of the previously discussed flavoprotein monooxygenases.

Regulation of Enzyme Activity by Posttranslational Modification

Phenylalanine hydroxylase catalyzes the first step in the complete oxidation of phenylalanine. Disorders of phenylalanine catabolism, due to a mutation of either phenylalanine hydroxylase or a tetrahydrobiopterin synthetic enzyme, result in phenylketonuria or hyperphenylalaninemia (85). The two related tyrosine and tryptophan hydroxylases catalyze the rate-limiting steps in catecholamine and serotonin synthesis, respectively, thereby playing key roles in neural transmission and hormone regulation (79, 95). These enzymes are subject to strict regulatory controls, which reflects the importance of roles they play.

Tyrosine hydroxylase is activated by phosphorylation, heparin, salts, and phospholipids (152). It is also feedback inhibited by the binding of catecholamines to the oxidized form of the active-site iron. The binding blocks the reduction of this Fe(III) by tetrahydrobiopterin. The dissociation of catecholamines from the enzyme is promoted by phosphorylation of serine 40, and the release of catecholamine may be a major mechanism of phosphorylation activation of this enzyme (40).

The activity of phenylalanine hydroxylase is regulated by its substrate, phenylalanine, which binds to two different sites: an effector-binding site (the activation site) at which no hydroxylation occurs, and a substrate-binding site at which it is hydroxylated (122). Reversible phosphorylation further regulates phenylalanine hydroxylase, apparently by affecting the dissociation constant of the amino acid at the activation site (33). Proteolytic cleavage of the 52-kDa phenylalanine hydroxylase with chymotrypsin yields three peptide fragments. Of these, a 36-kDa fragment, corresponding to the central region of the enzyme, is enzymatically active, having a 30-fold higher specific activity than the native enzyme. An 11-kDa amino-terminal fragment containing the regulatory phosphorylation site may constitute a domain that represses enzyme activity (57).

Reaction Mechanisms

Kinetic experiments have shown that tetrahydrobiopterin binds first to the free hydroxylases, followed by the binding of dioxygen and an amino acid (25). The binding of dioxygen to the aromatic amino acid hydroxylases may occur in two steps. Dioxygen binding to the catalytic Fe(II) may activate dioxygen, which is then followed by binding to tetrahydrobiopterin to form 4a-peroxytetrahydropterin, the rate-limiting intermediate (39). This intermediate resembles 4a-hydroperoxy-flavin, an intermediate of flavoprotein monooxygenases. The nucleophilic attack on 4a-peroxytetrahydropterin by an aromatic-ring carbon may result in cleavage of the oxygen-oxygen bond and in hydroxylation of the substrate aromatic ring (26). This reaction mechanism is very similar to that of *p*-hydroxybenzoate hydroxylase. The iron cofactor of phenylalanine hydroxylase not only interacts with tetrahydrobiopterin but also with the substrate (phenylalanine), although the significance of this interaction is not yet known (81).

Phenylalanine hydroxylase converts the substrate analog L-[2,5-H_2] phenylalanine to the corresponding 3,4-epoxide (86), although the natural substrate, phenylalanine, is not necessarily hydroxylated via an epoxide intermediate (124). The tetrahydrobiopterin-dependent phenylalanine hydroxylase from *Chromobacterium violaceum* requires Cu(II) instead of iron for enzymatic activity (104). Mechanistically, this enzyme may be different from its eukaryotic counterpart. In the *C. violaceum* enzyme, dioxygen binds to a single Cu(II) center as the first substrate. The Cu(II) center may also be the site of tetrahydrobiopterin binding.

Primary Structure of Aromatic Amino Acid Hydroxylases

Homologous sequences are unevenly distributed along the polypeptide chains of aromatic amino acid hydroxylases. The amino-terminal 104 amino acid residues of rabbit tryptophan hydroxylase exhibit only 15% sequence identity with those of rat tyrosine hydroxylase, but show about 40% identity with those of human and rat phenylalanine hydroxylases. The sequence identity between these aromatic amino acid hydroxylases exceeds 60% in the carboxyl-terminal regions of the proteins (35). This strongly conserved part of the amino acid sequences corresponds to the region of the enzymatically active 36-kDa proteolytic fragment of phenylalanine hydroxylase (57). A monoclonal antibody was used to localize an amino acid segment of phenylalanine hydroxylase, residues 263–289, responsible for the binding of tetrahydrobiopterin (59). In tryptophan and tyrosine hydroxylases, the corresponding regions are strongly conserved. The amino-terminal regions that contain the regulatory phosphorylation site vary in size from 104 amino acid residues in rabbit tryptophan hydroxylase to 204 residues in human tyrosine hydroxylase (35, 63). The sequence heterogeneity observed in this region may reflect the different regulatory specificities of the three enzymes. Clearly phenylalanine, tyrosine, and tryptophan hydroxylases stem from a common ancestor. The amino-terminal sequences, encoded by exons one and two in human tyrosine hydroxylase, however, may have been acquired at a later stage in their evolutionary divergence.

Mutant genes from patients with phenylketonuria allowed the identification of amino acid substitutions responsible for the phenylalanine hydroxylase deficiencies (10). In the near future, the three-dimensional structure of aromatic amino acid hydroxylases may be available (15), and crucial features of the aromatic amino acid hydroxylases will most likely then be elucidated.

ELECTRON-TRANSPORT COMPONENTS OF MULTICOMPONENT OXYGENASE SYSTEMS

With the exception of flavoprotein monooxygenases, oxygenases that introduce hydroxyl groups into hydrocarbon substrates require an electron-transport component. In this section, these electron-transport components are collectively discussed.

Dihydropteridine Reductase

As described in the previous section, tetrahydrobiopterin is a cofactor of phenylalanine, tyrosine, and tryptophan hydroxylases. Dihydropteridine reductase catalyzes the NADH-dependent reduction of quinonoid dihydrobiopterin to tetrahydrobiopterin, and, hence, is an essential component of the pterin-dependent aromatic amino acid hydroxylases. The reductases from human and rat are very similar 25-kDa polypeptides (73, 120). The preliminary X-ray diffraction pattern at 2.3-Å resolution was obtained (83). The amino-terminal region contains an ADP-binding $\beta\alpha\beta$-fold motif. This region may be the NADH-binding site because a mutation at Asp37 inside the motif reduced the affinity of the enzyme to NADH (82).

Electron-Transport Components of Cytochrome P-450s

Generally, a NADPH-P-450 reductase of approximately 80 kDa serves as the electron-transport component of microsomal cytochrome P-450s (9). This enzyme contains one mole each of FMN and FAD per subunit and is anchored to the endoplasmic reticulum by its hydrophobic amino-terminal region. The primary structures of these reductases from yeast, trout, rat, rabbit, and pig share significant homology (33%) (146). In these sequences, putative domains for the binding of FMN, FAD, and NADPH are conserved. Two separate regions that may be involved in FMN binding reside in the amino-terminal portion of the protein. Two possible FAD-binding domains, one for the binding of the pyrophosphate group of FAD, and the other for the binding of the isoalloxazine ring of FAD, are located in the middle part of the sequences, while the NADPH-binding site may reside at the carboxyl-terminal portion. Its 70-kDa hydrophilic domain is released from the membrane upon protease treatment. The soluble domain cannot interact with the hydroxylase components of cytochrome P-450s but does exhibit NADPH-cytochrome *c* reductase activity.

Some mitochondrial and bacterial P-450s utilize electron-transport systems consisting of two different polypeptide species, a [2Fe-2S] ferredoxin and a flavoprotein. In the P-450_{cam} system, these two polypeptides are called putidaredoxin and NADH-putidaredoxin reductase, respectively. Putidaredoxin shuttles electrons from NADH-putidaredoxin reductase to P-450_{cam}. This system is similar to that of the mitochondrial P-450 systems that are comprised of adrenodoxin and NADH-adrenodoxin reductase. NADH putidaredoxin reductase in the P-450_{cam} system is a member of the glutathione reductase family (19). This oxidoreductase family possesses two $\beta\alpha\beta$-type ADP-binding folds, one for FAD and the other for NAD(P)H binding (144).

P-$450_{Bm\text{-}3}$ from *Bacillus megaterium* catalyzes the hydroxylation of long-chain fatty acids and is inducible by barbiturates. This enzyme has a mass of 118 kDa and exhibits the activities of both P-450 hydroxylase and NADPH-P-450 reductase (68, 90). Such composite P-450 enzymes are not found in eukaryotes, but enzymatically active fusions between a eukaryotic P-450 and a NADPH-P-450 reductase have been constructed (121).

The electron-transport component of the P-450_{sca} system from *Streptomyces carbophilus,* NADH-P-450_{sca} reductase, is a single polypeptide containing FMN and FAD. This reductase thus resembles the eukaryotic reductases, but its size, 51 kDa, is smaller than the sizes of its eukaryotic counterparts (119).

In eukaryotic NADPH-P-450 reductases, reducing equivalents are transferred from NADPH, via the FAD and FMN redox centers, which are separated by about 20 Å (7), to P-450 hydroxylase components (91). In the absence of the P-450 hydroxylase components, the FAD and FMN redox centers of NADPH-P-450 reductase are reduced by NADPH to the semiquinone form of FAD and the fully reduced form of FMN, respectively. Full reduction is not accomplished because the NADP/NADPH couple is significantly more positive than the FADH/$FADH_2$ couple.

In the P-450_{cam} system, NADH reduces FAD to $FADH_2$, and $FADH_2$ subsequently reduces the [2Fe-2S] center of putidaredoxin. The reduction of the P-450_{cam} hydroxylase component is facilitated by the docking of reduced putidaredoxin on the surface of P-450_{cam}. This association is believed to be stabilized by electrostatic interactions between carboxylate groups on the surface of putidaredoxin and complementary arginine and lysine residues on the surface of the hydroxylase component (126). The two redox centers of putidaredoxin and of the P-450_{cam} hydroxylase cannot directly interact because the latter is buried in the hydrophobic core of the protein. The terminal tryptophan residue of putidaredoxin is assumed to be the first component of a long-range electron-transfer network (126).

Electron-Transport Components of Aromatic-Ring Dioxygenases, Aromatic-Ring Hydroxylases, and Alkyl-Group Hydroxylases

Each of the electron-transport components of benzene, toluene, and naphthalene dioxygenases and of alkane hydroxylase is composed of two different proteins (42, 43, 56, 154). The first protein is a flavoprotein exhibiting an NAD(P)H-acceptor reductase activity, whereas the second protein is a ferredoxin. The flavoproteins of benzene and toluene dioxygenases contain two regions of approximatly 30 amino acids in length, one in the amino-terminal region and the other in the middle of the sequences, which fit the NADH- or FAD-binding $\beta\alpha\beta$ motif. Although the sequence of the equivalent component of naphthalene dioxygenase is not presently available, biochemical studies of this component (NADH-ferredoxin$_{NAP}$ reductase) suggest that it is a flavoprotein containing a chloroplast-type [2Fe-2S] cluster, hence different from the corresponding proteins in benzene and toluene dioxygenases. The ferredoxin components of the toluene, benzene, and naphthalene dioxygenases are structurally similar and contain one [2Fe-2S] cluster. Conserved sequence patterns of these ferredoxins suggest that their iron-sulfur clusters are Rieske-type (94).

In benzoate and toluate 1,2-dioxygenases that catalyze the dihydroxylation of the aromatic ring, and in xylene monooxygenase, multicomponent phenol hydroxylase, and methane monooxygenase that catalyze monohydroxylation of substrates, a single protein mediates the electron transfer from NADH to its hydroxylase partner (94, 98, 108, 129). In addition, the second component of naphthalene dioxygenase, NADH-ferredoxin$_{NAP}$ reductase (described above), may be structurally similar to these reductases. The primary structures of the electron-transport components of these enzymes that constitute a new flavoprotein family are similar to each other. Their amino-terminal sequences are similar to chloroplast-type ferredoxins while their carboxyl-terminal regions are similar to those of ferredoxin-NADP reductases (46, 94, 129). The sequence alignment of the electron-transport components of these oxygenases with the ferredoxin-NADP reductase whose three-dimensional structure has been determined (61) showed that strongly conserved sequences are located in the NAD(P)H- and FAD-binding regions. A domain of NADPH-P-450 reductase also appears to be related to this flavoprotein family (94). In the photosynthetic reactions, electrons are transferred from ferredoxin to FAD on ferredoxin-NADP reductase, then from reduced FAD to NADP to yield NADPH. In the oxygenase systems, however, electrons are transferred in the opposite direction: from NAD(P)H to FAD, then from FAD to the [2Fe-2S] cluster.

The *P. putida* (*oleovorans*) alkane hydroxylase is a three-component enzyme. The electron transfer from NADH to the active site of the membrane-bound hydroxylase component, AlkB, is achieved by the 41-kDa NADH-rubredoxin reductase (19) and the 18-kDa rubredoxin (64). NADH-rubredoxin reductase bears sequence similarity to NADH-putidaredoxin oxidoreductase of the P-450_{cam} hydroxylase and to several flavoprotein oxidoreductases such as glutathione reductase, *p*-hydroxybenzoate hydroxylase, and lipoamide dehydrogenase (19). The local peptide sequence of NADH-rubredoxin reductase could be superimposed upon the available three-dimensional structure of glutathione reductase using the WHATIF algorithm developed by Vriend (139). The primary structure of rubredoxin is related to rubredoxins from anaerobic bacteria (64), but not similar to those of other ferredoxins.

General Properties of Electron-Transport Components

The oxidation of NAD(P)H generates two electrons, but the reduction of hydroxylase components requires two one-electron transfer steps. The electron-transport components of the hydroxylase systems can convert a single two-electron transfer into two one-electron transfers. Two electrons in these systems are transferred from NAD(P)H to FAD (or FAD plus FMN in the case of NADPH-P-450 reductase). These flavins provide a reservoir for electrons, and their $2e^-$ forms (e.g. $FADH_2$) can transfer one electron to the [2Fe-2S] center, thereby becoming semiquinones (e.g. FAD*). The semiquinones can further transfer one electron to [2Fe-2S] and thus return to the $0e^-$ state (e.g. FAD).

The electron transfer in the hydroxylase systems can occur most efficiently when the midpoint potentials of electron acceptors are higher than those of electron donors. Where midpoint potentials are known, however, such as in the two redox centers of methane monooxygenase component C, electron acceptor potentials are not always higher than those of electron donors (74, 75). It is not evident why the array of the redox centers may be less than optimal. One possibility is that the electron-transfer reaction may be more efficient than the rate-limiting hydroxylation step, and therefore, an inefficient electron-transport system would not constitute a selective disadvantage. Many hydroxylase components change the midpoint potentials of their redox centers upon binding their substrates. Such a change could be an important mechanism to prevent undesirable uncoupling reactions between NAD(P)H oxidation and substrate hydroxylation. Some electron-transport components also change their midpoint potentials upon the formation of a complex with another component (67).

Cloning in *Escherichia coli* of an incomplete set, lacking one of the electron-transport components, of naphthalene dioxygenase genes, resulted in functional expression of this dioxygenase (66). This result suggests that the missing electron-transport component can be complemented by some unidentified *E. coli* electron-transport component(s). This nonspecific electron transfer is rather unexpected because interactions between unrelated electron-transport systems may cause undesirable short circuits. To what degree biological systems necessitate the specificity of a redox partnership is an interesting question. If a low degree of discrimination between correct and incorrect redox partners can be tolerated, then the evolutionary exchange of partnerships between hydroxylase and electron-transport components may occur frequently and may be the scenario of the natural evolution of multicomponent oxygenase systems.

CONCLUSIONS

Oxygenases from widely divergent organisms share many properties, structures, and reaction mechanisms. During the past decade, the isolation of many hydrocarbon-utilizing organisms combined with advances in recombinant DNA techniques allowed the rapid accumulation of primary sequences of many oxygenases. Sequence comparisons have enabled the classification of oxygenases into distinct families. Such data coupled with biochemical studies have provided important information concerning the structure-function relationships of these enzymes.

The grouping of proteins into related families is useful in several respects. Biochemical knowledge obtained from one protein may be applicable to another protein in the same family. Amino acid residues that are highly conserved among members of a gene family indicate residues essential for function. Furthermore, comparisons of related proteins may allow the reconstruction of genetic rearrangements that have taken place during the course of evolution and therefore suggest mechanisms of protein evolution (2). The oxygenases discussed here have been grouped into several families. Usually, those considered to be related show global sequence homologies: long stretches of amino acid sequences can be aligned without the introduction of many gaps, and overall sequence identities are more than 20%. Some regions of the aligned sequences show particularly high degrees of identity, and such patterns of conserved amino acids are noted, for instance, in regions that bind important cofactors. As discussed above, residues that bind NAD(P)H and FAD, and iron-sulfur centers, are highly conserved (94, 144).

Local sequence similarities are also observed among oxygenases from different families, especially in regions involved in cofactor binding. It is not known how the broad distribution of these coenzyme-binding domains arose. Similarities in the cofactor-binding regions of unrelated enzymes may reflect constraints imposed by the functioning of these cofactors. If this is the case, the observed sequence similarities may have arisen as the result of convergent evolution. Alternatively, common patterns in different oxygenases may be formed by the exchange and association of different protein domains. In eukaryotes, exons may code for compact protein structures called modules, and recombination events between the intervening introns may allow exon shuffling and thus creation of novel proteins (31). In bacteria, where introns do not generally exist, the propagation of protein-module-encoding DNA segments, if it does occur, should take place by a mechanism other than exon shuffling.

The electron-transport components of benzene dioxygenase from *P. putida* and the comparable component of benzoate dioxygenase from *A. calcoaceticus* do not appear to be evolutionarily related. The NADH- and FAD-binding regions of the electron-transport component of benzene dioxygenase are characterized by the ADP-binding $\beta\alpha\beta$ fold, whereas the equivalent cofactor-binding regions in the electron-transport component of benzoate dioxygenase resemble those of spinach ferredoxin-NADP$^+$ reductase, a member of a separate flavoprotein family (94). However, a comparison of the three-dimensional structures of the cofactor-binding regions in these two families shows that they are topologically similar (61). Whether similar cofactor-binding domains have been established in a process of divergent evolution or of convergent evolution is not clear.

Investigators have noticed genetic exchanges creating apparent protein fusions in some electron-transport components of oxygenases. The mammalian NADPH-P-450 reductases appear to be mosaic proteins, with distinct domains involved in the binding of different cofactors. In bacteria, the electron-transport component of benzoate dioxygenase and its family appear to be a fusion product of a chloroplast-type ferredoxin at the amino-terminal region with a ferredoxin-NADP reductase. A homologous chloroplast-type ferredoxin segment is found in the carboxyl-terminal region, rather than the amino-terminal region of the electron-transport component of vanillate demethylase (13).

In bacteria, catabolic genes, especially those for xenobiotic degradation, are often plasmid-born (5), and in some cases, these genes are integrated into transposons (135, 138). Chromosomally encoded catabolic genes often form supraoperonic structures. It is not known whether these genetic arrangements reflect requirements for gene expression, beyond transcriptional unity, or whether they reflect mechanisms of evolutionary change (103). Alternatively, they may reflect modes of genetic exchange. In the environment, certainly, genetic transfer could be mediated by plasmids, chromosomal gene mobilization, transduction, transformation, and transposition. Such mechanisms have also been suggested to mediate genetic exchange between prokaryotic and eukaryotic cells (54).

Several mechanisms may lead to the divergence of enzymes. The accumulation of point mutations and the power of natural selection have long been discussed. More recently, however, other kinds of DNA rearrangements have been suggested that may be responsible for dynamic changes in the primary structure of DNA (46). A variety of mutations in combination with gene amplification may provide additional possibilities of genetic evolution: a silent gene copy may accumulate mutations without imposing selective disadvantage to the host organism. In the process of gene conversion caused by DNA slippage, one or several of the acquired mutations may be integrated into the actively transcribed copy of the gene (23).

In the evolution of catabolic pathways, selective advantage may not be conferred until an entire suite of enzymes needed for substrate degradation is present. This situation may have created different pathways for the degradation of the same compound, which may reflect differences in evolutionary history. An example is provided by different pathways for the mineralization of toluene: in the TOL plasmid–encoded pathway, the methyl side chain of toluene is attacked by a monooxygenase, while in a chromosomally encoded pathway, the toluene ring is cleaved without the processing of this methyl side chain (148).

Enzymes may evolve by altering substrate specificity without changing catalytic mechanism. The plasmid-encoded *P. putida* alkane hydroxylase and xylene monooxygenase provide a naturally occurring example of structurally related but functionally distinct enzymes. Both enzymes catalyze the same kind of reaction, but they do not have a substrate in common. Mutations provoked by single amino acid substitutions can dramatically affect the substrate specificity of catechol 2,3-dioxygenase (113, 141). A wide spectrum of substrate specificity occurs naturally in enzymes of related gene families (12). Mechanisms of coupling substrate specificity and catalysis are largely unknown; often there is a trade-off between substrate specificity and catalytic efficiency. Oxygenases that can accept many substrates may be less efficient and may generate fewer specific oxidation products than their homologues that are more narrowly substrate specific (96).

The ability to engineer modifications in oxygenase function or specificity may be useful in environmental-protection efforts. Recalcitrant pollutants often resemble the natural substrates of microbial enzymes. The custom design of oxygenases may also be used in certain biotechnology processes in which biotransformations prove to be advantageous in the synthesis of useful chemicals. The accelerated evolution of catabolic pathways has been achieved in the laboratory, shedding light on the ways in which substrate specificity and regulation of existing enzymes may change (1). The biochemical studies of oxygenases will be relevant to these applications as well as to our general understanding of the function and structure of an important and interesting class of enzymes.

Acknowledgments

This research was supported by the Swiss National Science Foundation.

Literature Cited

1. Abril, M.-A., Michan, C., Timmis, K. N., Ramos, J. L. 1989. Regulator and enzyme specificities of the TOL plasmid-encoded upper pathway for degradation of aromatic hydrocarbons and expansion of the substrate range of the pathway. *J. Bacteriol.* 171:6782–90
2. Alber, T. 1989. Mutational effects on protein stability. *Annu. Rev. Biochem.* 58:765–98
3. Aoyama, T., Nagata, K., Yamazoe, Y., Kato, R., Matsunaga, E., et al. 1990. Cytochrome *b5* potentiation of cytochrome P-450 catalytic activity demonstrated by a vaccinia virus-mediated in situ reconstitution system. *Proc. Natl. Acad. Sci. USA* 87:5425–29
4. Arciero D. M., Lipscomb, J. D. 1986. Binding of the ^{17}O-labeled substrate and inhibitors to protocatechuate 4,5-dioxygenase-nitrosyl complex. *J. Biol. Chem.* 261:2170–78
5. Assinder, S. J., Williams, P. A. 1990. The TOL plasmids: determinants of the catabolism of toluene and the xylenes. *Adv. Microbiol. Phys.* 31:1–69
6. Bartels I., Knackmuss H.-J., Reineke W. 1984. Suicide inactivation of catechol 2,3-dioxygenase from *Pseudomonas putida* mt-2 by 3-halocatechols. *Appl. Environ. Microbiol.* 47:550–55
7. Bastiaens, P. I., Bonants, P. J., Muller, F., Visser, A. J. 1989. Time-resolved fluorescence spectroscopy of NADPH-cytochrome P-450 reductase: demonstration of energy transfer between the two prosthetic groups. *Biochemistry* 28: 8416–25
8. Batie, C. J., LaHaie, E., Ballou, D. P. 1987. Purification and characterization of phthalate oxygenase and phthalate oxygenase reductase from *Pseudomonas cepacia*. *J. Biol. Chem.* 262:1510–18
9. Benveniste, I., Lesot, A., Hasenfratz, M. P., Kochs, G. Durst, F. 1991. Multiple forms of NADPH-cytochrome P450 reductase in higher plants. *Biochem. Biophys. Res. Commun.* 177:105–12
10. Berthelon, M., Caillaud, C., Rey, F., Labrune, P., Melle, D., et al. 1991. Spectrum of phenylketonuria mutations in Western Europe and North Africa, and their relation to polymorphic DNA haplotypes at the phenylalanine hydroxylase locus. *Hum. Genet.* 86:355–58
11. Bollinger, J. M., Edmondson, D. E., Huynh, B. H., Filley, J., Norton, J. R., et al. 1991. Mechanism of assembly of the tyrosyl radical-dinuclear iron cluster cofactor of ribonucleotide reductase. *Science* 253:292–98
12. Broderick, J. B., O'Halloran, T. V. 1991. Overproduction, purification, and characterization of chlorocatechol dioxygenase, a non-heme iron dioxygenase with broad substrate tolerance. *Biochemistry* 30:7349–58
13. Brunel, F., Davison, J. 1988. Cloning and sequencing of *Pseudomonas* genes encoding vanillate demethylase. *J. Bacteriol.* 170:4924–30
14. Burrows, K. J., Cornish, A., Scott, D., Hoggins, I. J. 1984. Substrate specificities of the soluble and particulate methane mono-oxygenases of *Methylosinus trichosporium* OB3b. *J. Gen. Microbiol.* 130:3327–33
15. Celikel, R., Davis, M. D., Dai, X. P., Kaufman, S., Xuong, N. H. 1991. Crystallization and preliminary X-ray analysis of phenylalanine hydroxylase from rat liver. *J. Mol. Biol.* 218:495–98
16. Davis, M. D., Kaufman, S. 1989. Evidence for the formation of the 4a-carbinolamine during the tyrosine-dependent oxidation of tetrahydrobiopterin by rat liver phenylalanine hydroxylase. *J. Biol. Chem.* 264:8585–96
17. Dagley, S. 1978. Pathways for the utilization of organic growth substrates. In *The Bacteria,* ed. L. N. Ornston, J. R. Sokatch, 6:305–88. New York: Academic
18. Edwards, R. J., Murray, B. P., Singleton, A. M., Boobis, A. R. 1991. Orientation of cytochromes P450 in the endoplasmic reticulum. *Biochemistry* 30:71–76
19. Eggink, G., Engel, H., Vriend, G., Terpstra, P., Witholt, B. 1990. Rubredoxin reductase of *Pseudomonas oleovorans*. Structural relationship to other flavoprotein oxidoreductases based on one NAD and two FAD fingerprints. *J. Mol. Biol.* 212:135–42
20. Eich, F., Geary, P. J., Bernhardt, F.-H. 1985. Protein-protein interactions and antigenic relationships between the components of 4-methoxybenzoate monooxygenase and of benzene 1,2-dioxygenase from *Pseudomonas putida*. *Eur. J. Biochem.* 153:407–12
21. Enseley, B. D., Gibson, D. T. 1983. Naphthalene dioxygenase: purification and properties of a terminal oxygenase component. *J. Bacteriol.* 155:505–11
22. Entsch, B., Palfey, B. A., Ballou, D. P., Massey, V. 1991. Catalytic function of tyrosine residues in para-hydroxybenzoate hydroxylase as determined by the study of site-directed mutants. *J. Biol. Chem.* 266:17341–49

23. Fitch, D. H., Mainone, C., Goodman, M., Slightom, J. L. 1990. Molecular history of gene conversions in the primate fetal gamma-globin genes. Nucleotide sequences from the common gibbon, *Hylobates lar*. *J. Biol. Chem.* 265:781–93
24. Fitzpatrick, P. F. 1989. The metal requirement of rat tyrosine hydroxylase. *Biochem. Biophys. Res. Comm.* 161: 211–15
25. Fitzpatrick, P. F. 1991. Steady-state kinetic mechanism of rat tyrosine hydroxylase. *Biochemistry* 30:3658–62
26. Fitzpatrick, P. F. 1991. Studies of the rate-limiting step in the tyrosine hydroxylase reaction: alternate substrates, solvent isotope effects, and transition-state analogues. *Biochemistry* 30:6386–91
27. Fox, B. G., Froland, W. A., Dege, J. E., Lipscomb, J. D. 1989. Methane monooxygenase from *Methylosinus trichosporium* OB3b. *J. Biol. Chem.* 264:10023–33
28. Fox, B. G., Liu, Y., Dege, J. E., Lipscomb, J. D. 1991. Complex formation between the protein components of methane monooxygenase from *Methylosinus trichosporium* OB3b. Identification of sites of component interaction. *J. Biol. Chem.* 266:540–50
29. Ghisla, S., Massey, V. 1989. Mechanisms of flavoprotein-catalyzed reactions. *Eur. J. Biochem.* 181:1–17
30. Gibson, D. T. 1987. Microbial catabolism of aromatic hydrocarbons and the carbon cycle. In *Microbial Metabolism and the Carbon Cycle*, ed. S. R. Hagedorn, R. S. Hanson, D. A. Kunz, pp. 33–58. Chur, Switzerland: Horwood Academic
31. Gilbert, W. 1987. The exon theory of genes. *Cold Spring Harbor. Symp. Quant. Biol.* 52:901–5
32. Graham-Lorence, S., Khalil, M. W., Lorence, M. C., Mendelson, C. R., Simpson, E. R. 1991. Structure-function relationships of human aromatase cytochrome P-450 using molecular modeling and site-directed mutagenesis. *J. Biol. Chem.* 266:11939–46
33. Green, A. K., Cotton, G. H., Jennings, I., Fisher, M. J. 1990. Experimental determination of the phosphorylation state of phenylalanine hydroxylase. *Biochem. J.* 265:563–68
34. Green, J., Dalton, H. 1989. Substrate specificity of soluble methane monooxygenase. *J. Biol. Chem.* 264:17698–17703
35. Grenett, H. E., Ledley, F. D., Reed, L. L., Woo, S. L. C. 1987. Full-length cDNA for rabbit tryptophan hydroxylase: functional domains and evolution of aromatic amino acid hydroxyases. *Proc. Natl. Acad. Sci. USA* 84:5530–34
36. Guengerich, F. P. 1991. Reactions and significance of cytochrome P-450 enzymes. *J. Biol. Chem.* 266:10019–22
37. Guengerich, F. P., Macdonald, T. L. 1990. Mechanisms of cytochrome P-450 catalysis. *FASEB Lett.* 4:2453–59
38. Gurbiel, R. J., Batie, C. J., Sivaraja, M., True, A. E., Fee, J. A., et al. 1989. Electron-nuclear double resonance spectroscopy of ^{15}N-enriched phthalate dioxygenase from *Pseudomonas cepacia* proves that two histidines are coordinated to [2Fe-2S] Rieske-type clusters. *Biochemistry* 28:4861–71
39. Haavik, J., Le Bourdelles, B., Martinez, A., Flatmark, T., Mallet, J. 1991. Recombinant human tyrosine hydroxylase isozymes. Reconstitution with iron and inhibitory effect of other metal ions. *Eur. J. Biochem.* 199:371–78
40. Haavik, J., Martinez, A., Flatmark, T. 1990. pH-dependent release of catecholamines from tyrosine hydroxylase and the effect of phosphorylation of Ser-40. *FEBS Lett.* 262:363–65
41. Hahn, D. R., Solenberg, P. J., Baltz, R. H. 1991. Tn*5099*, a *xylE* promoter probe transposon for *Streptomyces* spp. *J. Bacteriol.* 173:5573–77
42. Haigler, B. E., Gibson, D. T. 1990. Purification and properties of ferredoxins$_{NAP}$, a component of naphthalene dioxygenase from *Pseudomonas* sp. strain NCIB 9816. *J. Bacteriol.* 172:465–68
43. Haigler, B. E., Gibson, D. T. 1990. Purification and properties of NADH-ferredoxin$_{NAP}$ reductase, a component of naphthalene dioxygenase from *Pseudomonas* sp. strain NCIB 9816. *J. Bacteriol.* 172:457–64
44. Halliwell, B., Gutteridge, J. M. C. 1984. Oxygen toxicity, oxygen radicals, transition metals and disease. *Biochem. J.* 219:1–14
45. Harayama, S., Rekik, M. 1989. Bacterial aromatic ring-cleavage enzymes are classified into two different gene families. *J. Biol. Chem.* 264:15328–33
46. Harayama, S., Rekik, M., Bairoch, A., Neidle, E. L., Ornston, L. N. 1991. Potential DNA slippage structures acquired during evolutionary divergence of *Acinetobacter calcoaceticus* chromosomal *benABC* and *Pseudomonas putida* TOL pWW0 plasmid *xylXYZ*, genes encoding benzoate dioxygenases. *J. Bacteriol.* 173:7540–48

47. Harayama, S., Rekik, M., Timmis, K. N. 1986. Genetic analysis of a relaxed substrate specificity aromatic ring dioxygenase, toluate 1,2-dioxygenase, encoded by TOL plasmid pWW0 of *Pseudomonas putida*. *Mol. Gen. Genet.* 202:226–34
48. Harayama, S., Rekik, M., Wubbolts, M., Rose, K., Leppik, R. A., et al. 1989. Characterization of five genes in the upper-pathway operon of TOL plasmid pWW0 from *Pseudomonas putida* and identification of the gene products. *J. Bacteriol.* 171:5048–55
49. Harayama, S., Timmis, K. N. 1989. Catabolism of aromatic hydrocarbons by *Pseudomonas*. In *Genetics of Bacterial Diversity,* ed. D. A. Hopwood, K. Charter, pp. 151–74. New York: Academic
50. Harpel, M. R., Lipscomb, J. D. 1990. Gentisate 1,2-dioxygenase from *Pseudomonas*. Purification, characterization, and comparison of the enzymes from *Pseudomonas testosteroni* and *Pseudomonas acidovorans*. *J. Biol. Chem.* 265:6301–11
51. Harpel, M. R., Lipscomb, J. D. 1990. Gentisate 1,2-dioxygenase from *Pseudomonas*. Substrate coordination to active site Fe^{2+} and mechanism of turnover. *J. Biol. Chem.* 265:22187–96
52. Hartnett, C., Neidle, E. L., Ngai, K.-L., Ornston, L. N. 1990. DNA sequences of genes encoding *Acinetobacter calcoaceticus* protocatechuate 3,4-dioxygenase: evidence indicating shuffling of genes and of DNA sequences within genes during their evolutionary divergence. *J. Bacteriol.* 172:956–66
53. Higashi, Y., Hiromasa, T., Tanae, A., Miki, T., Nakura, J. et al. 1991. Effects of individual mutations in the P-450(C21) pseudogene on the P-450(C21) activity and their distribution in the patient genomes of congenital steroid 21-hydroxylase deficiency. *J. Biochem. (Tokyo)* 109:638–44
54. Horn, J. M., Harayama, S., Timmis, K. N. 1991. DNA sequence determination of the TOL pWW0 *xylGFJ* genes of *Pseudomonas putida:* implications for the evolution of aromatic catabolism. *Mol. Microbiol.* 5:2459–74
55. Imai, M., Shimada, H., Watanabe, Y., Matsushima-Hibiya, Y., Makino, R., et al. 1989. Uncoupling of the cytochrome P-450_{CAM} monooxygenase reaction by a single mutation, threonine-252 to alanine or valine: a possible role of the hydroxy amino acid in oxygen activation. *Proc. Natl. Acad. Sci. USA* 86:7823–27
56. Irie, S., Doi, S., Yorijuki, T., Takagi, M., Yano, K. 1987. Nucleotide sequencing and characterization of the genes encoding benzene oxidation enzymes of *Pseudomnas putida*. *J. Bacteriol.* 169:5174–79
57. Iwaki, M., Phillips, R. S., Kaufman, S. 1986. Proteolytic modification of the amino-terminal and carboxyl-terminal regions of rat hepatic phenylalanine hydroxylase. *J. Biol. Chem.* 261:2051–56
58. Iwasaki, M., Juvonen, R., Lindberg, R., Negishi, M. 1991. Alteration of high and low spin equilibrium by a single mutation of amino acid 209 in mouse cytochromes P450. *J. Biol. Chem.* 266:3380–82
59. Jennings, I. G., Kemp, B. E., Cotton, R. G. 1991. Localization of cofactor binding sites with monoclonal anti-idiotype antibodies: phenylalanine hydroxylase. *Proc. Natl. Acad. Sci. USA* 88:5734–38
60. Kabisch, M., Fortnagel, P. 1990. Nucleotide sequence of metapyrocatechase I (catechol 2,3-oxygenase I) gene *mpcI* from *Alcaligenes eutrophus* JMP222. *Nucleic Acids Res.* 18:3405–6
61. Karplus, P. A., Daniels, M. J., Herriott, J. R. 1991. Atomic structure of ferredoxin-$NADP^+$ reductase: prototype for a structurally novel flavoenzyme family. *Science* 251:60–66
62. Katopodis, A. G., Smith, H. A., May, S. W. 1988. New oxyfunctionalization capabilities for omega-hydroxylases: asymmetric aliphatic sulfoxidation and branched ether demethylation. *J. Am. Chem. Soc.* 110:897–99
63. Kobayashi, K., Kaneda, N., Ichinose, H., Kishi, F., Nakazawa, A., et al. 1988. Structure of the human tyrosine hydroxylase gene: alternative splicing from a single gene accounts for generation of four mRNA types. *J. Biochem. (Tokyo)* 103:907–12
64. Kok, M., Oldenhuis, R., van der Linden, M. P., Meulenberg, C. H., Kingma, J., et al. 1989. The *Pseudomonas oleovorans alkBAC* operon encodes two structurally related rubredoxins and an aldehyde dehydrogenase. *J. Biol. Chem.* 264:5442–51
65. Kok, M., Oldenhuis, R., van der Linden, M. P., Raatjes, P., Kingma, J., et al. 1989. The *Pseudomonas oleovorans* alkane hydroxylase gene. Sequence and expression. *J. Biol. Chem.* 264:5435–41
66. Kurkela, S., Lehväslaiho, H., Palva, E. T., Teeri, T. H. 1988. Cloning, nucleotide sequence and characterization of genes encoding naphthalene dioxy-

genase of *Pseudomonas putida* strain NCIB9816. *Gene* 73:355–62
67. Lamberth, J. D., Seybert, D. W., Kamin, H. 1979. Ionic effects on adrenal steroidogenic electron transport. The role of adrenodoxin as an electron shuttle. *J. Biol. Chem.* 254:7255–64
68. Li, H. Y., Darwish, K., Poulos, T. L. 1991. Characterization of recombinant *Bacillus megaterium* cytochrome P-450_{BM-3} and its two functional domains. *J. Biol. Chem.* 266:11909–14
69. Lindberg, R. L. P., Negishi, M. 1989. Alteration of mouse cytochrome $P450_{coh}$ substrate specificity by mutation of a single amino-acid residue. *Nature (London)* 339:632–34
70. Liu, K. E., Lippard, S. J. 1991. Redox properties of the hydroxylase component of methane monooxygenase from *Methylococcus capsulatus* (Bath). Effects of protein B, reductase, and substrate. *J. Biol. Chem.* 266:12836–39
71. Locher, H. H., Leisinger, T., Cook, A. M. 1991. 4-Sulphobenzoate 3,4-dioxygenase: purification and properties of a desulphonative two-component enzyme system from *Comamonas testosteroni* T-2. *Biochem. J.* 274:833–42
72. Locher, H. H., Leisinger, T., Cook, A. M. 1991. 4-Toluene sulfonate methylmonooxygenase from *Comamonas testosteroni* T-2: purification and some properties of the oxygenase component. *J. Bacteriol.* 173:3741–48
73. Lockyer, J. L., Cook, R. G., Milstien, S., Kaufman, S., Woo, S. L. C., et al. 1987. Structure and expression of human dihydropteridine reductase. *Proc. Natl. Acad. Sci. USA* 84:3329–33
74. Lund, J., Dalton, H. 1985. Further characterisation of the FAD and Fe_2S_2 redox centres of component C, the NADH:acceptor reductase of the soluble methane monooxygenase of *Methylococcus capsulatus* >(Bath). *Eur. J. Biochem.* 147:291–96
75. Lund, J., Woodland, M. P., Dalton, H. 1985. Electron transfer reactions in the soluble methane monooxygenase of *Methylococcus capsulatus* (Bath). *Eur. J. Biochem.* 147:297–305
76. Lynch, J. B., Juarez-Garcia, C., Münck, E., Que, L. Jr. 1989. Mössbauer and EPR studies of the binuclear iron center in ribonucleotide reductase from *Escherichia coli*. A new iron-to-protein stoichiometry. *J. Biol. Chem.* 264:8091–96
77. Mabrouk, P. A., Orville, A. M., Lipscomb, J. D., Solomon, E. I. 1991. Variable-temperature variable-field magnetic circular dichroism studies of the Fe(II) active site in metapyrocatechase: implications for the molecular mechanism of extradiol dioxygenases. *J. Am. Chem. Soc.* 113:4053–61
78. Majamaa, K., Volkmar, G., Hanavske-Abel, H. M., Myllyla, R., Kivivikko, K. I. 1986. Partial identity of the 2-oxoglutarate and ascorbate binding sites of prolyl-4-hydroxylase. *J. Biol. Chem.* 261:7819–23
79. Makita, Y., Okuno, S., Fujisawa, H. 1990. Involvement of activator protein in the activation of tryptophan hydroxylase by cAMP-dependent protein kinase. *FEBS Lett.* 268:185–88
80. Markus, A., Krekel, D., Lingens, F. 1986. Purification and some properties of component A of the 4-chlorophenylacetate 3,4-dioxygenase from *Pseudomonas* species strain CBS. *J. Biol. Chem.* 261:12883–88
81. Martinez, A., Andersson, K. K., Haavik, J., Flatmark, T. 1991. EPR and ^{1}H-NMR spectroscopic studies on the paramagnetic iron at the active site of phenylalanine hydroxylase and its interaction with substrates and inhibitors. *Eur. J. Biochem.* 198:675–82
82. Matthews, D. A., Varughese, K. I., Skinner, M., Xuong, N. H., Hoch, J., et al. 1991. Role of aspartate-37 in determining cofactor specificity and binding in rat liver dihydropteridine reductase. *Arch. Biochem. Biophys.* 287:234–39
83. Matthews, D. A., Webber, S., Whiteley, J. M. 1986. Preliminary X-ray diffraction characterization of crystalline rat liver dihydropteridine reductase. *J. Biol. Chem.* 261:3891–93
84. Mazliak, P. 1973. Lipid metabolism in plants. *Annu. Rev. Plant. Physiol.* 24:287–310
85. McDonald, J. D., Bode, V. C., Dove, W. F., Shedlovsky, A. 1990. Pah^{hph-5}: a mouse mutant deficient in phenylalanine hydroxylase. *Proc. Natl. Acad. Sci. USA* 87:1965–67
86. Miller, R. J., Benkovic, S. J. 1988. L-[2,5-H_2]phenylalanine, an alternate substrate for rat liver phenylalanine hydroxylase. *Biochemistry* 27:3658–63
87. Nakai, C., Horiike, K., Kuramitsu, S., Kagamiyama, H., Nozaki, M. 1990. Three isozymes of catechol 1,2-dioxygenase (pyrocatechase), $\alpha\alpha$, $\alpha\beta$, and $\beta\beta$, from *Pseudomonas arvilla* C-1. *J. Biol. Chem.* 265:660–65
88. Nakai, C., Kagamiyama, H., Nozaki, M., Nakazawa, T., Inouye, S., et al. 1983. Complete nucleotide sequence of

the metapyrocatechase gene on the TOL plasmid of *Pseudomonas putida* mt-2. *J. Biol. Chem.* 258:2923–28
89. Naqui, A., Chance, B. 1986. Reactive oxygen intermediates in biochemistry. *Annu. Rev. Biochem.* 55:137–66
90. Narhi, L. O., Fulco, A. J. 1987. Identification and characterization of two functional domains in cytochrome P-450_{BM-3}, a catalytically self-sufficient monooxygenase induced by barbiturates in *Bacillus megaterium*. *J. Biol. Chem.* 262:6683–90
91. Nebert, D. W., Gonzalez, F. J. 1987. P450 genes: structure, evolution, and regulation. *Annu. Rev. Biochem.* 56: 945–93
92. Nebert, D. W., Nelson, D. R., Coon, M. J., Estabrook, R. W., Feyereisen, R., et al. 1991. The P450 superfamily: update on new sequences, gene mapping, and recommended nomenclature. *DNA Cell Biol.* 10:1–14
93. Neidle, E., Hartnett, C., Ornston, L. N., Bairoch, A., Rekik, M., et al. 1992. *Cis*-diol dehydrogenases encoded by the TOL pWW0 plasmid *xylL* gene and the *Acinetobacter calcoaceticus* chromosomal *benD* gene are members of the short-chain alcohol dehydrogenase superfamily. *Eur. J. Biochem.* 204:113–20
94. Neidle, E. L., Hartnett, C., Ornston, L. N., Bairoch, A., Rekik, M., et al. 1991. Nucleotide sequences of the *Acinetobacter calcoaceticus benABC* genes for benzoate 1,2-dioxygenase reveal evolutionary relationships among multicomponent oxygenases. *J. Bacteriol.* 173:5385–95
95. Nestler, E. J., McMahon, A., Sabban, E. L., Tallman, J. F., Duman, R. S. 1990. Chronic antidepressant administration decreases the expression of tyrosine hydroxylase in the rat locus coeruleus. *Proc. Natl. Acad. Sci. USA* 87:7522–26
96. Ngai, K.-L., Ornston, L. N. 1988. Abundant expression of *Pseudomonas* genes for chlorocatechol metabolism. *J. Bacteriol.* 170:2412–13
97. Noda, Y., Nishikawa, S., Shiozuka, K., Kadokura, H., Nakajima, H., et al. 1990. Molecular cloning of the protocatechuate 4,5-dioxygenase genes of *Pseudomonas paucimobilis*. *J. Bacteriol.* 172:2704–9
98. Nordlund, I., Powlowski, J., Shingler, V. 1990. Complete nucleotide sequence and polypeptide analysis of multicomponent phenol hydroxylase from *Pseudomonas* sp. strain CF600. *J. Bacteriol.* 172:6826–33
99. Nozaki, M. 1979. Oxygenases and dioxgenases. *Top. Curr. Chem.* 78:145–86
100. Nurk, A., Kasak, L., Kivisaar, M. 1991. Sequence of the gene (*pheA*) encoding phenol monooxygenase from *Pseudomonas* sp. EST1001: expression in *Escherichia coli* and *Pseudomonas putida*. *Gene* 102:13–18
101. Ohkuma, M., Tanimoto, T., Yano, K., Takagi, M. 1991. CYP52 (cytochrome $P450_{alk}$) multigene family in *Candida maltosa:* molecular cloning and nucleotide sequence of the two tandemly arranged genes. *DNA Cell. Biol.* 10: 271–82
102. Ohlendorf, D. H., Lipscomb, J. D., Weber, P. C. 1988. Structure and assembly of protocatechuate 3,4-dioxygenase. *Nature* 336:403–5
103. Ornston, L. N., Neidle, E. L. 1991. Evolution of genes for the β-ketoadipate pathway in *Acinetobacter calcoaceticus*. In *The Biology of* Acinetobacter, ed. K. J. Towner, E. Bergogne-Bergin, C. A. Fewson, pp. 201–37. New York: Plenum
104. Pember, S. O., Johnson, K. A., Villafranca, J. J., Benkovic, S. J. 1989. Mechanistic studies on phenylalanine hydroxylase from *Chromobacterium violaceum*. Evidence for the formation of an enzyme-oxygen complex. *Biochemistry* 28:2124–30
105. Perkins, E. J., Gordon, M. P., Caceres, O., Lurquin, P. F. 1990. Organization and sequence analysis of the 2,4-dichlorophenol hydroxylase and dichlorocatechol oxidative operons of plasmid pJP4. *J. Bacteriol.* 172:2351–59
106. Porter, T. D., Coon, M. J. 1991. Cytochrome P-450. Multiplicity of isoforms, substrates, and catalytic and regulatory mechanisms. *J. Biol. Chem.* 266:13469–72
107. Poulos, T. L., Finzel, B. C., Gunsalus, I. C., Wagner, G. C., Kraut, J. 1985. The 2.6-Å crystal structure of *Pseudomonas putida* cytochrome P-450. *J. Biol. Chem.* 260:16122–30
108. Powlowski, J., Shingler, V. 1990. In vitro analysis of polypeptide requirements of multicomponent phenol hydroxylase from *Pseudomonas* sp. strain CF600. *J. Bacteriol.* 172:6834–40
109. Raag, R., Poulos, T. L. 1989. Crystal structure of the carbon monoxide-substrate-cytochrome P-450_{CAM} ternary complex. *Biochemistry* 28:7586–92
110. Raag, R., Poulos, T. L. 1991. Crystal structures of cytochrome P-450_{CAM}

complexed with camphane, thiocamphor, and adamantane: factors controlling P-450 substrate hydroxylation. *Biochemistry* 30:2674–84
111. Rajasekharan, S., Rajasekharan, R., Vaidyanathan, C. S. 1990. Substrate-mediated purification and characterization of a 3-hydroxybenzoic acid-6-hydroxylase from *Micrococcus*. *Arch. Biochem. Biophys.* 278:21–25
112. Raju, S. G., Kamath, A. V., Vaidyanathan, C. S. 1988. Purification and properties of 4-hydroxyphenylacetic acid 3-hydroxylase from *Pseudomonas putida*. *Biochem. Biophys. Res. Commun.* 154:537–43
113. Ramos, J. L., Wasserfallen, A., Rose, K., Timmis, K. N. 1987. Redesigning metabolic routes: manipulation of TOL plasmid pathway for catabolism of alkylbenzoates. *Science* 235:593–96
114. Roper, D. I., Cooper, R. A. 1990. Subcloning and nucleotide sequence of the 3,4-dihydroxyphenylacetate (homoprotocatechuate) 2,3-dioxygenase gene from *Escherichia coli* C. *FEBS Lett.* 275:53–57
115. Ruettinger, R. T., Olson, S. T., Boyer, R. F., Coon, J. M. 1974. Identification of the omega-hydroxylase of *Pseudomonas oleovorans* as a nonheme iron protein requiring phospholipid for catalytic activity. *Biochem. Biophys. Res. Commun.* 57:1011–17
116. Schreuder, H. A., Hol, W. G. J., Drenth, J. 1988. Molecular modeling reveals the possible importance of a carbonyl oxygen binding pocket for the catalytic mechanism of *p*-hydroxybenzoate hydroxylase. *J. Biol. Chem.* 263: 3131–36
117. Schreuder, H. A., Hol, W. G. J., Drenth, J. 1990. Analysis of the active site of the flavoprotein *p*-hydroxybenzoate hydroxylase and some ideas with respect to its reaction mechanism. *Biochemistry* 29:3101–8
118. Sejlitz, T., Neujahr, H. Y. 1991. Arginyl residues in the NADPH-binding sites of phenol hydroxylase. *J. Protein Chem.* 10:43–48
119. Serizawa, N., Matsuoka, T. 1991. A two component-type cytochrome P-450 monooxygenase system in a prokaryote that catalyzes hydroxylation of ML-236B to pravastin, a tissue-selective inhibitor of 3-hydroxy-3-methylglutaryl coenzyme A reductase. *Biochim. Biophys. Acta* 1084:35–40
120. Shahbaz, M., Hoch, J. A., Trach, K. A., Hural, J. A., Webber, S., et al. 1987. Structural studies and isolation of cDNA clones providing the complete sequence of rat liver dihydropteridine reductase. *J. Biol. Chem.* 34:16412–16
121. Shibuta, M., Sakaki, T., Yabusaki, Y., Murakami, M., Ohkawa, M. 1990. Genetically engineered P450 monooxygenases: construction of bovine P450 c17/yeast reductase fused enzymes. *DNA Cell Biol.* 9:27–36
122. Shiman, R., Jones, S. H., Gray, D. W. 1990. Mechanism of phenylalanine regulation of phenylalanine hydroxylase. *J. Biol. Chem.* 265:11633–42
123. Shimizu, T., Tateishi, T., Hatano, M., Fujii-Kuriyama, Y. 1991. Probing the role of lysines and arginines in the catalytic function of cytochrome P450d by site-directed mutagenesis. Interaction with NADPH-cytochrome P450 reductase. *J. Biol. Chem.* 266:3372–75
124. Siegmund, H.-U., Kaufman, S. 1991. Hydroxylation of 4-methylphenylalanine by rat liver phenylalanine hydroxylase. *J. Biol. Chem.* 266:2903–10
125. Stainthorpe, A. C., Lees, V., Salmond, G. P. C., Dalton, H., Murrell, J. C. 1990. The methane monooxygenase gene cluster of *Methylococcus capsulatus* (Bath). *Gene* 91:27–34
126. Stayton, P. S., Sligar, S. G. 1991. Structural microheterogeneity of a tryptophan residue required for efficient biological electron transfer between putidaredoxin and cytochrome P-450_{CAM}. *Biochemistry* 30:1845–51
127. Streber, W. R., Timmis, K. N., Zenk, M. H. 1987. Analysis, cloning, and high-level expression of 2,4-dichlorophenoxyacetate monooxygenase gene *tfdA* of *Alcaligenes eutrophus* JMP134. *J. Bacteriol.* 169:2950–55
128. Subramanian, V., Liu, T.-N., Yeh, W.-K., Serdar, C. M., Wackett, L. P., et al. 1985. Purification and properties of ferredoxin$_{TOL}$. *J. Biol. Chem.* 260:2355–63
129. Suzuki, M., Hayakawa, T., Shaw, J. P., Rekik, M., Harayama, S. 1991. Primary structure of xylene monooxygenase: similarities to and differences from the alkane hydroxylation system. *J. Bacteriol.* 173:1690–95
130. Taira, K., Hayase, N., Arimura, N., Yamashita, S., Miyazaki, T., Furukawa, K. 1988. Cloning and nucleotide sequence of the 2,3-dihydroxybiphenyl dioxygenase gene from the PCB-degrading strain of *Pseudomonas paucimobilis* Q1. *Biochemicstry* 27:3990–96
131. Taylor, M. G., Massey, V. 1990. Decay of the 4a-hydroxy-FAD intermediate of

phenol hydroxylase. *J. Biol. Chem.* 265: 13687–94
132. Taylor, M. G., Massey, V. 1991. 6-Mercapto-FAD and 6-thiocyanato-FAD as active site probes of phenol hydroxylase. *J. Biol. Chem.* 266:8281–90
133. Deleted in proof
134. True, A. E., Orville, A. M., Pearce, L. L., Lipscomb, J. D., Que, L. Jr. 1990. An EXAFS study of the interaction of substrate with the ferric active site of protocatechuate 3,4-dioxygenase. *Biochemistry* 29:10847–54
135. Tsuda, M., Minegishi, K.-I., Iino, T. 1989. Toluene transposons Tn*4651* and Tn*4653* are class II transposons. *J. Bacteriol.* 171:1386–93
136. Unger, B., Jollie, D., Atkins, W., Dabrowski, M., Sligar, S. 1986. Site directed mutagenesis of cytochrome $P450_{CAM}$. *Fed. Proc.* 45:2298
137. van der Meer, J. R., Eggen, R. I. L., Zehnder, A. J. B., de Vos, W. M. 1991. Sequence analysis of the *Pseudomonas* sp. strain P51 *tcb* gene cluster, which encodes metabolism of chlorinated catechols: evidence for specialization of catechol 1,2-dioxygenases for chlorinated substrates. *J. Bacteriol.* 173:2425–34
138. van der Meer, J. R., Zehnder, A. J. B., de Vos, W. M. 1991. Identification of a novel composite transposable element, Tn*5280*, carrying chlorobenzene dioxygenase genes of *Pseudomonas* sp. strain P51. *J. Bacteriol.* 173:7077–83
139. Vriend, G., Sander, C. 1991. Detection of common three-dimensional substructures in proteins. *Proteins* 11:52–58
140. Wang, L.-H., Hamzah, R. Y., Yu, Y., Tu, S.-C. 1987. *Pseudomonas cepacia* 3-hydroxybenzoate 6-hydroxylase: induction, purification, and characterization. *Biochemistry* 26:1099–1104
141. Wasserfallen, A., Rekik, M., Harayama, S. 1991. A *Pseudomonas putida* strain able to degrade *m*-toluate in the presence of 3-chlorocatechol. *Bio/Technology* 9:296–98
142. Wende, P., Bernhardt, F.-H., Pfleger, K. 1989. Substrate-modulated reactions of putidamonooxin. The nature of the active oxygen species formed and its reaction mechanism. *Eur. J. Biochem.* 181:189–97
143. Whited, G. M., Gibson, D. T. 1991. Toluene-4-monooxygenase, a three-component enzyme system that catalyzes the oxidation of toluene to *p*-cresol in *Pseudomonas mendocina* KR1. *J. Bacteriol.* 173:3010–16
144. Wierenga, R. K., Terpstra, P., Hol. W. G. J. 1986. Prediction of the occurrence of the ADP-binding $\beta\alpha\beta$-fold in proteins, using an amino acid sequence fingerprint. *J. Mol. Biol.* 187:101–7
145. Witholt, B., de Smet, M.-J., Kingma, J., van Beilen, J. B., Kok, M., et al. 1990. Bioconversions of aliphatic compounds by *Pseudomonas oleovorans* in multiphase bioreactors: background and economic potential. *TIBTECH* 8:46–52
146. Yabusaki, Y., Murakami, H., Ohkawa, H. 1988. Primary structure of *Saccharomyces cerevisiae* NADPH-cytochrome P450 reductase deduced from nucleotide sequence of its cloned gene. *J. Biochem.* 103:1004–10
147. Yamaguchi, M., Fujisawa, H. 1982. Subunit structure of oxygenase component in benzoate 1,2-dioxygenase system from *Pseudomonas arvilla* C-1. *J. Biol. Chem.* 257:12497–12502
148. Yen, K.-M., Karl, M. R., Blatt, L. M., Simon, M. J., Winter, R. B., et al. 1991. Cloning and characterization of a *Pseudomonas mendocina* KR1 gene cluster encoding toluene-4-monooxygenase. *J. Bacteriol.* 173:5315–27
149. You, I.-S., Ghosal, D., Gunsalus, I. C. 1991. Nucleotide sequence analysis of the *Pseudomonas putida* PpG7 salicylate hydroxylase gene (*nahG*) and its 3'-flanking region. *Biochemistry* 30:1635–41
150. You, I.-S., Murray, R. I., Jollie, D., Gunsalus, I. C. 1990. Purification and characterization of salicylate hydroxylase from *Pseudomonas putida* PpG7. *Biochem. Biophys. Res. Commun.* 169:1049–54
151. Zeyer, J., Kocher, H. P. 1988. Purification and characterization of a bacterial nitrophenol oxygenase which converts ortho-nitrophenol to catechol and nitrite. *J. Bacteriol.* 170:1789–94
152. Zigmond, R. E., Schwarzschild, M. A., Rittenhouse, A. R. 1989. Acute regulation of tyrosine hydroxylase by nerve activity and by neurotransmitters via phosphorylation. *Annu. Rev. Neurosci.* 12:415–61
153. Zvelebil, M. J. J. M., Wolf, C. R., Sternberg, M. J. E. 1991. A predicted three-dimensional structure of human cytochrome P450: implications for substrate specificity. *Protein Eng.* 4:271–82
154. Zylstra, G. J., Gibson, D. T. 1989. Toluene degradation by *Pseudomonas putida* F1. *J. Biol. Chem.* 264:14940–46

Annu. Rev. Microbiol. 1992. 46:603–33

ARREST OF BACTERIAL DNA REPLICATION

Thomas M. Hill

Department of Bioscience and Biotechnology, Drexel University, Philadelphia, Pennsylvania 19104

KEY WORDS: termination, prokaryotic

CONTENTS

Abstract

The chromosomes of both gram-positive and gram-negative bacteria contain sites that arrest the progression of DNA replication forks. These replication-arrest sites limit the end of the replication cycle to a particular region of the chromosome, called the terminus region. Replication arrest is mediated by protein-DNA complexes that show polarity of function: they arrest DNA replication from one direction only. This paper reviews our current knowledge of the replication-arrest complexes of *Bacillus subtilis* and *Escherichia coli* and examines possibilities for the function and mechanism of action of these complexes within the bacterial cell.

0066-4227/92/1001-0603$02.00

INTRODUCTION

In the four decades that have passed since Watson & Crick published the structure of DNA, the overwhelming majority of research on DNA replication has focused on events that occur at replication origins during initiation and on the characterization of established, elongating replication forks. In recent years, interest has shifted to include the process of termination of DNA replication, which is defined here as the sequence of events that occur when two converging replication forks meet and conclude a cycle of DNA replication. To a large degree, the increased interest in replication termination is due to significant advances in describing and characterizing sites in bacterial chromosomes that arrest the progression of replication forks. The discovery of these replication arrest systems, along with the identification of new topoisomerases in *Escherichia coli* that possibly play a role in chromosome decatenation or partitioning (18, 44, 63), the isolation of mutants in the partitioning apparatus (38), and the characterization of genes controlling cell division (reviewed in 16) have combined to make the events surrounding the conclusion of the DNA replication cycle and the onset of cell division one of the most exciting and active areas of prokaryotic research.

In the strictest sense, the identification and characterization of sites that arrest DNA replication in bacterial chromosomes do not provide us with information about the events that occur at termination of replication per se, but only about the interaction of the replication arrest components with the replication fork. This distinction is made because the word "termination" as described above is often used interchangeably with the phrase "replication arrest," which is defined here as the inhibition of replication-fork progression. In reality, the process of replication arrest is only one step of many during the termination process. That replication arrest is not synonymous with replication termination is evidenced by: (*a*) replication termination occurs in the *E. coli* chromosome at locations other than the replication arrest sites and (*b*) in both *E. coli* and *Bacillus subtilis,* inactivation of sites that arrest DNA replication has no apparent effect on cell growth, suggesting that replication termination proceeds normally in the absence of functional arrest sites. Consequently, this review is restricted to a discussion of replication arrest in prokaryotes and focuses exclusively on the DNA-binding proteins and specific DNA sequences of *B. subtilis* and *E. coli* that halt DNA replication.

BACILLUS SUBTILIS

The Replication Terminus

Many of the ground-breaking experiments that characterized bacterial replication arrest came in *Bacillus subtilis,* particularly from the laboratory of R. G.

Wake. These studies eventually led to the identification of two components of the *B. subtilis* replication arrest system, the inverted repeat region (IRR) of the chromosome (containing the DNA sequences IR I and IR II) and the DNA-binding replication terminator protein (RTP), which associates with IR I and IR II. Although many characteristics of the replication-arrest systems of *B. subtilis* and *E. coli* are shared, several striking dissimilarities suggest that the mechanics of replication inhibition in these two systems are decidedly different.

Bidirectional replication of the chromosome of the gram-positive, spore-forming eubacterium *B. subtilis* is initiated from a unique origin located near the *purA* gene and ends when the two replication forks meet on the opposite side of the circular chromosome in the vicinity of the *gltA* and *citK* loci (25, 40). The region where forks converge has been designated the chromosomal terminus, or *terC*. Evidence that this region of the chromosome was not simply the place where replication forks met most often came from genetic studies of *B. subtilis* strain GSY1127, which contains a nontandem duplication of 25% of the counterclockwise-replicated arm of the chromosome. This duplication results in an asymmetric chromosome, placing an *ilvC* marker directly opposite the origin of replication and shifting the *terC* region considerably closer to *oriC* on the clockwise-replicated arm of the chromosome. Using this strain, O'Sullivan & Anagnotopoulos (72) determined transformation frequencies for genetic markers on both arms of the chromosome during spore outgrowth. They observed that the *gltA/citK* region, or *terC*, was replicated after *ilvC*, even though the clockwise replication fork should have arrived at *terC* well before the counterclockwise fork reached *ilvC*. This result suggested that clockwise replication was specifically inhibited at the normal replication terminus near *gltA* rather than at a point halfway around the chromosome.

Further characterization of the interesting properties displayed by *terC* required an accurate physical map of the region. By taking advantage of certain properties of the sporulation process, investigators could radioactively label only terminus-region DNA, permitting an unambiguous restriction map of approximately 200 kb of the terminus region to be constructed (67, 84). In addition to providing a map of the terminus region, this study helped establish a more precise determination of the site of replication arrest. Careful analysis of the incorporation of label into the various restriction fragments showed that one side of the terminus region was labeled to a much greater extent than the other and that the level of incorporation shifted abruptly at one particular position. Based on these results, Weiss & Wake proposed that the clockwise-traveling replication fork arrived at *terC* five minutes in advance of the counterclockwise replication fork and that an impediment to the progression of the clockwise replication fork was located in a 24.8-kb *Bam*HI restriction fragment (84, 85). Furthermore, they speculated that arrest of replication at *terC* should transiently produce a Y-structure in the *terC* restriction fragment during the period between inhibition of the clockwise replication fork and arrival of the counterclockwise fork. This Y-structure should be distinguishable from the normal, linear *terC* restriction fragment on the basis of different mobilities during electrophoresis. This was indeed the case, as Southern blots of *Bam*HI-digested DNA from strains 168 (wild-type) and GSY1127 both showed two bands that hybridized to a *terC* probe: the 24.8-kb *terC* linear fragment and a second, more slowly migrating band (86). This slowly migrating band was later characterized as a Y-structure containing an arrested replication fork 15.4 kb from one end (26, 87). These results, in combination with studies localizing the position of the Y-junction to a 1.75-kb restriction fragment (81) demonstrated for the first time that replication arrest in a bacterial chromosome occurred at a very specific position.

The *terC* region does not appear to be necessary for vegetative growth or sporulation of *B. subtilis,* as determined by studying the growth characteristics of strains in which the terminus region was deleted. Several strains containing deletions that originated in the SPβ prophage and removed the *citK* and *gltA* loci (92) were characterized in order to define the endpoints of the deletions (42). The junction fragment of the smallest deletion was identified, demonstrating conclusively that 230 kb of terminus-region DNA, including *terC,* were removed in this strain. When the growth rate and morphology of this deletion strain was compared to wild-type cells, no difference was observed, indicating that vegetative growth was undisturbed by loss of *terC*. However, this deletion strain did not sporulate. The inability to sporulate may have been a direct result of *terC* loss, or could have resulted from elimination of other sporulation genes in the 230-kb deletion. To determine if loss of *terC* had a specific effect on sporulation, another deletion strain was constructed that only removed 11.2 kb, including the *terC* region. This strain showed normal vegetative growth characteristics and could sporulate as well, indicating that *terC* was dispensable for the sporulation process (42).

THE INVERTED REPEAT REGION AND RTP When the nucleotide sequence of the region in *B. subtilis* strain 168 containing *terC* was determined, the most striking feature was the presence of two large imperfect, inverted repeats of 47–48 nucleotides (12). These inverted repeats, designated IR I and IR II, shared 77% identity, were separated by 59 nucleotides, and were located between two open reading frames (ORFs) (see Figure 1). A more precise determination of the site where the clockwise replication fork was immobilized placed the Y-structure junction in the vicinity of IR II, suggesting that these inverted repeats played a major role in the inhibition of replication (12). This supposition was confirmed when a series of deletions in the *terC*

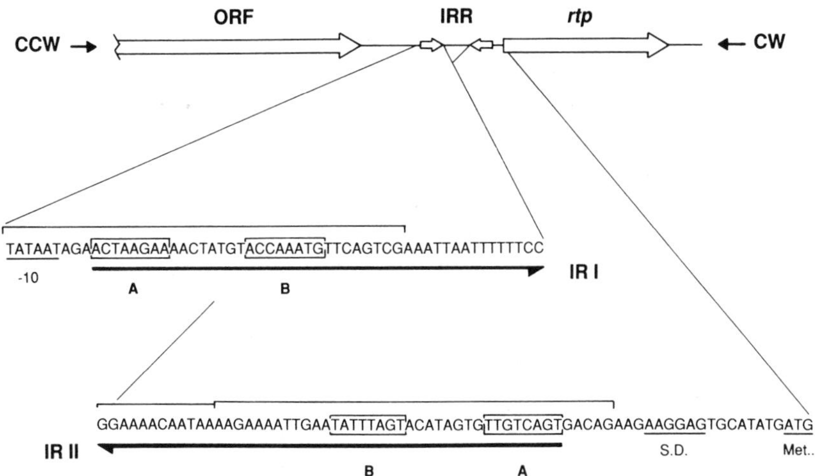

Figure 1 The *TerC* region of the *B. subtilis* chromosome. The direction of movement of clockwise (CW) and counterclockwise (CCW) replication forks as they approach the *TerC* region are indicated by the arrows. The *rtp* gene encodes a DNA-binding protein that binds to the inverted repeats IR I and IR II (denoted by the heavy bar beneath the two sequences), which are located within the inverted repeat region (IRR). The single bracket above the IR I sequence indicates the region protected from DNase I treatment by RTP binding. The double bracket above the IR II sequence indicates the regions protected from DNase I digestion at low and high concentrations of RTP. The bracket covering the 41 nucleotides on the right half of IR II indicates that they are protected at relatively low RTP concentrations (1.3×10^{-8} M); at higher concentrations of RTP (2.5×10^{-8} M), 12 additional nucleotides on the left half of IR II are protected by RTP binding. The boxed nucleotides labeled A and B are the proposed recognition sequences for RTP binding within the inverted repeats IR I and IR II. The underlined nucleotides labeled −10, S.D., and Met indicate promoter sequences, Shine-Delgarno sequences, and the start codon, respectively, for the *rtp* gene.

region were constructed and crossed into the chromosome of GSY1127 (82). Deletions starting on the left side of the *terC* region (as depicted in Figure 1) that removed the left ORF and extended up to, but did not include IR I, had no effect on replication arrest. However, deletions that removed IR I, but not IR II or the right ORF, abolished the impediment to replication, indicating that IR I was a necessary component of the *terC* fragment.

The IR elements alone, however, were not sufficient to halt replication forks. Deletions that started on the right side of the *terC* region and extended into the right ORF, but did not remove either IR II or IR I, also abolished replication arrest. This suggested that the protein product of this gene was required for inhibition of replication, possibly as a DNA-binding protein that associated with the IR elements (82). In an effort to confirm the importance of this gene, which has been designated *rtp* (replication terminator protein),

Lewis & Wake (57) sequenced the *terC* region of strain W23, whose genome has 67–89% DNA sequence homology to the chromosome of strain 168 (62, 78). Although there were 22 substitutions in the 366 bases of the W23 *rtp* gene, the amino acid sequence was identical to that of the *rtp* gene of strain 168, suggesting that function of the protein was conserved between the two strains. This study also identified a putative promoter for the *rtp* gene for both strains, with the -10 box positioned just three bases 5′ of IR I (Figure 1). IR I and IR II are therefore located between the promoter and the start codon of the *rtp* gene, raising the possibility that expression of this gene is autoregulated (82).

Based on the DNA sequence of the *rtp* gene, the replication terminator protein is a small, basic polypeptide of 122 amino acids with a calculated pI of 9.2. RTP shows homology to the DnaB protein of *B. subtilis* (49), which is involved in initiation of DNA replication and is not to be confused with the DnaB protein of *E. coli,* which is the major replicative helicase. The homology shared by RTP and DnaB may be part of a DNA-binding domain or could indicate a common ancestry for these replication-initiation and -termination proteins. No significant amino acid homology between RTP and the replication-arrest protein of *E. coli,* Tus, was detected (37). RTP has been overexpressed, purified to homogeneity, and migrates on SDS polyacrylamide gels with an apparent molecular mass of 14,500 (56), very similar to the predicted size of a RTP monomer. To determine its native configuration, sedimentation equilibrium studies of RTP were performed and demonstrated an apparent molecular weight of 29,000, consistent with a dimer (55). Dimerization depended upon RTP concentration, with an equilibrium association constant K_{12} for the monomer-to-dimer transition of 2×10^6 liters mol^{-1} and a K_{24} of 1×10^4 liters mol^{-1} for the dimer-to-tetramer transition. Thus, the dimeric form of RTP is the predominant species between 5×10^{-7} and 1×10^{-4} M in these buffer conditions. If these equilibrium constants reflect the true physiological values for the monomer-dimer transition, then approximately 70 molecules of the RTP monomer would be required intracellularly for 50% dimer formation.

Gel retardation assays were used to test the hypothesis that purified RTP would bind specifically to the IRR (inverted repeat region) containing IR I and IR II. Addition of low concentrations of purified RTP to a 209-bp fragment containing IRR (molar ratio of RTP:IRR = 7) produced four slowly migrating bands, presumably resulting from protein-DNA complexes containing one, two, three, or four molecules of RTP (56). As the RTP:IRR ratio was increased to 80, all of the IRR DNA was shifted into the slowest-migrating species. These results not only demonstrated a specific interaction between RTP and the IRR, but also suggested that each IR element contained two binding sites for RTP. This was confirmed later when DNA fragments containing only IR I or IR II were complexed with RTP and each IR fragment produced two slowly migrating bands on the gel (55).

A more detailed analysis of the binding interaction between RTP and the IRR allowed an estimation of the average equilibrium constant for RTP binding to a single site (55). RTP bound to the intact IRR (containing four binding sites) with an average K_D of 1×10^{-11} M for a single RTP:IRR interaction, indicating a strong affinity of RTP for its binding sites. Stoichiometric studies determined that the number of RTP monomers bound to a saturated RTP:IRR complex was seven to eight, suggesting that RTP bound as a dimer (55), a result consistent with the dimerization of unbound RTP. To measure the K_D, concentrations of RTP were used that were several orders of magnitude below the monomer-dimer equilibrium dissociation constant of 5×10^{-7} M, suggesting that the monomeric rather than the dimeric form of RTP was the predominant molecular species. However, the authors concluded that the high levels of glycerol (50%) used in binding and loading buffers for the K_D studies enhanced the formation and binding of the dimeric RTP molecules of such low protein concentrations. They demonstrated that the efficiency of RTP binding was greatly reduced in buffers containing less than 30% glycerol and suggested that the most likely explanation for the increased binding in the presence of glycerol was a simple excluded-volume effect or a glycerol-induced conformational change during the RTP/IRR interaction resulting in compaction of the protein-DNA complex.

DNase I footprinting studies of the RTP/IRR complex identified the RTP binding sites within the IRR (55). At the lowest concentrations of RTP tested (6.4×10^{-9} M), a 41-bp region that overlapped IR I was the only part of the IRR protected from digestion (Figure 1). As the concentration of RTP was increased to 1.3×10^{-8} M, a second protected region of 41 bp was observed that overlapped IR II. As the RTP concentration was increased further, to 2.5×10^{-8} M, an additional 12 bp of the IR II region was protected from DNase I digestion. This observation was consistent with the proposal that each IR contained two RTP binding sites. The concentration-dependent order of binding observed with the DNase I footprinting studies was also consistent with the equilibrium constant study described above, which indicated positive cooperativity amongst the sites for RTP binding and suggested that the sites were not equivalent.

The fully protected regions of both IR I and IR II were not positioned in the center of the homologous sequences, but instead covered about two-thirds of one end of the inverted repeats (Figure 1). When the protected regions from both IR elements of strain 168 were aligned with the equivalent sequences from the IRR of strain W23, the only potential candidate for the RTP binding sequence that was present twice in each protected region was an 8-bp segment with the consensus sequence ACYRARA/TR. The candidate sequences, designated A and B (Figure 1), were present as direct repeats separated by only 8 bp. They did not show the twofold rotational symmetry often associated with the recognition sites of dimeric DNA-binding proteins, suggesting that the IR elements might show polarity of function. Indirect evidence suggesting a role for this 8-bp sequence in RTP binding was obtained when a DNA fragment containing only the IR I-B box was tested for binding in the gel retardation assay and shown to produce only a single retarded species, even at high RTP concentrations (55).

By analogy with the observed polarity of replication arrest associated with the *Ter* sites of *E. coli* (described below), Lewis et al (55) suggested that IR I was responsible for arrest of the clockwise replication fork and that the clockwise fork passed through IR II. This possibility was tested by synthesizing a series of oligodeoxyribonucleotides corresponding to sequences at different positions upstream of IR I on the clockwise arm of the chromosome (89). These single-stranded probes were then hybridized to either the leading or lagging strand of the immobilized clockwise replication fork from strain GSY1127 to determine the location of replication arrest in the *terC* fragment. Although the majority of arrested leading strands were halted before reaching IR II, a small but significant number of leading strands traversed IR II and entered the 59-bp spacer region between IR I and IR II. No leading strands were detected beyond IR I. Probes specific for lagging-strand synthesis showed that no lagging strands passed through IR II. These results suggested that IR I was the primary impediment to at least a portion of the leading strands of the clockwise replication fork, consistent with the hypothesis that the IR elements were polar for replication arrest.

Recently, a similar result demonstrating that replication forks can traverse IR II but not IR I was obtained by inserting the IRR/*rtp* region into a plasmid (10). In this study, the IRR/*rtp* region was oriented such that the unidirectional replication fork initiated from the plasmid origin would encounter the *terC* region in the order *rtp*/IR II/IR I, the same order encountered by the clockwise replication fork in the chromosome. The products of leading-strand arrest were mapped at nucleotide resolution and shown to pass through *rtp* and IR II before ending within IR I, adjacent to the IR I-B binding site. The primary arrest sites of leading-strand synthesis were one and two nucleotides upstream of and one nucleotide within the IR I-B binding site. These results suggested that the polymerase replicated through the RTP-binding sites in IR II, but halted at the first properly oriented RTP bound in IR I. The arrest of leading-strand synthesis within a couple of nucleotides of the RTP-binding site is similar to the observation in *E. coli* that replication arrest occurs primarily within the binding site of the replication-arrest protein Tus (35, 53a).

RELOCATION OF *TERC* IN THE *B. SUBTILIS* CHROMOSOME Another approach to test the hypothesis that the IR elements function in a polar fashion was to relocate the *terC* fragment at different positions around the chromosome. Initially, a strain was constructed with a 1.75-kb IRR/*rtp* fragment inserted in the usual orientation into the chromosome 25 kb clockwise from the normal *terC* site, which had been deleted. The displaced *terC* functioned normally in this strain, indicating that the IRR/*rtp* fragment could be moved to different locations in the chromosome and still retain activity (41). Recently, Carrigan et al (13) built strains with the *terC* fragment relocated on the clockwise arm of the chromosome at positions 100° or 139° (on a 360° map) and in either orientation. These strains were tested to determine if the *terC* fragment retained the ability to arrest clockwise replication forks regardless of its orientation. Surprisingly, only the normal orientation of *terC* arrested the clockwise replication fork. When IR II was positioned to impede clockwise replication, it failed to do so, demonstrating that the IRR/*rtp* fragment as a unit showed polarity. Given the sequence similarities between IR II and IR I and the demonstration that RTP binds to both, the simplest explanation for these results is that IR II is unoccupied at normal intracellular concentrations of RTP (13). This postulation is consistent with the observation that RTP binds to both IR I-A and IR I-B sites before either of the IR II sites are bound. In addition, if occupation of IR I represses expression of the *rtp* gene (IR I-A overlaps the presumed transcription start), this will prevent further synthesis of RTP and keep the intracellular level of RTP below that necessary to bind both IR II sites.

Future Prospects

Analysis of replication arrest in *B. subtilis* will remain a fertile field for years to come. For instance, it is surprising that the recognition sequence for RTP binding is so small, because the 8-bp consensus binding site would be expected to be present approximately 2000 times in a 4-million-basepair genome. However, if replication arrest required the juxtaposition of two RTP-binding sites as a direct repeat separated by a fixed-length spacer, then the number expected per genome would drop drastically. This then raises the issue of how many RTP-binding sites are necessary for replication arrest. Can a single RTP-binding site function to halt replication, as has been shown for the *Ter* sites in *E. coli* (discussed below), or are two adjacent RTP binding sites necessary? If two RTP-binding sites are required, how do the two RTP dimers interact with the oncoming replication fork? Unless a DNA loop is formed, which was not detected in the DNase I footprinting studies (55), then the replication fork will encounter the two RTP dimers sequentially rather than simultaneously, perhaps suggesting that the replication fork contains two copies of the target of RTP, possibly one on each template strand. Also, if the IR II element is truly inactive or is rarely, if ever, occupied by RTP, why has it been conserved? Will additional functional IR elements be discovered on the two arms of the chromosome, as appears to be the case for the *E. coli Ter* sites, or are the only RTP binding sites in the chromosome found in *terC*? Finally, experiments directly addressing the molecular mechanism of replication arrest in *B. subtilis* will be of paramount importance, because these alone will determine if the replication-arrest systems of gram-positive and gram-negative bacteria are truly different.

ESCHERICHIA COLI

The concept of a replication terminus in *E. coli* was first advanced by Masters & Broda (65) and Bird et al (5), who concluded that bidirectional replication of the chromosome was initiated from a unique origin, *oriC* (min 84), and that the two replication forks met on the opposite side of the chromosome in the interval between the *trp* (min 28) and *his* (min 44) loci (Figure 2). In the late 1970s, investigators showed that the region where the converging replication forks met contained an impediment to replication-fork progression. Using *E. coli* strains with phage or plasmid replication origins integrated asymmetrically in the chromosome, the laboratories of Kuempel (50–52) and Louarn (59, 60) demonstrated that clockwise and counterclockwise replication forks were inhibited between the *trp* (min 28) and *manA* (min 36) loci. Thus, replication termination events were confined to this particular part of the chromosome, which was designated the terminus region.

Few genetic loci have been mapped to the terminus region compared with the remainder of the chromosome. The lack of genetic markers in the terminus was first observed during bacterial conjugation experiments, when approximately 8 min passed between transfer of the *trp* and *manA* markers. Two possible explanations could account for the genetic gap in the terminus region. One explanation was that *trp* and *manA* were in fact positioned very close to one another on the physical map, but the terminus region–replication impediment reduced the rate of conjugation by interfering with DNA transfer, causing a delay between the two markers (52). The other explanation was that the delay between transfer of the *trp* and *manA* markers simply reflected the physical distance between the two loci. This ambiguity regarding the actual physical structure of the terminus region was resolved in the early 1980s when a cotransduction map spanning the 8-min interval between *trp* and *manA* loci was published (6, 19) and when Bouché constructed a restriction map of 450 kb that included the entire terminus region, demonstrating that 380 kb separated *trp* and *manA* (8, 9). However, in spite of the advances in our understanding of the structure of the terminus, this region still has relatively few

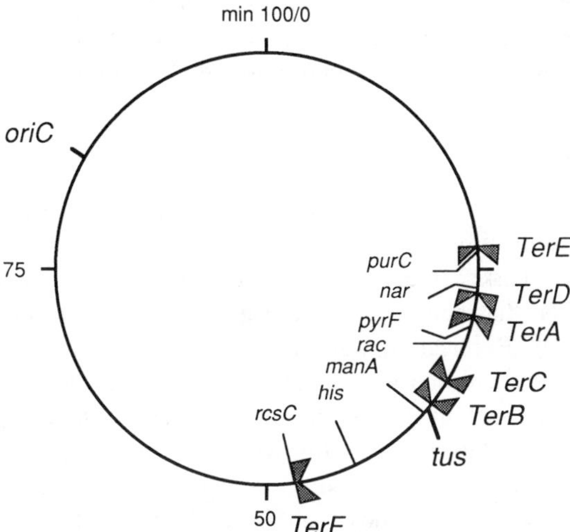

Figure 2 A map of the *Escherichia coli* chromosome showing the location of the termination sites (*TerA–TerF*) relative to the location of the origin of replication and other genetic markers. The sites *TerA*, *TerD*, and *TerE* arrest counterclockwise-traveling replication forks only, whereas *TerB*, *TerC*, and *TerF* arrest clockwise-traveling replication forks. The location of the *tus* gene, which produces the *Ter*-binding protein Tus, is also presented.

known genetic markers. Of the 1403 loci identified on the most recent map of the *E. coli* chromosome (1), only 40 or 2.9% reside within the terminus region, although this interval is equal to 7.7% of the total *E. coli* genome.

The *E. coli* terminus, like its counterpart in *B. subtilis,* is dispensable for cell viability, indicating that no essential genes are present in this part of the chromosome. An *E. coli* strain was isolated with a 340-kb deletion that removed almost all of the terminus region (27), and although the cells were filamentous, they were still viable. The barrier to replication forks was also removed in this strain, but this was not the reason for filamentation. Kuempel et al (53) recently reported that loss of the *dif* locus (min 33.5), a site for the RecA-independent recombination pathway mediated by XerC (7), caused the filamentation. Thus, for standard growth conditions in the laboratory, only deletion of the *dif* locus causes an observable phenotype; removal of the rest of the terminus region has no apparent effect on cell growth.

The feature of the terminus region that has received the most attention in recent years is the barrier to replication-fork progression, or the replication-arrest system. Two components have now been identified that are required for replication arrest in *E. coli*. These are the *Ter* sites, which are nonpalindromic DNA sequences of approximately 20 bp, and the Tus protein, which is a DNA-binding protein that recognizes and binds specifically to the *Ter* sequences. Although these components have been designated by different names in other publications, the nomenclature adopted by Bachmann in the latest genetic map of the *E. coli* chromosome (1) is used exclusively in this review.

The Ter *Sites*

IDENTIFICATION OF THE CHROMOSOMAL *TER* SITES The first studies that characterized replication arrest in the terminus region of the *E. coli* chromosome employed marker frequency assays to determine the relative amount of DNA replication that had occurred at a given position on the chromosome (50–52, 59, 60). Replication was initiated from plasmid or phage origins integrated into the chromosome at positions close to the terminus region, allowing an unambiguous assessment of the progress of the replication fork traversing the terminus region. However, the sensitivity of these marker frequency assays entirely depended on the number of DNA probes used for hybridization, and few chromosomal probes were available initially. As a result, the first replication assays only localized the replication block to broad areas of the terminus region. Clockwise replication forks were halted somewhere between *rac* (min 30) and *aroD* (min 37) and counterclockwise replication forks were inhibited between *rac* and *trp* (51, 52, 59, 60). In addition, clockwise replication forks could pass through the region where inhibition of counterclockwise replication occurred (51), the first hint that the impediment to replication functioned in a polar fashion.

Independent reports from Louarn's (17) and Kuempel's (33) laboratories separated the replication block into distinct loci that were specific for arrest of clockwise or counterclockwise replication forks and clearly demonstrated the polarity of the replication-arrest sites. Using an extensive battery of chromosomal probes in the marker frequency assay, both groups showed that two replication-arrest sites were situated at the extremities of the terminus region. Arrest of replication at these sites was not absolute; replication forks were eventually released and continued past the replication-arrest sites. One site, *TerA,* was located close to *pyrF* at 28.5 min and arrested only counterclockwise replication forks. A second site that only halted clockwise replication forks, *TerB,* was positioned on the opposite side of the terminus region. The map location of *TerB* by the two groups was different; deMassy et al (17) placed *TerB* at min 33.5, and Hill et al (33) placed it between min 34.5 and 35.7, near *manA*. This discrepancy was resolved when it was determined that two arrest sites (*TerC* and *TerB*) for clockwise replication were present on this side of the terminus (discussed below). The location and polar function of *TerA* and *TerB* suggested that, unlike *terC* in *B. subtilis,* the replication-arrest sites in the *E. coli* chromosome were widely separated and positioned at the edges of the terminus region to form a replication-fork trap, allowing replication forks to enter but not exit this region.

Even with a greater number of chromosomal probes, the marker frequency assays still lacked the sensitivity to accurately locate the sites of replication arrest. A more precise determination of the location of replication arrest was ultimately achieved by a series of techniques, including marker frequency analysis of deletion mutants (34), insertion of phage origins into the terminus region (20, 21), electrophoretic analysis of the Y-structures formed by arrested replication forks (73), and characterization of the replication products of plasmids carrying the chromosomal *Ter* sites (29, 74). These studies reduced the region of replication arrest to a size amenable to sequencing.

The *Ter* sites were initially identified as nonpalindromic 22- to 23-bp sequences (Figure 3) that resided within 100 bp of the site of replication arrest (29, 36). The lack of dyad symmetry conferred a directionality to the sequences, consistent with the observed polarity of replication arrest. Thus, the orientation of the *TerA* sequence in the chromosome was inverted with respect to the orientation of the *TerB* sequence. In addition to the *TerA* (min 28) and *TerB* (min 36) sites reported by Hill et al (36), Hidaka et al (29) also identified the sequences for *TerC* (min 34), which correspond to the replication-arrest site mapped by deMassy et al (17), and *TerD* (min 27). More recently, *TerE* (min 23) (30) and *TerF* (min 48) (B. Sharma & T. Hill, unpublished results) were reported (Figure 3). The six identified termination sites are asymmetrically distributed over approximately 25% of the total chromosome, with *TerA, TerD,* and *TerE* positioned to arrest counterclockwise replication, and *TerB, TerC,* and *TerF* positioned to arrest clockwise replication (Figure 2). These six *Ter* sites are probably not the only ones in the chromosome, since François et al (20) probed the *E. coli* genome with an oligomer containing the *TerA* site and identified 10 different genomic fragments containing homology to the *Ter* sites.

TER SITES IN PLASMIDS Replication-arrest sites have also been identified in several plasmids of gram-negative bacteria. The most well-characterized plasmid-replication terminus is in R6K. R6K replication is initiated unidirectionally from one of three origins (14, 43), and the replication fork proceeds to the terminus where it is arrested. A second replication fork is then initiated from the same origin that proceeds in the opposite direction and meets the arrested fork at the terminus region (61). Kolter & Helinski showed that the R6K terminus was active when inserted into a ColE1-type plasmid, which replicates unidirectionally, and that either orientation of the R6K terminus could arrest replication (48). The orientation-independent function of the R6K terminus was explained when the nucleotide sequence was deter-

	Sequence
Chromosomal	
TerA	A A T T A G T A T G T T G T A A C T A A A G T
TerB	A A T A A - T A - - - - - - - - - - - A A G T
TerC	A T A T A - G A - - - - - - - - - - - A T A T
TerD	C A T T A - T A - - - - - - - - - - - A A T G
TerE	T T A A A - T A - - - - - - - - - - - A G N N
TerF	C C T T C - T A - - - - - - - - - G - C G A T
Plasmid	
R6K*terR1*	C T C T T - T G - - - - - - - - - - - A A T C
R6K*terR2*	C T A T T - A G - - - - - - - - - - - C T A G
R100*terR1*	A T T A T - A A - - - - - - - - - - - C T T C
R100*terR2*	T G T C T - A G - - - - - - - - - - - A A G C
R1*terR1*	A T T A T - A A - - - - - - - - - - - C A T C
R1*terR2*	T T T T T - T G - - - - - - - - - - - A A G T
P307	A T T A T - A A - - - - - - - - - - - C A T T
Consensus	**N N A/T A/T A/T G A/T A/G T G T T G T A A C T A A/C N N N**
Position	5 (pos. 5), 10 (pos. 10), 15 (pos. 15), 20 (pos. 20)

Figure 3 Comparison of known *Ter* sequences from the *E. coli* chromosome and other replicons. As presented, these *Ter* sites will arrest replication forks approaching from the 5' side of the sequence and will allow replication forks approaching from the 3' side to pass unimpeded. The *TerC, TerD* and *TerE* sequences are from Hidaka et al (30), and the *TerF* sequence is from B. Sharma & T. Hill (unpublished data). All other sequences are from Hill et al (36). Nucleotides that represent the core sequences identical for all *Ter* sites are indicated by a dash.

mined (2) and two highly homologous inverted repeats were identified that were 20 bp in length and separated by 73 bp (12). Insertion of the individual sequences into ColE1 plasmids showed that each site could arrest plasmid replication in an orientation-dependent manner (39). Thus, the two polar R6K *Ter* sequences formed a replication-fork trap and halted replication from either direction. The extensive homology between the R6K *Ter* sites and the chromosomal *Ter* sites (29, 36, 39) suggested that the two operated by a similar mechanism. The demonstration that the R6K *Ter* sites, like the chromosomal *Ter* sites, formed a complex with Tus protein and depended on the presence of a functional *tus* gene (discussed below) confirmed this supposition (46, 74, 80).

The RepFIIA family of replicons also contains *Ter* sites, but unlike R6K, the *Ter* sites are located close to the plasmid origin of replication. The order of loci in the origin region of these plasmids is:

cw ⇒ *repA1* (5'→3') - *ori* (→) - *TerR2* (<) - *TerR1* (>) ⇐ ccw

where cw and ccw indicate the clockwise and counterclockwise directions, respectively. The *repA1* gene encodes a protein required for replication initiation (76). Replication occurs unidirectionally from the origin, *ori* (70), positioned approximately 200 bp from the 3' end of the *repA1* gene. The origin contains binding sites for the RepA protein (64), and replication is directed away from the *repA1* gene, as indicated by the arrow. In the plasmids R1 and R100, two *Ter* sites, *TerR1* and *TerR2*, are present as an inverted repeat and are positioned about 500 bp from the 3' end of the *repA1* gene. In the RepF1C origin of P307 (77), only the *TerR1* site is present, also located 500 bp 3' of the *repA1* gene. The *TerR1* site is oriented to arrest clockwise replication, and the *TerR2* site will arrest counterclockwise replication forks. An initial postulation was that the *TerR1* site might arrest the nascent replication fork if replication was actually initiated from the RepA-binding sites in the minimal origin region (36). However, because the *TerR1* site was 10-fold less efficient than the *TerR2* site in vivo, it was suggested that the newly synthesized replication fork would only be transiently arrested at this site and would ultimately pass through the block site to replicate the plasmid genome. More recent studies (66) have demonstrated that, although the minimal replication-origin region was adjacent to the *repA1* gene, the 5' end of the leading strand was initiated beyond both the *TerR1* and *TerR2* sites, approximately 800 bp from the 3' end of the *repA1* gene and bypassing the *TerR1* site altogether.

The role of the *Ter* sites in the replication cycle of plasmids remains obscure. Interestingly, deletion of the region containing *TerR1* in R100 results in the formation of plasmid multimers (71). In the absence of *TerR1*, rolling-circle replication may be initiated, leading to multimer formation. Alternately, the *Ter* sites may assist in the resolution of the plasmid multimers by an unknown mechanism. Inserting *Ter* sites into plasmids also reportedly affects plasmid maintenance. Introduction of a properly oriented *TerB* site into a hybrid plasmid containing a phage M13 origin of replication caused plasmid instability (83). Maintenance of the plasmid could be increased by overexpressing the phage replication protein gpII, and it was suggested that the observed instability resulted from replication arrest at the *TerB* site.

COMPARISON OF THE *TER* SEQUENCES Figure 3 lists the 13 known *Ter* sequences from plasmids and the *E. coli* chromosome, along with a derived consensus sequence. Although 22 to 23 basepairs were originally identified as the length for the *Ter* sites, the truly invariant nucleotides are restricted to the G residue at position 6 and the 11-bp sequence between residues 9 through 19. The one exception to this rule is *TerF*, which has a G substitution at position 18. Of the 32 interference and protection sites mapped in the Tus-*TerB* complex (24) (discussed below), 29 are located in the core region between positions 6 and 19, confirming the importance of these basepairs in the *Ter* sequence. The four nucleotides from positions 2 through 5, although not invariant nor protected by Tus binding, are generally A-T rich. The conservation of these A-T residues suggests that sequence context may influence Tus binding, although this has not yet been directly tested. Sista et al (79) have reported the effects of several mutations in the core region on Tus binding and replication-arrest activity. The *Ter* sites that contained single mutations at positions 10, 12, 13, and 14 did not bind Tus protein and consequently could no longer form productive replication-arrest complexes. However, Tus was able to bind to *Ter* sites containing single substitutions at positions 6, 8, 11, 16, and 18, and these mutated sequences still retained the ability to arrest DNA replication.

The Tus Protein

IDENTIFICATION OF A *TRANS*-ACTING TERMINATION FACTOR The first suggestion that a protein might be involved in replication arrest came in the early 1980s from Germino & Bastia, who examined cell-free replication of hybrid R6K/ColE1 plasmids. Replication of the hybrid plasmid substrates in vitro was arrested at the R6K terminus regardless of whether the crude cell extracts were prepared from cells that carried an R6K plasmid or not, suggesting that R6K did not encode a factor necessary for replication arrest (23). These results led the authors to speculate that any *trans*-acting factor needed for replication arrest must be supplied from the host cell.

The gene required for replication arrest in the *E. coli* chromosome was identified by Hill et al (34) in a genetic study using deletion mutations to map the location of the *TerA* and *TerB* sites. They observed that deletions that removed the *TerB* site also inactivated the *TerA* site, even though over 300 kb separated the two sites. The deletion mutants regained *TerA* activity if a plasmid carrying the *TerB* region was provided. It was postulated that a *trans*-acting factor encoded by a gene in the vicinity of *TerB* was required for replication arrest and that this gene encoded a DNA-binding protein that associated with the *Ter* sites. The gene was named *tus* (terminus utilization substance). Sequencing of the *tus* gene region revealed an open reading frame encoding a basic protein with a predicted molecular mass of 36,000 (32, 37). The *tus* gene produced a DNA-binding protein that associated with the *Ter* sites (37, 46, 80), and inactivation of the *tus* gene by insertional mutagenesis abolished DNA binding in vitro and replication arrest in vivo (37, 46). The predicted amino acid sequence of the Tus protein did not contain significant homology to the helix-turn-helix, leucine zipper, or zinc-finger motifs common to other known DNA-binding proteins (37).

EXPRESSION OF THE *TUS* GENE The *TerB* site was located immediately upstream of the *tus* gene, between the promoter and the start codon, suggesting that *tus* gene expression was regulated by Tus binding to the *TerB* site

(37). Both in vivo (32, 74a, 75) and in vitro (68) analysis of *tus* gene expression verified autoregulation. Hidaka et al (32) introduced a single mutation into the *TerB* site that inactivated Tus binding and subsequently increased *tus* gene expression, suggesting autoregulation. Natarajan et al (68) and Roecklein and coworkers (74a, 75) have mapped the 5' end of the *tus* mRNA and demonstrated that the primary transcriptional start for the *tus* gene lies within the *TerB* sequence. In the presence of Tus, the level of transcription was at least 30-fold lower than in the absence of Tus, suggesting that RNA polymerase was prevented from transcribing the gene. This hypothesis was confirmed in vitro by DNase I footprinting studies that showed that binding of Tus excluded binding of RNA polymerase (68). Thus, Tus binding at the *TerB* site occludes RNA polymerase and represses transcription of the *tus* gene.

The region containing the *tus* gene encodes several other genes as well (75). Immediately 3' of the *tus* gene is *fumC*, which encodes a fumarase whose function is unknown (91). The *tus* and *fumC* genes are transcribed convergently, and the stop codon of the *tus* gene overlaps the stop codon of the *fumC* gene. A potential rho-independent terminator for *tus* resides within the coding sequences of *fumC*, and the transcriptional terminator for *fumC* is positioned within *tus* (90). The effect, if any, of *fumC* transcription on the expression of *tus* has not been determined. Two genes encoding potential sensor/regulator proteins, dubbed *urpT* (unidentified regulatory protein/Terminus) and *uspT* (unidentified sensory protein/Terminus), were identified 5' of the *tus* gene. The 3' end of *uspT* was only 78 bp upstream of *tus*, with no obvious transcriptional terminator present in the intragenic region. Transcripts originating in these upstream genes did not extend into the *tus* gene in a tus^+ strain (74a). However, in a tus^- strain, readthrough transcription was observed, suggesting that Tus binding to *TerB* not only regulates *tus* expression by promoter occlusion, but also by impeding the progression of an actively transcribing RNA polymerase. In spite of the proximity of *urpT* and *uspT* to *tus*, they do not appear to contribute to replication arrest because insertional inactivation of *uspT* had no effect on the function of the *Ter* sites (37).

PROPERTIES OF THE *TUS* PROTEIN Several laboratories have purified Tus protein to homogeneity (32, 35, 45; E. Lee & A. Kornberg, unpublished results), and several of the physical and biochemical properties of purified Tus protein were recently determined (F. Coskun-Ari & T. Hill, unpublished results). Both gel filtration and sedimentation equilibrium studies have demonstrated that the soluble form of the Tus protein was a monomer, consistent with earlier reports using gel filtration alone (79). The Tus protein had a frictional coefficient of 1.06 and an axial ratio of 2, indicating that the polypeptide was nearly spherical. The stokes radius was 23 Å, suggesting that the protein would cover approximately 13 bp when bound to DNA, somewhat less than the 17 bp protected from hydroxyl-radical treatment (24). The ϵ_{280} was 39,400 $M^{-1}cm^{-1}$, very close to the predicted value of 38,800 $M^{-1}cm^{-1}$ calculated using Wetlaufer's method (88). The pI of the native protein was 7.5 (F. Coskun-Ari & T. Hill, unpublished results), although the calculated isoelectric point based on the nucleotide sequence of the *tus* gene was 10.1 (37). The difference in the observed and predicted pI may simply reflect the the folding of the native protein or may result from a posttranslational modification, such as phosphorylation or nucleotide binding. Tus binds to the *Ter* sites as a monomer (79), an observation that is consistent with the monomeric form of the soluble protein.

The Replication-Arrest Complex

The Tus-*Ter* complex is unique among DNA-protein complexes in its intrinsic ability to arrest DNA replication. Consequently, most of the recent investigations of this unusual protein-DNA complex have focused on the molecular mechanism of replication-fork arrest and the details of Tus binding to a *Ter* site. In these studies, two potential mechanisms have been considered for the activity of Tus-*Ter* complexes against DNA replication forks. The first of these is specific protein-protein interactions between Tus and some component of the approaching replication fork that impairs function of the replisome component and halts the replication fork. The second model is a general block to protein translocation along the DNA; that is, Tus binds to the *Ter* site in such a way as to impede protein translocation from one direction but not from the other.

EQUILIBRIUM AND KINETIC CONSTANTS If the Tus-*Ter* complex acts as a simple barrier to replication-fork progression, this property might be reflected by the affinity of Tus for the *Ter* sites and by the stability of the complex. Consequently, Gottlieb et al (24) used filter-binding studies of the Tus binding to the *Ter* sites to measure these parameters. Their work demonstrated a very high affinity of Tus for the chromosomal *TerB* site, with an equilibrium binding constant (K_D) of 3.4×10^{-13} M. The dissociation rate constant (k_d) was 2.1×10^{-5} sec^{-1} and the halflife was 550 min, indicating that the Tus-*TerB* complex was very stable. The association rate (k_a) of Tus-*TerB* complex formation was 1.4×10^8 $M^{-1}sec^{-1}$, in the range expected for simple diffusion-controlled bimolecular reactions. The interaction of Tus with the R6K*TerR2* site, which is a less efficient replication-arrest site (29, 74), revealed a K_D of 1×10^{-11} M, or an affinity 30 times lower than the binding of Tus to *TerB*. This difference was primarily the result of a faster dissociation rate of the Tus-*TerR2* complex, with a halflife of 43 min. In contrast, Sista et al (79) used a gel mobility-shift assay and reported a K_D of 5×10^{-9} M for the Tus-R6K*TerR* interaction. The 500-fold difference between the reported values for the Tus-R6K interaction probably results from the different buffers used in the binding studies.

The stability of the Tus-*TerB* complex, which is more than 100 times greater than the halflife of the *lac* repressor-operator complex, may cause cell cycle–dependent regulation of *tus* gene expression. When Tus binds to the *TerB* site, *tus* gene expression should be repressed until either the bound Tus protein dissociates or is displaced by a counterclockwise-traveling replication fork. Thus, in *E. coli* cells with doubling times of 30–40 minutes, *tus* gene expression should be restricted to the period of the cell cycle when replication forks reach the terminus region.

DNA FOOTPRINT A variety of footprinting studies have been used to probe the structure of the Tus-*Ter* complex. Footprinting of Tus-R6K*TerR* complexes by copper-phenanthroline (80) and DNase I (32) demonstrated that 15–24 bp of the conserved *Ter* sequence were protected from cleavage by bound Tus protein. Sista et al (79) employed a combination of hydroxyl radical footprinting, methylation interference, alkylation interference, and methylation protection studies to show that Tus protein contacted both major and minor grooves and was positioned asymmetrically on the R6K*TerR* sites, consistent with the observed polarity of function.

Gottlieb et al (24) used a similar collection of chemical-modification techniques to examine the Tus-*TerB* complex and also observed an asymmetric arrangement of protein-DNA contacts on the chromosomal *Ter* site (Figure 4). However, a greater number of protected residues were observed on the *TerB* site as compared to the R6K*TerR* sites, which probably reflects the lower equilibrium constant and greater stability of the Tus-*TerB* complex. The pattern of the protected residues indicated that the Tus protein was primarily positioned on one side of the double-helix, with protein fingers extending in the major groove to the other side of the helix. Interestingly, the side of the Tus-*TerB* complex that halts an approaching replication fork had a greater number of protected residues, and more importantly, these contacts extended across both the major and minor grooves on the front side of the complex and in the major groove on the back side of the DNA, suggesting that Tus was firmly clamped onto the *Ter* site in this region. By comparison, on the side of the Tus-*TerB* complex that is permissive for replication-fork progression, only a single strand was contacted by the Tus protein (Figure 4).

This observation, in conjunction with the stability of the Tus-*TerB* complex, suggested a possible model for generalized inhibition of DNA-unwinding proteins (24). A DNA-unwinding protein moving towards the Tus-*TerB* complex from the nonpermissive direction would encounter the

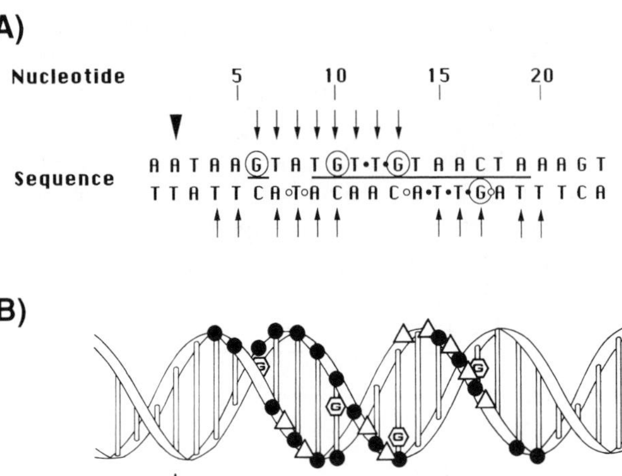

Figure 4 Location of protein-DNA contact sites in the Tus-*TerB* complex. (*A*) Circled nucleotides indicate guanine residues protected from DMS methylation. Filled-in circles between nucleotides indicate strong alkylation interference sites and open circles indicate weak alkylation interference sites. Small arrows denote hydroxyl radical protected sites. The large arrow shows the last nucleotide of leading-strand synthesis. The heavy bar between the strands denotes the core sequence common to all known *Ter* sequences. (*B*) Three-dimensional representation of DNA-protein contacts in the Tus-*Ter* complex. Closed circles show the position of hydroxyl radical–protected nucleotides; open triangles indicate positions of alkylation interference; and hexagons containing the letter "G" indicate DMS-protected guanine residues. Reproduced with permission from Ref. 24.

region of the *TerB* site where Tus contacted both major and minor grooves. The unwinding protein would not be able to pass through the stable barrier formed by the Tus-*TerB* complex. However, a DNA-unwinding protein moving towards the Tus-*TerB* complex from the permissive direction would encounter the protein-DNA complex where the Tus protein contacted only one strand of the *TerB* site (Figure 4). The DNA-unwinding protein would separate the strands of the *TerB* site, destabilizing the complex and allowing passage of the unwinding protein.

The pattern of protein-DNA contacts observed in the Tus-*TerB* complex could also accomodate the protein-protein interaction model equally well, because Tus contacts were positioned close to the last nucleotide synthesized by the leading strand (Figure 4). This observation suggested that a domain of the Tus protein could interact specifically with a helicase approaching from the nonpermissive direction, thereby halting its DNA unwinding activity. If the helicase approached from the permissive direction, it would pass unimpeded because the orientation of the Tus protein would not permit correct protein-protein interactions between the helicase and Tus.

THE MECHANISM OF REPLICATION ARREST In vitro studies of replication arrest have been employed to identify the component of the replisome that interacts with the Tus protein and the mechanism by which replication is halted. That the Tus-*Ter* complex alone was sufficient to arrest DNA replication was established by inhibition of a bona fide replication fork using an in vitro system composed entirely of purified proteins (35, 54). The arrest of DNA replication in vitro faithfully mimicked in vivo replication arrest, in that replication inhibition depended upon the orientation of the Tus-*Ter* complex with respect to the direction of replication-fork progression. In addition, these studies demonstrated that the primary arrest sites of the leading strand were the first and second nucleotides of the *Ter* site (Figure 3) (35, 53a).

A likely candidate for Tus-mediated arrest of replication is the replicative helicase, which separates the DNA strands in advance of the pol III holoenzyme. To examine the effect of the Tus-*Ter* complex on helicase activity, a relatively simple helicase assay has been employed to monitor strand displacement in the presence of Tus. In this assay, a partial DNA heteroduplex consisting of a short oligodeoxyribonucleic acid hybridized to a circular ssDNA molecule is incubated with a purified DNA helicase. The intrinsic unwinding activity of the helicase displaces the oligomer and the displaced oligomer can be easily separated from the heteroduplex substrate by electrophoresis. Since no other replication proteins are necessary for this assay, the direct action of Tus on the helicase can be determined. In addition, several different helicases can be tested using this simple protocol.

Lee et al (54) demonstrated that a Tus-*TerB* complex was capable of an orientation-dependent inhibition of unwinding activity of the DnaB (the major replicative helicase of *E. coli*), Rep, and UvrD helicases. The polarity of helicase inhibition was the same as observed for replication-fork inhibition in vivo. The orientation of the Tus-*TerB* complex that arrested helicase activity was independent of the direction of helicase movement; DnaB unwinds duplex DNA in the 5' to 3' direction, and Rep and UvrD unwind in the 3' to 5' direction. In a separate study, Khatri et al (45) tested the same three helicases, except that they used a R6K*TerR* site instead of the chromosomal *Ter* site. They reported that the Tus-R6K*TerR* complex only showed polar inhibition of DnaB activity; Rep and UvrD activity were unaffected. More recently, however, the results regarding UvrD have been reversed, and Bedorsian & Bastia (3) now report that the Tus-R6K*TerR* complex does halt UvrD in a polar manner.

These results suggested that the antihelicase activity of Tus was responsible for replication-fork arrest. More recent studies have strengthened this view of Tus as an antihelicase. The list of helicases and the direction of unwinding inhibited by the Tus-*TerB* complex in an orientation-dependent manner now include SV-40 T antigen (3' to 5') (31), helicase I from F plasmid, helicase B from mouse cells (both 5' to 3') (31), and *priA* (3' to 5') (28a, 53a). SV-40 T antigen was also shown to be inhibited by a Tus-R6K*TerR* complex (3), but in these studies the 3'-to-5' T-antigen helicase activity appeared to be inhibited by the orientation of the Tus-*Ter* complex that was normally permissive for replication. Because this inhibition seemed unlikely, the authors speculated that a 5'-to-3' activity of the T-antigen was preferentially inhibited by the usual orientation of the Tus-*Ter* complex.

In addition to its antihelicase activity, the Tus-*TerB* complex reportedly prevents strand displacement by a variety of polymerases, including T5, T7, and *E. coli* pol I large fragment (53a). Against these enzymes, however, inhibition of strand displacement occurred regardless of the orientation of the *Ter* site, suggesting that the strength of the protein-DNA interactions in the Tus-*TerB* complex was sufficient to impede the progress of the polymerases. The extent of polymerization into the *TerB* site on either side by T7 pol was also determined. From the nonpermissive side, T7 pol replicated up to the fourth nucleotide of the consensus sequence, and from the permissive side, replication occurred up to the sixth nucleotide (Figure 4). These endpoints of polymerization coincided almost exactly with the limits of the Tus footprint on the *TerB* site shown by Gottlieb et al (24), lending credence to the suggestion that the strength of the Tus-*TerB* complex alone is sufficient to halt these polymerase activities.

Although the multitude of replication enzymes inhibited by the Tus-*Ter* complex suggests a generalized mechanism for arrest rather than specific protein-protein interactions, some results are inconsistent with this model. For instance, the Dda helicase of T4 phage was not inhibited by either orientation of a Tus-R6K*terR* complex (3). More recently, results by Hasai & Marians (28a) have provided the most compelling evidence to date that protein-protein interactions may play a role in replication arrest and, in the process, have challenged the utility of helicase assays that use short oligomeric substrates for monitoring helicase activity. Using substrates with a short (60 base) *TerB* oligomer hybridized to a single-stranded circle, orientation-dependent inhibition of DnaB (5' to 3') and PriA (3' to 5') helicases was observed, consistent with reports from other laboratories. Also, strand displacement by a ϕx-type primosome, which contains both the DnaB and PriA helicases, was inhibited by both orientations of the Tus-*TerB* complex. However, when a *TerB* substrate containing an elongated duplex of approximately 250 bp was tested, the primosome was still halted in both directions as before, but PriA was only partially inhibited by the nonpermissive orientation, and DnaB passed through the Tus-*TerB* complex in both orientations. Thus, the effect of the Tus-*TerB* complex on helicase activity differed on short oligomeric substrates versus elongated duplex substrates. Also, in contrast to other published reports, Hasai & Marians did not observe inhibition of UvrD helicase with either orientation of the Tus-*TerB* complex on either short or elongated substrates. The authors postulated that studies assaying helicase activity using oligomeric substrates might only be observing strand-displacement activity rather than true helicase activity. On longer duplex substrates, strand-displacement activity gave way to true DNA unwinding, this transition occurring somewhere between 2 and 10 duplex DNA turns. Thus, the distribution of protein-DNA contacts and the stability of the Tus-*TerB* complex in the nonpermissive orientation could account for the generalized inhibition of helicase strand displacement on oligomeric substrates. However, these characteristics of the Tus-*TerB* complex were not sufficient to halt helicases once bona fide duplex unwinding had commenced, suggesting that protein-protein interactions must be an integral step in Tus-mediated arrest of DNA replication.

Why Do Ter *Sites Exist?*

Beyond their obvious role as endpoints of DNA replication, the biological function and selective advantage of *Ter* sites in the prokaryotic chromosome remain a mystery. The benefit of having chromosomal replication-arrest sites is attested to by the striking similarities in the organization of the replication-arrest systems of *B. subtilis,* a gram-positive bacterium, and *E. coli,* a gram-negative bacterium, and by the identification of a system in *Salmonella typhimurium* (75) that is homologous to the *Ter* sites and Tus protein of *E. coli*. The conservation of replication-arrest systems in bacteria and the identification of functionally similar systems in yeast (11), peas (28), and viruses (22) strongly support the argument that replication-arrest sites confer an evolutionary advantage to the cell. However, the exact nature of this advantage remains unclear, especially in light of the apparent dispensability of the replication-arrest system.

The absence of an observable phenotype in tus^{-} *E. coli* strains grown in normal laboratory conditions prompted Roecklein et al (75) to compare the growth characteristics of tus^{+} and *tus::*kan^{r} strains in a variety of other laboratory conditions. The two strain types were tested in rich vs minimal medium, after medium shifts, in anaerobic conditions, following UV irradiation, and following exposure to agents that inhibit DNA synthesis. The authors could not detect significant differences in growth between the wild-type strain and the strain carrying the *tus* null mutation in any of these conditions. These results indicated that the simple set of growth parameters available in the laboratory are not sufficient to produce an observable phenotype in tus^{-} cells. In a more natural environment, where the bacterium encounters a more complicated set of growth conditions, the selective advantage for a tus^{+} strain or a clear phenotype for a tus^{-} strain might be more readily observed.

Loss of the replication-arrest system appears to elicit an observable phenotype only in *E. coli* strains that replicate the chromosome unidirectionally (15). In these strains, called intR1 strains, *oriC* is inactivated and replaced by the unidirectional replication origin from plasmid R1 (69). As a result, the replication cycle of these strains is highly asymmetric. The chromosome is replicated predominately, if not entirely, in either the clockwise or counter-clockwise direction, depending upon the orientation of the inserted plasmid origin. Thus, to complete the chromosome, the single replication fork must pass through a series of active termination sites. When intR1 strains contain a functional *tus* gene, the cells are long filaments that pinch off. However, when the *tus* gene is inactivated, filamentation is significantly reduced, indicating that the abnormal phenotype is linked to replication arrest. A second characteristic of the tus^+ intR1 strains is that they require a functional *recA* gene for viability (S. Dasgupta & K. Nordström, unpublished results). The dependence of these strains on *recA* can be alleviated by inactivating the *tus* gene, suggesting that arrested replication forks in the chromosome of tus^+ intR1 strains produce structures that must be resolved by RecA. Alternately, stable DNA replication (*sdr*), which requires RecA (47), may be needed to complete the chromosome.

An intriguing observation reported recently by Louarn et al (58) was the increased frequency of RecA-mediated homologous recombination associated with the replication-arrest sites. In this study, the frequency of recombination at a given point on the chromosome was measured by monitoring the excision of a temperature-sensitive prophage that had been inserted into a transposon via homologous recombination. Excision of the prophage could be easily scored, allowing an accurate determination of the frequency of homologous recombination at many different points around the *E. coli* chromosome. The excision frequency was consistent (10^{-5}) for the majority of the chromosome (from min 44 clockwise to min 23), increased significantly (10- to 100-fold) at the edges of the terminus region near *TerA* and *TerB*, and reached a maximum of 10^{-2} at min 33.8, near the *TerC* locus. The increased frequency of recombination in the vicinity of the replication-arrest sites depended upon a functional *recA* gene, indicating that the machinery for homologous recombination is required to excise the prophage. The authors suggested that RecA activity, in conjunction with the presence of an arrested replication fork, could account for the observed level of hyperrecombination. Recent data from other laboratories also support the hypothesis that an increased recombination frequency is associated with the replication-arrest sites. T. Horiuchi and coworkers (unpublished results) found that recombinational hotspots were located in the terminus region and that in some cases hyperrecombination depended on the presence of a functional *tus* gene, suggesting that the replication-arrest sites were involved. Also, Bierne et al (4) have reported that the *TerB* locus can act as a deletion hotspot when inserted in the proper orientation into hybrid plasmids containing phage M13 origins of replication. Although the significance of these observations is unclear with regards to *Ter*-site function and the role of *Ter* sites in the physiology of *E. coli,* they suggest that these sites are associated with cellular activities in addition to replication arrest.

Although the frequency of *Ter*-site usage in exponentially growing *E. coli* is unknown, *TerC* is probably used most often in replication cycles that utilize a replication-arrest site. Louarn et al (58) pointed out that *TerC* is almost equidistant from *oriC* and that the region of the chromosome just upstream of *TerC* is replicated predominantly in the clockwise direction. Thus, the *TerC* region is the preferred region of replication termination. Clearly, however, other *Ter* sites are also utilized in a significant percentage of replication cycles. Pelletier et al (73) demonstrated that replication forks were arrested at both *TerA* and *TerB* in exponentially growing cells, indicating that these two sites were used relatively frequently. Surprisingly, strains containing chromosomal inversions that doubled the distance from *oriC* to *TerB* compared with the distance from *oriC* to *TerA* also showed significant use of both the *TerA* and *TerB* sites, even though replication of the clockwise arm of the chromosome should have taken twice as long as replication of the counterclockwise arm. The replication rates of the individual chromosomal arms might be adjusted to compensate for the difference in replication distance, allowing both replication forks to arrive at the terminus region at the same time.

Although the *Ter* sites are distributed and oriented to form a replication-fork trap and thus limit the end of the replication cycle to the terminus region, the recent identification of the replication-arrest sites *TerE* and *TerF* well outside of the traditional terminus region suggests that replication arrest by the *Ter* sites may serve other purposes as well. It is unlikely that a significant portion of replication forks originating from *oriC* would reach these arrest sites because the replication forks must first traverse the two arrest sites located closer to the terminus region before arriving at the outermost sites (Figure 2). However, if replication forks were initiated at origins other than *oriC,* then these nonterminus replication-arrest sites would limit replication along the terminus-to-origin axis while allowing replication to proceed normally along the origin-to-terminus axis. Examples of secondary replication origins that might utilize the replication-arrest sites outside of the terminus region are the *sdr* origins (47), prophage origins, or replication origins of integrated plasmids.

Future Prospects

The preceding section clearly shows that our understanding of the selective advantage of replication arrest is fragmentary at best. Unfortunately, the intriguing questions posed by the existence of such a highly conserved, yet apparently dispensable system of replication arrest are also the most difficult to address. Perhaps the recent association of hyperrecombination with replication-arrest sites is the first step in obtaining a broader understanding of the function of the replication-arrest systems of bacteria. From the biochemical standpoint, many questions regarding the ability of the Tus-*Ter* complex to arrest DNA replication also remain, but these appear to be more tractable. The mechanics of Tus-mediated arrest of replication using in vitro replication systems will continue to be explored in an effort to elucidate the action of Tus against replication-fork components. X-ray crystallography of the Tus-*Ter* complex should provide a wealth of information regarding the protein-DNA contacts and the possible implications of these contacts in Tus-mediated replication arrest. Also, a genetic approach involving the isolation of mutant Tus proteins would help clarify the mechanism of Tus action. If Tus uses protein-protein interactions to arrest DNA replication, the domain of the protein that interacts with the replication fork might be distinct from the DNA-binding domain of the protein. If this is so, it should be possible to introduce mutations that enable Tus to bind normally to a *Ter* site but impair the ability of the Tus-*Ter* complex to arrest DNA replication. Alternately, if all mutations in Tus that affect the efficiency of replication arrest also impair Tus binding, then this result would suggest that the replication arrest and DNA-binding functions are inseparable, a notion consistent with a generalized action against DNA-translocating proteins. Mutations in the *tus* gene may also help sort out the physiological advantage of the replication-arrest system.

Probably the most exciting applications of the study of replication-arrest systems will be the use of these systems as tools to examine other aspects of DNA replication and cell division. By inserting *Ter* sites at specific locations in a plasmid or chromosome, one can trap replication forks at defined points within the replicon. The products of the arrested replication fork can then be readily analyzed. The *E. coli Ter* sites have already been exploited to examine the initiation events of plasmid R1162 (93), and the *terC* region of *B. subtilis* was used to map the initiation site of plasmid pAMβ1 (10). Other potential uses of the replication-arrest system include dissection of the final steps of chromosome replication, analysis of the distribution of replisome proteins at the arrested replication fork, determination of membrane binding of the terminus region of *E. coli,* and possibly examination of the coordination of DNA replication with the onset of cell division. Clearly, the study of replication arrest and the use of replication-arrest systems to analyze the biology and biochemistry of DNA replication will continue to yield valuable information for many years to come.

ACKNOWLEDGMENTS

The author gratefully acknowledges Ann Flower, Peter Kuempel, Bryan Roecklein, Molly Schmid, and Gerry Wake for critically evaluating the contents of this manuscript.

Literature Cited

1. Bachmann, B. J. 1990. Linkage map of *Escherichia coli* K-12, Edition 8. *Microbiol. Rev.* 54:130–97
2. Bastia, D., Germino, J., Crosa, J. H., Ram, J. 1981. The nucleotide sequence surrounding the replication terminus of plasmid R6K. *Proc. Natl. Acad. Sci. USA* 78:2095–99
3. Bedrosian, C. L., Bastia, D. 1991. *Escherichia coli* replication terminator protein impedes simian virus 40 (SV40) DNA replication fork movement and SV40 large tumor antigen helicase activity in vitro at a prokaryotic terminus sequence. *Proc. Natl. Acad. Sci. USA* 88:2618–22
4. Bierne, H., Ehrlich, S. D., Michel, B. 1991. The replication termination signal *terB* of the *Escherichia coli* chromosome is a deletion hot spot. *EMBO J.* 10:2699–2705
5. Bird, R. E., Louarn, J., Martuscelli, J., Caro, L. 1972. Origin and sequence of chromosome replication in *Escherichia coli*. *J. Mol. Biol.* 70:549–66
6. Bitner, R. M., Kuempel, P. L. 1981. P1 transduction map spanning the replication terminus of *Escherichia coli* K12. *Mol. Gen. Genet.* 184:208–12
7. Blakely, G., Colloms, S., May, G., Burke, M., Sherratt, D. 1991. *Escherichia coli* XerC recombinase is required for chromosomal segregation at cell division. *New Biol.* 3:789–98
8. Bouché, J. P. 1982. Physical map of a 470 kilobase pair region flanking the terminus of DNA replication in the *Escherichia coli* K12 genome. *J. Mol. Biol.* 154:1–20
9. Bouché, J. P., Gélugne, J. P., Louarn, J., Louarn, J. M., Kaiser, K. 1982. Relationships between the physical and genetic maps of a 470×10^3 base-pair region around the terminus of *Escherichia coli* K12 DNA replication. *J. Mol. Biol.* 154:21–32
10. Braund, C., Ehrlich, S. D., Jannière, L. 1991. Unidirectional theta replication of the structurally stable *Enterococcus faecalis* plasmid pAMb1. *EMBO J.* 10: 2171–77
11. Brewer, B. J., Fangman, W. L. 1988. A replication fork barrier at the 3' end of yeast ribosomal RNA gene. *Cell* 55: 637–43
12. Carrigan, C. M., Haarsma, J. A., Smith, J. A., Smith, M. T., Wake, R. G. 1987. Sequence features of the replication terminus of the *Bacillus subtilis* chromosome. *Nucleic Acids Res.* 15:8501–9
13. Carrigan, C. M., Pack, R. A., Smith, M. T., Wake, R. G. 1991. Normal *terC*-region of the *Bacillus subtilis* chromosome acts in a polar manner to arrest the clockwise replication fork. *J. Mol. Biol.* 222:197–207
14. Crosa, J. H. 1980. Three origins of replication are active in vivo in the R plasmid RSF1040. *J. Biol. Chem.* 255: 11075–77
15. Dasgupta, S., Bernander, R., Nordström, K. 1991. In vivo effect of the *tus* gene on cell division in an *Escherichia coli* strain where chromosome replication is under control of plasmid R1. *Res. Microbiol.* 142:177–80
16. de Boer, P. A. J., Cook, W. R., Rothfield, L. I. 1990. Bacterial cell division. *Annu. Rev. Genet.* 24:249–74
17. deMassy, B., Bejar, S., Louarn, J., Louarn, J. M., Bouché, J. P. 1987. Inhibition of replication forks exiting the terminus region of the *Escherichia coli* chromosome occurs at two loci separated by 5 min. *Proc. Natl. Acad. Sci. USA* 84:1759–63
18. DiGate, R. J., Marians, K. J. 1988. Identification of a potent decatenating enzyme from *Escherichia coli*. *J. Biol. Chem.* 263:13366–73
19. Fouts, K. E., Barbour, S. D. 1982. Insertions of transposons through the major cotransduction gap of *Escherichia coli* K12. *J. Bacteriol.* 149:106–13
20. François, V., Louarn, J., Louarn, J.-M. 1989. The terminus region of the *Escherichia coli* chromosome is flanked by several polar replication pause sites. *Mol. Microbiol.* 3:995–1002
21. François, V., Louarn, J., Rebollo, J.-E., Louarn, J.-M. 1990. Replication ter-

mination, nondivisible zones, and structure of the *Escherichia coli* chromosome. In *The Bacterial Chromosome,* ed. K. Drlica, M. Riley, pp. 351–59. Washington, DC: Am. Soc. Microbiol.

22. Gahn, T. A., Schildkraut, C. L. 1989. The Epstein-Barr virus origin of plasmid replication, oriP, contains both the initiation and termination sites of DNA replication. *Cell* 58:527–35
23. Germino, J., Bastia, D. 1981. Termination of DNA replication in vitro at a sequence-specific replication terminus. *Cell* 23:681–87
24. Gottlieb, P. A., Wu S., Zhang, X., Tecklenburg, M., Kuempel, P., Hill, T. M. 1992. Equilibrium, kinetic, and footprinting studies of the Tus-Ter protein-DNA interaction. *J. Biol. Chem.* 267:7434–43
25. Harford, N. 1975. Bidirectional chromosome replication in *Bacillus subtilis. J. Bacteriol.* 121:835–47
26. Hanley, P. J. B., Carrigan, C. M., Rowe, D. B., Wake, R. G. 1987. Breakdown and quatitation of the forked termination of replication intermediate of *Bacillus subtilis. J. Mol. Biol.* 196: 721–27
27. Henson, J. M., Kuempel, P. L. 1985. Deletion of the terminus region (340 kilobase pairs) from the chromosome of *Escherichia coli. Proc. Natl. Acad. Sci. USA* 82:3766–70
28. Hernández, P., Lamm, S. S., Bjerknes, C. A., Van't hof, J. 1988. Replication termini in the rDNA of synchronized pea root cells (*Pisum sativum*). *EMBO J.* 7:303–8

28a. Hasai, H., Marians, K. J. 1992. Differential inhibition of the DNA translocation and DNA unwinding activities of DNA helicases by the *Escherichia coli* Tus protein. *J. Biol. Chem.* In press

29. Hidaka, M., Akiyama, M., Horiuchi, T. 1988. A consensus sequence of three DNA replication terminus sites on the E. coli chromosome is highly homologous to the terR sites of the R6K plasmid. *Cell* 55:467–75
30. Hidaka, M., Kobayashi, T., Horiuchi, T. 1991. A newly identified DNA replication terminus site, *TerE,* on the *Escherichia coli* chromosome. *J. Bacteriol.* 173:391–93
31. Hidaka, M., Kobayashi, T., Ishimi, Y., Seki, M., Enomoto, T., et al. 1992. Termination complex in *E. coli* inhibits SV40 DNA replication in vitro by impeding the action of T antigen helicase. *J. Biol. Chem.* 267:5361–65
32. Hidaka, M., Kobayashi, T., Takenaka, S., Takeya, H., Horiuchi, T. 1989. Purification of a DNA replication terminus (*ter*) site-binding protein in *Escherichia coli* and identification of the structural gene. *J. Biol. Chem.* 264: 21031–37
33. Hill, T. M., Henson, J. M., Kuempel, P. K. 1987. The terminus region of the *Escherichia coli* chromosome contains two separate loci that exhibit polar inhibition of replication. *Proc. Natl. Acad. Sci. USA* 84:1754–58
34. Hill, T. M., Kopp, B. J., Kuempel, P. K. 1988. Termination of DNA replication in *Escherichia coli* requires a trans-acting factor. *J. Bacteriol.* 170:662–68
35. Hill, T. M., Marians, K. J. 1990. *Escherichia coli* Tus protein acts to arrest the progression of DNA replication forks in vitro. *Proc. Natl. Acad. Sci. USA* 87:2481–85
36. Hill, T. M., Pelletier, A. J., Tecklenburg, M., Kuempel, P. L. 1988. Identification of the DNA sequence from the E. coli terminus region that halts replication forks. *Cell* 55:459–66
37. Hill, T. M., Tecklenburg, M., Pelletier, A. J., Kuempel, P. L. 1989. *tus,* the trans-acting gene required for termination of DNA replication in *Escherichia coli,* encodes a DNA-binding protein. *Proc. Natl. Acad. Sci. USA* 86:1593–97
38. Hiraga, S., Niki, H., Ogura, T., Ichinose, C., Mori, H., et al. 1989. Chromosome partitioning in *Escherichia coli:* novel mutants producing anucleate cells. *J. Bacteriol.* 171:1496–1505
39. Horiuchi, T., Hidaka, M. 1988. Core sequence of two separable terminus sites of the R6K plasmid that exhibit polar inhibition of replication is a 20 bp inverted repeat. *Cell* 54:515–23
40. Hye, R. J., O'Sullivan, M. A., Howard, K., Sueoka, N. 1976. Membrane association of origin, terminus, and replication fork in *Bacillus subtilis.* In *Microbiology—1976,* ed. D. Schlesinger, pp. 83–90. Washington, DC: Am. Soc. Microbiol.
41. Iismaa, T. P., Carrigan, C. M., Wake, R. G. 1988. Relocation of the replication terminus, *terC,* of *Bacillus subtilis* to a new chromosomal site. *Gene* 67:183–91
42. Iismaa, T. P. Wake, R. G. 1987. The normal replication terminus of the *Bacillus subtilis* chromosome, *terC,* is dispensable for vegetative growth and sporulation. *J. Mol. Biol.* 195:299–310
43. Inuzuka, N., Inuzuka, M., Helinski, D.

R. 1980. Activity in vitro of three replication origins of the antibiotic resistance plasmid RSF1040. *J. Biol. Chem.* 255:11071–74

44. Kato, J., Nishimura, Y., Imamura, R., Niki, H., Hiraga, S., Suzuki, H. 1990. New topoisomerase essential for chromosome segregation in E. coli. *Cell* 63:393–404
45. Khatri, G. S., MacAllister, T., Sista, P. R., Bastia, D. 1989. The replication terminator protein of E. coli is a DNA sequence-specific contra-helicase. *Cell* 59:667–74
46. Kobayashi, T., Hidaka, M., Horiuchi, T. 1989. Evidence of a *ter* specific binding protein essential for the termination reaction of DNA replication in *Escherichia coli. EMBO J.* 8:2435–41
47. Kogoma, T. 1985. RNase H-defective mutants of *Escherichia coli. J. Bacteriol.* 166:361–63
48. Kolter, R., Helinski, D. 1978. Activity of the replication terminus of plasmid R6K in hybrid replicons in *Escherichia coli. J. Mol. Biol.* 124:425–41
49. Kralicek, A. V., Day, A. J., Wake, R. G., King, G. F. 1991. A sequence similarity between proteins involved in intiation and termination of bacterial chromosome replication. *Biochem. J.* 275:823–23
50. Kuempel, P. L., Duerr, S. A. 1979. Chromosome replication in *Escherichia coli* is inhibited in the terminus region near the *rac* locus. *Cold Spring Harbor Symp. Quant. Biol.* 43:563–67
51. Kuempel, P. L., Duerr, S. A., Maglothin, P. D. 1978. Chromosome replication in an *Escherichia coli dnaA* mutant integratively suppressed by prophage P2. *J. Bacteriol.* 134:902–12
52. Kuempel, P. L., Duerr, S. A., Seeley, N. R. 1977. Terminus region of the chromosome in *Escherichia coli* inhibits replication forks. *Proc. Natl. Acad. Sci. USA* 74:3927–31
53. Kuempel, P. L., Henson, J. M., Dircks, L., Tecklenburg, M., Lim, D. F. 1991. *dif,* a *recA*-independent recombination site in the terminus region of the chromosome of *Escherichia coli. New Biol.* 3:799–811

53a. Lee, E. H., Kornberg, A. 1992. Features of replication fork blockage by the *Escherichia coli* termination-binding protein. *J. Biol. Chem.* 267:8778–84

54. Lee, E. H., Kornberg, A., Hidaka, M., Kobayashi, T., Horiuchi, T. 1989. *Escherichia coli* replication termination protein impedes the action of helicases. *Proc. Natl. Acad. Sci. USA* 86:9104–8
55. Lewis, P. J., Ralston, G. B., Christopherson, R. I., Wake, R. G. 1990. Identification of the replication terminator protein binding sites in the terminus region of the *Bacillus subtilis* chromosome and stoichiometry of the binding. *J. Mol. Biol.* 241:73–84
56. Lewis, P. J., Smith, M. T., Wake, R. G. 1989. A protein involved in termination of chromosome replication in *Bacillus subtilis* binds specifically to the *terC* site. *J. Bacteriol.* 171:3564–67
57. Lewis, P. J., Wake, R. G. 1989. DNA and protein sequence conservation at the replication terminus in *Bacillus subtilis* 168 and W23. *J. Bacteriol.* 171:1402–8
58. Louarn, J.-M., Louarn, J., François, V., Patte, J. 1991. Analysis and possible role of hyperrecombination in the termination region of the *Escherichia coli* chromosome. *J. Bacteriol.* 173:5097–5104
59. Louarn, J., Patte, J., Louarn, J.-M. 1977. Evidence for a fixed termination site of chromosome replication in *Escherichia coli* K12. *J. Mol. Biol.* 115: 295–314
60. Louarn, J., Patte, J., Louarn, J.-M., 1979. Map position of the replication terminus on the *Escherichia coli* chromosome. *Mol. Gen. Genet.* 172:7–11
61. Lovett, M. A., Sparks, R. B., Helinski, D. R. 1975. Bidirectional replication of plasmid R6K DNA in *Escherichia coli;* correspondence between origin of replication and position of single-strand break in relaxed complex. *Proc. Natl. Acad. Sci. USA* 72:2905–9
62. Lovett, P. S., Young, F. E. 1969. Identification of *Bacillus subtilis* NRRL B-3275 as a strain of *Bacillus pumilus. J. Bacteriol.* 100:658–61
63. Luttinger, A. L., Springer, A. L., Schmid, M. B. 1991. A cluster of genes that affects nucleoid segregation in *Salmonella typhimurium. New Biol.* 3:687–97
64. Masai, H., Arai, K.-I. 1987. RepA and DnaA proteins are required for initiation of R1 plasmid replication in vitro and interact with the *oriR* sequence. *Proc. Natl. Acad. Sci. USA* 84:4781–85
65. Masters, M., Broda, P. 1971. Evidence for the bidirectional replication of the *Escherichia coli* chromosome. *Nature (London) New Biol.* 232:137–40

66. Miyazaki, C., Kawai, Y., Ohtsubo, H., Ohtsubo, E. 1988. Unidirectional replication of plasmid R100. *J. Mol. Biol.* 204:331–43
67. Monterio, M. J., Sargent, M. G., Piggot, P. J. 1984. Characterization of the replication terminus of the *Bacillus subtilis* chromosome. *J. Gen. Microbiol.* 130:2403–14
68. Natarajan, S., Kelly, W. L., Bastia, D. 1991. Replication terminator protein of *Escherichia coli* is a transcriptional repressor of its own synthesis. *Proc. Natl. Acad. Sci. USA* 88:3867–71
69. Nordström, K., Bernander, S., Dasgupta, S. 1991. Analysis of bacterial cell cycle using strains in which chromosome replication is controlled by plasmid R1. *Res. Microbiol.* 142:181–88
70. Ohtsubo, E., Feingold, J., Ohtsubo, H., Mickel, S., Bauer, W. 1977. Unidirectional replication of three small plasmids derived from R factor R12 in *Escherichia coli. Plasmid* 1:8–18
71. Ohtsubo, H., Vassino, B., Ryder, T., Ohtsubo, E. 1982. A simple method for shortening a plasmid genome using a system of plasmid cointegration mediated by a Tn*3* mutant. *Gene* 20:245–54
72. O'Sullivan, M. A., Anagnostopoulos, C. 1982. Replication terminus of the *Bacillus subtilis* chromosome. *J. Bacteriol.* 151:135–43
73. Pelletier, A. J., Hill T. M., Kuempel, P. L. 1988. Location of sites that inhibit progression of replication forks in the terminus region of *Escherichia coli. J. Bacteriol.* 170:4293–98
74. Pelletier, A. J., Hill T. M., Kuempel, P. L. 1989. Termination sites T1 and T2 from the *Escherichia coli* chromosome inhibit DNA replication in ColE1-derived plasmids. *J. Bacteriol.* 171: 1739–41
74a. Roecklein, B., Kuempel, P. L. 1992. Autoregulation of *tus* gene expression in *Escherichia coli* by inhibition of transcription initiation and elongation. *Mol. Microbiol.* In press
75. Roecklein, B., Pelletier, A. J., Kuempel, P. L. 1991. The *tus* gene of *Escherichia coli:* autoregulation, analysis of flanking sequences, and identification of a complementary system in *Salmonella typhimurium. Res. Microbiol.* 142:169–76
76. Ryder, T. B., Rosen, J., Armstrong, K., Davidson, D., Ohtsubo, E., Ohtsubo, H. 1981. Dissection of the replication region controlling incompatibility, copy number, and initiation of DNA synthesis in the resistance plasmids, R100 and R1. In *The Initiation of DNA Replication, ICN-UCLA Symposia on Molecular and Cellular Biology,* ed. D. S. Ray, 22:91–111. New York: Academic
77. Saadi, S., Maas, W. K., Hill, D. F., Berquist, P. L. 1987. Nucleotide sequence analysis of RepFIC, a basic replicon present in IncFI plasmids P307 and F, and its relation to the RepA replicon of IncFII plasmids. *J. Bacteriol.* 169:1836–46
78. Seki, T., Oshima, T., Oshima, Y. 1975. Taxonomic study of *Bacillus* by deoxyribonucleic acid–deoxyribonucleic acid hybridization and interspecific transformation. *Int. J. Syst. Bacteriol.* 25:258–70
79. Sista, P. R., Hutchison, C. A., Bastia, D. 1991. DNA-protein interaction at the replication termini of plasmid R6K. *Genes Dev.* 5:74–82
80. Sista, P. R., Mukherjee, S., Patel, P., Khatri, G. S., Bastia, D. 1989. A host-encoded DNA-binding protein promotes termination of plasmid replication at a sequence-specific replication terminus. *Proc. Natl. Acad. Sci. USA* 86:3026–30
81. Smith, M. T., Aynsley, C., Wake, R. G. 1985. Cloning and localization of the *Bacillus subtilis* chromosome replication terminus, *terC. Gene* 38:9–17
82. Smith, M. T., Wake, R. G. 1988. DNA sequence requirements for replication fork arrest at *terC* in *Bacillus subtilis. J. Bacteriol.* 170:4083–90
83. Uzest, M., Ehrlich, S. D., Michel, B. 1991. The *Escherichia coli TerB* sequence affects maintenance of a plasmid with the M13 phage replication origin. *J. Bacteriol.* 173:7695–97
84. Weiss, A. S., Wake, R. G. 1983. Restriction map of DNA spanning the replication terminus of the *Bacillus subtilis* chromosome. *J. Mol. Biol.* 171: 119–37
85. Weiss, A. S., Wake, R. G. 1984. Impediment to replication fork movement in the terminus region of the *Bacillus subtilis* chromosome. *J. Mol. Biol.* 179:745–50
86. Weiss, A. S., Wake, R. G. 1984. A unique DNA intermediate associated with termination of chromosome replication in Bacillus subtilis. *Cell* 39:683–89
87. Weiss, A. S., Wake, R. G., Inman, R. B. 1986. An immobilized fork as a termination of replication intermediate in *Bacillus subtilis. J. Mol. Biol.* 188:199–205

88. Wetlaufer, D. B. 1962. Ultraviolet spectra of proteins and amino acids. *Adv. Protein Chem.* 17:303–90
89. Williams, N. K., Wake, R. G. 1989. Sequence limits of DNA strands in the arrested replication at the *Bacillus subtilis* chromosome terminus. *Nucleic Acids Res.* 17:9947–56
90. Woods, S. A., Miles, J. S., Roberts, R. E., Guest, J. R. 1986. Structural and functional relationships between fumarase and aspartase. *Biochem. J.* 237:547–57
91. Woods, S. A., Schwartzbach, S. D., Guest, J. R. 1988. Two chemically distinct classes of fumarase in *Escherichia coli*. *Biochim. Biophys. Acta* 954:14–26
92. Zahler, S. A. 1982. Specialized transduction in *Bacillus subtilis*. In *The Molecular Biology of the Bacilli*, Vol. 1, Bacillus subtilis, ed. D. Dubnau, pp. 269–305. New York: Academic
93. Zhou, H., Byrd, C., Meyer, R. J. 1991. Probing the activation of the replicative origin of broad host-range plasmid R1162 with Tus, the *E. coli* anti-helicase protein. *Nucleic Acids Res.* 19:5379–83

Annu. Rev. Microbiol. 1992. 46:635–54

TREATMENT OF THE PICORNAVIRUS COMMON COLD BY INHIBITORS OF VIRAL UNCOATING AND ATTACHMENT

Mark A. McKinlay and Daniel C. Pevear

Department of Virology, Sterling Winthrop Pharmaceuticals Research Division, Rensselaer, New York 12144

Michael G. Rossmann

Department of Biological Sciences, Purdue University, West Lafayette, Indiana 47907

KEY WORDS: rhinovirus, antiviral, rational drug design, enterovirus, WIN compounds

---

CONTENTS

0066-4227/92/1001-0635$02.00

Abstract

The human rhinoviruses are the leading cause of the ubiquitous, mild, and self-limiting infections generally referred to as the common cold. Considerable research effort has been expended in the search for well tolerated antiviral agents capable of preventing and treating the common cold. Although no anitrhinovirus drug is yet commercially available, considerable progress has been made in the discovery and development of novel, viral specific inhibitors of rhinovirus replication. This report reviews the history and current status of the research that has focused on inhibitors of the early steps in the virus life cycle: attachment to the cellular receptor and uncoating of the viral RNA. Molecules directed at these targets currently possess the greatest potential for generating a safe and efficacious treatment for the rhinovirus common cold.

INTRODUCTION

The common cold remains one of the most frequent causes of virus-induced morbidity in humans. Although the disease is mild and self-limiting, in the United States alone the common cold is responsible for 17% of all acute illnesses requiring medical attention or restricted activity, 27 million visits to physicians, and 161 million days of restricted activity (70). Despite the best efforts of several pharmaceutical research groups, no effective agent has yet been developed for the treatment of the common cold. The obstacles to success are many; one of the most significant is the number of different viruses that can cause the disease.

Members of the largest class of the synthetic antirhinoviral agents that have been evaluated clinically share common mechanisms of action as a result of binding in a specific hydrophobic pocket in the virion capsid. Binding in this pocket results in the inhibition of attachment to the host cell receptor and/or uncoating of the virion RNA. The results of clinical trials with compounds of this mechanistic class have been largely disappointing to date. However, much has been learned by studying the inadequacies of these molecules, which will hopefully lead to the discovery of agents with greater potential to safely and effectively treat this common viral disease.

DRUG DISCOVERY

When faced with the challenge of finding a safe and effective treatment for the common cold, one must consider several factors before selecting the approach with the greatest likelihood of success. Choosing the family or families of viruses to target for therapy is the first factor to be considered. The two families of viruses responsible for the majority of colds are the picornaviruses

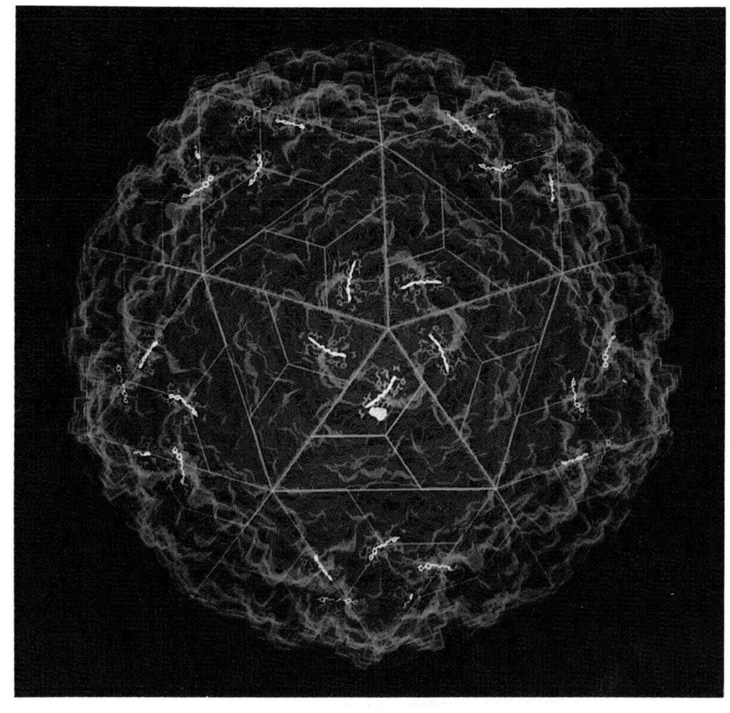

B

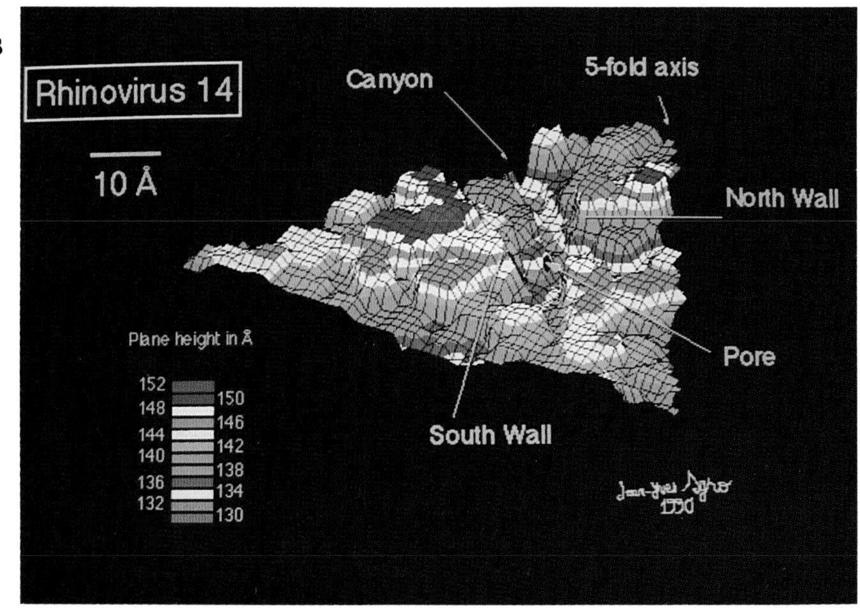
Rhinovirus 14
10 Å
Canyon
5-fold axis
North Wall
Pore
South Wall
Plane height in Å
152
150
148
146
144
142
140
138
136
134
132
130

C

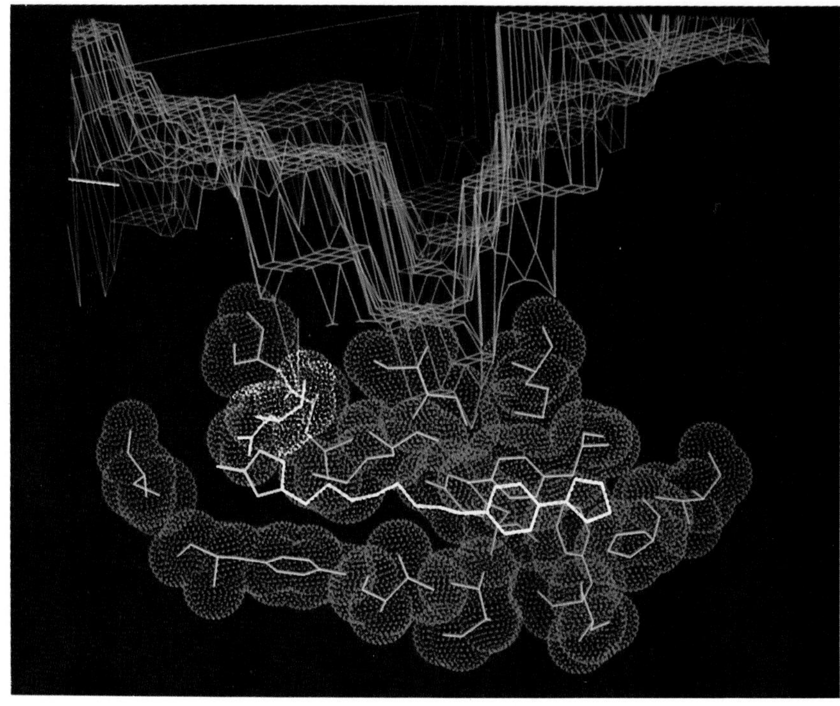

Figure 1 Structure of human rhinovirus type 14. (*A*) Surface topography of the virion with an antiviral WIN compound bound in the 60 VP 1 sites that surround the five-fold axis of symmetry (center of pentameter). (*B*) Topographical representation of the surface of one asymmetric unit (1/60) of the virus, showing the canyon or receptor-binding site, and the pore through which the capsid-binding agents gain access to the hydrophobic pocket beneath the canyon. (*C*) Hydrophobic drug-binding pocket beneath the floor of the canyon. A WIN compound is shown bound in the pocket (*yellow*). The van der Waals radii of the hydrophobic side chains lining the pocket are highlighted (*blue*), with asparagine 219, the only potential hydrogen-binding site, highlighted in red.

and coronaviruses (42). Of the two, the rhinovirus members of the picornavirus family are responsible for up to 50% of the total (28, 63), and the coronaviruses up to only 20% (70). The remaining 30% or more of colds are caused by the adenoviruses, parainfluenza, respiratory syncytial virus, influenza virus, and another group of picornaviruses, the enteroviruses. Because of the predominance of the rhinoviruses in the etiology of common colds, the majority of drug discovery efforts are directed towards the picornavirus family.

Infection by at least 100 distinct serotypes of rhinoviruses can result in a cold (29). For this reason, a vaccine approach to preventing infection does not appear to be practical. The multitude of serotypes also presents a considerable challenge to the discovery of antiviral agents because the molecular target of any broad spectrum agent would necessarily need to be reasonably well conserved among serotypes.

The next factor to be considered is the molecular target for antiviral attack. The life cycle of the human rhinoviruses contains within it a number of viral-specific molecular targets for chemotherapeutic intervention (60). This review focuses on the early replication steps of attachment to the cellular receptor and uncoating of the virion RNA. Much progress has been made in the discovery of inhibitors that block these steps, although in the case of the uncoating inhibitors, the mechanism of action was not determined until after the agents had been discovered. While the research strategy varied between research groups, most uncoating blockers were found during the screening of compound libraries for their ability to inhibit the rhinovirus-induced cytopathic effect in cell culture. Despite the serendipitous nature of the discovery of this class of inhibitors, the following discussion makes clear that the search for the cure for the common cold is progressing rapidly and is at the forefront of the emerging field of rational drug design.

INHIBITORS OF VIRION ATTACHMENT

Cellular Receptors

The first step in infection by rhinoviruses involves attachment to a cellular receptor(s) on the surface of a host cell. Ninety percent of the human rhinoviruses use the cellular protein intracellular adhesion molecule-1 (ICAM-1) to initiate infection (27, 72, 75). The rhinoviruses binding to ICAM-1 have been termed the major receptor binding group. With one exception, the remaining 10% (minor receptor group) all bind to another unidentified receptor (76).

The normal function of ICAM-1 is to bind to the integrin, lymphocyte-function associated molecule-1 (LFA-1), on the surface of lymphocytes to promote adhesion between lymphocytes. ICAM-1 is found on the surface of nasal epithelium and most cells of the body, and the receptor density can be regulated through several inflammatory mediators including interleukin-1, gamma interferon, and tumor necrosis factor (72). The roles of these inflammatory mediators and receptor density in determining whether an infection is initiated, how rapidly it will spread once it is underway, and the severity of symptoms are not known. However, one could speculate that nasal irritation/inflammation resulting from a number of environmental causes such as low humidity could be a significant predisposing factor to infection.

Receptor Blockade

Because 90% of all rhinoviruses use ICAM-1 as a mechanism to gain entry to the host cell, antagonism of the virus-receptor interaction would appear to be an effective way to inhibit a broad spectrum of rhinoviruses. In this regard, several approaches to inhibiting attachment have been shown to prevent rhinovirus replication in cell culture. Colonno and coworkers (16) were first to demonstrate the effectiveness of this approach utilizing a murine monoclonal antibody directed at the major cellular receptor. This high affinity antibody was effective in competing off bound virus and effectively blocked virus replication in cell culture in the absence of cellular toxicity. Hayden et al (30) subsequently evaluated the ability of this antibody to prevent rhinovirus infection in experimentally infected human volunteers. Intranasal inoculation of the antibody in a prophylaxis regimen delayed the onset of symptoms by one to two days but had no effect on the overall incidence of infection. It is not clear whether higher doses of antibody or prolonged dosing would have had a more pronounced and lasting effect.

Soluble Receptor

ICAM-1 is comprised of five immunoglobulin-like, glycosylated extracellular domains, a hydrophobic transmembrane region, and a small intracellular domain. Marlin et al (47) have shown that the soluble form of ICAM-1 (extracellular domain) produced in CHO cells can also effectively block rhinovirus replication in cell culture. Although this result is encouraging, the likely cost of production of soluble ICAM-1 would not make this a viable approach to treating a cold. With this in mind, a number of groups are working to identify a smaller fragment of ICAM-1 that would mimic the function of the entire molecule. Recent findings suggest that domain 1 and possibly 2 are predominantly involved in binding to rhinovirus and LFA-1 (43, 49, 50).

One potential advantage of the ICAM-1 approach to inhibiting rhinovirus infection is that the virus would not be expected to easily develop resistance to the inhibitor. This hypothesis is supported by the observation that the major receptor group viruses do not appear able to mutate to a phenotype capable of binding to an alternate receptor (16). Mutations affecting the action of an ICAM-1 mimic would likely not bind or would bind with a lower affinity to the cell surface, resulting in a virus that is potentially less virulent or is nonviable. Recent findings with poliovirus (a member of the picornavirus family) mutants indicate that unlike rhinovirus mutants, these mutants can attach to cells in the presence of soluble receptor concentrations that inhibit wild-type virus replication (37). Whether reduced potential for resistance is or is not an advantage of the ICAM-1 mimic approach to therapy will only be established in clinical trials.

Capsid-Binding Agents

In 1985, Rossmann and coworkers published the first atomic structure of a human rhinovirus, namely that of human rhinovirus type 14 (66). The elucidation of this structure (Figure 1*A*, color insert) revealed several significant structural features of particular relevance to this review. A 25-Å deep depression, or canyon, was found to encircle each five-fold axis of symmetry on the surface of the virion (Figure 1*B*, color insert). Rossmann proposed that this canyon was the site of attachment to the cellular receptor (ICAM- 1) based on the narrowness of the canyon, which would likely make the floor of the canyon inaccessible to neutralizing antibodies (65, 66). By locating the critical site of attachment in a region inaccessible to immune surveillance, the virus can maintain a conserved binding site. Evidence for the "canyon hypothesis" was provided by site-directed mutations in the canyon floor that resulted in altered receptor-binding affinity (17). Additional evidence was provided by Pevear et al, who showed that antiviral WIN compounds that raise the floor of the canyon as a consequence of binding in a hydrophobic pocket beneath the canyon (Figures 1*B* and 1*C*, color insert) also inhibit attachment (61). These capsid-binding antiviral agents affected the attachment of all of the major receptor-group viruses evaluated while having no effect on the attachment of the minor receptor-group viruses. This difference may be attributed to the absence of conformational changes in the canyon upon compound binding to minor receptor-group viruses, as was the case for rhinovirus 1A (41).

Inhibition of attachment is likely only one mechanism by which the WIN compounds and other capsid-binding agents act to inhibit the replication of the major receptor group viruses. Because these agents inhibit the uncoating of the minor group rhinoviruses and polioviruses (26, 78) they presumably would also prevent the uncoating of any major receptor-group virus that happens to enter the cell. Uncoating inhibition by these compounds is discussed in depth in the following section.

UNCOATING INHIBITORS

Inhibitors of Virion Uncoating

Following attachment to the cellular receptor, rhinoviruses are thought to enter cells by a process of endocytosis (44). After penetrating the cell membrane, the eclipsed virus becomes acidified within the endocytic vesicle (47). During the acidification process, an internal polypeptide, VP_4, is released from the virion capsid (46), possibly through a channel at the center of the pentameric unit (Figure 1*A*) (V. L. Giranda, B. A. Heinz, M. A. Oliviera, I. Minor, M. G. Rossmann & R. R. Rueckert, submitted). The loss of VP_4 may result in the exposure of hydrophobic domains that fuse with the endosomal membrane prior to release of the viral RNA into the cytoplasm (45, 48). This process of disassembly or uncoating requires that the virion be sufficiently stable to survive the conditions that exist in the extracellular environment but not so stable as to withstand the receptor and pH-induced conformational changes that must occur to allow for efficient RNA release. This stability balance is critical and forms the basis for the mechanism of action of the capsid-binding uncoating inhibitors discussed below. By increasing the stability of the virion to the point at which the virion resists the receptor and pH-induced conformational changes, the uncoating-blocking agents retain the viral RNA in the encapsidated state and the infection cycle is effectively blocked.

Many of the chemical classes of uncoating inhibitors (Figure 2) were discovered by screening the chemical files of various pharmaceutical companies for compounds capable of preventing the rhinovirus-induced cytopathic effect in cell cultures. Subsequent to the discovery of the antiviral activity, researchers demonstrated that these compounds acted by binding directly to the virion capsid and thereby inhibiting the uncoating process (2, 9, 18, 26, 34, 40, 56, 57, 69, 73, 74, 78) as well as attachment, as discussed earlier (61).

Structural Diversity

One feature of the capsid-binding compounds shown in Figure 2 is the striking diversity of chemical structures capable of binding to rhinovirus capsids. Included in this group are WIN 54954 (77), WIN 51711 or disoxaril (25, 36, 51, 52, 54, 58), R77975 (9), R61837 (6), BW683C or dichloroflavan (15), Ro 09-0410 (34, 35, 56), Ro 09-0696 (57), RMI 15,731 (12), 44,081 R.P. (2, 79), BW4294 (38), MDL 20,610 (39, 40), and SCH-38057 (67). Despite the structural variations in terms of differing heterocycles and alkyl chain lengths between compounds, they all share a common physical property: hydrophobicity. From this observation, one might have predicted that the binding site within the virus would be predominately hydrophobic.

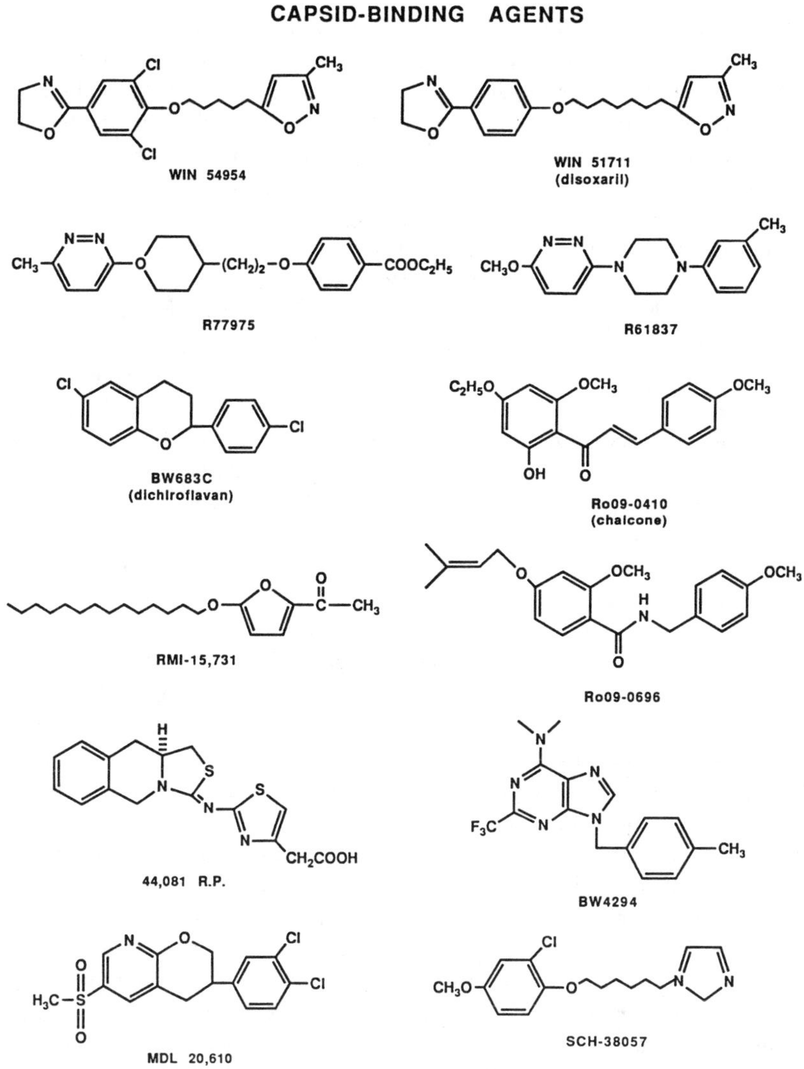

CAPSID-BINDING AGENTS

WIN 54954

WIN 51711 (disoxaril)

R77975

R61837

BW683C (dichiroflavan)

Ro09-0410 (chalcone)

RMI-15,731

Ro09-0696

44,081 R.P.

BW4294

MDL 20,610

SCH-38057

Figure 2 Structures of compounds known to, or presumed to, bind to the capsids of picornaviruses.

Site of Compound Binding

At the time of the discovery of most of the capsid-binding agents, nothing was known about the binding site within the virus other than what could be inferred from the common features of the inhibitors known to act directly on the virion. The solving of the atomic structure of a human rhinovirus (66) enabled investigators to determine for the first time whether these molecules were binding to a specific site, and if so, where the binding was occurring. The breakthrough came in 1986 when Smith et al pinpointed the binding site of the WIN compounds in the capsid of rhinovirus-14 (69). The binding site was located beneath the floor of the canyon that surrounds each five-fold symmetry axis, and access to the binding site was through a pore near the canyon floor (Figure 1*B*). The binding pocket was found to be lined predominately by hydrophobic amino acid side chains from virion capsid protein VP_1 (Figure 1*C*). Because the virion consists of 60 copies of each of 4 virion proteins, 60 drug-binding sites were present within each virion.

Two notable features of the virus were found to change upon drug binding. In the native state, the WIN pocket, or drug-binding pocket, is occluded by a polypeptide strand of VP_1-containing methionine 221. Upon drug binding, this region is pushed upward by up to 5 Å, causing a corresponding upheaval in the floor of the canyon (13). This deformation in the floor of the canyon is thought to be responsible for the inhibition of attachment to ICAM-1 discussed earlier (61).

Another feature that was modified following drug binding was the apparent increase in occupancy of a divalent cation at the vertex of each pentameric unit (Figure 1*A*) (69). The significance of this may relate directly to the mechanism by which the compounds block the release of the RNA from the capsid. Because the first step in the uncoating process appears to be loss of VP_4 and because VP_4 is clustered beneath the center of the pentamer (33, 66), the divalent cation probably serves as a closed gate at the end of the channel to the center of the virion. Support for this notion has been provided by studies of the effect of low pH on the structure of rhinovirus 14 (V. L. Giranda, B. A. Heinz, M. A. Oliviera, I. Minor, M. G. Rossmann & R. R. Rueckert, submitted). It is not clear how the binding of the WIN compounds affects the center of the pentameric unit, which is about 46 Å away.

One of the great challenges of developing a clinically useful antirhinovirus agent is finding a compound with activity against a sufficiently broad spectrum of the 100 distinct serotypes. Because the residues lining the pocket vary from serotype to serotype (59, 69), and technology does not exist to confidently model the binding sites of a virus with the precision necessary for drug design, the X-ray structures of additional serotypes are necessary. Kim et al (41) solved the structure of the minor receptor group rhinovirus, 1A, and found that the pocket was shorter than that of rhinovirus 14. The dimensions of the pocket were consistent with the observation that shorter molecules possessed relatively greater activity against 1A than 14. Another major difference in the drug-binding site of rhinovirus-1A was the absence of conformational changes in the canyon floor induced by drug binding. Like poliovirus (24), rhinovirus-1A binds a cofactor, possibly a lipid, in the drug-binding pocket. The antiviral agents appear to displace this cofactor in HRV 1A without inducing additional changes in the already open pocket configuration. The absence of change in the shape of the pocket and floor of the canyon is the likely explanation for why the antiviral agents have no effect on attachment of HRV 1A.

Mechanism of Uncoating Inhibition

How does the binding of a small molecule within a hydrophobic pocket of the virus prevent the disassembly of the virion? One explanation may be the effect on the divalent cation occupancy at the center of the pentamer as discussed above. Another likely explanation is that filling space within the capsid would increase the stability of the virus in a way similar to that seen when larger hydrophobic amino acid side chains are substituted for smaller side chains in enzymes from thermophiles (11). By increasing the stability of the virion, drug binding would be expected to enable the virion to resist the pH and receptor-induced pressure to disassemble. When Andries et al (7) looked at one compound against five rhinovirus serotypes, they demonstrated that there was no correlation between the concentration of a compound necessary to inhibit replication and the concentration necessary to prevent inactivation of the virus by heat or low pH. These authors concluded that while both antiviral and stabilizing activities are the result of drug binding, antiviral activity was not the result of an increase in virion stability. Whether there is a causal relationship remains an open question, however. Drug-requiring mutants of rhinovirus 1A have been isolated that are more thermolabile in the absence of drug than wild-type virus (D. C. Pevear, in preparation). Presumably, drug binding returns these mutants to the infectious wild-type stability phenotype. Further work needs to be done to better understand the relationship of antiviral activity and virion stabilization.

Yet another possible mechanism whereby uncoating can be blocked by drug binding is based on the assumption that the empty pocket found in native rhinovirus 14 is empty for a reason. Smith et al (69) proposed that the empty pocket is necessary to allow for the extensive conformational shifts that must occur in order for the virion to disassemble. The presence of the capsid-binding agent would prevent this necessary collapse of the pocket, thereby keeping the virion intact. Which mechanism(s) are operating to prevent uncoating following drug binding is still open to speculation.

Relationship of Structure and Activity of Antiviral Agents

Much has been published on the relationship of structure and activity for the WIN capsid-binding agents (20–23). Unlike drug discovery in most areas, the design of a clinically useful antirhinovirus agent must take into account the multitude of serotypes, and therefore, distinct targets against which an

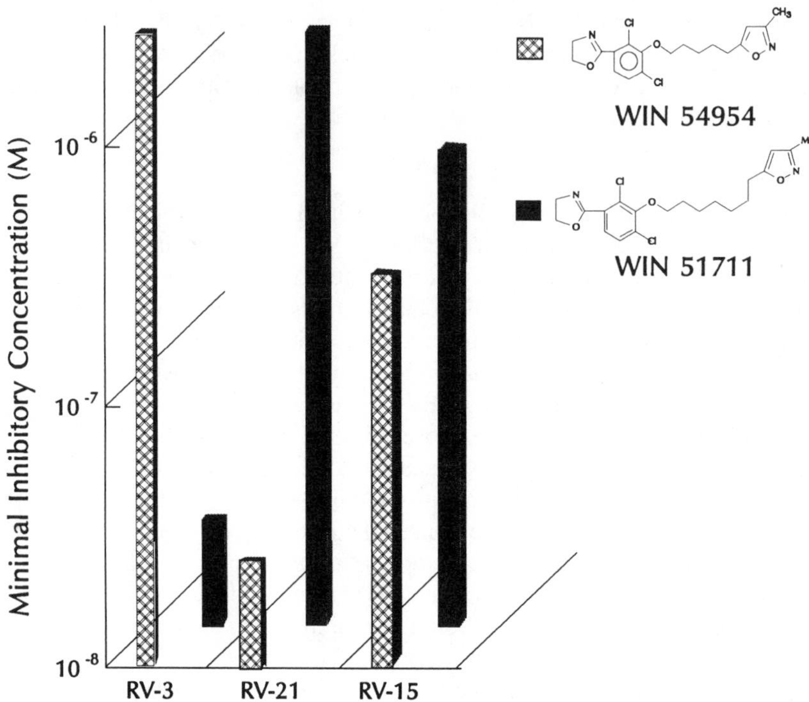

Figure 3 Inhibitory activity of two closely related antiviral compounds, WIN 51711 and WIN 54954, against rhinovirus 3, 15, and 21. The results demonstrate the significant effect small structural changes in the agent can have on the antiviral activity versus certain serotypes.

agent must be active. In essence, the synthesizing chemist must design compounds with specificity for binding to the rhinovirus pocket in VP_1 while recognizing that the pocket varies from serotype to serotype. Figure 3 shows an example of the range of sensitivity to the antiviral effects of two compounds. Some serotypes behave like rhinovirus 15 and are equally sensitive to both agents, but many serotypes such as rhinovirus 3 or 21 will be more sensitive to one compound than another. In essence, the structure-activity relationships for each serotype must be combined in order to yield a true picture of the overall structural requirements for ultimate potency and spectrum of activity.

Prior to the availability of the atomic-resolution structure of rhinovirus 14 and 1A (41, 66), drug design had to be based solely on the activity of synthesized compounds against a panel of serotypes thought to be representative of all 100 serotypes. This intensive approach revealed patterns of sensitivity to chemical modifications such as compound length and enabled

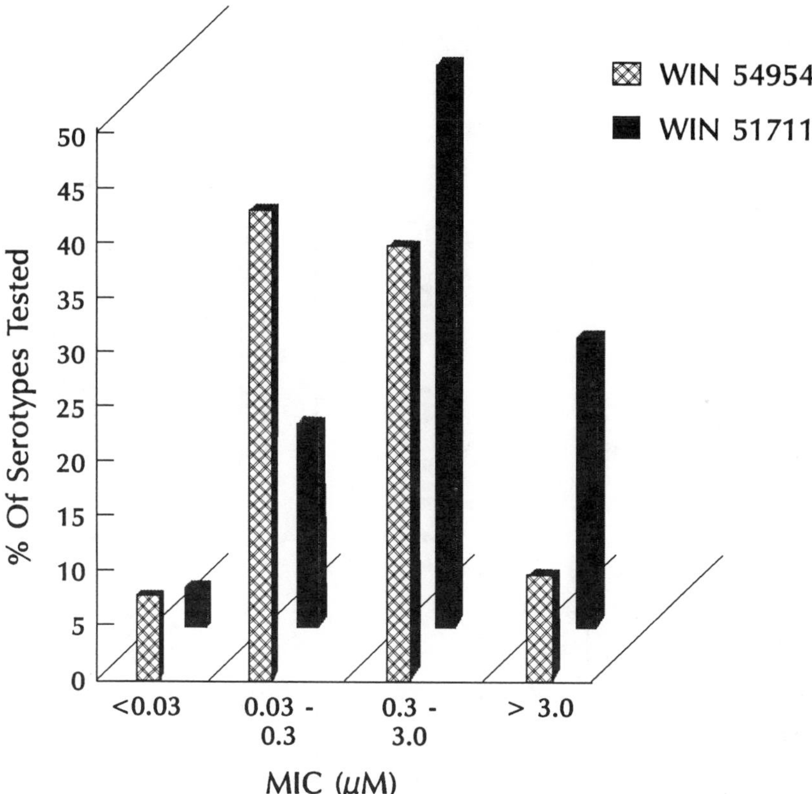

Figure 4 Spectrum of antiviral activity of two closely related antiviral agents, WIN 51711 (n = 33 serotypes tested) and WIN 54954 (n = 51 serotypes tested).

researchers to more rationally select a representative group of serotypes for routine screening (8, 10), and thereby increase the potency against a spectrum of serotypes (Figure 4).

In addition to the variability in sensitivity between serotypes, one must also evaluate the sensitivity of various isolates of a given serotype. Since a serotype is determined by surface antigenic determinants, the drug-binding pocket and drug sensitivity may differ between isolates of the same serotypes. This was found to be the case when clinical isolates of coxsackievirus B1–5 and A9 were evaluated for their sensitivity to WIN 54954 (Figure 5). The sensitivity of these isolates varied over 100-fold in some cases, suggesting that differences must exist in the binding site between isolates. Despite the range in sensitivities, WIN 54954 was effective against the vast majority of

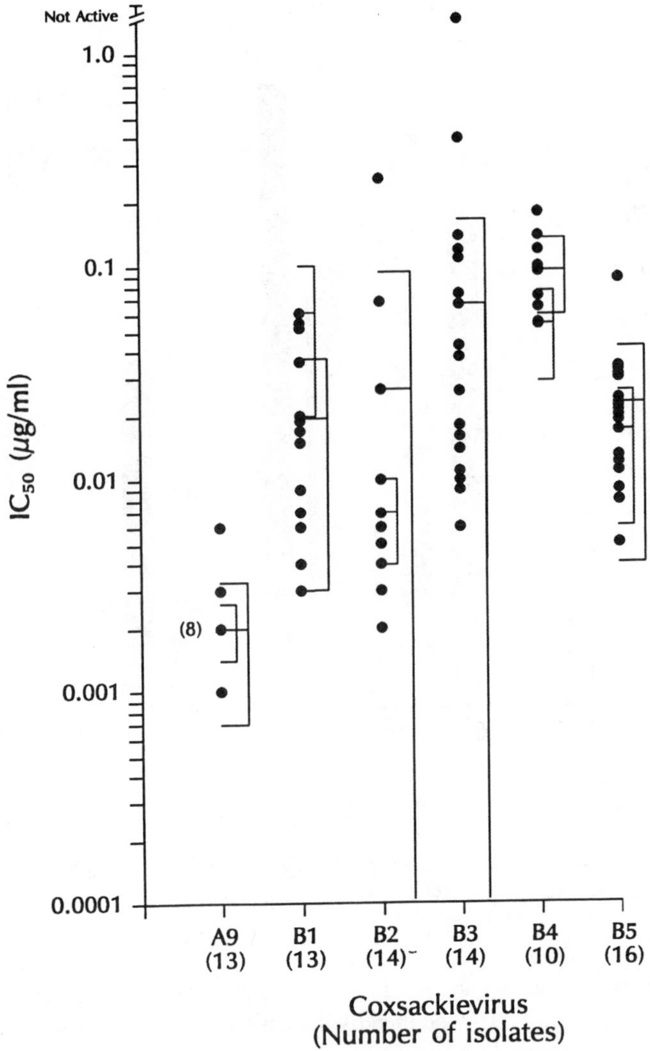

Figure 5 Activity of WIN 54954 against several clinical isolates of six different coxsackievirus serotypes.

serotypes. Various isolates of the closely related rhinoviruses should behave in a similar manner.

Rational Drug Design

The availability of the atomic structures of the rhinoviruses has made it possible to more rationally design compounds with greater potency (53).

Rational drug design is still in the early stages of evolution, however, and many of the early observations of X-ray structures of drug-virus complexes were unexpected and indicate the current state of the science.

One of the first unexpected observations made with the WIN compounds was the ability of closely related homologues to bind in opposite orientation within the pocket (69). Although certain rules were predicted to govern compound orientation (13), this result demonstrated the importance of confirming orientation with crystallographic data.

The extensive drug-induced conformational change discussed earlier was also unexpected. The magnitude of the conformational changes a given compound will induce in a given rhinovirus serotype is difficult to predict with the necessary precision for drug design, again making it necessary to obtain crystallographic data on drug-bound virus in addition to native virus. The extent of the conformational changes in human rhinovirus 14 was found to be similar for nine structurally related WIN compounds (13) but was significantly different when R77975 was bound (M. G. Rossmann, in preparation). The relationship of the movement of polypeptide chains to the antiviral potency of a given compound is not clear. The results with the WIN compounds, however, demonstrated that a 120-fold change in antiviral potency was possible in the absence of any significant differences in conformation detectable at the limits of resolution (13). The extent of hydrophobic interactions, or snugness, of the fit within the pocket is probably the most important determinant of binding affinity and antiviral activity.

Designing antirhinovirus compounds with greater potency requires that there be a good correlation between binding affinity (K_d) and in vitro antiviral activity (MIC, or minimal inhibitory concentration). Such a correlation was shown to be the case for a limited series of closely related WIN compounds (25). Several reasons, however, such as chemical instability in cell culture or inability of the compounds to penetrate the cell membrane, could account for the K_d not correctly predicting the MIC. The problem obviously becomes infinitely greater if one attempts to correlate in vivo activity with K_d.

Dissolution, absorption, excretion, metabolism, and plasma protein binding must also be addressed in order for a high affinity binding agent to be active orally against a rhinovirus infection. Because of the hydrophobic nature of the binding pocket, active molecules will likely be hydrophobic. Unfortunately, hydrophobic molecules tend to be more extensively metabolized, more slowly dissolved, and more tightly bound to plasma proteins, making an orally active agent with high binding affinity to the virus the exception and not the rule.

Drug Resistance

Single-stranded RNA viruses have high mutation frequencies in the range of 10^{-4} to 10^{-5}, as was found when rhinovirus 14 mutants resistant to high concentrations of a WIN compound were selected (32). All of the high resistance mutations were found in 2 of 16 amino acid residues lining the drug-binding pocket. In both positions, the mutations were to bulkier side chains, suggesting that resistance resulted from steric blocking of drug binding. Since all capsid-binding agents appear to bind to the same site, high level drug resistance to one compound is generally seen for all others (7, 57, 74). Human rhinoviruses resistant to lower concentrations of a WIN compound were observed at a yet higher frequency in tissue culture (32). The mutations responsible for the low resistance phenotype all mapped to regions near the canyon floor that are disrupted upon drug binding. Some of these mutations partially reversed the ability of the capsid-binding agents to block attachment to ICAM-1.

A completely different pattern of drug-resistance has been seen when rhinovirus 1A–resistant mutants were isolated. In this case, all of the independently derived mutants isolated, regardless of the drug employed, grew to significantly higher titer in the presence of the antiviral agent (D. C. Pevear, in preparation). This drug-enhanced or drug-dependent phenotype is similar to that described by Schrom et al (68) for arildone (55), a predecessor of the antirhinovirus WIN compounds, when it was used to select drug-resistant mutants of poliovirus. Drug-dependent mutants of rhinovirus 9 have also been reported (1). Many of the drug-enhanced mutants are less stable in the absence of drug, which suggests that the compounds may act to restore wild-type stability to the virion, whereas in the absence of drug the virions degrade or prematurely uncoat, resulting in an inefficient infection.

What is the likely clinical significance of drug resistance given the high rate of mutation observed in vitro? While the answer to this question must await clinical use of an effective agent, preliminary evidence suggests that resistant virus can be readily isolated from infected volunteers being treated with R61837 (19). The important question, however, is whether or not resistant virus will have the same propensity to cause symptomatic disease as does wild-type virus. At least one study suggests that these mutants are less virulent (1).

CLINICAL TRIALS

One of the attractive aspects of developing antirhinovirus drugs is that their clinical efficacy can be evaluated in a well-controlled environment where human volunteers are experimentally infected with a rhinovirus of known sensitivity to the test drug. Because of the significant placebo effect known to exist for rhinoviruses, such studies must be done with sufficient blinding and with as many objective measures of efficacy as possible (70). In addition to the subjective assessment of cold symptoms, clinical trials have used such objective measurements as nasal mucous discharge weights, viral shedding, cough counts, number of tissues used, etc.

Dichoroflavan (5, 64), Ro 09-0410 (4), Ro 09-0415 (62), and 44,081 R.P. (79) have all been evaluated for their ability to prevent experimental colds with largely disappointing results. In cases where dichloroflavan and Ro 09-415 were delivered orally, activity was not demonstrated despite the fact that blood levels in excess of the MIC for the infecting virus were achieved (62, 64). Neither compound could be found in nasal secretions, raising the question of whether they reached the site of infection. Intranasal administration of 44,081 R.P. (79), Ro 09-410 (4), and dichloroflavan (5) was also shown to be ineffective in preventing colds. Whether the compounds were cleared rapidly by ciliary action and/or did not reach inhibitory concentration within the nasal epithelium is not clear because the intranasal route of delivery has been shown to be effective in trials of R61837 (3, 14). In this study, six-time-daily dosing with R61837 beginning 4–28 hours prior to infection and continued for 4–6 days resulted in significant reductions in nasal secretion rates and mean clinical scores. Prolonged dosing appeared to be necessary to prevent the appearance of symptoms following cessation of dosing. No efficacy was observed when dosing was initiated after infection. Efficacy has also been demonstrated for R77975, a more potent analog of R61837 (30). This compound was effective when administered in a prophylaxis regimen six times daily but was ineffective when the dosing frequency was reduced to three times daily. While the six-times-per-day dosing regimen required for efficacy in this study clearly demonstrates the efficacy of R77975, it would probably not be acceptable to consumers.

Efficacy trials have recently been completed with WIN 54954. In experimental infections with rhinovirus T39 and rhinovirus 23, in which WIN 54954 was administered orally in a three-times-daily prophylaxis regimen, no significant effect was observed on symptom scores or objective measures of efficacy (R. Turner & F. Hayden, abstract submitted to meeting of the 1992 International Society for Antiviral Research, Vancouver). Indeed, attack rates and symptom scores appeared to be higher in the WIN 54954–treated group in the rhinovirus T39 study. In both studies, blood levels in excess of the MIC for the infecting virus were maintained throughout the dosing period, although nasal-secretion concentrations were generally low or undetectable.

In marked contrast to the negative results in the two rhinovirus studies, a prophylaxis trial with WIN 54954 against coxsackievirus A21 clearly demonstrated the oral efficacy of the drug (G. M. Schiff, J. R. Sherwood, E. C. Young, L. J. Mason, abstract submitted to the meeting of the 1992 International Society for Antiviral Research, Vancouver). Coxsackievirus A21, while not a rhinovirus, does bind to ICAM-1 and causes an illness in volunteers similar to the common cold, although the virus has been associated with more severe systemic illness (71). In this trial, WIN 54954 was administered orally at 600 mg three times daily in a prophylaxis mode similar to that used in the rhinovirus studies. Subjects receiving WIN 54954 showed significant reduction in symptomatic attack rate ($p = 0.0001$), illness severity ($p = 0.0009$), and viral shedding ($p = 0.016$). The striking difference in efficacy between coxsackievirus A21 and rhinovirus 23 could not be attributed to antiviral sensitivity because the viruses were roughly equisensitive to WIN 54954. This represents the first demonstration of oral efficacy of an antipicornavirus agent.

DEVELOPMENT CHALLENGES

A number of significant hurdles must be overcome before a clinically useful and commercially viable drug for the prevention and treatment of the rhinovirus common cold is available. Some of the more significant hurdles are:

1. The drug must be effective against the vast majority of rhinovirus serotypes since no particular serotypes appear to predominate. This point is particularly relevant to the capsid-binding agents with their dramatic range of potency against various serotypes.
2. A successful agent for oral delivery must have a safe toxicity and side-effect profile consistent with the disease being treated, and must have a pharmacokinetic profile that results in sufficient drug reaching the site of infection.
3. For intranasally delivered agents, the drug should be sufficiently stable in nasal tissue, and should be capable of achieving and maintaining inhibitory concentrations at the site of infection when given at a reasonable frequency (three times daily).
4. The drug must be capable of reducing the duration and severity of infection by a clinically significant margin (about 50%?) when administered no later than 24–36 hours after the appearance of symptoms. This goal will only be achievable if virus replication does not initiate an irreversible cascade of events prior to the onset of symptoms.
5. The marketed drug must be available at a cost that is not prohibitive. This would be a challenge for the protein-based ICAM-mimics that are currently being studied.

Despite the magnitude of the challenges facing researchers involved in the discovery and development of antirhinovirus agents, significant progress has been made toward the goals of preventing and treating this most common of human virus infections. The vast majority of the work cited in this review has been published within the past 10 years. Much of the literature from the past 5 provides some sense of the rate of progress in this area. Although other molecular targets certainly exist for rhinovirus therapy (60), most of the emphasis has been placed on inhibitors of attachment and uncoating. Given this emphasis, if a "cure" for the common cold is developed within the next decade, it is likely to come from this mechanistic class. For the optimists among us, however, the question is not if a cure can be found, but when.

ACKNOWLEDGMENTS

We thank Marjorie Celentano and Mary Verity for their patience and assistance in the preparation of this manuscript and acknowledge Jean-Yves Sgro for the creative and artistic way he illustrated the drug-virus interactions (Figure 1).

Literature Cited

1. Ahmad, A. L. M., Dowsett, A. B., Tyrrell, D. A. J. 1987. Studies of rhinovirus resistant to an antiviral chalcone. *Antivir. Res.* 8:27–39
2. Alarcon, B., Zerial, A., Dupiol, C., Carrasco, L. 1986. Antirhinoviruses compound 44081 R. P. inhibits virus uncoating. *Antimicrob. Agents Chemother.* 30:31–34
3. Al-Nakib, W., Higgins, P. G., Barrow, G. I., Tyrrell, D. A. J., Andries, K., et al. 1989. Suppression of colds in human volunteers challenged with rhinovirus by a new synthetic drug (R61837). *Antimicrob. Agents Chemother.* 33:522–25
4. Al-Nakib, W., Higgins, P. G., Barrow, I., Tyrrell, D. A. J., Lenox-Smith, I., et al. 1987. Intranasal chalcone, Ro 09-0410, as prophylaxis against rhinovirus infection in human volunteers. *J. Antimicrob. Chemother.* 20:887–92
5. Al-Nakib, W. J., Willman, J., Higgins, P. G., Tyrrell, D. A. J., Shepherd, W. M., et al. 1987. Failure of intranasally administered 4',5-dichloroflavan to protect against rhinovirus in man. *Arch. Virol.* 92:255–60
6. Andries, K., DeWindt, B., DeBrabander, M., Stokbroekx, R., Janssen, P. A. J. 1988. In vitro activity of R61837, a new antirhinovirus compounds. *Arch. Virol.* 101:155–67
7. Andries, K., Dewindt, B., Snoeks, J., Willebrords, R. 1989. Lack of quantitative correlation between inhibition of replication of rhinoviruses by an antiviral drug and their stabilization. *Arch. Virol.* 106:56–61
8. Andries, K., DeWindt, B., Snoeks, J., Willebrords, R., Stokbroekx, R., et al. 1991. A comparative test of fifteen compounds against all known human rhinovirus serotypes as a basis for a more rational screening program. *Antivir. Res.* 16:213–25
9. Andries, K., DeWindt, B., Snoeks, J., Willebrords, R., VanEemeren, K. 1991. Antirhinovirus spectrum and mechanism of action of R77975. *Antivir. Res.* 1:98 (Suppl.)
10. Andries, K., DeWindt, B., Snoeks, J., Wouters, L., Moereels, H., et al. 1990. Two groups of rhinoviruses revealed by a panel of antiviral compounds present sequence divergence and differential pathogenicity. *J. Virol.* 64:1117–23
11. Argos, P., Rossmann, M. G., Gran, U. M., Zuber, H., Frank, G., Tratschin, J. D. 1979. Thermal stability and protein structure. *Biochemistry* 18:5698–5703
12. Ash, R. J., Parker, R. A., Hagan, A. C., Mayer, G. D. 1979. RMI 15,731 (1-[5-Tetradecyloxy-2-furanyl]ethanone), a new antirhinovirus compound. *Antimicrob. Agents Chemother.* 16:301–5
13. Badger, J., Minor, I., Kremer, M. J., Oliviera, M. A., Smith, T. J., et al. 1988. Structural analysis of a series of antiviral agents complexed with human rhinovirus 14. *Proc. Natl. Acad. Sci. USA* 85:3304–8
14. Barrow, G. I., Higgins, P. G., Tyrrell, D. A. J., Andries, K. 1990. An appraisal of the efficacy of the antiviral R 61837 in rhinovirus infections in human volunteers. *Antivir. Chem. Chemother.* 1(5):279–83
15. Bauer, D. J., Selway, J. W. T., Batchelor, J. F., Tisdale, M., Caldwell, I. C., et al. 1981. 4',6-Dichloroflavan (BW683C), a new anti-rhinovirus compound. *Nature* 292:369–70

16. Colonno, R. J., Calahan, P. L., Long, W. J. 1986. Isolation of a monoclonal antibody that blocks attachment of the major group of rhinoviruses. *J. Virol.* 57:7–12
17. Colonno, R. J., Condra, J. H., Mizutani, S., Callahan, P. L., Davies, M. E., et al. 1988. Evidence for the direct involvement of the rhinovirus canyon in receptor binding. *Proc. Natl. Acad. Sci. USA* 85:5449–53
18. Conti, C., Orsi, N., Stein, M. L. 1988. Effect of isoflavans and isoflavenes on rhinovirus 1B and its replication in HeLa cells. *Antivir. Res.* 10:117–27
19. Dearden, C., Al-Nakib, W., Andries, K., Woestenborghs, R., Tyrell, D. A. J., et al. 1989. Drug resistant rhinoviruses from the nose of experimentally treated volunteers. *Arch. Virol.* 109:71–81
20. Diana, G. D., Cutcliffe, D., Oglesby, R. C., Otto, M. J., Mallamo, J. P., et al. 1989. Synthesis and structure-activity of some disubstituted phenylisoxazoles against human picornaviruses. *J. Med. Chem.* 32:450–55
21. Diana, G. D., McKinlay, M. A., Otto, M. J., Akullian V., Oglesby, C. 1985. (4,5,-dihydro-2-oxazoyl) phenoxyalkylisoxazoles, inhibitors of picornavirus uncoating. *J. Med. Chem.* 28:1906–10
22. Diana, G. D., Ogelsby, C., Akullian, V., Carabateas, P. M., Cutcliffe, D., et al. 1987. Structure activity studies of 5-[4-(4,5,dihydro-2-oxazoyl)phenoxy]-alkyl]-3-methyl isoxazoles: Inhibitors of picornavirus uncoating. *J. Med. Chem.* 30:383–88
23. Diana, G. D., Otto, M. J., Treasurywala, A. M., McKinlay, M. A., Ogelsby, R. C., et al. 1988. Enantiomeric effects of homologues of disoxaril on the inhibitory activity against human rhinovirus-14. *J. Med. Chem.* 31:540–44
24. Filman, D. J., Syed, R., Chow, M., Macadam, A. J., Minor, P. D., et al. 1989. Structural factors that control conformational transitions and serotype specificity in type 3 poliovirus. *EMBO J.* 8:1567–79
25. Fox, M. P., McKinlay, M. A., Diana, G. D., Dutko, F. J. 1991. Binding affinities of structurally related human rhinovirus capsid-binding compounds are related to their activities against human rhinovirus type 14. *Antimicrob. Agents Chemother.* 35:1040–47
26. Fox, M. P., Otto, M. J., McKinlay, M. A. 1986. The prevention of rhinovirus and poliovirus uncoating by WIN 51711: A new antiviral drug. *Antimicrob. Agents Chemother.* 30:110–16
27. Greve, J. M., Davis, G., Meyer, A. M., Forte, C. P., Connolly-Yost, S., et al. 1989. The major human rhinovirus receptor is ICAM-1. *Cell* 56:839–47
28. Gwaltney, J. M. 1985. The common cold. In *Principles and Practices of Infectious Disease,* ed. L. Mandel, R. G. Douglas, J. E. Bennett, pp. 351–55. New York: Wiley. 2nd ed.
29. Hamparian, V. V., Colonno, R. J., Cooney, M. K., Dick, E. C., Gwaltney, J. M., et al. 1987. A collaborative report: rhinoviruses—extension of the numbering system from 89 to 100. *Virology* 159:191–92
30. Hayden, F. G., Andries, K., Janssen, P. A. J. 1990. Efficacy of intranasal R77975 for the prevention of experimentally induced rhinovirus infection and illness. *Antivir. Res.* Suppl. 1, 103 pp.
31. Hayden, F. G., Gwaltney, J. M., Colonno, R. J. 1988. Modification of experimental rhinovirus colds by receptor blockade. *Antivir. Res.* 9:233–47
32. Heinz, B. A., Rueckert, R. R., Shepard, D. A., Dutko, F. J., McKinlay, M. A., et al. 1989. Genetic and molecular analysis of spontaneous mutants of human rhinovirus 14 resistant to an antiviral compound. *J. Virol.* 63:2476–85
33. Hogle, J. M., Chow, M., Filman, D. J. 1985. Three-dimensional structure of poliovirus at 2.9 Å resolution. *Science* 229:1358–65
34. Ishitsuka, H., Ninomiya, Y. T., Ohsawa, C., Fujiu, M., Suhara, Y. 1982. Direct and specific inactivation of rhinovirus by chalcone. *Antimicrob. Agents Chemother.* 22:617–21
35. Ishitsuka, H., Ninomiya, Y., Suhara, Y. 1986. Molecular basis of drug resistance to new antirhinovirus agents. *J. Antimicrob. Chemother.* 18B:11–18
36. Jubelt, J. A., Wilson, K., Ropka, S. L., Guidinger, P. L., McKinlay, M. A., et al. 1989. Clearance of a persistant human enterovirus infection of the mouse central nervous system by the antiviral agent disoxaril. *J. Infect. Dis.* 159:866–71
37. Kaplan, G., Peters, D., Racaniello, V. R. 1990. Poliovirus mutants resistant to neutralization with soluble receptors. *Science* 250:1596–99
38. Kelley, J. L., Linn, J. A., Selway, J. W. T. 1990. Antirhinovirus activity of 6-anilino-9-benzyl-2-chloro-9H-purines. *J. Med. Chem.* 33:1360–63
39. Kenny, M. T., Dulworth, J. K., Bargar, T. M., Torney, H. L., Graham, M. C., et al. 1986. In vitro antiviral activity of

the 6-substituted 2-(3',4''-dichlorophenoxy) 2H-pyranol(2,3-b)pyridines MDL 20,610,MDL 20,646 and MDL 20,957. *Antimicrob. Agents Chemother.* 30:516–18
40. Kenny, M. T., Torney, H. L., Dulworth, J. K. 1988. Mechanism of action of the antiviral compound MDL 20,610. *Antivir. Res.* 9:249–61
41. Kim, S., Smith, T. J., Chapman, M. S., Rossmann, M. G., Pevear, D. C., et al. 1989. Crystal structure of human rhinovirus serotype 1A (HRV 1A). *J. Mol. Biol.* 210:91–111
42. Larson, H. E., Reed, S. E., Tyrell, D. A. J. 1980. Isolation of rhinoviruses and coronaviruses from 38 colds in adults. *J. Med. Virol.* 5:221–29
43. Lineberger, D. W., Graham, D. J., Tomassini, J. E., Colonno, R. J., 1990. Antibodies that block rhinovirus attachment map to domain 1 of the major receptor group. *J. Virol.* 64:2582–87
44. Lonberg-Holm, K. 1975. The effects of conconavalin A on the early events of infection by rhinovirus type 2 and poliovirus type 2. *J. Gen. Virol.* 28: 313–27
45. Lonberg-Holm, K., Gosser, L. B., Shimshick, E. J. 1976. Interaction of liposomes with subviral particles of poliovirus type 2 and rhinovirus type 2. *J. Virol.* 19:746–49
46. Lonberg-Holm, K., Korant, B. D. 1972. Early interaction of rhinoviruses with host cells. *J. Virol.* 9:29–40
47. Madshus, I. H., Olsnes, S., Sandvig, K. 1984. Different pH requirements for entry of the two picornaviruses, human rhinovirus 2 and murine encephalomyocarditis virus. *Virology* 139:346–57
48. Madshus, I. H., Olsnes, S., Sandvig, K. 1984. Mechanism of entry into the cytosol of poliovirus type 1: requirement for low pH. *J. Cell Biol.* 98:1194–1200
49. Marlin, S. D., Staunton, D. E., Springer, T. A., Stratowa, C., Sommergruber, W., et al. 1990. A soluble form of intracellular adhesion molecule-1 inhibits rhinovirus infection. *Nature* 344: 70–72
50. McClelland, A., DeBear, J., Connolly Yost, S., Meyer, A. M., Marlor, C. W., et al. 1991. Identification of monoclonal antibody epitopes and critical residues for rhinovirus binding in domain 1 of intracellular adhesion molecule 1. *Proc. Natl. Acad. Sci. USA* 88:7993–97
51. McKinlay, M. A. 1985. WIN 51711: a new systemically active broad-spectrum antipicornavirus agent. *J. Antimicrob. Chemother.* 16:284–86
52. McKinlay, M. A., Frank, J. A., Steinberg, B. A. 1986. Use of WIN 51711 to prevent echovirus type 9-induced paralysis in suckling mice. *J. Infect. Dis.* 154: 676–81
53. McKinlay, M. A., Rossmann, M. G. 1989. Rational design of antiviral agents. *Annu. Rev. Pharmacol. Toxicol.* 29:111–22
54. McKinlay, M. A., Steinberg, B. A. 1986. Oral efficacy of WIN 51711 in mice infected with human poliovirus. *Antimicrob. Agents Chemother.* 29:30–32
55. McSharry, J. J., Caliguiri, L. A., Eggers, H. J. 1979. Inhibition of uncoating of poliovirus by arildone, a new antiviral agent. *Virology* 97:307–15
56. Ninomiya, Y., Ohsawa, C., Aoyama, M., Umeda, I., Suhara, Y., et al. 1984. Antivirus agent, R009–0410, binds to rhinovirus specifically and stabilizes the virus conformation. *Virology* 134:269–76
57. Ninomiya, Y., Shimma, N., Ishitsuka, H. 1990. Comparative studies on the antirhinovirus activity and mode of action of the rhinovirus capsid binding agents, chalcone amides. *Antivir. Res.* 13:61–74
58. Otto, M. J., Fox, M. P., Fancher, M. J., Kuhrt, M. F., Diana, G. D., et al. 1985. In vitro activity of WIN 51711, a new broad-spectrum antipicornavirus drug. *Antimicrob. Agents Chemother.* 27:883–86
59. Palmenberg, A. C. 1989. Sequence alignments of picornaviral capsid proteins. In *Molecular Aspects of Picornavirus Infections and Detection,* ed. B. L. Semler, E. Ghrenfeld, pp. 211–41. Washington, DC: Am. Soc. Microbiol.
60. Pevear, D. C., Diana, G. D., McKinlay, M. A. 1992. Molecular targets for antiviral therapy of the common cold. *Sem. Virol.* In press
61. Pevear, D. C., Fancher, M. J., Felock, P. J., Rossmann, M. G., Miller, M. S., et al. 1989. Conformational changes in the floor of the human rhinovirus canyon blocks adsorption to HeLa cell receptors. *J. Virol.* 63:2002–7
62. Phillpotts, R. J., Higgins, P. G., Willman, J. S., Tyrrell, D. A. J., Lenox-Smith, I. 1984. Evaluation of the antirhinovirus chalcone Ro 09–0415 given orally to volunteers. *J. Antimicrob. Chemother.* 14:403–9
63. Phillpotts, R. J., Tyrrell, D. A. S. 1985. Rhinovirus colds. *Br. Med. Bull.* 41:386–90
64. Phillpotts, R. J., Wallace, J., Tyrrell,

D. A. J., Freestone, D. S., Shepherd, W. M., et al. 1983. Failure of oral 4',6-dichloroflavan to protect against rhinovirus infection in man. *Arch. Virol.* 75:115–21
65. Rossmann, M. G. 1989. The canyon hypothesis. *J. Biol. Chem.* 264:14587–90
66. Rossmann, M. G., Arnold, E., Erickson, J. W., Frankenberger, E. A., Griffith, J. P., et al. 1985. Structure of a common cold virus and functional relationship to other picornaviruses. *Nature* 317:145–53
67. Rozhon, E., Cox, S., Buontempo, J., O'Connell, J., Schwartz, J., et al. 1990. A new molecule with broad spectrum inhibitory activity against picornaviruses. *Antivir. Res.* Suppl. 1:62
68. Schrom, M., Laffin, J. A., Evans, B., McSharry, J. J., Caliguiri, L. A. 1982. Isolation of poliovirus variants resistant to and dependent on arildone. *Virology* 122:492–97
69. Smith, T. J., Kremer, M. J., Luo, M., Vriend, G., Arnold, E., et al. 1986. The site of attachment in human rhinovirus 14 for antiviral agents that inhibit uncoating. *Science* 233:1286–93
70. Sperber, S. J., Hayden, F. G. 1988. Chemotherapy of rhinovirus colds. *Antimicrob. Agents Chemother.* 32:409–19
71. Spickard, A., Evans, H., Knight, V., Johnson, K. 1963. Acute respiratory disease in normal volunteers associated with coxsackie A-21 viral infection. III. Response to nasopharyngeal and enteric inoculation. *J. Clin. Invest.* 42:840–52
72. Staunton, D. E., Merluzzi, V. J., Rothlein, R., Barton, R., Marlin, S. O., et al. 1989. A cell adhesion molecule, ICAM-1, is the major receptor for rhinoviruses. *Cell* 56:849–53
73. Tisdale, M., Selway, J. W. T. 1983. Inhibition of an early stage of rhinovirus replication by dichloroflavan (BW683C). *J. Gen. Virol.* 64:795–803
74. Tisdale, M., Selway, J. W. T. 1984. Effect of dichloroflavan (BW683C) on the stability and uncoating of rhinovirus type 1B. *J. Antimicrob. Chemother.* 14:97–105 (Suppl.)
75. Tomassini, J. E., Graham, D., DeWitt, C. M., Lineberger, D. W., Rodkey, J. A., et al. 1989. cDNA cloning reveals that the major group rhinovirus receptor on HeLa cells is intracellular adhesion molecule-1. *Proc. Natl. Acad. Sci. USA* 86:4907–11
76. Uncapher, C. R., DeWitt, C. M., Colonno, R. J. 1991. The major and minor group receptor families contain all but one human rhinovirus serotype. *Virology* 180:814–17
77. Woods, M. G., Diana, G. D., Rogge, M. C., Otto, M. J., Dutko, F. J., et al. 1989. In vitro and in vivo activities of WIN 54954, a new broad-spectrum antipicornavirus drug. *Antimicrob. Agents Chemother.* 33:2069–74
78. Zeichhardt, H., Otto, M. J., McKinlay, M. A., Willingham, P., Habermehl, K.-O. 1987. Inhibition of poliovirus uncoating by disoxaril (WIN 51711). *Virology* 160:281–85
79. Zerial, A., Werner, G. H., Phillpotts, R. J., Willmann, J. S., Higgins, P. G., et al. 1985. Studies on 44081 R. P., a new antirhinovirus compound, in cell cultures and in volunteers. *Antimicrob. Agents Chemother.* 27:846–50

Annu. Rev. Microbiol. 1992. 46:655–93

HUMAN IMMUNODEFICIENCY VIRUS AND THE CENTRAL NERVOUS SYSTEM

David C. Spencer and Richard W. Price

Department of Neurology, University of Minnesota, Minneapolis, Minnesota 55455

KEY WORDS: dementia, AIDS, AIDS dementia complex, brain, retroviruses

CONTENTS

Abstract

Neurological disease frequently complicates HIV-1 infection. In addition to opportunistic infections, a syndrome of combined cognitive and motor impairment, referred to as the AIDS dementia complex, has been recognized. While presumed to relate to HIV-1 itself, the pathogenesis of this syndrome remains uncertain. Because of the limited extent of productive brain HIV-1

0066-4227/92/1001-0655$02.00

infection in many cases, and because such infection involves macrophages and microglia rather than cells of neuroectodermal origin, current speculation centers on indirect mechanisms of brain injury including virus- or cell-coded neurotoxins. We review clinical and laboratory studies and also describe models of the interaction of HIV-1 and immune responses that might account for brain injury.

INTRODUCTION

Among the clinically important and biologically intriguing aspects of human immunodeficiency virus type one (HIV-1) are its effects on the central nervous system (CNS). This clinical importance stems from the frequency of neurological disease over the course of infection, particularly during its late phase of severe immunodeficiency (for general reviews see 16, 100, 172). The CNS of HIV-1 infected individuals is not only vulnerable to opportunistic infections (e.g. cerebral toxoplasmosis, cryptococcal meningitis, and progressive multifocal leukoencephalopathy) and opportunistic neoplasms (e.g. primary CNS lymphoma), but also to a unique syndrome of neurological impairment referred to here as the AIDS dementia complex (ADC), which appears to result from a more fundamental effect of HIV-1 itself, rather than simply from an opportunistic infection by another pathogen. However, the relation between the AIDS retrovirus and clinical neurological dysfunction noted in this syndrome remains poorly understood. In this review, we examine both clinical and laboratory observations bearing on the AIDS dementia complex and attempt to provide a guide to emerging hypotheses of pathogenesis provoked by these observations.

Although slow to achieve clear definition, the AIDS dementia complex was recognized early as outside of the spectrum of neurological complications typically seen in other immunosuppressed patients (56, 71, 151). Once the AIDS virus was identified, investigators speculated that the AIDS dementia complex might result from direct CNS infection by HIV-1 (then referred to as LAV or HTLV-III). Additional impetus for this speculation came from earlier studies showing that retroviruses cause neurological diseases in animals (62). Particularly provocative in this regard was the observation that HIV-1 was similar to the lentivirus, visna, which causes slow infection of the nervous system in sheep (55). A retroviral etiology of the AIDS dementia complex soon gained more direct support from studies identifying HIV-1 in the brain, first by Southern blot detection of proviral DNA (148) and soon thereafter by detection of viral mRNA, antigens, and particles, as well as direct isolation of virus (7, 45, 51, 53, 70, 80, 134, 160, 169).

However, as more detailed information about HIV-1 brain infection accrued, it became increasingly evident that the relationship between HIV-1 infection of the CNS and the neurological symptoms and signs of the dementia complex was not straightforward and could not be explained by simple cytolytic brain infection (129). Thus, while HIV-1 clearly can infect the CNS, this infection does not appear to involve the neuroectodermally derived functional elements of the brain (i.e. neurons, oligodendrocytes, and astrocytes), but rather cells of bone-marrow origin that populate or are attracted to the CNS (macrophages and microglia). Consequently, recent speculation centers on indirect mechanisms of brain damage and dysfunction resulting from infection of these cells (91, 128, 132).

The relation between brain infection and disease is also confounded by several additional observations. The brain appears to be exposed to the virus early in the course of systemic infection, and HIV-1 may be detected in the cerebrospinal fluid (CSF) during the clinically latent period of asymptomatic seropositivity, yet the AIDS dementia complex, particularly in its severe form, is generally a late development seen in only a subset of patients. Additionally, studies anatomically mapping the distribution and extent of productive HIV-1 infection of the CNS indicate that in many cases infection appears more limited than might be anticipated from the clinical-disease severity. Indeed, some AIDS-dementia-complex cases exhibit little or no detectable productive brain infection.

In reviewing the data bearing upon these issues, a simple diagram can be employed to highlight the three principal factors that appear to be involved in pathogenesis (Figure 1). Two of these, the virus and the immune system, can be considered the primary agonists that sustain a dynamic and changing interaction as disease evolves. The third factor is the target: the CNS. After reviewing the salient clinical, pathological, and virological aspects of the AIDS dementia complex and CNS HIV-1 infection, we return to this model and examine how these interactions might answer some of the following fundamental questions regarding how the brain is injured by HIV-1:

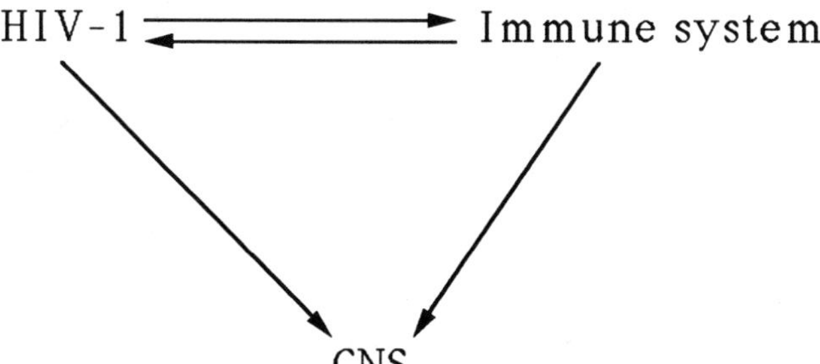

Figure 1 Interrelations of the major pathogenetic factors in the AIDS dementia complex.

1. What is the relation between HIV-1 infection and the AIDS dementia complex?
2. What is the nature and course of CNS HIV-1 infection?
   a. When and how does the virus enter the CNS?
   b. What becomes of it after entry?
      i. Does it persist in the brain, and if so, in what molecular form?
      ii. What cells are infected?
      iii. What controls the subsequent profile of HIV-1 infection in various cell types—latent, indolent, or active productive infection?
   c. What determines the tissue and regional localization of brain HIV-1 infection?
   d. What directs the emergence of neuropathic viral strains?
3. What role does the immune system play?
   a. In protecting the CNS?
   b. In damaging the CNS?
4. How is the CNS injured?
   a. What are the principal pathogenic agonists?
   b. What are the cellular targets?
   c. What are the mechanisms of injury?

THE CLINICAL PROBLEM: CNS HIV-1 INFECTION AND THE AIDS DEMENTIA COMPLEX

Neurological abnormalities have been described in all phases of HIV-1 infection, beginning with the primary infection associated with seroconversion, during which a number of encephalitic syndromes, usually monophasic and self-limiting in character, may occur (15, 157). Whether these relate to acute brain infection, parainfectious autoimmune reactions, or components of both is still uncertain. In the clinically latent or asymptomatic seropositive phase (generally associated with helper or $CD4^+$ T-lymphocyte counts $> 500/mm^3$), studies of the CSF have documented frequent asymptomatic infection (see below). The transitional phase, often marked by the onset of constitutional symptoms, including those of the AIDS-related complex (generally with $CD4^+$ counts between 200 and 500 cells/mm^3), can be complicated by aseptic meningitis, possibly related to HIV-1 infection, as well as by the appearance of the AIDS dementia complex, usually in a mild form. This syndrome, however, is more characteristic of the late phase, when when CNS opportunistic infections and neoplasms also manifest ($CD4^+$ counts < 200 cells/mm^3). For simplicity, throughout this essay we refer to the phases of infection by these terms as if they are biologically uniform, while recognizing that the relation between immunosuppression and disease is more complex.

The AIDS Dementia Complex: A Subcortical Dementia

The term AIDS dementia complex was introduced to designate a clinical syndrome, i.e. a cohesive constellation of symptoms and signs, rather than a clear etiopathogenetic disease entity (106, 107). Whether one prefers this or another term, one should distinguish between the concept of this syndrome and HIV-1 brain infection, which refers to a pathobiological process. Thus, CNS HIV-1 infection ranges from early asymptomatic, and seemingly innocent exposure and perhaps colonization to late productive, pathogenic parenchymal infection. Although infection of the CNS likely accounts in some way for at least one of the pathological subtypes of the AIDS dementia complex, it is less clearly implicated in milder forms and other pathological subtypes (132).

The AIDS dementia complex has been classified with the subcortical dementias on the basis of the character of cognitive dysfunction as revealed by both bedside examination and more formal and quantitative neuropsychological testing (107, 131, 132, 159). This term is used to group the cognitive impairment found in Parkinson's disease, Huntington's disease, hydrocephalus, progressive supranuclear palsy, and other conditions in which psychomotor slowing and difficulty with attention and concentration are salient clinical deficits (6, 37). This grouping is contradistinct from the cortical dementias, such as Alzheimer's disease and Creutzfeldt-Jacob disease, in which focal deficits in memory retrieval, language (aphasia), and the like predominate. Although the anatomical justification for this dichotomy remains open to question, it is a clinically useful distinction and emphasizes the similarities of the AIDS dementia complex to the other subcortical dementias mentioned above. It also provides a clue to the pathophysiology of cerebral dysfunction in these patients.

Because the AIDS dementia complex is defined as a syndrome rather than a disease, one can legitimately ask whether the full spectrum of the condition actually relates to a single pathogenetic process. Arguing for a single process is the observation that the character of the dementia is similar over the full spectrum of severity: a combination of cognitive and motor slowing that worsens with disease progression (131, 159). This review focuses on disease in adults, but a parallel condition affects children with AIDS and shares the same general features, albeit involving the developing brain (5, 46). It begins with loss of previously acquired developmental milestones and may progress to include dementia, quadriplegia, and rigidity. Acquired microcephaly is almost universal.

Staging, Epidemiology, and Natural History of the AIDS Dementia Complex

In order to provide a common vocabulary of functional severity of the AIDS dementia complex for both clinical definition and comparative studies correlating clinical features with various laboratory findings, we devised a simple staging scheme (127). Based on evaluation of cognitive and motor performance, this system uses simple functional criteria (130) to classify patients as unaffected (Stage 0), subclinically or equivocally affected (Stage 0.5), or suffering mild (Stage 1), moderate (Stage 2), severe (Stage 3), and end-stage disease (Stage 4). This staging can also be roughly translated into newer World Health Organization (WHO) and American Academy of Neurology (AAN) classifications, which introduce the term *HIV-1 associated cognitive/motor complex* to encompass the full constellation of the AIDS dementia complex and add subcategories to refer to patients with predominantly cognitive (HIV-1–associated dementia) or myelopathic (HIV-1–associated myelopathy) presentations of sufficient severity to interfere with work or activities of daily living (roughly equivalent to AIDS dementia complex Stages 2–4) (2a, 171). The term *HIV-1 associated minor cognitive/motor disorder* was introduced to designate patients with mild but definite symptoms and signs and only minimal functional impairment of work or activities of daily living (roughly equivalent to Stage 1). In addition to attempting to separate the patients with predominant myelopathy from those with cognitive changes, this terminology restricts the term dementia to patients with a level of cognitive impairment consistent with that used in other formal definitions. The WHO classification also does not include the implicit assumption that the disorder is a single disease entity differing only in severity. Whether research criteria proposed by the AAN will prove workable for clinical diagnosis and patient management remains to be determined (14).

Although the epidemiology and natural history of the AIDS dementia complex have not been precisely defined, both the prevalence and severity of AIDS dementia complex clearly increase as the immune system fails and the other complications that define AIDS accumulate (129). Available information derives chiefly from clinical-case series rather than prospective epidemiological studies, although several studies of the latter type have characterized some of the neurological and neuropsychological aspects of asymptomatic HIV-1 seropositive subjects (103, 130, 144). During the clinically latent period, when patients are constitutionally well, functionally significant abnormalities are rare, although a small percentage of asymptomatic seropositive patients may exhibit mild signs. In the transitional phase of HIV-1 infection, subclinical abnormalities (Stage 0.5) may be present in one third or more patients, though probably less than one in ten exhibits features of Stage 1 or higher. When patients first present with AIDS-defining opportunistic infections (e.g. *Pneumocystis carinii* pneumonia), mild (Stage 1) AIDS dementia complex may be detected in perhaps one fifth of patients, while more severe neurological impairment is present in a lesser number (J. J. Sidtis & R. W. Price, unpublished data). Finally, the AIDS dementia complex is a more frequent and significant clinical problem in the late stages of systemic HIV-1 infection, and prior to death a majority of untreated AIDS patients may exhibit neurological symptoms and signs, with perhaps one-half of these showing functional (Stage 2 or greater) disability (107).

This epidemiological and natural history profile can be interpreted as indicating that immunosuppression, as measured by $CD4^+$ T-lymphocyte counts, has a permissive effect on the development of more severe AIDS dementia complex; advanced neurological dysfunction (Stages 3 and 4) occurs almost exclusively in individuals with severely depressed immune function. However, the relationship between neurological severity and immune suppression is highly variable, and hence, immune status is not the only determinant of the condition. Exceptional patients may progress to Stage 2–4 AIDS dementia complex without experiencing major systemic complications of HIV-1 (109). At the other extreme are patients who remain neurologically intact and continue to function well despite low $CD4^+$ lymphocyte counts and severe episodes of opportunistic infection. Also, within-group analyses of patients with low $CD4^+$ T-lymphocyte counts have generally noted a poor correlation between CD4 counts and neurological- or neuropsychological-test impairment. Thus, the link between severe progressive AIDS dementia complex and systemic immunosuppression is not absolute and other factors clearly must impact on its development.

Morphological Substrate of the AIDS Dementia Complex

Both neuroimaging and autopsy studies have provided important information regarding the anatomical and physiological pathology of the AIDS dementia complex and HIV-1 brain infection, and both point to a common theme of subcortical involvement.

NEUROIMAGING Anatomical imaging techniques, including both computed tomography (CT) and magnetic resonance imaging (MRI), have revealed that the AIDS dementia complex is almost always accompanied by brain atrophy, although the regional and cellular substrate of tissue loss has not been defined (73, 105, 126). The most consistent finding of these studies is an increase in the size of both the cortical sulci and the ventricles. In some patients, MRI also frequently reveals altered signal in the cerebral white matter and, at times, in the basal ganglia, consistent with focal or diffuse increased water content in these subcortical structures (74). In children, mineralization of the basal ganglia very often accompanies atrophy (4).

Physiological imaging using positron emission tomography (PET) or single photon emission tomography (SPECT) have also shown abnormalities. The most extensive PET study suggested that hypermetabolism of the diencephalic nuclei may be among the earliest abnormalities in mild AIDS dementia complex, perhaps related to the selective vulnerability of these brain regions to infection or injury (140). Later stages of the syndrome may be accompanied by either global or patchy alteration in glucose metabolism or blood flow (124, 140).

HISTOPATHOLOGY The pathology of the AIDS dementia complex and HIV-1 infection is reviewed elsewhere (19, 21, 22, 106, 120, 138, 151). In simplified terms, the major neuropathological findings fall into three overlapping subsets: (*a*) central gliosis and white-matter pallor, (*b*) multinucleated-cell encephalitis, and (*c*) vacuolar myelopathy. All three of these preferentially involve subcortical regions (or spinal cord) rather than cortex.

The most common of these is the diffuse white-matter pallor accompanied by astrocytic reaction. This involves particularly the central white matter (except the subcortical U fibers) and diencephalic nuclei. The process that underlies this common finding is unknown, and it has not been defined at the ultrastructural level whether this poor staining of myelin relates to local edema, true alteration of myelin sheaths, or other effects. When simple pallor is present without multinucleated cells, inflammation is most often scant on routine pathological examination, although both astrocyte and microglia cell numbers are increased.

Multinucleated cells, derived from infected macrophages and likely microglia, are found in a subgroup of patients who usually have suffered more severe clinical disease. In these brains, reactive infiltrates are more prominent and consist of perivascular and parenchymal macrophages and microglia, along with the multinucleated cells. Because of the association of these findings with direct HIV-1 brain infection as described below, this pathological subset may be legitimately termed HIV-1 encephalitis (120). The characteristic multinucleated cell and macrophage infiltrates are most often concentrated in white matter and deep gray structures. In the white matter, they may be surrounded by focal rarefaction, and less commonly by frank demyelination.

Vacuolar myelopathy pathologically resembles subacute combined degeneration caused by vitamin B_{12} deficiency (121), with vacuolization of the spinal cord and intramyelin edema, but serum levels of this vitamin, as well as serum, red blood cell, and CSF folate, are generally normal in patients with this condition (77). However, metabolic abnormalities in the CSF methylation pathway have been noted in some AIDS patients and may provide a clue to this pathology (77).

The cellular pathology of HIV-1 infection is only beginning to be examined in more detail. Although the AIDS dementia complex is clearly not accompanied by conspicuous neuronal loss, recent quantitative studies have shown that there is indeed a loss of neurons from the frontal cortex of AIDS patients ranging from 18–38% (49, 78) and a loss of up to 50% of large neurons in the neocortex of patients with HIV-1 encephalitis (168). One study found that this loss of neurons was consistent at various disease stages and was not correlated with the presence of HIV-1 encephalitis (49). Additionally, changes in the synaptic density of the frontal cortex and in the complexity of dendritic arborization have been described (168). Because subcortical areas have been defined as the regions supporting productive infection, the role of ascending pathways in this proposed cortical pathology will need to be examined. A preliminary report of decreases in the enzyme that synthesizes acetylcholine (choline acetyl transferase or ChAT) might reflect such an ascending pathway loss (108).

VIRUS LOCALIZATION AND CORRELATION WITH HISTOPATHOLOGY AND CLINICAL STATUS Productive HIV-1 infection, likely limited to macrophages and microglia as noted earlier, is focally distributed in the brain and is highly correlated with the histopathological finding of multinucleated-cell encephalitis. Detectable infection appears to preferentially involve diencephalic structures, particularly the globus pallidus, but also other basal ganglia nuclei, the substantia nigra, and deeper white matter, with less frequent infection of the cerebral cortex (84, 139).

Some general statements regarding the clinical correlations of both pathological and virological findings are possible (Table 1), although one can find clear exceptions to these generalities. In patients with only mild AIDS dementia complex (Stages 0.5–1 and some with Stage 2) who present with predominantly cognitive and general motor slowing, pathological findings are usually restricted to central gliosis and white-matter pallor with limited inflammatory response and little or no evidence of productive HIV-1 infection (138). In those with more severe and progressive AIDS dementia complex (Stages 3 and 4 and some with Stage 2) multinucleated-cell encephalitis, with HIV-1 infected macrophages, microglia, and derivative multinucleated cells, is commonly superimposed on the central gliosis and pallor. There is therefore a general, but not absolute, correlation between severe AIDS dementia complex and HIV-1 encephalitis, but even in these cases the degree of clinical dysfunction often exceeds what might be anticipated on the basis of the extent of infection, and occasionally, severe clinical AIDS dementia complex may be accompanied by very limited or no evidence of viral replication. This evidence that infection is restricted both in quantity and cell type (macrophages and microglia) underlies the current search for mechanisms of brain injury that are less direct, in which either virus-coded products or cell-coded cytokines released by infected or reactive cells might exert toxic effects on uninfected brain cells, as discussed below.

Table 1 General clinical correlations with neuropathology and virology

Clinical	Pathology	Virology
Mild ADC	Gliosis/pallor	Negative for productive infection
Severe ADC	Multinucleated-cell encephalitis	Productive infection
Myelopathy	Vacuolar myelopathy	Negative for productive infection

In patients with prominent spinal cord signs and symptoms, usually spastic-ataxic paraparesis, vacuolar myelopathy is most frequently noted. Although multinucleated-cell encephalitis and vacuolar myelopathy may coexist, the vacuolar spinal cord pathology itself does not appear to relate to productive infection, at least not in the same way as the process that generates multi-nucleated cells (139). The similarity between this pathology and that of subacute combined degeneration suggests that an altogether different pathological process may be at work.

PATHOGENESIS OF HIV-1 INFECTION AND CNS INJURY

The presumption that the AIDS dementia complex relates in some way to HIV-1 itself is based on the uniqueness of the syndrome to HIV-1 infected patients in contrast to other immunosuppressed patients, the lack of another identified etiology, the identification of HIV-1 in the brain of some patients, and the role of other retroviruses in animal models of neurological disease. But how the virus causes this clinical spectrum is not yet clear, given the lack of productive infection of functional elements and the seeming paucity of infection in some brains. Earlier, we outlined a general model of the interaction of the the virus and immune reactions and their effects on the CNS (128). In this section, we review pertinent information related to each of the major pathogenetic components outlined in Figure 1.

HIV-1 Infection of the CNS

The control of HIV-1 replication is a complex process. In simplified terms, factors modulating replication can be considered in two categories: (*a*) those relating to the virus itself, including the actions of regulatory genes and genetic variation in the capacity to replicate under different conditions, and (*b*) those related to the host cell, including both cell type, which determines the potential capacity to be infected (by availability of cell-surface receptors, but also the complex internal cell machinery necessary for viral replication), as well as the cell state, which may vary according to epigenetic influences and hence with the environmental milieu. These factors have been studied predominantly in nonneural cells in vitro (principally lymphocytes and lymphocyte cell lines, but also monocyte-macrophages or derivative cells). However, some studies have been carried out involving cells from the CNS, and lessons from nonneural cells can be extrapolated, at least in principle, to considerations of nervous-system infection.

Outside of the nervous system, infection by HIV-1 is principally confined to cells expressing the CD4 receptor, helper T-lymphocytes, and cells of the monocyte-macrophage lineage. A discussion of factors within these cells that determine susceptibility to and kinetics of infection, for example the action of transcription factors like NFκB, are beyond the scope of this paper and are reviewed elsewhere (36). Here, it suffices to emphasize that cells may vary in their capacity to support HIV-1 replication according to the cell state, which may in turn be altered by exogenous signals, including cytokines; in this way, cells of the immune system may exert considerable effect on the course of infection. Among the cytokines influencing HIV-1 replication are tumor necrosis factor alpha (TNF-α), tumor necrosis factor beta (TNF-β), granulocyte/macrophage colony-stimulating factor (GM-CSF) (upregulation), and interferon gamma (IFN-γ) and interferon beta (IFN-β) (downregulation) (for review see 99). Such influences may help determine the permissiveness of cells to infection in vitro and in vivo and may contribute to the marked variability in the course of infection and viral replication in vivo.

In this section, we consider the evidence regarding the susceptibility and role of HIV-1 infection of the CNS, separately reviewing evidence regarding infection of cells of neuroectodermal origin and infection of CNS macrophages. We also review the issue of neurotropism and current understanding of the course of CNS infection.

INFECTION OF CELLS OF NEUROECTODERMAL ORIGIN This category comprises cells of neuronal and (macro)glial origin. Several studies have now shown that both primary cell cultures and cell lines of neuroectodermal origin can be infected in the laboratory. Kunsch & Wigdahl (82) recently reviewed these studies. Although research has not shown that primary neurons can be infected, several neuroblastoma and other cells lines with neuronal phenotypes have been shown to support replication of HIV-1. Similarly, several astroglial cell lines have been infected, and in some studies, primary cells of glial origin have proved infectible as well (30, 33, 38, 47, 79, 153).

In vitro infection of these various cell types exhibits a number of consistent differences from infections of lymphocytes and lymphocytic cell lines. In contrast to infection of lymphocytes, infection of neuroglial cells is characterized by low yields of free virus, often requiring cocultivation or other amplification techniques to rescue infectious virus. Also, at any given time only a small percentage of cells appear to bear productive infection. Infection is thus low-grade and indolent and cannot be induced to higher output.

Additionally, infection of these cell types does not appear to involve the CD4 receptor. These cells do not express CD4 and infection is not blocked by anti-CD4 antibodies that prevent lymphocyte infection. Thus, the surface of these neuroectodermal cells appears to have an alternative HIV-1 receptor that Gonzalez-Scarano and colleagues have recently suggested might be galactosyl ceramide (GalC), a glycolipid found in particularly high concentration on the surface of oligodendrocytes (63). Specific antibodies raised against GalC blocked infection of glioma and neuroblastoma cell lines, but not of HeLa T4 positive cells. These investigators have also presented kinetic analysis of interaction of HIV-1 gp120 with GalC. The significance of these observations and particularly their bearing on in vivo events remains to be fully demonstrated.

Is infection of these neuroectodermal cells restricted in nature because of the limiting effect of this alternative receptor, or because of other characteristics of these cells? Viral entry is a complex process, and entry via an alternative receptor might restrict replication. In many cells, once the cell membrane barrier to infection is overcome (e.g. by transfection of proviral DNA or pseudotyping with other viruses), the cells are permissive to replication (1, 25, 87, 94, 112, 152). Other cells seem to severely restrict viral expression regardless of the mode of viral entry, suggesting that lack of a high affinity receptor might not be the limiting factor (95). Generally, cultured astroglial cells transfected with infectious clones of HIV-1 support replication (39, 40, 158), and a neuroglioma cell line genetically engineered to express the CD4 receptor exhibited high-level virus production comparable to infected T-cells (161). The importance of transcriptional factors or other variables in these cell types has not been fully discerned.

The in vivo relevance of these cell-culture studies is currently uncertain. Early reports suggested that neurons, oligodendrocytes, and astrocytes might be infected in vivo, but subsequent work has cast doubt on some of these findings. Suspicion that neurons might be infected was based on studies identifying cell type by morphology (134, 155, 169) that have not been substantiated by later work. Prominent in vivo white-matter changes suggested infection of oligodendrocytes, the cells that synthesize and maintain central myelin. Several groups (61, 134, 155) reported such infection, although some investigators interpreted an early electron microscopic study purporting to demonstrate HIV-1 replication in oligodendrocytes (61) as a case of misidentified cell type (23, 146). However, a recent study by Esiri has confirmed infection (or at least antigen positivity) of occasional oligodendrocytes (48). Nonetheless, productive infection of these cells is likely rare, and Esiri interprets the most conspicuous morphological change in oligodendrocytes as reactive changes, presumably directed at myelin regeneration. Likewise, in vitro findings of infected astrocyte cell lines have not been extended to the in vivo situation. Rare infected astrocytes have been reported in vivo (45, 61, 104, 134, 155, 163), yet an extensive double labeling study demonstrated that cells of all morphologic types, even cells with bipolar or multipolar processes, were positive for macrophage-microglial markers but negative for astrocyte as well as neuron or endothelial markers (83).

These negative results must still be tempered by the fact that studies of infected cells in AIDS brain tissue have used relatively insensitive tools capable of detecting only high levels of mRNA transcription or structural protein production. The in vitro studies of indolent, limited infection raise the question of whether a similar profile of infection might occur in vivo and underlie persistent infection. Such infection might act as a reservoir of virus in the CNS, resulting in continued immune reaction in the brain and a source of virus that might later be amplified by more permissive cells, such as macrophages, just as neural cell culture virus is rescued by cocultivated indicator cells. Ongoing studies exploring the combined use of polymerase chain reaction (PCR) and in situ hybridization may resolve this issue in the near future.

INFECTION OF MACROPHAGES AND MICROGLIA AND TROPISM FOR THESE CELLS Studies using both in situ hybridization to detect viral mRNA and immunocytochemistry to detect viral proteins have quelled some of the controversy regarding the cell type infected in the CNS and suggest that macrophages and microglia are the principal, if not exclusive, cells supporting productive infection in vivo (45, 51, 80, 102, 122, 134, 155, 160, 169). Macrophages are bone marrow–derived monocytes that have migrated to populate tissues and are terminally differentiated. Microglia are almost certainly of bone-marrow rather than glial origin but differ from macrophages in the timing of their migration to the CNS (41). Although the issue is still unsettled, microglia appear to populate the CNS antenatally and distribute into both grey and white matter (58, 118). They appear identical to cells of monocyte-macrophage lineage in both antigen profile and functional properties [e.g. phagocytosis, antigen presentation, interleukin (IL)-1 synthesis, superoxide anion production, etc] (58). Like macrophages, these cells are activated [e.g. to express major histocompatability complex (MHC) class II antigens and cytokines] by infection and other stimuli. Limited in vitro study of microglial infection has not revealed any difference between these cells and macrophages with respect to HIV-1 replication. Supporting in vitro findings, one study showed that only the microglial elements in human primary brain cultures supported demonstrable productive infection (164). Additionally, in this same system, HIV-1 strains known to replicate well in macrophages, the so-called macrophage-tropic strains, showed a similar affinity for microglia. Infection of microglia was determined to be CD4 dependent (75). Further definition of issues regarding the origin of microglia, the timing of their migration to the CNS (pre- or post-HIV-1 infection), and the means by which they are replenished may have implications for the pathogenesis of HIV-1 infection of the CNS.

If macrophages and microglia infected in brain indeed support HIV-1 replication in the same way as macrophages in cell culture, they may: (*a*) be susceptible to infection by virtue of the CD4 receptor on their surface; (*b*) support protracted productive infection without cell lysis; (*c*) respond to local stimuli including cytokines that up- or down-regulate replication; and (*d*) undergo cell fusion related to the interactions of the viral glycoproteins and the CD4 receptor. They could possibly serve as reservoirs of persistent infection, amplifiers of infection, and sources of toxins.

Neurotropic variants of HIV-1 HIV-1 strains that infect the CNS have been referred to as neurotropic strains. This term is appropriate in a broad definition of neurotropism indicating ability to replicate in the CNS, but is less fitting if a more restricted definition of neurotropism as infection of neuroectodermally derived cells is used. Because most evidence now suggests that strains of HIV-1 that favorably replicate in the brain are principally macrophage-tropic (see below), the term neurotropic may be confusing and should probably be avoided in this context. Work of several investigators suggested that there might be a propensity for brain-derived strains to replicate more favorably in cells of glial origin, but this has not been a general finding (30). Another term used to refer to HIV-1 strains is *neuropathic*. This term implies that particular strains cause nervous system disease without specifying a pathogenic mechanism. Thus, it is possible that certain strains might not only replicate better in brain but might also be more likely to cause disease by processes other than replication, such as secretion or induction of neurotoxins (see below).

The genetic instability of HIV-1 results in marked variability of the genome, so that not only do different patients possess distinct strains, but individual patients are populated by a related array, or quasispecies, of viral variants. The presence of quasispecies is evident from the time of initial infection throughout the course of infection, generated by the error-prone viral reverse transcriptase, with drift perhaps driven and directed by immune selection (88, 162, 170). Distinct populations of HIV-1 have been demonstrated between the brain and blood or brain and other tissues of individual patients (44, 114, 154). Hot spots for mutation and immune escape are found in the region of the genome coding for the viral envelope glycoprotein (170), resulting in changes that may lead to heterogeneity of biological behavior, including differences in range of host cells infected and replicative ability, cytopathicity, susceptibility to neutralization by antibodies, and ability to modulate the CD4 receptor on host cells (27).

Given this variability, might genetic variation contribute to the pathogenesis of HIV-1 brain infection and the AIDS dementia complex? Variability in the appearance of this syndrome and of frank HIV-1 encephalitis among patients also provokes this question. If immunosuppression has only a permissive influence, as discussed earlier, then the emergence of viral variants that are more likely to cause CNS infection is an attractive second control mechanism explaining the incidence of the disorder.

Macrophage tropism Several strains of HIV-1 have now been identified in brain that share the character of macrophage tropism, or a propensity to replicate well in primary macrophages (for review see 28). These strains also replicate in peripheral blood mononuclear cells (PBMCs) routinely used to isolate HIV-1 from blood, but not in transformed lymphocyte cell lines. They differ from the so-called T-cell tropic HIV-1 strains, which are unable to productively infect primary macrophages, but do infect lymphoblastoid cell lines as well as PBMCs.

Macrophage tropism is a characteristic of most brain-derived isolates studied carefully to date (81, 89, 92). These isolates have generally come from patients with demonstrable productive brain infection and multinucleated-cell encephalitis. Studies of brain isolates derived from culture have shown that such isolates replicate to high titer in cultures of primary macrophages. One study demonstrates that brain isolates share a similar tropism in brain-derived macrophages (146a). There are some concerns about the possibility of in vitro selection in these cultured isolates; do they accurately represent the CNS strains in vivo? Recently, however, macrophage tropism was noted in virus cloned directly from brain, establishing that this characteristic is not an artefact of selection in culture (89).

Studies using chimeric reconstructions created by exchange of DNA fragments between culture-derived macrophage-tropic and macrophage nonreplicative strains have shown that regions coding for the gp120 portion of the viral *env* glycoprotein are both necessary and sufficient to confer macrophage tropism (26, 29, 31, 32, 92, 111, 149, 166, 167, for review see 28). These exchanged fragments do not include regions coding for the viral LTR or viral regulatory genes. Further studies fine-tuning the definition of the locus for macrophage tropism have pointed to the V3 loop region of the gp120 *env* glycoprotein (72), a domain distinct from the putative CD4 binding region. Studies by Ratner and colleagues using these chimeric reconstructions with direct viral isolates affirm the importance of the V3 loop and also suggest that alterations in another region of gp120, upstream from the V3 loop, result in increased efficiency of viral replication in macrophages (167). The importance of the V3 loop in determining tropism in other cell types has been demonstrated by Takeuchi et al, who found that a single nucleotide change confers tropism for brain fibroblast-like cells on a T-cell tropic clone (156). The finding that specific proteases can recognize amino acid residues at the crown of the V3 loop has suggested a possible mechanism for V3 loop determination of tropism (35, 65). This model proposes that following the binding of virus to cell by the well-defined gp120-CD4 interaction, specific cell-surface proteases might cleave the V3 loop, driving a conformational change in the *env* glycoprotein that leads to virus-cell fusion and subsequent entry. This hypothesis has been supported by the finding that specific protease inhibitors can block viral infectivity, and by the fact that the majority of neutralizing antibodies are directed at the V3 loop, perhaps making amino acid residues unavailable to proteases. In addition, mutations at the relatively conserved tip of the V3 loop abolish the ability of the *env* protein to induce cell fusion (50, 113). Although further work is needed regarding these observations, the finding of consistent macrophage tropism is consonant both with the identification of macrophages as the predominant cell productively infected in brain and the role of infection of these cells in pathogenesis.

Unresolved is the issue of how such strains arise and why they might emerge in some patients and not in others. Most efforts at isolation of virus from blood using macrophages are successful, implying that variants with the capacity to replicate in macrophages are probably common throughout infection. One recent study has suggested that these strains predominate early in the course of infection (143). How then can pathogenicity of these strains in only some patients be explained? Does long-term infection gradually exhaust the supply of $CD4^+$ T-lymphocytes and drive more efficient replication in the remaining susceptible cells, or is there some more fundamental change that causes strong selection and rapid proliferation of virus in brain? The restricted genetic polymorphism of viral strains isolated from brain (114) suggests that a more radical genetic shift, rather than gradual and multiple-strain genetic drift, leads to HIV-1 encephalitis, at least in some patients. Thus far, analysis of viral strains has been confined largely to patients with florid and widespread productive infection. It will be of interest to examine the genotype, genetic heterogeneity, and macrophage tropism of strains derived from milder cases of productive CNS infection or from genomes of earlier, asymptomatic cases amplified using PCR.

OTHER CELL TYPES INFECTED BY HIV-1 An additional cell type reported to support HIV-1 infection in brain is the vascular endothelial cell. However, these findings were based on light microscopic identification of elongated cells in or near the vessel wall that could not be confirmed as endothelial cells by double-labeling experiments; they might also be perivascular cells of microglial-macrophage lineage (20, 86). Choroid plexus cells have been shown to harbor latent infection in vitro, but in vivo evidence is lacking (64).

THE COURSE OF CNS HIV-1 INFECTION IN VIVO Current understanding of the course of HIV-1 infection of the CNS derives from a reconstruction of information described above, the results of studies of CSF, as well as the character of infection in cell culture and studies of simian immunodeficiency virus (SIV) and other lentiviruses. Some fundamental questions regarding the overall picture of CNS HIV-1 infection are:

1. When does virus enter and at what rate and significance?
   a. During initial virema?
   b. Throughout the asympotomatic phase?
   c. Later, during secondary virema?
2. How does the virus enter?
   a. As free virus or within cells?
   b. Across brain capillaries or choroid plexus?
3. What happens to entering virus?
   a. Circulation (egress of infected cells)?
   b. Latency?
   c. Persistent low-grade infection?
4. Where is the virus found at each stage of infection?
   a. What cell type?
   b. What brain regions?
5. What causes subsequent escape and amplification?
   a. Loss of immune control?
   b. Emergence of neuropathic variants?
   c. Cofactor activation?

Clues from the study of CSF Analysis of CSF has provided some of the most interesting information regarding the interaction of HIV-1 and the CNS, particularly concerning the early events of infection (3, 43, 57, 96, 98, 136). Indeed, in the absence of serial sampling of brain, and particularly in the absence of anatomic sampling of brain early in disease, the CSF has provided a unique window into events in the CNS that relate to both the presence of the virus in the CNS and immune responses within or adjacent to this compartment. Studies of asymptomatic HIV-1 seropositives have demonstrated abnormalities in the CSF that reflect nonspecific host responses, including increased cell counts and protein, as well as specific reactions to HIV-1 with intrathecal production of antibodies against viral antigens (24, 34, 136). The virus can be isolated from the CSF of between 10 and 30% of asymptomatic but serologically positive subjects. These observations indicate that HIV-1 reaches the CNS early in systemic infection and elicits local responses, yet brain function remains normal under these conditions.

Late in the course of infection, HIV-1 p24 antigen may be detected by immunoassay in the CSF in a minority of ADC patients (116). Although a direct correlation of antemortem CSF findings and brain analysis has not been done, combined observations are compatible with the interpretation that detectable p24 in the CSF may derive from the brain or at least correlate with the presence of virus replication in the brain and with multinucleated-cell encephalitis.

We must extrapolate from these fragments to understand the course of CNS HIV-1 infection in vivo. The CNS is apparently exposed early in the course of systemic infection, but it is uncertain how the virus enters the nervous system, how long it persists, and whether cultures remain positive because of persistence or continued reentry of virus. Ability to isolate virus in the CSF does not appear to correlate with disease stage or $CD4^+$ T-lymphocyte counts, but is significantly associated with CSF cell count (24). Hence, isolation from CSF may imply either continued traffic of infected cells with only relatively short passage, or could imply chronic local persistence.

Entry of HIV-1 into the CNS Of the possible modes of entry of HIV into the CNS, the most favored is entry within infected monocytes, by the so-called Trojan horse mechanism postulated for visna virus (117). In this scheme, monocytes infected by HIV-1 in the peripheral blood are subsequently recruited to the brain or spinal cord by some inflammatory or other signal, and enter the CNS, bringing the virus with them. Studies of early simian immunodeficiency virus (SIV) infection support this idea; entry occurs through the choroid plexus rather than through the brain capillaries (145). Whether infected lymphocytes may also play a role in viral entry is unknown, though peripheral activation of lymphocytes, without a specific CNS inflammatory signal, is sufficient to increase traffic of these cells in the CNS (68). It is unclear whether this early seeding of the CNS has special implications for subsequent disease. In fact, later events might overshadow early seeding, because viremia may take place throughout the course of infection and increase with immunosuppression and worsening systemic condition, and hence viral entry may be less important than viral amplification. Again, the issue is whether immune release or emergence of new strains with capacity to replicate in brain are primarily responsible for productive brain infection.

Cell, tissue, and regional localization of infection A full understanding of HIV-1 infection requires explanation of the localization of infection at the cell

Table 2 Distribution of infection in brain: hypotheses explaining cell, tissue, and brain region distribution

	Primary determinant	Secondary determinant
Cell (macrophage-microglia)	CD4 receptor	(a) Activation to permissive state (b) Presence of macrophage tropic viral strains
Tissue (perivascular)	Local amplification (perivascular influx of permissive macrophages)	Arrival of virus (a) Hematogenous entry (b) Local virus presence (CSF→ choroid plexus, latent infection)
Regional (deep nuclei)	(a) CSF→ centrifugal spread	Amplification
	(b) Selective vulnerability of certain cell populations to injury	Amplification
	(c) Selective hematogenous viral entry	Amplification

(macrophages and microglia), tissue (perivascular predilection), and regional (deep grey and white matter) levels (Table 2).

The infected cell types have been discussed above. The predilection for infection to be perivascular likely involves at least two steps: (*a*) arrival of the virus and (*b*) local amplification of infection. The local seeding might be the result of hematogenous infection with virus or, more likely, virus-infected cells transgressing the local capillaries. Alternatively, virus might be present within the parenchyma (by spread from the CSF or activation of latent or indolent infection). Once the virus arrives, local amplification may be the critical process. This process, initiated by local and blood-borne monocyte-macrophages attracted to the site, may be perpetuated by further activation of these cells by the elaboration of cytokines, creating a positive-feedback loop.

The regional distribution of infection to the deeper brain regions is somewhat more difficult to explain. Simple hematogenous dissemination does not satisfactorily explain this distribution, because blood flow to the cortex is higher than that to the white matter. Three alternative hypotheses can be offered. First, virus (or infected cells) might primarily reach the brain parenchyma via the choroid plexus and CSF, and move centrifugally outward to the deep brain regions. Secondary amplification as described above would then explain the perivascular distribution in these regions. This explains some aspects of regional distribution, but not all. For example, certain nuclei, notably the globus pallidus, appear to be infected to an extent greater than might be explained by the gradient distance from the lateral ventricles. A second hypothesis that might account for deep distribution relates to possible selective vulnerability of certain cells (e.g. globus pallidus or substantia nigra neurons) to the toxic effects of infection (see below). Thus, if certain cells are selectively vulnerable to neurotoxins, death of these cells might result in local cell responses with secondary amplification as a result of activation and attraction of local microglia and blood-borne monocytes, along with altered capillary access. This also would create a positive-feedback loop, resulting in focal regional damage. Finally, as a third explanation, hematogenous viral entry could be reconciled with regional distribution if local differences exist in the permeability or permissiveness of the blood brain barrier (e.g. differential distribution of cellular adhesion molecules) that result in different interactions between infected or activated cells and the perivascular barrier. Cytokines regulate the expression of adhesion molecules on endothelial cells and leukocytes, and so might mediate these local differences (85).

The Immune System and the AIDS Dementia Complex

Immune reactions and the cells of the immune system appear to play major, yet dual, roles in the development and pathogenesis of the AIDS dementia complex and CNS HIV-1 infection. These two roles relate to (*a*) protection and (*b*) cell injury, with the relative importance or balance of these roles changing during the course of HIV-1 infection (Table 3).

Table 3 Some interactions of the immune system and HIV-1

	Primary	Sequelae
Effects of HIV-1 on the immune system	Infection of helper T lymphocytes Infection of macrophages and microglia	Depression of cell-mediated and other aspects of immunity Eventual loss of immune response to emerging viral variants Upregulation of some cell responses
	Defense	Pathology
Effects of the immune system on HIV-1 infection	Suppression of replication and spread of HIV-1	Supply susceptible cells (macrophages/microglia) Activation of cells to permissive state Upregulation of replication Selection of macrophage-tropic strains

PROTECTIVE ROLE OF THE IMMUNE SYSTEM Although the specific mechanisms involved are only partially characterized, host immune defenses are clearly involved in inhibiting HIV-1 replication, yet cannot eliminate the virus. Perhaps the most compelling empirical evidence for effective host responses in vivo derives from observations that viremia after initial exposure to HIV-1 is self-limiting and that initial high titers of virus noted in the blood of some patients are reduced and do not recur to similar levels until the late stages of infection years later. Presumably, HIV-1 replication is kept in check through the stage of asymptomatic seropositivity, but subsequently breaks through when immune defenses are debilitated by the action of the virus itself. If this is the case for systemic infection, it likely occurs in the brain as well. Hence, CNS HIV-1 infection can be considered an opportunistic infection, but one in which systemic infection by the same agent predisposes the host by damaging defenses that would inhibit the agent's free replication. The relative importance of the individual arms of the immune system in inhibiting HIV-1 replication is uncertain. HIV induces strong humoral (neutralizing antibody), antibody-dependent cellular cytotoxicity (ADCC), and cytotoxic T-lymphocyte responses directed against HIV-1 antigens (93, 110, 137).

The immune system may also play a role in the selection of viral variants. Thus, as the host responds to the virus, antigenically altered escape mutants emerge. The host response to these drives further change, until later in infection the host capacity to mount an immune response to novel antigenic variants may be exhausted and new, more virulent, and truly immune-resistant strains emerge. The V3 loop of the *env* glycoprotein is the immunodominant site for neutralizing antibodies, and thus may be a common component to emergence of both macrophage tropism and strains resistant to host defenses.

However, for the present consideration of brain injury, the specific mechanisms of defense are perhaps less critical than the broad concept that the efficiency and effectiveness of anti-HIV-1 responses probably wanes with progression of systemic disease and falling $CD4^+$ T-lymphocyte counts. Studies of CSF immune responses early in disease show evidence of local reactions to the virus in the form of mildly elevated protein levels and pleocytosis, along with intrathecal anti-HIV-1 antibody synthesis. Early in the asymptomatic phase of HIV-1 infection, either host responses are effective or the virus is not damaging to the CNS. Cell-mediated defenses in the CNS or CSF compartments early in infection have not yet been studied. Later in infection, anti-p24 antibody levels in the CSF may decrease as a result of either reduced production or complexing with locally synthesized virus in the context of productive infection, or a combination of both. Antibodies against the viral-envelope glycoproteins have not yet been studied.

IMMUNOPATHOGENETIC ROLE OF THE IMMUNE SYSTEM The limited presence and restricted (both regionally and cellularly) degree of productive brain HIV-1 infection in relation to the clinical severity of the AIDS dementia complex have led to the search for involvement of other pathogenic processes. Paradoxically, the prime candidate for a second pathogenic agonist is the immune system. This, of course, is not a novel concept, and certainly immunopathological mechanisms have been suggested to operate in varying degrees in a broad range of viral infections, including visna infections (62). However, the AIDS dementia complex is distinctly different from many of these other infections. In other infections in which immunopathology is implicated, inflammatory reactions are characteristically prominent, while in the AIDS dementia complex, such responses may be relatively inconspicuous but perhaps pathogenetically important because of their chronicity as well as failure of modulatory influences.

In addition to the specific and general systemic host responses to HIV-1 infection mentioned above, upregulation of the markers of immune activation have been documented in studies of CSF and brain in AIDS patients, indicating that despite profound depression of some immune defenses, other reactions are actually enhanced. Among the surrogate markers indicating activated immune reactions are beta-2-microglobulin (β_2M), a noncovalently bound portion of the class 1 major histocompatibility complex (MHC) expressed on the surface of all nucleated cells (though to a limited extent in the CNS). Levels of β_2M are relatively constant in healthy individuals; elevated concentrations often result from induction of MHC expression on the surface of activated immune cells, particularly lymphocytes, and have been noted to be elevated in the CSF of patients infected with HIV (11, 13). Neopterin, a by-product of tryptophan metabolism, is a second marker that likely is an indirect measure of cytokine (e.g. INF-γ) activity (12). Quinolinic acid is similarly elevated (66, 67); this product of tryptophan metabolism can be induced by INF-γ and perhaps other cytokines. In addition to being a marker of immune activation, quinolinic acid has a potential pathogenetic role; it can act at the N-methyl-D-aspartate (NMDA) receptor as an endogenous excitotoxin or excitatory amino acid capable of producing toxic effects. Heyes and colleagues have demonstrated CSF concentrations of quinolinic acid as high as those shown to cause neurotoxicity in vitro (66).

Elevation of β_2M, neopterin, and quinolinic acid in the blood, and independently in the CSF, indicate that, although AIDS patients are immunosuppressed, certain immune-cell responses are augmented by systemic or CNS disease, respectively. The concentrations of these three markers of immune-cell activation in the CSF correlate, to some extent, with AIDS dementia complex severity as well as with the presence of opportunistic infections in the brain. More particularly, since these markers may be increased by the action of cytokines and because some are macrophage products, these observations raise the question of whether cytokine-related reactions involving macrophages might actually be involved in the production of CNS injury.

We have previously postulated that the high levels of surrogate markers might result from the disturbance of a negative-feedback loop that normally limits immune responses (128). The inability of the weakened immune system to suppress viral replication could result in further stimulation of the now ineffective responses. These responses come with a cost to the host: increased immunopathologic cell injury. We suggested that HIV-1 infection stimulates reactions in the brain involving macrophages, microglia, and perhaps other cells (endothelial cells or astrocytes) to elaborate cytokines, which in turn results in upregulation of these markers. The regulation of these cytokine markers in AIDS dementia complex is uncertain, but the pattern of associated elevations of β_2M, neopterin, and quinolinic acid could be explained by the pleotropic effects of INF-γ, which induces the synthetic pathways for neopterin and quinolinic acid and also induces β_2M expression on the cell surface. Recently, one group reported detecting tumor necrosis factor alpha (TNF-α) in CSF in late HIV-1 infection (60), although levels were not correlated with AIDS dementia complex stage and, indeed, TNF-α detection has not been confirmed by others (52, 147). If either INF-γ, TNF-α, or other cytokines are induced within the brain, we must then ask, what processes lead to their induction and what is their pathogenetic role.

This process could be set into motion by either (*a*) immune response to the virus (by uninfected cells) or (*b*) dysregulation of infected cells by viral genes and gene products. The first of these involves the normal orchestra of responses to a foreign antigen and to tissue injury. Previously, we have suggested that these immunopathological responses are minimal when defenses are effective in reducing the virus load, but that when responses become less effective, they are induced to still higher levels leading to predominantly destructive processes (128). The second mechanism hypothesizes that viral gene expression, perhaps most likely expression of regulatory genes, causes upregulation of host genes within infected cells and hence the synthesis of cellular products. The effects of activation by viral genes or gene products on cellular transcription or translation would lead to increased output of cytokines by infected cells, leading to local or (if hematogenously disseminated) more diffuse cell dysfunction. Cell-culture studies bearing on this issue conflict, but in most cases, infected macrophages do not appear to produce excessive amounts of the specific markers or potential effectors that have been measured, though some yet unidentified macrophage product might still be involved (54, 133) (see below). Lack of in vitro evidence for these processes is not definitive because of the variability of macrophage state and response depending upon the milieu.

Whatever the mechanism of initial induction, the effector may then secondarily trigger other processes, leading to increases in certain detectable markers but, more importantly, to cell injury as discussed below. However, these increases may also affect the process of viral infection. Cytokines may have either stimulatory or inhibitory effects on infection. Stimulatory effects, as discussed earlier in relation to amplification of infection, occur both by attracting and enlisting permissive cells (e.g. macrophages and lymphocytes) and by increasing the output of virus from infected cells (e.g. the colony-stimulating factors).

Overall, observations of increased markers of immune activation in the CSF indicate complex processes in which the virus and both effective and ineffective immune reactions might be linked to brain injury. These same processes are involved in the symptomatology of many acute brain infections, but the difference with HIV-1 infection is the chronicity of the infection and the paradox of profound immunosuppression yet active immunopathology.

EVOLVING INTERACTION OF HIV-1 AND THE IMMUNE SYSTEM Let us now view these combined mechanisms of immune protection and virus- and immune-induced injury as an evolving, dynamic process (Table 4). As pre-

Table 4 Hypothetical evolving relationship between HIV-1, immune responses, and the CNS

Phase	Immune reactions	HIV-1	CNS
Latent	Effective immune control of HIV-1 replication.	Limited replication.	No clinical dysfunction.
Transitional	Less efficient anti-HIV-1 defenses. Reduced inhibition, increased viral stimulation, increasing gain of cell reactions. Upregulation of effective and ineffective immune reactions. Mild immunopathologic reactions.	Slight increase in replication insufficient to cause virus-coded cytotoxicity. Slow evolution of viral strains.	Mild dysfunction related to immune reactions.
Late	Ineffective immune control of HIV-1 replication. Reduced inhibition, increased viral stimulation, increasing gain of cell reactions. Upregulation of ineffective immune reactions, resultant cytotoxicity. Moderate to marked immunopathological reactions.	Accelerated replication. Evolution of macrophage-tropic pathogenic strains. Virus coded toxicity. Virus activation of cell-coded toxins.	Moderate to severe dysfunction related to both virus-coded toxins and immune reactions.

viously argued, during the clinically asymptomatic phase, immune defenses are effective in suppressing much of viral replication at low cost to host CNS tissue. Neurological symptoms and signs are not seen despite evidence of virus exposure and cell reactions. Later, during the transitional phase, the virus load (detectable productive brain infection) is still largely kept in check, but as the gain of the defensive systems increases, the cost increases in the form of immunopathology (128). One can speculate that diffuse gliosis and white-matter pallor may be the morphological substrate of this global reaction, either resulting from systemic infection or smoldering CNS infection. Finally, in the late stage, defenses are ineffective and increasing immunopathology is accompanied by viral replication with combined cell-coded and virus-coded toxin release. Either required for or augmenting this final stage is the emergence of macrophage-tropic viral strains, which may evolve more rapidly in the absence of general host-mediated viral suppression, at a time when response to novel antigens is crippled, resulting in true HIV-1 multinucleated-cell encephalitis.

The Targets and Mechanisms of Injury in the AIDS Dementia Complex

In this section, we consider both the targets of pathology and the possible mechanisms of injury of these targets. We approach this subject by first considering three conceptual models of virus-related injury that emerge from the previous discussion. Once these models are introduced, we then consider the cells participating in the pathogenesis according to a functional pathogenic classification before considering the types of effects that these processes may have on the major functional targets. Finally, we consider the mechanisms of the effects on these targets. This discussion, as did previous discussions, outlines broad conceptual models rather than specific mechanisms.

Figure 2 outlines three models of virus-induced injury, with increasing levels of complexity. Each simply schematizes the events triggered by the viral genome [the genome is cited rather than the virion in order to distinguish at this initiating level the difference between genetic effects on the infected cell (the first cell in each of these models) and the effects of the structural components of the virus (e.g. gp120)]. Functional targets are restricted to neural cells involved directly in signaling and include for simplicity the neuron and oligodendrocyte. One might include the astrocyte as well, but its involvement is less likely and hence is omitted here.

Before examining these models by category, let us consider the possible fates of the principal target cells, which are the final step in each of the three models. The most drastic outcome is cell death. Less severe is alteration of functional capacity without or preceding cell death. In the case of neurons, abnormal function might affect reception, modulation, conduction, and

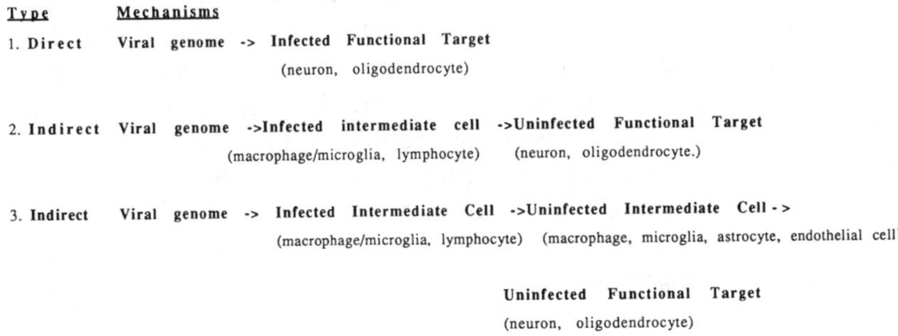

Figure 2 Three models of HIV-1 induced CNS injury.

transmission of signals. These changes might result from a discrete alteration of a luxury function or a more general process affecting neuronal metabolism. For oligodendrocytes, cell death results in frank demyelination whereas cell injury might cause changes in myelin resulting in altered conduction.

THE TARGET-CELL-INFECTION MODEL Let us now turn to the first model, which proposes a direct effect of the viral genome on the functional target. The lack of clear evidence of productive infection of neurons argues against a simple productive-lytic infection paradigm such as poliovirus infection of an anterior horn cell but does not rule out nonproductive infection with partial viral gene expression resulting in either cell death or dysfunction. This possibility would entail partial gene expression that disrupts normal cell metabolism without viral replication, robust transcription, or detectable production of viral structural gene products. Such processes could theoretically cause selective or indolent neuronal loss that would result in no discernable neuronal changes by light microscopy, or could cause luxury changes accounting for the alterations of synapses and dendritic complexity recently reported.

However, although infection of functional elements is possible, there is no clear evidence that it occurs. In vitro studies suggest that neurons are resistant to productive infection, and no existing evidence indicates that this is different in vivo. Whether resistance to infection is wholly at the receptor level, or whether the need for cells to divide in order to replicate the retrovirus and differences in transcription factors are predominant remains to be explored. Likewise, there is no evidence that neurons either harbor the latent HIV-1 genome or, more importantly, transcribe viral genes. Hence, for neurons, the first model appears unlikely even with respect to nonproductive infection.

The case of oligodendrocytes may be less clear. Esiri recently showed viral antigen within an occasional oligodendrocyte, and in vitro work suggesting that GalC might serve as an alternative receptor also points to the potential infectibility of this cell type. However, this infection must be somewhat restricted, otherwise one would anticipate focal demyelination as occurs in progressive multifocal leukoencephalopathy in which JC virus destroys oligodendrocytes (7), rather than white-matter pallor or local edema. Possible demyelination is confined to areas with large numbers of macrophages. Entry of HIV-1 via the alternative receptor might perturb oligodendrocyte function and reduce the capacity to maintain myelin without cell death. This idea is highly speculative and is based on only fragmentary evidence.

Overall, current evidence favors the other two models, both of which involve primary infection of a nonneuroectodermal cell with secondary events leading to indirect injury of critical functional targets. Both of these indirect models involve an infected intermediate cell, the most important candidate being the macrophage-microglial cell as discussed earlier. The $CD4^+$ T-lymphocyte could also be involved, but probably plays at most a secondary role, because these cells are severely depleted during the late phase of HIV-1 infection. The two models differ in the path from the infected cell to the targets, which in both cases are uninfected and altered by toxins secreted by other cells.

THE TWO-CELL MODEL The second model views the infected intermediate cell as the source of toxins. These might be of two classes: (*a*) viral gene products secreted or released in the course of infection and (*b*) cell gene products that are upregulated by infection, e.g. products transactivated by the HIV-1 *tat* gene or similar mechanism.

Several cell-culture studies have now suggested that viral gene products might be toxic to neurons. The surface glycoprotein gp120 is shed by the virus and by HIV-1 infected cells. Brenneman et al (10) suggested that this viral product might be toxic to neurons and demonstrated in vitro that picomolar concentrations of gp120 were lethal to rat hippocampal neurons. They have also presented in vivo evidence of dystrophic neurites in hippocampal pyramidal cells and behavioral deficits in rat pups following intraventricular injection of gp120 (8, 115, 119). Lipton et al (42, 76) reported similar in vitro findings in rat retinal and hippocampal neurons; both native and recombinant gp120 were neurotoxic, and the toxicity was blocked by anti-gp120 antibodies. This group also presented evidence that the gp120 toxicity was mediated via entry or redistribution of Ca^{2+}, resulting in increased free intracellular Ca^{2+}. They demonstrated a protective effect of the dihydropyridine class of Ca^{2+}-channel blockers (nifedipine, nimodipine), suggesting a role of the L-type Ca^{2+} channel in this pathology (42). Whether this results from a direct effect on ion channels or from action at a receptor coupled to an ion channel is unclear.

The possible role of the CD4 receptor in this neurotoxicity model is not well defined. It is generally accepted that most neurons do not express the CD4 receptor; the predominant form of mRNA for CD4 in the mammalian CNS is truncated and apparently not expressed in neurons (97). Brenneman et al (10) showed that anti-CD4 antibodies protected against gp120 induced neurotoxicity, while Kaiser et al (76) showed that they did not protect against either cell death or rise in intracellular Ca^{2+}. The sequence homology of gp120 with the peptide neurotransmitter vasoactive intestinal peptide (VIP) and peptide T (a five–amino acid peptide with sequence homology to VIP) suggested that gp120 might compete at receptors for this endogenous peptide (10). Brenneman et al (10) support this idea with the finding that both VIP and peptide T protect neurons from gp120 toxicity. Because neurotoxicity mediated by the NMDA (glutamate subtype) receptor, a common mechanism of neurotoxicity in many disease states, shows a similar profile of early rise in intracellular Ca^{2+} and delayed neuronal death, NMDA antagonists were studied in an in vitro model. This study showed that these antagonists also have a protective effect against gp120-induced neurotoxicity, though NMDA agonists did not kill neurons, even at high concentrations. Cytoprotection by a particular antagonist does not imply a specific pathogenetic mechanism. For example, the rise in intracellular Ca^{2+} may be limited in several ways that may be remote from the actual signaling mechanism.

Lipton (90) proposes a synergistic effect of gp120 and NMDA agonists. His group found that gp120 was not toxic to rat retinal ganglion cells in the absence of endogenous glutamate, and as stated earlier, NMDA agonists alone could not cause neurotoxicity. They propose a model in which both a glutamate agonist (e.g. quinolinate) and gp120 are necessary for neurotoxicity, but neither is sufficient alone.

Other viral gene products have also been reported to be neurotoxic, including both the *tat* and *nef* gene products. Werner et al (165) noted sequence homology between the *nef* regulatory gene product and scorpion peptides targeted to K^+ channels. In patch-clamp experiments, they demonstrated a reversible increase in K^+ current in chick dorsal root ganglion neurons exposed to recombinant Nef. They propose that such an extracellular source of Nef might come from lysis of infected cells or the production of a secreted form by infected cells and lead to neuronal dysfunction. Infected astrocyte cell lines have been shown to produce high levels of Nef, and antibodies to this protein have been demonstrated in vivo (2, 101). Sabatier et al (141) found that Tat, the 86–amino acid viral regulatory gene product, manifested cytotoxic activity in nerve-cell lines and induced lethal neurotoxicity at comparable doses to snake toxins when injected into the cerebral ventricles of mice. Binding assays by this group found that Tat specifically binds to rat synaptic membranes, rat glioma cells, and murine neuroblastoma cells. Electrophysiological studies revealed that the protein can generate membrane depolarization by modifying cell-membrane permeability in vertebrate and invertebrate excitable cell preparations.

Evidence for a role of toxic cellular gene products in this model derives from studies of Pulliam and Guillian. Pulliam (133) has shown that infected macrophages release a small [<2 kilodaltons (kDa)], protease resistant, heat-stable toxin that causes death and vacuolation of neurons and other cells in human brain-cell aggregates. Because HIV-1-infected lymphocytes do not exert a similar effect, this toxin is probably a cellular, and more specifically a macrophage, gene product, rather than a viral product. Likewise, the studies of Guillian show that the products of infected macrophages cause NMDA-mediated neuronal toxicity and may involve a similar mechanism (54). As discussed earlier, the results of studies attempting to characterize increases in specific macrophage products have been ambiguous and require further definition.

THE THREE-CELL MODEL The importance of cell gene products and the lack of evidence for direct induction of toxin in infected cells raises the question of whether this third model, in which an uninfected intermediate cell is involved, might be more apt. The uninfected cell (or networks of uninfected cells) would be induced to produce a neurotoxin either directly by a viral gene product that alters the uninfected cell, or by a less specific reaction of the intermediate cell to infection or damage of the infected cell. The uninfected intermediate cell might be considered an amplifier and provide a mechanism whereby limited infection might lead to toxin production by a larger number of uninfected cells. This cell might also induce the uninfected intermediate cells to become permissive, and hence provide positive feedback for infection. This model can also be reconciled with some of the cell-culture studies already cited. Thus, Pulliam's and Guillian's systems include uninfected macrophages as well as infected cells, and it may be the uninfected cells reacting to infected cells that produces the toxin described in their studies. This hypothesis might explain the failure of some infected macrophages to produce known cytokines or toxins. Likewise, Lipton's suggestion that the NMDA receptor might be involved could mean that another cell is acted upon by gp120 and secondarily releases an endogenous excitotoxin. This hypothesis might fit best with the concept that quinolinic acid is involved in HIV-1-related neurotoxicity, with INF-γ, TNF-α, or some other primary agonist inducing quinolinic acid production in the uninfected cell (66). The candidate cells for the uninfected intermediate include principally the macrophage-microglia and the astrocyte. The vascular endothelial cell might also be involved in a similar series of steps whereby products of the infected cell intermediate (or even another uninfected cell intermediate) might lead to altered permeability of the blood-brain barrier, and hence exposure of target cells to blood-borne toxins.

In conclusion, the indirect hypotheses have in common an external toxic mechanism rather than a direct effect of the viral genome on the target. Several putative toxins have been investigated, some identified, and some only localized to a likely cell source.

RESULTS OF THERAPY: IMPLICATIONS AND PROSPECTS

Treatment of the AIDS dementia complex has received disproportionately little direct study. We focus on what the results of therapy have suggested about the pathogenesis of disease, how the ideas about pathogenesis discussed above might bear on treatment strategies, and the implications for future therapies for this and other neurological diseases.

Implications for the Model

The results of therapy appear to support the contention that HIV-1 is the prime mover in the pathogenesis of the AIDS dementia complex. Several studies now suggest that use of the antiretroviral zidovudine (ZDV, also azidothymidine or AZT), which interferes with virus-directed reverse transcription, is beneficial for both therapy and prophylaxis in this syndrome (18, 123, 142, 173; J. J. Sidtis, C. Gatsonis, R. W. Price, E. J. Singer & A. C. Collier, submitted). Therapeutic efficacy has been documented by case reports (142, 173) and controlled study (J. J. Sidtis, C. Gatsonis, R. W. Price, E. J. Singer & A. C. Collier, submitted), although the optimal dose, the limits of therapy, and other more refined questions remain to be addressed. Neuropathic evidence has also supported this contention; one study found HIV-1 encephalitis was less common in zidovudine-treated patients (59). Better definition of the early events of CNS infection may help answer some questions regarding therapy. For example, higher antiviral drug levels might be required if the virus sets up local long-term persistent infection, whereas in a model of continuous reinfection, lower doses might be adequate. A preventive role for antiviral therapy is suggested by a study documenting an abrupt fall in the incidence of the AIDS dementia complex in the Netherlands coincident with the introduction of zidovudine (125). The mode of efficacy is uncertain, but may involve: (*a*) reduction of systemic infection and hence of seeding (initial or continued) of the CNS, (*b*) direct inhibition of infection in the CNS, or even (*c*) some reversal of immunosuppression thereby reducing permissiveness in the CNS for direct HIV-1 infection. The mode is important because of dosing implications: systemic effects likely require lower doses than do direct

CNS effects, and antiretroviral drugs vary in their ability to penetrate the blood-brain barrier.

Some additional observations are of interest in suggesting that immunopathology is driven by HIV-1 and not by self-perpetuating autoimmune reactions. Particularly intriguing in this regard are observations of reductions in the CSF levels of surrogate markers, including β_2M, neopterin, and quinolinic acid with zidovudine therapy (13, 66). The simplest and most parsimonious explanation for the substantial reductions noted in these markers is that they are stimulated by local infection, and that zidovudine reduces this stimulation by an effect on infected cells (as opposed to a nonspecific metabolic effect). Observations of decreased blood-brain barrier permeability in CSF and of decreased edema using MRI and PET with zidovudine are also consistent with this picture (B. J. Brew & R. W. Price, unpublished data).

Implications of the Model

Perhaps the salient implication of the model, also supported by results of treatment to date, is that the primary approach to the AIDS dementia complex and brain HIV-1 infection is to attack the prime mover—to inhibit HIV-1 replication using antiviral strategies. However, there are other potential implications if one considers the secondary processes as pathogenetically important. One can envision interventions that target the secondary processes of neurotoxicity. Indeed, this is the underlying strategy of a proposed use of nimodipine (to reduce Ca^{2+} toxicity) or pentoxyphylline (to oppose TNF-α toxicity) in these patients. The AIDS Clinical Trials Group is currently embarking on a trial of nimodipine. Blocking the indirect effects of HIV-1 infection is also the rationale for the use of peptide T, which was initially advanced as an antiretroviral on the basis of homology with a portion of gp120 (a portion that turned out to be variable) but is now promoted on the basis of its neuronal protective effect related to homology with vasointestinal peptide (VIP), an endogenous neurotransmitter (9, 17). Phase I testing has demonstrated the safety of peptide T and preliminary results suggest possible improved cognitive and neuromotor function in the context of stable immune function (17). One could also contemplate the cautious use of corticosteroids with such an antiimmune strategy.

Implications Beyond HIV-1 and the AIDS Dementia Complex

None of the mechanisms discussed here are unique to HIV-1 infection. What is perhaps special about HIV-1 is its chronic course, which allows these mechanisms to induce toxicity not seen in more acute, rapidly reversible conditions, and the coexistence of severe immunosuppression and dysregulated immune response.

Hence, lessons are here for other neurological conditions, ranging from acute infections where these mechanisms likely contribute to other chronic conditions of known etiology, such as HTLV-I infection, which causes a neurological condition [tropical spastic paraparesis/HTLV-1 associated myelopathy (TSP/HAM)] (69), or of unknown cause, such as multiple sclerosis. These all likely involve important interactions of immune reactions, particularly cytokine networks, which participate in tissue injury and perpetuation of disease. Therefore, pursuing these issues regarding HIV-1 and the AIDS dementia complex will be important in order to find clues not only to its own prevention and treatment, but also to open doors to the management of other conditions.

ACKNOWLEDGMENTS

Our research program studying the AIDS Dementia Complex is supported by U.S. Public Health Service grant NS-25701. David Spencer is a Howard Hughes Medical Institute Medical Student Research Training Fellow.

Literature Cited

1. Adachi, A., Gendelman, H. E., Koenig, S., Folks, T., Willey, R., et al. 1986. Production of acquired immunodeficiency syndrome-associated retrovirus in human and non-human cells transfected with an infectious molecular clone. *J. Virol.* 59:284–91
2. Ameisen, J.-C., Guy, B., Chamaret, S., Loche, M., Mouton, Y., et al. 1989. Antibodies to the *nef* protein and to *nef* peptides in HIV-1 infected seronegative individuals. *AIDS Res. Hum. Retrovirus* 5:279–91
2a. American Academy of Neurology AIDS Task Force. 1991. Nomenclature and research case definitions for neurologic manifestations of human immunodeficiency virus-type 1 (HIV-1) infection. *Neurology* 41:778–85
3. Appleman, M. E., Marshall, D. W., Brey, R. L., Houk, R. W., Beatty, D. C., et al. 1988. Cerebrospinal fluid abnormalities in patients without AIDS who are seropositive for the human immunodeficiency virus. *J. Infect. Dis.* 158:193–99
4. Belman, A. L., Lantos, G., Horoupian, D., Novick, B. E., Ultmann, M. H., et al. 1986. Calcification of the basal ganglia in infants and children. *Neurology* 36:1192–99
5. Belman, A. L., Ultmann, M. H., Horoupian, D., Novick, B., Sprio, A. J., et al. 1985. Neurological complications in infants and children with acquired immune deficiency syndrome. *Ann. Neurol.* 18:560–66
6. Benson, D. F. 1987. The spectrum of dementia: a comparison of the clinical features of AIDS dementia and dementia of the Alzheimer's type. *Alzheimer Dis. Assoc. Disord.* 1(4):217–20
7. Berger, J. R. Kaszovitz, B., Post, M. J., Dickinson, G. 1987. Progressive multifocal leukoencephalopathy associated with human immunodeficiency virus infection. *Ann. Intern. Med.* 107:78–87
8. Brenneman, D. E., Hill, J. M. 1990. Gp120 induced retardation of behavioral development in neonatal rats: prevention by peptide T. *Soc. Neurosci. Abstr.* 16:615
9. Brenneman, D., Westerbrook, G., Fitzgerald, S., Ennis, D., Elkins, K., et al. 1988. Neuronal killing by the envelope protein of HIV and its prevention by vasoactive intestinal peptide. *Nature* 335:639–42
10. Brenneman, D. E., Westerbrook, G. L., Fitzgerald, S. P., Ennist, D. L., Elkins, K. L., et al. 1988. Neuronal cell killing by the envelope glycoprotein of HIV and its prevention by vasoactive intestinal peptide. *Nature* 335:639–42
11. Brew, B. J., Bhalla, R. B., Fleisher, M., Paul, M., Khan, A., et al. 1989. Cerebrospinal fluid $\beta 2$ microglobulin in patients infected with human im-

munodeficiency virus. *Neurology* 39: 830–34
12. Brew, B. J., Bhalla, R. B., Paul, M., Gallardo, H., McArthur, J. C., et al. 1990. Cerebrospinal fluid neopterin in human immunodeficiency virus type 1 infection. *Ann. Neurol.* 28:556–60
13. Brew, B. J., Bhalla, R., Paul, M., Sidtis, J. J., Keilp, J. J., et al. 1992. Cerebrospinal fluid B2-microglobulin in patients with AIDS dementia complex: an expanded series including response to zidovudine treatment. *AIDS* In press
14. Brew, B. J., Perdices, M. 1992. HIV nomenclature. *Neurology* 42:265
15. Brew, B. J., Perdices, M., Darveniza, P., Edwards, P., Whyte, B., et al. 1989. The neurological features of early and "latent" human immunodeficiency virus infection. *Aust. N. Z. Med.* 19:700–5
16. Brew, B., Sidtis, J., Petito, C. K., Price, R. W. 1988. The neurological complications of AIDS and human immunodeficiency virus infection. In *Advances in Contemporary Neurology*, ed. F. Plum, pp. 1–49. Philadelphia: F. A. Davis Co
17. Bridge, T. P., Heseltine, P. N., Parker, E. S., Eator, E. M., Ingraham, L. J., et al. 1991. Results of extended Peptide T administration in AIDS and ARC patients. *Psychopharm. Bull.* 27(3): 237–245
18. Brouwers, P., Moss, H., Wolters, P., Eddy, J., Balis, F., Poplack, D. G. 1990. Effect of continuous-infusion zidovudine therapy on neuropsychologic functioning in children with symptomatic human immunodeficiency virus infection. *J. Pediatr.* 116:980–85
19. Budka, H. 1989. Human immunodeficiency virus (HIV)-induced disease of the central nervous system: pathology and implications for pathogenesis. *Acta Neuropathol. (Berlin)* 77:225–36
20. Budka, H. 1990. Human immunodeficiency virus (HIV) envelope and core proteins in CNS tissues of patients with the acquired immune deficiency syndrome (AIDS). *Acta Neuropathol.* 79:611–19
21. Budka, H. 1991. Neuropathology of HIV encephalitis. *Brain Pathol.* 1:163–75
22. Budka, H., Costanzi, G., Cristina, S., Lechi, A., Parravicini, C., et al. 1987. Brain pathology induced by infection with the human immunodeficiency virus (HIV). A histological, immunocytochemical and electron microscopical study of 100 autopsy cases. *Acta Neuropathol. (Berlin)* 75:185–98
23. Budka, H., Lassman, H. 1988. Human immunodeficiency virus in glial cells? *J. Infect. Dis.* 157:203
24. Buffet, R., Agut, H., Chieze, R., Katlama, C., Bolgert, F., et al. 1991. Virological markers in the cerebrospinal fluid from HIV-1 infected individuals. *AIDS* 5:1419–24
25. Canivet, M., Hoffman, A. D., Hardy, D., Sernatinger, J., Levy, J. A. 1990. Replication of HIV-1 in a wide variety of animal cells following phenotypic mixing with murine retroviruses. *Virology* 178:543–51
26. Cann, A. J., Churcher, M. J., Boyd, M., O'Brien, W., Zhao, J.-Q., et al. 1992. The region of the envelope gene of human immunodeficiency virus type 1 responsible for determination of cell tropism. *J. Virol.* 66(1):305–9
27. Castro, B. A., Cheng-Mayer, C., Evans, L. A., Levy, J. A. 1988. HIV heterogeneity and viral pathogenesis. *AIDS* 2(1):S17–S27
28. Cheng-Mayer, C. 1990. Biological and molecular features of HIV-1 related to tissue tropism. *AIDS* 4(1):S49–S56
29. Cheng-Mayer, C., Quiroga, M., Tung, J. W., Dina, D., Levy, J. A. 1990. Viral determinants of human immunodeficiency virus type 1 T-cell or macrophage tropism, cytopathogenicity, and CD4 antigen modulation. *J. Virol.* 64(9): 4390–98
30. Cheng-Mayer, C., Rutka, J. T., Rosenblum, M. L., McHugh, T., Stites, D. P., Levy, J. A. 1987. Human immunodeficiency virus can productively infect cultured human glial cells. *Proc. Natl. Acad. Sci. USA* 84:3523–39
31. Cheng-Mayer, C., Shioda, T., Levy, J. A. 1991. Host range, replicative, and cytopathic properties of human immunodeficiency virus type 1 are determined by very few amino acid changes in *tat* and gp120. *J. Virol.* 65(12):6931–41
32. Chesebro, B., Nishio, J., Perryman, S., Cann, A., O'Brien, W., et al. 1991. Identification of human immunodeficiency virus envelope gene sequences influencing viral entry into CD4-positive HeLa cells, T-leukemia cells, and macrophages. *J. Virol.* 65(11):5782–89
33. Chiodi, F., Fuerstenberg, S., Gidlund, M., Åsjö, B, Fenyö, E. M. 1987. Infection of brain derived cells with human immunodeficiency virus. *J. Virol.* 61(4): 1244–47
34. Chiodi, F., Norkrans, G., Hagberg, L., Sonnerborg, A., Gaines, H., et al. 1988.

Human immunodeficiency virus infection of the brain. II. Detection of intrathecally synthesized antibodies by enzyme linked immunosorbent assay and imprint immunofixation. *J. Neurol. Sci.* 87:37–48

35. Clements, G. J., Price-Jones, M. J., Stephens, P. E., Sutton, C., Schulz, T. F., et al. 1991. The V3 loops of HIV-1 and HIV-2 surface glycoproteins contain proteolytic cleavage sites: a possible function in viral fusion? *AIDS Res. Hum. Retroviruses* 7(1):3–16
36. Cullen, B. R. 1991. Regulation of HIV-1 gene expression. *FASEB J.* 5:2361–68
37. Cummings, J. L., Benson, D. F. 1984. Subcortical dementia: review of an emerging concept. *Arch. Neurol.* 41:874–79
38. Dewhurst, S., Bresser, J., Stevenson, M., Sakai, K., Evinger-Hodges, M. J., Volsky, D. J. 1987. Susceptibility of human glial cells to infection with human immunodeficiency virus (HIV). *FEBS Lett.* 213:138–43
39. Dewhurst, S., Sakai, D., Bresser, J., Stevenson, M., Evinger-Hodges, M. J., Volsky, D. J. 1987. Persistent productive infection of human glial cells by human immunodeficiency virus (HIV) and by infectious molecular clones of HIV. *J. Virol.* 61:3774–82
40. Dewhurst, S., Sakai, K., Zhang, X. H., Wasiak, A., Volsky, D. J. 1988. Establishment of human glial cell lines chronically infected with the human immunodeficiency virus. *Virology* 162(1): 151–9
41. Dickson, D. W., Mattiace, L. A., Kure, K., Hutchins, K., Lyman, W. D., Brosnan, C. F. 1991. Biology of disease: microglia in human disease, with an emphasis on acquired immune deficiency syndrome. *Lab. Invest.* 64:135–56
42. Dreyer, E. B., Kaiser, P. K., Offermann, J. T., Lipton, S. A. 1990. HIV-1 coat protein neurotoxicity prevented by calcium channel antagonists. *Science* 248:364–67
43. Elovaara, I., Iivanainen, M., Valle, S. L., Suni, J., Tervo, T., Lahdevirta, J. 1987. CSF protein and cellular profiles in various stages of HIV infection related to neurological manifestations. *J. Neurol. Sci.* 78:331–42
44. Epstein, L. G., Kuiken, C., Blumberg, B. M., Hartman, S., Sharer, L. R., et al. 1991. HIV-1 V3 domain variation in brain and spleen of children with AIDS: tissue specific evolution within host-determined quasispecies. *Virology* 180: 583–90
45. Epstein, L. G., Sharer, L. R., Cho, E.-S., Meyenhofer, M., Navia, B. A., Price, R. W. 1985. HTLVIII/LAV-like retrovirus particles in the brains of patients with AIDS encephalopathy. *AIDS Res.* 1:447–54
46. Epstein, L. G., Sharer, L. R., Joshi, V. V., Fojas, M. M., Koenigsberger, M. R., Oleske, J. M. 1985. Progressive encephalopathy in children with acquired immune deficiency syndrome. *Ann. Neurol.* 17:488–96
47. Erfle, V., Stoeckbauer, P., Kleinschmidt, A., Kohleisen, B., Mellert, W., et al. 1991. Target cells for HIV in the central nervous system: macrophages or glial cells? *Res. Virol.* 142:139–44
48. Esiri, M. M., Morris, C. S., Millard, P. R. 1991. Fate of oligodendrocytes in HIV-1 infection. *AIDS* 5:1081–88
49. Everall, I. P., Luthert, P. J., Lantos, P. L. 1991. Neuronal loss in the frontal cortex in HIV infection. *Lancet* 337: 1119–21
50. Freed, E. O., Myers, D. J., Risser, R. 1991. Identification of the principal neutralizing determinant of human immunodeficiency virus type 1 as a fusion domain. *J. Virol.* 65:1190–94
51. Gabuzda, D. H., Ho, D. D., DeLaMonte, S. M., Hirsch, M. S., Rota, T. R., Sobel, R. A. 1986. Immunohistochemical identification of HTLV-III antigen in brains of patients with AIDS. *Ann. Neurol* 20:289–95
52. Gallo, P., Piccinno, M. G., Krzalic, L., Tavolato, B. 1989. Tumor necrosis factor α (TNFα) and neurological diseases. Failure in detecting TNFα in the cerebrospinal fluid from patients with multiple sclerosis, AIDS dementia complex, and brain tumors. *J. Neuroimmunol.* 23(1):41–4
53. Gartner, S., Markovits, P., Markovits, D. M., Betts, R. F., Popovic, M. 1986. Virus isolation from and identification of HTLV-III/LAV producing cells in brain tissue from a patient with AIDS. *J. Am. Med. Assoc.* 256:2365–71
54. Giulian, D., Vaca, K., Noonan, C. A. 1990. Secretion of neurotoxins by mononuclear phagocytes infected with HIV-1. *Science* 250:1593–96
55. Gonda, M. A., Wong-Staal, F., Gallo, R. C., Clements, J. G., Narayan, O., Gilden, R. V. 1985. Sequence homology and morphologic similarity of HTLV III and visna virus, a pathogenic lentivirus. *Science* 227:173–77
56. Gopinathan, G., Laubenstein, L. J., Mondale, B., Krigel, R. G. 1983. Central nervous system manifestations of the

acquired immunodeficiency syndrome in homosexual men. *Neurology* 33(2):105
57. Goudsmit, J., DeWolf, F., Paul, D. A., Epstein, L. G., Lange, J. M., et al. 1986. Expression of human immunodeficiency virus antigen (HIV-Ag) in serum and cerebrospinal fluid during acute and chronic infection. *Lancet* 2(8500):177–80
58. Graeber, M. B., Streit, W. J. 1990. Microglia: immune network in the CNS. *Brain Pathol.* 1:2–5
59. Gray, F., Geny, C., Dournon, E., Fenelon, G., Lionnet, F., Gherardi, R. 1991. Neuropathological evidence that zidovudine reduces the incidence of HIV infection of brain. *Lancet* 337:852–3
60. Grimaldi, L. M. E., Martino, G. V., Franciotta, D. M., Brustia, R., Castagna, A., et al. 1991. Elevated α-tumor necrosis factor levels in spinal fluid from HIV-1 infected patients with central nervous system involvement. *Ann. Neurol.* 29:21–5
61. Györkey, F., Melnick, J. L., Györkey, P. 1987. Human immunodeficiency virus in brain biopsies of patients with AIDS and progressive encephalopathy. *J. Infect. Dis.* 155:870–76
62. Haase, A. T. 1986. Pathogenesis of lentivirus infections. *Nature* 322:130–36
63. Harouse, J. M., Bhat, S., Spitalnik, S. L., Laughlin, M., Stefano, K., et al. 1991. Inhibition of entry of HIV-1 in neural cell lines by antibodies against galactosyl ceramide. *Science* 253:320–22
64. Harouse, J. M., Wroblewska, Z., Laughlin, M. A., Hickey, W. F., Schonwetter, B. S., Gonzalez-Scarano, F. 1989. Human choroid plexus cells can be latently infected with human immunodeficiency virus. *Ann. Neurol.* 25:406–11
65. Hattori, T., Koito, A., Takatsuki, K., Kido, H., Katunuma, N. 1989. Involvement of trypase-related cellular protease(s) in human immunodeficiency type 1 infection. *FEBS Lett.* 248:48–52
66. Heyes, M. P., Brew, B. J., Martin, A. 1991. Quinolinic acid in cerebrospinal fluid and serum in HIV-1 infection: relationship to clinical and neurologic status. *Ann. Neurol.* 29:202–9
67. Heyes, M. P., Brew, B., Martin, A., Markey, S. P., Price, R. W., et al. 1991. Cerebrospinal fluid quinolinic acid concentrations are increased in acquired immune deficiency syndrome. *Adv. Exp. Med. Biol.* 294:687–90
68. Hickey, W. F. 1991. Migration of hematogenous cells through the blood-brain barrier and the initiation of CNS inflammation. *Brain Pathol.* 1:97–105
69. Hjelle, B. 1991. Human T-cell leukemia/lymphoma viruses: life cycle, pathogenicity, epidemiology, and diagnosis. *Arch. Pathol. Lab. Med.* 115:440–50
70. Ho, D. D., Rota, T. R., Schooley, R. T., Kaplan, J. C., Allan, J. D., et al. 1985. Isolation of HTLV-III from cerebrospinal fluid and neural tissues of patients with neurologic syndromes related to the acquired immunodeficiency syndrome. *New Engl. J. Med.* 313: 1493–97
71. Horowitz, S. L., Benson, D. F., Gottleib, M. S., Davos, I., Bentson, J. R. 1982. Neurological complications of gay-related immunodeficiency disorder. *Ann. Neurol.* 12:80
72. Hwang, S. S., Boyle, T. J., Lyerly, H. K., Cullen, B. R. 1991. Identification of the envelope V3 loop as the primary determinant of cell tropism in HIV-1. *Science* 253:71–74
73. Jakobsen, J., Gyldensted, C., Brun, B., Bruhn, P., Helweg-Larsen, S., Arlien-Soborg, P. 1989. Cerebral ventricular enlargement relates to neuropsychological measures in unselected AIDS patients. *Acta Neurol. Scand.* 79:59–62
74. Jarvik, J. G., Hesselink, J. R., Kennedy, C., Teschke, R., Wiley, C., et al. 1988. Acquired immunodeficiency syndrome. Magnetic resonance patterns of brain involvement with pathologic correlation. *Arch. Neurol.* 45(7):731–36
75. Jordan, C. A., Watkins, B. A., Kufta, C., Dubois-Dalcq, M. 1991. Infection of brain microglial cells by human immunodeficiency virus type 1 is CD4 dependent. *J. Virol.* 65:2736–42
76. Kaiser, P. K., Offerman, J. T., Lipton, S. A. 1990. Neuronal injury due to HIV-1 envelope protein is blocked by anti-gp120 antibodies but not by anti-CD4 antibodies. *Neurology* 40:1757–61
77. Keating, J. N., Trimble, K. C., Mulcahy, F., Scott, J. M., Weir, D. G. 1991. Evidence of brain methyltransferase inhibition and early brain involvement in HIV+ patients. *Lancet* 337:935–39
78. Ketzler, S., Weis, S., Haug, H., Budka, H. 1990. Loss of neurons in the frontal cortex in AIDS brains. *Acta Neuropathol.* 80:92–94
79. Keys, B., Albert, J., Kövamees, J., Chiodi, F. 1991. Brain-derived cells can be infected with HIV isolates derived from both blood and brain. *Virology* 183:834–39

80. Koenig, S., Gendelman, H. E., Orenstein, J. M., Dal Canto, M. C., Pezeshkpour, G. H., et al. 1986. Detection of AIDS virus in macrophages in brain tissue from AIDS patients with encephalopathy. *Science* 233:1089–93
81. Koyangi, Y., Miles, S., Mitsuyasu, R. T., Merrill, J. E., Vinters, H. V., Chen, I. S. Y. 1987. Dual infection of the central nervous system by AIDS viruses with distinct cellular tropisms. *Science* 236:819–22
82. Kunsch, C., Wigdahl, B. 1989. Role of HIV in human nervous system dysfunction. *AIDS Res. Hum. Retroviruses* 5:369–74
83. Kure, K., Lyman, W. D., Weidenheim, K. M., Dickson, D. W., 1990. Cellular localization of an HIV-1 antigen in subacute AIDS encephalitis using an improved double-labeling immunohistochemical method. *Am. J. Pathol.* 136(5):1085–92
84. Kure, K., Weidenhiem, K. M., Lyman, W. D., Dickson, D. W. 1990. Morphology and distribution of HIV-1 gp41-positive microglia in subacute AIDS encephalitis. *Acta Neuropath.* 80:393–400
85. Lassman, H., Rössler, K., Zimprich, F., Vass, K. 1991. Expression of adhesion molecules and histocompatability antigens at the blood-brain barrier. *Brain Pathol.* 1:115–23
86. Lassmann, H., Zimprich, F., Vass, K., Hickey, W. F. 1991. Microglial cells are a component of the perivascular glial limitans. *J. Neurosci. Res.* 28(2):236–43
87. Levy, J. A., Cheng-Mayer, C., Dina, D., Luciw, P. A. 1986. AIDS retrovirus (ARV-2) clone replicates in transfected human and animal fibroblasts. *Science* 232:998–1001
88. Li, W. H., Tanimura, M., Sharp, P. M. 1988. Rates and dates of divergence between AIDS virus nucleotide sequences. *Mol. Biol. Evol.* 5:313–30
89. Li, Y., Kappes, J. C., Conway, J. A., Price, R. W., Shaw, G. M., Hahn, B. H. 1991. Molecular characterization of human immunodeficiency virus type 1 cloned directly from uncultured brain tissue: identification of replication-competent and -defective viral genomes. *J. Virol.* 65(8):3973–85
90. Lipton, S. A. 1991. HIV-related neurotoxicity. *Brain Pathol.* 1:193–99
91. Lipton, S. A. 1992. Models of neuronal injury in AIDS: another role for the NMDA receptor? *Trends Neurosci.* 15(3):75–79
92. Liu, Z.-Q., Wood, C., Levy, J. A., Cheng-Meyer, C. 1990. The viral envelope gene is involved in macrophage tropism of a HIV-1 strain isolated from brain tissue. *J. Virol.* 64(12):6148–53
93. Ljungrenn, K., Biberfeld, G., Jondal, M., Fenyo, E. M. 1989. Antibody dependant cellular cytotoxicity detects type and strain specific antigens among HIV types 1 and 2 and SIV mac isolates. *J. Virol.* 63:3376–80
94. Lusso, P., Veronese, F., Ensoli, B., Franchini, G., Jemma, C., et al. 1990. Expanded HIV-1 cellular tropism by phenotypic mixing with murine endogenous retroviruses. *Science* 247:848–52
95. Ma, X., Sakai, K., Sinangil, F., Golub, E., Volsky, D. 1990. Interaction of a noncytopathic human immunodeficiency virus type 1 (HIV-1) with target cells: efficient viral entry followed by delayed expression of its RNA and protein. *Virology* 176:184–94
96. McArthur, J. C., Cohen, B. A., Farzadegan, H., Cornblath, D. R., Selnes, O. A., et al. 1988. Cerebrospinal fluid abnormalities in homosexual men with and without neuropsychiatric findings. *Ann. Neurol.* 23(Suppl):S34-S37
97. Maddon, P. J., Dalgleish, A. G., McDougal, J. S., Clapham, P. R., Weiss, R. A., Axel, R. 1986. The T4 gene encodes the AIDS virus receptor and is expressed in the immune system and the brain. *Cell* 47:333–48
98. Marshall, D. W., Brey, R. L., Cahill, W. T., Houk, R. W., Zajac, R. A., Boswell, R. N. 1988. Spectrum of cerebrospinal fluid findings in various stages of human immunodeficiency virus infection. *Arch. Neurol.* 45:954–58
99. Matsuyama, T., Kobayashi, N., Yamamoto, N. 1991. Cytokines and HIV infection: is AIDS a tumor necrosis factor disease? *AIDS* 5:1405–17
100. McArthur, J. C. 1987. Neurologic manifestations of AIDS. *Medicine* 66(6): 407–37
101. Mellert, W., Festl, H., Kleinschmidt, A., Muller-Lantzsch, N., Erfle, V. 1990. Cell specific control of HIV production in chronically infected glial cells. *Int. Conf. on AIDS, 5th, San Francisco, June 1990*. Abstr. 607
102. Michaels, J., Price, R. W., Rosenblum, M. K. 1988. Microglia in the human immunodeficiency virus encephalitis of acquired immunodeficiency syndrome: proliferation, infection, and fusion. *Acta Neuropathol.* 76:373–79
103. Miller, E. B., Selnes, O. A., McArthur, J. C., Satz, P., Becker, J. T., et al.

1990. Neuropsychological performance in HIV-1-infected homosexual men: the multi-center AIDS cohort study (MACS). *Neurology* 40:197–203
104. Mirra, S. S., del Rio, C. 1989. The fine structure of acquired immune deficiency syndrome encephalopathy. *Arch. Pathol. Lab. Med.* 113:858–65
105. Moeller, A. A., Backmund, H. C. 1990. Ventricle brain ratio in the clinical course of HIV infection. *Acta Neurol. Scand.* 81:512–15
106. Navia, B. A., Cho, E.-W., Petito, C. K., Price, R. W. 1986. The AIDS dementia complex: II. Neuropathology. *Ann. Neurol.* 19:525–35
107. Navia, B. A., Jordan, B. D., Price, R. W. 1986. The AIDS dementia complex: I. Clinical features. *Ann. Neurol.* 19:517–24
108. Navia, B. A., Khan, A., Pumarola-Sune, T. Price, R. W. 1986. Choline acetyltransferase activity is reduced in AIDS dementia complex. *Ann. Neurol.* 20:142
109. Navia, B. A., Price, R. W. 1987. The acquired immunodeficiency syndrome dementia complex as the presenting or sole manifestation of human immunodeficiency virus infection. *Arch Neurol* 44:65–69
110. Nixon, D. F., McMichael, A. J. 1991. Cytotoxic T-cell recognition of HIV proteins and peptides. *AIDS* 5:1049–59
111. O'Brien, W. A., Koyanagi, Y., Namazie, A., Zhao, J.-Q., Diagne, A., et al. 1990. HIV-1 tropism for mononuclear phagocytes can be determined by regions of gp120 outside the CD4-binding domain. *Nature* 348:69–73
112. Page, K., Landau, N. R., Littman, D. 1990. Construction and use of a human immunodeficiency virus vector for analysis of virus infectivity. *J. Virol.* 64:5270–76
113. Page, K. A., Stearns, S. M., Littman, D. R. 1992. Analysis of mutations in the V3 domain of gp160 that affect fusion and infectivity. *J. Virol* 66(1):524–33
114. Pang, S., Vinters, H. V., Akashi, T., O'Brien, W. A., Chen, I. S. Y. 1991. HIV-1 *env* sequence variation in brain tissue of patients with AIDS-related neurological disease. *J. AIDS* 4:1082–92
115. Panlilio, L. V., Hill, J. M., Brenneman, D. E. 1990. Gp120 and VIP receptor antagonists impair Morris water maze performance in rats. *Soc. Neurosci. Abstr.* 16:1330
116. Paul, M. O., Brew, B. J., Khan, A., Gallardo, M., Price, R. W. 1989. Detection of HIV-1 in cerebrospinal fluid (CSF): correlation with presence and severity of the AIDS dementia complex. *Int. Conf. on AIDS, 5th, San Francisco, June 1990.* Abstr. 238
117. Peluso, R., Haase, A., Stowring, L., Edwards, M., Ventura, P. 1985. A Trojan Horse mechanism for the spread of visna virus in monocytes. *Virology* 147:231–36
118. Perry, V. H., Simon, G. 1988. Macrophages and microglia in the nervous system. *Trends Neurosci.* 11:273–77
119. Pert, C. B., Mervis, R. F., Hill, J. M., Brenneman, D. E., Ruff, M. R., et al. 1989. Morphological and behavioral response to acute and subacute i.c.v. gp120 administration in the rat. *Soc. Neurosci. Abstr.* 15:1387
120. Petito, C. K., Cho, E.-S., Lemann, W., Navia, B. A., Price, R. W .1986. Neuropathology of acquired immunodeficiency syndrome (AIDS): an autopsy review. *J. Neuropathol. Exp. Neurol.* 45:635–46
121. Petito, C. K., Navia, B. A., Cho, E.-S., Jordan, B. D., George, D. C., Price, R. W. 1985. Vacuolar myelopathy pathologically resembling subacute combined degeneration in patients with acquired immune deficiency syndrome (AIDS). *New Engl. J. Med.* 312:874–79
122. Peudenier, S., Hery, C., Montagnier, L., Tardieu, M. 1991. Human microglial cells: characterization in cerebral tissue and in primary culture, and study of their susceptibility to HIV-1 infection. *Ann. Neurol.* 29:152–61
123. Pizzo, P. A., Eddy, J., Falloon, J., Balis, F. M., Murphy, R. F., et al. 1988. Effect of continuous intravenous infusion of zidovudine (AZT) in children with symptomatic HIV infection. *New Engl. J. Med.* 319:889–96
124. Pohl, P., Vogl, G., Fill, H., Rossler, H., Zangerle, R., Gerstenbrand, F. 1988. Single photon emission computed tomography in AIDS dementia complex. *J. Nucl. Med.* 29(8):1382–86
125. Portegies, P., de Gans, J., Lange, J. M., Derix, M. M., Speelman, H., et al. 1989. Declining incidence of AIDS dementia complex after introduction of zidovudine treatment. *Br. Med. J.* 299:819–21
126. Post, M. J., Tate, L. G., Quencer, R. M., Hensley, G. T., Berger, J. R., et al. 1988. CT, MR, and pathology in HIV encephalitis and meningitis. *Am. J. Roentgenol.* 151:373–80
127. Price, R. W., Brew, B. J. 1988. The AIDS dementia complex. *J. Infect. Dis.* 158:1079–83

128. Price, R. W., Brew, B. J., Rosenblum, M. 1990. The AIDS dementia complex and HIV-1 brain infection: a pathogenetic model of virus-immune interaction. In *Immunologic Mechanisms in Neurologic and Psychiatric Disease,* ed. B. H. Waksman, pp. 269–90. New York: Raven

129. Price, R. W., Brew, B. J., Sidtis, J., Rosenblum, M., Scheck, A. C., Cleary, P. 1988. The brain in AIDS: central nervous system HIV-1 infection and AIDS dementia complex. *Science* 239:586–92

130. Price, R. W., Sidtis, J. J. 1990. Early HIV infection and the AIDS dementia complex. *Neurology* 40:323–26

131. Price, R. W., Sidtis, J. J. 1992. The AIDS dementia complex. In *AIDS and Other Manifestations of HIV Infection,* ed. G. P. Wormser, pp. 373–82. New York: Raven

132. Price, R. W., Sidtis, J. J., Brew, B. J. 1991. AIDS dementia complex and HIV-1 infection: a view from the clinic. *Brain Pathol.* 1:155–62

133. Pulliam, L., Herndler, B. G., Tanf, N. M., McGrath, M. S. 1991. Human immunodeficiency virus–infected macrophages produce soluble factors that cause histological and neurochemical alterations in cultured human brains. *J. Clin. Invest.* 87:503–12

134. Pumarole-Sune, T., Navia, B. A., Cordon-Cardo, C., Cho, E.-S., Price, R. W. 1987. HIV antigen in the brains of patients with the AIDS dementia complex. *Ann. Neurol.* 21:490–96

135. Deleted in proof

136. Resnick, L., Berger, J. R., Shapshak, P., Tourtellotte, W. W. 1988. Early penetration of the blood-brain-barrier by HIV. *Neurology* 38:9–14

137. Robert-Guroff, M., Brown, M., Gallo, R. C. 1985. HTLV-III neutralizing antibodies in patients with AIDS and AIDS-related complex. *Nature* 316:72–74

138. Rosenblum, M. K. 1990. Infection of the central nervous system by the human immunodeficiency virus type 1: morphology and relation to syndromes of progressive encephalopathy and myelopathy in patients with AIDS. *Pathol. Ann.* 25:117–69

139. Rosenblum, M., Scheck, A. C., Cronin, K., Brew, B. J., Khan, A., et al. 1989. Dissociation of AIDS-related vacuolar myelopathy and productive human immunodeficiency virus type 1 (HIV-1) infection of the spinal cord. *Neurology* 39:892–96

140. Rottenberg, D. A., Moeller, J. R., Strother, S. C., Sidtis, J. J., Navia, B. A., et al. 1987. The metabolic pathology of the AIDS dementia complex. *Ann. Neurol.* 22:700–6

141. Sabatier, J.-M., Vives, E., Marbrouk, K., Benjouad, A., Rochat, H., et al. 1991. Evidence for neurotoxic activity of *tat* from human immunodificiency virus type 1. *J. Virol.* 65(2):961–67

142. Schmitt, F. A., Bigley, J. W., McKinnis, R., Logue, P. E., Evans, R. W., Drucker, J. L. 1988. Neuropsychological outcome of zidovudine (AZT) treatment of patients with AIDS and AIDS-related complex. *New Engl. J. Med.* 319:1573–78

143. Schuitemaker, H., Koot, M., Kootstra, N. A., Wouter Dercksen, M., et al. 1992. Biological phenotype of human immunodeficiency virus type 1 clones at different stages of infection: progression of disease is associated with a shift from monocytotropic to T-cell tropic virus populations. *J. Virol.* 66(3):1354–60

144. Selnes, O. A., Miller, E., McArthur, J., Gordon, B., Munoz, A., et al. 1990. HIV-1 infection: no evidence of cognitive decline during the asymptomatic stages. *Neurology* 40:204–8

145. Sharer, L. R., Michaels, J., Murphey-Corb, M., Hu, F.-S., Kuebler, D. J., et al. 1991. Serial pathogenesis study of SIV brain infection. *J. Med. Primatol.* 20:211–17

146. Sharer, L. R., Prineas, J. W. 1988. Human immunodeficiency virus in glial cells, continued. *J. Infect. Dis.* 157:204

146a. Sharpless, N. E., O'Brien, W. A., Verdin, E., Kufta, C. V., Chen, I. S. Y., et al. 1992. Human immunodeficiency virus type 1 tropism for brain microglial cells is determined by a region of the env glycoprotein that also controls macrophage tropism. *J. Virol.* 66(4):2588–93

147. Shaskan, E. G., Thompson, R. M., Price, R. W. 1992. Undetectable tumor necrosis factor-alpha in spinal fluid from HIV-1 infected patients. *Ann. Neurol.* In press

148. Shaw, G. M., Harper, M. E., Hahn, B. H., Epstein, L. G., Gajdusek, D. C., et al. 1985. HTLV-III infection in brains of children and adults with AIDS encephalopathy. *Science* 227:177–82

149. Shioda, T., Levy, J. A., Cheng-Mayer, C. 1991. Macrophage and T cell-line tropisms of HIV-1 are determined by specific regions of the envelope gp120 gene. *Nature* 349:167–69

150. Deleted in proof

151. Snider, W. D., Simpson, D. M., Neilsen, S., Gold, J. W., Metroka, C. E., Posner, J. B. 1983. Neurological complications of acquired immune de-

ficiency syndrome: analysis of 50 patients. *Ann. Neurol.* 14:403–18
152. Spector, D. A., Wade, E., Wright, D. A., Koval, V., Clark, C., et al. 1990. Human immunodeficiency virus pseudotypes with expanded cellular and species tropism. *J. Virol.* 64:2298–2308
153. Stavrou, D., Mehraein, P., Mellert, W., Bise, K., Schmidtke, K., et al. 1989. Evaluation of intracerebral lesions in patients with acquired immunodeficiency syndrome. Neuropathological findings and experimental data. *Neuropathol. Appl. Neurobiol.* 15:207–22
154. Steular, H., Storch-Hagenlocher, B., Wildemann, B. 1992. Distinct populations of human immunodeficiency virus type 1 in blood and cerebrospinal fluid. *AIDS Res. Hum. Retroviruses* 8(1):53–59
155. Stoler, M. H., Eskin, T. A., Benn, S., Angerer, R. C., Angerer, L. M. 1986. Human T-cell lymphotropic virus type III infection of the central nervous system. A preliminary in situ analysis. *J. Am. Med. Assoc.* 256:2360–64
156. Takeuchi, Y., Akutsu, M., Murayama, K., Shimizu, N., Hoshino, H. 1991. Host range mutant of human immunodeficiency virus type 1: modification of cell tropism by a single point mutation at the neutralization epitope in the *env* gene. *J. Virol.* 65(4):1710–18
157. Tindall, B., Cooper, D. A. 1991. Primary HIV infection: host responses and intervention strategies. *AIDS* 5:1–14
158. Tornatore, C., Nath, A., Amemiya, K., Major, E. O. 1991. Persistent human immunodeficiency virus type 1 infection in human fetal glial cells reactivated by T-cell factor(s) or by the cytokines tumor necrosis factor alpha and interleukin-1 beta. *J. Virol.* 65(11):6094–6100
159. Tross, S., Price, R. W., Navia, B. A., Thaler, H. T., Gold, J., et al. 1988. Neuropsychological characterization of the AIDS dementia complex: a preliminary report. *AIDS* 2(2):81–88
160. Vazeux, R., Brousse, N., Jarry, A., Henin, D., Marche, C., et al. 1987. AIDS subacute encephalitis: identification of HIV-infected cells. *Am. J. Pathol.* 126:403–10
161. Volsky, B., Sakai, K., Reddy, M. M., Volsky, D. J. 1992. A system for the high efficiency replication of HIV-1 in neural cells and its application to anti-viral evaluation. *Virology* 186(1): 303–8
162. Wain-Hobson, S. 1989. HIV genome variability in vivo. *AIDS* 3(1):S13–S18
163. Ward, J. M., O'Leary, T. J., Baskin, G. B., Benveniste, R., Harris, C. A., et al. 1987. Immunohistochemical localization of human and simian immunodeficiency viral antigens in fixed tissue sections. *Am. J. Pathol.* 127:199–205
164. Watkins, B. A., Dorn, H. H., Kelly, W. B., Armstrong, R. C., Potts, B. J., et al. 1990. Specific tropism of HIV-1 for microglial cells in primary human brain cultures. *Science* 249:549–53
165. Werner, T., Ferroni, S., Saermark, T., Brack-Werner, R., Banati, R. B., et al. 1991. HIV-1 *nef* protein exhibits structural and functional similarity to scorpion peptides interacting with K^+ channels. *AIDS* 5:1301–8
166. Westervelt, P., Gendelman, H. E., Ratner, L. 1991. Identification of a determinant within the human immunodeficiency virus 1 surface envelope glycoprotein critical for productive infection of primary monocytes. *Proc. Natl. Acad. Sci. USA* 88:3097–3101
167. Westervelt, P., Trowbridge, D. B., Epstein, L. G., Blumberg, B. M., Li, Y., et al. 1992. Macrophage tropism determinants of human immunodeficiency virus type 1 in vivo. *J. Virol.* 66(4):2577–82
168. Wiley, C. A., Masliah, E., Morey, M., Lemere, C., DeTeresa, R., et al. 1991. Neocortical damage during HIV infection. *Ann. Neurol.* 29:651–57
169. Wiley, C. A., Schrier, R. D., Nelson, J. A., Lampert, P. W., Oldstone, M. B. A. 1986. Cellular localization of human immunodeficiency virus infection within the brains of acquired immune deficiency patients. *Proc. Natl. Acad. Sci. USA* 83:7089–93
170. Wolfs, T. F. W., De Jong, J.-J., Van Den Berg, H., Tunagel, J. M., Krone, W. J. A., Goudsmit, J. 1990. Evolution of sequences encoding the principal neutralization epitope of HIV-1 is host-dependent, rapid, and continuous. *Proc. Natl. Acad. Sci. USA* 87(24):9938–42
171. World Health Organization. 1990. World Health Organization consultation on the neuropsychiatric aspects of HIV-1 infection. *AIDS* 4(9):935–6
172. Worley, J., Price, R. W. 1992. Management of neurologic complications of HIV-1 infection and AIDS. In *The Medical Management of AIDS*, ed. M. A. Sande, P. A. Volberding. Philadelphia: W. B. Saunders Co. 4th ed. In press
173. Yarchoan, R., Berg, G., Brouwers, P., Fischl, M. A., Spetzer, A. R., et al. 1987. Response of human immunodeficiency virus associated neurological disease to 3'-azido-3'-deoxythymidine. *Lancet* 1:132–35

Annu. Rev. Microbiol. 1992. 46:695–729

METABOLISM AND FUNCTIONS OF TRYPANOTHIONE IN THE KINETOPLASTIDA

Alan H. Fairlamb

Department of Medical Parasitology, London School of Hygiene and Tropical Medicine, Keppel Street, London WC1E 7HT, United Kingdom

Anthony Cerami

The Picower Institute for Medical Research, 350 Community Drive, Manhasset, New York 11034

KEY WORDS: glutathione, polyamines, flavoprotein disulfide oxidoreductases, chemotherapy, protozoan parasites

---

CONTENTS

0066-4227/92/1001-0695$02.00

Abstract

Trypanosomatids differ from all other organisms in their ability to conjugate the sulfur-containing tripeptide, glutathione, and the polyamine, spermidine, to form trypanothione [N^1,N^8-bis(glutathionyl)spermidine]. Together with the NADPH-dependent flavoprotein, trypanothione reductase, the dithiol form of trypanothione provides an intracellular reducing environment in these parasites, substituting for glutathione and glutathione reductase found in the mammalian host. Trypanothione and its related enzymes are involved in defense against damage by oxidants, certain heavy metals, and possibly xenobiotics. Trypanothione and its metabolic precursor, glutathionylspermidine, are also implicated in the modulation of spermidine levels during growth. Several existing trypanocidal drugs interact with the trypanothione system, suggesting that trypanothione metabolism may be a good target for the development of new drugs. The purification and properties of three key enzymes (glutathionylspermidine synthetase, trypanothione synthetase, and trypanothione reductase) are discussed, and the catalytic mechanism, substrate-specificity, and the three-dimensional structure of trypanothione reductase are compared to that of glutathione reductase.

INTRODUCTION

Without exception, all living organisms contain high levels of two classes of low-molecular weight compounds: aliphatic nitrogenous bases, collectively known as polyamines, and thiol-containing compounds, of which glutathione is the most ubiquitous. Both polyamines and glutathione have been implicated in a wide range of cellular functions too numerous to elaborate here; the reader is referred elsewhere for a detailed treatment of the history (34, 145) and metabolism and functions of these important compounds (48, 89, 143, 146, 156, 157, 182, 195, 212).

Trypanothione[1] was discovered as a result of studies on an apparently unusual glutathione reductase activity in the African trypanosome, *Trypanosoma brucei brucei* (63). Dialyzed cell-free extracts of this organism cannot reduce glutathione disulfide (GSSG) with NADPH unless a low molecular weight cofactor is added. Studies demonstrated that the cofactor contains an enzymatically reducible disulfide group and that the cofactor is present in extracts from representatives of the Kinetoplastida, but absent from a variety of other biological materials (63). Using the dialyzed crude enzyme as a biological assay, Fairlamb et al (61) subsequently purified the cofactor and

[1]In accordance with the general accepted nomenclature for glutathione, the term trypanothione ($T[SH]_2$) is defined as bis(L-γ-glutamyl-L-cysteinylglycyl)spermidine and its disulfide as trypanothione disulfide ($T[S]_2$). Total trypanothione is defined as the sum of these in trypanothione equivalents.

determined by amino acid analysis, mass-spectrometry, and chemical synthesis that the structure is N^1, N^8-bis(glutathionyl)spermidine. As it was uniquely found in parasitic protozoans of the suborder Trypanosomatina, the cofactor was christened trypanothione.

Interest in trypanothione metabolism is twofold. First, what possible advantage can this group of flagellated protozoa gain by the evolution of this compound? And, second, on a more practical note, how can this unique and potentially vulnerable area of metabolism be exploited for rational drug design (53, 57, 58, 100, 126, 187, 203)? The remainder of this review addresses both of these questions.

PROPERTIES OF TRYPANOTHIONE

Figure 1 shows the structure of trypanothione disulfide (T[S]2) and its two-electron reduced form, dihydrotrypanothione (T[SH]2). At physiological pH values, both compounds are zwitterions with a net charge of +1. In contrast, glutathione (GSH) and glutathione disulfide (GSSG) have a net charge of −2 (Figure 1). This may form the basis for the substrate-discrim-

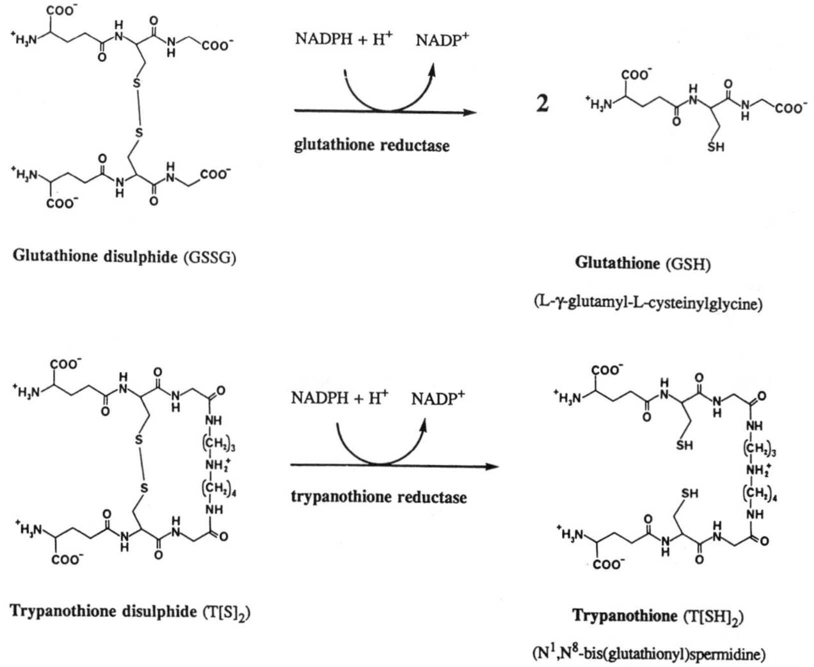

NADPH + H+ NADP+

glutathione reductase

Glutathione disulphide (GSSG)

Glutathione (GSH)

(L-γ-glutamyl-L-cysteinylglycine)

NADPH + H+ NADP+

trypanothione reductase

Trypanothione disulphide ($T[S]_2$)

Trypanothione ($T[SH]_2$)

(N^1,N^8-bis(glutathionyl)spermidine)

Figure 1 Reactions catalyzed by human glutathione reductase and parasite trypanothione reductase.

inatory properties of trypanothione reductase (TR) and human glutathione reductase (GR) discussed below. $T[S]_2$ has been synthesized by conventional liquid phase peptide synthesis using either a hexahydropyrimidine derivative of spermidine (61, 96) or N^1,N^8-bis(glycyl)N^4-boc spermidine (94, 96) as protected intermediates. $T[S]_2$ is not particularly easy to synthesize by either route and, although commercially available from Bachem, is expensive to buy. The facile synthesis of an alternative substrate for TR, *N*-benzyloxy-carbonyl-L-cysteinyl-glycyl-3-dimethylaminopropylamine disulfide was recently published (49, 50). Enzymatic synthesis from glutathione, spermidine, and ATP is also feasible using partially purified glutathionylspermidine synthetase and trypanothione synthetase (98), or $T[S]_2$ can be purified from *Crithidia fasciculata* (61). Both of these latter methods are laborious, yielding only small amounts of trypanothione.

Space-filling models of the 24-membered ring macrocycle, trypanothione disulfide, and the linear dithiol, dihydrotrypanothione, indicate that both forms might assume the conformation of a β-pleated sheet (61). However, NMR studies indicated that both peptides showed considerable flexibility in aqueous solution (94). Neither form has been crystallized in a form suitable for X-ray crystallography.

Knowledge of the redox potential of the $T[SH]_2/T[S]_2$ couple is important for an understanding of the role of these compounds in the cell. As measured by an equilibrium displacement technique, the standard redox potential of E_0' is -0.242 ± 0.002 V (64). This is slightly more electronegative than glutathione ($E_0' = -0.230$ V), but less than other physiological dithiols such as thioredoxin ($E_0' = -0.260$ V) or lipoic acid ($E_0' = -0.288$ V). Moreover, $T[SH]_2$ cannot be regarded as a biological equivalent of dithiothreitol since the latter has an E_0' of -0.330 V. The kinetics of nonenzymatic thiol-disulfide exchange reactions with $T[SH]_2$ have not been studied in any detail.

$T[SH]_2$, like other dithiols (190), forms stable complexes with trivalent organic arsenicals (67). However, the association constant for the 25-membered ring formed between trypanothione and the trypanocidal drug, melarsen oxide, ($K_a = 1.21 \times 10^7\ M^{-1}$) is much less than the five-membered dithioarsenite complexes of the therapeutic compounds, melarsoprol ($7.93 \times 10^{10}\ M^{-1}$) and trimelarsen ($4.50 \times 10^{10}\ M^{-1}$) (67). Interestingly, the six-membered complexes formed between melarsen oxide and dihydrolipoate ($4.51 \times 10^9\ M^{-1}$) and dihydrolipoamide ($5.47 \times 10^9\ M^{-1}$) show intermediate stability (68), which may be relevant to either the mechanism of uptake or the trypanocidal action of these compounds, in view of the likely association of these compounds with dihydrolipoamide dehydrogenase in the plasma membrane of the African trypanosome (38, 110).

One can readily determine cellular $T[SH]_2$ by derivatization with the thiol-specific fluorescent agent, monobromobimane (123), and by separation with HPLC (181). Table 1 lists thiol levels for a range of different trypanosomatids, including the biosynthetic precursors GSH and N^1-glutathionylspermidine (GSH-SPD). Of the total intracellular trypanothione, ≥98% is $T[SH]_2$ (65).

BIOSYNTHESIS OF TRYPANOTHIONE

Polyamines

In African trypanosomes, spermidine is synthesized from ornithine and methionine via the same route that operates in mammalian cells. All of the enzymes involved in this pathway have been identified: ornithine decarboxylase (ODC) (19, 162, 163), *S*-adenosylmethionine synthetase (209), *S*-adenosylmethionine decarboxylase (24, 30), and spermidine synthase (25). Metabolic labeling (11, 12) and inhibitor studies (8, 65, 80, 209) also support the pathway outlined in Figure 2. Current evidence also suggests that a similar pathway operates in *Leishmania* spp. (3, 13, 120) and in *C. fasciculata* (66, 105, 106). The initial step of putrescine formation in *Trypanosoma cruzi* is less clear, because growth of these organisms is not sensitive to inhibition by D,L-α-difluoromethylornithine (DFMO) (122). Moreover, growth of *T. cruzi* in macrophages is affected by high concentrations of D,L-α-difluoromethylarginine (DFMA) (122), an inhibitor of putrescine biosynthesis from arginine via agmatine (22), which suggests that *T. cruzi* could contain arginine decarboxylase rather than ODC. However, neither enzyme activity could be detected in vitro (122), nor could intact cells form radiolabeled putrescine from either [^{14}C]-ornithine or [^{14}C]-arginine (3). A related study found that DFMA inhibited the growth of *C. fasciculata,* but this was attributed to metabolism of DFMA to DFMO and subsequent inhibition of ODC (106) as

Table 1 Levels of glutathione, glutathionylspermidine, and dihydrotrypanothione in various trypanosomatids

Organism	Total GSH[a] nmol/10^8 cells	% GSH recovered as GSH	GSH-SPD	$T[SH]_2$	References
T. brucei bloodstream forms	5.6	24	5	71	65
T. brucei procyclic forms	11.4	27	4	69	16
C. fasciculata (exponential phase)	51.1	12	20	68	181
C. fasciculata (stationary phase)	38.7	10	68	22	181
L. braziliensis guyanensis promastigotes	13.7	14	4	82	120
L. donovani promastigotes	15.6	14	4	82	54
T. cruzi epimastigotes	9.5	19	2	79	Unpublished data[b]

[a] Total GSH = [GSH] + [GSH-SPD] + $2[T(SH)_2]$.
[b] T. Sharpington & A. H. Fairlamb, unpublished data.

found in tobacco and mammalian cells (186) rather than to inhibition of arginine decarboxylase.

Glutathione

All trypanosomatids contain 10–30% of their total GSH as the free peptide (see Table 1). Although the biosynthetic pathway has not been investigated in any detail, buthionine sulfoximine, a specific inhibitor of L-γ-glutamyl-L-cysteine synthetase in mammalian cells (82), inhibits glutathione biosynthesis in *T. brucei* (4), as well as in *T. cruzi* and *C. fasciculata* (K. Smith & A. H. Fairlamb, unpublished results), suggesting that a similar pathway is present in these parasites (see Figure 2).

Trypanothione

The subsequent steps from glutathione and spermidine are not found in mammalian cells. Metabolic labeling studies with intact cells and cell-free

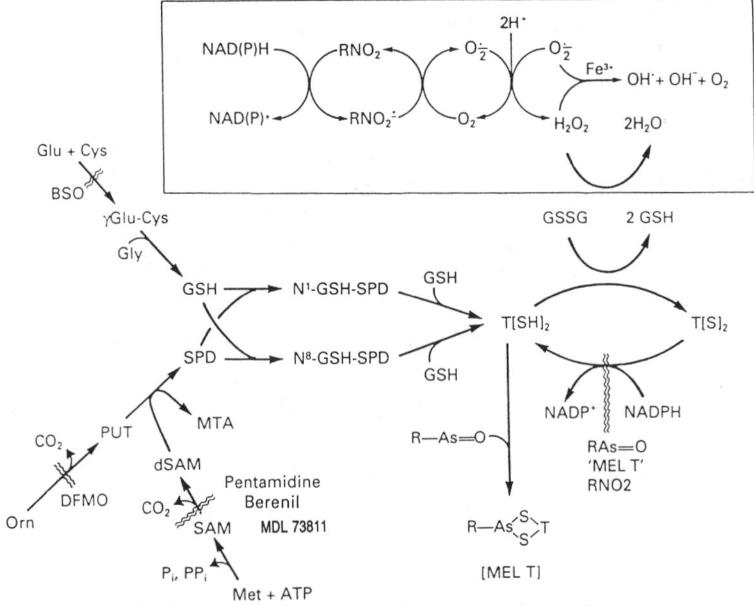

Figure 2 Metabolism and functions of trypanothione, showing possible sites of action of trypanocidal compounds. The boxed insert illustrates "futile redox-cycling" by nitro compounds (RNO_2) to form hydrogen peroxide (H_2O_2) and hydroxyl radical (OH•). Abbreviations: BSO, buthionine sulfoximine; DFMO, difluoromethylornithine; R-As=O, melarsen oxide; MelT, trypanothione:melarsen adduct; PUT, putrescine; SPD, spermidine; GSH-SPD, glutathionyl-spermidine; SAM, *S*-adenosylmethionine; dSAM, decarboxylated *S*-adenosylmethionine; MTA, methylthioadenosine. Modified from Ref. 55.

lysates have shown that glutathione and spermidine are conjugated in two ATP-dependent reactions forming both N^1- and N^8-isomers of glutathionylspermidine as intermediates (66). Until recently, a single enzyme was thought to be responsible for the conversion of glutathione and spermidine to trypanothione (98). However, this enzyme activity has now been resolved into two distinct proteins: a glutathionylspermidine synthetase of 90 kilodaltons (kDa) and a trypanothione synthetase of 82 kDa. (188). Both enzymes are highly specific for their respective substrates and, under conditions of low to moderate ionic strength, copurify as a functional complex. As discussed below, the differences in pH optima of glutathionylspermidine synthetase (pH 6.5) and trypanothione synthetase (pH 7.5) could possibly account for the relative differences in concentration of glutathionylspermidine and trypanothione between logarithmic and stationary phases of growth. Both these enzymes represent important and potentially specific targets for drug development.

Inhibition of Trypanothione Biosynthesis

Several experimental and therapeutic drugs inhibit various steps in the biosynthetic pathway (Figure 2). DFMO, the first new drug to be licensed for the treatment of human African trypanosomiasis in the past 40 years, inhibits ODC, the first enzyme in polyamine biosynthesis (10, 11, 142, 171, 172), resulting in a complete loss of putrescine and in a marked decrease in spermidine and trypanothione (16, 65). At the other end of the pathway, trivalent arsenical drugs sequester $T[SH]_2$ as a dithioarsane adduct, MelT (67), which in turn is a competitive inhibitor of TR (62, 67). Thus the marked synergistic effect between these two compounds could result from their combined action in lowering trypanothione levels (114–116, 142).

In the absence of arsenicals, the basis for the selective action of DFMO against the African trypanosome is the subject of much debate. Undoubtedly, the primary target of DFMO is ODC since coadministration of polyamines abrogates the curative effect of the drug in vivo (151). However, the trypanocidal effect is not due to selective inhibition of ODC, since DFMO is an equally potent inhibitor of both host and parasite enzymes in vitro and because both enzymes are kinetically similar in other respects (19, 147, 163). Likewise, selective uptake by the parasite is an unlikely explanation because DFMO enters both host and parasite by simple diffusion (20). A more subtle proposal invokes the slow turnover of trypanosome ODC relative to mammalian ODC (162, 205). This hypothesis assumes that pharmacokinetic fluctuations in the extracellular and intracellular levels of DFMO in vivo is coupled to the continuous rapid synthesis of new ODC molecules in the host, but not in the parasite, allowing host cells to transiently escape from the drug's inhibitory effect on polyamine biosynthesis. There is little or no evidence for or against this intriguing hypothesis at present.

The secondary effects of inhibition of ODC by DFMO in the trypanosome are clear, namely: a general depletion of polyamines and trypanothione, and a marked accumulation of *S*-adenosylmethionine and decarboxylated *S*-adenosylmethionine (8, 65, 76, 209). However, opinion is divided as to which of these secondary effects is most important for the trypanocidal effect. Depletion of polyamines is most likely responsible for the observed marked decrease in macromolecular synthesis (8), especially of variant surface glycoprotein (23), which is essential for evasion of the host immune response. One school of thought emphasizes that DFMO has a cytostatic effect and maintains that an antibody response is necessary for complete clearance of the parasites (26, 43). However, this hypothesis does not explain the transient decrease in parasitaemia observed in immunosuppressed mice (26, 43), nor does it account for the slow trypanocidal effect of DFMO observed in vitro (76, 164). Another school of thought maintains that irreversible differentiation to nondividing intermediate or short stumpy forms is an essential feature of DFMO's action (8, 42, 79, 80). Presumably these forms would have a limited life span in the host before being eliminated. This effect of DFMO on differentiation was recently disputed (15). Finally, a reduction in $T[SH]_2$ content could contribute by seriously compromising the antioxidant defenses of the trypanosome (65).

The other effect of DFMO and several inhibitors of *S*-adenosylmethionine decarboxylase, including the trypanocidal drugs berenil and pentamidine (24) and the experimental compound MDL 73811 (21, 30), is to promote the accumulation of millimolar levels of *S*-adenosylmethionine (see Figure 2). One study reported that the antitrypanosomal effects of these inhibitors correlate better with the observed increase in *S*-adenosylmethionine than with the decrease in polyamines, suggesting that aberrant methylation could be important for the trypanocidal effect of DFMO (30). Thus, although the primary target of DFMO is undoubtedly ODC, the precise mechanism by which trypanosomes are killed remains the subject of great debate and merits further investigation.

FUNCTIONS OF TRYPANOTHIONE

Maintenance of Thiol Redox

Without exception, all organisms contain high concentrations of at least one low-molecular-weight thiol for maintenance of an intracellular reducing environment. Although this is usually achieved by means of GSH, some organisms employ analogs of GSH such as homoglutathione [mung beans (33)], N^1-glutathionylspermidine [stationary phase *Escherichia coli* (196) and stationary phase *C. fasciculata* (181)], trypanothione (trypanosomatids listed in Table 1), and L-γ-glutamyl-L-cysteine [Halobacteria (152)]. Other prokaryotic organisms dispense with GSH entirely and use other thiol-containing compounds such as cysteine, coenzyme M, lipoic acid, hydrogen sulfide, or thiosulfate for this purpose (51).

The activity of many enzymes can be affected markedly by modification of protein thiols either by the formation of mixed disulfides (Equation 1) or by the formation of intramolecular disulfides (Equation 2).

$$\text{EnzSH} + \text{RSSR} \leftrightarrow \text{EnzSSR} + \text{RSH} \qquad 1.$$

$$\text{Enz(SH)}_2 + \text{RSSR} \leftrightarrow \text{Enz(S)}_2 + 2\text{RSH} \qquad 2.$$

Mammalian cells have two general mechanisms for the reduction of intracellular protein disulfides and other disulfides: GR, glutathione, and glutaredoxin (also known as thioltransferase) and thioredoxin reductase and thioredoxin (Figure 3). The reader is referred to recent reviews for further details of these systems (74, 101–103, 136). As thioredoxin reductase has not been detected

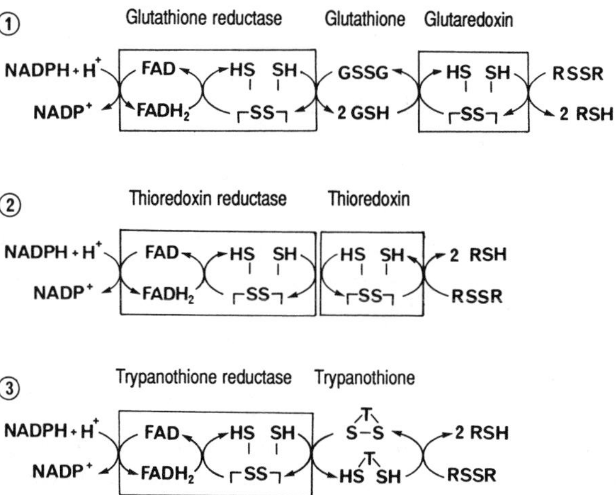

Figure 3 General purpose NADPH-dependent disulfide-reducing systems: (scheme 1) glutathione reductase, glutathione, and glutaredoxin; (scheme 2) thioredoxin reductase and thioredoxin; (scheme 3) trypanothione reductase and trypanothione. Boxes represent proteins. Modified from Ref. 136.

in *Crithidia* spp. (G. B. Henderson & A. H. Fairlamb, unpublished results), it may be that the TR system (Figure 3, scheme 3) substitutes for both the GR and thioredoxin reductase systems found in most other cells (Figure 3, schemes 1 and 2). If so, one should note that thioredoxin and glutaredoxin also serve as essential cofactors for the reduction of ribonucleotides by ribonucleotide reductases (74, 191). However, the enzyme and cofactors involved in ribonucleotide metabolism in trypanosomatids have not been studied.

Numerous studies have shown that chemically induced oxidation of intracellular thiols can deleteriously affect many complex cellular processes, including carbohydrate metabolism, calcium homeostasis, tubulin polymerization, mitosis, membrane integrity, and sensitivity to oxidant and chemical damage (see 109, 213 for further details). However, with the one notable exception of the oxidative burst in macrophages, large perturbations in intracellular thiol redox do not normally occur. Although numerous authors have suggested that the thiol-disulfide status of many enzymes plays a role in metabolic regulation (109), a recent critical review concludes that the hypothesis, though attractive, remains unproven (213).

Protein Folding

Disulfide bond formation, breakage, and rearrangement can be a useful tool for identifying intermediates in protein folding pathways (36, 37). Moreover, intramolecular disulfide bonds can contribute significantly to the overall stability of some proteins, especially those destined for excretion by the cell (72, 75). Clearly such reactions depend on the redox nature of the environment in which the folding events are taking place and are likely to be quantitatively more important in an extracellular oxidizing environment. Whether trypanothione has a physiological role in disulfide formation and thiol-disulfide exchange in proteins has not been investigated so far.

Formation of disulfide bonds is an early step in the posttranslational modification of secretory proteins, which may be catalyzed by protein-disulfide isomerase, a membrane-bound enzyme located on the luminal side of the endoplasmic reticulum (72). Protein-disulfide isomerase, which is also known as glutathione-insulin transhydrogenase (202), catalyzes thiol-disulfide exchange reactions and therefore promotes protein-disulfide formation, isomerization, or reduction, depending on the imposed redox potential (75). A developmentally regulated gene, expressed in bloodstream *T. brucei,* encodes a protein with homology to protein-disulfide isomerase and to mammalian phosphoinositide-specific phospholipase C (104). Its function remains to be determined.

Oxidant Defenses

Like other organisms living in an aerobic environment, trypanosomatids are exposed to reactive oxygen intermediates such as superoxide anion (O_2^-), hydrogen peroxide (H_2O_2) and hydroxyl radical (OH•). These compounds are generated internally from cofactors (reduced flavins, quinones, thiols, etc) and drug metabolism (reduction of nitroheterocyclics, quinones, bleomycins, etc; see Figure 2) and externally by the hosts' immune defense system. The interested reader is referred to books (84, 170, 183) and reviews (46, 55, 87, 132) for further information. The most reactive species, the hydroxyl radical, is thought to be generated from O_2^- and H_2O_2 by means of the transition-metal ion–catalyzed Haber-Weiss reaction (6):

$$O_2^- + Fe^{3+} \rightarrow O_2 + Fe^{2+} \quad 3.$$

$$H_2O_2 + Fe^{2+} \rightarrow OH\cdot + OH^- + Fe^{3+} \quad 4.$$

$$\text{Sum: } O_2^- + H_2O_2 \rightarrow OH\cdot + OH^- + O_2. \quad 5.$$

The hydroxyl radical (or a related species) can then cause lethal damage by reacting with cellular components such as DNA and membrane lipids. Once the radical forms, enzymatic removal is not possible. Instead, OH• is trapped nonenzymatically by low-molecular-weight radical scavengers such as vitamins A, C, and E (β-carotene, ascorbate, and tocopherol) and other metabolites (uric acid, bilirubin, etc) and by thiols, notably glutathione and trypanothione (46, 56, 59, 134, 138, 169). The disulfide end products of the latter reaction are then converted to their free thiols by the appropriate NADPH disulfide oxidoreductase. These mechanisms of trapping OH• and other free-radicals can be regarded as the last line of defense, because the general cellular strategy is to minimize OH• formation by keeping the levels of the precursors O_2^- and H_2O_2 as low as possible by enzymatic means.

The first line of defense involves removal of superoxide anion by superoxide dismutase (SOD) (see reviews 14, 73) (Equation 6):

$$2O_2^- + 2H^+ \rightarrow H_2O_2 + O_2. \quad 6.$$

Trypanosomatids contain a cytosolic Fe-containing SOD (133), which is distinct from the Cu,Zn- and Mn-containing SODs of mammalian cells, but similar to one found in certain plants, fungi, and bacteria. The enzyme has been purified to homogeneity from *C. fasciculata* and has been characterized (133). Another trypanosomatid, *T. cruzi,* also reportedly contains a cyanide-sensitive SOD, presumably of the Cu,Zn-type found in mammalian cells, but this enzyme has not been purified to homogeneity (28).

The second line of defense involves the removal of H_2O_2, which is achieved in mammalian cells by catalases and various peroxidases: glutathione peroxidase (cytosol and mitochondrion), catalase (peroxisomes), and other peroxidases such as cytochrome *c* peroxidase (mitochondrion) and ascorbate peroxidase. In general, trypanosomatids contain little or no catalase or glutathione peroxidase (see reviews 31, 46, 47, 132). The trace amounts of glutathione peroxidase and catalase detected in amastigotes of *Leishmania* spp. are probably contaminants of the host cell. Likewise, glutathione peroxidase has been detected in crude extracts of *T. cruzi* (27), but this report should be interpreted with caution because these authors also found GR activity, which can be ascribed to the presence of trypanothione and TR in their extracts. Glutathione peroxidase is clearly absent from *C. fasciculata* and *T. brucei*, in which a trypanothione-dependent peroxidase activity has been found in crude dialyzed cell extracts (92). The enzyme will use *t*-butyl hydroperoxide and cumene hydroperoxide as well as H_2O_2 as substrate but has not been purified. Attempts to demonstrate that the enzyme is a selenoprotein, like other H_2O_2-dependent glutathione peroxidases (71), were unsuccessful. Other indirect evidence has shown that representatives of the Trypanosomatidae can metabolize exogenously added H_2O_2 and that this process could be inhibited by sulfydryl reagents, suggesting that trypanothione is involved in this process (158–160). Thus, trypanothione and its ancillary enzymes TR and trypanothione peroxidase (where present) constitute an antioxidant system analogous to that of glutathione, GR, and glutathione peroxidase of mammalian cells. This observation may be important in relation to the mode of action of redox-cycling drugs, such as the nitrofuran nifurtimox, which are thought to act by forming free radicals that swamp the parasite's antioxidant defenses (see boxed insert in Figure 2).

The glutathione antioxidant system may have evolved primarily to protect the mitochondrion from oxidant damage (52). Much evidence supports this hypothesis: (*a*) *Entamoeba histolytica,* which lacks mitochondria, is completely devoid of glutathione (52). [*Trichomonas vaginalis*, another parasitic protozoan that lacks mitochondria, also lacks glutathione (A. H. Fairlamb, unpublished observation).] (*b*) A significant proportion of the glutathione, GR, and glutathione peroxidase in mammalian cells is localized in mitochondria (83, 141, 197). (*c*) Mitochondria possess a high affinity transport system for the uptake of cytosolically synthesized glutathione (83, 129, 139). (*d*) Glutathione deficiency induced by administration of buthionine sulfoximine causes mitochondrial damage in the brains of neonatal rats (111). However, if the glutathione system were essential for mitochondrial function in all eukaryotes, then one would expect that the trypanothione system would show a similar distribution pattern. This was not found to be the case—TR is exclusively localized in the cytosolic compartment of the African trypanosome in two different stages of its life cycle (189).

GSH has also been proposed to play an important role in protection of DNA from radical-mediated damage induced by ionizing radiation (167). Recent studies suggest that $T[SH]_2$ is a much better DNA radioprotectant than GSH or spermidine, presumably because the polyamine moiety of $T[SH]_2$ allows a greater local concentration of SH groups near the DNA (7).

Defense Against Xenobiotics

Higher organisms contain a remarkably diverse family of glutathione *S*-transferases that catalyze the nucleophilic attack of glutathione on a variety of hydrophobic electrophilic xenobiotics to form *S*-conjugated products that are often less toxic and more easily excreted than the parent compounds. The nomenclature and general properties of these enzymes has been reviewed elsewhere (35, 137, 165). Some members of this multigene family of enzymes may also participate in the repair of oxidative damage to DNA and membrane lipids (121, 149). Numerous studies have also implicated glutathione *S*-transferases in resistance to drugs, toxic heavy metals, herbicides, and insecticides (88, 99, 130, 131, 149, 206).

Surprisingly little is known about glutathione *S*-transferases in trypanosomatids. One enzyme has been purified from *T. cruzi;* it is a heterodimer comprising subunits of 20 and 17 kDa and has low activity with methyl iodide or 1-chloro-2,4-dinitrobenzene as substrate (211). Its physiological substrate and function are unknown, and its isolation predates the discovery of trypanothione, so the specificity for this dithiol has not been investigated. We recently identified and partially purified an apparent trypanothione *S*-transferase activity in extracts of *C. fasciculata* (E. Oben Etah, K. Smith & A. H. Fairlamb, unpublished data). The enzyme is highly specific for trypanothione and shows negligible activity with glutathione. Its relationship to the enzyme isolated from *T. cruzi* is not known at present.

Heavy Metals

Protein thiol groups are the targets for heavy metal poisons such as mercury, cadmium, lead, silver, gold, and arsenic. Some metals, such as zinc and copper, are essential trace nutrients that participate at low concentrations in a variety of enzyme reactions but are toxic at higher concentrations. Sequestration of these metals frequently involves sulfydryl groups, for example: glutathione in mammalian cells (185) or its poly(γ-glutamylcysteinyl)glycine analogs, the phytochelatins in plants and fungi (166), and low-molecular-weight, cysteine-rich proteins, the metallothioneins (85). Thiol groups are also essential features of many of the protein components that confer resistance to inorganic mercuric salts in bacteria, including the scavenger protein (MerP), the transport protein (MerT), the regulatory protein (MerR), and the detoxifying enzyme mercuric ion reductase (MerA), which catalyzes the reduction of Hg^+ to nontoxic elemental mercury, Hg^0 (90, 173, 204).

Because trypanothione forms a stable complex with trivalent organic arsenical drugs (67), we have investigated whether trypanothione, like glutathione (185), could play a wider role in defense against heavy metals. Exposure of *C. fasciculata* to sublethal concentrations of Cd^{2+} or Hg^{+}, but not other heavy-metal ions, causes a marked increase in trypanothione (10-fold), a moderate increase in glutathionylspermidine (4-fold), and a slight increase in glutathione levels (2.2-fold) (K. Smith & A. H. Fairlamb, unpublished results). Following inhibition of glutathione and trypanothione biosynthesis by buthionine sulfoximine, the cells become hypersensitive to the toxic effects of cadmium. These results indicate that trypanothione is of paramount importance in defense against cadmium toxicity in *C. fasciculata*. There are no reports of metallothionein-like proteins in the Trypanosomatidae.

Coenzyme Functions

Glutathione can also serve as a coenzyme in isomerization reactions, notably in the glyoxalase system where methylglyoxal, formed from dihydroxyacetone phosphate, is converted to D-lactate (200). Of the trypanosomatids, four species of *Leishmania* and *Trypanosoma lewisi* can produce D-lactate from glucose (39, 41). Metabolism of methylglyoxal by these organisms requires glutathione (40, 78). Whether glutathionylspermidine or trypanothione can replace glutathione as cofactor has not been reported. The possible role of trypanothione in other glutathione-dependent reactions such as the metabolism of formaldehyde (199) or in the *cis*-to-*trans* isomerization of maleylacetoacetate (178) deserves further investigation.

Regulation of Polyamine Levels

Elevated levels of polyamines are essential for growth and proliferation in mammalian cells (see reviews 113, 156, 157, 195). Consequently, the biosynthesis and back-conversion of polyamines is highly regulated by alterations in the activity of three enzymes, namely ODC, *S*-adenosylmethionine decarboxylase, and polyamine N^1-acetyltransferase. Control is mediated primarily by alterations in the amount of enzyme protein by modulation of the rate of enzyme synthesis and degradation. These three enzymes all have half-lives of less than one hour and are rapidly degraded by a mechanism that is not clearly understood at present. Both ODC and *S*-adenosylmethionine decarboxylase contain sequences rich in the amino acids proline, glutamate, serine, threonine, and aspartate (PEST sequences) that may be important in the degradation of many proteins with extremely short half lives (168). In contrast, ODC and *S*-adenosylmethionine decarboxylase from trypanosomatids do not turn over rapidly (9, 162, 205), even when expressed in mammalian cell lines (77). Indeed, *T. brucei* ODC lacks the PEST sequences found in the mammalian enzyme, despite high sequence homology in other regions (162). Nevertheless, levels of polyamines are elevated during the exponential phase of growth in trypanosomatids (3, 9, 13, 86, 120, 181). In *C. fasciculata*, free spermidine can be sequestered as N^1-glutathionylspermidine at the expense of trypanothione in stationary phase of growth, when the medium becomes acidic (181). When transferred to fresh growth medium, N^1-glutathionylspermidine is rapidly converted back into free spermidine and trypanothione, without the need for the de novo synthesis of glutathione or spermidine. This effect might be mediated by changes in intracellular pH, affecting the relative catalytic activity of glutathionylspermidine synthetase and trypanothione synthetase (66). *E. coli* employs a similar mechanism for sequestration of spermidine as glutathionylspermidine (196), although this mechanism does not involve trypanothione (A. Tranguch & A. H. Fairlamb, unpublished data). Whether the accumulation of glutathionylspermidine triggers growth arrest or is merely a secondary phenomenon is not known for either of these organisms.

No studies report that trypanosomatids can catabolize polyamines via the interconversion or terminal-oxidative pathways found in mammalian cells (177).

TRYPANOTHIONE REDUCTASE

As discussed above, many of the antioxidant functions of trypanothione result in the formation of $T[S]_2$. Thus, TR is central to the regeneration of $T[SH]_2$ and is a principal target for rational drug design. A thorough understanding of the structural and mechanistic properties of the target enzyme and how it differs from the host GR is an essential part of this approach.

Isolation and Characterization

TR has been identified in all species of trypanosomatids examined so far (63). The enzyme was first isolated in pure form from *C. fasciculata* (179) and subsequently from *T. cruzi* (117). TR has also been partially purified from *T. brucei* (62, 97) and isolated as a recombinant protein from *Trypanosoma congolense* (192) and *T. cruzi* (194). TRs from all of these sources are very similar to GRs in that they are homodimeric proteins of approximately 50 kDa per subunit, contain FAD as the coenzyme, are specific for NADPH as electron donor, and contain a redox-active cysteine disulfide in the active site (see reviews 60, 76, 91, 125, 126, 148, 174, 187, 203) (Table 2).

The gene encoding TR has been cloned and sequenced from *T. congolense* (180), *T. cruzi* (194), *C. fasciculata* (1, 70), *T. brucei* (1), and *Leishmania donovani* (198). However, despite an amino acid identity of about 40% between TRs and human GR, these enzyme classes possess a high degree of specificity for their respective disulfide substrates (see Table 2) (93, 117, 179). GR from *E. coli* is an interesting exception in that it displays marked enzymatic activity with $T[S]_2$ as well as GSSG (95). This observation could

Table 2 Properties of trypanothione reductases and glutathione reductases

Source[a]	Trypanothione reductase			Glutathione reductase	
	C. fasciculata	*T. cruzi*	*T. congolense*	*E. coli*	Human
Flavin	FAD	FAD	FAD	FAD	FAD
Pyridine dinucleotide	NADPH	NADPH	NADPH	NADPH	NADPH
Subunit M_r[b]	54000[d]	53900[e]	53400[f]	48700[g]	52500
Amino acids/subunit	491[d]	492[e]	492[f]	450[g]	478
Oligomeric structure	Dimer	Dimer	Dimer	Dimer	Dimer
E_{ox}, λ_{max}(nm)	464	461	464	462[h]	460
ϵ_0 at λ_{max} (mM^{-1} cm^{-1})	11.3	11.2	10.6	—	11.3
Charge transfer in EH_2	Yes	Yes	Yes	Yes	Yes
ϵ_0 at λ_{530} (mM^{-1} cm^{-1})	3.63	4.9	3.7	—	4.5
K_m (μM)[c]					
Trypanothione	53 (51,[i] 58[j])	45 (55,[j] 50[e])	31 (18[k])	2000[l]	
Glutathione				66 (61,[l] 70[m])	65
k_{cat} (min^{-1})					
Trypanothione	31000 (28600[j])	14200	9600	6100[l]	
Glutathione				44000[l]	12600
k_{cat}/K_m (M^{-1} sec^{-1})	9.8×10^6	5.3×10^6	5.2×10^6	6.2×10^6	3.1×10^6
K_M^{app}(NADPH)	7	5	5	16 (25[m])	9

[a] Unless stated otherwise, the data are obtained from the following references: *C. fasciculata* (179); *T. cruzi* (124); recombinant *T. congolense* (192); *E. coli* (140). References to the human erythrocyte GR can be found by consulting Ref. 174.

[b] Calculated from the derived amino acid sequence, not including FAD.

[c] Absolute values vary due to different experimental conditions. In particular, the catalytic parameters for TR from *T. cruzi* are particularly sensitive to changes in pH and ionic strength (117).

[d–m] These footnotes denote supplementary data from references: d (70); e (194); f (180); g (81); h (45); i (93); j (49); k (193); l (95); m (176).

be related to the fact that *E. coli* can synthesize glutathionylspermidine in stationary phase of growth as mentioned above.

Reaction Mechanism

The reaction mechanism of GR has been extensively studied using kinetic analyses including chemical modification and UV spectroscopy (5, 18, 32, 135, 201, 207, 208, 210), by site-directed mutagenesis (17, 44, 175), and by X-ray crystallography (118, 119, 154, 155). All available evidence supports a ping-pong mechanism in which electrons are transferred from NADPH to GSSG via the isoalloxazine ring of FAD and the redox active cysteine disulfide bridge (see Figure 4). The first half of the catalytic cycle involves reduction of the enzyme to the EH_2 form, characterized by the charge-transfer complex between the thiolate anion of Cys63 in human GR and the isoalloxazine ring of FAD. The second half of the catalytic cycle involves reoxidation of the EH_2 form by GSSG via a transient covalent disulfide intermediate between Cys58 of GR and Cys-I of GSSG [GS-I designates the half of GSSG that is covalently bound during catalysis (118)]. The second stage also involves His467′ acting as a proton donor/acceptor in the reaction. Interestingly, mutagenesis of the equivalent histidine residue in *E. coli* GR to glutamine or alanine does not completely abolish activity, and the mutant enzymes are thought to recruit an alternative proton donor, thereby allowing catalysis to proceed at about 1% or less of that of the native enzyme (44).

The reaction mechanism of TR has not been studied in such great detail. However, alignment of all TRs sequenced so far suggests conservation of the redox active cysteines, the histidine base together with the glutamate that holds the histidine imidazolium side chain in the correct orientation, and the tyrosine that stacks between the two tripeptide moieties of GSSG (Table 3). The importance of other conserved residues in relation to structure and function have been reviewed elsewhere (1, 70). Spectroscopic studies show the formation of a charge-transfer complex on reduction of the enzyme (124, 179), suggesting that the reaction is mediated via a two-electron reduced intermediate. Alkylation of the NADPH-reduced enzyme with iodoacetamide indicates an essential catalytic role for the redox-active cysteines (124, 179). Thus, currently available evidence suggests that the reaction mechanisms of TR and GR are similar, if not identical.

Substrate Specificity

Prior to any primary sequence information, the disulfide-binding site of the enzyme was extensively mapped by means of chemically synthesized structural analogs of $T[S]_2$ (93). Modifications of the spermidine portion of the molecule established that: (*a*) a cyclic macrocyclic structure was not essential

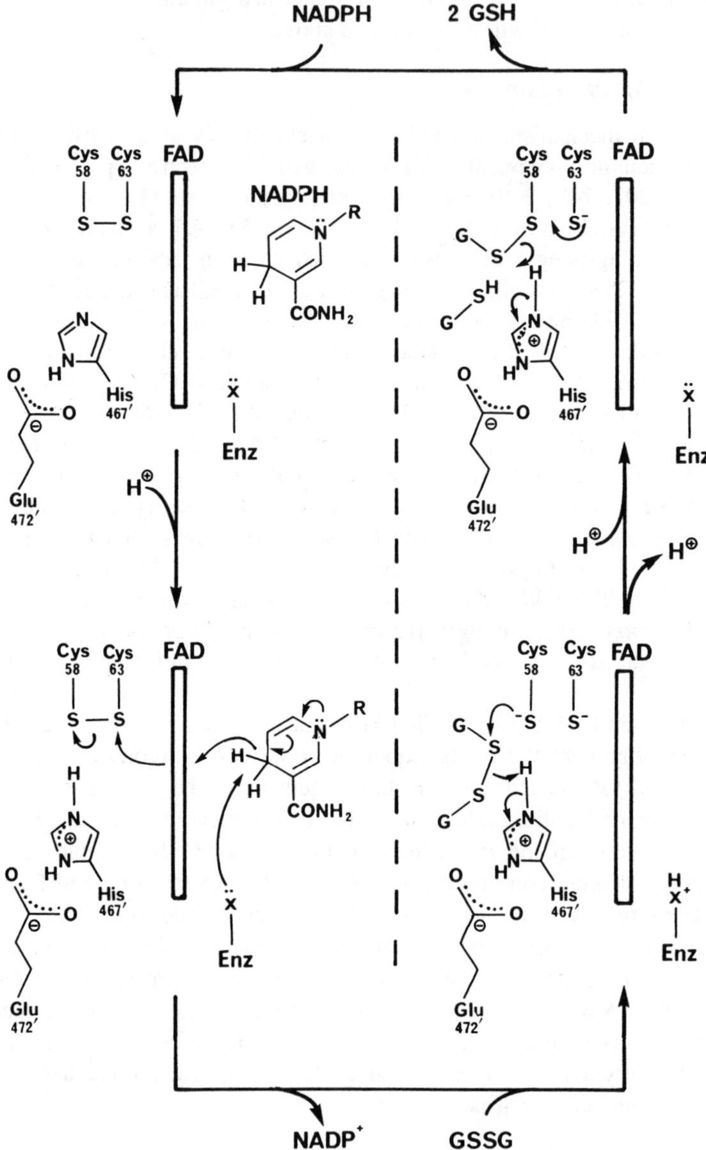

Figure 4 Postulated reaction mechanism for human glutathione reductase, based on Refs. 155, 207. The base X has been tentatively identified as Lys66 (155).

Table 3 Catalytic residues in the disulfide-binding site of human glutathione reductase and their homologues in trypanothione reductase

Human GR[a,b]	Function in human glutathione reductase[b]	Trypanothione reductase				
		T. congolense[c]	*T. cruzi*[d]	*C. fasciculata*[e]	*T. brucei*[f]	*L. donovani*[g]
Cys58	Redox-active disulfide	Cys52	Cys53	Cys52	Cys52	Cys52
Cys63	Redox-active disulfide	Cys57	Cys58	Cys57	Cys57	Cys57
Tyr114	Stacks between GS moieties in GSSG	Tyr110	Tyr111	Tyr110	Tyr110	Tyr110
His467'	Active-site base	His461'	His462'	His461'	His461'	His461'
Glu472'	Hydrogen bonds to active-site histidine	Glu466'	Glu467'	Glu466'	Glu466'	Glu466'

[a] Residues marked with a prime are from the opposite subunit in the dimer.
[b] Data taken from Ref. 118.
[c] Data taken from Ref. 180.
[d] Data taken from Ref. 194
[e] Data taken from Ref. 70.
[f] Data taken from Ref. 1.
[g] Data taken from Ref. 198.

for catalysis because N^1-glutathionylspermidine disulfide was an efficient substrate, (*b*) simply eliminating the negative charge on the glycine carboxylates by amidation was insufficient for substrate recognition, and (*c*) a hydrophobic component, preferably with a positively charged secondary or tertiary amino function, was important (93). These observations accord with the discovery of a hydrophobic and negatively charged region in the active site of TR, described below (127).

More recently, the γ-glutamyl binding site was examined using eight chemically synthesized analogs of L-γ-glutamyl-L-cysteinylglycyl-dimethylaminopropylamide (50; A. F. El-Waer, K. Smith, J. H. McKie, T. Benson, A. H. Fairlamb & K. T. Douglas, in preparation). These studies showed that the α-amino group is more important than the α-carboxyl group for binding and catalysis by TR, in agreement with similar studies on human GR (112). Unexpectedly, two intermediates in the synthesis of these analogs (the γ-glutamyl residue is replaced by either a benzyloxycarbonyl or *t*-butyloxycarbonyl moiety) were found to be excellent substrates, one of which is now used as a cheap and easy-to-synthesize alternative to trypanothione for monitoring enzyme purification (49). None of the above analogs had any activity with human GR.

All of the physiological substrates for TR and GR have a common γ-glutamylcysteinyl disulfide moiety (Figure 1) and therefore might be expected to have conserved the amino acid side chains involved in recognition of this portion of the substrate. Human GR contains 19 residues that interact with GSSG (118). Table 4 lists these residues and compares them with the equivalent residues in other members of this group as determined by sequence alignments. All the residues that interact with Cys-I, Cys-II, and Glu-II are highly conserved; only minor homologous substitutions are permitted. The most striking difference occurs at the Glu-I binding-site, where Arg347[2] in GR is replaced by an alanine in all TRs (or a methionine in GR from *Pseudomonas aeruginosa*). Mutating this alanine to an arginine in *T. congolense* TR (A434R) does not markedly affect catalytic efficiency (k_{cat}/K_m) for either $T[S]_2$ or GSSG, indicating that it is not important for substrate discrimination (193). These results therefore suggest that the orientation of the γ-glutamylcysteinyl disulfide moiety is likely to be similar in GRs and TRs.

Because $T[S]_2$ and GSSG differ markedly at the glycyl carboxylate region (see Figure 1), one should not be surprised that four out of five amino acid residues that interact with this region are different in GR and TRs (see Table 4; note that the conserved Tyr114 also interacts with both Cys-I and Cys-II).

[2]Some confusion has arisen in the literature that Arg347 interacts with the Gly-I carboxylate of GSSG rather than Glu-I (127, 180).

Attempts to engineer a reversal of substrate specificity between GR and TR, and vice versa, have been partially successful, particularly when Ala34 and Arg37, the residues in human GR that interact with the Gly-I carboxylate of GSSG, have been mutated in combination to Glu and Trp, respectively (see Tables 4 and 5). The equivalent mutations in *E. coli* GR (together with R22N) only cause a modest (fivefold) increase in catalytic efficiency toward $T[S]_2$, probably because the native enzyme already possesses significant activity with this substrate. However, the Ala → Glu and Asn or Arg → Trp mutations do have a pronounced effect on the ability of the mutant *E. coli* enzyme to process GSSG. The reason that *E. coli* GR can utilize $T[S]_2$ as substrate with such relative efficiency has not been investigated. Also, all TRs and *E. coli* GR have a serine residue conserved at the Gly-II site, which replaces the more bulky isoleucine in other GRs (Table 4). In addition, all TRs and *E. coli* GR have a hydrophobic Met or Val instead of the hydrophilic Asn in human GR at the Gly-II site. When this Met is replaced by Asn as part of a quadruple mutant, specificity for $T[S]_2$ drops markedly compared with the triple mutant (203). A mutational analysis of human GR at the Gly-II site in combination with the Gly-I mutations noted above would therefore be of interest. Although the combinational mutations produced so far have resulted in quite spectacular alterations in catalytic activity with GSSG or $T[S]_2$, these genetically engineered proteins are nonetheless considerably less efficient than the native proteins. Presumably this is the result of more subtle changes in the relative disposition of the amino acid segments that comprise the active-site region, as discussed below.

Structural Studies

Two groups have published models of *T. congolense* TR (150, 187) based on the crystal structure of human erythrocyte GR (119). Both groups observed a high degree of homology in the disulfide-binding active-site region, particularly of the redox-active cysteines (Cys52, Cys57) and the active-site base, His461′. Henderson's group suggested that Asp112 and Asp116 could represent a recognition site for the spermidine moiety of $T[S]_2$ (150). However, subsequent mutagenesis (95) or crystallographic analysis (127) did not support this supposition. In contrast, Smith et al correctly predicted that Trp21, Met113, and Leu17 could form a cluster of hydrophobic residues in the active site and proposed that these amino acids could interact with the hydrophobic regions of the spermidine part of $T[S]_2$ (187). The mutagenesis experiments described above indicate that Trp21 is important for this purpose, but Met113 much less so.

TR has been crystallized from *T. cruzi* (124, 194) and *C. fasciculata* (108, 128), but only the latter enzyme has yielded suitable X-ray diffraction data. In one case, *C. fasciculata* TR was further resolved by anion-exchange

Table 4 Residues in human GR interacting with GSSG and their homologues in other glutathione and trypanothione reductases[a]

	Glutathione reductase			Trypanothione reductase				
GSSG site	Human	*E. coli*	*P. aeruginosa*	*C. fasciculata*	*T. congolense*	*T. cruzi*	*T. brucei*	*L. donovani*
Glu-I	T_{339}	*	*	*	*	*	*	*
	I_{343}	V	L	*	*	*	*	*
	R_{347}	*	**M**	**A**	**A**	**A**	**A**	**A**
	$H_{467'}$	*	*	*	*	*	*	*
	$T_{476'}$	*	*	S	S	S	S	S
Glu-II	M_{406}	*	*	L	L	L	L	L
	$H_{467'}$	*	*	*	*	*	*	*
	$P_{468'}$	*	*	*	*	*	*	*
	$T_{469'}$	*	*	*	*	*	*	*
	$E_{472'}$	*	*	*	*	*	*	*
	$E_{473'}$	*	*	*	*	*	*	*
Cys-I	S_{30}	*	*	*	*	*	*	*
	V_{59}	*	*	*	*	*	*	*
	V_{64}	*	*	*	*	*	*	*
	Y_{114}	*	*	*	*	*	*	*
	$H_{467'}$	*	*	*	*	*	*	*
Cys-II	L_{110}	I	*	I	I	I	I	I
	Y_{114}	*	*	*	*	*	*	*
	$H_{467'}$	*	*	*	*	*	*	*
Gly-I	A_{34}	*	**R**	**E**	**E**	**E**	**E**	**E**
	R_{37}	**N**	*	**W**	**W**	**W**	**W**	**W**
	Y_{114}	*	*	*	*	*	*	*
Gly-II	I_{113}	**S**	*	**S**	**S**	**S**	**S**	**S**
	Y_{114}	*	*	*	*	*	*	*
	N_{117}	**V**	**L**	**M**	**M**	**M**	**M**	**M**

[a] Residues that are identical to those in human GR are marked *. Nonconservative substitutions are marked in bold. Glu-I, Cys-I, and Gly-I refer to the residues of GSSG that are covalently linked to the enzyme during catalysis. Residues marked with a prime are from the opposite subunit in the dimer. References to the sequence data are the same as those given in Table 3. The sequence for *Pseudomonas aeruginosa* is from Perry et al (161).

Table 5 Kinetic parameters for wild-type and mutant trypanothione reductase and glutathione reductases[a]

Enzyme	GSSG			$T[S]_2$			k_{cat}/K_m relative to TR (%)
	K_m (μM)	k_{cat} (min^{-1})	k_{cat}/K_m (min^{-1} M^{-1})	K_m (μM)	k_{cat} (min^{-1})	K_{cat}/K_m (min^{-1} M^{-1})	
T. congolense WT TR	—	—	0.84	18	9,100	5.1×10^8	100
E18A/W21R	>100,000	—	4.4×10^3	5,200	3,000	5.7×10^5	0.11
E. coli WT GR	61	44,000	7.2×10^8	2,000	6,100	3.1×10^6	0.60
A18E/N21W/R22N	>20,000	—	—	660	9,900	1.5×10^7	3.0
H. sapiens WT GR	66	8,200	1.2×10^8	—	—	1.4×10^4	0.0028
A34E/R37W	—	—	9.0×10^3	120	730	6.1×10^6	1.2

[a] Kinetic data for wild-type (WT) and corresponding mutant enzymes for *T. congolense* (193), *E. coli* (95), and human (29).

chromatography into three separate peaks, all of which yielded well-ordered crystals belonging to space group $P2_1$ (127), which could be caused by a glutamate substituting for a glutamine in the C-terminal extension of one of the gene sequences (70). Another crystal form, $P4_1$, has been obtained using a different cloned line of *C. fasciculata* (108). Curiously, each type of *C. fasciculata* TR can only be crystallized under one set of conditions (W. N. Hunter, personal communication). The X-ray structure of the $P2_1$ form has been determined at 2.4-Å resolution (127) and the $P4_1$ form at 2.8-Å resolution (107). Both structures closely resemble that of GR; each monomer consists of four domains: a N-terminal FAD-binding domain, a NADPH-binding domain, a central domain that also provides part of the FAD-binding site, and the interface domain. Each disulfide-binding site is a crevice formed by contributions from the FAD and central domains on one side and from the interface domain from the opposing subunit on the other side. The redox-active cysteines are located at the base of the active site, with the isoalloxazine ring of the FAD lying below. The overall structure of TR superimposes well on GR, except that the binding site for $T[S]_2$ is more open in TR due to rotations of two helical domains forming part of the active site (127). Presumably this allows TR to accommodate the bulkier substrate, $T[S]_2$, and may account for the failure of the protein engineering studies described above to achieve complete reversal of substrate specificity.

A comparison of the disulfide-binding sites of human GR and *C. fasciculata* TR indicates that the bottom and one side of the active site containing the catalytic components (Cys, Cys, His, and Glu) is remarkably similar. However, the region where the glycine carboxylates of GSSG interact with GR are markedly different in TR (127). Overall, the highly positively charged and hydrophilic region of GR is replaced by a hydrophobic and negatively charged pocket in TR, which is formed from the terminal methyl group of Met112 and the side chains of Leu16, Tyr109, Trp20, and Glu17. Glu17 could either hydrogen bond with one of the amide linkages between the spermidine and glycine carboxylate (127) or could interact with the positively charged secondary amine of spermidine (203). Although $T[S]_2$ has been modeled in the active site of the crystal structure of TR (127), one should remember that $T[S]_2$ is asymmetric about the N^4-secondary amine and that therefore two orientations are possible. The ultimate resolution to these questions awaits the high resolution X-ray structure of the enzyme-ligand complex.

Inhibitors

The trivalent arsenical drugs, such as melarsen oxide, are potent inhibitors of the NADPH-reduced TR in vitro, presumably by reaction with the cysteine residues in the active site of the enzyme (62; M. Cunningham, K. Smith & A. H. Fairlamb, unpublished data). However, when intact *T. brucei* are exposed to melarsen oxide in vitro, the arsenical reacts primarily with intracellular $T[SH]_2$ to form a stable dithioarsane adduct, MelT (see Figure 2) (67). This compound is a competitive inhibitor of TR (K_i = 9 μM), but whether this effect is sufficient to account for the trypanocidal activity of the arsenicals remains an open question (62, 67). However, dihydrolipoamide dehydrogenase from *T. brucei* is much less sensitive to inactivation by melarsen oxide than TR and is not inhibited by the equivalent lipoamide:melarsen adduct (68). Other trypanocidal compounds—pentamidine, berenil, and suramin—show less than 10% inhibition of TR when tested at 10^{-4} M (A. H. Fairlamb, unpublished data).

A series of suitably substituted naphthoquinones and nitrofuran derivatives inhibit reduction of $T[S]_2$ by TR and also undergo one-electron redox-cycling in the presence of molecular oxygen (97) (see Figure 2). Structure-activity relationships showed that the most effective inhibitors contained basic functional groups in the side-chain residues, in accord with what is now known about the recognition determinants for the disulfide-binding site of TR (see above). Because these compounds promote the formation of reactive oxygen species, they effectively subvert the antioxidant function of TR and have been termed "subversive substrates" (97) or "turn-coat inhibitors" (117). Now that the crystal structure of TR is known, it should be possible to further enhance selectivity of these subversive substrates toward TR and against GR (117).

The cytostatic agent 1,3-*bis*(2-chloroethyl)-1-nitrosourea (BCNU), which covalently inhibits glutathione reductase and dihydrolipoamide dehydrogenase (2), also inhibits TR (117). As above, it should be possible to produce a carbamoylating agent with specificity towards TR by judicious modification of the structure. Other potential approaches to selective-inhibitor design include: interface peptide mimetics that would interfere with subunit-subunit interactions promoting the formation of inactive monomers, and chimaeric active site inhibitors containing two ligands covalently linked by an appropriate spacer. The former approach has already been successfully developed for ribonucleotide reductase of Herpes simplex (144) and the latter with adenylate kinase from erythrocytes (69).

AN EVOLUTIONARY PERSPECTIVE

The Kinetoplastida have several features uniquely different from other organisms. The most striking of these are: (*a*) the unusual organization of their mitochondrial DNA into a single large concatenated mass of maxi- and minicircles, known as the kinetoplast (184); (*b*) the localization of many of the enzymes of glycolysis (and other enzymes) into a microbody-like organelle, the glycosome (153); and (*c*) the glutathionylspermidine conjugate, trypanothione. Why these organisms should possess these features, whereas their nearest taxonomic neighbor *Euglena* apparently does not, is anyone's guess. However, the evolution of trypanothione is not unprecedented because *E. coli* can synthesize N^1-glutathionylspermidine, and *E. coli* GR appears to have evolved to be able to reduce glutathionylspermidine disulfides as well as GSSG. At present, one could speculate that trypanothione has evolved as a convenient means of regulating spermidine levels during growth, as a more effective radioprotector of DNA than GSH, as a better sequestering agent for toxic heavy metals than GSH, and possibly as a convenient replacement for the thioredoxin and glutaredoxin system of other organisms. More research is required before these ideas can be regarded as more than speculation.

CONCLUSION AND OUTLOOK

Less than ten years have passed since the discovery of trypanothione, yet considerable progress has been made in delineating the metabolism and functions of this novel polyamine-containing peptide. Its potential importance as a target for drug development is clearly underlined by the fact that several existing trypanocidal agents appear to act on the trypanothione system. Already some novel targets have been identified and one of them, trypanothione reductase, characterized in great detail at the molecular level. The future prospects for the design of safer, more effective treatments for these important tropical diseases indeed looks promising.

ACKNOWLEDGMENTS

This review is dedicated to the memory of our colleague Dr. Graeme Henderson who was actively involved in many aspects of the work reviewed here. We thank the Wellcome Trust (A. H. F.) and the National Institutes of Health (AI 19428 to A. C.) for their financial support. We also thank Dr. K. Smith for communicating his unpublished results and for critically reading the manuscript.

Literature Cited

1. Aboagye-Kwarteng, T., Smith, K., Fairlamb, A. H. 1992. Molecular characterization of the trypanothione reductase gene from *Crithidia fasciculata* and *Trypanosoma brucei:* comparison with other flavoprotein disulphide oxidoreductases with respect to substrate specificity and catalytic mechanism. *Mol. Microbiol*. In press
2. Ahmad, T., Frischer, H. 1985. Active site-specific inhibition by 1,3-*bis*(2-chloroethyl)-1-nitrosourea of two genetically homologous flavoenzymes: glutathione reductase and lipoamide dehydrogenase. *J. Lab. Clin. Med.* 105: 464–71
3. Algranati, I. D., Sanchez, C., Gonzales, N. S. 1989. Polyamines in *Trypanosoma cruzi* and *Leishmania mexicana*. In *The Biology and Chemistry of Polyamines,* ed. S. H. Goldenberg, I. D. Algranati, pp. 137–46. Oxford: Oxford Univ. Press
4. Arrick, B. A., Griffith, O. W., Cerami, A. 1981. Inhibition of glutathione synthesis as a chemotherapeutic strategy for trypanosomiasis. *J. Exp. Med.* 153: 720–25

5. Arscott, L. D., Thorpe, C., Williams, C. H. Jr. 1981. Glutathione reductase from yeast. Differential reactivity of the nascent thiols in two-electron reduced enzyme and properties of a monoalkylated derivative. *Biochemistry* 20: 1513–20
6. Aust, S. D., Morehouse, L. A., Thomas, C. E. 1985. Role of metals in oxygen radical reactions. *J. Free Radic. Biol. Med.* 1:3–25
7. Awad, S., Henderson, G. B., Cerami, A., Held, K. D. 1992. Effects of trypanothione on the biological activity of irradiated transforming DNA. *Int. J. Radiat. Biol.* In press
8. Bacchi, C. J., Garofalo, J., Mockenhaupt, D., McCann, P. P., Diekema, K. A., et al. 1983. In vivo effects of α-DL-difluoromethylornithine on the metabolism and morphology of *Trypanosoma brucei brucei*. *Mol. Biochem. Parasitol.* 7:209–25
9. Bacchi, C. J., Garofalo, J., Santana, A., Hannan, J. C., Bitonti, A. J., McCann, P. P. 1989. *Trypanosoma brucei brucei:* regulation of ornithine decarboxylase in procyclic forms and trypomastigotes. *Exp. Parasitol.* 68: 392–402
10. Bacchi, C. J., McCann, P. P. 1987. Parasitic protozoa and polyamines. See Ref. 143, pp. 317–44
11. Bacchi, C. J., Nathan, H. C., Hutner, S. H., McCann, P. P., Sjoerdsma, A. 1980. Polyamine metabolism: a potential therapeutic target in trypanosomes. *Science* 210:332–34
12. Bacchi, C. J., Vergara, C., Garofalo, J., Lipschik, G. Y., Hutner, S. H. 1979. Synthesis and content of polyamines in bloodstream *Trypanosoma brucei*. *J. Protozool.* 26:484–88
13. Bachrach, U., Brem, S., Wertman, S. B., Schnur, L. F., Greenblatt, C. L. 1979. *Leishmania* spp.: cellular levels and synthesis of polyamines during growth cycles. *Exp. Parasitol.* 48:457–63
14. Bannister, J. V., Bannister, W. H., Rotillo, G. 1987. Aspects of the structure, function, and applications of superoxide dismutase. *Crit. Rev. Biochem.* 22:111–80
15. Bass, K. E., Wang, C. C. 1991. The in vitro differentiation of pleomorphic *Trypanosoma brucei* from bloodstream into procyclic form requires neither intermediary nor short-stumpy stage. *Mol. Biochem. Parasitol.* 44:261–70
16. Bellofatto, V., Fairlamb, A. H., Henderson, G. B., Cross, G. A. 1987. Biochemical changes associated with α-difluoromethylornithine uptake and resistance in *Trypanosoma brucei*. *Mol. Biochem. Parasitol.* 25:227–38
17. Berry, A., Scrutton, N. S., Perham, R. N. 1989. Switching kinetic mechanism and putative proton donor by directed mutagenesis of glutathione reductase. *Biochemistry* 28:1264–69
18. Bilzer, M., Krauth-Siegel, R. L., Schirmer, R. H., Akerboom, T. P., Sies, H., Schulz, G. E. 1984. Interaction of a glutathione *S*-conjugate with glutathione reductase. Kinetic and X-ray crystallographic studies. *Eur. J. Biochem.* 138:373–78
19. Bitonti, A. J., Bacchi, C. J., McCann, P. P., Sjoerdsma, A. 1985. Catalytic irreversible inhibition of *Trypanosoma brucei brucei* ornithine decarboxylase by substrate and product analogs and their effect in murine trypanosomiasis. *Biochem. Pharmacol.* 34:1773–77
20. Bitonti, A. J., Bacchi, C. J., McCann, P. P., Sjoerdsma, A. 1986. Uptake of α-difluoromethylornithine by *Trypanosoma brucei brucei*. *Biochem. Pharmacol.* 35:351–54
21. Bitonti, A. J., Byers, T. L., Bush, T. L., Casara, P. J., Bacchi, C. J., et al. 1990. Cure of *Trypanosoma brucei brucei* and *Trypanosoma brucei rhodesiense* infections in mice with an irreversible inhibitor of *S*-adenosylmethionine decarboxylase. *Antimicrob. Agents Chemother.* 34:1485–90
22. Bitonti, A. J., Casara, P. J., McCann, P. P., Bey, P. 1987. Catalytic irreversible inhibition of bacterial and plant arginine decarboxylase activities by novel substrate and product analogues. *Biochem. J.* 242:69–74
23. Bitonti, A. J., Cross-Doersen, E., McCann, P. P. 1988. Effects of α-difluoromethylornithine on protein synthesis and synthesis of the variant-specific glycoprotein (VSG) in *Trypanosoma brucei brucei*. *Biochem. J.* 250:295–98
24. Bitonti, A. J., Dumont, J. A., McCann, P. P. 1986. Characterization of *Trypanosoma brucei brucei S*-adenosyl-L-methionine decarboxylase and its inhibition by Berenil, pentamidine and methylglyoxal bis(guanylhydrazone). *Biochem. J.* 237:685–89
25. Bitonti, A. J., Kelly, S. E., McCann, P. P. 1984. Characterization of spermidine synthase from *Trypanosoma brucei brucei*. *Mol. Biochem. Parasitol.* 13:21–28
26. Bitonti, A. J., McCann, P. P., Sjoerdsma, A. 1986. Necessity of antibody response in the treatment of African

trypanosomiasis with α-difluoromethylornithine. *Biochem. Pharmacol.* 35: 331–34

27. Boveris, A., Sies, H., Martino, E. E., Docampo, R., Turrens, J. F., Stoppani, A. O. 1980. Deficient metabolic utilization of hydrogen peroxide in *Trypanosoma cruzi*. *Biochem. J.* 188: 643–48
28. Boveris, A., Stoppani, A. O. 1977. Hydrogen peroxide generation in *Trypanosoma cruzi*. *Experientia* 33:1306–8
29. Bradley, M., Bucheler, U. S., Walsh, C. T. 1991. Redox enzyme engineering: conversion of human glutathione reductase into a trypanothione reductase. *Biochemistry* 30:6124–27
30. Byers, T. L., Bush, T. L., McCann, P. P., Bitonti, A. J. 1991. Antitrypanosomal effects of polyamine biosynthesis inhibitors correlate with increases in *Trypanosoma brucei brucei S*-adenosyl-L-methionine. *Biochem. J.* 274:527–33
31. Callahan, H. L., Crouch, R. K., James, E. R. 1988. Helminth anti-oxidant enzymes: a protective mechanism against host oxidants? *Parasitol. Today* 4:218–25
32. Carlberg, I., Mannervik, B. 1986. Reduction of 2,4,6-trinitrobenzenesulfonate by glutathione reductase and the effect of $NADP^+$ on the electron transfer. *J. Biol. Chem.* 261:1629–35
33. Carnegie, P. R. 1963. Structure and properties of a homologue of glutathione. *Biochem. J.* 89:471–78
34. Cohen, S. S. 1971. *Introduction to the Polyamines*. Englewood Cliffs, NJ: Prentice-Hall
35. Coles, B., Ketterer, B. 1990. The role of glutathione and glutathione transferases in chemical carcinogenesis. *Crit. Rev. Biochem. Mol. Biol.* 25:47–70

35a. Coombs, G. H., North, M. J., eds. 1991. *Biochemical Protozoology*. London: Taylor and Francis

36. Creighton, T. E. 1986. Disulphide bonds as probes of protein folding pathways. *Methods Enzymol.* 131:83–106
37. Creighton, T. E. 1990. Understanding protein folding pathways and mechanisms. In *Protein Folding: Deciphering the Second Half of the Genetic Code*, ed. L. Gierasch, J. King, pp. 157–70. Washington, DC: Am. Assoc. Adv. Sci.
38. Danson, M. J., Conroy, K., McQuattie, A., Stevenson, K. J. 1987. Dihydrolipoamide dehydrogenase from *Trypanosoma brucei*. Characterisation and cellular location. *Biochem. J.* 243:661–65
39. Darling, T. N., Balber, A. E., Blum, J. J. 1988. A comparative study of D-lactate, L-lactate and glycerol formation by four species of *Leishmania* and by *Trypanosoma lewisi* and *Trypanosoma brucei gambiense*. *Mol. Biochem. Parasitol.* 30:253–57
40. Darling, T. N., Blum, J. J. 1988. D-lactate production by *Leishmania braziliensis* through the glyoxalase pathway. *Mol. Biochem. Parasitol.* 28: 121–27
41. Darling, T. N., Davis, D. G., London, R. E., Blum, J. J. 1987. Products of *Leishmania braziliensis* glucose catabolism: release of D-lactate and, under anaerobic conditions, glycerol. *Proc. Natl. Acad. Sci. USA* 84:7129–33
42. de Gee, A. L. W., Carstens, P. H., McCann, P. P., Mansfield, J. M. 1984. Morphological changes in *Trypanosoma brucei rhodesiense* following inhibition of polyamine biosynthesis *in vivo*. *Tissue & Cell* 16:731–38
43. de Gee, A. L. W., McCann, P. P., Mansfield, J. M. 1983. Role of antibody in the elimination of trypanosomes after D,L-α-difluoromethylornithine chemotherapy. *J. Parasitol.* 69:818–22
44. Deonarain, M. P., Berry, A., Scrutton, N. S., Perham, R. N. 1989. Alternative proton donors/acceptors in the catalytic mechanism of the glutathione reductase of *Escherichia coli:* the role of histidine-439 and tyrosine-99. *Biochemistry* 28:9602–7
45. Deonarain, M. P., Scrutton, N. S., Berry, A., Perham, R. N. 1990. Directed mutagenesis of the redox-active disulphide bridge in glutathione reductase from *Escherichia coli*. *Proc. R. Soc. London Ser. B* 241:179–86
46. Docampo, R. 1990. Sensitivity of parasites to free radical damage by antiparasitic drugs. *Chem.-Biol. Interact.* 73:1–27
47. Docampo, R., Moreno, S. N. J. 1984. Free-radical intermediates in the antiparasitic action of drugs and phagocytic cells. In *Free Radicals in Biology*, ed. W. A. Pryor, pp. 243–88. New York: Academic. 6th ed.
48. Dolphin, D., Poulson, R., Avramovic, O., eds. 1989. *Glutathione: Chemical, Biochemical and Medical Aspects. Parts A and B*. New York: Wiley Interscience
49. El-Waer, A., Douglas, K. T., Smith, K., Fairlamb, A. H. 1991. Synthesis of *N*-benzyloxycarbonyl-L-cysteinylglycine 3-dimethylaminopropylamide disulphide: a cheap and convenient new assay for trypanothione reductase. *Anal. Biochem.* 198:212–16
50. El-Waer, A. F. 1991. *Substrate spec-*

ificity of trypanothione reductase. PhD thesis. Univ. Manchester, Manchester, UK
51. Fahey, R. C., Newton, G. L. 1983. Occurrence of low molecular weight thiols in biological systems. See Ref. 130a, pp. 251–60
52. Fahey, R. C., Newton, G. L., Arrick, B., Overdank-Bogart, T., Aley, S. B. 1984. *Entamoeba histolytica:* a eukaryote without glutathione metabolism. *Science* 224:70–72
53. Fairlamb, A. H. 1982. Biochemistry of trypanosomiasis and rational approaches to chemotherapy. *TIBS* 7: 249–53
54. Fairlamb, A. H. 1988. The role of glutathionylspermidine and trypanothione in regulation of intracellular spermidine levels during growth of *Crithidia fasciculata*. See Ref. 212, pp. 667–74
55. Fairlamb, A. H. 1989. Novel biochemical pathways in parasitic protozoa. *Parasitology* 99S:93–112
56. Fairlamb, A. H. 1989. Metabolism and functions of trypanothione with special reference to leishmaniasis. See Ref. 86a, pp. 487–94
57. Fairlamb, A. H. 1990. Trypanothione metabolism and rational approaches to drug design. *Biochem. Soc. Trans.* 18: 717–20
58. Fairlamb, A. H. 1990. Future prospects for the chemotherapy of human trypanosomiasis. *Trans. R. Soc. Trop. Med. Hyg.* 84:613–17
59. Fairlamb, A. H. 1990. The interaction of trypanocidal drugs with the metabolism and functions of trypanothione. In *Chemotherapy for Trypanosomiasis: Proceedings of a Workshop Held at ILRAD, Nairobi, Kenya, 21–24 August 1989*, ed. A. S. Peregrine, pp. 25–31. Nairobi: Int. Lab. Res. Animal Diseases
60. Fairlamb, A. H. 1991. Trypanothione metabolism in the chemotherapy of leishmaniasis and trypanosomiasis. See Ref. 205a, pp. 107–21
61. Fairlamb, A. H., Blackburn, P., Ulrich, P., Chait, B. T., Cerami, A. 1985. Trypanothione: a novel bis(glutathionyl)spermidine cofactor for glutathione reductase in trypanosomatids. *Science* 227:1485–87
62. Fairlamb, A. H., Carter, N. S., Cunningham, M., Smith, K. 1992. Characterisation of melarsen-resistant *Trypanosoma brucei brucei* with respect to cross-resistance to other drugs and trypanothione metabolism. *Mol. Biochem. Parasitol.* 53:213–22
63. Fairlamb, A. H., Cerami, A. 1985. Identification of a novel, thiol-containing co-factor essential for glutathione reductase enzyme activity in trypanosomatids. *Mol. Biochem. Parasitol.* 14:187–98
64. Fairlamb, A. H., Henderson, G. B. 1987. Metabolism of trypanothione and glutathionylspermidine in trypanosomatids. In *Host-Parasite Cellular and Molecular Interactions in Protozoal Infections*, ed. K. -P. Chang, D. Snary, pp. 29–40. Berlin: Springer-Verlag/ NATO ASI Series, Vol. H11
65. Fairlamb, A. H., Henderson, G. B., Bacchi, C. J., Cerami, A. 1987. In vivo effects of difluoromethylornithine on trypanothione and polyamine levels in bloodstream forms of *Trypanosoma brucei*. *Mol. Biochem. Parasitol.* 24: 185–91
66. Fairlamb, A. H., Henderson, G. B., Cerami, A. 1986. The biosynthesis of trypanothione and N^1-glutathionylspermidine in *Crithidia fasciculata*. *Mol. Biochem. Parasitol.* 21:247–57
67. Fairlamb, A. H., Henderson, G. B., Cerami, A. 1989. Trypanothione is the primary target for arsenical drugs against African trypanosomes. *Proc. Natl. Acad. Sci. USA* 86:2607–11
68. Fairlamb, A. H., Smith, K., Hunter, K. J. 1992. The interaction of arsenical drugs with dihydrolipoamide and dihydrolipoamide dehydrogenase from arsenical resistant and sensitive strains of *Trypanosoma brucei brucei*. *Mol. Biochem. Parasitol.* 53:223–32
69. Feldhaus, P., Frohlich, T., Goody, R. S., Isakov, M., Schirmer, R. H. 1975. Synthetic inhibitors of adenylate kinases in the assays for ATPases and phosphokinases. *Eur. J. Biochem.* 57: 197–204
70. Field, H., Cerami, A., Henderson, G. B. 1992. Cloning, sequencing, and demonstration of polymorphism in trypanothione reductase from *Crithidia fasciculata*. *Mol. Biochem. Parasitol.* 50:47–56
71. Flohe, L. 1989. The selenoprotein glutathione peroxidase. See Ref. 48, Part A, pp. 643–731
72. Freedman, R. B. 1984. Native disulphide bond formation in protein biosynthesis: evidence for the role of protein disulphide isomerase. *TIBS* 9:438–41
73. Fridovich, I. 1989. Superoxide dismutases: an adaptation to a paramagnetic gas. *J. Biol. Chem.* 264:7761–64
74. Fuchs, J. A. 1989. Glutaredoxin. See Ref. 48, Part B, pp. 551–70
75. Gething, M. -J., Sambrook, J. 1992.

Protein folding in the cell. *Nature* 355:33–45

76. Ghisla, S., Massey, V. 1989. Mechanisms of flavoprotein-catalyzed reactions. *Eur. J. Biochem.* 181:1–17
77. Ghoda, L., Phillips, M. A., Bass, K. E., Wang, C. C., Coffino, P. 1990. Trypanosome ornithine decarboxylase is stable because it lacks sequences found in the carboxyl terminus of the mouse enzyme which target the latter for intracellular degradation. *J. Biol. Chem.* 265:11823–26
78. Ghoshal, K., Banerjee, A. B., Ray, S. 1989. Methylglyoxal-catabolizing enzymes of *Leishmania donovani* promastigotes. *Mol. Biochem. Parasitol.* 35:21–29
79. Giffin, B. F., McCann, P. P. 1989. Physiological activation of the mitochondrion and the transformation capacity of DFMO induced intermediate and short stumpy bloodstream form trypanosomes. *Am. J. Trop. Med. Hyg.* 40:487–93
80. Giffin, B. F., McCann, P. P., Bitonti, A. J., Bacchi, C. J. 1986. Polyamine depletion following exposure to DL-α-difluoromethylornithine both in vivo and in vitro initiates morphological alterations and mitochondrial activation in a monomorphic strain of *Trypanosoma brucei brucei*. *J. Protozool.* 33:238–43
81. Greer, S., Perham, R. N. 1986. Glutathione reductase from *Escherichia coli:* cloning and sequence analysis of the gene and relationship to other flavoprotein disulfide oxidoreductases. *Biochemistry* 25:2736–42
82. Griffith, O. W., Meister, A. 1979. Potent and specific inhibition of glutathione synthesis by buthionine sulfoximine (*S-N*-butyl homocysteine sulfoximine). *J. Biol. Chem.* 254: 7558–60
83. Griffith, O. W., Meister, A. 1985. Origin and turnover of mitochondrial glutathione. *Proc. Natl. Acad. Sci. USA* 82:4668–72
84. Halliwell, B., Gutteridge, J. M. C. 1985. *Free Radicals in Biology and Medicine*. Oxford: Clarendon Press
85. Hamer, D. H. 1986. Metallothionein. *Annu. Rev. Biochem.* 55:913–51
86. Hannan, J. C., Bacchi, C. J., McCann, P. P. 1984. Induction of ornithine decarboxylase in *Leptomonas seymouri*. *Mol. Biochem. Parasitol.* 12:117–24

86a. Hart, D. T., ed. 1989. *Leishmaniasis: The Current Status and New Strategies for Control*. New York: Plenum Press/NATO ASI Series Vol. 163.

87. Hassett, D. J., Cohen, M. S. 1989. Bacterial adaptation to oxidative stress: implications for pathogenesis and interaction with phagocytic cells. *FASEB J.* 3:2574–82
88. Hayes, J. D., Wolf, C. R. 1988. Role of glutathione transferase in drug resistance. In *Glutathione Conjugation: Mechanisms and Biological Significance*, ed. H. Sies, B. Ketterer, pp. 315–55. London: Academic
89. Heby, O., Persson, L. 1990. Molecular genetics of polyamine synthesis in eukaryotic cells. *Trends. Biochem. Sci.* 15:153–58
90. Helmann, J. D., Ballard, B. T., Walsh, C. T. 1990. The MerR metalloregulatory protein binds mercuric ion as a tricoordinate, metal-bridged dimer. *Science* 247:946–48
91. Henderson, G. B., Fairlamb, A. H. 1987. Trypanothione metabolism: a chemotherapeutic target in trypanosomatids. *Parasitol. Today* 3:312–15
92. Henderson, G. B., Fairlamb, A. H., Cerami, A. 1987. Trypanothione dependent peroxide metabolism in *Crithidia fasciculata* and *Trypanosoma brucei*. *Mol. Biochem. Parasitol.* 24: 39–45
93. Henderson, G. B., Fairlamb, A. H., Ulrich, P., Cerami, A. 1987. Substrate specificity of the flavoprotein trypanothione disulfide reductase from *Crithidia fasciculata*. *Biochemistry* 26:3023–27
94. Henderson, G. B., Glushka, J., Cowburn, D., Cerami, A. 1990. Synthesis and NMR characterization of the trypanosomatid metabolite, N^1,N^8-bis (glutathionyl)spermidine disulphide (trypanothione disulphide). *J. Chem. Soc. Perkin Trans. 1* pp. 911–14
95. Henderson, G. B., Murgolo, N. J., Kuriyan, J., Osapay, K., Kominos, D., et al. 1991. Engineering the substrate-specificity of glutathione reductase toward that of trypanothione reduction. *Proc. Natl. Acad. Sci. USA* 88:8769–73
96. Henderson, G. B., Ulrich, P., Fairlamb, A. H., Cerami, A. 1986. Synthesis of the trypanosomatid metabolites trypanothione, and N^1-mono- and N^8-mono-glutathionylspermidine. *J. Chem. Soc. Chem. Commun.* pp. 593–94
97. Henderson, G. B., Ulrich, P., Fairlamb, A. H., Rosenberg, I., Pereira, M., et al. 1988. "Subversive" substrates for the enzyme trypanothione disulfide reductase: alternative approach to chemotherapy of Chagas' disease. *Proc. Natl. Acad. Sci. USA* 85:5374–78

98. Henderson, G. B., Yamaguchi, M., Novoa, L., Fairlamb, A. H., Cerami, A. 1990. Biosynthesis of the trypanosomatid metabolite trypanothione: purification and characterization of trypanothione synthetase from *Crithidia fasciculata*. *Biochemistry* 29:3924–29

99. Hinson, J. A., Kadlubar, F. F. 1988. Glutathione conjugation: mechanisms and biological significance. In *Glutathione and Glutathione Transferases in the Detoxification of Drug and Carcinogen Metabolites*, ed. H. Sies, B. Ketterer, pp. 235–80. London: Academic

100. Hol, W. G. J. 1986. Protein crystallography and computer graphics—towards rational drug design. *Angew. Chem. Int. Ed. Eng.* 25:767–78

101. Holmgren, A. 1985. Thioredoxin. *Annu. Rev. Biochem.* 54:237–71

102. Holmgren, A. 1989. Thioredoxin and glutaredoxin systems. *J. Biol. Chem.* 264:13963–66

103. Holmgren, A., Brändén, C.-I., Jornvall, H., Sjoberg, B. M., eds. 1986. *Thioredoxin and Glutaredoxin Systems: Structure and Function*. New York: Raven

104. Hsu, M., Muhich, M. L., Boothroyd, J. C. 1989. A developmentally regulated gene of trypanosomes encodes a homologue of rat protein-disulfide isomerase and phosphoinositol-phospholipase C. *Biochemistry* 28:6440–46

105. Hunter, K. J., Strobos, C. A. M., Fairlamb, A. H. 1990. The interaction of trypanocidal drugs with polyamine and trypanothione metabolism. *Biochem. Soc. Trans.* 18:1094–96

106. Hunter, K. J., Strobos, C. A. M., Fairlamb, A. H. 1991. Inhibition of polyamine biosynthesis in *Crithidia fasciculata* by D,L-α-difluoromethyl-ornithine and D,L-α-difluoromethylarginine. *Mol. Biochem. Parasitol.* 46:35–44

107. Hunter, W. N., Bailey, S., Habash, J., Harrop, S. J., Helliwell, J. R., et al. 1992. The active site of trypanothione reductase: a target for rational drug design. *J. Mol. Biol.* In press

108. Hunter, W. N., Smith, K., Derewenda, Z., Harrop, S. J., Habash, J., et al. 1990. Initiating a crystallographic study of trypanothione reductase. *J. Mol. Biol.* 216:235–37

109. Inoue, M. 1989. Glutathione; dynamic aspects of protein mixed disulfide formation. See Ref. 48, Part B, pp. 613–44

110. Jackman, S. A., Hough, D. W., Danson, M. J., Stevenson, K. J., Opperdoes, F. R. 1990. Subcellular localisation of dihydrolipoamide dehydrogenase and detection of lipoic acid in bloodstream forms of *Trypanosoma brucei*. *Eur. J. Biochem.* 193:91–95

111. Jain, A., Martensson, J., Stole, E., Auld, P. A. M., Meister, A. 1991. Glutathione deficiency leads to mitochondrial damage in brain. *Proc. Natl. Acad. Sci. USA* 88:1913–17

112. Janes, W., Schulz, G. E. 1990. Role of the charged groups of glutathione disulfide in the catalysis of glutathione reductase: crystallographic and kinetic studies with synthetic analogues. *Biochemistry* 29:4022–30

113. Janne, J., Alhonen, L., Leinonen, P. 1991. Polyamines: from molecular biology to clinical applications. *Ann. Med.* 23:241–59

114. Jennings, F. W. 1988. Chemotherapy of trypanosomiasis: the potentiation of melarsoprol by concurrent difluoromethylornithine (DFMO) treatment. *Trans. R. Soc. Trop. Med. Hyg.* 82: 572–73

115. Jennings, F. W. 1988. The potentiation of arsenicals with difluoromethylornithine (DFMO): experimental studies in murine trypanosomiasis. *Bull. Soc. Pathol. Exot. Filiales* 81:595–607

116. Jennings, F. W. 1990. Future prospects for the chemotherapy of human trypanosomiasis. 2. Combination chemotherapy and African trypanosomiasis. *Trans. R. Soc. Trop. Med. Hyg.* 84: 618–21

117. Jockers-Scherubl, M. C., Schirmer, R. H., Krauth-Siegel, R. L. 1989. Trypanothione reductase from *Trypanosoma cruzi:* catalytic properties of the enzyme and inhibition studies with trypanocidal compounds. *Eur. J. Biochem.* 180:267–72

118. Karplus, P. A., Pai, E. F., Schulz, G. E. 1989. A crystallographic study of the glutathione binding site of glutathione reductase at 0.3-nm resolution. *Eur. J. Biochem.* 178:693–703

119. Karplus, P. A., Schulz, G. E. 1987. Refined structure of glutathione reductase at 1.54 Angstrom resolution. *J. Mol. Biol.* 195:701–29

120. Keithly, J. S., Fairlamb, A. H. 1989. Inhibition of *Leishmania* species by α-difluoromethylornithine. See Ref. 86a, pp. 749–56

121. Ketterer, B., Meyer, D. J., Clark, A. G. 1988. Soluble glutathione transferase isozymes. See Ref. 182, pp. 73–135

122. Kierszenbaum, F., Wirth, J. J., McCann, P. P., Sjoerdsma, A. 1987. Arginine decarboxylase inhibitors reduce the capacity of *Trypanosoma cruzi* to infect

and multiply in mammalian host cells. *Proc. Natl. Acad. Sci. USA* 84:4278–82

123. Kosower, E. M., Kosower, N. S., Radkowsky, A. 1983. Fluorescent thiol labeling and other reactions with bromobimanes; glutathione sulfide. See Ref. 130a, pp. 243–50
124. Krauth-Siegel, R. L., Enders, B., Henderson, G. B., Fairlamb, A. H., Schirmer, R. H. 1987. Trypanothione reductase from *Trypanosoma cruzi:* purification and characterization of the crystalline enzyme. *Eur. J. Biochem.* 164:123–28
125. Krauth-Siegel, R. L., Jockers-Scherubl, M. C., Becker, K., Schirmer, R. H. 1989. NADP-dependent disulphide reductases. *Biochem. Soc. Trans.* 17: 315–17
126. Krauth-Siegel, R. L., Lohrer, H., Bucheler, U. S., Schirmer, R. H. 1991. The antioxidant enzymes glutathione reductase and trypanothione reductase as drug targets. See Ref. 35a, pp. 493–505
127. Kuriyan, J., Kong, X.-P., Krishna, T. S. R., Sweet, R. M., Murgolo, N. J., et al. 1991. X-ray structure of trypanothione reductase from *Crithidia fasciculata* at 2.4-Å resolution. *Proc. Natl. Acad. Sci. USA* 88:8764–67
128. Kuriyan, J., Wong, L., Guenther, B. D., Murgolo, N. J., Cerami, A., Henderson, G. B. 1990. Preliminary crystallographic analysis of trypanothione reductase from *Crithidia fasciculata*. *J. Mol. Biol.* 215:335–37
129. Kurosawa, K., Hayashi, N., Sato, N., Kamada, T., Tagawa, K. 1990. Transport of glutathione across the mitochondrial membranes. *Biochem. Biophys. Res. Commun.* 167:367–72
130. Lamoureux, G. L., Rusness, D. G. 1989. The role of glutathione and glutathione-S-transferases in pesticide metabolism, selectivity, and mode of action in plants and insects. See Ref. 48, Part B, pp. 153–96
130a. Larsson, A., Orrenius, S., Holmgren, A., Mannervik, B., eds. 1983. *Functions of Glutathione: Biochemical, Physiological, Toxicological and Clinical Aspects*. New York: Raven
131. Lee, T. C., Wei, M. L., Chang, W. J., Ho, I. C., Lo, J. F., et al. 1989. Elevation of glutathione levels and glutathione *S*-transferase activity in arsenic-resistant chinese hamster ovary cells. *In Vitro Cell. Dev. Biol.* 25:442–48
132. Leid, R. W., Suquet, C. M., Tanigoshi, L. 1989. Oxygen detoxifying enzymes in parasites: a review. *Acta Leiden.* 57:107–14
133. Le Trang, N., Meshnick, S. R., Kitchener, K., Eaton, J. W., Cerami, A. 1983. Iron-containing superoxide dismutase from *Crithidia fasciculata:* purification, characterization, and similarity to leishmanial and trypanosomal enzymes. *J. Biol. Chem.* 258:125–30
134. Machlin, L. J., Bendich, A. 1987. Free radical tissue damage: protective role of antioxidant nutrients. *FASEB J.* 1: 441–45
135. Mannervick, B., Boggaram, V., Carlberg, I., Larson, K. 1980. The catalytic mechanism of glutathione reductase. In *Flavins and Flavoproteins,* ed. K. Yagi, T. Yamono, pp. 173–87. Tokyo: Jpn. Sci. Soc.
136. Mannervick, B., Carlberg, I., Larson, K. 1989. Glutathione: general review of mechanism of action. See Ref. 48, Part A, pp. 475–516
137. Mannervick, B., Danielson, U. H. 1988. Glutathione transferases—structure and catalytic activity. *Crit. Rev. Biochem.* 23:283–337
138. Mannervik, B. 1986. Glutathione and the evolution of enzymes for detoxication of products of oxygen metabolism. *Chem. Scr.* 26B:281–84
139. Mårtensson, J., Lai, J. C. K., Meister, A. 1990. High-affinity transport of glutathione is part of a multicomponent system essential for mitochondrial function. *Proc. Natl. Acad. Sci. USA* 87: 7185–89
140. Mata, A. M., Pinto, M. C., Lopez-Barea, J. 1984. Purification by affinity chromatography of glutathione reductase (EC 1.6.4.2) from *Escherichia coli* and characterization of such enzyme. *Z. Naturforsch.* 39c:908–15
141. Mbemba, F., Houbion, A., Raes, M., Remacle, J. 1985. Subcellular localization and modification with ageing of glutathione, glutathione peroxidase and glutathione reductase activities in human fibroblasts. *Biochim. Biophys. Acta* 838:211–20
142. McCann, P. P., Bacchi, C. J., Nathan, H. C., Sjoerdsma, A. 1983. Difluoromethylornithine and the rational development of polyamine antagonists for the cure of protozoan infection. In *Mechanisms of Drug Action,* ed. T. P. Singer, R. N. Ondarza, pp. 159–73. New York: Academic
143. McCann, P. P., Pegg, A. E., Sjoerdsma, A., eds. 1987. *Inhibition of Polyamine Metabolism: Biological Significance and Basis for New Therapies*. Orlando, FL: Academic

144. McClements, W., Yamanaka, G., Garsky, V., Perry, H., Bacchetti, S., et al. 1988. Oligopeptides inhibit the ribonucleotide reductase of Herpes simplex virus by causing subunit separation. *Virology* 162:270–73
145. Meister, A. 1988. On the discovery of glutathione. *TIBS* 13:185–88
146. Meister, A., Anderson, M. E. 1983. Glutathione. *Annu. Rev. Biochem.* 52: 711–60
147. Metcalf, B. W., Bey, P., Danzin, C., Jung, M. J., Casara, P., Vevert, J. P. 1978. Catalytic irreversible inhibition of mammalian ornithine decarboxylase (E. C. 4.1.1.17) by substrate and product analogues. *J. Am. Chem. Soc.* 100:2551–53
148. Milhous, W. K., Weatherly, N. F., Bowdre, J. H., Desjardins, R. E. 1985. In vitro activities of and mechanisms of resistance to antifol antimalarial drugs. *Antimicrob. Agents Chemother.* 27:525–30
149. Morrow, C. S., Cowan, K. H. 1990. Glutathione *S*-transferases and drug resistance. *Cancer Cells* 2:15–22
150. Murgolo, N. J., Cerami, A., Henderson, G. B. 1989. Trypanothione reductase. *Ann. N. Y. Acad. Sci.* 569:193–200
151. Nathan, H. C., Bacchi, C. J., Hutner, S. H., Rescigno, D., Sjoerdsma, A. 1981. Antagonism by polyamines of the curative effects of α-difluoromethylornithine in *Trypanosoma brucei brucei* infections. *Biochem. Pharmacol.* 30:3010–13
152. Newton, G. L., Javor, B. 1985. γ-Glutamylcysteine and thiosulfate are the major low-molecular-weight thiols in Halobacteria. *J. Bacteriol.* 161:438–41
153. Opperdoes, F. R. 1987. Compartmentation of carbohydrate metabolism in trypanosomes. *Annu. Rev. Microbiol.* 41:127–51
154. Pai, E. F. 1988. Crystallographic analysis of the binding of NADPH, NADPH fragments and NADPH analogues to glutathione reductase. *Biochemistry* 27: 4465–74
155. Pai, E. F., Schulz, G. E. 1983. The catalytic mechanism of glutathione reductase as derived from X-ray diffraction analyses of reaction intermediates. *J. Biol. Chem.* 258:1752–57
156. Pegg, A. E. 1986. Recent advances in the biochemistry of polyamines in eukaryotes. *Biochem. J.* 234:249–62
157. Pegg, A. E., McCann, P. P. 1988. Polyamine metabolism and function in mammalian cells and protozoans. In *ISI Atlas of Science: Biochemistry*, pp. 11–18. Philadelphia, PA: ISI
158. Penketh, P. G., Kennedy, W. D. K., Patton, C. L., Sartorelli, A. C. 1987. Competent metabolic utilization of hydrogen peroxide by trypanosomes. *Acta Trop.* 44:461–62
159. Penketh, P. G., Kennedy, W. P. K., Patton, C. L., Sartorelli, A. C. 1987. Trypanosomatid hydrogen peroxide metabolism. *FEBS Lett.* 221:427–31
160. Penketh, P. G., Klein, R. A. 1986. Hydrogen peroxide metabolism in *Trypanosoma brucei. Mol. Biochem. Parasitol.* 20:111–21
161. Perry, A. C. F., Ni Bhriain, N., Brown, N. L., Rouch, D. A. 1991. Molecular characterization of the *gor* gene encoding glutathione reductase from *Pseudomonas aeruginosa:* determinants of substrate specificity among pyridine nucleotide-disulphide oxidoreductases. *Mol. Microbiol.* 5(1):163–71
162. Phillips, M. A., Coffino, P., Wang, C. C. 1987. Cloning and sequencing of the ornithine decarboxylase gene from *Trypanosoma brucei:* implications for enzyme turnover and selective difluoromethylornithine inhibition. *J. Biol. Chem.* 262:8721–27
163. Phillips, M. A., Coffino, P., Wang, C. C. 1988. *Trypanosoma brucei* ornithine decarboxylase: enzyme purification, characterization, and expression in *Escherichia coli. J. Biol. Chem.* 263: 17933–41
164. Phillips, M. A., Wang, C. C. 1987. A *Trypanosoma brucei* mutant resistant to α-difluoromethylornithine. *Mol. Biochem. Parasitol.* 22:9–17
165. Pickett, C. B., Lu, A. Y. H. 1989. Glutathione *S*-transferases: gene structure, regulation, and biological function. *Annu. Rev. Biochem.* 58:743–64
166. Rauser, W. E. 1990. Phytochelatins. *Annu. Rev. Biochem.* 59:61–86
167. Revesz, L. 1985. The role of endogenous thiols in intrinsic radioprotection. *Int. J. Radiat. Biol.* 47:361–68
168. Rogers, S., Wells, R., Rechsteiner, M. 1986. Amino acid sequences common to rapidly degraded proteins: the PEST hypothesis. *Science* 234:364–68
169. Ross, D. 1988. Glutathione, free radicals and chemotherapeutic agents: mechanisms of free-radical induced toxicity and glutathione-dependent protection. *Pharmacol. Ther.* 37:231–49
170. Rotilio, G., ed. 1986. *Superoxide and Superoxide Dismutase in Chemistry, Biology and Medicine.* Amsterdam: Elsevier
171. Schechter, P. J., Barlow, J. L. R.,

Sjoerdsma, A. 1987. Clinical aspects of inhibition of ornithine decarboxylase with emphasis on therapeutic trials of Eflornithine (DFMO) in cancer and protozoan diseases. See Ref. 143, pp. 345–64

172. Schechter, P. J., Sjoerdsma, A. 1986. Difluoromethylornithine in the treatment of African trypanosomiasis. *Parasitol. Today* 2:223–24

173. Schiering, N., Kabsch, W., Moore, M. J., Distefano, M. D., Walsh, C. T., Pai, E. F. 1991. Structure of the detoxification catalyst mercuric ion reductase from *Bacillus* sp. strain RC607. *Nature* 352:168–72

174. Schirmer, R. H., Krauth-Siegel, R. L., Schulz, G. E. 1989. Glutathione reductase. See Ref. 48, Part A, pp. 553–96

175. Scrutton, N. S., Berry, A., Deonarain, M. P., Perham, R. N. 1990. Active site complementation in engineered heterodimers of *Escherichia coli* glutathione reductase created in vivo. *Proc. R. Soc. London Ser. B* 242:217–24

176. Scrutton, N. S., Berry, A., Perham, R. N. 1987. Purification and characterization of glutathione reductase encoded by a cloned and over-expressed gene in *Escherichia coli*. *Biochem. J.* 245: 875–80

177. Seiler, N. 1987. Inhibition of enzymes oxidizing polyamines. See Ref. 143, 49–77

178. Seltzer, S. 1989. Maleylacetoacetate *cis-trans* isomerase. See Ref. 48, Part A, pp. 733–51

179. Shames, S. L., Fairlamb, A. H., Cerami, A., Walsh, C. T. 1986. Purification and characterization of trypanothione reductase from *Crithidia fasciculata*, a newly discovered member of the family of disulphide-containing flavoprotein reductases. *Biochemistry* 25:3519–26

180. Shames, S. L., Kimmel, B. E., Peoples, O. P., Agabian, N., Walsh, C. T. 1988. Trypanothione reductase of *Trypanosoma congolense:* gene isolation, primary sequence determination, and comparison to glutathione reductase. *Biochemistry* 27:5014–19

181. Shim, H., Fairlamb, A. H. 1988. Levels of polyamines, glutathione and glutathione-spermidine conjugates during growth of the insect trypanosomatid *Crithidia fasciculata*. *J. Gen. Microbiol.* 134:807–17

182. Sies, H., Ketterer, B., ed. 1988. *Glutathione Conjugation: Mechanisms and Biological Significance*. London: Academic

183. Sies, H., ed. 1985. *Oxidative Stress*. London: Academic

184. Simpson, L. 1987. The mitochondrial genome of kinetoplastid protozoa: genomic organization, transcription, replication, and evolution. *Annu. Rev. Microbiol.* 41:363–82

185. Singhal, R. K., Anderson, M. E., Meister, A. 1987. Glutathione, a first line of defense against cadmium toxicity. *FASEB J.* 1:220–23

186. Slocum, R. D., Bitonti, A. J., McCann, P. P., Feirer, R. P. 1988. D,L-α-Difluoromethyl[3,4-^{3}H]arginine metabolism in tobacco and mammalian cells. *Biochem. J.* 255:197–202

187. Smith, K., Mills, A., Thornton, J. M., Fairlamb, A. H. 1991. Trypanothione metabolism as a target for drug design: molecular modelling of trypanothione reductase. See Ref. 35a, pp. 482–92

188. Smith, K., Nadeau, K., Walsh, C. T., Fairlamb, A. H. 1992. Purification of glutathionylspermidine and trypanothione synthetases from *Crithidia fasciculata*. *Protein Sci.* 1:874–83

189. Smith, K., Opperdoes, F. R., Fairlamb, A. H. 1991. Subcellular distribution of trypanothione reductase in bloodstream and procyclic forms of *Trypanosoma brucei*. *Mol. Biochem. Parasitol.* 48: 109–12

190. Stocken, L. A., Thompson, R. H. S. 1946. Dithiol compounds as antidotes for arsenic. *Biochem. J.* 40:535–54

191. Stubbe, J. 1990. Ribonucleotide reductases: amazing and confusing. *J. Biol. Chem.* 265:5329–32

192. Sullivan, F. X., Shames, S. L., Walsh, C. T. 1989. Expression of *Trypanosoma congolense* trypanothione reductase in *Escherichia coli:* overproduction, purification, and characterization. *Biochemistry* 28:4986–92

193. Sullivan, F. X., Sobolov, S. B., Bradley, M., Walsh, C. T. 1991. Mutational analysis of parasite trypanothione reductase: acquisition of glutathione reductase activity in a triple mutant. *Biochemistry* 30:2761–67

194. Sullivan, F. X., Walsh, C. T. 1991. Cloning, sequencing, overproduction and purification of trypanothione reductase from *Trypanosoma cruzi*. *Mol. Biochem. Parasitol.* 44:145–48

195. Tabor, C. W., Tabor, H. 1984. Polyamines. *Annu. Rev. Biochem.* 53:749–90

196. Tabor, H., Tabor, C. W. 1975. Isolation, characterization, and turnover of glutathionylspermidine from *Escherichia coli*. *J. Biol. Chem.* 250:2648–54

197. Taniguchi, M., Hara, T., Honda, H. 1986. Similarities between rat liver mitochondrial and cytosolic glutathione reductases and their apoenzyme accumulation in riboflavin deficiency. *Biochem. Int.* 1330:447–54
198. Taylor, M. C. 1992. *The trypanothione reductase gene of* Leishmania donovani. PhD thesis. Univ. London, London
199. Uotila, L., Koivusalo, M. 1989. Glutathione-dependent oxidoreductases: formaldehyde dehydrogenase. See Ref. 48, Part A, pp. 517–51
200. Vander Jagt, D. L. 1989. The glyoxalase system. See Ref. 48, Part A, pp. 597–641
201. Vanoni, M. A., Wong, K. K., Ballou, D. P., Blanchard, J. S. 1990. Glutathione reductase: comparison of steady-state and rapid reaction primary kinetic isotope effects exhibited by the yeast, spinach, and *Escherichia coli* enzymes. *Biochemistry* 29:5790–96
202. Varandani, P. T. 1989. Glutathione-insulin transhydrogenase (protein-disulphide interchange enzyme). See Ref. 48, Part A, pp. 753–65
203. Walsh, C. T., Bradley, M., Nadeau, K. 1991. Molecular studies on trypanothione reductase, a target for antiparasitic drugs. *TIBS* 16:305–9
204. Walsh, C. T., Distefano, M. D., Moore, M. J., Shewchuk, L. M., Verdine, G. L. 1988. Molecular basis of bacterial resistance to organomercurial and inorganic mercuric salts. *FASEB J.* 2: 124–30
205. Wang, C. C. 1991. The ornithine decarboxylase of *Trypanosoma brucei:* a stable protein in vivo. See Ref. 205a, pp. 87–94
205a. Wang, C. C., ed. 1991. *Molecular and Immunological Aspects of Parasitism.* Washington, DC: Am. Assoc. Adv. Sci.
206. Waxman, D. J. 1990. Glutathione *S*-transferases: role in alkylating agent resistance and possible target for modulation chemotherapy—a review. *Cancer Res.* 50:6449–54
207. Wong, K. K., Vanoni, M. A., Blanchard, J. S. 1988. Glutathione reductase: solvent equilibrium and kinetic isotope effects. *Biochemistry* 27:7091–96
208. Worthington, D. J., Rosemeyer, M. A. 1974. Human glutathione reductase: purification of the crystalline enzyme from erythrocytes. *Eur. J. Biochem.* 48:167–77
209. Yarlett, N., Bacchi, C. J. 1988. Effect of D,L-α-difluoromethylornithine on methionine cycle intermediates in *Trypanosoma brucei brucei. Mol. Biochem. Parasitol.* 27:1–10
210. Yarlett, N., Goldberg, B., Nathan, H. C., Garofalo, J., Bacchi, C. J. 1991. Differential susceptibility of *Trypanosoma brucei rhodesiense* isolates to *in vitro* lysis by arsenicals. *Exp. Parasitol.* 72:205–15
211. Yawetz, A., Agosin, M. 1981. Purification of the glutathione-*S*-transferase of *Trypanosoma cruzi. Comp. Biochem. Physiol.* 68B:237–43
212. Zappia, V., Pegg, A. E., ed. 1988. *Progress in Polyamine Research: Novel Biochemical, Pharmacological, and Clinical Aspects.* New York: Plenum
213. Ziegler, D. M. 1985. Role of reversible oxidation-reduction of enzyme thiols-disulfides in metabolic regulation. *Annu. Rev. Microbiol.* 54:305–29

SUBJECT INDEX

E

H

Q

R

S

CUMULATIVE INDEXES

CONTRIBUTING AUTHORS, VOLUMES 42–46

CHAPTER TITLES, VOLUMES 42–46

APPLIED MICROBIOLOGY AND ECOLOGY